THIRD EDITION

SYSTEMS ENGINEERING AND ANALYSIS

BENJAMIN S. BLANCHARD
Virginia Polytechnic Institute and State University

WOLTER J. FABRYCKY
Virginia Polytechnic Institute and State University

PRENTICE HALL, Upper Saddle River, New Jersey 07458

Library of Congress Cataloging-in-Publication Data

Blanchard, Benjamin S.
Systems engineering and analysis / Benjamin S. Blanchard, ard.
Wolter J. Fabrycky.—3rd ed.
p. cm.
Includes bibliographical references and index.
ISBN 0-13-135047-1
1. Systems engineering. 2. System analysis. I. Fabrycky, W. J.
(Wolter J.), 1932- . II. Title.
TA168.B53 1997
620'.001'1—dc21 97-44423
CIP

Acquisitions editor: ALICE DWORKIN
Production editor: RHODORA V. PENARANDA
Editor-in-chief: MARCIA HORTON
Managing editor: BAYANI MENDOZA DE LEON
Director of production and manufacturing: DAVID W. RICCARDI
Copy editor: ANDREA HAMMER
Cover designer: BRUCE KENSELAAR
Manufacturing buyer: DONNA SULLIVAN
Editorial assistant: NANCY GARCIA
Compositor: PREPARÉ/EMILCOMP

Printed in the United States of America

10 9 8 7 6 5 4

ISBN 0-13-135047-1

Prentice-Hall International (UK) Limited, *London*
Prentice-Hall of Australia Pty. Limited, *Sydney*
Prentice-Hall Canada Inc., *Toronto*
Prentice-Hall Hispanoamericana, S.A., *Mexico*
Prentice-Hall of India Private Limited, *New Delhi*
Prentice-Hall of Japan, Inc., *Tokyo*
Simon & Schuster Asia Pte. Ltd., *Singapore*
Editora Prentice-Hall do Brasil, Ltda., *Rio de Janeiro*

Dedication

This third edition of Systems Engineering and Analysis
is dedicated to the more than 350 individuals
who received interdisciplinary masters degrees
in Systems Engineering from the College of Engineering
at Virginia Polytechnic Institute and State University
over the past 25 years and to systems engineering
students and practicing systems engineers worldwide

Benjamin S. Blanchard
Wolter J. Fabrycky

PRENTICE HALL INTERNATIONAL SERIES IN INDUSTRIAL AND SYSTEMS ENGINEERING

W. J. Fabrycky and J. H. Mize, Editors

AMOS AND SARCHET *Management for Engineers*
AMRINE, RITCHEY, MOODIE AND KMEC *Manufacturing Organization and Management, 6/E*
ASFAHL *Industrial Safety and Health Management, 3/E*
BABCOCK *Managing Engineering and Technology, 2/E*
BADIRU *Comprehensive Project Management*
BADIRU *Expert Systems Applications in Engineering and Manufacturing*
BANKS, CARSON AND NELSON *Discrete Event System Simulation, 2/E*
BLANCHARD *Logistics Engineering and Management, 5/E*
BLANCHARD AND FABRYCKY *Systems Engineering and Analysis, 3/E*
BROWN *Technimanagement: The Human Side of the Technical Organization*
BURTON AND MORAN *The Future Focused Organization*
BUSSEY AND ESCHENBACH *The Economic Analysis of Industrial Projects, 2/E*
BUZACOTT AND SHANTHIKUMAR *Stochastic Models of Manufacturing Systems*
CANADA AND SULLIVAN *Economic and Multi-Attribute Evaluation of Advanced Manufacturing Systems*
CANADA, SULLIVAN AND WHITE *Capital Investment Analysis for Engineering and Management, 2/E*
CHANG AND WYSK *An Introduction to Automated Process Planning Systems*
CHANG, WYSK AND WANG *Computer-Aided Manufacturing, 2/E*
DELAVIGNE AND ROBERTSON *Deming's Profound Changes*
EAGER *The Information Payoff: The Manager's Concise Guide to Making PC Communications Work*
EBERTS *User Interface Design*
EBERTS AND EBERTS *Myths of Japanese Quality*
ELSAYED AND BOUCHER *Analysis and Control of Production Systems, 2/E*
FABRYCKY AND BLANCHARD *Life-Cycle Cost and Economic Analysis*
FABRYCKY, THUESEN AND VERMA *Economic Decision Analysis, 3/E*
FISHWICK *Simulation Model Design and Execution: Building Digital Worlds*
FRANCIS, MCGINNIS AND WHITE *Facility Layout and Location: An Analytical Approach, 2/E*
GIBSON *Modern Management of the High-Technology Enterprise*
GORDON *Systematic Training Program Design*
GRAEDEL AND ALLENBY *Industrial Ecology*
HALL *Queuing Methods: For Services and Manufacturing*
HANSEN *Automating Business Process Reengineering*
HAMMER *Occupational Safety Management and Engineering, 4/E*
HAZELRIGG *Systems Engineering*
HUTCHINSON *An Integrated Approach to Logistics Management*
IGNIZIO *Linear Programming in Single- and Multiple-Objective Systems*
IGNIZIO AND CAVALIER *Linear Programming*
KROEMER, KROEMER AND KROEMER-ELBERT *Ergonomics: How to Design for Ease and Efficiency*
KUSIAK *Intelligent Manufacturing Systems*
LAMB *Availability Engineering and Management for Manufacturing Plant Performance*
LANDERS, BROWN, FANT, MALSTROM AND SCHMITT *Electronic Manufacturing Processes*
LEEMIS *Reliability: Probabilistic Models and Statistical Methods*
MICHAELS *Technical Risk Management*
MOODY, CHAPMAN, VAN VOORHEES AND BAHILL *Metrics and Case Studies for Evaluating Engineering Designs*
MUNDEL AND DANNER *Motion and Time Study: Improving Productivity, 7/E*
OSTWALD *Engineering Cost Estimating, 3/E*
PINEDO *Scheduling: Theory, Algorithms, and Systems*
PRASAD *Concurrent Engineering Fundamentals, Vol. I: Integrated Product and Process Organization*
PRASAD *Concurrent Engineering Fundamentals, Vol. II: Integrated Product Development*
PULAT *Fundamentals of Industrial Ergonomics*
SHTUB, BARD AND GLOBERSON *Project Management: Engineering Technology and Implementation*
TAHA *Simulation Modeling and SIMNET*
THUESEN AND FABRYCKY *Engineering Economy, 8/E*
TURNER, MIZE, CASE AND NAZEMETZ *Introduction to Industrial and Systems Engineering, 3/E*
TURTLE *Implementing Concurrent Project Management*
VON BRAUN *The Innovation War*
WALESH *Engineering Your Future*
WOLFF *Stochastic Modeling and the Theory of Queues*

CONTENTS

PREFACE

This book is about *systems*. It focuses on the *engineering* of systems and on systems *analysis*. In the first case, emphasis is on the process of bringing systems into being, beginning with the definition of need and extending through requirements analysis, functional analysis and allocation, design synthesis, design evaluation, and system validation. In the second case, concern is with the improvement of systems already in existence. Through the iterative steps of analysis, evaluation, feedback, modification and control, many systems in use today can be improved in their effectiveness, output quality, ownership cost, and user satisfaction. Systems analysis methods and techniques are integrated within the *systems engineering process*, which remains the overarching theme for this book.

Systems may be classified as either "natural" or "human-made." Natural systems are those which came into existence by natural processes. Human-made, or technical systems, are those where people have intervened in the natural order by applying pervasive technologies through system components, attributes, and relationships. The types and variety of human-made systems are numerous and encompass the domains of communications, defense, education, healthcare, manufacturing, transportation, and others. Only human-made systems are the central focus in this book.

Experience in recent decades indicates that properly coordinated and functioning human-made systems, with a minimum of undesirable side effects, require the application of an integrated, life-cycle oriented "systems" approach. The consequences of not applying systems engineering in the design and development and/or reengineering of systems have been disruptive and costly. Accordingly, the main objective of this book is to provide engineers, systems analysts, technical personnel, and managers with the essential systems concepts, methodologies, models, and tools needed to understand and apply *systems engineering* to all types of human-made systems.

The topics presented in this book have been organized into 6 parts and 19 chapters. Part I presents an introduction to systems and systems engineering in the context of system science and good engineering practice. Part II addresses the system design process as a series of evolutionary steps, progressing from the identification of a need through conceptual design, preliminary design, detail design and development, and test and evaluation. Part III derives some of the most useful mathematical models and tools for systems analysis. Emphasis is placed upon the application of modeling and analysis techniques as an integral part of the systems engineering process. Part IV addresses "design for operational feasibility " by discussing those characteristics of design found to be most significant for successful system operation and customer satisfaction. Separate chapters are devoted to reliability, maintainability, usability (human factors), supportability (serviceability), producibility, disposability, and affordability (life-cycle cost). Part V presents an overview of systems engineering management, with planning and organization discussed in one chapter and program management and control in another. Part VI contains a set of comprehensive appendices providing supporting topics, checklists, tables, references, and related resource materials.

This third edition is more comprehensive than earlier editions. While the overall organization of material has not changed, a morphology for design synthesis, analysis, and evaluation has been integrated with the system life-cycle concept. The entire life cycle is now covered more comprehensively with the expansion of chapters to include contemporary design tools. Part II has been strengthened in the areas of requirements analysis, the identification and prioritization of technical performance measures, and functional analysis and allocation. Part III is now oriented more toward design evaluation as an essential activity within the systems engineering process. A design evaluation display for multiple criteria is the central unifying construct. Part IV has been expanded to include new chapters dealing with design for producibility, disposability, and affordability.

This book is intended for use in the classroom at either the undergraduate or graduate level, or by the practicing professional in business, industry, or government. The text material includes over 320 illustrations and 430 problem exercises arranged in such a manner as to guide the engineer or analyst through the entire system life cycle. The concepts and techniques presented are applicable to any type of system, and the topics discussed may be "tailored" for both large- and small-scale systems. Much of the material has been developed from industrial and research experience in systems engineering over three decades.

Six other Prentice Hall books by the authors provided some of the raw material from which this text was fashioned. These books cover the subject areas of engineering organization and management, logistics engineering and management, procurement and inventory systems, applied operations research and management science, engineering economy, and life-cycle cost and economic analysis.

Benjamin S. Blanchard
Wolter J. Fabrycky

PART I
INTRODUCTION TO SYSTEMS

1 SYSTEM DEFINITIONS AND CONCEPTS

Systems are as pervasive as the universe in which we live. At one extreme, they are as grand as the universe itself. At the other, they are as infinitesimal as the atom. Systems appeared first in natural forms, but with the appearance of human beings, a variety of human-made systems have come into existence. Only recently have we come to understand the underlying structure and characteristics of natural and human-made systems in a scientific way.

In this first chapter, some system definitions and system science concepts are presented that provide a basis for the study of systems engineering and analysis. This includes definitions of system characteristics, a classification of systems into various types, a discussion of the current state of system science, and a discussion of the transition to the systems age now under way. Finally, this chapter presents technology and the nature of engineering in the systems age.

1.1 SYSTEM DEFINITIONS AND ELEMENTS

A *system* is an assemblage or combination of elements or parts forming a complex or unitary whole, such as a river system or a transportation system; any assemblage or set of correlated members, such as a system of currency; an ordered and comprehensive assemblage of facts, principles, or doctrines in a particular field of knowledge or thought, such as a system of philosophy; a coordinated body of methods or a complex scheme or plan of procedure, such as a system of organization and management; any regular or special method of plan of procedure, such as a system of marking, numbering, or measuring.[1] Not every set of items, facts, methods, or procedures is a system. A

[1] This definition was adapted from *The Random House Dictionary of the English Language*. 2nd ed. (New York: Random House, Inc., 1994).

random group of items in a room would constitute a set with definite relationships between the items, but it would not qualify as a system because of the absence of unity, functional relationship, and useful purpose.

The Elements of a System

Systems are composed of components, attributes, and relationships. These are described as follows:

1. *Components* are the operating parts of a system consisting of input, process, and output. Each system component may assume a variety of values to describe a system state as set by some control action and one or more restrictions.
2. *Attributes* are the properties or discernible manifestations of the components of a system. These attributes characterize the system.
3. *Relationships* are the links between components and attributes.

A system is a set of interrelated components working together toward some common objective or purpose. The set of components has the following properties:

1. The properties and behavior of each component of the set has an effect on the properties and behavior of the set as a whole.
2. The properties and behavior of each component of the set depends on the properties and behavior of at least one other component in the set.
3. Each possible subset of components has the two properties listed previously; the components cannot be divided into independent subsets.

The properties listed earlier ensure that the set of components constituting a system always has some characteristic or behavior pattern that cannot be exhibited by any of its subsets. A system is more than the sum of its component parts. However, the components of a system may themselves be systems, and every system may be part of a larger system in a hierarchy.

The objective or purpose of a system must be explicitly defined and understood so that system components may be selected to provide the desired output for each given set of inputs. Once defined, the objective or purpose makes it possible to establish a measure of effectiveness indicating how well the system performs. Establishing the purpose of a human-made system and defining its measure of effectiveness is often a challenging task.

The purposeful action performed by a system is its *function*. A common system function is that of altering material, energy, or information. This alteration embraces input, process, and output. Some examples are the materials processing in a manufacturing system or a digestive system, the conversion of coal to electricity in a power plant system, and the information processing in a computer system.

Systems that alter material, energy, or information are composed of structural components, operating components, and flow components. *Structural components* are the static parts; *operating components* are the parts that perform the processing; and *flow components* are the material, energy, or information being altered. A motive force must be present to provide the alteration within the restrictions set by structural and operating components.

Structural, operating, and flow components have various attributes that affect their influence on the system. The attributes of an electrical system may be described in terms of inductance, capacitance, impedance, and so on. The system may change its condition over time in only certain ways, as in the *on* or *off* state of an electrical switching system. A system, condition, situation, or state is set forth to describe a set of components, attributes, and relationships.

A systems view is only one way of understanding complexity. Another is that of a *relational view*. Three major differences exist between a relation and a system. First, a relation exists between two and only two components, whereas a system is described by the interaction between many components. Second, a relation is formed out of the imminent qualities of the components, whereas a system is created by the particular position and spatial distribution of its components. The components of a relation are separated spatially, whereas a system is made up of the interacting distribution of its components. Third, the connection between the components of a relation is direct, whereas the connection in a system depends on a common reference to the entire set of components making up the system.

Relationships that are functionally necessary to each other may be characterized as *first order*. An example is symbiosis, the necessary relationship of dissimilar organisms, such as an animal and a parasite. *Second-order* relationships, called *synergistic*, are those that are complementary and add to system performance. *Redundance* in a system exists when duplicate components are present for the purpose of assuring continuation of the system function.

Systems and Subsystems

The definition of a system is not complete without consideration for its position in the hierarchy of systems. Every system is made up of *components*, and any component can be broken down into smaller components. If two hierarchial levels are involved in a given system, the lower is conveniently called a *subsystem*. For example, in an air transportation system, the aircraft, terminals, ground support equipment, and controls are subsystems. Equipment items, people, and information are components. Clearly, the designations of system, subsystem, and component are relative because the system at one level in the hierarchy is the component at another.

In any particular situation, it is important to define the system under consideration by specifying its limits, boundaries, or scope. Everything that remains outside the boundaries of the system is considered to be the *environment*. However, no system is completely isolated from its environment. Material, energy, and/or information must often pass through the boundaries as *input* to the system. In reverse, material, energy, and/or information that passes from the system to the environment is called *output*.

That which enters the system in one form and leaves the system in another form is usually called *throughput.*

The total system, at whatever level in the hierarchy, consists of all components, attributes, and relationships needed to accomplish an objective. Each system has an objective, providing a purpose for which all system components, attributes, and relationships have been organized. Constraints placed on the system limit its operation and define the boundary within which it is intended to operate. Similarly, the system places boundaries and constraints on its subsystems.

An example of a total system is a fire department. The components of this "fire control system" are the building, the fire engines, the firefighters and small equipment, the communication equipment, and the maintenance facilities. These components are the major subsystems of the fire department. Each of these subsystems has several contributing subsystems or components. At each level in the hierarchy, the description must include all components, all attributes of these components, and all relationships.

The systems viewpoint looks at a system from the top down rather than from the bottom up. Attention is first directed to the system as a black box that interacts with its environment. Next, attention is focused on how the smaller black boxes (subsystems) combine to achieve the system objective. The lowest level of concern is then with individual components.

The process of bringing systems into being, and of improving systems already in existence, in a wholistic sense is receiving continued attention. By bounding the total system for study purposes, the systems engineer or systems analyst will be more likely to achieve a satisfactory result. Focusing on systems, subsystems, and components in a hierarchy forces consideration of all pertinent functional relationships. Components and attributes are important, but only to the end that the purpose of the whole system is achieved through the functional relationships linking them.

1.2 A CLASSIFICATION OF SYSTEMS[2]

Systems may be classified for convenience and to provide insight into their wide range. This will be accomplished by several dichotomies conceptually contrasting system similarities and dissimilarities. In this section, descriptions are given of natural and human-made systems, physical and conceptual systems, static and dynamic systems, and closed and open systems.

Natural and Human-Made Systems

The origin of systems gives a most important classification opportunity. *Natural systems* are those that came into being by natural processes. *Human-made systems* are those in which human beings have intervened through components, attributes, or relationships.

[2] The classifications in this section are only some of those that could be presented. All system types have embedded information flow components and, therefore, information systems are not included as a separate classification.

All human-made systems, when brought into being, are embedded into the natural world. Important interfaces often exist between human-made systems and natural systems. Each affects the other in some way. The effect of human-made systems on the natural world has only recently become a keen subject for study by concerned people, especially in those instances where the effect is undesirable.

Natural systems exhibit a high degree of order and equilibrium. This is evidenced in the seasons, the food chain, the water cycle, and so on. Organisms and plant life adapt themselves to maintain an equilibrium with the environment. Every event in nature is accompanied by an appropriate adaptation, one of the most important being that material flows are cyclic. In the natural environment there are no dead ends, no wastes, only continual recirculation.

Only recently have significant human-made systems appeared. These systems make up the human-made world, their chief engineer being human. The rapid evolution of human beings is not adequately understood, but their coming upon the scene has significantly affected the natural world, often in undesirable ways. Primitive beings had little impact on the natural world, for they had not yet developed a potent and pervasive technology.

A good example of the impact of human-made systems on natural systems is the set of problems that arose from building the Aswan Dam on the Nile River. Construction of this massive dam ensures that the Nile will never flood again, solving an age-old problem. However, several new problems arose. The food chain was broken in the eastern Mediterranean, thereby reducing the fishing industry. Rapid erosion of the Nile Delta took place, introducing soil salinity into upper Egypt. No longer limited by periodic dryness, the population of bilharzia (a water-borne snail parasite) has produced an epidemic of intestinal disease along the Nile. These side effects were not adequately considered by those responsible for the project. A systems view encompassing both natural and human-made elements might have led to a better solution to the problem of flooding.

Physical and Conceptual Systems

Physical systems are those that manifest themselves in physical form. They are composed of real components and may be contrasted with *conceptual systems*, where symbols represent the attributes of components. Ideas, plans, concepts, and hypotheses are examples of conceptual systems.

A physical system consumes physical space, whereas conceptual systems are organizations of ideas. One type of conceptual system is the set of plans and specifications for a physical system before it is actually brought into being. A proposed physical system may be simulated in the abstract by a mathematical or other conceptual model. Conceptual systems often play an essential role in the operation of physical systems in the real world.

The totality of elements encompassed by all components, attributes, and relationships focused on a given result employ a process in guiding the state of a system. A process may be mental (thinking, planning, and learning), mental-motor (writing, drawing, and testing), or mechanical (operating, functioning, and producing). Processes exist equally in both physical and conceptual systems.

Process occurs at many different levels within systems. The subordinate process essential to the operation of a total system is provided by the subsystem. The subsystem may, in turn, be dependent on more detailed subsystems. System complexity is the feature that defines the number of subsystems present and, consequently, the number of processes involved. A system may be bounded for the purpose of study at any process or subsystem level.

Static and Dynamic Systems

Another system dichotomy is the distinction of static and dynamic systems. A *static system* is one having structure without activity, as exemplified by a bridge. A *dynamic system* combines structural components with activity. An example is a school, combining a building, students, teachers, books, and curricula.

For centuries people have viewed the universe of phenomena as unchanging. A mental habit of dealing with certainties and constants developed. The substitution of a process-oriented description for the static description of the world is one of the major characteristics separating modern science from earlier thinking.

A dynamic conception of the world has become a necessity. Yet, a general definition of a system as an ongoing process is incomplete. Many systems would not be included under this broad definition because they lack motion in the usual sense. A highway system is static, yet it contains the system elements of components, attributes, and relationships.

It is recognized that a system is static only in a limited frame of reference. A bridge is constructed over a period, and this is a dynamic process. It is then maintained and perhaps altered to serve its intended purpose more fully. Even a crystal passes through several forms during its period of growth.

Systems may be characterized as having random properties. In almost all systems in both the natural and human-made categories, the inputs, process, and output can only be described in statistical terms. Uncertainty often occurs in both the number of inputs and the distribution of these inputs over time. For example, it is difficult to predict exactly the number of passengers that will check in for a flight, or the exact time each will arrive at the airport. However, these factors can be described in terms of probability distributions, and system operation is said to be *probabilistic*.

Closed and Open Systems

A *closed system* is one that does not interact significantly with its environment. The environment provides only a context for the system. Closed systems exhibit the characteristic of equilibrium resulting from internal rigidity that maintains the system in spite of influences from the environment. An example is the chemical equilibrium eventually reached in a closed vessel when various reactants are mixed together. The reaction can be predicted from a set of initial conditions. Closed systems involve deterministic interactions, with a one-to-one correspondence between initial and final states.

An *open system* allows information, energy, and matter to cross its boundaries. Open systems interact with their environment, examples being plants, ecological systems, and business organizations. They exhibit the characteristics of *steady state*, wherein a dynamic interaction of system elements adjusts to changes in the environment. Because of this steady state, open systems are self-regulatory and often self-adaptive.

It is not always easy to classify a system as either open or closed. Open systems are typical of those that have come into being by natural processes. People-made systems have characteristics of open and closed systems. They may reproduce natural conditions not manageable in the natural world. They are closed when designed for invariant input and statistically predictable output, as in the case of an aircraft in flight.

Both closed and open systems exhibit the property of entropy. *Entropy* is defined here as the degree of disorganization in a system and is analogous to the use of the term in thermodynamics. In the thermodynamic usage, entropy is the energy unavailable for work resulting from energy transformation from one form to another.

In systems, increased entropy means increased disorganization. A decrease in entropy occurs as order occurs. Life represents a transition from disorder to order. Atoms of carbon, hydrogen, oxygen, and other elements become arranged in a complex and orderly fashion to produce a living organism. A conscious decrease in entropy must occur to create a human-made system. All human-made systems, from the most primitive to the most complex, consume entropy—the creation of more orderly states from less orderly states.

1.3 SCIENCE AND SYSTEMS SCIENCE

The significant accumulation of scientific knowledge, which began in the eighteenth century and rapidly expanded in the twentieth, made it necessary to classify what was discovered into scientific disciplines. Science began its separation from philosophy almost two centuries ago. It then proliferated into more than 100 distinct disciplines. A relatively recent unifying development is the idea that systems have general characteristics, independent of the area of science to which they belong. In this section, the evolution of a science of systems is presented through an examination of cybernetics, general systems theory, and systemology.

Cybernetics

The word *cybernetics* was first used in 1947 by Norbert Wiener, but it is not explicitly defined in his classical book.[3] Cybernetics comes from the Greek word meaning "steersman" and is a cognate of "governor." In its narrow view, cybernetics is equivalent to servo theory in engineering. In its broad view, it may encompass much of natural science. Cybernetics has to do with self-regulation, whether mechanical, electromechanical, electrical, or biological.

The concept of feedback is central to cybernetic theory. All goal-seeking behavior is controlled by the feedback of corrective information about deviation from a

[3] N. Wiener, *Cybernetics* (New York: John Wiley & Sons, Inc., 1948).

desired state. The best known and most easily explained illustration of feedback is the action of a thermostat. The thermometer component of a thermostat senses temperature. When the actual temperature falls below that set into the thermostat, an internal contact is made, activating the heating system. When the temperature rises above that set into the thermostat, the contact is broken, shutting the heating system off.

Biological organisms are endowed with the capacity for self-regulation, called *homeostasis*. The biological organism and the physical world are both very complex. Instructive analogies exist between them and human-made systems. Through these analogies humans have learned some things about their properties that might have not been learned from the study of natural systems alone. As people develop even more complex systems, we will gain a better understanding of how to control them and our environment.

The science of cybernetics has made three important contributions to the area of regulation and control. First, it stresses the concept of information flow as a distinct system component and clarifies the distinction between the activating power and the information signal. Second, it recognizes that similarities in the action of control mechanisms involve principles that are fundamentally identical. Third, the basic principles of feedback control are given mathematical treatment.

A practical application of cybernetics has been the tremendous development of automatic equipment and processes, most controlled by microcomputers. However, its significance is greater than this technological contribution. The science of cybernetics is important not only for the control engineer but also for the purest of scientists. Cybernetics is a new science of purposeful and optimal control applicable to complex processes in nature, society, and business organizations.

General Systems Theory

An even broader unifying concept than cybernetics took shape during the late 1940s. It was the idea that basic principles common to all systems could be found that went beyond the concept of control and self-regulation. A unifying principle for science and a common ground for interdisciplinary relationships needed in the study of complex systems was being sought. Ludwig von Bertalanffy used the phrase *general systems theory* around 1950 to describe this endeavor.[4]

General systems theory is concerned with developing a systematic framework for describing general relationships in the natural and the human-made world. The need for a general theory of systems arises out of the problem of communication among the various disciplines. Although the scientific method brings similarity between the methods of approach, the results are often difficult to communicate across disciplinary boundaries. Concepts and hypotheses formulated in one area seldom carry over to another where they could lead to significant forward progress. The difficulties are greatest among the various disciplines, the physical and life sciences, the social and behavioral sciences, and the humanities.

[4] L. von Bertalanffy, "General System Theory: A New Approach to Unity of Science," *Human Biology*, December 1951. A related contribution was by K. Boulding, "General Systems Theory: The Skeleton of Science," *Management Science*, April 1956.

One approach to an orderly framework is the structuring of a hierarchy of levels of complexity for basic units of behavior in the various fields of inquiry. A *hierarchy of levels* can lead to a systematic approach to systems that has broad application. Such a hierarchy was formulated by Boulding approximately as follows[5]:

1. The level of static structure or *frameworks*, encompassing the geography and anatomy of the universe.
2. The level of the simple dynamic system of *clockworks*, encompassing a significant segment of chemistry, physics, and engineering science.
3. The level of the *thermostat* or cybernetic system, encompassing the transmission and interpretation of information.
4. The level of the *cell*, the self-maintaining structure or open system where life begins to be evident.
5. The level of the *plant*, with genetic-societal structure making up the world of botany.
6. The level of the *animal*, encompassing mobility, teleological behavior, and self-awareness.
7. The level of the *human*, encompassing self-consciousness and the ability to produce, absorb, and interpret symbols.
8. The level of *social organization*, where the content and meaning of messages, value systems, transcription of images into historical record, art, music, and poetry, and complex human emotion are of concern.
9. The level of the *unknowables*, where structure and relationship may be postulated but where answers are not yet available.

The first level in Boulding's hierarchy is the most pervasive. Static systems are everywhere, and this category provides a basis for analysis and synthesis of systems at higher levels. Dynamic systems with predetermined outcomes are predominant in the natural sciences. At higher levels cybernetic models are available, mostly in closed-loop form. Open-loop systems are currently receiving scientific attention, but modeling difficulties arise regarding their self-regulating properties. Beyond this level, there is little systematic knowledge available. However, general systems theory provides science with a useful framework within which each specialized discipline may contribute. It allows scientists to compare concepts and similar findings, its greatest benefit being that of communication across disciplines.

Systemology

The science of systems or their formation is called *systemology*. As system science is pushed forward by the formation of interdisciplines, humankind should benefit from a more appropriate application of the discipline. The problems and problem complexes faced by human beings are not organized along disciplinary lines. It is only through a new organization of scientific and professional effort based on the common attributes and characteristics of problems that beneficial progress will be made.

[5] Boulding, "General Systems Theory: The Skeleton of Science," *Management Science*, April 1956.

Disciplines in science and the humanities developed largely by what society permitted scientists and humanists to investigate. Areas that provided the least challenge to cultural, social, and moral beliefs were given priority. The survival of science was also of concern in the progress of the discipline, and this is still so. However, recent developments have added to the respectability of most areas. Much credit for this can be given to the new respectability of interdisciplinary inquiry.

During the 1940s, scientists of established reputation accepted the challenge of attempting to understand a host of common processes in military operations. Their team effort was called *operations research*, and the focus of their attention was the science of military systems. After the war, this interdisciplinary area began to take on the attributes of a discipline and a profession. Today a body of systematic knowledge exists for military and commercial operations. But operations research is not the only science of systems available today. Cybernetics, general systems research, organizational and policy sciences, management science, and the information sciences are others.

Formation of interdisciplines in the past 60 years has brought about an evolu tionary synthesis of knowledge. This has occurred not only within science, but between science and technology and between science and the humanities. The forward progress of systemology in the study of large-scale complex systems requires a synthesis of science and the humanities as well as a synthesis of science and technology. One of the most important contributions of systemology is that it offers a single vocabulary and a unified set of concepts applicable to many types of systems.

1.4 TRANSITION TO THE SYSTEMS AGE[6]

There is considerable evidence to suggest that the advanced nations of the world are leaving one technological age and entering another. It appears that this transition is bringing about a change in the conception of the world in which we live. This conception is both a realization of the complexity of natural and human-made systems and a basis for improvement in people's position relative to these systems.

The Machine Age

Two ideas have been dominant in the way people seek to understand the world around them. The first is called *reductionism*. It consists of the belief that everything can be reduced, decomposed, or disassembled to simple indivisible parts. These were taken to be atoms in physics, simple substances in chemistry, cells in biology, and monads, instincts, drives, motives, and needs in psychology.

Reductionism gives rise to an analytical way of thinking about the world, a way of seeking explanations and understanding. Analysis consists, first, of taking apart what is to be explained, disassembling it, if possible, down to the independent and indivisible parts of which it is composed; second, of explaining the behavior of these parts; and, finally, of aggregating these partial explanations into an explanation of the whole.

[6] This section was adapted from R. L. Ackoff, *Redesigning the Future* (New York: John Wiley & Sons, Inc., 1974).

For example, the analysis of a problem consists of breaking it down into a set of as simple problems as possible, solving each, and assembling their solutions into a solution of the whole. If the analyst succeeds in decomposing a problem into simpler problems that are independent of each other, aggregating the partial solutions is not required because the solution to the whole is the sum of the solutions to its independent parts. In the *Machine Age*, understanding the world was taking to be the sum, or result, of an understanding of its parts, which were conceptualized as independently of each other as was possible.

The second basic idea was that of *mechanism*. All phenomena were believed to be explainable by using only one ultimately simple relation, cause and effect. One thing or event was taken to be the *cause* of another (its *effect*) if it was both necessary and sufficient for the other. Because a cause was taken to be sufficient for its effect, nothing was required to explain the effect other than the cause. Consequently, the search for causes was environment free. It employed what is now called "closed-system" thinking. Laws such as that of freely falling bodies were formulated so as to exclude environmental effects. Specially designed environments, called *laboratories*, were used so as to exclude environmental effects on phenomena under study.

Causal laws permit no exceptions. Effects are completely determined by causes. Hence, the prevailing view of the world was deterministic. It was also mechanistic because science found no need for teleological concepts (such as functions, goals, purposes, choice, and free will) in explaining any natural phenomenon; they considered such concepts to be unnecessary, illusory, or meaningless. The commitment to causal thinking yielded a conception of the world as a machine; it was taken to be like a hermetically sealed clock—a self-contained mechanism whose behavior was completely determined by its own structure.

The *Industrial Revolution* brought about *mechanization*, the substitution of machines for people as a source of physical work. This process affected the nature of work left for people to do. They no longer did all the things necessary to make a product; they repeatedly performed a simple operation in the production process. Consequently, the more machines were used as a substitute for people at work, the more workers were made to behave like machines. The dehumanization of work was an irony of the Industrial Revolution and the Machine Age.

The Systems Age

Although eras do not have precise beginnings and endings, the 1940s can be said to have contained the beginning of the end of the Machine Age and the beginning of the *Systems Age*. This new age is the product of a new intellectual framework in which the doctrines of reductionism and mechanism and the analytical mode of thought are being supplemented by the doctrines of expansionism, teleology, and a new synthetic (or systems) mode of thought.

Expansionism is a doctrine that considers all objects and events, and all experiences of them, are parts of larger wholes. It does not deny that they have parts, but it focuses on the wholes of which they are part. It provides another way of viewing things, a way that is different from, but compatible with, reductionism. It turns attention from ultimate elements to a whole with interrelated parts—to systems.

Preoccupation with systems brings with it the synthetic mode of thought. In the *analytic* mode, an explanation of the whole was derived from explanations of its parts. In *synthetic* thinking, something to be explained is viewed as part of a larger system and is explained in terms of its role in that larger system. The Systems Age is more interested in putting things together than in taking them apart.

Analytic thinking is outside-in thinking; synthetic thinking is inside-out thinking. Neither negates the value of the other, but by synthetic thinking one can gain understanding that cannot be obtained through analysis, particularly of collective phenomena.

The synthetic mode of thought, when applied to systems problems is called the *systems aproach*. This way of thinking is based on the observation that, when each part of a system performs as well as possible, the system as a whole may not perform as well as possible. This follows from the fact that the sum of the functioning of the parts is seldom equal to the functioning of the whole. Accordingly, the synthetic mode seeks to overcome the often observed predisposition to perfect details and ignore system outcomes.

Because the Systems Age is *teleologically oriented*, it is preoccupied with systems that are goal seeking or purposeful; that is, systems that can display choice of either means or ends, or both. It is interested in purely mechanical systems only insofar as they can be used as instruments of purposeful systems. Furthermore, the Systems Age is largely concerned with purposeful systems, some of whose parts are purposeful; these are called *social groups*. The most important class of social groups is the one containing systems whose parts perform different functions that have a division of functional labor; these are called *organizations*.

In the Systems Age, attention is focused on groups and on organizations as parts of larger purposeful societal systems. Participative management, collaboration, group decision making, and total quality management are new working arrangements within the organization. Among organizations is now found a keen concern for social and environmental factors, whereas competition continues to increase worldwide.

1.5 TECHNOLOGY AND TECHNICAL SYSTEMS

Technology is defined broadly as the branch of knowledge that deals with industrial arts, applied science, and engineering, or the sum of the ways in which social groups provide themselves with the material objects of their civilization.[7] A social group possessing inherent abilities and the knowledge to maintain its stock of technology is said to have a civilization. Modern civilizations possess pervasive and potent technical systems that provide needed products, systems, structures, and services.

Technology and Society

Human society is characterized by its culture. Each human culture manifests itself through the media of technology. In turn, the manifestation of culture is an important indicator of the degree to which a society is technologically advanced.

[7] This definition was adapted from *The Random House Dictionary of the English Language*, 2nd ed. (New York: Random House, Inc., 1994).

The entire history of humankind is closely related to the progress of technology. But, technological progress is often stressful on people and organizations alike. This need not be. The challenge should be to find ways for people to live better lives as a result of new technological capability and social organizational structure.

In general, the complexity of systems desired by societies is increasing. As new technologies become available, the pull of "want" is augmented by a push to incorporate these new capabilities into both new and existing systems. The desire for bigger and better systems produces an ever-changing set of requirements. The identification of the "true" need and the elicitation of "real" requirements is, in itself, a technological challenge. There has been a tendency to "design now and fix later," with a negative impact on the worth of many systems to intended users and society.

To transition from the past, to present and future technological states, is not a one-step process. Continuing cascades of technical advances are available to society as time unfolds. Societal response is often to make one transition and then to adopt a static pattern of behavior. A better response would be to seek new but well-thought-out possibilities for advancement continuously.

Improvement in technological literacy should increase the population of individuals capable of participating in this desirable activity. One key to imparting this literacy are the communication technologies now expanding as never before. Thus, technology in this sphere may act favorably to aid the understanding and subsequent evaluation by society of technologies in other spheres.

Technical Systems[8]

The phrase *technical system* may be used to represent all types of human-made artifacts, including technical products and processes. Therefore, the technical system is the subject of the collection of activities that are preformed by engineers within the processes of engineering design, including generating, retrieving, processing, and transmitting information about products. It is also the subject of various tasks in the production process, including work preparation and planning, and in many economic considerations, both internal and social.

In museums we see thousands of technical objects and recognize them as products of technology. Their variety of functions, form, size, and so forth, tends to obscure common properties and features. But vast variety also exists in nature, and in those circumstances clearly defined kingdoms of natural objects have been defined for study in the natural sciences. Likewise, attempts have been made to define a term that conceptually describes classes of technical objects.

Technical objects can be referred to as simply objects, products, things, machines, implements, or technical works. The results of a manufacturing activity, as the conceptual content of technology, can be termed artifacts or *instrumentum*. Such definitions are meant to include all manner of machines, appliances, implements, structures, weapons, and ships that represent the technical means by which humans achieve their ends. But, to be complete, this definition must recognize the hierarchical nature of sys-

[8] This section was adapted from V. Hubka and W. E. Eder, *Theory of Technical Systems* (Berlin: Springer-Verlag, 1988).

tems and the interactions that occur between levels in the hierarchy. For example, the "system" of interest may be a transportation system, an airline system within the transportation system, or an aircraft system contained within the airline system.

Little difficulty exists in the classification of systems as either natural or technical (human-made). But it is difficult to classify technical systems accurately. One approach is to classify in accordance with the well-established subdivisions of technology in industry: e.g., civil engineering, electrical engineering, and mechanical engineering. However, from a practical and organizational viewpoint, this does not permit a precise definition of a machine system or electrical system because no firm boundary can be drawn by describing these systems as products of mechanical or electrical engineering. Modern developments of technical systems have generally blurred the boundaries. Electrical and computer products, especially software, are increasingly used together with mechanical and human interfaces. Each acts as a subsystem to a system of greater complexity and purpose. Most systems in use today are hybrids of the simple systems of the past.

1.6 ENGINEERING IN THE SYSTEMS AGE

Engineering activities of analysis and design for human-made or technical systems are not an end in themselves but are a means for satisfying human wants. Thus, modern engineering has two aspects. One aspect concerns itself with the materials and forces of nature; the other is concerned with the needs of people.

In the Systems Age, successful accomplishment of engineering objectives requires a combination of technical specialties and expertise. Engineering in the Systems Age must be a team activity where various individuals involved are cognizant of the important relationships between specialties and between economic factors, ecological factors, political factors, and societal factors. Engineering decisions of today require consideration of these factors in the early stage of system design and development, and the results of such decisions have a definite impact on these factors. Conversely, these factors usually impose constraints on the design process. Thus, technical expertise must include not only the basic knowledge of individual specialty fields of engineering but a knowledge of the context of the system being brought into being.

System Complexity and Scope

The world is increasing in complexity because of human intervention. Through the advent of advanced technologies, transportation times have been reduced, and vastly more efficient means of communication have been introduced. Every aspect of human existence has become more intimate and interactive. The need for integration and conflict resolution becomes more important. At the same time, increasing populations and the desire for larger and better systems is leading to the accelerated exploitation of resources and increased environmental impact. A variety of technically literate specialists is needed.

Although relatively small products, such as a wireless telephone, an electrical household appliance, or even an automobile may employ a limited amount of direct engineering personnel and supporting resources, there are many large-scale systems that require the combined input of specialists representing a wide variety of engineering disciplines. An example is that of a ground mass-transit system.

Civil engineers are required for the layout and/or design of railroad tracks, tunnels, bridges, cables, and facilities. Electrical engineers are involved in the design of automatic train control provisions, traction power, substations for power distribution, automatic fare collection, digital data systems, and so on. Mechanical engineers are necessary in the design of passenger vehicles and related mechanical equipment. Architectural engineers provide support in the construction of passenger terminals. Reliability and maintainability engineers are involved in the design for system availability and the incorporation of supportability characteristics. Industrial engineers deal with the production aspects of passenger vehicles and vehicle components. Test engineers evaluate the system to ensure that all performance, effectiveness, and system support requirements are met. Engineers in the planning and marketing areas are required to keep the public informed and to promote the technical aspects of the system (i.e., to keep the politicians and local citizens informed). General systems engineers are required to ensure that all aspects of the system are properly integrated and function as a single entity.

Although the preceding example is not all inclusive, it is evident that many different engineering disciplines are directly involved. In fact, there are some large projects, such as the development of a new airplane, where the number of engineers assigned to perform engineering functions is in the thousands. In addition, the different engineering types often range in the hundreds. These engineers, forming a part of a large organization, must not only be able to communicate with each other but must be conversant with such interface areas as purchasing, accounting, personnel, and legal.

Another major factor associated with large projects is that much system development, production, evaluation, and support is often accomplished at supplier (sometimes known as *subcontractor*) facilities located throughout the world. Often there is a prime producer or contractor who is ultimately responsible for the development and production of the total system as an entity, and there are numerous suppliers providing different system components. Thus, much of the project work and many of the associated engineering functions may be accomplished at dispersed locations, often worldwide.

Technological Growth and Change

Technological growth and change is occurring continuously and is stimulated by an attempt to respond to some unmet current need and/or by attempting to perform ongoing activities in a more effective and efficient manner. In addition, changes are being stimulated by social factors, political objectives, and ecological constraints.

Generally, people are not satisfied with the impact of the human-made or technical systems on the natural world and on ourselves. Because engineering and the applied sciences are largely responsible for bringing technical systems into being, it is not surprising that there is some dissatisfaction with these fields of endeavor. Accordingly, technical and economic feasibility can no longer be the sole determinants of what engineers do. Ecological, political, social, cultural, and even psychological influences are equally important considerations. The number of factors in any given engineering project has multiplied. Because of the shifts in social attitudes toward moral responsibility, the ethics of personal decisions are becoming a major professional concern. Engineering is not alone in facing up to these considerations.

Some examples of these considerations may be cited. For instance, environmental concerns have resulted in recent legislation and regulations requiring new methods for crop protection from insects, new means for the disposal of medical waste, and new methods for treating solid waste. Concern for shortages of fossil fuel sources as well as ecological impacts brought about a great focus on energy conservation and alternative energy sources. These and other comparable situations were created through both properly planned programs and as a result of panic situations. All have stimulated beneficial technological innovation.

The response in fulfilling technology requirements is dependent on the scientists and engineers available in the needed fields of expertise and whether they are up to date and creative in their respective specialty areas. In some fields, such as electronics and the medical profession (from the standpoint of engineering innovations), technological growth is rapid. Engineers in these fields have an extremely difficult time maintaining their skills.

The continuing trend in technological advances has created an increasing demand for engineers in many fields. There will always be a demand for engineers who can synthesize and adapt. Certain technical specialties will become obsolete with time. The astute engineer should be able to detect trends and plan accordingly for satisfactory transition by acquiring knowledge to broaden his or her horizons. To help in this process is one aim of this book.

QUESTIONS AND PROBLEMS

1. Identify and describe an example system for each system category named in the dictionary definition of Section 1.1.
2. Pick a system that alters material and identify its structural components, operating components, and flow components.
3. Pick a natural system and describe it in terms of components, attributes, and relationships; repeat for a human-made system.
4. Select a complex system and discuss it in terms of the hierarchy of systems.
5. Identify and contrast a physical and a conceptual system.
6. Identify and contrast a static and a dynamic system.
7. Identify and contrast a closed and an open system.
8. Describe cybernetics through an example of your choice.
9. Give a system example at each level in Boulding's hierarchy.
10. Give an example of a problem requiring an interdisciplinary appoach and identify the needed disciplines.
11. Identify the attributes of the Machine Age and the Systems Age.
12. What are the special engineering requirements in the Systems Age?
13. What benefits could result from increasing societies' technological literacy?
14. What is the difficulty encountered in attempting to classify technical systems?
15. Why is technological and economic feasibility no longer the sole determinant of success in engineering undertakings?

2 Bringing Systems Into Being

The world in which we live may be divided into the natural world and the human-made world. Included in the former are all elements of world that came into being by natural processes. The human-made world is made up of all products, systems, and structures made by people for the use of people. But we are not satisfied with the impact of the human-made world on the natural world and ourselves. Never before have there been greater improvement possibilities from the proper application of the concepts and principles of systems engineering.

Systems engineering and analysis, when coupled with new and emerging technologies, reveals unexpected opportunities for bringing new and improved systems and products into being that will be more competitive in the world economy. These technologies are acting to expand physically realizable design options and to enhance capabilities for developing more cost-effective entities for human use. This chapter introduces a technologically based process encompassing an extension of engineering through all phases of the system life cycle—design and development, production or construction, utilization and support, and phaseout and disposal.

2.1 ENGINEERING FOR PRODUCT COMPETITIVENESS

Product competitiveness is desired by both commercial and public-sector producers worldwide. It is the product, or consumer good, that must meet customer expectations. Accordingly, the systems engineering challenge is to bring products and systems into being that meet these expectations cost-effectively.

Because of intensifying international competition, producers are seeking ways to gain a sustainable competitive advantage in the marketplace. Acquisitions, mergers, and extensive advertising seem unable to create the intrinsic wealth and good will so essential for the long-term health of the organization. Economic competitiveness is essential. Engineering with an emphasis on economic competitiveness must become coequal with concerns for advertising, finance, production, and the like.

Available resources are dwindling. The industrial base is expanding, and international competition is increasing at a rapid pace. Many organizations are downsizing, looking to improve their operations, and seeking international partners. Competition has also reduced the number of suppliers and subcontractors able to respond. This is occurring at a time when the number of qualified team members required for complex system development is increasing. Also, needed new systems are being deferred in favor of extending the life of existing systems.

Engineering has always been concerned with the economical use of limited resources for the benefit of people.[1] The purpose of engineering activities of design and analysis is to determine how physical factors may be altered to create the most utility for the least cost, in terms of product cost, product service cost, and social cost. Viewed in this context, engineering must be practiced in an expanded way, with engineering of the system placed ahead of concern for components thereof. Specifically, emphasis must be placed on the following:

1. Improving methods for defining product and system requirements as they relate to true customer needs. This should be done early in the design phase, along with a determination of performance, effectiveness, and essential system characteristics.
2. Addressing the total system with all of its elements from a life-cycle perspective, and from the product or prime equipment to its elements of support. This means defining the system in functional terms before identifying hardware, software, people, facilities, information, or combinations thereof.
3. Considering the overall system hierarchy and interactions between various levels in the hierarchy. This includes intra-relationships among system elements and interrelationships between higher and lower levels within the system.
4. Organizing and integrating the necessary engineering and related disciplines into the main systems engineering effort in a timely concurrent manner.
5. Establishing a disciplined approach with appropriate review, evaluation, and feedback provisions to insure orderly and efficient progress from the initial identification of need through phaseout and disposal.

Products, systems, and structures are designed and developed in accordance with processes that are not as well understood as they might be. The cost-effectiveness of the resulting technical entities can be enhanced by giving more attention to what they are to do, before addressing what they are composed of. Simply stated, form should follow function.

[1] According to the definition of engineering adapted by the Accreditation Board for Engineering and Technology (ABET), "Engineering is the profession in which a knowledge of the mathematical and natural sciences gained by study, experience, and practice is applied with judgment to develop ways to utilize economically, the materials and forces of nature for the benefit of mankind."

All other factors being equal, people will meet their needs by purchasing goods and services that offer the highest value-cost ratio, subjectively evaluated. This ratio can be increased by giving more attention to the resource-constrained world within which engineering is practiced. To ensure economic competitiveness regarding the end item, engineering must become more closely associated with economics and economic feasibility. This is best accomplished through a life-cycle approach to engineering.

2.2 SYSTEM LIFE-CYCLE ENGINEERING

In general, classical engineering has focused mainly on product performance as the main objective rather than on development of the overall system of which the product is a part.[2] Experience in recent decades indicates that a properly functioning system that is competitive cannot be achieved through efforts applied largely after it comes into being. Accordingly, it is essential that engineers be sensitive to utilization outcomes during the early stages of system design and development, and that they assume the responsibility for *life-cycle engineering* that has been largely neglected in the past.

The System Life Cycle

Fundamental to the application of systems engineering is an understanding of the system life-cycle process illustrated for the product in Figure 2.1. The life cycle begins with the identification of a need and extends through conceptual and preliminary design, detail design and development, production and/or construction, product use, phaseout, and disposal. The program phases are classified as *acquisition* and *utilization* to recognize producer and customer activities.[3]

A detailed presentation of the elaborate technological activities and interactions that must be integrated over the system life-cycle process is given in Figure 2.2. The

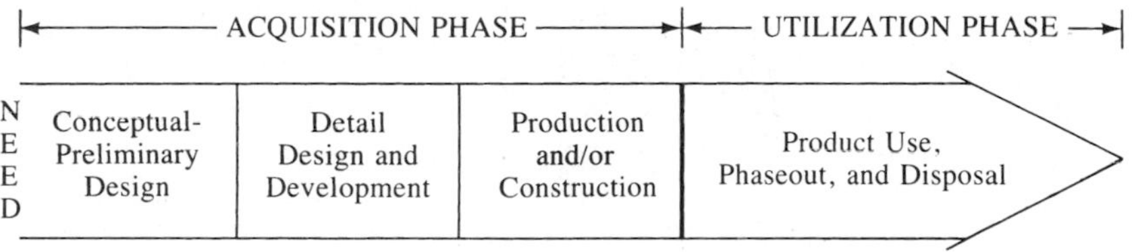

Figure 2.1 The product life cycle.

[2] A *product*, for example, may be a television set, an automobile, or an appliance. The product cannot come into being without a manufacturing capability, a support capability, and so on. Accordingly, in dealing with systems, one must not only consider the product, but its manufacturing process, utilization, maintenance and support, and retirement and disposal.

[3] This classification represents a generic approach. Sometimes the *acquisition* process may involve both the customer (or procurring agency) and the producer (or contractor), whereas utilization may include a combination of contractor and consumer (or ultimate user) activities. In some instances, the customer may not be the ultimate consumer (as is the case in the defense sector) but must represent the consumer's interests in the acquisition process. Additionally, the product phaseout and material disposal may involve contractor support. The situation will vary depending on the type and nature of the system.

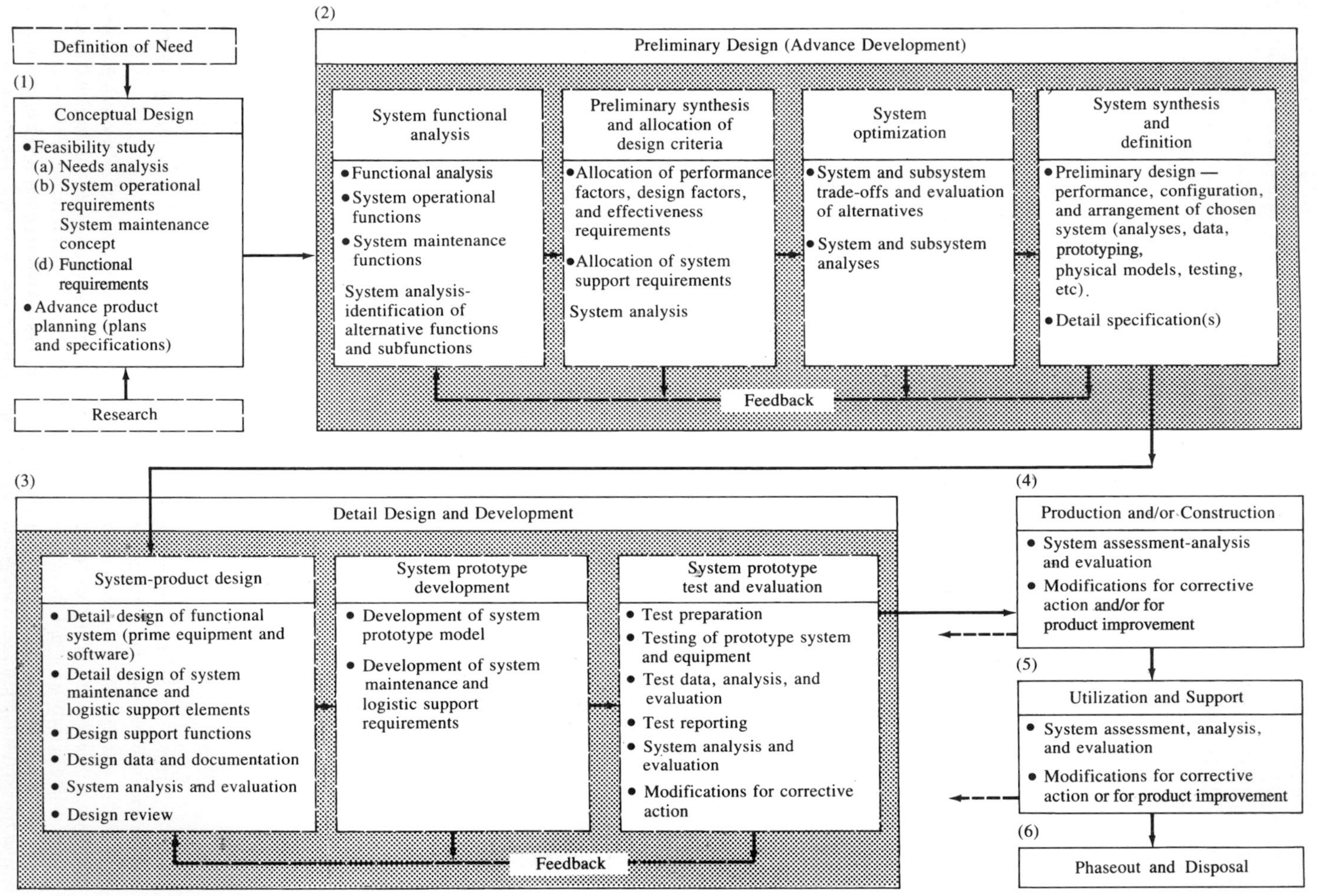
Definition of Need
(1)
Conceptual Design
• Feasibility study
(a) Needs analysis
(b) System operational requirements
System maintenance concept
(d) Functional requirements
• Advance product planning (plans and specifications)
Research
(2)
Preliminary Design (Advance Development)
System functional analysis
• Functional analysis
• System operational functions
• System maintenance functions
System analysis-identification of alternative functions and subfunctions
Preliminary synthesis and allocation of design criteria
• Allocation of performance factors, design factors, and effectiveness requirements
• Allocation of system support requirements
System analysis
System optimization
• System and subsystem trade-offs and evaluation of alternatives
• System and subsystem analyses
System synthesis and definition
• Preliminary design — performance, configuration, and arrangement of chosen system (analyses, data, prototyping, physical models, testing, etc).
• Detail specification(s)
Feedback
(3)
Detail Design and Development
System-product design
• Detail design of functional system (prime equipment and software)
• Detail design of system maintenance and logistic support elements
• Design support functions
• Design data and documentation
• System analysis and evaluation
• Design review
System prototype development
• Development of system prototype model
• Development of system maintenance and logistic support requirements
System prototype test and evaluation
• Test preparation
• Testing of prototype system and equipment
• Test data, analysis, and evaluation
• Test reporting
• System analysis and evaluation
• Modifications for corrective action
Feedback
(4)
Production and/or Construction
• System assessment-analysis and evaluation
• Modifications for corrective action and/or for product improvement
(5)
Utilization and Support
• System assessment, analysis, and evaluation
• Modifications for corrective action or for product improvement
(6)
Phaseout and Disposal

Figure 2.2 The system life-cycle process.

progression is iterative from left to right, and not serial in nature as might be implied. Although the level of activity and detail may vary, the life-cycle functions described and illustrated are generic. They are applicable whenever a new need or changed requirement is identified, with the process being common to large as well as small-scale systems. It is essential that this process be implemented completely, not only in the acquisition of new systems but also in the re-engineering of existing or legacy systems.

Designing for the Life Cycle

Design within the system life-cycle context is different from design in the ordinary sense. Life-cycle focused design is simultaneously responsive to customer needs (i.e., to requirements expressed in functional terms) and to life-cycle outcomes. Design should not only transform a need into a definitive product and system configuration, but should ensure the design's compatibility with related physical and functional requirements. Further, it should consider operational outcomes expressed as producibility, reliability, maintainability, supportability, serviceability, disposability, and others, as well as performance, effectiveness, and affordability.

Major technical functions performed during the acquisition and utilization phases of the life cycle are summarized in Figure 2.2. When a new customer need is identified, a planning function is initiated followed by conceptual, preliminary, and detail design activities. Producing and/or constructing the system is the function that completes the acquisition phase. System operation and support functions constitute the utilization phase of the life cycle. Phaseout and disposal are important final functions to be considered in design for the life cycle.

System life-cycle engineering goes beyond the product life cycle. It must simultaneously embrace the life cycle of the manufacturing process as well as the life cycle of the product support and service capability. Actually, there are three concurrent life cycles progressing in parallel as is illustrated in Figure 2.3. This is the basis for *concurrent engineering*.[4]

The need for the product comes into focus first. This recognition initiates conceptual design to meet the need. Then, during conceptual design of the product, consideration should simultaneously be given to its production. This gives rise to a parallel life cycle for bringing a manufacturing capability into being. It requires many production-related activities to become ready for manufacturing.

Also shown in Figure 2.3 is another life cycle of great importance which is often neglected until product and production design is completed. This is the life cycle for the maintenance and logistic support activities needed to service the product during use and to support the manufacturing capability during its duty cycle. Logistic and maintenance requirements planning should begin during product conceptual design in a coordinated manner.

The communication and coordination needed to develop the product, the manufacturing process, and the support capability in a coordinated manner is not easy to

[4] *Concurrent Engineering* is defined as a systematic approach to creating a product design that simultaneously considers all elements of the product life cycle, from conception through disposal, to include consideration of manufacturing processes, transportation processes, maintenance processes, and so on.

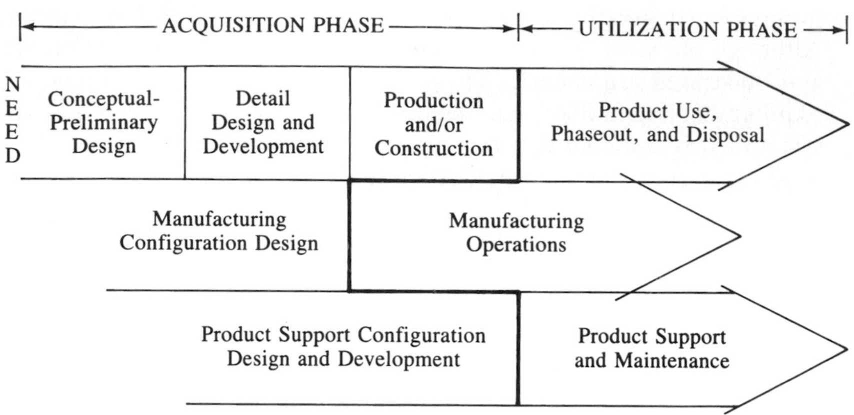

Figure 2.3 Product, manufacturing, and support life cycles.

achieve. Progress in this area is facilitated by new technologies that make more timely acquisition and the use of design information possible. Computer-Aided Design (CAD) and Computer-Aided Manufacturing (CAM) technology are only two of these. Others are being developed which can integrate relevant design and development activities over the entire life cycle of the system.

Concern for the entire *life cycle* is strong within the Department of Defense (DOD). This may be attributed to the fact that acquired defense systems are owned, operated, and maintained by the DOD. This is unlike the situation most often encountered in the private sector, where the consumer or user is usually not the producer. Those private firms serving as defense contractors are obliged to design and develop in accordance with DOD directives, specifications, and standards. Because the DOD is the customer and also the user of the resulting system, considerable intervention occurs during the acquisition phase.[5]

Many firms that produce for private sector markets have chosen to design with the life cycle in mind. For example, design for energy efficiency is now common in appliances like water heaters and air conditioners. Fuel efficiency is a required design characteristic for automobiles. Some truck manufacturers promise that life-cycle maintenance requirements will be within stated limits. These developments are commendable, but they do not go far enough. When the producer is not the consumer, it is less likely that potential operational problems will be addressed during development. Undesirable outcomes too often end up as problems of the user of the product instead of the producer.

[5] This intervention is guided by DODR 5000.2, "Mandatory Procedures for Major Defense Acquisition Programs (MDAPs) and Major Automated Information System (MAIS) Acquisition Programs" Washington, D.C.: (Department of Defense, 1996). This is supported by a host of directives, specifications, and standards.

2.3 SYSTEMS ENGINEERING DEFINITIONS

There is no commonly accepted definition of *systems engineering* in the literature. Both the definition and approach seem to be based on the background and experience of the individual or organization.[6] Variety is evident from three published definitions given subsequently.

1. "The application of scientific and engineering efforts to (a) transform an operational need into a description of system performance parameters and a system configuration through the use of an iterative process of definition, synthesis, analysis, design, test, and evaluation; (b) integrate related technical parameters and ensure compatibility of all physical, functional, and program interfaces in a manner that optimizes the total system definition and design; and (c) integrate reliability, maintainability, safety, survivability, human engineering, and other such factors into the total engineering effort to meet cost, schedule, supportability, and technical performance objectives."[7]
2. "An interdisciplinary approach encompassing the entire technical effort to evolve and verify an integrated and life-cycle balanced set of system, people, product, and process solutions that satisfy customer needs. Systems engineering encompasses (a) the technical efforts related to the development, manufacturing, verification, deployment, operations, support, disposal of, and user training for, system products and processes; (b) the definition and management of the system configuration; (b) the translation of the system definition into work breakdown structures; and (d) development of information for management decision making."[8]
3. "An interdisciplinary collaborative approach to derive, evolve, and verify a life-cycle balanced system solution which satisfies customer expectations and meets public acceptability."[9]

Although the definitions vary, there are some common threads. Basically, systems engineering is good engineering with special areas of emphasis. Some of these are the following:

1. A *top-down* approach that views the system as a whole. Although engineering activities in the past have adequately covered the design of various system components (representing a bottom-up approach), the necessary overview and understanding of how these components effectively fit together is frequently overlooked.

[6] An overwiew of systems engineering technology, approach, challenges, and related issues is presented in the Inaugural Issue of *Systems Engineering*. The Journal of the International Council on Systems Engineering (INCOSE), Volume 1, Number 1, July/September, 1994.

[7] DSMC, *Systems Engineering Management Guide*, Defense Systems Management College, Superintendent of Documents, U.S. Government Printing Office, Washington, D.C. 1990, pp. 12.

[8] EIA/IS 632, "Systems Engineering," Electronic Industries Association, 2001 Pennsylvania Avenue N.W., Washington, D.C., 1994, pp. 43.

[9] IEEE P1220, "Standard for Application and Management of the Systems Engineering Process," Institute of Electrical and Electronics Engineers, 345 East 47th Street, New York, N.Y., 1994, pp. 11.

2. A *life-cycle* orientation that addresses all phases to include system design and development, production and/or construction, distribution, operation, maintenance and support, retirement, phaseout, and disposal. Emphasis in the past has been placed primarily on design and system acquisition activities, with little (if any) consideration given to their impact on production, operations, maintenance, support, and disposal. If one is to adequately identify risks associated with the up-front decision-making process, then such decisions must be based on life-cycle considerations.
3. A better and more complete effort is required regarding the initial *definition of system requirements*, relating these requirements to specific design criteria and the follow-on analysis effort to ensure the effectiveness of early decision making in the design process. The true system requirements need to be well defined and specified, and the traceability of these requirements from the system level downward needs to be visible. In the past, the early "front-end" analysis as applied to many new systems has been minimal. The lack of defining an early "baseline" has resulted in greater individual design efforts downstream. Many of these are not well integrated with other design activities, causing costly modifications later on.
4. An *interdisciplinary* or team approach throughout the system design and development process to ensure that all design objectives are addressed in an effective and efficient manner. This requires a complete understanding of many different design disciplines and their interrelationships, together with the methods, techniques, and tools that can be applied to facilitate implementation of the system engineering process.

Systems engineering is not a traditional engineering discipline in the same sense as civil engineering, electrical engineering, industrial engineering, mechanical engineering, reliability engineering, or any of the other engineering specialties. It should not be organized in a similar manner, nor does the implementation of systems engineering (or its methods) require extensive resources. However, a well-planned and highly disciplined approach must be followed. The systems engineering process involves the use of appropriate technologies and management principles in a synergetic manner. Its application requires synthesis and a focus on process, along with a new "thought process" that should lead to a change in "culture."

2.4 THE SYSTEMS ENGINEERING PROCESS

Although there is general agreement regarding the principles and objectives of systems engineering, its actual implementation will vary from one program to the next. The process approach and steps used will depend on the backgrounds and experiences of the individuals involved. To establish a common frame of reference for improving communication and understanding, it is important that a "baseline" be defined that describes the systems engineering process, along with the essential life-cycle phases and steps within that process. Augmenting this common frame of reference are top-down and bottom-up approaches. Also, there are other process models that have attracted various degrees of attention. Each of these topics are presented in this section.

Life-Cycle Process Phases and Steps

Figure 2.4 illustrates the major life-cycle process phases and selected milestones for a typical system. This is the "model" that will serve as a frame of reference for material presented in subsequent chapters. Included are the basic steps in the systems engineering process (i.e., requirements analysis, functional analysis and allocation, synthesis, trade-off studies, evaluation, and so on). A newly identified need, as well as an evolving need, reveals a new system requirement. The basic phases of conceptual design and onward through system retirement and phaseout are then applicable, as described in the paragraphs that follow.

Program phases described in Figure 2.4 are not intended to convey specific periods or levels of funding. Individual program requirements will vary from one application to the next. The figure reflects an overall *process* that needs to be followed in system acquisition. Regardless of the type, size and complexity of the system, there is a conceptual design requirement (i.e., to include requirements analysis), a preliminary design requirement, and so on. Also, to ensure maximum effectiveness the concepts presented in Figure 2.4 must be properly "tailored" to the particular system being addressed.[10]

Figure 2.4 (Blocks 0.1 to 0.8) shows the basic steps in the system engineering process to be iterative in nature, providing a top-down definition of the system, and then proceeding down to the subsystem level (and below as necessary). On completion of Block 0.2, the system is defined in *functional* terms (having identified the "whats" from a requirements perspective). These "whats" are translated into an applicable set of "hows" through the iterative activities of functional partitioning, requirements allocation, trade-off studies, synthesis, and evaluation. This is where the initial requirements for the system configuration (or system architecture) are defined.

The functional definition of the system serves as the baseline for the identification of resource requirements (i.e., hardware, software, people, facilities, data, elements of support, or a combination thereof). Depending on the degree of definition required, these steps may often be accomplished concurrently (i.e., functional requirements are described for certain complex elements of the system while, at the same time, the requirements for hardware, software, and the like, are being identified for other elements of the system).

As one proceeds from the top-down in the early phases of system design and development, there is a follow-on "bottom-up" procedure. During the latter phases of preliminary design (and in the detail design and development phase) components are combined, assembled, and integrated into the specified system configuration This, in turn, leads to the iterative process of system evaluation. Inherent within the systems engineering process must be a provision for continuous feedback and corrective action. The system engineering process is continuous, iterative, and must incorporate feedback actions to ensure convergence as is illustrated in Figure 2.5.

[10] B. S. Blanchard, "The System Engineering Process: An Application for the Identification of Resource Requirements," *Systems Engineering*, Journal of the International Council on Systems Engineering (INCOSE), Volume 1, Number 1, July/September 1994.

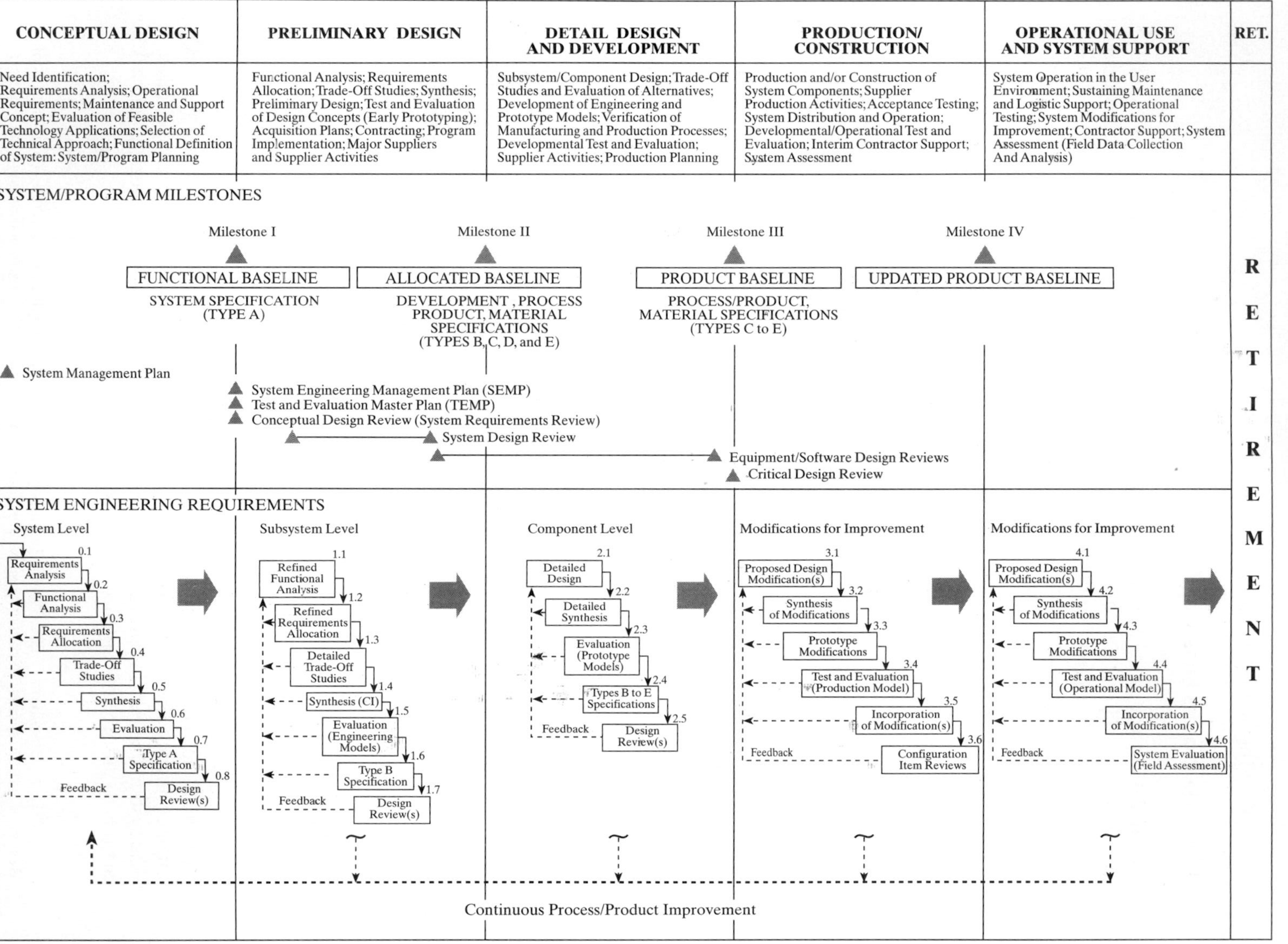

Figure 2.4 The system acquisition process ("baseline").

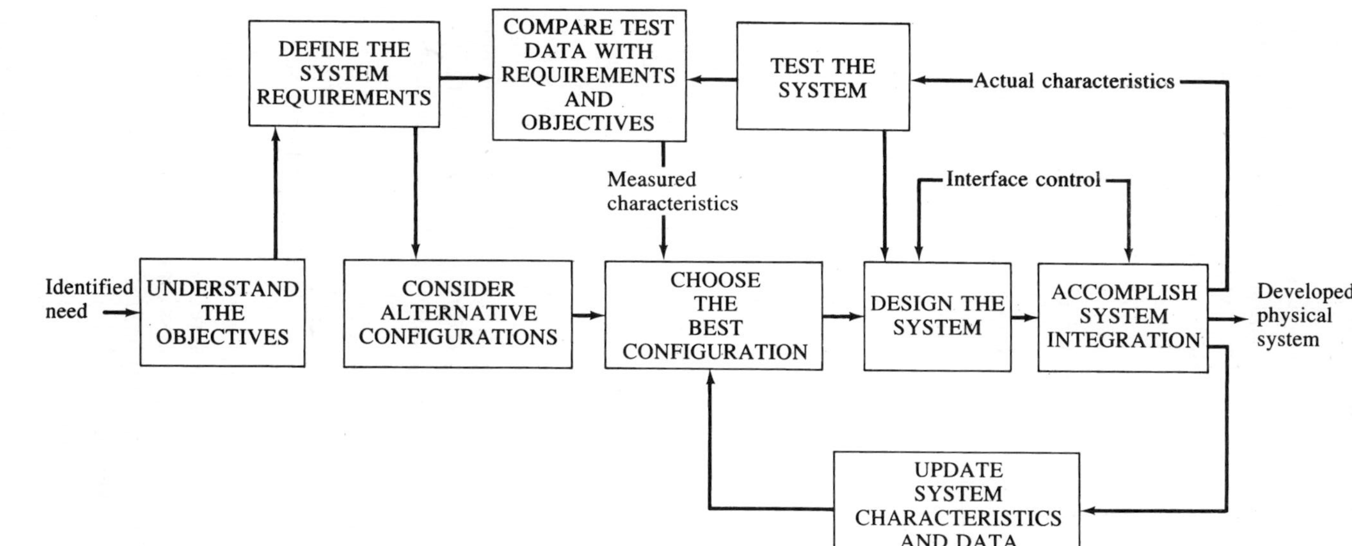

Figure 2.5 Feedback in the systems engineering process.

Top-Down and Bottom-Up Approaches

Traditional engineering design methods are based on a bottom-up approach. Starting with a set of known elements, design engineers create the product or system by synthesizing a combination of system elements. However, it is unlikely that the functional need will be met on the first attempt unless the system is simple. After determining the product's performance and deviation from what is required, the elements and their combination are altered and the performance determined again. This bottom-up process is iterative, with the number of iterations (and design efficiency) determined by the experience and creativity of the designer, as well as by the complexity of the product or system.

A more directed methodology is evoked by the systems engineering process, which is based on a top-down approach to design. Starting with requirements about the external behavior of any part of the system (expressed in terms of the function provided by that part), that behavior is analyzed to identify its functional characteristics. These functional behaviors are then described in more detail and made specific through refinement. Finally, the appropriateness of this choice of functional components is verified by synthesizing the original system part.

There are two main characteristics of the top-down process. First, the process is applicable to any part of the system. Starting with the system as a whole, repeated application of this process will result in partitioning of the system into smaller and smaller elements. Second, the process is self-consistent. External properties of the whole system, as described by the inputs and outputs and relations between parts, must be reproduced by the external properties of the set of interacting elements.

The beginning of the top-down approach recognizes that general functions are involved in transforming inputs into outputs. A designer must abstract from the particular case to the underlying generic case, and represent the generic case by several interacting functional elements. The use of functional elements is the essential difference in systems engineering methodology compared with systems integration. A particular functional element is applicable to a whole class of systems. Consequently, only a few such elements are needed to represent many real systems. Functional elements allow one to engage in system design before physical manifestations have been defined. This contrasts with designing a system by using the bottom-up methodology, where one starts out with a defined set of real elements (components) and synthesizes a system out of members from the set.

Two main differences occur in the bottom-up and top-down approaches. In the case of bottom-up design, physical realizability in terms of known elements is assured, whereas the top-down design process ends with the system elements as functional entities. Their physical realizability is not guaranteed. In the top-down approach, the requirements are always satisfied through every step of the design process because it is an inherent part of the methodology, whereas in the bottom-up approach the methodology provides no assurance that this will occur.

Systems engineering is not likely to replace bottom-up design completely. Every end product incorporates physical objects working together to meet the need. At some point in the design process there must be a transition from the functional (or abstract) to the physical. Accordingly, most projects will employ both methodologies: first systems engineering to reduce the complexity by partitioning the system into its elements and then bottom-up design to realize the elements for that system.

Other Systems Engineering Process Models

The systems engineering process, and the steps illustrated in Figure 2.4, is developed and described in detail in Part II of this text (Chapter 3 to 6). Again, the objective is to describe a *process* (as a frame of reference) that must be "tailored" to be specific program need. The illustration presented in Figure 2.4 is not intended to represent any particular model, such as the "waterfall" model, the "spiral" model, the "Vee" model, or equivalent. These are other well-known process models are illustrated and briefly described in Figure 2.6, with references to the literature found in Appendix F.

2.5 SYSTEM DESIGN EVALUATION

The systems engineering process is suggested as the best approach for bringing products, systems, and structures into being that will be cost-effective and competitive. An essential technical activity within the process is that of system design evaluation. But it should not be pursued in isolation. System design evaluation should be inherent within the systems engineering process and invoked regularly as the system design evolves. It is the assurance of continuous design improvement.[11]

To design is to synthesize (i e., to put known elements together into a new combination). Thus, a design alternative is a projection of what could be. Evaluation is a prediction of how good the design alternative might be if it is chosen for implementation. System design evaluation is preceded by systems analysis, which, in turn, is preceded by synthesis. These relationships are shown conceptually in Figure 2.7.

Development of Design Criteria[12]

An initial step in a design evaluation effort is to establish the *baseline* against which a given design configuration is to be evaluated. This baseline is described through the iterative process of requirements analysis (i.e., the identification of need, the conduct of feasibility analysis, definition of system operational requirements and the maintenance concept). The functions that the system must perform to satisfy a specific consumer need should be described along with the expectations in terms of time, frequency, effectiveness, costs, and other such factors. These functional requirements,

[11] W. J. Fabrycky, "Modeling and Indirect Experimentation in System Design Evaluation," *Systems Engineering*, Journal of the International Council on Systems Engineering (INCOSE), Volume 2, Number 1, July/September 1994.

[12] Design *criteria* constitute a set of "design-to" requirements, which can be expressed in both qualitative and quantitative terms. These requirements represent the bounds within which the designer must "operate" when engaged in the iterative process of synthesis, analysis, and evaluation. Design criteria may be established for each level in the system hierarchical structure.

(a)

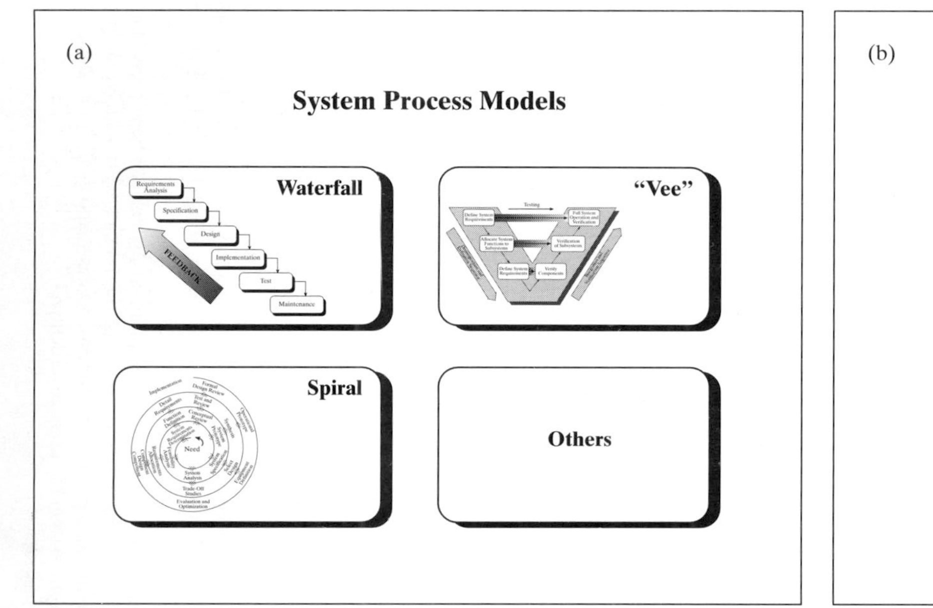

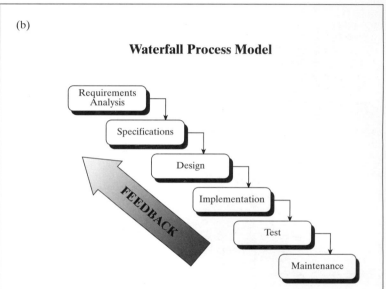

- It is observed that the preference expressed by individuals and groups for one of the system models is subjective.
- A study of the literature and current practice is needed to identify which model fits a specific situation best. Refer to Appendix F.

(b)

Waterfall Process Model

Requirements Analysis → Specifications → Design → Implementation → Test → Maintenance

FEEDBACK

- The waterfall model, introduced by Royce in 1970, initially was used for software development. This model usually consists of five to seven series of steps or phases for systems engineering or software development. Boehm expanded this into an eight-step series of activities in 1981.
- A similar model splits the hardware and software into two distinct efforts. Ideally, each phase is carried out to completion in sequence until the product is delivered. However, this rarely is the case because when deficiencies are found, phases must be repeated until the product is correct.

Figure 2.6 System process models (examples—sheet 1).

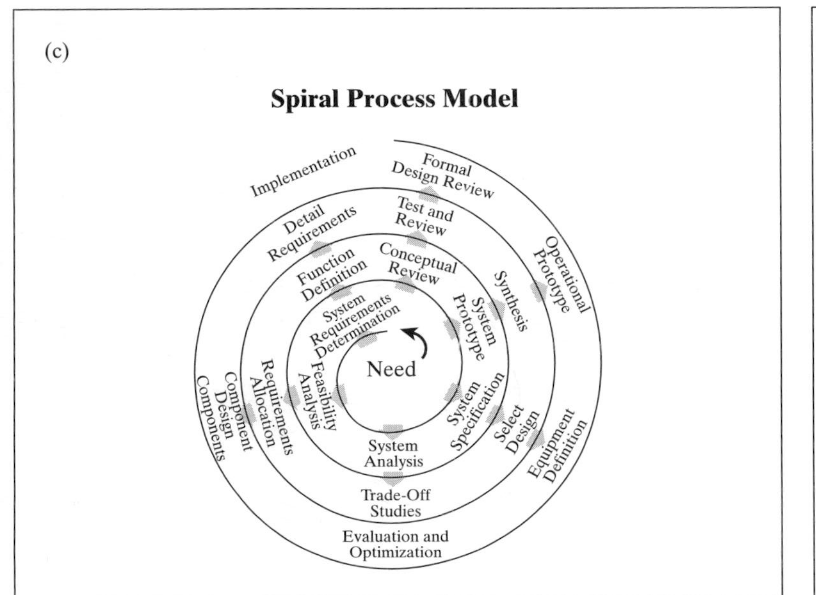

- The spiral process model of the development life cycle (developed by Boehm in 1986 using Hall's work in systems engineering from 1969) is intended to introduce a risk-driven approach for the development of products or systems.
- This model is an adaption of the waterfall model, which does not mandate the use of prototypes. The spiral model incorporates features from other models, such as feedback, etc.
- Application of the spiral model is iterative and proceeds through the several phases each time that a different type of prototype is developed. It allows for an evaluation of risk before proceeding to a subsequent phase.

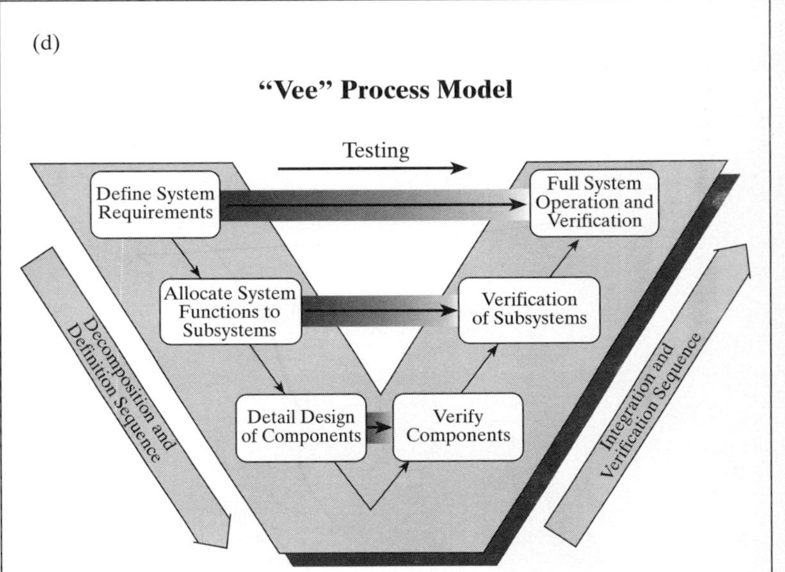

- Forsberg and Mooz describe what they call "the technical aspect of the project cycle" by the "Vee" process model. This model starts with user needs on the upper left and ends with a user-validated systems on the upper right.
- On the left side, decomposition and definition activities resolve the system architecture, creating details of the design. Integration and verification flows up and to the right as successively higher levels of subsystems are verified, culminating at the system level.
- Verification and validation progress from the component level to the validation of the operational system. At each level of testing, the originating specifications and requirements documents are consulted to ensure that component/subsystems/system meet all specifications.

Figure 2.6 System process models (examples—sheet 2).

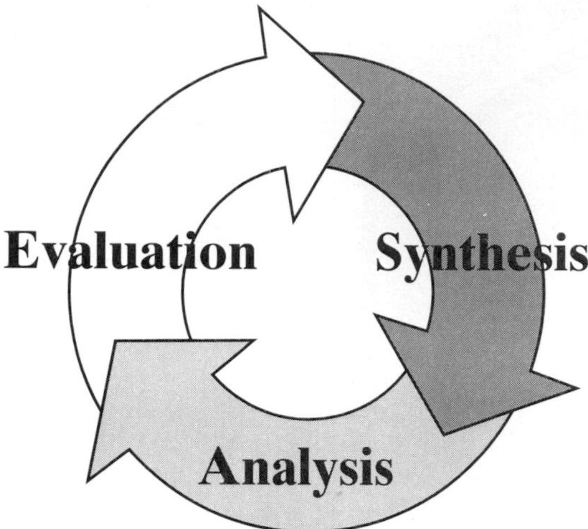

Figure 2.7 System design, analysis, and evaluation.

starting at the *system level*, will ultimately lead to determining the characteristics that should be incorporated within the design of the system and its elements.

Referring to Figure 2.8, the requirements for the system as an entity are established by describing the functions that must be performed. Both operational functions (i.e., those required to accomplish a specified mission scenario, or series of missions) and maintenance and support functions (i.e., those required to ensure that the system is operational when required) must be described at the top level. As part of this process, it is necessary to establish some system "metrics" related to performance, effectiveness, cost, and other such quantitative factors as required to meet customer expectations. For instance, what functions must the system perform, where are these to be accomplished, and at what frequency, with what degree of reliability, and at what cost? Some of these factors may be considered to be more important than others by the customer, which will, in turn, influence the design process in placing different levels of emphasis on the selection of design criteria. The result is the identification and prioritization of *technical performance measures* (TPMs) for the system overall. The output from the requirements analysis process will establish an initial baseline for system evaluation (refer to Figure 2.4, Block 0.1).

With the applicable TPMs identified at the system level, the next step is to determine the specific characteristics that must be incorporated and be made inherent within the design itself. *Design-dependent parameters* (DDPs) are identified, analysis and trade-off studies are conducted by considering various possible design alternatives, design synthesis is performed, and the iterative process of design evaluation takes place. This process is accomplished at the system level, then the subsystem level, and on down to the level necessary to ensure that the ultimate system configuration will fully meet the expectations of the consumer. This process of design synthesis, analysis, and evaluation (as conveyed in Figure 2.7) is inherent within the overall systems engineering process (illustrated in Figure 2.4).

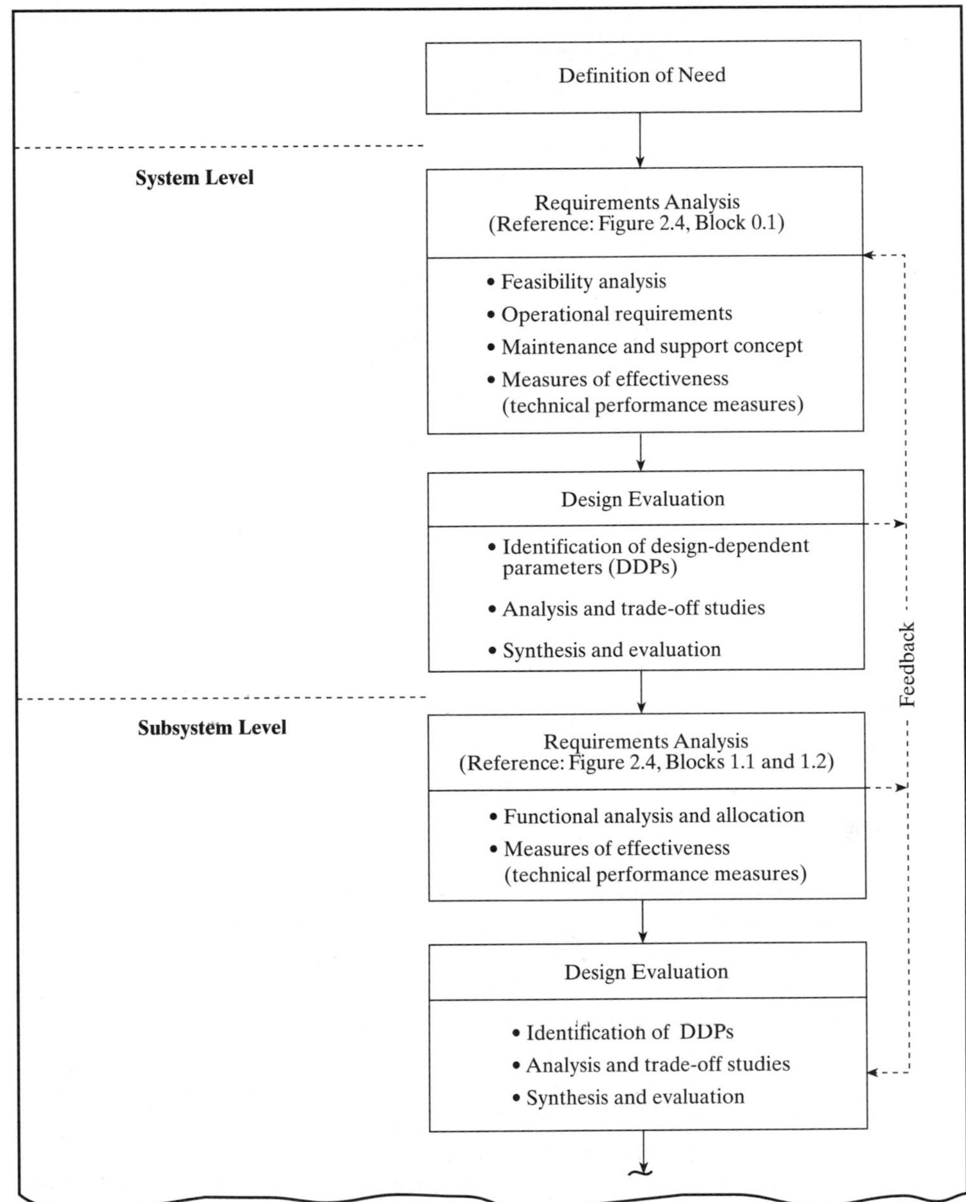

Figure 2.8 Decomposition of design criteria.

Design Evaluation and Multiple Criteria

In Figure 2.8, the prioritized TPMs at the top level reflect the overall performance characteristics of the system as it accomplishes its mission objectives in response to the needs of the customer. There may be numerous factors, such as system size and weight, range and accuracy, speed of performance, capacity, operational availability, reliability and maintainability, supportability, cost, and so on. These *measures of effectiveness* (MOEs) must be specified in terms of some level of importance, as determined by the customer and the criticality of the functions to be performed. For instance, there may be certain mission scenarios where system availability is critical, with reliability being less important as long as there are maintainability considerations built into the system that will allow for ease of repair. Conversely, for missions where the accomplishment of maintenance is not feasible, reliability becomes more important. Thus, the nature and criticality of the mission(s) to be accomplished will lead to the identification of specific requirements and the relative levels of importance of the applicable TPMs.[13]

Given the requirements at the top level, it may be appropriate to develop a "design-objectives" tree similar to that presented in Figure 2.9. First-, second-, and third-order (and lower-level) considerations are noted. Based on the established MOEs for the system, a top-down breakout of requirements will lead to the identification of characteristics that should be included and inherent within the design (i.e., the identification of *design-dependent parameters* (DDPs). For instance, a first-order consideration may be *system value*, which, in turn, may be broken down into *economic* factors and *technical* factors.

Technical factors may be expressed in terms of *system effectiveness*, which is a function of performance, operational availability, dependability, and so on. This leads to the consideration of such features as speed of performance, reliability and maintainability, size and weight, and flexibility. Assuming that maintainability represents a high priority in design, then such features as accessibility, diagnostic aids, packaging, mounting, and interchangeability should be stressed in the design. Thus, the *criteria* for design and the associated DDPs may be established early during conceptual design and then carried through the entire design cycle. The DDP's establish the extent and scope of the design space within which *trade-off* decisions may be made. During the process of making trade-off decisions, requirements must be related to the appropriate hierarchical level in the system structure (i.e., system, subsystem, and configuration item) as in Figure 2.8.

Generating and Evaluating Design Alternatives

As system development evolves, the design evaluation activity is iterative and continues through the steps illustrated in Figure 2.4 (i.e., from system-level design, to subsystem design, and down to the component level). For the purposes of illustration, the morphology (or "structure") shown in Figure 2.10 is presented as a frame of reference. Inherent within this structure are the elements of synthesis, analysis, and evaluation.

Referring to the figure, the top block (Block 0) reflects the set of requirements, or bounds, within which different design alternatives are synthesized. Referring to Figure 2.8, the results of the needs analysis, feasibility analysis, operational requirements, maintenance and support concept, and the applicable technical performance measures

[13] Establishing the appropriate levels of importance and the prioritization of TPMs is critical as it may be necessary to make compromises and "trade off" certain features during the design evaluation process (i.e., eliminate certain desired features in favor of others with a greater degree of importance).

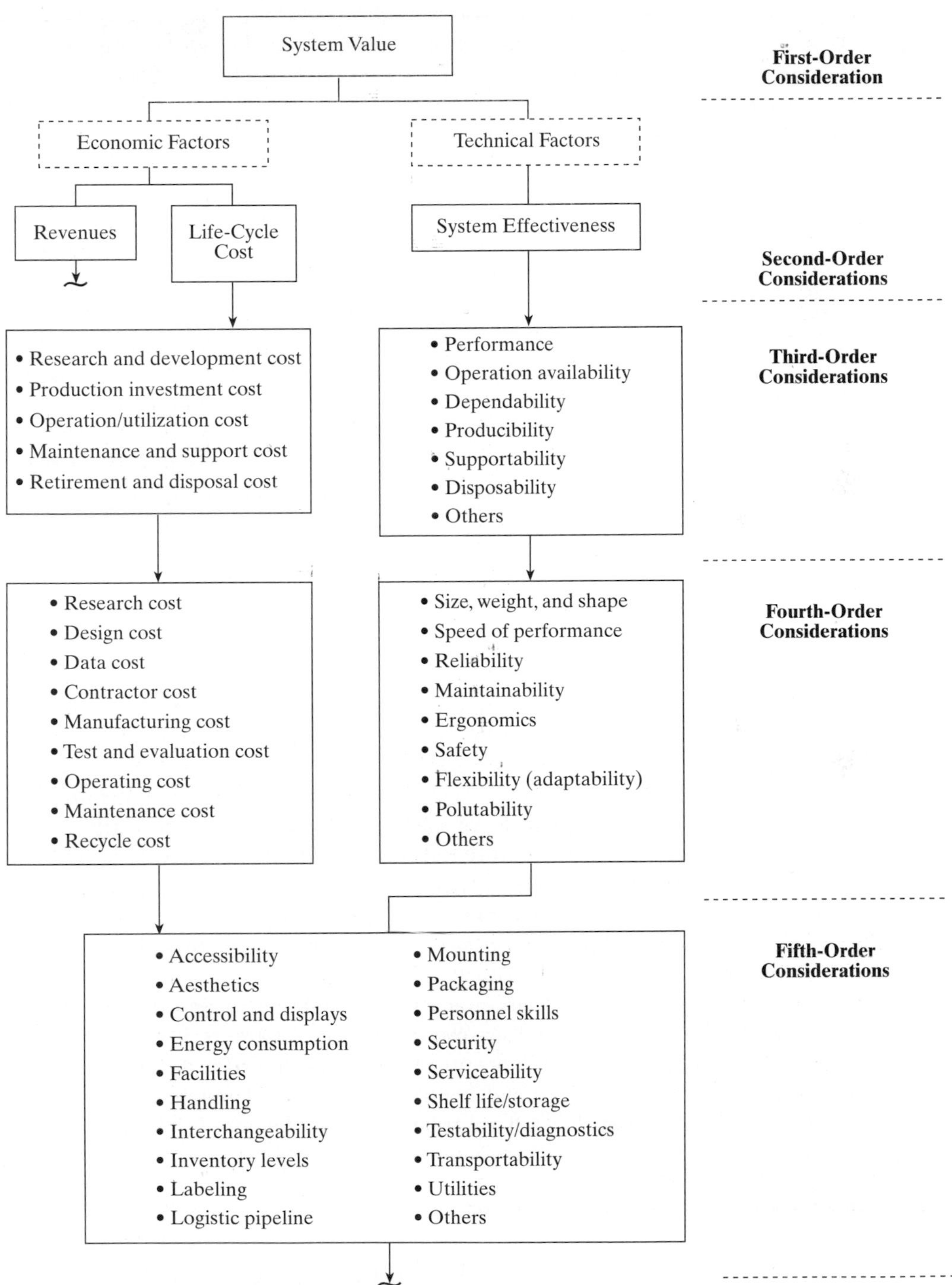

Figure 2.9 Design consideration hierarchy.

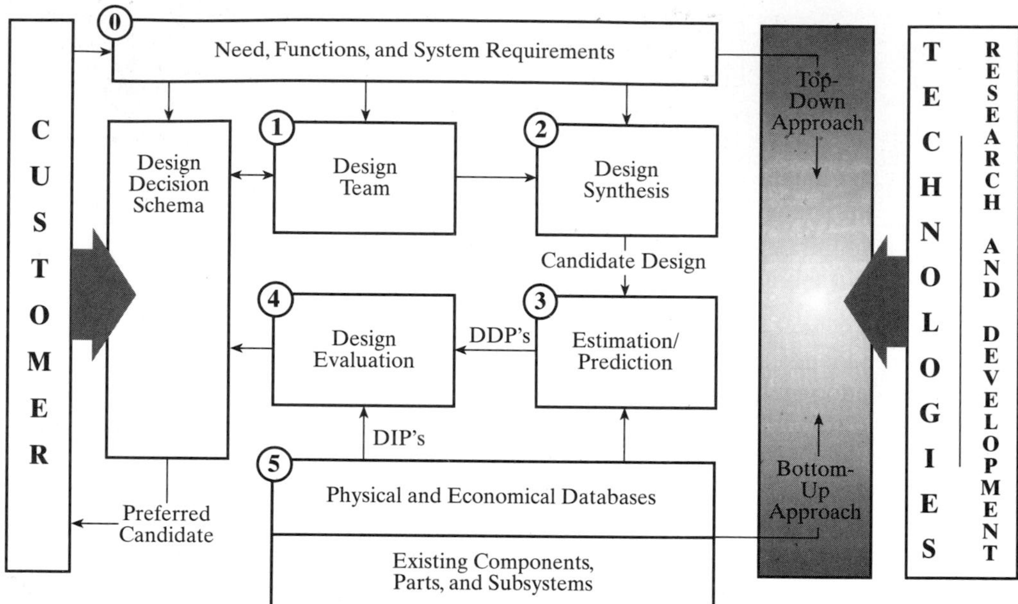

Figure 2.10 Morphology for design synthesis, analysis, and evaluation.

(TPMs) are specified, and the various design alternatives are evaluated in terms of these requirements. As design progresses, the individual designer or design team (Block 1), through creative activity, identifes the various feasible alternative candidates that may be evaluated in arriving at the best design solution. Through the use of various computer-aided design (CAD) or computer-aided engineering (CAE) tools, design synthesis can be accomplished (Block 2). This leads to the generation, through prediction and/or some form of estimation, of the appropriate metrics associated with each of the design alternatives being considered for evaluation (Block 3). The TPMs (Figure 2.8) and the appropriate DDPs (Figure 2.9) must be estimated or predicted and related to the applicable design alternative. The evaluation effort is then accomplished (Block 4), using DDP values and design independent parameters (DIPs) and other resource information as required (Block 5).

The subject of design evaluation and its applications, as an inherent part of the systems engineering process, are discussed throughout Chapters 3 to 6 (Part II) as one proceeds through the phases of conceptual design, preliminary system design, detail design and development, and test and evaluation. Analytical and modeling approaches for systems analysis and evaluation are treated in Chapter 7 to 11 (Part III).

2.6 IMPLEMENTING SYSTEMS ENGINEERING

Current trends indicate that, in general, the complexity of systems is increasing, and many of those systems in use today are not meeting the needs of the customer in terms of performance, effectiveness, and overall cost. New technologies are being introduced on a continuing basis, while the life cycles for many systems are being extended. The length of time that it takes to develop and acquire a new system needs to be reduced, the costs of modifiying existing systems are getting higher, and available resources are dwindling. At the same time, there is a greater degree on international cooperation, and competition is increasing worldwide. These factors, when combined, create a need to address "systems" from a different perspective.

When evaluating past experiences regarding the development of systems, most of the problems noted have been the direct result of not applying a *disciplined* top-down "systems approach" in meeting the desired objectives. The overall requirements for the system were not defined well from the beginning; the perspective in terms of meeting a need has been relatively "short term" in nature; and, in many instances, the approach followed has been to "deliver it now and fix it later," using strictly a bottom-up approach to design. In essence, the systems design and development process has suffered from the lack of good early planning and the subsequent definition and allocation of requirements in a complete and methodical manner. Yet, it is at this early stage in the life cycle when decisions are made that can have a large impact on the overall effectiveness and cost of the system.

Referring to Figure 2.11 experience has indicated that there can be a large commitment in terms of technology applications, the establishment of a system configuration and its performance characteristics, the obligation of resources, and potential

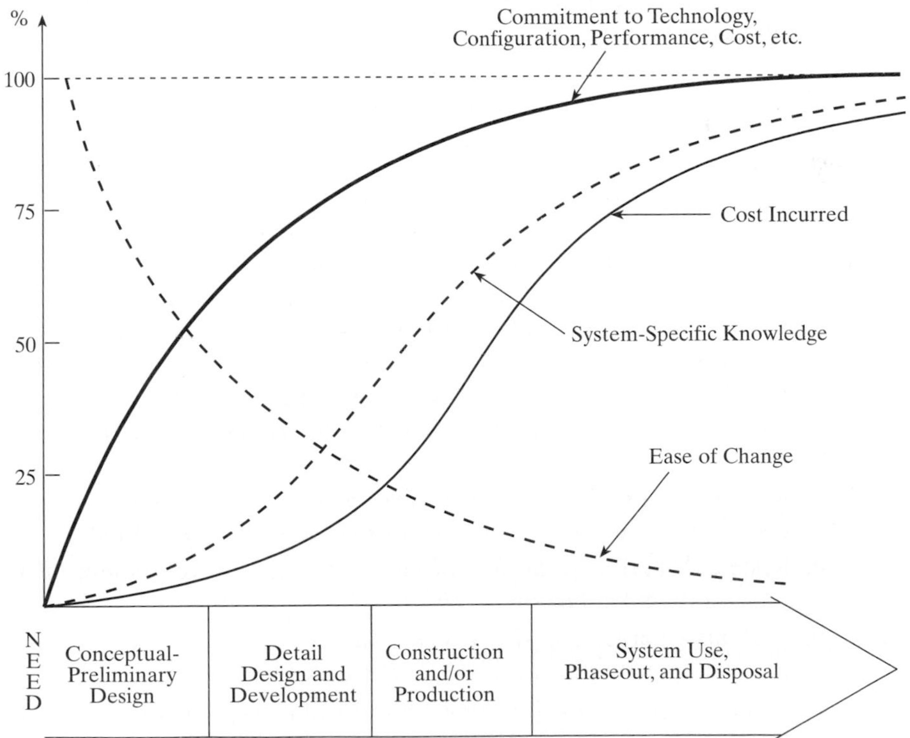

Figure 2.11 Commitment, system-specific knowledge, and cost.

life-cycle cost at the early stages of a program. It is at this point when system-specific knowledge is limited, but when major decisions are made pertaining to the selection of technologies, the selection of materials and potential sources of supply, equipment packaging schemes and levels of diagnostics, the selection of a manufacturing process, the establishment of a maintenance approach, and so on. It is estimated that from 50% to 75% of the projected life-cycle cost for a given system can be commited (i.e.,"locked in") based on engineering design and management decisions made during the early stages of conceptual and preliminary design. Thus, it is at this stage where the implementation of systems engineering concepts and principles is critical. It is essential that one start off with a good understanding of the customer need and a definition of system requirements, evolving through the process conveyed in Figure 2.2 and described further in Part II of this text.

Applications for Systems Engineering

There are many categories of human-made systems, and there are many applications where the concepts and principles of systems engineering can be effectively implemented. Every time that there is a newly identified need to accomplish some function, a new *system* requirement is established. In each instance, there is a new design and development effort that must be accomplished at the *system* level. This, in turn, may lead to a variety of approaches at the subsystem level and below (i.e., the design and development of new equipment and software, the selection and integration of new commercial off-the-shelf [COTS] items, the modification of existing items already in use, or combinations thereof). In any event, for every new consumer requirement, there is a required design effort for the system overall, and the steps described in Section 2.4 are applicable. Although the extent and depth of effort will vary, the concepts and principles for bringing a system into being are basically the same. Some specific examples of application are highlighted in Figure 2.12, and some application categories include the following:

1. Large-scale systems with many components, such as a space-based system, an urban transportation system, or a hydroelectric power-generating system.
2. Small-scale systems with relatively few components, such as a local area communications system, a computer system, a hydraulic system, or a mechanical braking system.
3. Manufacturing or production systems where there are input-output relationships, processes, processors, control software, facilities, and people.
4. Systems where a great deal of new design and development effort is required (e.g., in the introduction of advanced technologies).
5. Systems where the design is based largely on the use of existing COTS equipment, commercial software, or existing facilities.
6. Systems that are highly equipment, software, facilities, or data intensive.
7. Systems where there are several suppliers involved in the design and development process at the national and possibly international level.

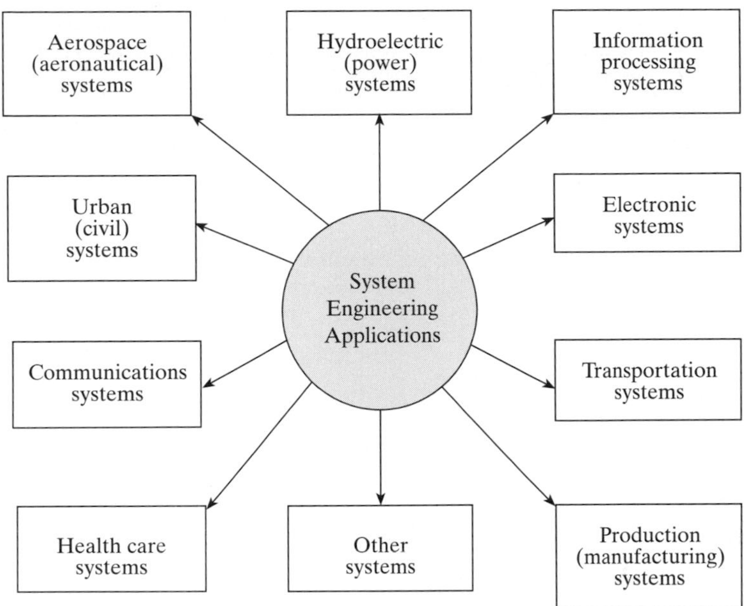

Figure 2.12 Application areas for systems engineering.

8. Systems being designed and developed for use in the defense, civilian, commercial, or private sectors separately or jointly.

The Management of Systems Engineering

The systems engineering process is applicable in all phases of the life cycle, with the greatest benefit being derived from emphasis in the early stages as illustrated in Figure 2.11. The objective is to influence design early, in an effective and efficient manner, through a comprehensive needs analysis, requirements definition, functional analysis and allocation, and then to address the follow-on activities in a logical and progressive manner with the appropriate feedback provisions. As conveyed in Figure 2.13, the overall objective is to influence design in the early phases of acquisition, leading to the identification of individual design disciplinary needs as one evolves from system-level requirements to design of the various subsystems and below.[14]

As the system design and development activity evolves, the objective is to ensure the proper integration of design requirements at the appropriate level in the system hierarchical structure. Having initially established specific design requirements as described in Section 2.5 (Figures 2.8 and 2.9), the goal from hereon is to ensure that these requirements are properly balanced and integrated as conveyed in Figure 2.14.

[14] It should be noted that in Figure 2.13 the intent is to convey the degree of "design influence" from the application of the systems engineering process, and not to imply levels of human effort or cost. A single individual with the appropriate experience and technical expertise can exert a great deal of influence on design, whereas, conversely, the establishment of a new organization and the assignment of many people to a project may have little influence.

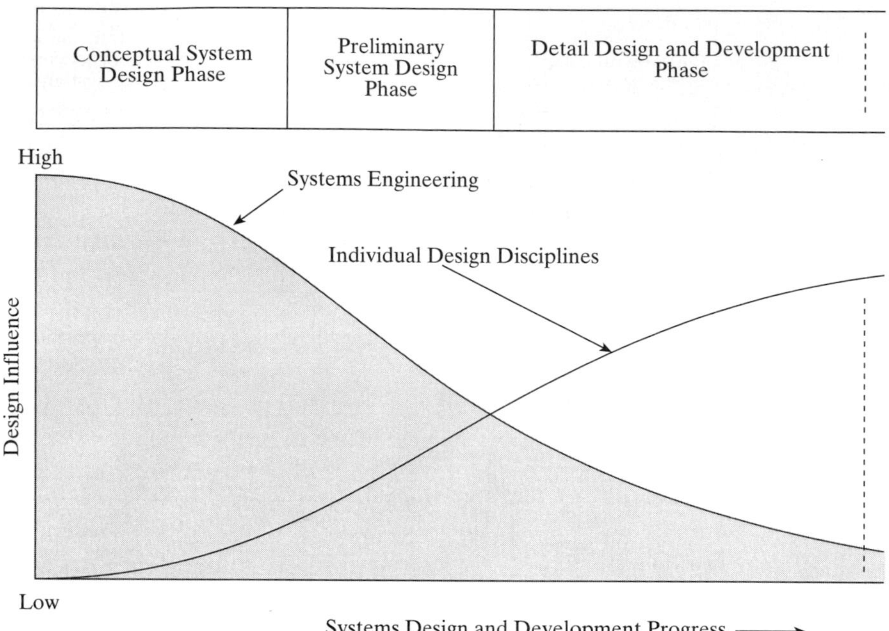

Figure 2.13 Systems engineering influence on design.

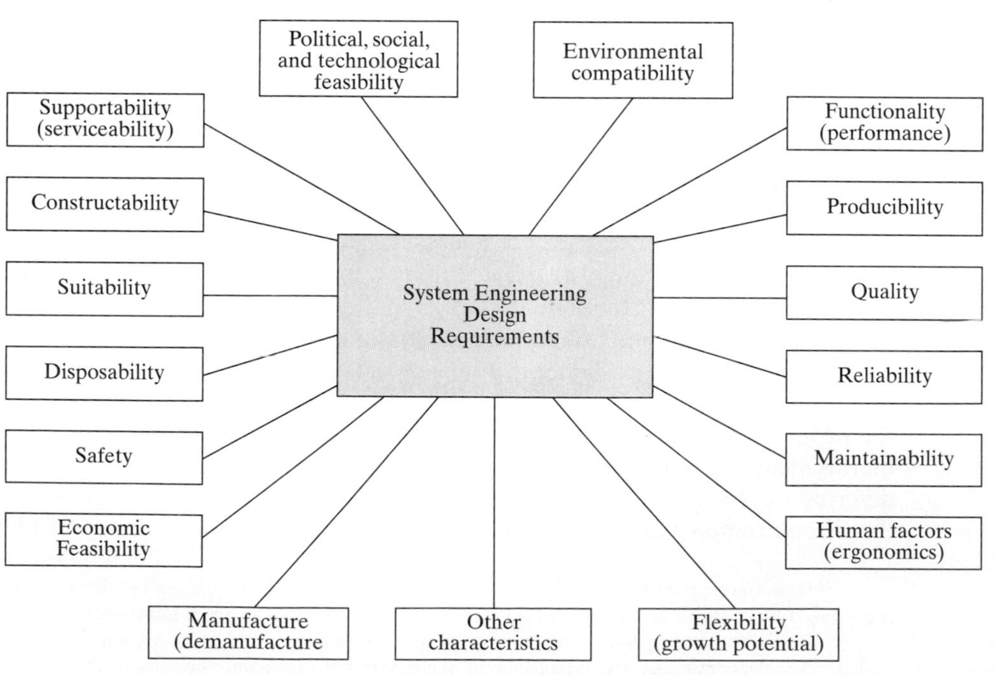

Figure 2.14 The integration of design considerations.

Referring to the figure, an understanding of the interrelationships among these various factors is critical. Given this, there is a requirement to ensure that the applicable engineering disciplines responsible for design of the individual system elements are properly integrated. Implementing the concepts and principles of concurrent engineering and the necessity to promote and establish a good communications network involving all critical project personnel is essential. Often, the various design disciplines (or "domains") are remotely located, and there is a major challenge in promoting the proper level of communications across the board. This goal has been facilitated to some extent through the use of CAD tools and the establishment of a network, such as the one conveyed in Figure 2.15. It is the role of systems engineering to first establish the requirements for design and then to ensure that the proper *design integration* occurs throughout the life cycle as required.

Successful implementation of the systems engineering process is dependent, not only on the availability and application of the appropriate technologies and tools, but on the planning and management of the activities required to accomplish the overall objective. Although the steps described in Section 2.4 may be specified for a given program, successful implementation (and the benefits to be derived) will not be realized unless the proper organizational environment is established that will encourage it to happen. There have been numerous instances where a project organization included a "systems engineering" function but where the impact on design has been almost nonexistent, with the objectives not being met.

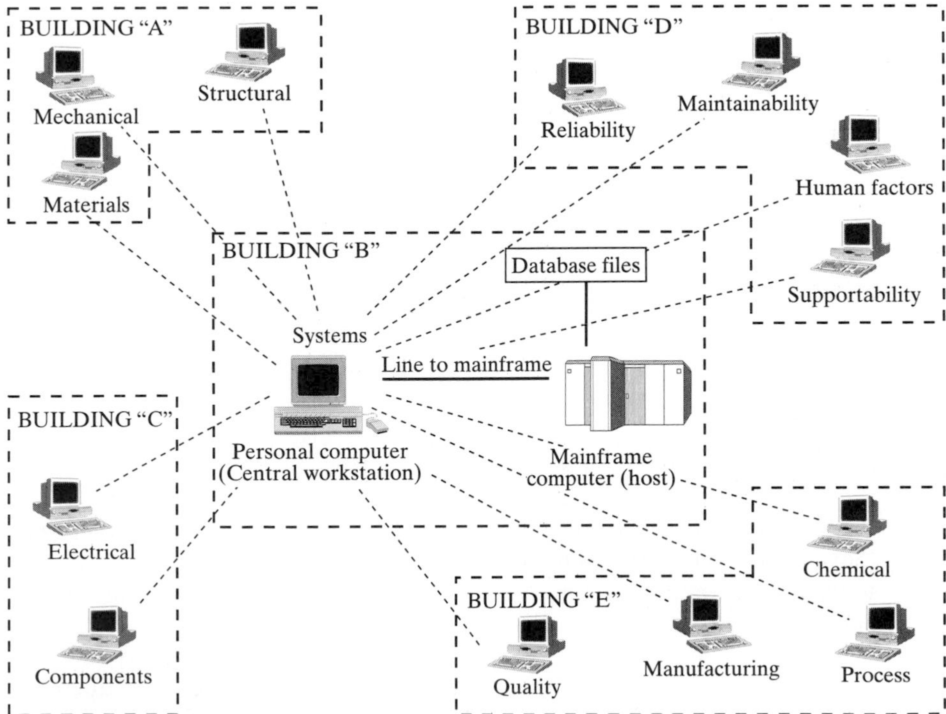

Figure 2.15 System design communications network (example).

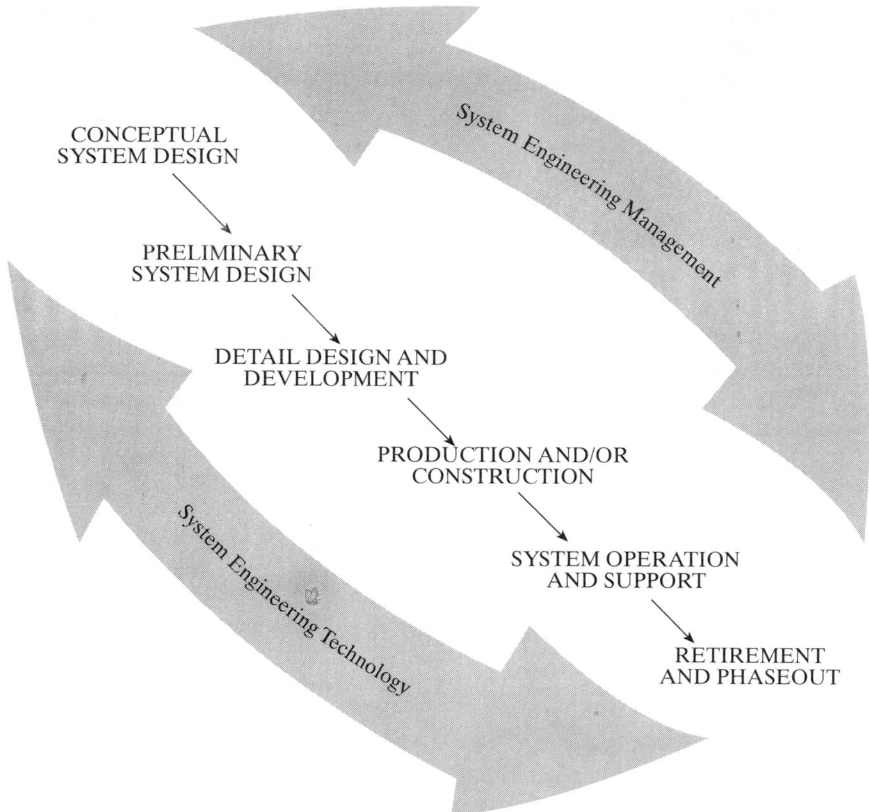

Figure 2.16 Application of technology and management activities to systems engineering.

Without the proper organizational emphasis from the top down, the establishment of an environment that will allow for creativity and innovation, a leadership style that will promote a "team" approach to design, and so on, implementation of the concepts and methodologies described herein will not occur. Thus, systems engineering must be addressed in terms of both *technology* and *management* as conveyed in Figure 2.16

Potential Benefits From Systems Engineering

Some of the benefits associated with application of the concepts and principles of systems engineering have been implied throughout this chapter, but it may be worthwhile to provide a compact summary for reference. Accordingly, application of the systems engineering process can lead to the following benefits:

1. Reduction in the cost of system design and development, production and/or construction, system operation and support, system retirement and material disposal; hence, a reduction in life-cycle cost should occur. Often it is perceived that the

implementation of systems engineering will increase the cost of system acquisition. Although there may be a few more steps to perform during the early (conceptual and preliminary) system design phases, this investment could significantly reduce the requirements in the integration, test, and evaluation efforts accomplished late in the detail design and development phase. The bottom-up approach involved in making the system work can be simplified if a good engineering effort is initiated from the beginning. Additionally, experience indicates that the early emphasis in systems engineering can result in a cost savings later on in the production, operations and support, and retirement phases of the life cycle.

2. Reduction in system acquisition time (or the time from the initial identification of a customer need to the delivery of a system to the customer). Evaluation of all feasible alternative approaches to design early in the life cycle (with the support of available design technologies, such as the use of CAD methods) should help to promote greater design maturity earlier. Changes can be incorporated at an early stage before the design is "fixed" and more costly to modify. Further, the results should enable a reduction in the time that it takes for final system integration, tests, and evaluation.
3. More visibility and a reduction in the risks associated with the design decision-making process. Increased visibility is provided through viewing the system from a long-term and life-cycle perspective. The long-term impacts as a result of early design decisions and "cause-and-effect" relationships can be assessed at an early stage. This should cause a reduction in potential risks, resulting in greater customer satisfaction.

The implementation of systems engineering is the responsibility of systems engineering management. Part V of this text is devoted to this important activity.

QUESTIONS AND PROBLEMS

1. Describe some of the interfaces between the natural world and the human-made world as they pertain to the process of bringing systems, products, and structures into being.
2. What are some of the problems for producers in today's competitive economic environment? How can the application of system engineering preclude some of these problems in the future?
3. What does system life-cycle thinking add to engineering as practiced traditionally? What are the expected benefits to be gained (if any)?
4. Various phases of the system life cycle are shown in Figure 2.1. Describe how the system engineering process can be effectively applied to each phase (if at all).
5. Refer to Figure 2.3 and describe some of the interrelationships between the three life cycles.
6. Select a system of your choice and describe the system life cycle and the activities in each phase as appropriate (the description should be "tailored" for the system that you have selected).
7. Describe some of the main goals and objectives for implementing the system engineering process in the design and development of the system you selected in Question 6.
8. One of the first steps in the system engineering process is the definition, of systems requirements. Based on your own experience, how is this accomplished, and why is it important?

9. What are the major systems engineering functions in conceptual design, preliminary system design, detail design and development, production and/or construction, system operation, maintenance and support, system retirement, and material disposal/recycling?
10. What is the significance of the "feedback" process illustrated in Figure 2.5? Relate it to feedback in Figure 2.2.
11. What is meant by "tailoring," and why is it important in the engineering of a system?
12. Describe your understanding of the interfaces between synthesis, analysis, and evaluation.
13. Relate the activities of Figure 2.7 to the elements of Figure 2.8; that is, map one on the other.
14. Design decisions at the system level are multicriteria in nature. Pick a hypothetical system, and classify the essential cost and effectiveness measures to be considered.
15. Describe some of the application categories of the system engineering process. Give specific examples from your own experience.
16. What are some of the impediments (as you see them) to the successful implementation of systems engineering?
17. Describe some of the management requirements necessary for the successful implementation of the system engineering process.
18. What are the differences (or similarities) between "systems engineering" and some of the more traditional disciplines, such as civil engineering, electrical engineering, industrial engineering, or mechanical engineering?
19. Describe some of the benefits that may result from the proper implementation of the system engineering process.

PART II THE SYSTEM DESIGN PROCESS

3 CONCEPTUAL SYSTEM DESIGN

Conceptual design is the first step in the overall process of system design and development, as introduced in Chapter 2. The system engineering process evolves from an identified need to the definition of system requirements in functional terms, establishment of design criteria, and development of a system to meet customer needs in an effective and efficient manner. Thus, systems engineering is concerned with the transition from a need and requirements definition to a fully defined system configuration ready for production and subsequent use.

Conceptual design is the foundation on which the life-cycle phases of preliminary system design, detail design and development, and so forth are based (refer to Figure 2.2, Block 1). Within the conceptual design phase are activities related to the identification of customer need and the several steps involved in the definition of system design requirements (refer to Figure 2.4, Block 0.1). This chapter primarily addresses customer needs and requirements and their conversion to design criteria as a basis for the topics presented in Chapters 4 to 6.

3.1 IDENTIFICATION OF NEED

The systems engineering process begins with the identification of a "need," "want," or "desire" for one or more new entities, or for a new or improved capability. It should be based on a real (or perceived) deficiency. For example, a current system may not be adequate in meeting certain performance goals, may not be available when needed, cannot be properly supported, is too costly to operate, and so on. Or there is a lack of capability to communicate between point A and point B, at a desired bit rate X, with a reliability of Y, and within a specified cost of Z. Accordingly, a new system requirement may be defined along with the priority for its realization, the date when the new capability is to be introduced for use, and an estimate of the resources needed for its acquisition. The statement of need should be presented in specific qualitative and quantitative terms and in enough detail to justify proceeding to the next step.

To identify the need seems to be basic or self-evident. However, a design project is often initiated as a result of a personal interest or a political whim, without first having adequately defined the requirement. Defining the problem is the most difficult part of the system engineering process. However, a complete description of the current deficiency and the real need is necessary. It is essential that the results truly reflect a requirement. This objective can be met best by involving the customer, or ultimate "user," in the process from the beginning.[1]

3.2 ACCOMPLISHMENT OF FEASIBILITY ANALYSIS

Having a definition of need, it is then necessary to (1) identify possible system-level design approaches that can be pursued to meet that need; (2) evaluate the most likely approaches in terms of performance, effectiveness, maintenance and logistic support, and economic criteria; and (3) recommend a preferred course of action. There may be many possible alternatives; however, the number must be narrowed down to a few feasible ones, consistent with the availability of resources (i.e. personnel, materials, and money).

In considering alternative design approaches, different technology applications are investigated. For instance, in the design of a communications system, should one use fiber optics technology or the conventional twisted-wire approach? In aircraft design, to what extent should the use of composite materials be considered? In automobile design, should one incorporate high-speed electronic circuitry in a certain control application, or should an electromechanical approach be taken? In the design of a data communications capability, should a digital or conventional format be used? In the design of a process, to what extent should embedded computer capabilities be incorporated, or artificial intelligence be used? Included in the evaluation process are considerations pertaining to the type and maturity of the technology, its stability and growth potential, the anticipated life of the technology, the number of supplier sources, and so forth.

It is at this early stage in the life cycle (i.e., the conceptual design phase) that major decisions are made relative to adapting a specific design approach, and it is at this stage that the results of such decisions can have a great impact on the ultimate characteristics and life-cycle cost of a system. Technology applications are evaluated and, in some instances where there is not enough information available, research may

[1] The application of Quality Function Deployment (QFD), or equivalent methods, can be useful in helping to estabilish necessary communications between the customer and producer and in identifying and ranking design goals. Two good references are Y. Akao, (ed.), *Quality Function Deployment: Integrating Customer Requirements Into Product Design*, translated into English (Portland: Productivity Press, 1990); and L. Cohen, *Quality Function Deployment: How To Make QFD Work for You* (Reading, Mass.: Addison-Wesley Publishing Co., 1995). The QFD method was developed at the Kobe Shipyard, Mitsubishi Heavy Industries, Japan, in the late 1960s and has been applied in many situations since.

be initiated with the objective of developing new knowledge for specific applications. Finally, the "need" should drive the "technology" and not vice versa.

The results of a feasibility analysis will significantly impact the operational characteristics of the system including its producibility, supportability, disposability, and similar characteristics. The selection (and application) of a given technology has reliability and maintainability implications, may affect manufacturing operations in terms of the processes required, may significantly impact the requirements for test equipment and spare parts, and will certainly affect life-cycle cost. Thus, it is essential that life-cycle considerations be an inherent aspect of the feasibility analysis activity.

3.3 ADVANCE SYSTEM PLANNING

Given an identified need for a feasible new system, advance planning activities are initiated that will establish a project for the conceptual design of that system. Advance planning activities most often included in the conceptual design phase are highlighted in Figure 3.1. These include (1) communication with the customer (consumer) to

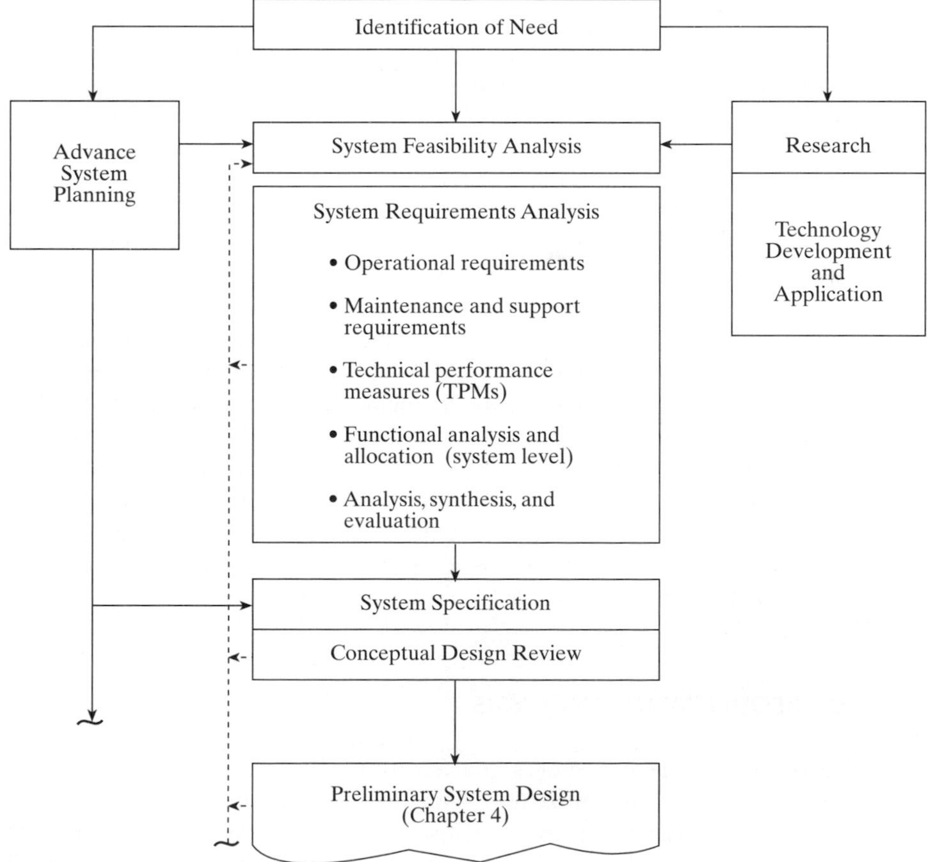

Figure 3.1 System requirements definition process (example).

obtain an in-depth delineation of the need, (2) completion of a feasibility analysis to determine the availability of technologies applicable to the need (i.e., the "technical" approach to design), (3) definition of system operational requirements, (4) development of the system maintenance and support concept, (5) identification and prioritization of technical performance measures (and other design dependent "metrics"), (6) completion of a top-level functional analysis, (7) preparation of a system specification, and (8) conduct of a conceptual design review.

The results from advance system planning may be classified in terms of the *technical* requirements, included in specifications, and the *management* requirements provided in program planning documentation. Of particular significance are the System Specification (type A) and the System Engineering Management Plan (SEMP), both shown in Figure 2.4. It is these documents that initiate the requirements for the implementation of a given program. The relationships between these documents are shown in Figure 3.2, and it is essential that they complement each other.

The system specification provides the design requirements for the system as an entity. It usually includes such information as a general description of the system, operational requirements and the maintenance concept, top-level functional analysis and the allocation of requirements to the subsystem level, performance and effectiveness factors (technical performance measures), physical characteristics, design characteristics, design data requirements, materials and manufacturing processes, logistic support provisions, test and evaluation requirements, and quality assurance provisions. The System Specification leads to lower-level development, product, process, or material specifications as illustrated in Figure 3.2.

In preparing specifications, it is important to recognize that requirements should be stated in terms of the "whats" versus the "hows." There has been a tendency to "overspecify" by telling suppliers how to do something versus defining the overall performance-related requirements in functional terms. This can be costly, particularly if the specifications are not written well in the first place.

Systems engineering requirements, from a program management perspective, are initiated through the SEMP. This plan includes a description of the tasks to be accomplished, the program organizational structure showing the interfaces with various design and supporting disciplines, customer-supplier relationships, the Work Breakdown Structure (WBS), task schedules, cost projections, data and documentation needs, program reporting requirements, design reviews, a risk management plan, and so on. A sample outline and discussion pertaining to systems engineering planning and organization is presented in Part V (Chapter 18).

3.4 SYSTEM REQUIREMENTS ANALYSIS

The activities of requirements analysis, functional analysis, requirements allocation, and so on are iterative in nature. They lead from an identified need to the definition of a system in functional terms, as shown in Figure 2.4. Within each block, there is some degree of iteration. For example, trade-off studies are accomplished at all levels. But it is difficult to convey graphically every iteration that occurs in system engineering.

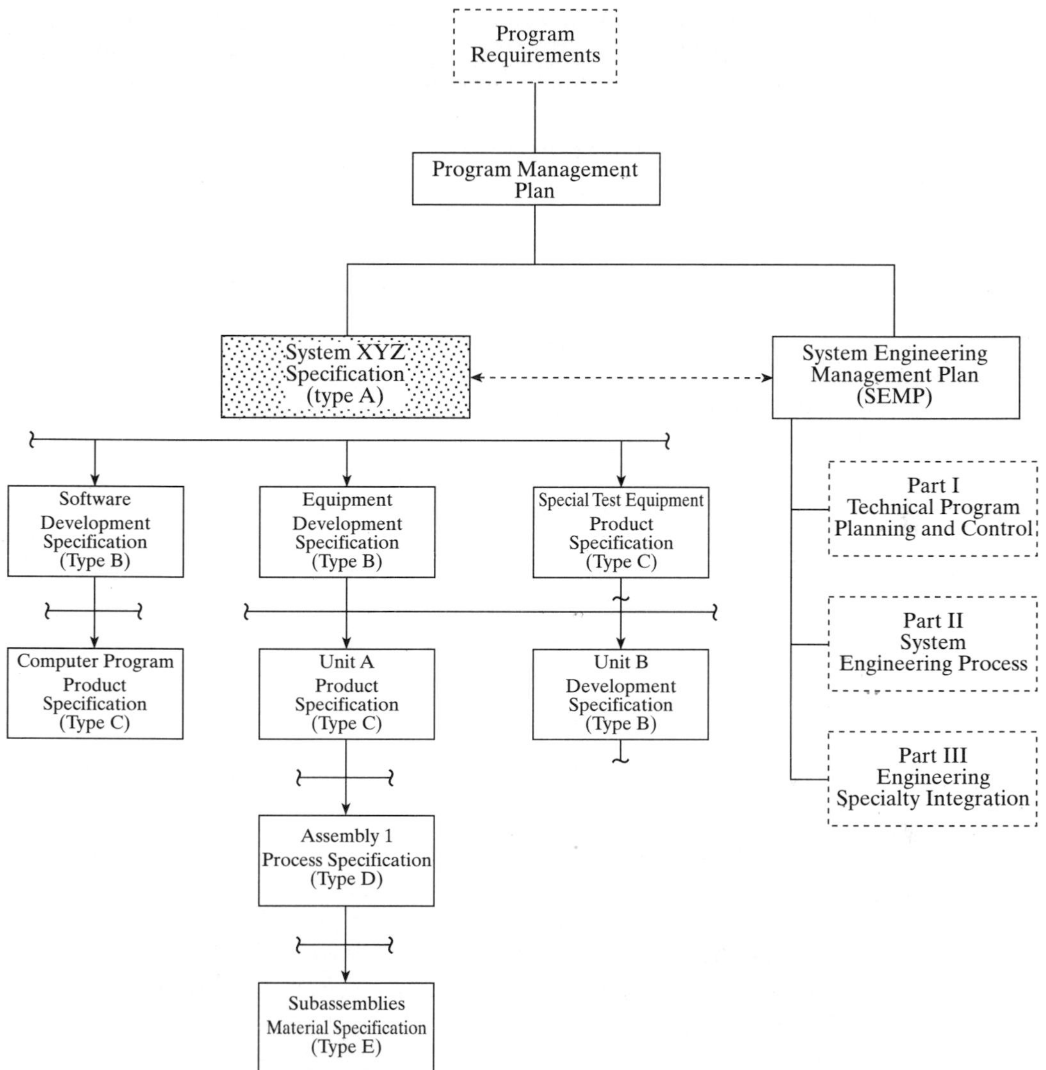

Figure 3.2 Program documentation tree.

Thus, the format presented in Figure 2.4 serves only as a basis for highlighting some of the key activities that must occur.

In any event, the process is initially top down and evolutionary in nature, leading from definition at the system level, to the subsystem level, and to major components of the system. The objective is to describe *requirements* at each level in the system hierarchy; i.e., to describe the "whats (not the "hows" in terms of specific hardware, software, facilities, people, data, etc.). The resources supporting the "hows" will ultimately evolve from the functional analysis and allocation process. Requirements must be complete and fully describe the user's need; they should be objective and designable into the system, must be measurable and demonstrable, and so on.

The important ingredient in the implementation of the requirements analysis and definition process (which is also iterative in nature), is the necessary communications that must exist between the customer and the producer or contractor.

Definition of Operational Requirements

After the needs analysis is combined with the selection of a feasible technical approach, one is ready to project the relevant information to derive anticipated operational requirements. These requirements include the following considerations:

1. *Operational distribution or deployment*—the number of customer sites where the system will be used, the geographical distribution and deployment schedule, and the type and number of system components at each location. This responds to the question—where is the system to be used? Figure 3.3 presents a hypothetical worldwide distribution pattern.
2. *Mission profile or scenario*—identification of the prime mission for the system, and its alternative or secondary missions. What is the system to accomplish and what functions must be performed in responding to the need? This may be defined through a series of operational profiles, illustrating the "dynamic" aspects required in accomplishing a mission. An aircraft flight path between two cities, an automobile or a shipping route, and the number of products to be produced in a factory are examples. Figure 3.4 presents a simple illustration of possible profiles.
3. *Performance and related parameters*—definition of the basic operating characteristics or functions of the system. This refers to parameters, such as range, accuracy, rate, capacity, throughput, power output, size, and weight. What are the

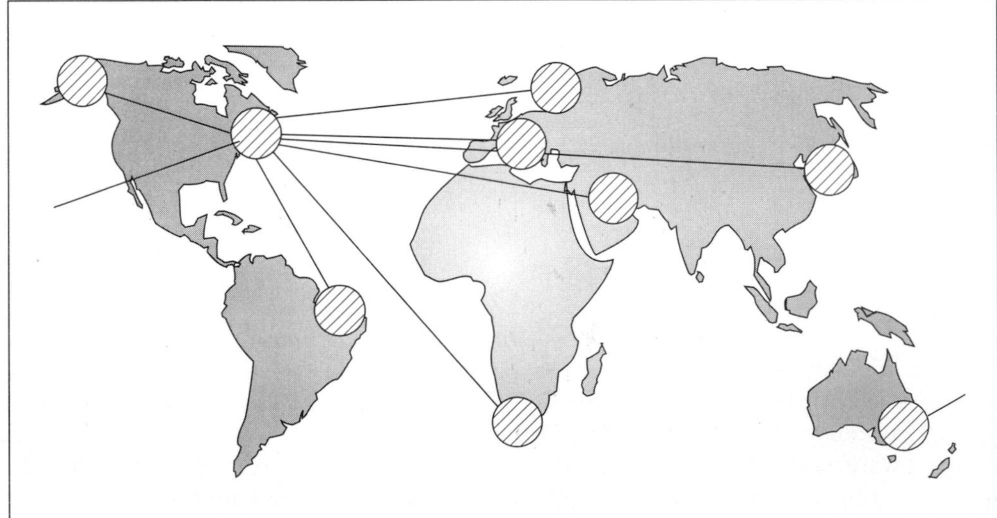

Figure 3.3 Operational requirements (geographical distribution).

There is usually little attention given to system maintenance and support. In general, emphasis in the past has been directed to only part of the system and not the entire system as defined earlier.

To meet the overall objectives of systems engineering, it is essential that all aspects of the system be considered on an integrated basis from the beginning. This includes not only the prime mission-oriented elements of the system but the support capability as well. The prime system elements must be designed in such a way that they can be effectively and efficiently supported throughout the planned life cycle. The overall support capability must be responsive to this requirement. This, in turn, means that one should also address the characteristics of design as they pertain to the supply support network (i.e., spares, repair parts, and operating inventories), test and support equipment, transportation and handling equipment, computer resources (i.e., software), personnel and training, facilities, and technical data. It is essential that a system "maintenance and support concept" be developed during the conceptual design phase.[2]

The maintenance concept evolves from the definition of system operational requirements and includes the activities illustrated in Figure 3.5. It constitutes a before-the-fact series of illustrations and statements on how the system is to be designed for supportability, whereas the "maintenance plan" defines the follow-on requirements for system support based on the results of the supportability or logistic support analysis, or some equivalent analysis of a given configuration. The maintenance concept ultimately evolves into a detailed maintenance plan.

Referring to Figure 3.5, one must deal with the flow of activities and materials from design, through production, and to the operational sites for customer use. A maintenance flow exists when items are returned from the operational site to the intermediate and suppliers or depot levels of maintenance. By reviewing the network in total, one should address issues, such as the levels of maintenance, the responsibilities and functions to be performed at each level, design criteria pertaining to the various elements of support (e g., type of spares and levels of inventory, reliability of the test equipment, personnel numbers and skill levels), and the effectiveness factors for the overall support capability. Although design of the prime elements of a system may appear to be adequate, the overall ability of the system to successfully fulfill its mission objective is highly dependent on the effectiveness of the support infrastructure.

While there are some variations that arise as a function of the nature and type of system, the maintenance concept generally includes the following information:

1. *Levels of maintenance*—corrective and preventive maintenance may be accomplished on the system itself (or an element thereof) at the site where the system is being used, in an intermediate shop near the consumer/user site, or at a depot or manufacturer's plant facility. Maintenance level pertains to the division of functions and tasks for each area where maintenance is performed. Anticipated frequency of maintenance, task complexity, personnel skill-level requirements, special facility needs, and so on, dictate, to a great extent, the specific maintenance functions to be accomplished at each level. Depending on the nature and

[2] B. S. Blanchard, *Logistics Engineering and Management*, 5th ed., (Upper Saddle River, N. J.: Prentice Hall, 1998).

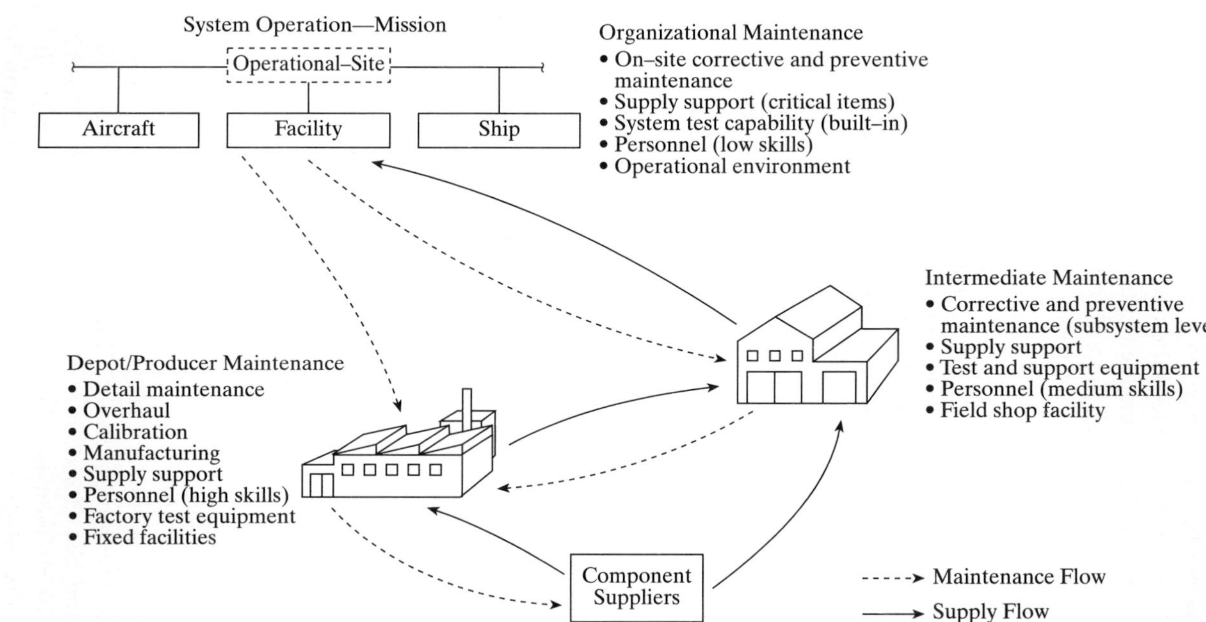

Figure 3.5 System operational and maintenance flow.

Criteria	Organizational Maintenance	Intermediate Maintenance		Supplier/Manufacturer/Depot Maintenance
Done where?	At the operational site or wherever the prime elements of the system are located	Mobile or semimobile units	Fixed units	Supplier/manufacturer/depot facility
		Truck, van, portable shelter, or equivalent	Fixed field shop	Specialized repair activity or manufacturer's plant
Done by whom?	System/equipment operating personnel (low maintenance skills)	Personnel assigned to mobile, semimobile, or fixed units (intermediate maintenance skills)		Depot facility personnel or manufacturer's production personnel (high maintenance skills)
On whose equipment?	Using organization's equipment	Equipment owned by using organization		
Type of work accomplished?	Visual inspection Operational checkout Minor servicing External adjustments Removal and replacement of some components	Detailed inspection and system checkout Major servicing Major equipment repair and modifications Complicated adjustments Limited calibration Overload from organizational level of maintenance		Complicated factory adjustments Complex equipments repairs and modifications Overhaul and rebuild Detailed calibration Supply support Overload from intermediate level of maintenance

Figure 3.6 Major levels of maintenance.

mission of the system, there may be two to four levels of maintenance. However, for the purposes of further discussion, maintenance may be classified as "organizational," "intermediate," and "supplier/depot." Figure 3.6 describes the basic differences between these levels.

2. *Repair policies*—within the constraints illustrated in Figures 3.5 and 3.6, there may be several possible policies specifying the extent to which repair of a system component will be accomplished (if at all). A repair policy may dictate that an item should be designed to be nonrepairable, partially repairable, or fully repairable. Repair policies are initially established (through accomplishing a repair level analysis), criteria are developed, and system design progresses within the bounds of the repair policy that is selected. An example of a repair policy for System XYZ, developed as part of the maintenance concept during conceptual design, is illustrated in Figure 3.7.
3. *Organizational responsibilities*—the accomplishment of maintenance may be the responsibility of the customer, the producer (or supplier), a third party, or a combination thereof. The responsibilities may vary, not only with different components of the system but as one progresses in time through the operational use and system support phase. Decisions pertaining to organizational responsibilities may impact system design from the diagnostic and packaging standpoint, as well as dictating repair policies, contract warranty provisions, and the like. Although conditions may change, some initial assumptions are required.
4. *Logistic support elements*—as part of the definition of the initial maintenance concept, criteria must be established relating to the various elements of logistic support. These elements include supply support (spare and repair parts, associated inventories, provisioning data), test and support equipment, personnel and

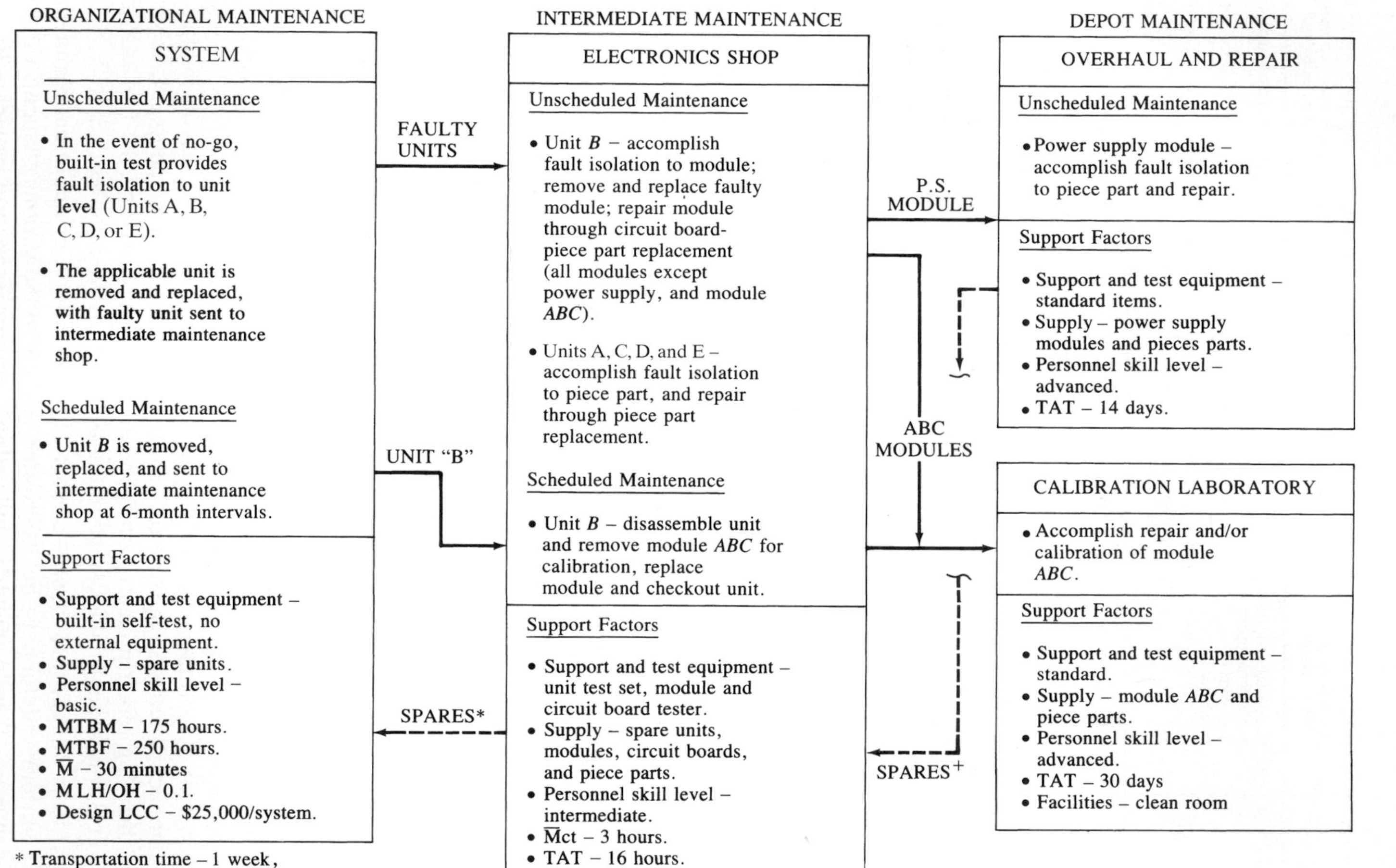

Figure 3.7 System maintenance and repair policy.

training, transportation and handling equipment, facilities, data, and computer resources. Such criteria, as an input to design, may involve self-test provisions, built-in versus external test requirements, packaging and standardization factors, personnel quantities and skill levels, transportation and handling factors and constraints, and so on. The maintenance concept provides some initial system design criteria pertaining to the activities illustrated in Figure 3.5. The final determination of specific logistic, support requirements will occur through the completion of the maintenance or supportability analysis, to be accomplished as design progresses.

5. *Effectiveness requirements*—this constitutes the effectiveness factors associated with the support capability. In the supply support area, this may include a spare part demand rate, the probability of a spare part being available when required, the probability of mission success given a designated quantity of spares, and the economic order quantity as related to inventory procurement. For test equipment, the length of the queue while waiting for test, the test station processing time, and the test equipment reliability are key factors. In transportation, transport rates, transportation times, the reliability of transportation, and transportation costs are of significance. For personnel and training, one should be interested in personnel numbers and skill levels, training rates, training times, and training equipment reliability. In software, the number of errors per module or hundred lines of code may be an important measure. These factors, as related to a specific system-level requirement, must be addressed. It is meaningless to specify a tight quantitative requirement applicable to the repair of prime elements of the system when it takes months to acquire a needed spare part. The effectiveness requirements applicable to the support capability must complement the requirements for the overall system.
6. *Environment*—definition of the environment as it pertains to maintenance and support. This includes temperature, shock and vibration, humidity, noise, arctic versus tropical, mountainous versus flat terrain, shipboard versus ground, and so on, as applicable to maintenance activities and related transportation, handling, and storage functions.

In summary, the maintenance concept provides the basis for the establishment of maintenance and supportability requirements in system design. Not only do these requirements impact the prime mission-oriented elements of the system, but they provide guidance in the design or procurement of the necessary elements of maintenance and logistic support. Additionally, the maintenance concept forms the baseline for the development of the detailed maintenance plan, to be prepared during detail design and development.

3.5 TECHNICAL PERFORMANCE MEASURES (TPMs)

Evolving from the operational requirements and the maintenance and support concept is the development of qualitative and quantitative design-to *criteria*, as conveyed in Figure 2.8. Of particular interest are the quantitative factors or *metrics* associated with the system being developed. These metrics, or *technical performance measures* (TPMs), lead to the identification of *design-dependent parameters* (DDPs) and the desired characteristics that should be incorporated into the design (refer to Figure 2.9), and must be established initially as part of the requirements definition process during conceptual design.

In defining the quantitative requirements for a particular system, it is essential that the correct factors be established along with their respective priorities in terms of degrees of importance as viewed by the customer. These prioritized factors must then be translated into the appropriate design criteria and reflected by the DDPs as an input into the design process. Accomplishing this objective requires a good continuous communications link between the consumer (customer) and the producer (contractor).

A good method for facilitating the early consumer-producer communications process is through the use of the Quality Function Deployment (QFD) technique. QFD constitutes a "team" approach to help ensure that the "voice of the customer" is reflected in the ultimate design. The purpose is to establish the necessary *requirements* and to translate those requirements into technical solutions. Customer requirements and preferences are categorized as *attributes*, which are then weighted based on the degree of importance. The QFD method provides the design team an understanding of customer desires, forces the customer to prioritize those desires, and enables a relatively complete comparison of one design approach with another. Each customer attribute is then satisfied by a technical solution.[3]

The QFD process involves constructing one or more matrices, the first of which is often referred to as the "House of Quality" (HOQ).[4] A modified version of the HOQ is presented in Figure 3.8. Starting on the left side of the structure is the identification of customer needs and the ranking of those needs in terms of priority, the levels of importance being specified quantitatively. This reflects the "whats" that must be addressed. A team, with representation from both customer and design organizations, determines the priorities through an iterative process of review, evaluation, revision, re-evaluation, and so on. The top part of the HOQ identifies the designer's *technical* response relative to the attributes that must be incorporated into the design to respond to the needs (i.e., the "voice of the customer"). This constitutes the "hows," and there should be at least one technical solution for each identified customer need. The interrelationships among attributes (or technical correlations) are identified, as well as possible areas of conflict. The center part of the HOQ conveys the strength or impact of the proposed technical response on the identified requirement. The bottom part allows for a comparison between possible alternatives, and the right side of the HOQ is used for planning purposes.

The QFD method is used to facilitate the translation of a prioritized set of subjective customer requirements into a set of *system-level* requirements during conceptual design. A similar approach may be used to subsequently translate system-level requirements into a more detailed set of requirements at each stage in the design and development process. Referring to Figure 3.9, the "hows" from one house become the

[3] See footnote 1 for Akao and Cohen. Also, refer to the bibliography in Appendix F.

[4] J. R. Hauser, and D. Clausing, "The House Of Quality," *Harvard Business Review*, May-June 1988.

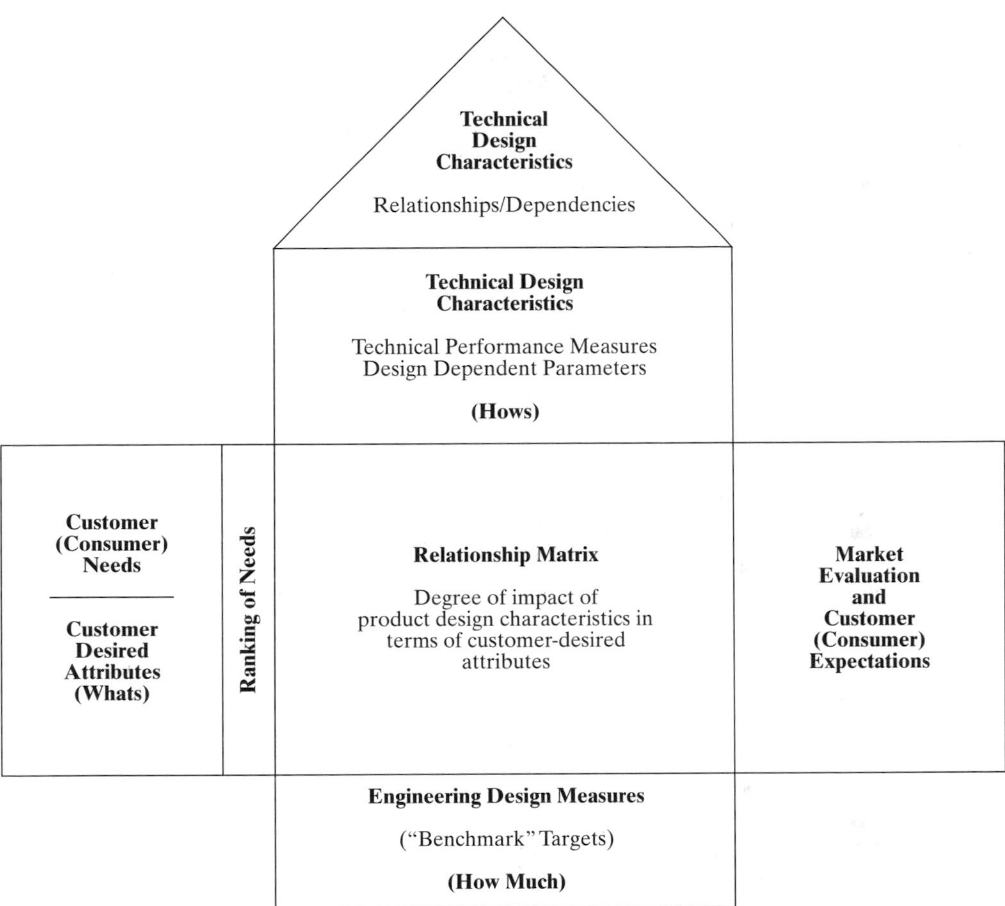

Figure 3.8 Modified "house of quality."

"whats" for a succeeding house. Requirements may be developed for the system, subsystem, component, the manufacturing process, the support infrastructure, and so on. The objective is to ensure the required justification and traceability of requirements from the top down. Further, requirements should be stated in *functional* terms.

Although the QFD method may not be the only approach used in helping to define the requirements for system design, it does constitute an excellent tool for creating the necessary visibility from the beginning. One of the largest contributors to "risk" is the lack of a good set of requirements and an adequate system specification. Inherent within the system specification should be the identification and prioritization of TPMs, as illustrated in Figure 3.10. The TPM, its associated measure (i.e., "metric"), its relative importance, and "benchmark" objective and of what is currently available will provide the designer with the necessary guidance for accomplishing his or her task. This is essential for establishing the appropriate levels of design emphasis, defining the criteria as an input to the design, and identifiying the levels of possible risk should the requirements not be met.

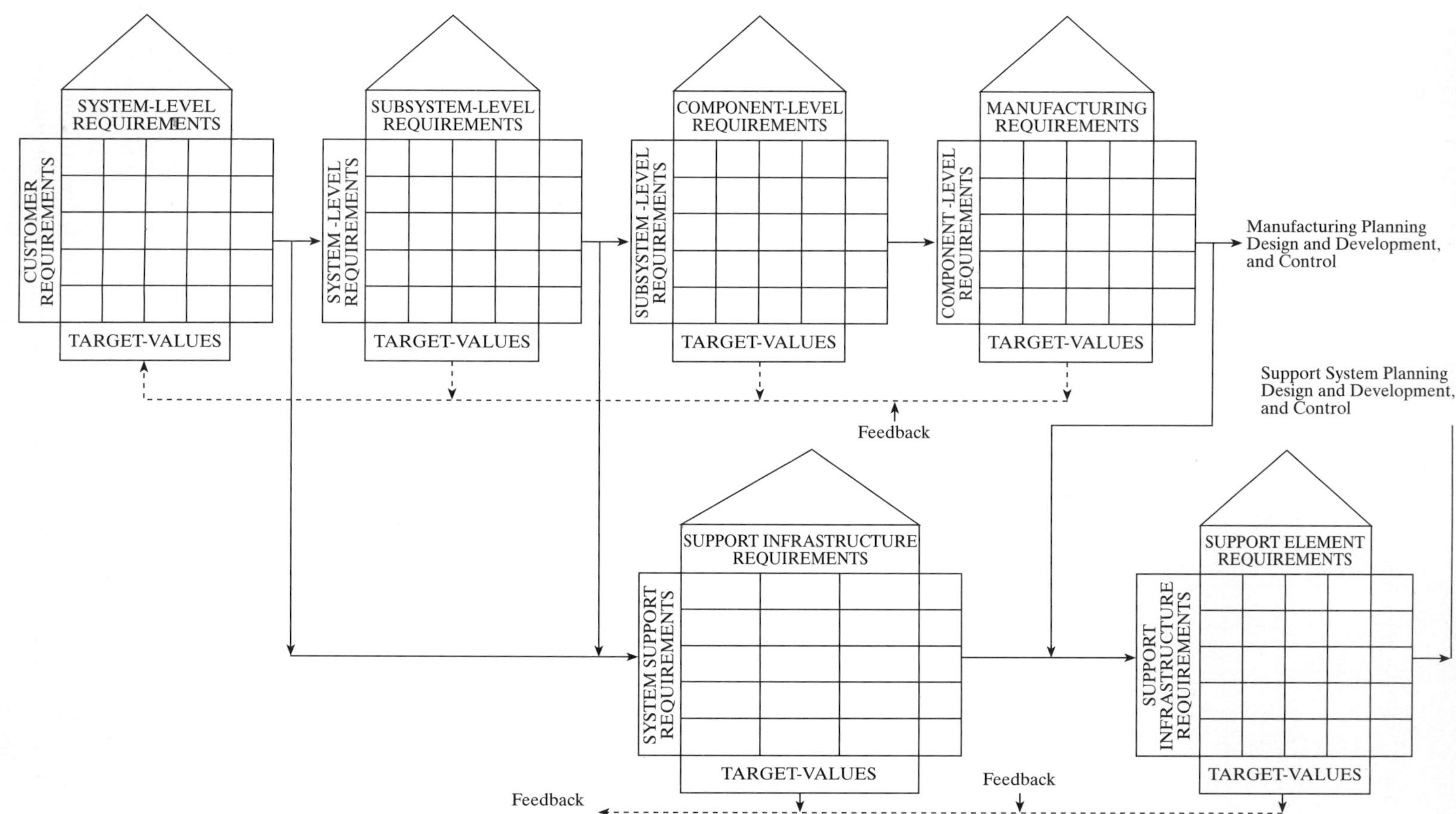

Figure 3.9 Family of houses (traceability of requirements). Source: Blanchard, B.S., *System Engineering Management*, 2nd ed. New York: John Wiley & Sons, 1998.

Technical performance measure	Quantitative requirement ("metric")	Current "benchmark" (competing systems)	Relative importance (customer desires) (%)
Process time (days)	30 days (maximum)	45 days (system M)	10
Velocity (MPH)	100 mph (minimum)	115 mph (system B)	32
Availability (operational)	98.5% (minimum)	98.9% (system H)	21
Size (feet)	10 feet long 6 feet wide 4 feet high (maximum)	9 feet long 8 feet wide 4 feet high (system M)	17
Human factors	Less than 1% error rate per year	2% per year (system B)	5
Weight (pounds)	600 pounds (maximum)	650 pounds (system H)	6
Maintainability (MTBM)	300 miles (minimum)	275 miles (system H)	9
			100

Figure 3.10 Prioritization of technical performance measures.

3.6 FUNCTIONAL ANALYSIS AND ALLOCATION

Functional analysis is the process of translating system requirements into detailed design criteria, along with the identification of specific resource requirements at the subsystem level and below. One starts with an abstraction of the needs of the customer and works down to identify the requirements for hardware, software, people, facilities, data, or combinations thereof. The initial step in this process results in the definition of system requirements (refer to Figure 2.4, Block 0.1). The results may also be specified as part of the overall *system architecture*, a term sometimes used the literature to address "requirements" and "structure" at a top level.[5]

[5] E. Rechtin, *System Architecting: Creating and Building Complex Systems* (Upper Saddle River, N. J.: Prentice Hall, 1991).

Functional Analysis

Although Figure 2.4 appears to show a sequential approach, the functional analysis actually begins with the initial identification of the "functions" that the proposed system is to accomplish (defined as part of the definition of need as explained in Section 3.1). A function refers to a specific or discrete action that is necessary to achieve a given objective (e.g., an operation that the system must perform to accomplish its purpose or a maintenance action that is necessary to restore the system to operational use).

Functions may ultimately be accomplished through the actions of equipment, software, people, and the like. However, the objective is to specify the "whats" and not the "hows" (i.e., *what* is needed to be accomplished versus *how* it should be done). No piece of equipment, item of software, data item or element of logistic support should be identified and purchased without first being justified through a functional analysis. As obvious as this may seem, the design practice often followed is to acquire an item of equipment or software and then attempt to justify its need.

Functional analysis is the iterative process of breaking down, or decomposing, requirements from the system level, to the subsystem level, and as far down the hierarchical structure as necessary to identify specific resources and components of the system. The result represents a definition of the system in *functional* terms and includes system design functions, production functions, distribution and transportation functions, operating functions, maintenance and support functions, disposal functions, and so on.

The functional analysis (as it evolves from the system to the subsystem level and below and applied during the preliminary system design phase) is covered further in Chapter 4, Section 4.1. The specific mechanics for the development of functional block diagrams are discussed in Appendix A, Section A.1.

Functional Flow Block Diagrams

Accomplishment of the functional analysis is often facilitated through the use of functional flow block diagrams. Figure 3.11 shows a simplified flow diagram with some decomposition. Top-level functions are broken down into second-level functions, operational functions lead to maintenance functions, and block numbering is used for the purposes of initially providing top-down traceability of requirements (and later a bottom-up justification of resources in terms of these requirements).[6]

Figure 3.12 shows an expansion of the functional block diagram, identifying several levels and showing operational functions leading into maintenance functions. Note that the words in each block are "action-oriented." Each block can be expanded (through a further downward iteration) and then evaluated in terms of inputs, out-

[6] The preparation of functional block diagrams may be accomplished through the use of any one of a number of graphical methods to include the Integrated DEFinition (IDEF) modeling method, the Behavioral Diagram method, and the N-Squared Charting method. Although the graphical descriptions are different, the ultimate objective is similar. These methods are compared and discussed (within the context of the functional analysis) in B. S. Blanchard, D. Verma, and E. L. Peterson, *Maintainability: A Key to Effective Serviceability and Maintenance Management* (New York: John Wiley & Sons, Inc., 1995).

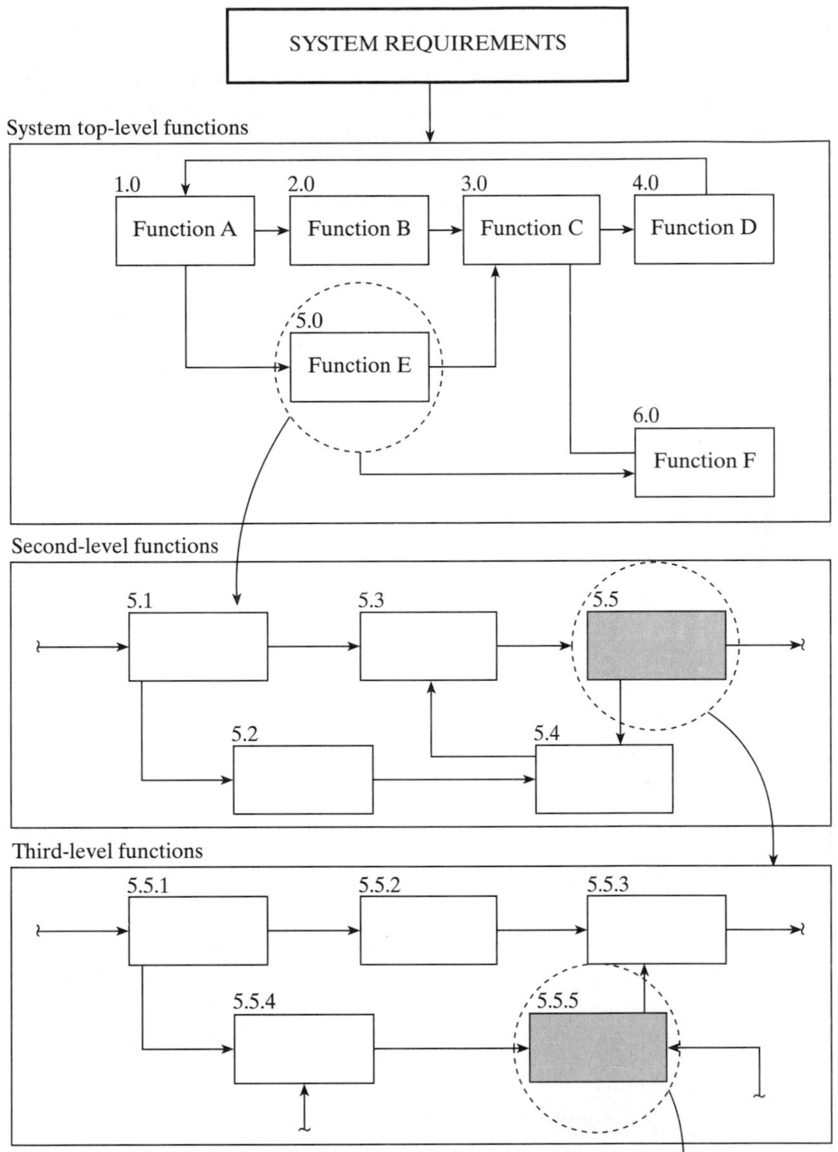

Figure 3.11 System functional breakdown.

puts, controls and/or constraints, and enabling mechanisms. Basically, the "mechanisms" lead to identification of the specific resources necessary to accomplish the function; i.e., equipment, an item of software, a design analysis tool, a facility, people, data, and so forth.

Functional block diagrams are developed for the primary purpose of structuring system requirements into functional terms. They serve to illustrate system organization and to identify major functional interfaces. The functional analysis is initiated during

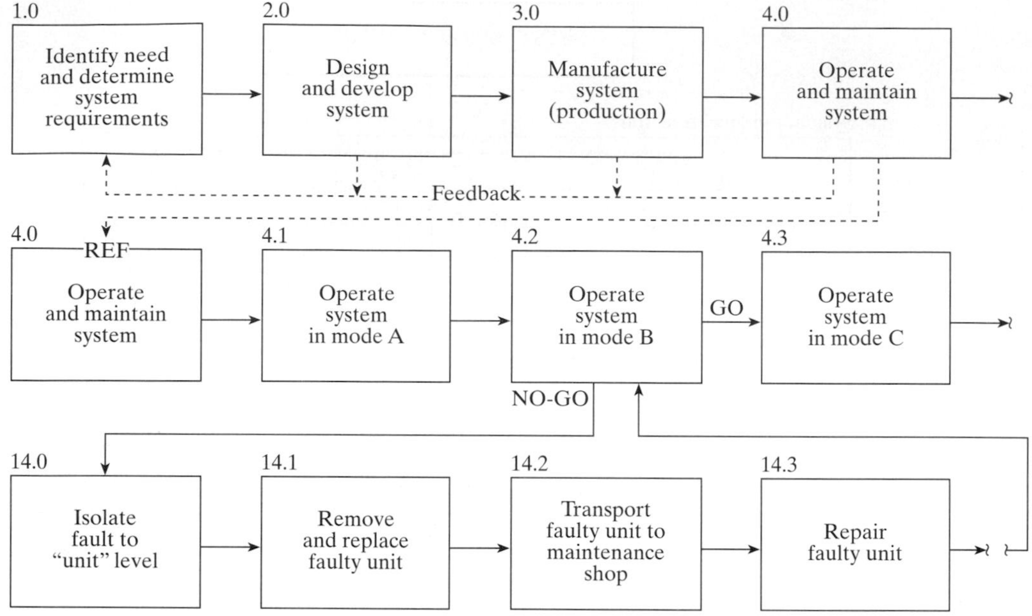

Figure 3.12 Functional block diagram expansion (partial).

the latter stages of conceptual design and is intended to enable the completion of the system design and development process in a comprehensive and logical manner. More specifically, the functional approach helps to ensure that

1. All facets of system design and development, production, operation, and support are considered (i.e., all significant activities within the system life cycle).
2. All elements of the system are fully recognized and defined (i.e., prime equipment, spare/repair parts, test and support equipment, facilities, personnel, data, and software).
3. A means is provided for relating system packaging concepts and support requirements to specific system functions (i.e., satisfying the requirements of good functional design).
4. The proper sequences of activity and design relationships are established along with critical design interfaces.

Functional Allocation

Given a top-level description of the system in functional terms, the next step is to combine, or group, similar functions into logical subdivisions, identifying major subsystems and lower-level elements of the overall system (i.e., the development of a functional packaging scheme for the system). An "open-system architecture" approach is used where the functions (and the functional interfaces) are well defined. At the same time, the "whats" are converted into "hows," and the system is broken down into components as illustrated in Figure 3.13.

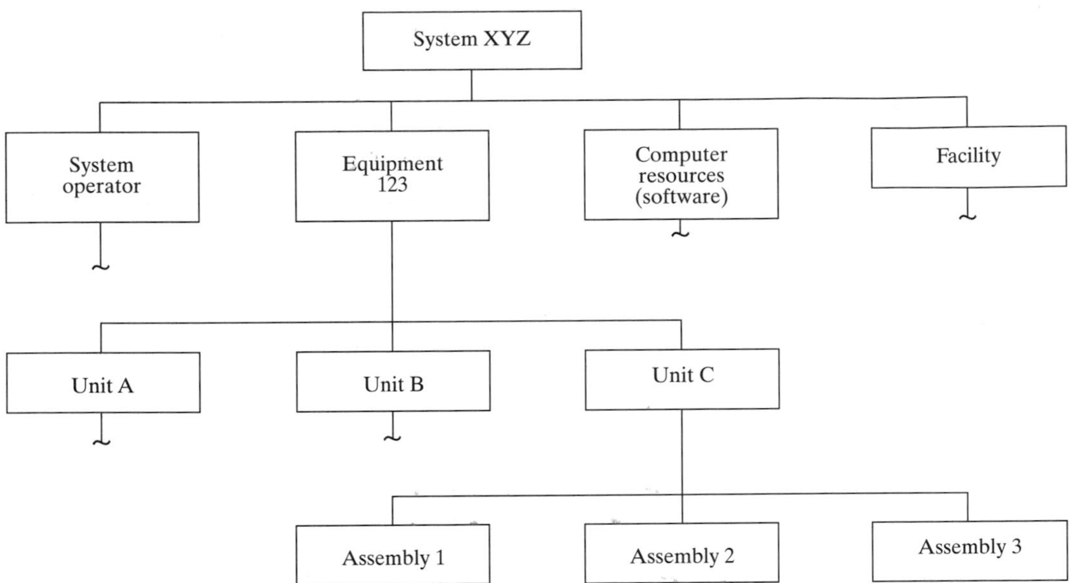

Figure 3.13 The function breakdown of the system into components.

With this structure as a framework for preliminary system design, the qualitative and quantitative requirements evolving from development of the TPMs and DDPs can be allocated to the appropriate system element (i.e., the criteria dictating the design requirements for "Equipment 123," or "Computer Resources (Software)," or "System Operator," or "Facility"). These requirements, developed through the allocation process identified in Block 0.3 of Figure 2.4, may be included in the appropriate lower-level *development, product, process,* or *material* specifications (refer to Figures 2.4 and 3.2).

Through the process of functional packaging and allocation, trade-off studies are conducted in evaluating the different design approaches that can be followed in responding to a given functional requirement (i.e., the "how!"). It may be appropriate to accomplish a designated function through the use of equipment (hardware), software, facilities, people, or combinations thereof. The proper mix is established, and the respective system elements are either developed or procured as required. At this stage, a plan is generated covering the steps necessary for the acquisition of the system elements that have been justified. Figure 3.14 presents an example of the evolution of hardware, software, and human requirements from the functional analysis and the steps required for acquisition. From a systems engineering perspective, it is essential that the requirements for equipment, software, people, facilities, and so on, be justified by responding to some *functional need.* Further, it is critical that the appropriate coordination and integration of the activities shown in Figure 3.14 be initiated from the beginning.[7]

[7] Referring to Figure 3.14, it is not uncommon for hardware engineers, software engineers, facility engineers, and other specialists to each proceed through a given series of steps without the appropriate level of communications between these activities. Further, the human element is often left out. This often results in significant problems being detected and numerous modifications being required during the final system

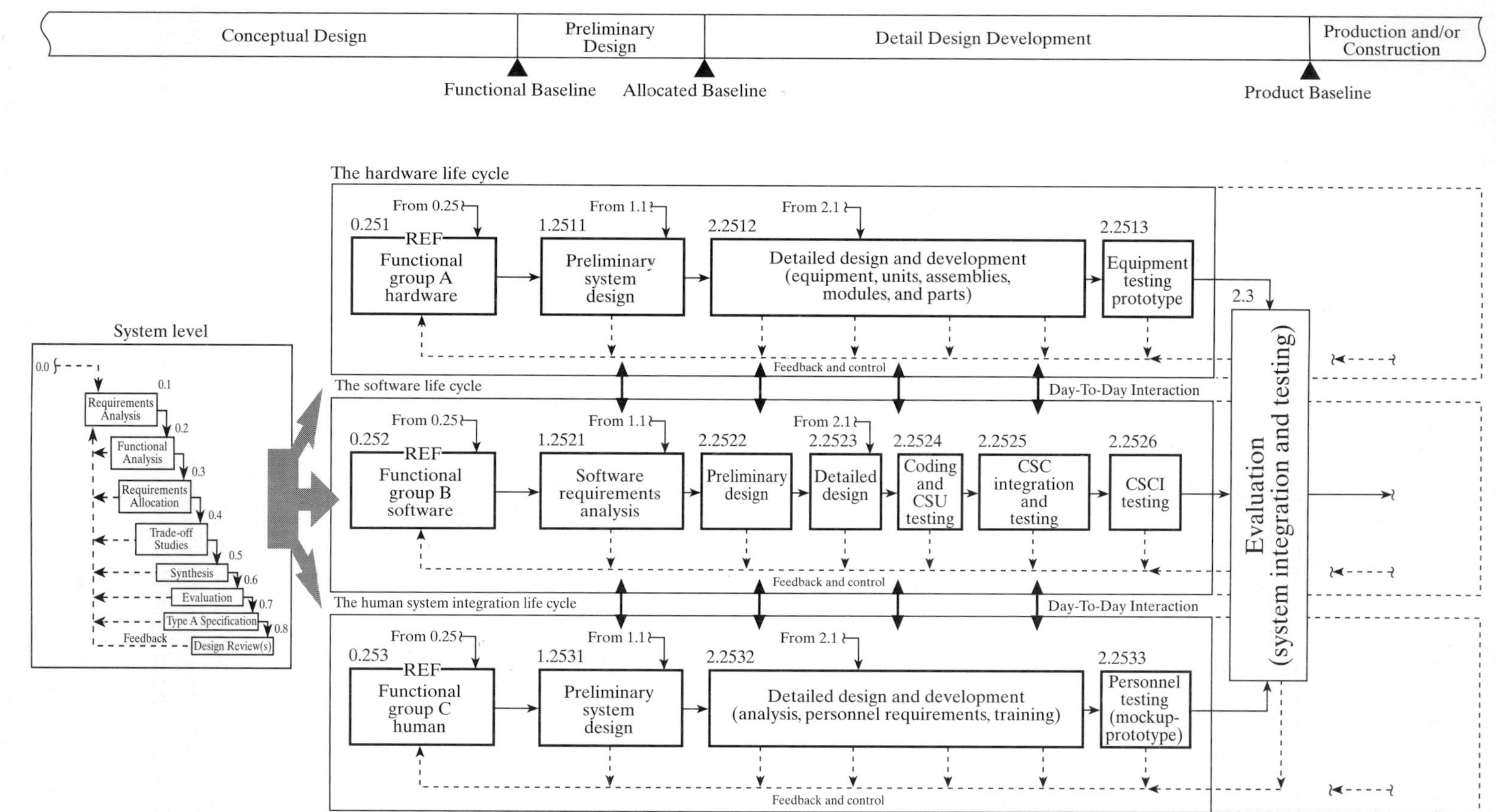

Figure 3.14 The evolution of hardware, software, and human requirements from the functional analysis (refer to Figure 2.4).

3.7 SYNTHESIS, ANALYSIS, AND EVALUATION

As the system design progresses, there are many possible trade-offs involving such issues as the evaluation and selection of different technologies, different materials, alternative system packaging schemes, alternative diagnostic routines, the evaluation and selection of COTS items, alternative maintenance and support policies, the incorporation of automation versus the accomplishment of functions by human means, and so on. Later, there may be alternative manufacturing processes, detailed maintenance plans, alternative logistic support structures, and/or alternative methods of material disposal that require evaluation. In general, the approach followed in the accomplishment of almost any trade-off study (or evaluation) is illustrated in Figure 3.15. One must first define the problem, identify the specific design criteria or measures against which the various alternatives will be evaluated (i.e., the applicable TPMs and DDPs), select the appropriate evaluation techniques, select or develop a model to facilitate the evaluation process, acquire the necessary input data, evaluate each of the candidates being considered, perform a sensitivity analysis and identify the potential areas of risk and finally recommend a course of action. This process can be "tailored" and applied at any point in the life cycle as illustrated in Figures 2.2 and 2.4. Inherent within this process is the application of the design evaluation morphology illustrated in Figure 2.10. Only the depth of the analysis and evaluation effort will vary, depending on the nature of the problem at hand.

The trade-off analysis leads into *synthesis*. Synthesis refers to the combining and structuring of components in such a way as to represent a feasible system configuration. The basic requirements have been established, some trade-off studies have been completed, and a baseline configuration is developed to demonstrate the concepts presented earlier.

Synthesis is *design*. Initially, synthesis is used in the development of preliminary concepts and to establish relationships among the various components of the system. Later, when sufficient functional definition and decomposition have occurred, synthesis is used to further define the "hows" at a lower level. Synthesis involves the creation of a configuration which could be representative of the form that the system will ultimately take (although a final configuration is certainly not to be assumed at this point in the process).

Given a synthesized configuration, one needs to evaluate the characteristics of that configuration in terms of the requirements initially specified. Changes are incorporated as required, leading to a preferred design configuration. This iterative process of analysis, synthesis, evaluation, and design refinement leads initially to the establishment of the "functional baseline," then the "allocated baseline," and finally the "product baseline" (refer to Figure 2.4). A good description of these configuration baselines, combined with a *disciplined* approach to baseline management, is essential for the successful implementation of the system engineering process.

integration and test task when it is discovered that the various elements do not fit (refer to Block 2.3)! The objective is to provide the proper day-to-day coordination (with the appropriate feedback) as one proceeds through Blocks 0.251, 0.252, 0.253, and so on, identifying potential problems as early in the life cycle as possible.

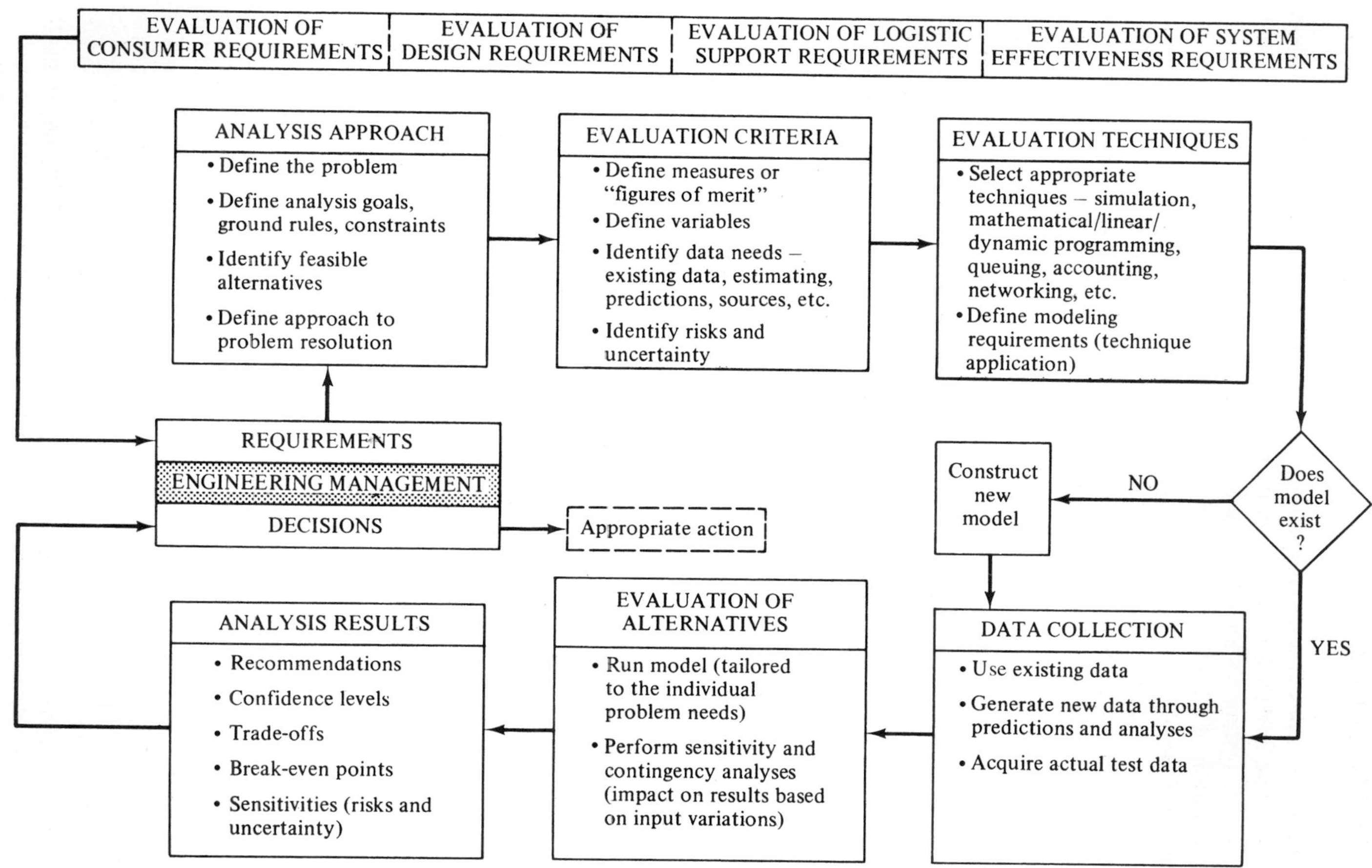

Figure 3.15 A generic systems analysis process.

3.8 SYSTEM SPECIFICATION

The *technical* requirements for the system and its elements are documented through a series of specifications, as conveyed in Figure 2.4. Of significance at this stage in the life cycle is the *System Specification* (type A). This is the single most important engineering *design* document, defining the system functional baseline and including the results from the needs analysis, feasibility analysis, operational requirements and the maintenance concept, top-level functional analysis, and identifying the critical TPMs and DDPs. This top-level specification leads into one or more subordinate specifications, as shown in Figure 3.2, covering applicable subsystems, configuration items, equipment, software, and other components of the system. Although the individual specifications for a given program may assume a different set of designations, a generic approach is used here. Further, the different categories of specifications are described subsequently.[8]

1. *System specification* (type A)—includes the technical, performance, operational and support characteristics for the system as an entity. It includes the allocation of requirements of functional areas, and it defines the various functional-area interfaces. The information derived from the feasibility analysis, operational requirements, maintenance concept, and the functional analysis is covered.
2. *Development specification* (type B)—includes the technical requirements for any item below the system level where research, design, and development are accomplished. This may cover an equipment item, assembly, computer program, facility, critical item of support, and so on. Each specification must include the performance, effectiveness, and support characteristics that are required in the evolving of design from the system level and down.
3. *Product specification* (type C)—includes the technical requirements for any item below the top system level that is currently in the inventory and can be procured "off the shelf." This may cover standard system components (equipment, assemblies, units, cables), a specific computer program, a spare part, a tool, and so on.
4. *Process specification* (type D)—includes the technical requirements that cover a service that is performed on any component of the system (e.g., machining, bending, welding, plating, heat treating, sanding, marking, packing, and processing).
5. *Material specification* (type E)—includes the technical requirements that pertain to raw materials, mixtures (e.g., paints, chemical compounds), or semifabricated materials (e.g., electrical cable, piping) that are used in the fabrication of a product.

The System Specification (type A) provides the *technical baseline* for the system as an entity, must be written in "performance-related" terms, and must describe design requirements in terms of the "whats" (i.e., the functions that the system is to perform and the associated metrics). Although the individual content should be "tailored" to the particular system in question, Figure 3.16 shows an example of what might be

[8] These categories were initially developed for application in the defense sector and are used in a "generic" sense throughout this text.

System Specification

1.0 Scope
2.0 Applicable Documents
3.0 Requirements
  3.1 System Definition
    3.1.1 General Description
    3.1.2 Operational Requirements (Need, Mission, Use Profile, Distribution, Life Cycle)
    3.1.3 Maintenance Concept
    3.1.4 Functional Analysis and System Definition
    3.1.5 Allocation of Requirements
    3.1.6 Functional Interfaces and Criteria
  3.2 System Characteristics
    3.2.1 Performance Characteristics
    3.2.2 Physical Characteristics
    3.2.3 Effectiveness Requirements
    3.2.4 Reliability
    3.2.5 Maintainability
    3.2.6 Usability (Human Factors)
    3.2.7 Supportability
    3.2.8 Transportability/Mobility
    3.2.9 Flexibility
    3.2.10 Other
  3.3 Design and Construction
    3.3.1 CAD/CAM Requirements
    3.3.2 Materials, Processes, and Parts
    3.3.3 Mounting and Labeling
    3.3.4 Electromagnetic Radiation
    3.3.5 Safety
    3.3.6 Interchangeability
    3.3.7 Workmanship
    3.3.8 Testability
    3.3.9 Economic Feasibility
  3.4 Documentation/Data
  3.5 Logistics
    3.5.1 Maintenance Requirements
    3.5.2 Supply Support
    3.5.3 Test and Support Equipment
    3.5.4 Personnel and Training
    3.5.5 Facilities and Equipment
    3.5.6 Packaging, Handling, Storage, and Transportation
    3.5.7 Computer Resources (Software)
    3.5.8 Technical Data
    3.5.9 Customer Services
  3.6 Producibility
  3.7 Disposability
  3.8 Affordability
4.0 Test and Evaluation
5.0 Quality Assurance Provisions
6.0 Distribution and Customer Service

Figure 3.16 Type A system specification format (example).

included for a large system. It is essential that a good comprehensive well-written specification be prepared from the beginning.[9]

3.9 CONCEPTUAL DESIGN REVIEW

Design progresses from an abstract notion to something that has form and function, is fixed, and can ultimately be reproduced in designated quantities to satisfy a need. Initially, a need is identified. From this point, a design evolves through a series of stages (i.e., conceptual design, preliminary system design, detail design and development). In each major stage of the design process, an evaluative function is accomplished to ensure that the design is correct at that point before proceeding with the next stage. The evaluative function includes both the informal day-to-day project coordination and data review, and the formal design review. Design information is released and reviewed for compliance with the basic system-equipment requirements (i.e., performance, reliability, maintainability, usability, etc., as defined by the system specification). If the requirements are satisfied, the design is approved as is. If not, recommendations for corrective action are initiated and discussed as part of the formal design review.

The formal design review constitutes a coordinated activity (including a meeting or series of meetings) directed to satisfy the interests of the design engineer and the technical discipline support areas (reliability, maintainability, human factors, logistics, manufacturing, industrial engineering, quality assurance, and program management). The purpose of the design review is to formally and logically cover the proposed design from the total system standpoint in the most effective and economical manner through a combined integrated review effort. The formal design review serves a number of purposes.

1. It provides a formalized check (audit) of the proposed system/subsystem design with respect to specification requirements. Major problem areas are discussed and corrective action is taken.
2. It provides a common baseline for all project personnel. The design engineer is provided the opportunity to explain and justify his or her design approach, and representatives from the various supporting organizations (e.g., maintainability, logistic support) are provided the opportunity to hear the design engineer's problems. This serves as an excellent communication medium and creates a better understanding among design and support personnel.

[9] It is not uncommon for the *system specification* to be written in very vague terms, with the objective of keeping the requirements "loose" in the beginning in order to provide for as much flexibility in design as possible (i.e., to allow for innovation on the part of the designer). Without a good foundation on which to build, the follow-on *B*, *C*, *D*, and *E* specifications, which are usually stated in rather *specific* terms, may not properly reflect what is ultimately desired by the customer in terms of system performance. Further, these lower-level specifications may not be mutually supportive or compatible if there is a poor description of the system upon which to base requirements. Thus, it is essential that a well-written *performance-based* specification be prepared in the beginning.

3. It provides a means for solving interface problems and promotes the assurance that all system elements will be compatible.
4. It provides a formalized record of what design decisions were made and the reasons for making them. Analyses, predictions, and trade-off study reports are noted and are available to support design decisions. Compromises to performance, reliability, maintainability, human factors, cost, and logistic support are documented and included in the trade-off study reports.
5. It promotes a higher probability of mature design, as well as the incorporation of the latest techniques (where appropriate). Group review may identify new ideas, possibly resulting in simplified processes and ultimate cost savings.

The formal design review, when appropriately scheduled and conducted in an effective manner, causes a reduction in the producer's risk relative to meeting specification requirements and results in improvement of the producer's methods of operation. Also, the customer often benefits through receipt of a better product.

Design reviews are generally scheduled before each major evolutionary step in the design process, as illustrated in Figure 2.4. In some instances, this may entail a single review toward the end of each stage (i.e., conceptual, preliminary system design, and detail design and development). For the other projects, where a large system is involved and the amount of new design is extensive, a series of formal reviews may be conducted on designated elements of the system. This may be desirable to allow for the early processing of some items while concentrating on the more complex, high-risk items.

Although the number and type of design reviews scheduled may vary from program to program, four basic types are readily identifiable and are common to most programs. They include the conceptual design review (i.e., system requirements review), the system design review, the equipment/software design review, and the critical design review. Of particular interest relative to the activities discussed in this chapter is the *conceptual design review*. The conceptual design review may be scheduled during the early part of a program (preferably not more than 4 to 8 weeks after program start) when operational requirements and the maintenance concept have been defined. Feasibility studies justifying preliminary design concepts should be reviewed. Logistic support requirements at this point are generally included in the maintenance-concept definition.

QUESTIONS AND PROBLEMS

1. In accomplishing a needs analysis in response to a given deficiency, what type of information would you include? Describe the process that you would use in developing the necessary information.
2. What is the purpose of the feasibility analysis? What considerations should be addressed in the completion of such an analysis?
3. Through a review of the literature, describe the QFD approach and how it could be applied in helping to define the requirements for a given system design.
4. Why is the definition of system operational requirements important? What type of information is included?

5. Why is it important to define specific mission scenarios (or operational profiles) within the context of the system operational requirements?
6. What information should be included in the system maintenance concept? How is it developed (describe the steps), and at what point in the system life cycle should it be developed?
7. How do system operational requirements influence the maintenance concept (if at all)?
8. How does the maintenance concept affect system/product design? Give some specific examples.
9. Select a system of your choice and develop the operational requirements for that system. Based on the results, develop the maintenance concept for the system. Construct the necessary operational and maintenance flows, identify repair policies, and apply quantitative effectiveness factors as appropriate.
10. In evaluating whether or not a two or three-level maintenance concept should be specified, what factors would you consider in the evaluation process?
11. In developing the maintenance concept, it is essential that all levels of maintenance be considered on an integrated basis. Why?
12. Why is the development of technical performance measures (TPMs) important?
13. Refer to Figure 3.10. Describe the steps that you would complete in developing the information included in the figure. Be specific.
14. Given the information provided in Figure 3.10, how would you apply this information in the system design process (if at all)?
15. What is meant by functional analysis? When in the system life cycle is it accomplished? What purpose does it serve? Identify some of the benefits derived. Can a functional analysis be accomplished for any system?
16. How does the functional analysis lead into the definition of specific resource requirements in the form of hardware, software, people, data, facilities, and so on? Briefly describe the steps in the process, and include an example. What is the purpose of the block numbering shown in Figures 3.11 and 3.12?
17. What is the purpose of allocation? To what depth in the system hierarchical structure should allocation be accomplished? How does it impact system design (if at all)?
18. Briefly define "synthesis," "analysis," and "evaluation." How are they related (if at all)?
19. Describe what is meant by a "design dependent parameter" (DDP). How are DDP's determined?
20. What is the purpose of the formal design review? What are some of the benefits derived from the conduct of design reviews? Describe some of the negative aspects.

4 Preliminary System Design

Preliminary system design, also known as advance development, begins by establishing a functional baseline for the system (refer to Figure 2.2, Block 2). It extends the translation of system-level requirements into design requirements for the subsystem level, configuration-item level, and below. Included is an extension of functional analysis and requirements allocation from the baseline described in Section 3.6, to the depth needed to identify specific requirements for hardware, software, people, facilities, logistic support elements, and related resources. Trade-off studies are conducted and the results described in the form of an allocated baseline (i.e., development, product, process, and material specifications).

This chapter continues the top-down approach to overall system design and development. It presents refined functional analysis and allocation (refer to Figure 2.4, Blocks 1.1 and 1.2), detailed trade-off studies (Block 1.3), synthesis and evaluation at the subsystem level (Blocks 1.4 and 1.5), and a final description of the major elements of the system. Results from these activities support detail design and development to be addressed in Chapter 5.

4.1 SUBSYSTEM FUNCTIONAL ANALYSIS

As described in Section 3.6, functional analysis is the iterative process of breaking down (or decomposing) requirements from the system level, to the subsystem level, and as far down the hierarchical system structure as necessary to describe functional interfaces and identify resource needs adequately. These resources are in the form of hardware, software, people, facilities, data, or combinations thereof. The functional analysis evolves from an identified need, the results of the feasibility analysis, the definition of system operational requirements, and the maintenance concept.

The Functional Analysis Process

The system is defined in functional terms through the generation of a series of block diagrams, as illustrated in Figure 3.12. This block diagramming approach aids in the establishment of activity sequences and series-parallel relationships, and in defining major functional interfaces that exist between various system components. Appendix A.1 gives a description of the development of functional block diagrams, along with selected illustrations.

The functional analysis reflects the "whats" that must be accomplished to achieve desired objectives. Conversion to "hows" is accomplished by evaluating each individual block, defining the necessary "inputs" and expected "outputs," describing external controls and constraints, and determining the mechanisms by which the function could be accomplished. Specific quantitative effectiveness factors (i.e., metrics) may be established for each functional block. By defining the mechanisms or the resources required for accomplishing the function, one is led to the identification of hardware, software, people, facilities, data, or combinations thereof. The example in Figure 4.1 shows a breakout of the design function with the need for people, materials, facilities, and design aids.[1]

Figure 4.2 presents a documentation format that may be used to record specific equipment needs, software needs, etc., for each function, beginning with the early stages of system design and extending through system retirement and disposal. In some cases, the same item of equipment, or item of software, or computer-based model may be identified for more than one function. A timeline analysis may be used to identify *when* the specific "resource" will be required and for *how long*. If there are parallel and conflicting requirements, two or more of the same item may be required. Conversely, combining and integrating some resources to accomplish multiple functions may be possible, leading to economy in resource consumption.

The general sequence of steps leading from the identification of a need to the functional analysis is illustrated in Figure 4.3. Although the functional block diagram covers events through the entire life cycle, Block 9 has been expanded to cover the operational scenario illustrated (i.e., fly aircraft from point A to point B). This scenario may be decomposed into segments, indicating the "whats" at a lower level. Eventually, one needs to convert to the "hows" resulting in the identification of a specific UHF communication subsystem (refer to Block 9.5.1). Actually, this shows only one of the many subsystems required to meet an airborne transportation need. Other subsystems must be defined in a similar manner.[2]

[1] With the proper application of performance and effectiveness factors, the need for design aids may lead to the identification of a specific design work station, computer-aided design tools, rapid prototyping capability, simulation, and/or other analytical design evaluation aids.

[2] *Operational functions* are mission related, covering various modes of system operation. Typical gross operating functions might be: "prepare aircraft for flight," "fly aircraft from City A to City B," and "recycle aircraft for the next flight." In the case of a communication system, typical operating functions might include:

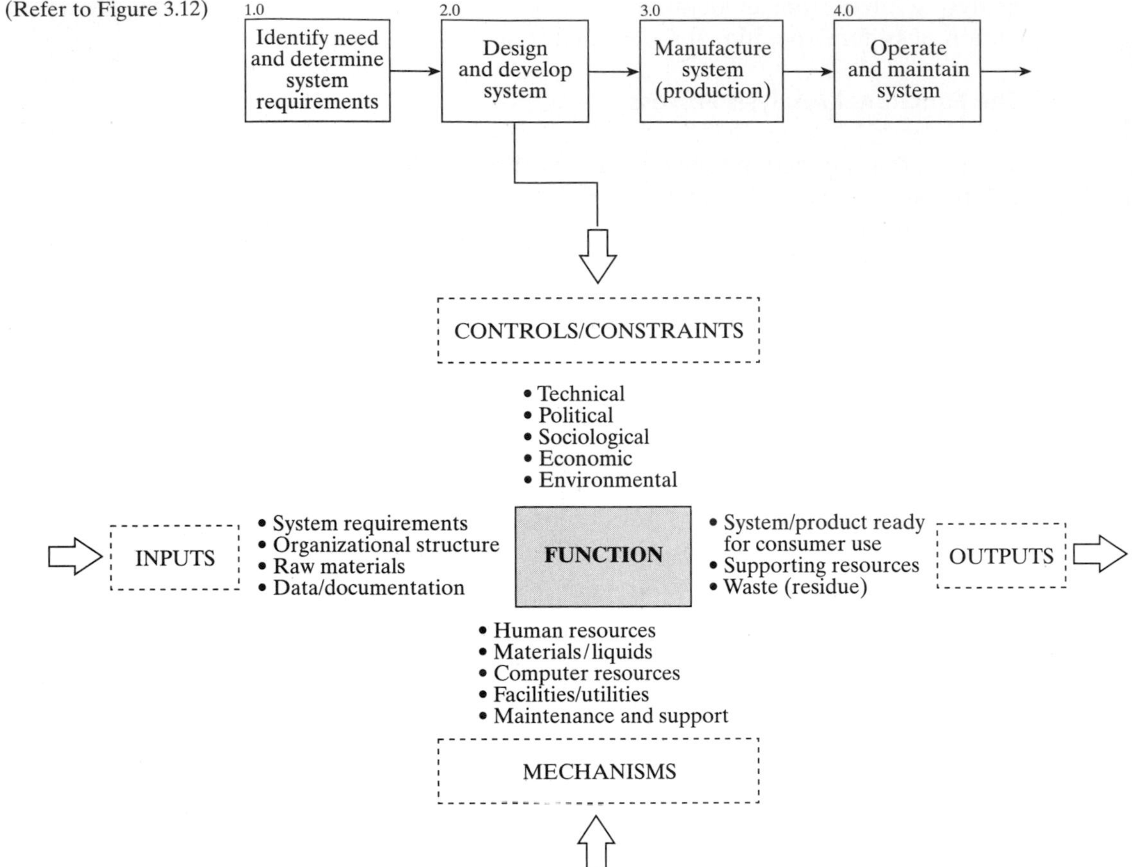

Figure 4.1 Identification of resource requirements (mechanisms).

Maintenance Functional Flows

Once operational functions are defined, the system description should lead to the identification of *maintenance functions*. For example, there are specified performance requirements for each operational function (signal level, tolerances accuracies, unit dimensions, output capacity, effectiveness measures, etc.). A check of the applicable function will indicate either a "go" or a "no-go" decision (represented by "G" and "$\overline{G}$" respectively). A "go" decision leads to a check of the next operational function. A "no-go" decision (constituting a symptom of failure) provides a starting point for the development of detailed maintenance functional flows and logic troubleshooting diagrams. A top-level functional flow diagram is illustrated in Figure 4.4.

"develop a communications system for an urban area of a given geographical size and population", "produce and install a communication network", and "accomplish the communication of certain designated information throughout the urban area seven days per week, six hours per day, with a specified degree of reliability." Operating functions cover the mission(s) to be accomplished in response to a customer need.

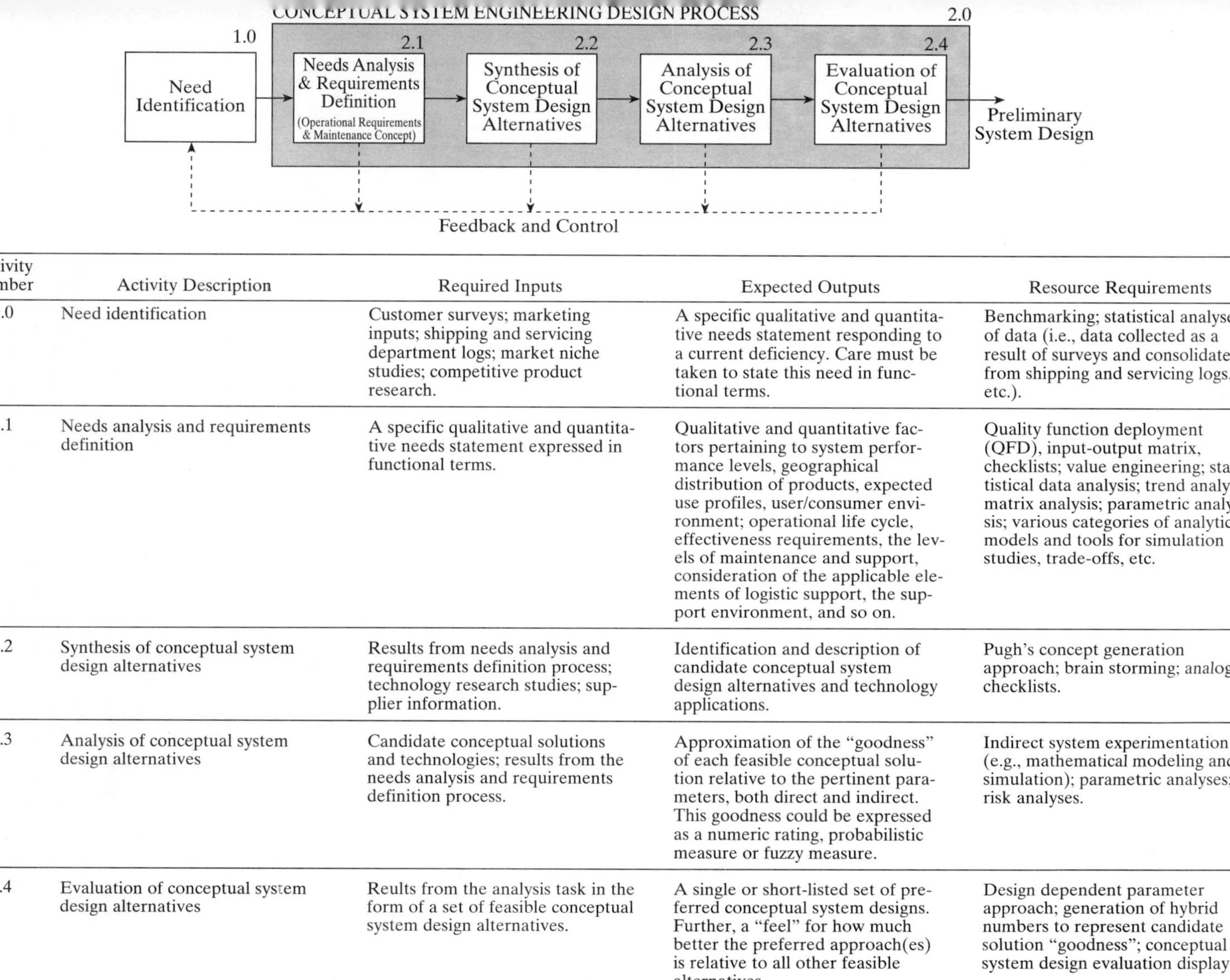

Activity Number	Activity Description	Required Inputs	Expected Outputs	Resource Requirements
1.0	Need identification	Customer surveys; marketing inputs; shipping and servicing department logs; market niche studies; competitive product research.	A specific qualitative and quantitative needs statement responding to a current deficiency. Care must be taken to state this need in functional terms.	Benchmarking; statistical analyses of data (i.e., data collected as a result of surveys and consolidated from shipping and servicing logs, etc.).
2.1	Needs analysis and requirements definition	A specific qualitative and quantitative needs statement expressed in functional terms.	Qualitative and quantitative factors pertaining to system performance levels, geographical distribution of products, expected use profiles, user/consumer environment; operational life cycle, effectiveness requirements, the levels of maintenance and support, consideration of the applicable elements of logistic support, the support environment, and so on.	Quality function deployment (QFD), input-output matrix, checklists; value engineering; statistical data analysis; trend analysis; matrix analysis; parametric analysis; various categories of analytical models and tools for simulation studies, trade-offs, etc.
2.2	Synthesis of conceptual system design alternatives	Results from needs analysis and requirements definition process; technology research studies; supplier information.	Identification and description of candidate conceptual system design alternatives and technology applications.	Pugh's concept generation approach; brain storming; analogy; checklists.
2.3	Analysis of conceptual system design alternatives	Candidate conceptual solutions and technologies; results from the needs analysis and requirements definition process.	Approximation of the "goodness" of each feasible conceptual solution relative to the pertinent parameters, both direct and indirect. This goodness could be expressed as a numeric rating, probabilistic measure or fuzzy measure.	Indirect system experimentation (e.g., mathematical modeling and simulation); parametric analyses; risk analyses.
2.4	Evaluation of conceptual system design alternatives	Reults from the analysis task in the form of a set of feasible conceptual system design alternatives.	A single or short-listed set of preferred conceptual system designs. Further, a "feel" for how much better the preferred approach(es) is relative to all other feasible alternatives.	Design dependent parameter approach; generation of hybrid numbers to represent candidate solution "goodness"; conceptual system design evaluation display.

Figure 4.2 Documentation format reflecting functions, input-output requirements, and resource requirements (mechanisms).

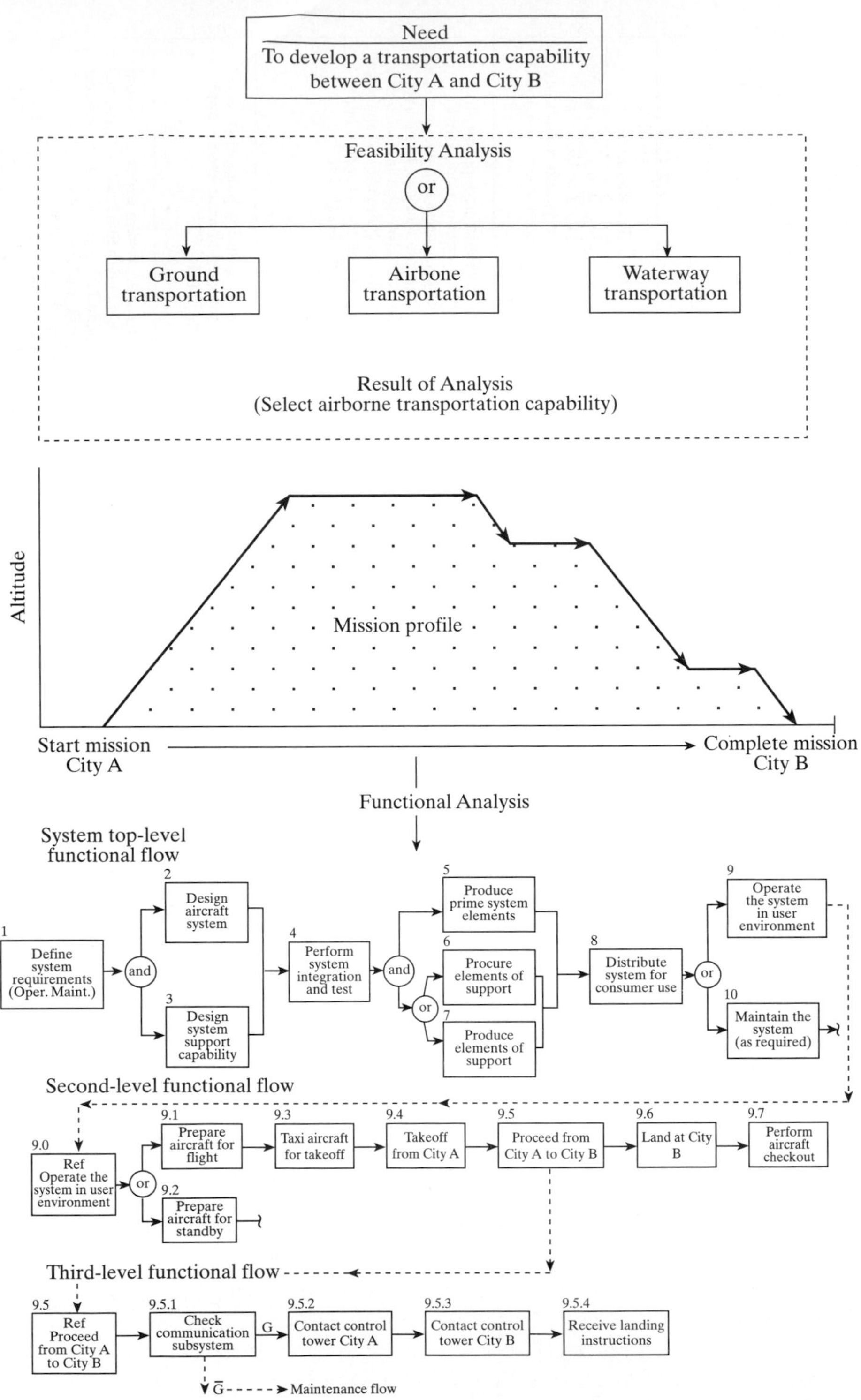

Figure 4.3 Progression from the "need" to the functional analysis.

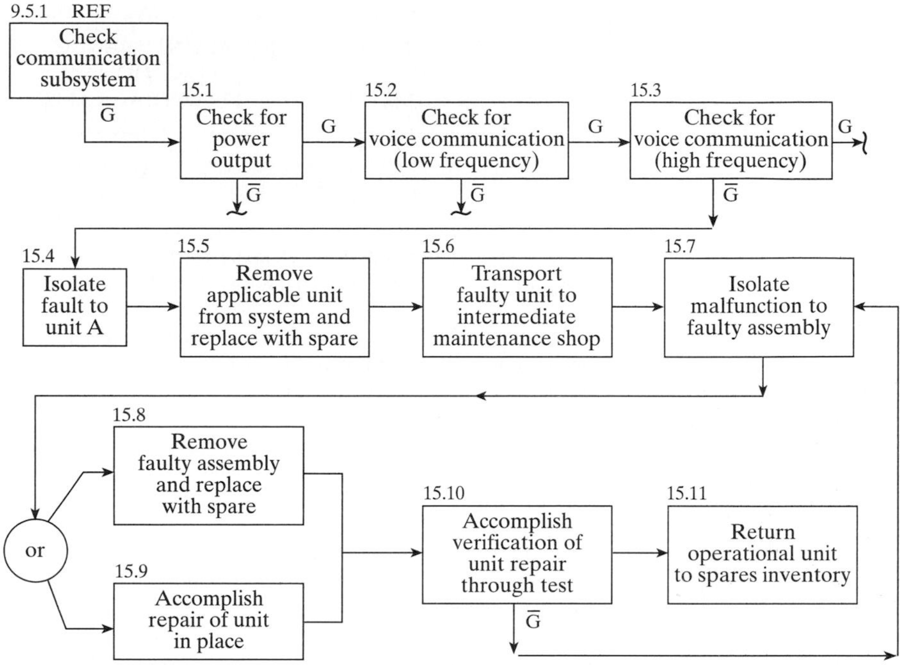

Figure 4.4 Maintenance functional block diagram.

Maintenance functional flow diagrams are prepared for the purpose of establishing supportability design criteria for both the prime elements of the system and the maintenance and logistic support infrastructure. These flow diagrams should be prepared for corrective maintenance, preventive maintenance, inspection, servicing, transportation and handling, support equipment maintenance, and so on. Further, they are developed and used to supplement and/or update the system maintenance concept.

Functional Analysis Summary

Functional analysis provides an initial description of the system. Accordingly, its applicability is extensive. It serves as a "baseline" for definition of equipment requirements, software requirements, whether to support operational functions or maintenance functions, and so on. In the past, these requirements have often been derived independently. The results can conflict because of different baselines having been followed. Often, the electrical design has evolved from one baseline, the mechanical design from another, the reliability and maintainability requirements from yet another, the logistics requirements from an independent source, and so on. The various design disciplines often do not "track" the same baseline, leading to unsatisfactory results.

A major objective of system engineering is to develop *one* set of requirements and to define a *single* baseline from which all lower-level requirements may evolve. Figure 4.5 identifies some areas where a functional analysis is required as an input. It is essential that all such activities evolve from the *same* functional definition.

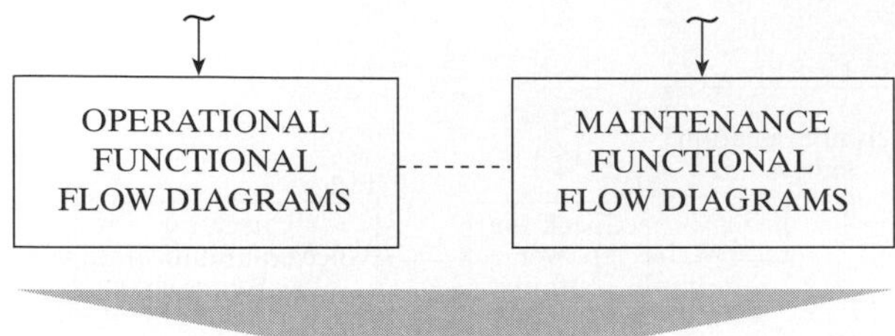

- Functional packaging of system elements (technology integration; equipment packaging; software modularization).
- Reliability analysis (reliability models and block diagrams; failure mode, effects, and criticality analysis; fault-tree analysis; reliability-centered maintenance).
- Maintainability analysis (maintainability analysis; maintenance task analysis; diagnostic and testability analysis; level of repair analysis).
- Human factors and safety (operator task analysis; operational sequence diagrams; system safety/hazard analysis; personnel training requirements).
- Maintenance and logistic support analysis (spares/repair parts and associated inventory requirements; test and support equipment requirements; personnel requirements; facility layout and design; technical data packaging; computer resources).
- Economical analysis (life-cycle cost analysis) (functional costing; activity-based costing by function).

Note: Many of these activities, which evolve from the functional analysis, are discussed in Part IV (Chapters 12–17).

Figure 4.5 Activity flow evolving from the functional requirements.

4.2 REQUIREMENTS ALLOCATION

Lower-level elements of the system are defined by combining (or grouping) similar functions into logical subdivisions, identifying major subsystems, configuration units, and so forth (refer to Figure 3.13). Figure 4.6 presents a summary of this process, evolving from a functional definition of System XYZ to the packaging of the system into three units (i.e., Unit A, Unit B, and Unit C).[3]

[3] The system may be broken down into subsystems, subsystems may be broken down into configuration items or units, and so on. The specific packaging concept (and nomenclature) may vary depending on the complexity and make-up of the system. However, the objective is to divide the system into discrete elements and to allocate specific design criteria as an input for the development and/or procurement of that element.

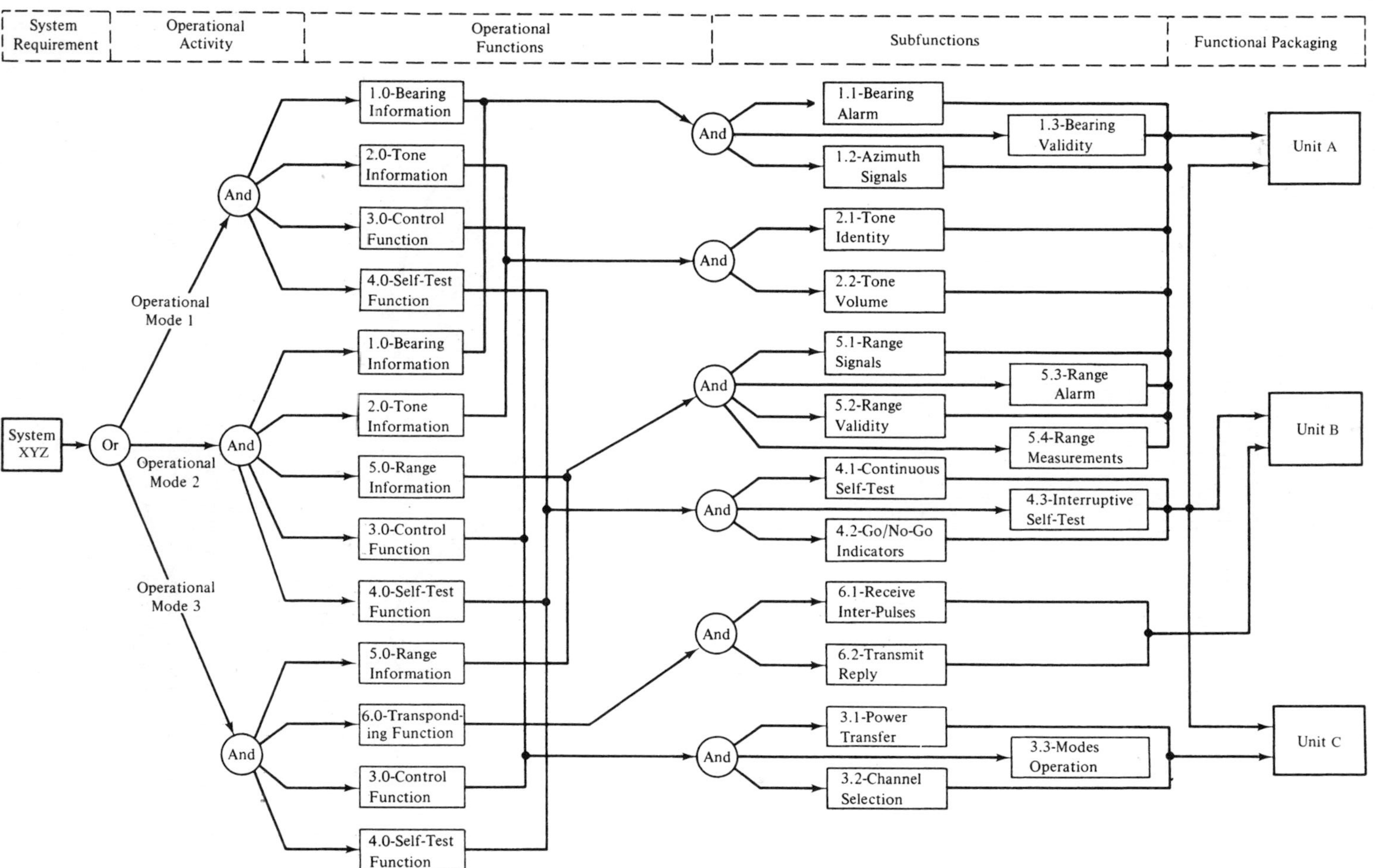

Figure 4.6 The functional packaging of the system into major elements.

Given the "packaging" concept shown in Figure 4.6, it is now appropriate to apply the technical performance measures (TPMs) to the units. The objective is to allocate, or apportion, the system-level "design-to" requirements down to Unit A, Unit B, and Unit C, and to develop specific qualitative and quantitative "design-to" requirements for each of the units (i.e., DDPs). This is accomplished for the purpose of including the necessary criteria in the lower-tier specifications (refer to Figure 3.2). It leads to either the development of new units, or the procurement of off-the-shelf commercial standard items that will satisfactorily perform the required functions. The goal is to influence the design from the top down.

The approach often followed results in designing or procuring the units first and then hoping that, when integrated later, the overall system requirements will be met. If the proper top-down requirements are established from the beginning, there is a much better chance that the various elements of the system will be compatible with system functional requirements and with each other. Accordingly, the need for downstream modifications will be minimized.

In the development of design goals at the unit level, priorities established for the system must be addressed (illustrated in Figure 3.10), and the desired characteristics of design must be identified. These design characteristics should be "tailored" in response to the relative importance of each of the planned system measures. Design characteristics, presented in a hierarchy, are as shown in Figure 2.9. These parameters, initially viewed form the top down, are compared, trade-off studies are conducted to evaluate the interaction affects, and those characteristics most significant in meeting the overall objectives are selected. An example of some trade-off approaches is shown in Figure 4.7. In each illustration, the allowable "design space" is shown by the shaded area.

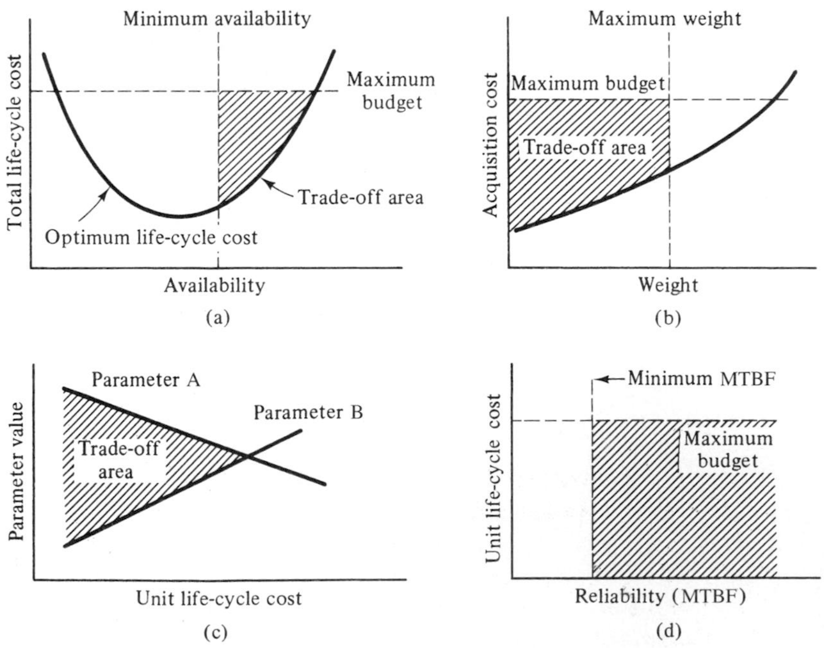

Figure 4.7 Trade-off areas for optimization (example).

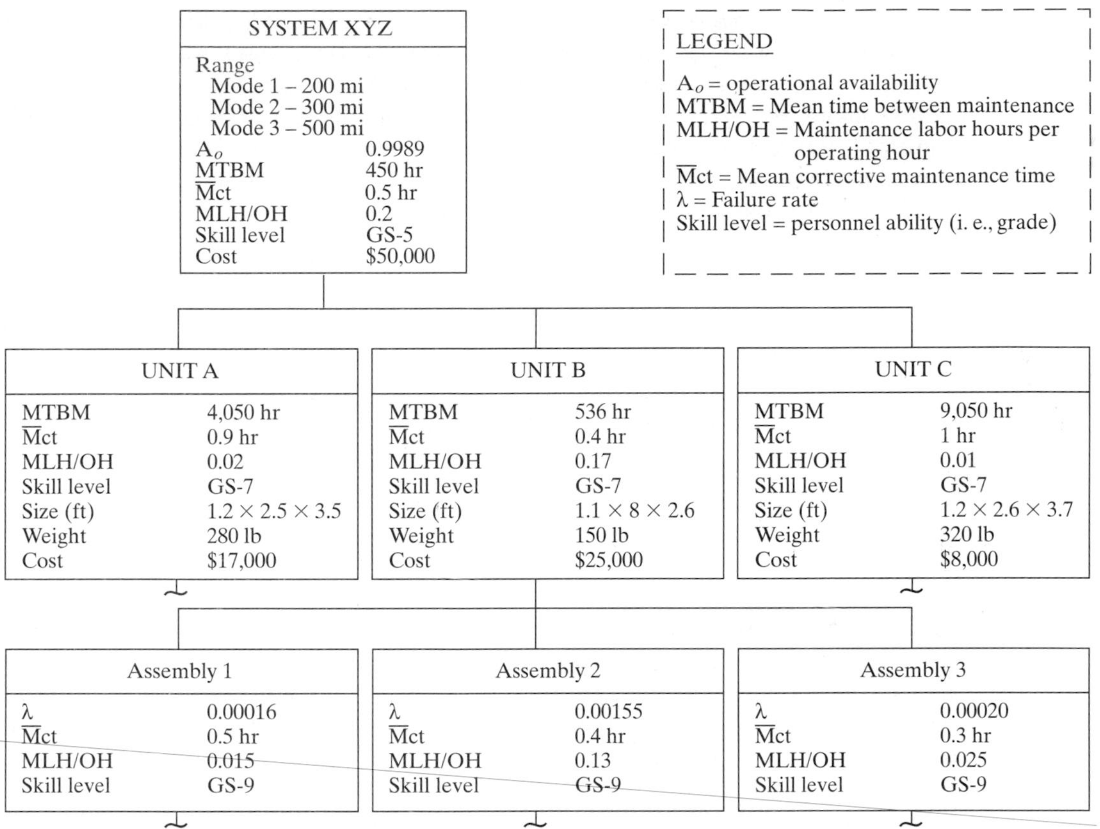

Figure 4.8 System XYZ requirements allocation.

Figure 4.8 shows the results of an allocation of System XYZ requirements to the unit level and the assembly level. The objective is to establish criteria or goals for the design or procurement of the three units and the assemblies. Allocation should consider all significant parameters stated in the form of maximum or minimum requirements (with tolerance bands where appropriate). These include the following:

1. System performance and physical parameters, such as range, accuracy, speed, capacity, output rate, power rating, size, and volume.
2. System effectiveness factors, such as operational availability, readiness, reliability, maintainability, producibility, dependability, supportability, and disposability.
3. System support capability factors, such as the transportation or supply times between levels of maintenance, spare parts availability, turnaround times for maintenance, test equipment use and reliability, facility use, information transfer rates, and personnel effectiveness factors.
4. System life-cycle cost factors including research and development cost, investment or production cost, operation and maintenance support cost, and system retirement and disposal cost.

The purpose of allocation is to provide, as an input, some guidelines to assist design engineers develop or select a product (whether it be equipment, an item of software, personnel, data, etc.) that will ultimately be compatible with, and contribute to, meeting system-level requirements. That is, when combining and integrating the units in Figure 4.8, the results must be in compliance with system requirements. This usually requires an iterative process because it may not always be possible to meet the allocated parameter values with the available technology exactly. Trade-offs are made with other elements of the system, the allocations are revised as necessary, and the best approach is defined as a model. In any event, a solid *baseline* must be established before proceeding.[4]

4.3 DESIGN REQUIREMENTS (PARAMETERS)

The basic design objective(s) for the system and its elements must (1) be compatible with the system operational requirements and the maintenance concept discussed in Section 3.4, (2) comply with the allocated design criteria established by the procedure described in Section 4.2, and (3) meet all specification requirements. Specific design characteristics to be incorporated will vary from one instance to the next, depending on the type and complexity of the system and the mission it is intended to accomplish. In all cases, design activity must consider downstream outcomes having to do with a primary purpose as well as with production, operation, maintenance, phaseout, and disposal. These design dependent parameters (and TPMs) are briefly described subsequently:

1. *Design for functional capability*—Functional capability derives from the characteristics of design that relate to the technical performance of the system. This includes such factors as size, weight, volume, shape, accuracy, capacity, flow rate, throughput, units per time period, speed of travel, power output, and all of the technical and physical characteristics that the system (when operating) must exhibit to accomplish its intended mission. Functional capability is the main concern of aeronautical design, chemical design, electrical design, mechanical design, structural design, and others.
2. *Design for reliability*—Reliability is that characteristic of design and installation concerned with the successful operation of the system throughout its planned mission. Reliability (R) is often expressed as the probability of success, or is measured in terms of MTBF, MTTF, λ, or a combination of these. An objective is to maximize operational reliability while minimizing system failure. Chapter 12 addresses design for reliability.
3. *Design for maintainability*—Maintainability is that characteristic of design and installation that reflects the ease, accuracy, safety, and economy of performing maintenance actions. Maintainability can be measured in terms of maintenance

[4] Various forms of allocation are covered in Part IV of this text. Specifically, reliability allocation is described in Chapter 12 and maintainability allocation is covered in Chapter 13.

times (mean corrective maintenance time or $\overline{M}ct$, mean preventive maintenance time or $\overline{M}pt$, active maintenance time or $\overline{M}$, maximum maintenance time or Mmax, logistics delay time or LDT, administrative delay time or ADL, and total maintenance downtime or MDT); maintenance frequency (mean time between maintenance or MTBM); maintenance labor hours (MLH); and/or maintenance cost. The objective is to minimize maintenance times and labor hours while maximizing supportability characteristics in design (e.g., accessibility, diagnostic provisions, standardization, interchangeability), minimize the logistic support resources required in the performance of maintenance (e.g., spares and supporting inventories, test equipment, maintenance personnel), and minimize maintenance cost. Chapter 13 addresses design for maintainability.

4. *Design for usability and safety*—Usability is that characteristic of design concerned with the interfaces between the human and hardware, the human and software, the human and facilities, and so on (i.e., ensuring the compatibility between, and safety of, system physical and functional design features and the human element in the operation, maintenance, and support of the system). Considerations must include anthropometric, human sensory, physiological, and psychological factors as they apply to operability and supportability. An objective is to minimize the number of people and skill-level requirements, minimize training requirements, and minimize human errors while maximizing productivity and safety. Chapter 14 addresses human factors and safety in design.
5. *Design for supportability and serviceability*—Supportability and Serviceability are characteristics of design that ensure that the system can ultimately be serviced and supported effectively and efficiently throughout its planned life cycle. An objective is to consider both the internal characteristics of the prime mission-related elements of the system (e.g., prime equipment, software, and personnel used in system operation) and the design of the maintenance and logistic support infrastructure. Supportability includes the consideration and application of reliability, maintainability, and human factors in design. Design for supportability is addressed in Chapter 15.
6. *Design for producibility and disposability*—Producibility is that characteristic of design that pertains to the ease and economy of producing a system or product. The objective is to design an item that can be produced (in multiple quantities) easily and economically, using conventional and flexible manufacturing methods and processes without sacrificing function, performance, effectiveness, or quality. Considerations include the utilization of standard parts, the construction and packaging for the ease of assembly (and disassembly), the use of standard equipment and tools in manufacturing, and so on. Disposability is that characteristic of design that allows for the disassembly and disposal of elements or components of the system easily, rapidly, and economically without causing environmental degradation. Design for producibility and disposability are addressed in Chapter 16.
7. *Design for affordability*—Economic feasibility or affordability are characteristics of design and installation that impact budget constraints. An objective is to base design decisions on *life-cycle* cost, and not just system acquisition cost (or purchase price). Economic feasibility is dependent on the balanced incorporation of

reliability, maintainability, human factors, supportability, and other related characteristics of design. Design for economic feasibility is addressed in Chapter 17.

Many design considerations interact with each other. Accordingly, the degree of emphasis from the standpoint of incorporation into the ultimate configuration depends on the prioritization and relative importance of the TPMs as described in Section 3.5. Given a specific set of quantitative "design-to" parameters, the specific characteristics that need to be incorporated and built into the design must directly support the specified TPMs (refer to Figure 2.9). Thus, some features may be more important than others.

Accomplishing the overall objective requires an appropriate balance between performance features, reliability and maintainability characteristics, human factors, supportability factors, life-cycle cost, and so on. This balance is difficult to attain, since some stated objectives appear to be in opposition to others. For example, incorporating high performance and highly sophisticated functional design may reduce the reliability and increase logistic support requirements and life-cycle cost; incorporating too much reliability may require the use of costly system components and thus increase both acquisition cost and life-cycle cost; the incorporation of too much accessibility for improving maintainability may cause a significant increase in equipment size. The goal is to conduct the necessary trade-off studies and to incorporate only those features considered to be essential to meet the requirements—not too many or too few. One must be careful not to overdesign or underdesign.

Figure 4.9 is an extension of the concept in Figures 2.4 and 3.14. It emphasizes three individual life cycles (hardware, software, and the human), and shows the evolution of design from the system level, to the subsystem level, and so on. As one proceeds through the design process, the specific characteristics of design must be defined for each level (refer to Figure 2.9). Reliability considerations are important at each level as are maintainability considerations, and so forth, and the various characteristics must be complementary as one proceeds downward. These characteristics must support the allocated requirements identified in Figure 4.8. Finally, appropriate design characteristics identified at each level must be included in the applicable development, product, process, or material specification(s).

4.4 ENGINEERING DESIGN TECHNOLOGIES

To design is to project and evaluate ideas for bringing a needed system or product into being. A design is an abstraction of what could be. For the design process to yield a good result, numerous abstractions (design alternatives) must be formulated and evaluated. This must be accomplished accurately, with relative ease, and in a limited period. The design engineer or the design team needs to converge to a final design which meets the need effectively and efficiently. Several computer-based technologies are emerging to facilitate this design assurance.

INTEGRATION OF THE HARDWARE, SOFTWARE, AND HUMAN LIFE CYCLES

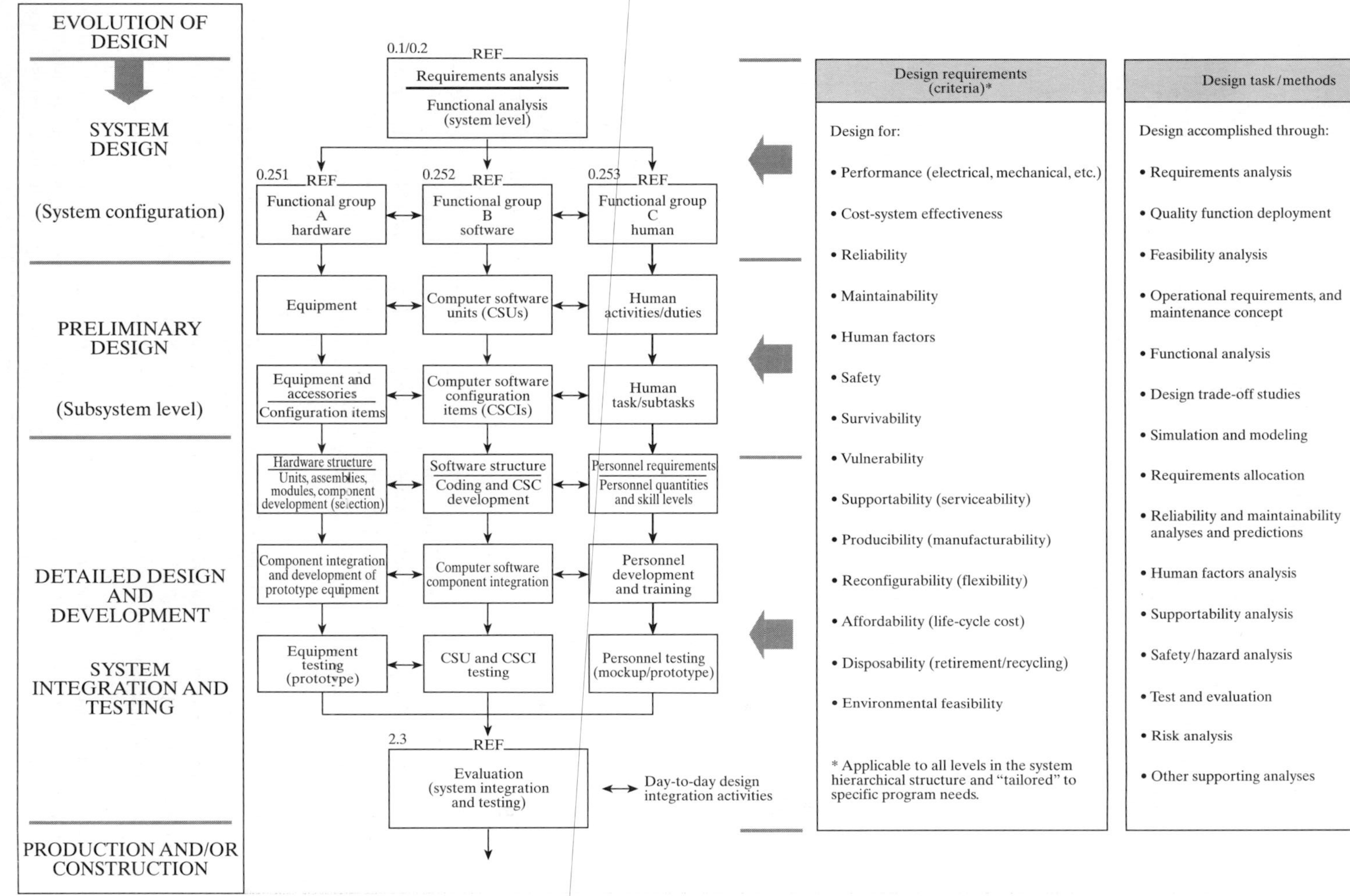

Figure 4.9 The breakout of design requirements by system indenture level (refer to Figure 3.14).

Computer-Aided Engineering Design (CAED)

The designer must solve a variety of problems and evaluate numerous alternatives in a limited amount of time, and still must define a total system or product that will meet all requirements effectively and efficiently in a highly competitive international environment. To meet this challenge requires use of a wide range of computer-based design aids. Generic categories of computer-aided methods include CAD, CAED, computer-aided design/computer-aided manufacturing (CAD/CAM), and computer-aided acquisition and logistic support (CALS).[5]

Computer usage in the design process is not new. It is used to generate drawings and graphic displays, to facilitate the accomplishment of analyses, to develop material lists, and so on. However, these various applications have been accomplished primarily on an individual basis. One computer program has been developed to produce design drawings, another program has been designed to aid in accomplishing a reliability analysis, a third program is used to generate component part lists, and so on. In essence, the expression relating to "islands of automation" addresses a reality. There are many different independent approaches intended to help solve a wide variety of problems. Further, the language requirements and program formats are different, and there is limited (if any) integration of these design methods. This approach can be quite costly overall.

Application of CAED Technologies

The application of CAD, CAM, CALS, and related tools promotes the integration of these various methods into a single entity. These design-related aids must be readily available and easy to use, compatible with each other facilitating the flow of information in a timely manner, and adaptable to the system or product design configuration at hand. The overall concept illustrated in Figure 4.10 reflects some of the major interfaces that must be addressed.

Referring to the figure, the objective of CAD is to allow for the accomplishment of design analyses, the projection of various design configurations in terms of three-dimensional graphical presentations, the accomplishment of reliability and maintainability predictions, the generation of design material lists, etc. There are many individual problems that require resolution and, through CAD, are integrated into the overall design process. The results are projected through the use of electronic data bases and, subsequently, a digital format, or in the form of hard drawings using CAD.

The results of CAD, in terms of components that must be manufactured, should feed directly into CAM. In other words, components that have been designed through CAD and must be produced should be addressed using CAM. Computer-aided process planning (CAPP) is used in responding to the CAD output with the objective of determining specific process requirements. Again, the information flow should be continu-

[5] The Department of Defense (DOD) currently uses the term, "continuous acquisition and life-cycle support," in addressing computer-aided applications to the system maintenance and support activities. However, "computer-aided acquisition and logistic support" is more descriptive. In any event, there are many different areas where the application of computer-aided methods is appropriate.

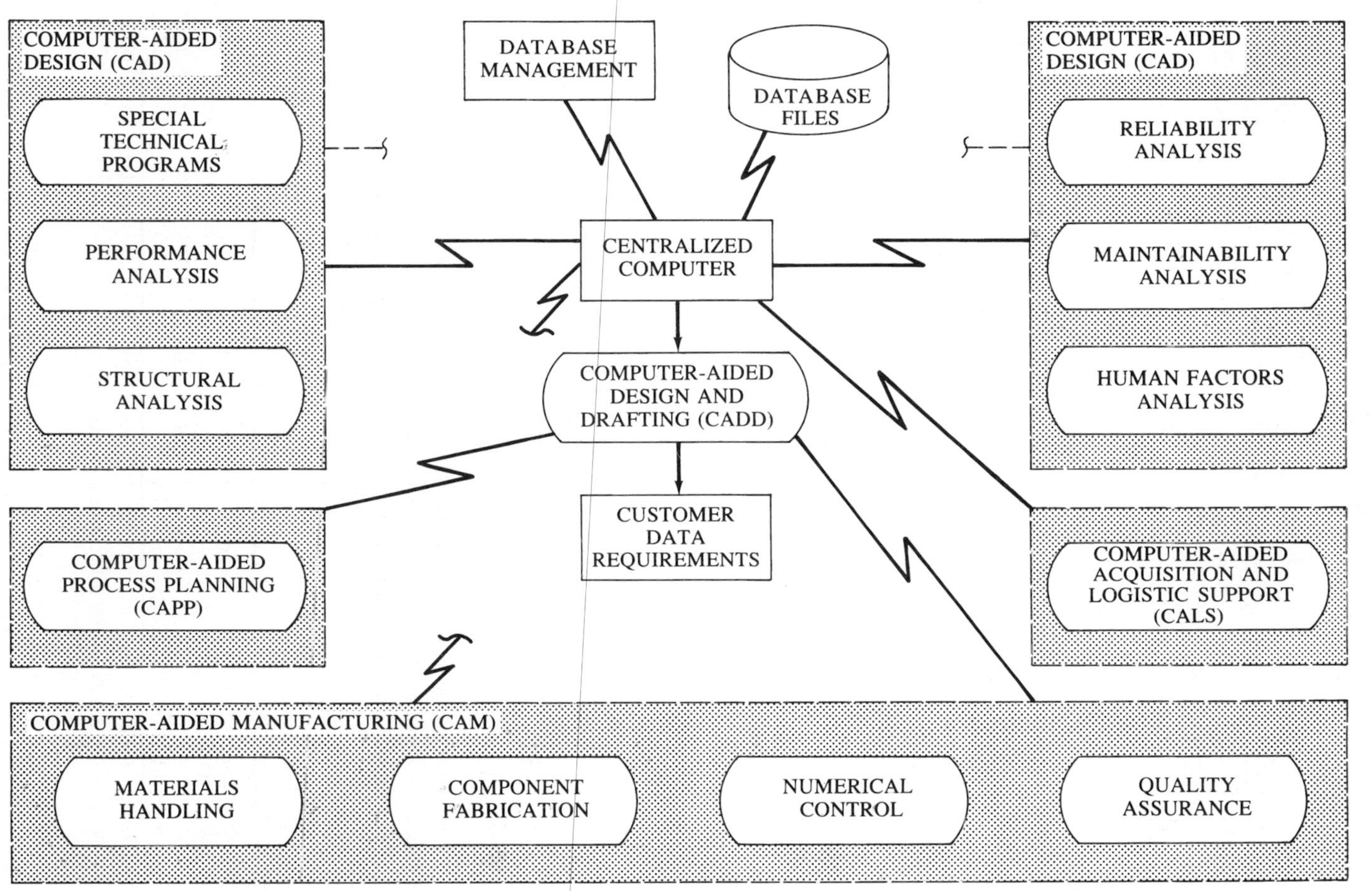

Figure 4.10 An example of the CAD, CAM, and CALS interface.

ous, providing the necessary continuity between design activities and production activities.

CALS, which primarily addresses the logistic support aspects of the system, is oriented both to system design and production. The objective of CALS is to provide a vehicle for design evaluation, as well as a tool for the processing of logistics data. CALS is addressed further in Chapter 15.

The use of CAD, CAM, and CALS methods, implemented on an integrated basis, offers many advantages. A few of these are the following:

1. The designer can address many different alternatives in a relatively short time frame. With the capability of evaluating a greater number of possible options, the risks associated with design decision making are reduced.
2. The designer is able to simulate, and verify design, for a greater number of configurations by using three-dimensional projections. Employing an electronic data base to facilitate this objective may eliminate the need to build a physical model later on, thus reducing cost.
3. The ability to incorporate design changes is enhanced, both in terms of the reduced time for accomplishing them and in the accuracy of data presentation. If the information flow is properly integrated, changes will be reflected throughout the data base as applicable. Also, the vehicle for incorporating a change is greatly simplified.
4. The quality of design data is improved, both in terms of the methods for data presentation and in the reproduction of individual data elements.
5. The availability of an improved database earlier in the system life cycle (with better methods for presenting design configurations) facilitates the training of personnel assigned to the project. Not only is it possible to better describe the design through graphical means, but a common database will help to ensure that all design activities are tracking the same baseline.

An integrated life-cycle approach to the application of computer-aided methods is required. One way of reflecting such an approach is through the concept of MacroCAD illustrated in Figure 4.11. The objective of MacroCAD is to link early life-cycle

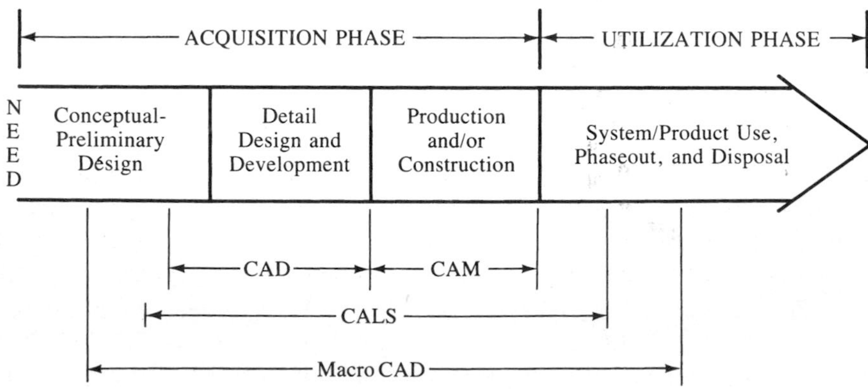

Figure 4.11 The relationship of CAD, CAM, CALS, and MacroCAD.

design decisions with operational outcomes, using simulation methods and mathematical modeling. This approach is feasible because of the large body of knowledge available about operational modeling and simulation. Key is the identification of design dependent system parameters and optimization over design variables. For each set of design dependent system parameters, one can identify an optimal value for the chosen measure of evaluation. By comparing these optimal values for a set of alternatives, a preferred configuration can be selected. The mathematical and algorithmic basis for MacroCAD is developed further in Part III of this text.

Analytical Models and Tools

The design evaluation process may be facilitated further through the use of various analytical models or tools in support of the MacroCAD objective. A model, in this context, is a simplified representation of the real world which abstracts features of the situation relative to the problem being analyzed. It is a tool employed by an analyst to assess the likely consequences of various alternative courses of action being examined. The model must be adapted to the problem at hand and the output must be oriented to the selected evaluation criteria. The model, in itself, is not the decision maker but is a tool that provides the necessary data in a timely manner in support of the decision-making process.

The extensiveness of the model will depend on the nature of the problem, the number of variables, input parameter relationships, number of alternatives being evaluated, and the complexity of operation. The ultimate objective in the selection and development of a model is simplicity and usefulness. The model used should incorporate the following features.

1. The model should represent the dynamics of the system configuration being evaluated in a way that is simple enough to understand and manipulate, and yet close enough to the operating reality to yield successful results.
2. The model should highlight those factors that are most relevant to the problem at hand, and suppress (with discretion) those that are not as important.
3. The model should be comprehensive by including *all* relevant factors and be reliable in terms of repeatability of results.
4. Model design should be simple enough to allow for timely implementation in problem solving. Unless the tool can be used in a timely and efficient manner by the analyst or the manager, it is of little value. If the model is large and highly complex, it may be appropriate to develop a series of models where the output of one can be tied to the input of another. Also, it may be desirable to evaluate a specific element of the system independently from other elements.
5. Model design should incorporate provisions for ease of modification or expansion to permit the evaluation of additional factors as required. Successful model development often includes a series of trials before the overall objective is met. Initial attempts may suggest information gaps which are not immediately apparent and consequently may suggest beneficial changes.

The use of mathematical models offers significant benefits. In terms of system application, several considerations exist—operational considerations, design considerations, product/construction considerations, testing considerations, and logistic support considerations. There many interrelated elements that must be integrated as a system and not treated on an individual basis. The mathematical model makes it possible to deal with the problem as an entity and allows consideration of all major variables of the problem on a simultaneous basis.

1. The mathematical model will uncover relations between the various aspects of a problem that are not apparent in the verbal description.
2. The mathematical model enables a comparison of *many* possible solutions and aids in selecting the best among them rapidly and efficiently.
3. The mathematical model often explains situations that have been left unexplained in the past by indicating cause-and-effect relationships.
4. The mathematical model readily indicates the type of data that should be collected to deal with the problem in a quantitative manner.
5. The mathematical model facilitates the prediction of future events, such as effectiveness factors, reliability and maintainability parameters, logistics requirements, and so on.
6. The mathematical model aids in identifying areas of risk and uncertainty.

When analyzing a problem in terms of selecting a mathematical model for evaluation purposes, it is desirable to first investigate the tools that are currently available. If a model already exists and is proven, then it may be feasible to adopt that model. However, extreme care must be exercised to relate the right technique with the problem being addressed and to apply it to the depth necessary to provide the sensitivity required in arriving at a solution. Improper application may not provide the results desired, and the result may be costly.

Conversely, it might be necessary to construct a new model. In accomplishing such, one should generate a comprehensive list of system parameters that will describe the situation being simulated. Next, it is necessary to develop a matrix showing parameter relationships, each parameter being analyzed with respect to every other parameter to determine the magnitude of relationship. Model input-output factors and parameter feedback relationships must be established. The model is constructed by combining the various factors and then testing for validity. Testing is difficult to do because the problems addressed primarily deal with actions in the future which are impossible to verify. However, it may be possible to select a known system or equipment item that has been in existence for several years and exercise the model using established parameters. Data and relationships are known and can be compared with historical experience. In any event, the analyst might attempt to answer the following questions:

1. Can the model describe known facts and situations sufficiently well?
2. When major input parameters are varied, do the results remain consistent and are they realistic?

3. Relative to system application, is the model sensitive to changes in operational requirements, design, production/construction, and logistic support?
4. Can cause-and-effect relationships be established?

Models and appropriate systems analysis techniques are presented further in Part III. Trade-off and system optimization relies heavily on these techniques and their application. A fundamental knowledge of probability and statistics, economic analysis techniques, modeling and optimization, simulation, queuing theory, control techniques, and the other analytical techniques is essential to accomplishing effective systems analysis.

Finally, the results of the overall system evaluation effort will lead to the selection of a preferred approach. If the configuration selected is significantly better than other options, the decision may be clear-cut and the systems engineer may proceed toward greater system definition. However, in most cases the engineer will perform some kind of a self-analysis to ensure the correct decision. The following questions will usually arise.

1. How much better is the selected approach than the next-best alternative? Is there a significant difference between the results of the comparative evaluation?
2. Have all feasible alternatives been considered?
3. What are the areas of risk and uncertainty?

These and other comparable questions should be addressed, and to obtain some of the answers may require one or more iterations of the process illustrated in Figure 3.15. The objective is to arrive at a decision where the selected approach is clearly the best among the alternatives evaluated with the associated risk and uncertainty minimized.

4.5 SYNTHESIS AND DESIGN DEFINITION

Synthesis refers to the combining and structuring of parts and elements in such a manner so as to form a functional entity, as discussed in Section 3.7. System synthesis is achieved when sufficient preliminary design progress has been made and trade-off studies have been accomplished to confirm and assure the completeness of system performance and other design requirements (as allocated for detail design). The performance, configuration and arrangement of a chosen system and its elements are portrayed along with the techniques for their test, operation, and life-cycle support. These portrayals cover intrasystem and intersystem and item interfaces, and provide enough information forming a definitive baseline that can be presented in the form of detail specifications. Synthesis may be accomplished by analytical means or through the development and preliminary test of physical models.

When synthesizing a specific system design configuration employing analytical means, it may be feasible to accomplish a system availability analysis, an evaluation of consumer operations, a shop turnaround time analysis, a level of repair analysis, and/or a spares inventory policy evaluation as part of verifying design adequacy in terms of compliance with system requirements. One may also wish to determine the impact of one system parameter on another, or the interaction effects between two elements of logistic support. In accomplishing such, problem resolution may require the utilization of a number of different models combined in such a manner to provide a variety of output factors.

The analysis of a system (to include the prime elements and associated elements of logistic support) is sometimes fairly complex; however, this process can be simplified by applying the right techniques and developing the proper analytical tools. The combining of techniques to facilitate the analysis task can be accomplished by following the steps illustrated in Figure 3.15 (i.e., defining the problem, identifying feasible alternatives, selecting evaluation criteria, etc.) The output may appear as illustrated in Figure 4.12.

Referring to the figure, the analyst may wish to develop a series of individual models as illustrated. Each model may be used separately to solve a specific detailed problem or the models may be combined to solve a higher-level problem. The models may be used to varying degrees, depending on the depth of analysis required. In any event, the overall analytical task must be approached in an organized methodical manner and applied judiciously to provide the results desired.

The approach conveyed in Figure 4.12 may appear to be obvious. However, the results of a recent survey indicated that there are at least 350 computer-based models available in the commercial market. Most of the identified models were developed on a relatively "independent" or "isolated" basis in terms of selected platform, context or computer language, input data requirements, degrees or "user friendliness," and so on. In general, the models do not "talk to each other" are complex, require too much input data, and can only be used in "downstream" phases of the life cycle (during detail design and development). From a systems engineering perspective, a good objective is to select or develop an integrated design workstation (incorporating CAE/CAD and the appropriate supporting analytical tools) that can be utilized in all phases of system acquisition, and one that can be adapted to address the varying levels of design definition as one progresses from the requirements definition stage on down through detail design and development (refer to Figure 4.10).[6]

4.6 SYSTEM DESIGN REVIEWS

The basic objectives of formal design reviews are introduced in Section 3.9. Design is a progression from a defined need to an entity that will perform a useful function in a satisfactory manner. Design evolves through a series of stages: conceptual design, preliminary system design, and detail design and development. In each major stage of the design process, an evaluative function is accomplished to ensure that the design is adequate at that point prior to proceeding with the next stage.

[6] B. S. Blanchard, W. J. Fabrycky, and D. Verma (eds.), *Application of the Systems Engineering Process to Define Requirements for Computer-Based Design Tools*, Monograph (Hyattsville, Md.: Society of Logistics Engineers, 1994).

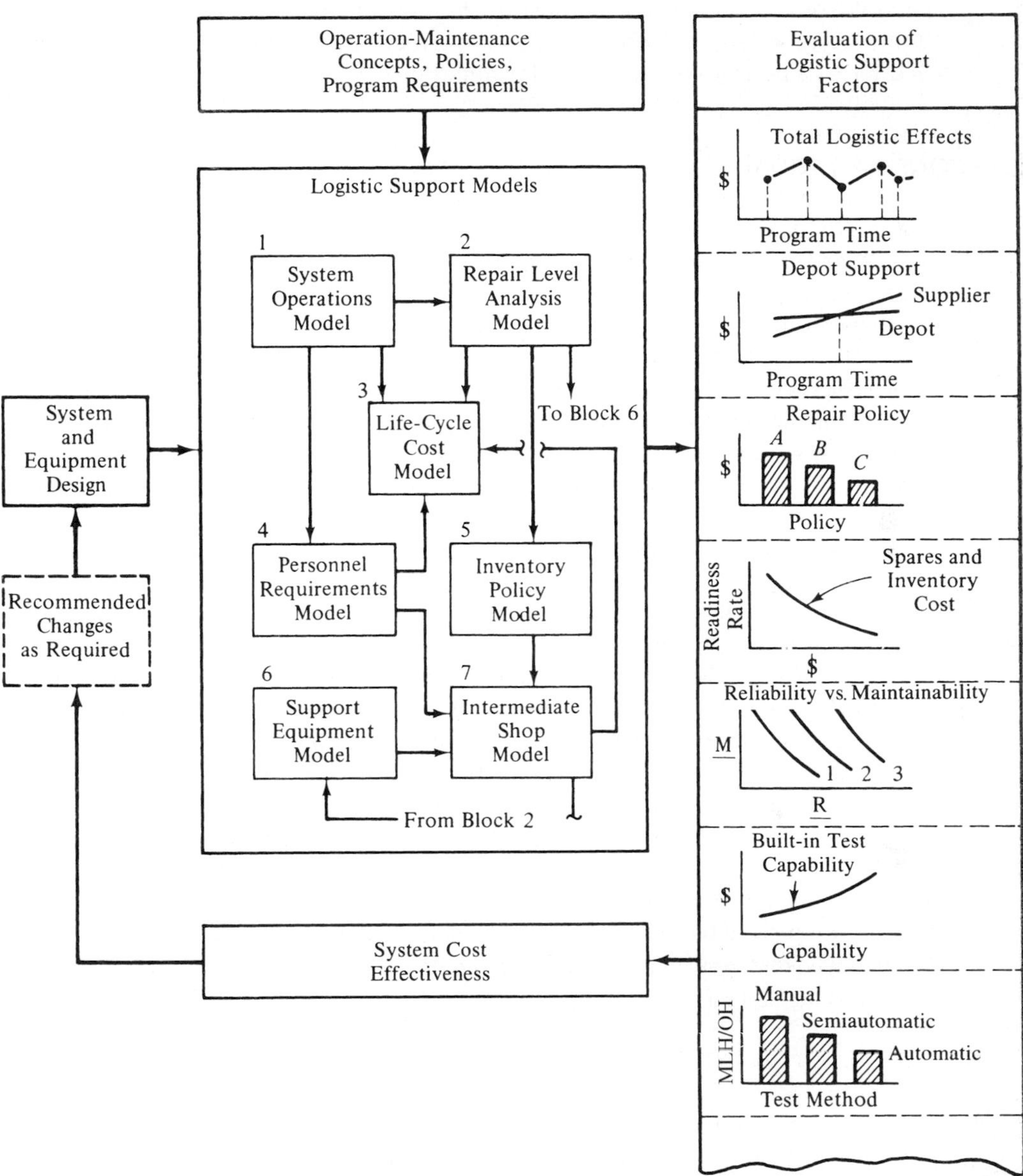

Figure 4.12 Application of models (example).

As the major subsystems (or configuration items) are defined throughout the preliminary system design stage, one or more formal design reviews may be scheduled to verify that overall requirements are being met. The results of the functional analysis and allocation process, the trade-off studies accomplished, the design approach selected (and the justification for it), etc., are reviewed for compliance with initially specified requirements. Deviations are noted, and the necessary corrective action should be initiated as appropriate. This corrective-action procedure is described in Chapter 6 as part of system test and evaluation.

QUESTIONS AND PROBLEMS

1. Describe how specific resource requirements (i.e., hardware, software, people, facilities, data, and elements of support) are derived from the functional analysis.
2. Describe the steps involved in transitioning from the functional analysis to a "packaging scheme" for the system (refer to Figure 4.6). Provide an example from your own experience and show by illustration.
3. Select a system of your choice and develop operational functional flow diagrams to the third level and maintenance functional flow diagrams to the second level. Show how the maintenance functional flow diagrams evolve from the operational flows.
4. Select a system of your choice and assign some top-level TPMs. Allocate these requirements as appropriate to the second and third levels.
5. What is meant by "design criteria?" How are design criteria developed? How are these criteria applied in the design process?
6. Refer to Figure 4.8. Describe how these factors were developed. How do these parameters relate to the goals specified in Figure 3.10 (if at all)?
7. Define CAD, CAM, CALS, and the interrelationships. When should these methods be applied in the system life cycle? What benefits can be gained through their application.
8. Assume that you are a design engineer and are looking for some analytical models/tools to aid you in the synthesis, analysis, and evaluation process. Develop the criteria that you would apply in selecting the most appropriate tools for your application.
9. List some of the benefits that can be derived through the use of mathematical models in system analysis.
10. How can the use of analytical models facilitate the design synthesis process.
11. What are some of the concerns associated with mathematical models and their application?
12. What type of information is usually covered through the system design review(s)?
13. What is the desired output of the preliminary system design stage?

5 DETAIL DESIGN AND DEVELOPMENT

Detail design begins with the configuration and the allocated baseline derived during preliminary design. With definition of the overall system and its major subsystems in hand, one may proceed to the realization of specific system components. Realization includes the technical activities of (1) describing subsystems, units, assemblies, lower-level components, software, people, and elements of logistic support that make up the system (and addressing their interrelationships); (2) preparing specifications and design data for all system components; (3) procuring COTS items or doing detail design for unique items; (4) developing an engineering model, a service test model, or a prototype model of the system and its components for test and evaluation; (5) integrating system components and testing the integration to verify compliance with the specified requirements; and (6) redesigning, reengineering, and retesting the system or an element thereof as required.

This chapter primarily addresses items 1 through 4 per Block 3 in Figure 2.2 (also see Figure 2.4, Blocks 2.1 through 2.5). Chapter 6 covers system test and evaluation.

5.1 DETAIL DESIGN REQUIREMENTS

Specific design requirements at this stage in the system design process are derived from the System Specification (type A) and evolve through applicable lower-level specifications, as illustrated in Figure 3 2. Included within these specifications are appropriate design-dependent parameters, technical performance measures and associated design-to criteria for characteristics that must be incorporated into the design of system elements and components. This is facilitated through the requirements allocation process illustrated in Figure 4.8.

Given this top-down approach for establishing requirements at each level in the hierarchical structure, the design process evolves through the iterative steps of synthesis, analysis, evaluation, to the definition of system components, and to the establishment of a product baseline (refer to Figure 2.4). At this point, the procurement and acquisition of system components begins, components are combined and integrated into a next higher assembly, and a physical model of the system is constructed for test and evaluation. The integration, test, and evaluation steps constitute a bottom-up activity, and should result in a configuration that can be assessed for compliance with initially specified customer requirements.[1]

Progressing through the system design and development process in an expeditious manner is essential in today's competitive environment. Minimizing the time that it takes from the initial identification of need to the ultimate delivery of the system for customer use is critical. This requires that certain design and development activities be accomplished on a *concurrent* basis. Figure 5.1 (an extension of Figure 2.3) illustrates this concurrency. The designer must think in terms of the four life cycles (and their interrelationships) at the same time and in an integrated manner, in lieu of the sequential approach to design often followed in the past. Note that the time to system use should be is shorter in the concurrency approach. Further, within each of the blocks (i.e., system/product design, production/construction, etc.) are activities that must also

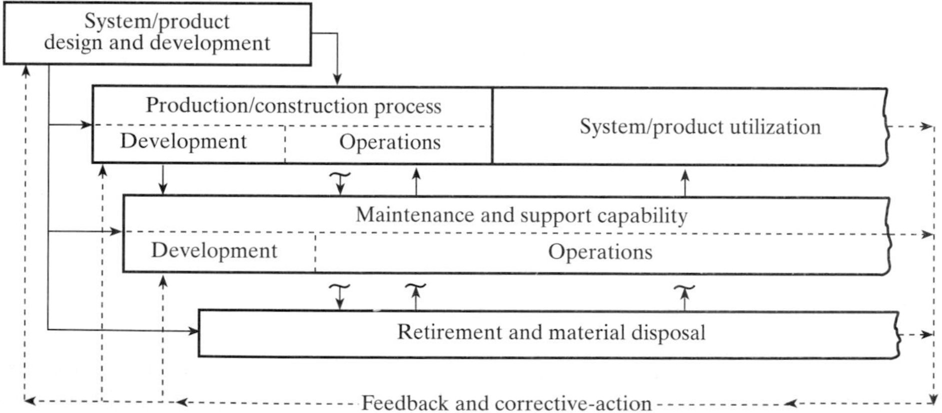

Figure 5.1 Sequential versus concurrent approaches in system design.

[1] The "Vee" process model best illustrates the relationship between top-down and bottom-up activities of system design (refer to Figure 2.6, Sheet 2). Also, refer to Section 2.4 for discussion of the top-down/bottom-up process.

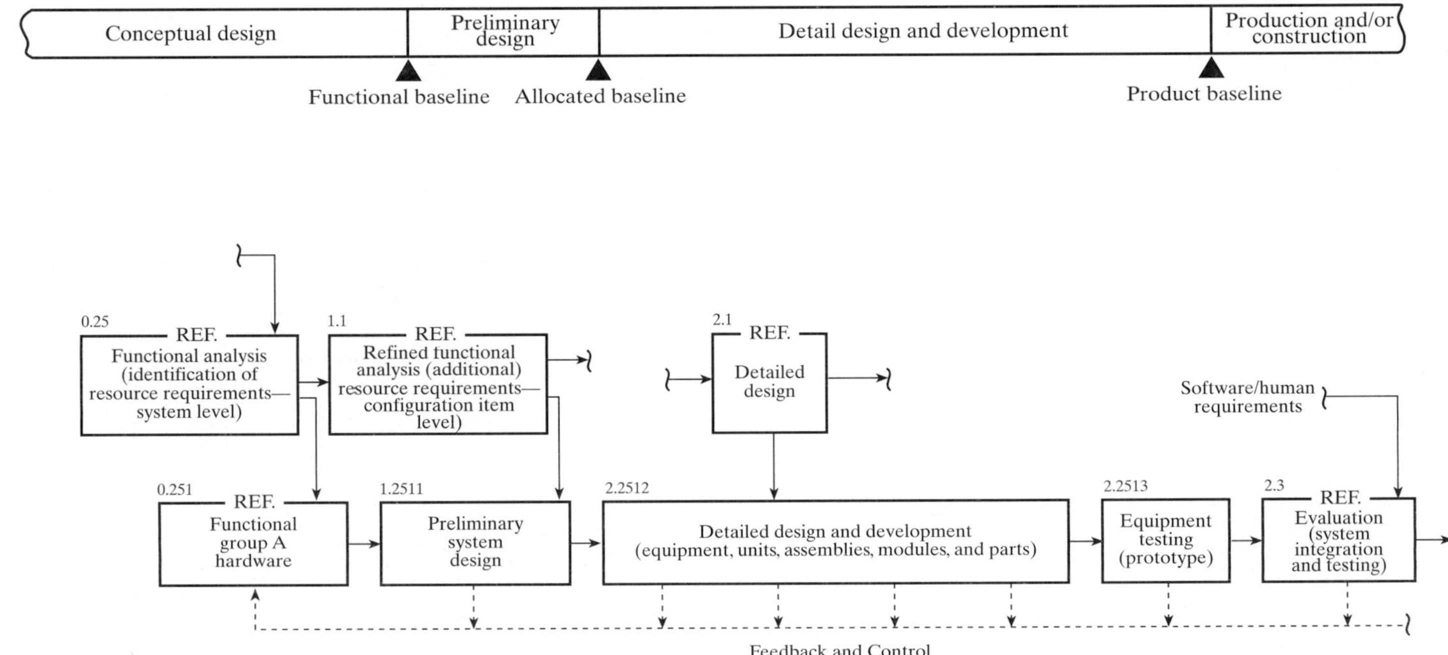

Figure 5.3 The hardware acquisition process (refer to Figure 2.4).

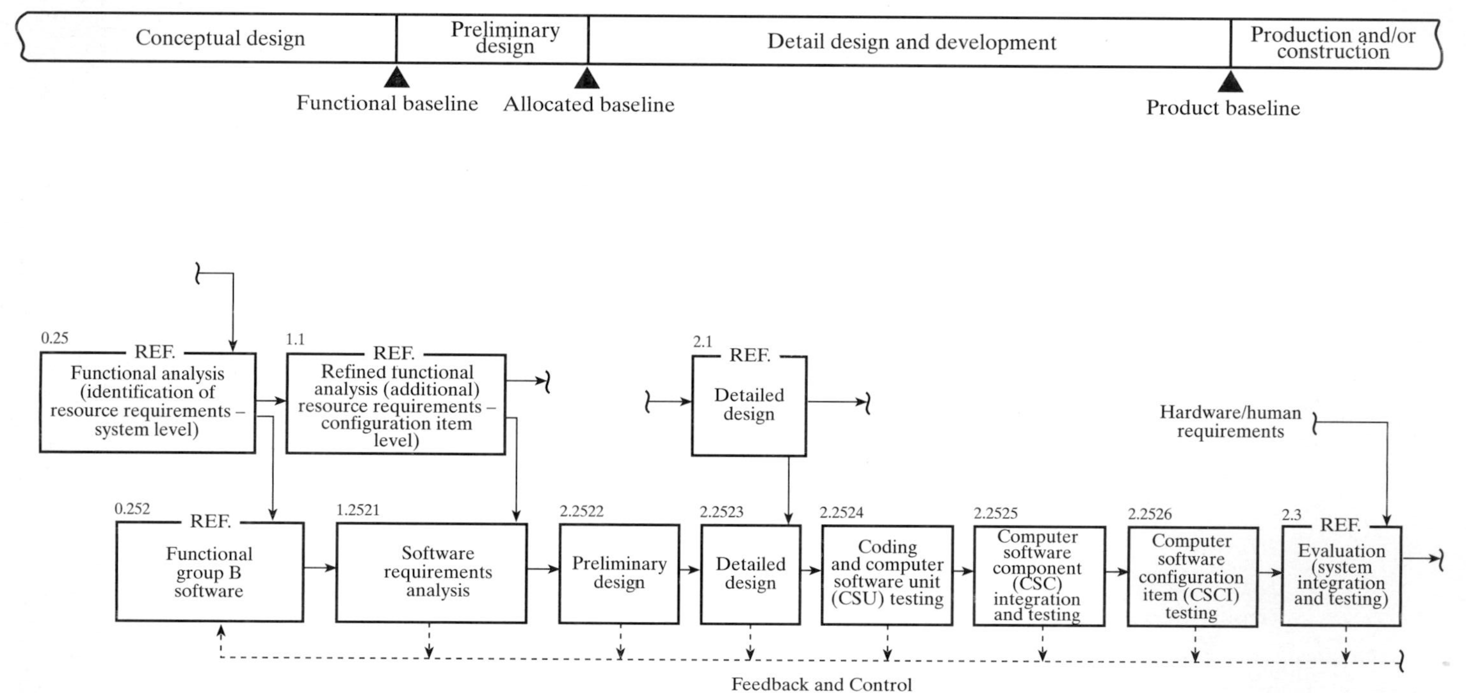

Figure 5.4 The software acquisition process (refer to Figure 2.4).

gration and testing. Sometimes, the technology of "rapid prototyping" is applied where software is developed and verified (with customer involvement) on an evolutionary basis.[5] Software engineers, like hardware engineers, often operate strictly within their own domain without considering necessary interfaces with the hardware development process.

There are a variety of approaches utilized in the development of software. The "functional approach" involves the identification of major functions and their associated relationships. These functions are then broken down into subfunctions, and so on, and continue down to the module at the lowest level. The objective is to provide a structure for a number of modules which, in turn, facilitates coding, testing, and the maintenance or replacement of software later.

Conversely, an "object-oriented" approach views the system as a "collection of objects rather than functions with messages passed between objects, and each object possesses its own operational characteristics."[6] Object characteristics are called attributes, and operations are defined as dynamic entities that may change over time. Although the object-oriented approach to software design and development has gained considerable visibility and popularity, it does not ensure functional compatibility and compliance with system requirements.

Figures 5.3 and 5.4 illustrate the major steps in the acquisition of only two categories of system elements (i.e., hardware and software). There may be several additional elements including people, facilities, data, test and support equipment, and so on. Figure 3.14 showed the steps relating to the development of humanability requirements, which also need to be addressed. Similar illustrations can be presented for facilities, data, and the development of various elements of maintenance and support. In any event, these acquisition cycles must be dealt with in a coordinated way. Proceeding down each domain-oriented path on an independent basis, without providing proper day-to-day coordination until final system integration and test, can result in numerous incompatibilities, many modifications, and the expenditure of resources that could have been avoided.

A primary objective of system engineering is to ensure the proper coordination and timely integration of all system elements (and the activities associated with each) from the beginning. This can be accomplished with a good definition of requirements through the preparation of well-written specifications, followed by a structured and disciplined approach to design. This includes the scheduling of an appropriate number of formal design reviews to ensure proper communications across the project and to check that all elements of the system are compatible. In the event of incompatibilities, the hardware design effort should not be allowed to proceed without first ensuring that the software is compatible, the software development effort should not be allowed to proceed without first ensuring compatibility with hardware, and so on.

[5] The concept of *rapid prototyping*, its advantages and difficulties, is discussed in B. W. Boehm, *Software Engineering Economics* (Upper Saddle River N. J.: Prentice Hall, Inc. 1981), pp. 655–657.

[6] For an in depth discussion of object-oriented design, refer to A. P. Sage and J. D. Palmer, *Software Systems Engineering* (New York: John Wiley and Sons, Inc., 1990).

5.3 DESIGN ENGINEERING ACTIVITIES

The day-to-day design activities begin with the implementation of the appropriate planning in the conceptual design phase (i.e., the program management and system engineering management plans identified in Figures 2.4 and 3.2). These functions usually include the establishment of the design team and initiation of specific design tasks, the development of design data, the conduct of design reviews, and the initiation of corrective action as necessary.

Establishing the Design Team

Success in system engineering derives from the realization that design activity requires a "team" approach. As one proceeds from conceptual design into the subsequent phases of the life cycle, the actual team "make-up" will vary in terms of the specific expertise required and the number of project personnel assigned. Early in the conceptual and preliminary design phases, there is a need for a *few* highly qualified individuals with broad technical knowledge and having and understanding of the customer and the user's operational environment, the major functional elements of the system and their interface relationships, and the general process for bringing a system into being. Emphasis is on a feel key individuals who understand and believe in the "systems approach," and know when to call on the appropriate disciplinary expertise.

As the design progresses, representation from the various individual design disciplines increases as illustrated in Figure 2.14. Depending on the project, there may be relatively few individuals assigned, or there may be hundreds of people involved. Further, some of the expertise desired may be located within the same physical facility, whereas other members of the project team may be remotely located in supplier organizations.

A major goal in the implementation of the system engineering process is first to understand system requirements and the expectations of the customer, and then to provide the *technical* guidance necessary to ensure that the ultimate system configuration will meet the need. Realization of this goal is dependent on providing the right personnel and material resources in a timely manner. Such resources may include a combination of the following:

1. Engineering technical expertise (e.g., electrical engineers, mechanical engineers, software engineers, reliability engineers).
2. Engineering technical support (e.g., technicians, component part specialists, model builders, computer programmers, test technicians).
3. Nontechnical support (marketing, purchasing and procurement, contracts, budgeting and accounting, industrial relations and human resources).

Whatever the case, the objective of systems engineering is to promote the "team" approach, and to create the proper environment for the necessary communications and the on-going exchange of information on a daily basis. Figure 5.5 illustrates this essen-

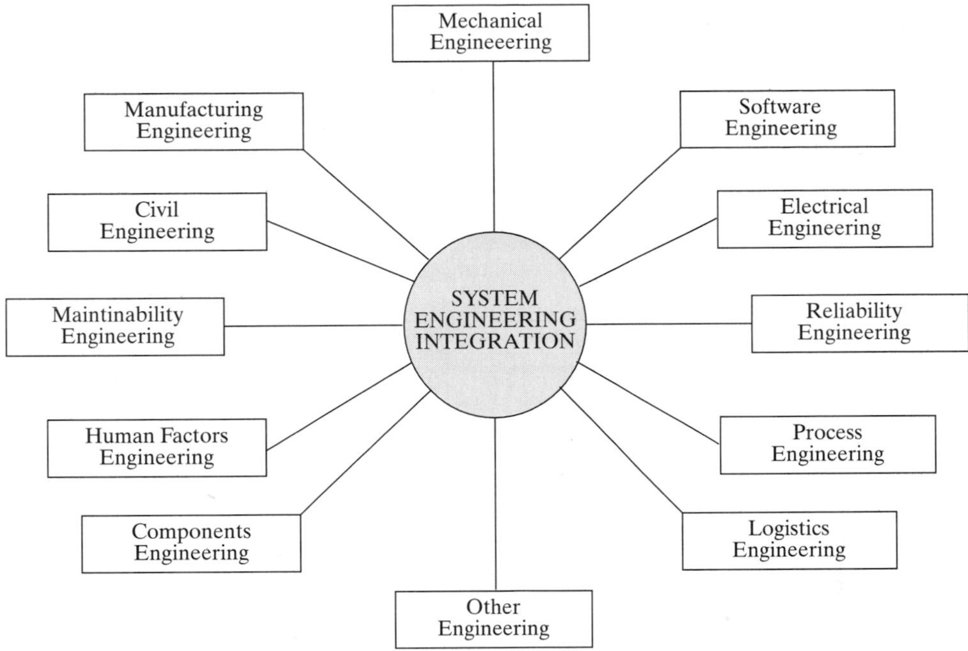

Figure 5.5 The integration of engineering disciplines.

tial system integration function. The organization for systems engineering is discussed further in Chapter 18.

Evolution of Detail Design

Detail design evolution follows the basic sequence of activities shown in Figure 5.6. The process is iterative, proceeding from system-level definition to a product configuration that can be produced in multiple quantities. There are "checks and balances" in the form of reviews at each stage of design progression, and a feedback loop allows for corrective action as necessary. These reviews may be relatively informal and occur continuously compared with the formal design reviews scheduled at specific milestones. In this respect, the process is similar to the synthesis, analysis, evaluation, and product definition, accomplished in the preliminary design stage, except that the requirements are at a lower level (i.e., units, assemblies, subassemblies, etc.).

As the level of detail increases, actual definition is accomplished through the development of data describing the item being designed. These data may be presented in the form of a digital description of the item in an electronic database format, design drawings, material and part lists, reports and analyses, computer programs, and so on.

The actual process of design iteration may occur through the use of a local area network (LAN), the use of telecommunications and compressed video conferences, or some equivalent form of communication. The design configuration selected may be the best possible in the eyes of the responsible designer. However, the results are practically useless unless properly documented, so that others can first understand what is

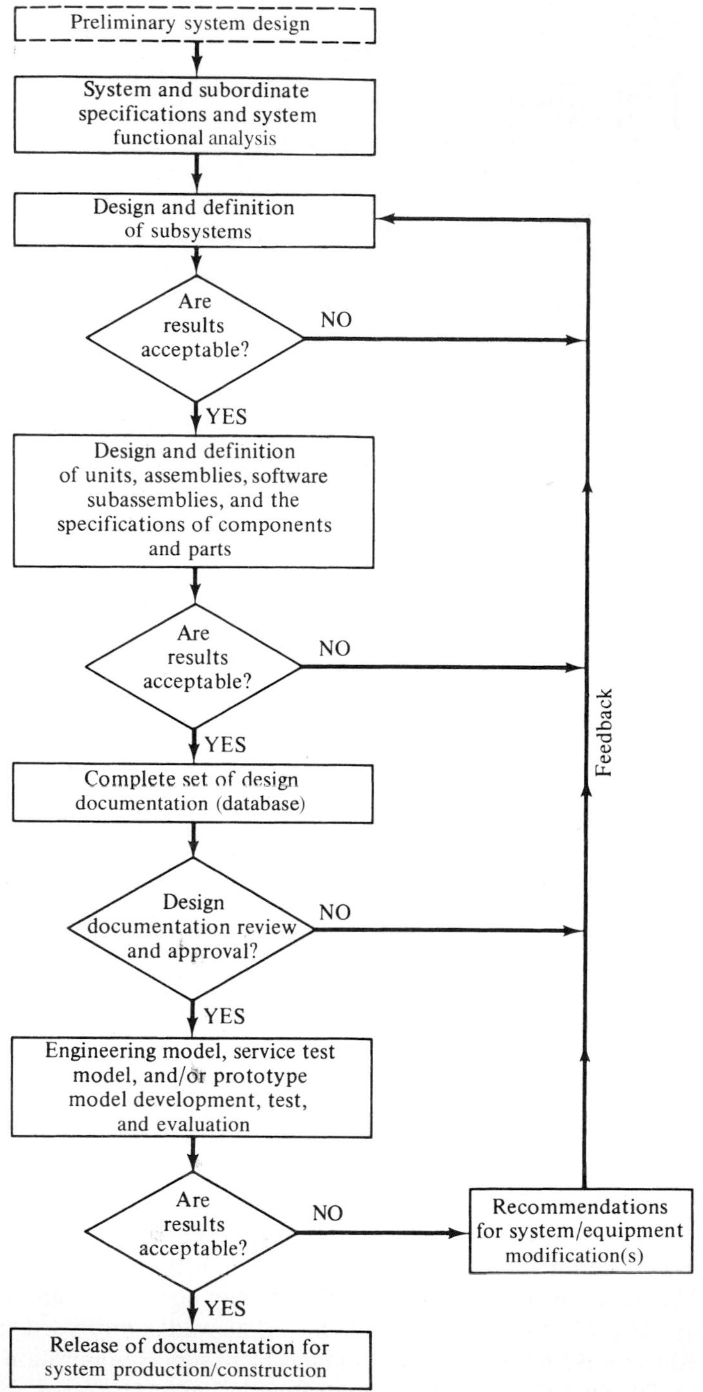

Figure 5.6 Basic design sequence (refer to Figure 5.8 for design data review cycle).

being conveyed and then be able to translate the output into an entity that can be produced or constructed.

5.4 DETAIL DESIGN AIDS

Successful completion of the design process is dependent on the availability of appropriate tools that will aid the design team in accomplishing its objectives in an effective and efficient manner. The application of CAE/CAD aids enables the projection and evaluation of *many* design alternatives early in the life cycle. It is important that this be done when requirements are being defined so that the results can have a beneficial impact on all downstream activities. Through the proper use of design technologies, the designer is able to produce a more robust design more quickly, while reducing the overall risk.

Additionally, the proper selection and application of analytical computer-based models can facilitate the accomplishment of synthesis, analysis, and evaluation at each level in the system hierarchy. Referring to Figure 4.9, there are analysis functions that can be applied in the design and definition of the system, at the subsystem level, and down to the module and the component level. These models/tools must be viewed on an integrated basis (as shown in Figures 4.10 and 4.12), and must support the application of CAE, CAD, CAM, and CALS methods as applicable. The design evaluation methodology described in Sections 2.5 and 4.4 illustrates how some of these models/tools can be applied.

As the design of an item progresses, it is often difficult to visualize it in its true perspective. Simulating of a three-dimensional view of an item within the system is possible through the use of CAD. As an additional aid to the designer, three-dimensional scale physical models or mock-ups are sometimes constructed to provide a realistic simulation of a proposed final system or equivalent configuration. Savings will result if this is done at an early point in the program, prior to the development of final design data or actual prototype hardware.

Models or mock-ups can be developed to any desired scale and to varying depths of detail depending on the level of emphasis required. Mock-ups can be developed for large systems as well as small systems and may be constructed of heavy cardboard, wood, metal, or a combination of materials. Mock-ups can be developed on a relatively inexpensive basis and in a short period of time when employing the right materials and personnel services (industrial design, human factors, or model shop personnel are usually well oriented to this area and should be used to the greatest extent possible). The uses and values of a mock-up are numerous.

1. It provides the design engineer with the opportunity of experimenting with different facility layouts, packaging schemes, panel displays, cable runs, and so on, before the preparation of final design data.
2. It provides the reliability-maintainability-human factors engineer with the opportunity to accomplish a more effective review of a proposed design configuration for the incorporation of supportability characteristics. Problem areas readily become evident.

3. It provides the maintainability-human factors engineer with a tool for use in the accomplishment of predictions and detailed task analyses. It is often possible to simulate operator and maintenance tasks to acquire task sequence and time data.
4. It provides the design engineer with an excellent tool for conveying his or her final design approach during a formal design review.
5. It serves as an excellent marketing tool.
6. It can be employed to facilitate the training of system operator and maintenance personnel.
7. It is utilized by production and industrial engineering personnel in developing fabrication and assembly processes and procedures and in the design of factory tooling and associated test fixtures.
8. At a later stage in the system life cycle, it may serve as a tool for the verification of a modification kit design prior to the preparation of formal data and the development of kit hardware.

In the domain of software development, in particular, designers are oriented toward the building of "one-of-a-kind" software packages. The issues in software development differ from those in other areas of engineering in that mass production is not the normal objective. Instead, the goal is to develop software that accurately portrays the features that are desired by the user (customer). For instance, in the design of a complex workstation display, the user may not at first comprehend the implications of the proposed command routines and data format on the screen. When the system is ultimately delivered, problems occur, and the "user interface" is not acceptable for one reason or another. Changes are then recommended, implemented, and the costs of modification and rework are usually high.

The alternative is to develop a "protoype" early in the system design process, design the applicable software, involve the user in the operation of the prototype, identify areas that need improvement, incorporate the necessary changes, involve the user once again, and so on. This iterative and evolutionary process of software development, accomplished throughout the preliminary and detail design phases, is referred to as *rapid prototyping*. Rapid prototyping is a practice that is often implemented and is inherent within the system engineering process, particularly in the development of large software-intensive systems.

5.5 DETAIL DESIGN DOCUMENTATION

The methods for documenting design are changing rapidly as a result of advances in information systems technology. These advances have promoted the use of electronic databases for the purposes of information processing, storage, and retrieval. Through the use of CAD techniques, information can be stored in the form of three-dimensiona1 representations, regular two-dimensional line drawings, in digital format, or combinations of these. By using computer graphics, word processing capability, e-mail communications, video-disc technology, and the like, design can be presented faster, in more detail, and in easily modified format.

Although many advances have been made in the application of computerized methods to data acquisition, storage, and retrieval, there are still needs for some of the more conventional methods of design documentation. These include a combination of the following:

1. *Design drawings*—assembly drawings, control drawings, logic diagrams, installation drawings, schematics, and so on.
2. *Material and part lists*—part lists, material lists, long-lead-item lists, bulk-item lists, provisioning lists, and so on.
3. *Analyses and reports*—trade-off study reports supporting design decisions, reliability and maintainability analyses and predictions, human factors analyses, safety reports, supportability analyses, configuration identification reports, computer documentation, installation and assembly procedures, and so on.[7]

Design drawings, constituting a primary source of definition, may vary in form and function depending on the design objective; that is, the type of equipment being developed, the extent of development required, whether the design is to be subcontracted, and so on. Some typical types of drawings are specified in Figure 5.7.

During the process of detail design, engineering documentation is rather preliminary, and then gradually progresses to the depth and extent of definition necessary to enable product manufacture. The responsible designer, using appropriate design aids, produces a functional diagram of the overall system. The system functions are analyzed and initial packaging concepts are assumed. With the aid of specialists representing various disciplines (i.e., electrical, mechanical, components, reliability, maintainability, etc.) and supplier data, detail design layouts are prepared for subsystems, units, assemblies, and subassemblies. The results are analyzed and evaluated in terms of functional capability, reliability, maintainability, human factors, safety, producibility, and other design parameters to assure compliance with the allocated requirements and the initially established design criteria. This review and evaluation occurs at each stage in the basic design sequence and generally follows the steps presented in Figure 5.8.

Engineering data are reviewed against design standards and checklist criteria. Throughout the industrial and governmental sectors are design standards manuals and handbooks developed to cover preferred component parts and supplier data, preferred design and manufacturing practices, designated levels of quality for specified products, requirements for safety, and the like. These standards, as applicable, may serve as a basis for design review and evaluation.[8]

[7] Trade-offs and analyses are accomplished throughout the design process as discussed in Chapters 3 and 4. It is important that the results of these analyses be adequately documented to support design decisions. The reports identified here may be newly generated, or reports prepared during conceptual and preliminary system design that have been updated to reflect new information.

[8] Such standards may include ISO or ANSI standards or those prepared by professional societies (e.g., IEEE, ASQC, SME, ASCE), or industry associations (e.g., Electronic Industries Association).

1. *Arrangement drawing*—shows in any projection or perspective, with or without controlling dimensions, the relationship of major units of the item covered.
2. *Assembly drawing*—depicts the assembled relationship of (a) two or more parts, (b) a combination of parts and subassemblies, or (c) a group of assemblies required to form the next higher indenture level of the equipment.
3. *Connection diagram*—shows the electrical connections of an installation or of its component devices or parts.
4. *Construction drawing*—delineates the design of buildings, structures, or related construction (including architectural and civil engineering operations).
5. *Control drawing*—an engineering drawing that discloses configuration and configuration limitations, performance and test requirements, weight and space limitations, access clearances, pipe and cable attachments, support requirements, etc., to the extent necessary that an item can be developed or procured on the commercial market to meet the stated requirements. Control drawings are identified as envelope control (i.e., configuration limitations), specification control, source control, interface control, and installation control.
6. *Detail drawing*—depicts complete end item requirements for the part(s) delineated on the drawing.
7. *Elevation drawing*—depicts vertical projections of buildings and structures or profiles of equipment.
8. *Engineering drawing*—an engineering document that discloses by means of pictorial or textual presentations, or a combination of both, the physical and functional end product requirements of an item.
9. *Installation drawing*—shows general configuration and complete information necessary to install an item relative to its supporting structure or to associated items.
10. *Logic diagram*—shows by means of graphic symbols the sequence and function of logic circuitry.
11. *Numerical control drawing*—depicts complete physical and functional engineering and product requirements of an item to facilitate production by tape control means.
12. *Piping diagram*—depicts the interconnection of components by piping, tubing or hose, and when desired, the sequential flow of hydraulic fluids or pneumatic air in the system.
13. *Running (wire) list*—a book-form drawing consisting of tabular data and instructions required to establish wiring connections within or between items.
14. *Schematic diagram*—shows, by means of graphical symbols, the electrical connections and functions of a specific circuit arrangement.
15. *Software diagrams*—functional flow diagrams, process flows, and coding drawings.
16. *Wiring and cable harness drawing*—shows the path of a group of wires laced together in a specified configuration, so formed to simplify installation.

Figure 5.7 Typical engineering drawing classifications.

Checklists may be developed and "tailored" for a specific system application. For example, the requirements definition process aids in the initial identification of technical performance measures (TPMs). These TPMs are then prioritized to reflect levels of importance (refer to Figure 3.10). Top-level qualitative and quantitative factors are allocated to establish the appropriate design criteria at each level in the system hier-

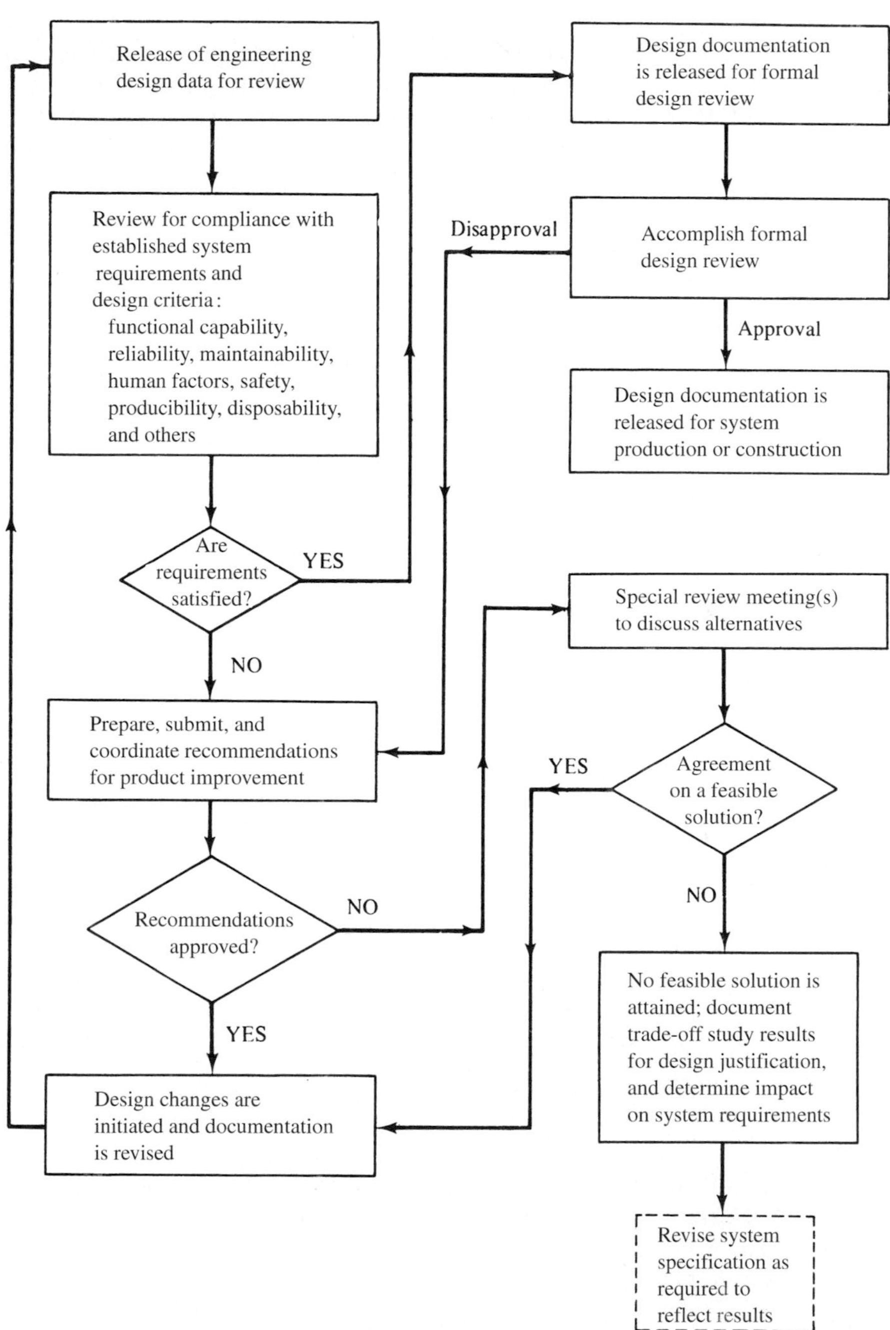

Figure 5.8 Design data review cycle (refer to Figure 5.6).

archical structure (refer to Figures 4.8 and 4.9). A design parameter tree may be prepared to capture critical top-down relationships. Finally, questions are developed in support of each parameter and included in a comprehensive checklist. Each question should be traceable back to a given TPM. The resulting checklist is then used for the purpose of design evaluation. Through a review of the available design data, the degree of compliance with the checklist criteria is assessed, and a recommendation is made as to whether to accept the item as is, to modify it, or to reject it.

Figure 5.9 constitutes an abbreviated checklist, which is generic in nature, but which can serve as a baseline for the purposes of "tailoring." Each of the categories identified is supported by detailed questions (refer to Appendix A.2). The questions, in turn, should be based on specific qualitative and quantitative criteria, with the appropriate weighting factors to indicate level of importance. Checklists may be developed and tailored to any level of detail desired. They serve as a reminder and aid to the design engineer in achieving a complete and robust design.

5.6 SYSTEM PROTOTYPE DEVELOPMENT

Thus far, Part II has primarily covered system design from the standpoint of a theoretical entity supported by concepts, analyses, drawings, and related documentation. The design function has been able to establish specific system objectives and subsequently assess a given design configuration relative to compliance with these objectives. The assessment up to this point has been analytical in nature, providing a certain level of confidence that all qualitative and quantitative requirements have been met. Although the analytical approach fulfills a definite need throughout the total design proccss, there is also a need to verify one's concepts and design configuration through the use of actual system components as soon as practicable. In other words, an objective in the detail design phase is to develop hardware, software, and the appropriate elements of maintenance and logistic support; combine and integrate these elements into the proper system configuration (to the extent possible); and accomplish the necessary test and evaluation to physically demonstrate that system requirements have been met (refer to Figure 2.2, Block 3).

Given that design has progressed to the point where definition can be attained through formalized specifications and supporting documentation, the applicable design may be evaluated through the construction and use of physical working models of one type or another. Categories of models are noted subsequently.

1. *Engineering model* represents a working system, or an element of the system, that will exhibit the functional performance characteristics defined in the specification. An engineering model may be developed in either the preliminary system design stage or in the system detail design stage, and is employed primarily to verify the technical feasibility of an item. It does not necessarily represent the system in terms of physical dimensions (i.e., form or fit).
2. *Service test model* represents a working system, or an element of the system, that reflects the end product in terms of functional performance and physical dimen-

System Design Review Checklist

General

1. System operational requirements defined
2. Effectiveness factors established
3. System maintenance concept defined
4. Functional analysis and allocation accomplished
5. System trade-off studies documented
6. System specification and supporting specifications completed
7. System engineering management plan completed
8. Design documentation completed
9. Logistic support requirements defined
10. Ecological requirements met
11. Societal requirements met
12. Economic feasibility determined

Design Features—Does the design reflect adequate consideration of

1. Accessibility
2. Adjustments and alignments
3. Cables and connectors
4. Calibration
5. Disposability
6. Environment
7. Fasteners
8. Handling
9. Human factors
10. Interchangeability
11. Maintainability
12. Packaging and mounting
13. Panel displays and controls
14. Producibility
15. Reliability
16. Safety
17. Selection of parts/materials
18. Servicing and lubrication
19. Software
20. Standardization
21. Supportability
22. Testability

When reviewing design (layouts, drawings, parts lists, computer graphics, engineering reports, program plans), this checklist may prove beneficial in covering various program functions and design features. The items listed can be supported with more detailed criteria as discussed in Appendix A.2. The response to each item listed should be YES.

Figure 5.9 Sample design review checklist.

sions. It is like the operational configuration except that the same component parts may not be incorporated (i.e., substitute components may be used as long as the system functions properly). The service test model may be developed in the detail design stage to verify the functional performance-physical configuration interface.

3. *Prototype model* represents the production configuration of a system in all aspects of form, fit, and function except that it has not been fully qualified in

terms of operational and environmental testing. A prototype may evolve through a series of working hardware configurations such as an engineering model and a service test model, or may evolve directly from a mock-up. A prototype of the prime equipment, software, and associated elements of logistic support is produced in the detail design stage for the purpose of final system/equipment test and evaluation before entering into the production and/or construction phase. The intent is to verify design adequacy to the maximum extent considered appropriate at this stage in the life cycle.

As part of the overall design process, engineering evaluation functions include the use of engineering models, service test models, or prototype models to assist in the verification of technical concepts and various system design approaches. Areas of noncompliance with the specified requirements are identified and correction action is initiated as required.

5.7 DETAIL DESIGN REVIEWS

The nature and scheduling of formal design reviews may vary from program to program because of system size and complexity, the number of personnel and organizations involved, the number and location of suppliers, etc. Four basic levels of design review are common in most programs. These include the conceptual design review discussed in Section 3.9, the system design review discussed in Section 4.6, and the equipment/software and critical design review discussed here.

Figure 5.10 (an extension of Figure 2.4) identifies design reviews in terms of life-cycle phase. Specific points in time in the acquisition process are inferred, but the actual date to be set for the conduct of such reviews should be based on the maturity of design and its overall progress toward meeting project objectives. Formal design reviews, usually scheduled during the detail design and development phase, are discussed in the sections that follow.

Equipment/Software Design Review

Equipment and software design reviews are scheduled to cover items of equipment, software, and other elements of the system below the subsystem and configuration-item levels (i.e., those items not previously covered by system design reviews). Refer to Section 4.6 for the nature of the system design reviews.

Mechanical and electrical drawings, layouts, functional and logic diagrams, computer programs, and material and component parts lists are needed to support the design review of equipmment and software. Trade-off study reports, reliability and maintainability analyses and prediction reports, and any documentation that supports and justifies design decisions are also made available on an as-required basis. In addition, engineering laboratory models, service test models, and prototypes may be available to support this review process.

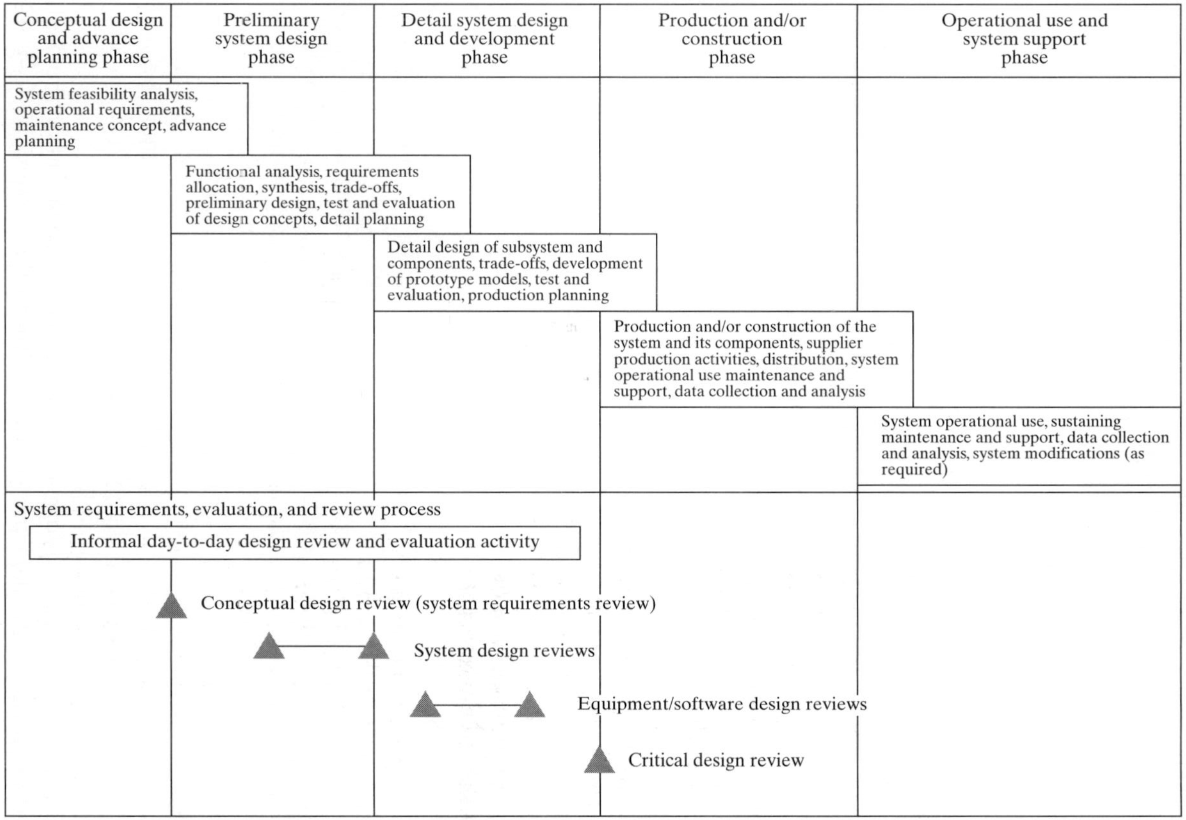

Conceptual design and advance planning phase

Preliminary system design phase

Detail system design and development phase

Production and/or construction phase

Operational use and system support phase

System feasibility analysis, operational requirements, maintenance concept, advance planning

Functional analysis, requirements allocation, synthesis, trade-offs, preliminary design, test and evaluation of design concepts, detail planning

Detail design of subsystem and components, trade-offs, development of prototype models, test and evaluation, production planning

Production and/or construction of the system and its components, supplier production activities, distribution, system operational use maintenance and support, data collection and analysis

System operational use, sustaining maintenance and support, data collection and analysis, system modifications (as required)

System requirements, evaluation, and review process

Informal day-to-day design review and evaluation activity

Conceptual design review (system requirements review)

System design reviews

Equipment/software design reviews

Critical design review

Figure 5.10 Formal design reviews (example).

Critical Design Review

The critical design review is generally scheduled after detail design has been completed, but before the release of firm design data to initiate production and/or construction. The objective is to establish a good "product baseline" as shown in Figure 2.4. Such a review is conducted to verify the adequacy and producibility (or constructability) of the design. Design is essentially "frozen" at this point, and manufacturing (or construction) methods, schedules, and costs are reevaluated for final approval.

The critical design review covers all design efforts accomplished after completion of the equipment/software reviews. This includes any changes resulting from recommendations for corrective action stemming from the equipment/software reviews and the prototype testing. Data requirements include applicable CAM data, manufacturing drawings, material and parts lists, final analysis and prediction reports, test reports, a production/construction plan, a system evaluation plan, a system utilization and support plan, and a retirement and material disposal plan. Before beginning production of a multiple quantity of items, or the construction of a single entity, the product baseline must be evaluated in terms of environmental impact, social acceptance, and even political factors.

Design Review Goal

The purpose of conducting any type of design review is the need to assess just how well the design configuration (as it is described at a particular time) may reflect desired product characteristics. Technical performance measures (TPMs) established and prioritized in the conceptual design phase (refer to Section 3.5) must be measured and "tracked" as one proceeds from one review to the next. Figure 5.11, an extension of Figure 3.10, reflects the TPMs for a given system which should be weighted in terms of degrees of importance. As the review process evolves, different organizational disciplines may be assigned to take the lead in monitoring the current design status relative to compliance with the initially specified requirements. The figure indicates the probable degrees of organizational interest based on potential impact should a given requirement not be met. Figure 5.12 identifies five typical parameters, the measures associated with each, the desired maximum or minimum bounds, and the current trends (whether upward or downward). If the predictions (or measures) at that time are within the prescribe limits, or the trends show design improvement toward meeting the requirements, no action may be taken. Conversely, if it appears that the specific design requirement will not be met, associated risks must be assessed and corrective-action taken accordingly. The various aspects of evaluation and risk management are discussed further in Part V.

Finally, the success of a formal design review is dependent on the depth of planning, organization, and data preparation prior to the review itself. A tremendous amount of coordination is required involving:

1. Identification of the items to be reviewed.
2. A selected date for the review.

Technical Performance Measures (TPMs) \ Engineering Design Functions	Aeronautical Engineering	Components Engineering	Cost Engineering	Electrical Engineering	Human Factors Engineering	Logistics Engineering	Maintainability Engineering	Manufacturing Engineering	Materials Engineering	Mechanical Engineering	Reliability Engineering	Structural Engineering	Systems Engineering
Availability (90%)	H	L	L	M	M	H	M	L	M	M	M	M	H
Diagnostics (95%)	L	M	L	H	L	M	H	M	M	H	M	L	M
Interchangeability (99%)	M	H	M	H	M	H	H	H	M	H	H	M	M
Life cycle cost ($350K, unit)	M	M	H	M	M	H	H	L	M	M	H	M	H
$\overline{M}ct$ (30 min)	L	L	L	M	M	H	H	M	M	M	M	M	M
MDT (24 hr)	L	M	M	L	L	H	M	M	L	L	M	L	H
MLH/OH (15)	L	L	M	L	M	M	H	L	L	L	M	L	H
MTBF (300 hr)	L	H	L	M	L	L	M	H	H	M	H	M	M
MTBM (250 hr)	L	L	L	L	L	M	H	L	L	L	M	L	H
Personnel skill levels	M	L	M	M	H	M	H	L	L	L	L	L	H
Size (150 ft by 75 ft)	H	H	M	M	M	M	M	H	H	H	M	H	M
Speed (450 mph)	H	L	L	L	L	L	L	L	L	L	L	M	H
System effectiveness (80%)	M	L	L	M	L	M	M	L	L	M	M	M	H
Weight (150K lb)	H	H	M	M	M	M	M	H	H	H	L	H	M

Figure 5.11 The relationship between TPMs and responsible design disciplines (refer to Figure 3.10). H = high interest; M = medium interest; L = low interest.

3. The location or facility where the review is to be conducted.
4. An agenda for the review (including a definition of the basic objectives).
5. A design review board representing the organizational elements and disciplines affected by the review. Basic design functions, reliability, maintainability, human factors, quality control, manufacturing, and logistic support representation are included. Individual organization responsibilities should be identified. Depending on the type of review, the customer and/or individual equipment suppliers may be included.

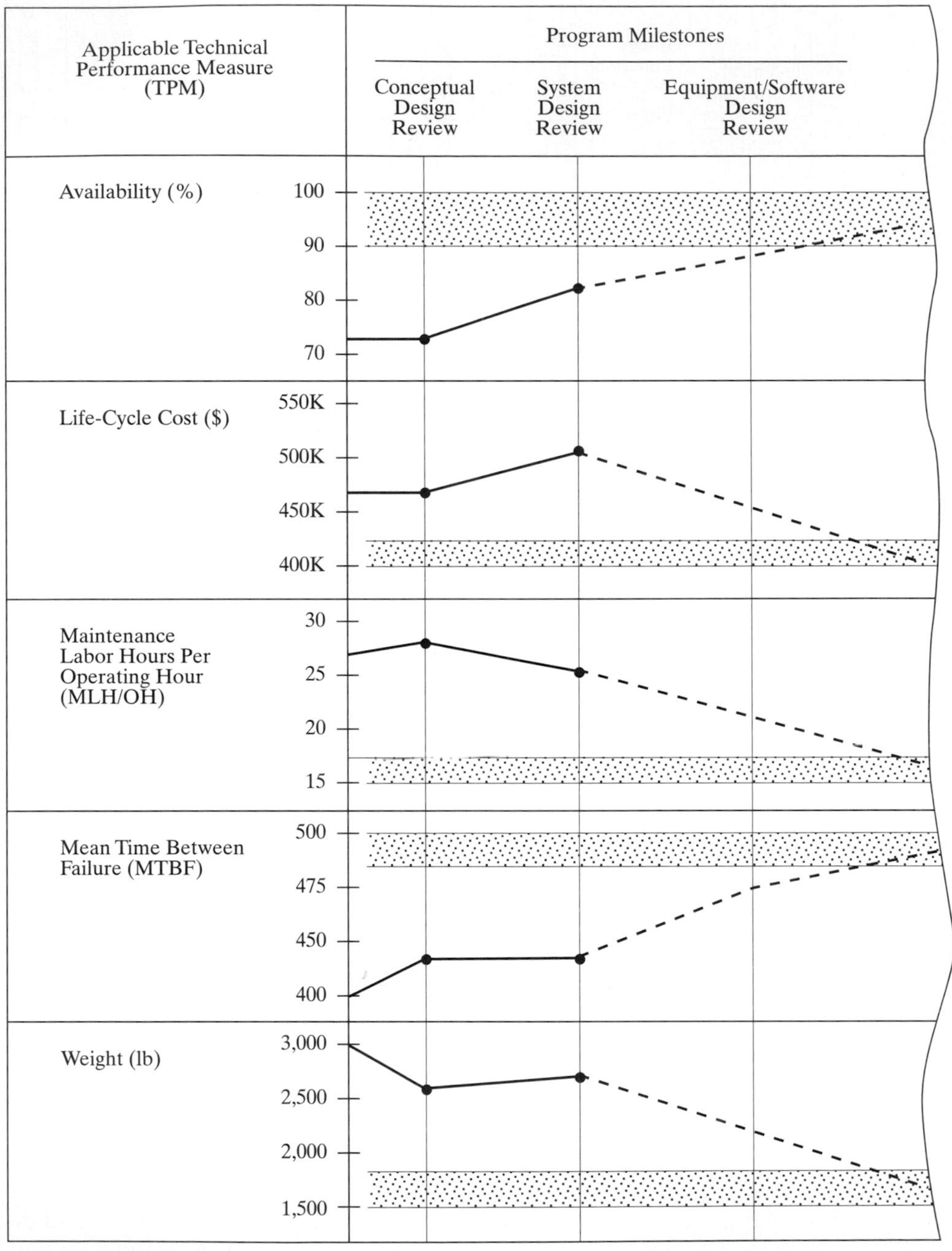

Figure 5.12 System parameter measurement and evaluation at design review.

6. Equipment (hardware) and/or software requirements for the review. Engineering prototypes and/or mock-ups may be required to facilitate the review process.
7. Design data requirements for the review. This may include all applicable specifications, lists, drawings, predictions and analyses, logistic data, computer data, and special reports.
8. Funding requirements. Planning is necessary in identifying sources and a means for providing the funds for conducting the review.
9. Reporting requirements and the mechanism for accomplishing the necessary follow-up actions stemming from design review recommendations. Responsibilities and action-item time limits must be established.

The design review involves a number of different discipline areas and covers a wide variety of design data and in some instances hardware and/or software. In order to fulfill its objective expeditiously (i.e., review the design to ensure that all system requirements are met in an optimum manner), the design review must be well organized and firmly controlled by the design review board chairman. Design review meetings should be brief and to the point, and must not be allowed to drift away from the topics on the agenda. Attendance should be limited to those who have a direct interest and can contribute to the subject matter being presented. Specialists who participate should be authorized to speak and make decisions concerning their area of specialty. Finally, the design review must make provisions for the identification, recording, scheduling, and monitoring of corrective actions. Specific responsibility for follow-up action must be designated by the chairman of the design review board.

QUESTIONS AND PROBLEMS

1. What are the basic differences between conceptual design, preliminary system design, and detail design? Are these stages of design applicable in the acquisition of all systems?
2. Refer to Figure 5.1. What are some of the advantages of the concurrent approach to design? Identify some of the problems that could occur in its implementation.
3. Refer to Figures 5.3 and 5.4. What commonalties exist between the two life cycles ? Describe some of the major differences.
4. Assume that you have just been assigned as the system engineering manager for the acquisition of a new system and that the system includes the development of both hardware and software. What specific steps would you take to ensure that the proper integration occurs from the beginning and on a day-by-day basis?
5. Describe what is meant by "rapid prototyping" as it is applied in the development of software.
6. Describe what is meant by "object-oriented design." What are some of the advantages in applying this approach from a system engineering perspective? Describe some of the potential problems (if any).
7. Identify some of the problems in the past relative to the development of software. Describe some of your own experiences in this area.
8. Design is a "team" effort. True or false? Why?

9. Identify some of the organizational characteristics that need to be inherent within the "design team" to meet the desired system engineering objectives.
10. Why are design standards important?
11. Describe the "checks and balances" in the design process as you see them.
12. Why is engineering documentation necessary?
13. Define what is meant by system synthesis, system analysis, and system evaluation. How are they related (if at all), and how does each fit into the design process?
14. Describe what is meant by CAD, CAM, and CALS. How are they related (if at all)? Provide some examples based on your own experience.
15. Select a system (or an element of that system) of your choice and develop a design review checklist that you can use for evaluation purposes.
16. Describe some of the benefits of physical models and mock-ups.
17. What are the basic differences between an engineering model, a service test model, and a prototype model?
18. What benefits can be acquired through implementation of the formal design review process?
19. Describe the input requirements for the conceptual design review, system design reviews, equipment/software design reviews, and critical design review.
20. What determines the success of a design review?
21. Why is good "baseline management" so important in the implementation of the system engineering process?
22. Refer to the definition of systems engineering in Chapter 2. How can the principles and concepts be applied in the detail and development phase? If you were appointed as the new systems engineering manager, what steps would you take to help ensure a successful outcome?

6 SYSTEM TEST AND EVALUATION

Evaluation refers to the examination and assessment of a system (or an element of a system) in terms of relative worth, quality of performance, degree of effectiveness, anticipated cost, and so on. Evaluation should be a continuous process which begins during conceptual design, extends through the operational use and support phase, and concludes when the system is retired. The purpose of evaluation is to determine (through a combination of prediction, analysis, and measurement activities) *true* system characteristics and to ensure that the system will successfully fulfill its intended purpose or mission.

The primary objective of this chapter is to address the measurement and evaluation aspect of the design process as depicted by Blocks 0.6, 1.5, and 2.3 of Figure 2.4. Referring to the figure, one initially defines requirements, develops a system, and then measures the results. If the results are in compliance with the initially specified requirements, then production or construction may begin; if not, corrective action is required. System evaluation and its relationship(s) with system requirements definition and with system development is illustrated in Figure 6.1.

6.1 REQUIREMENTS FOR TEST AND EVALUATION

Specific needs for test and evaluation are initially defined during conceptual design, when the overall requirements for the system are established. As system-level requirements are identified through feasibility studies, the definition of operational requirements and the maintenance concept, the identification and prioritization of technical performance measures (TPMs) and DDP's, a method must be established for evaluation and the subsequent determination of whether these requirements will be met. As

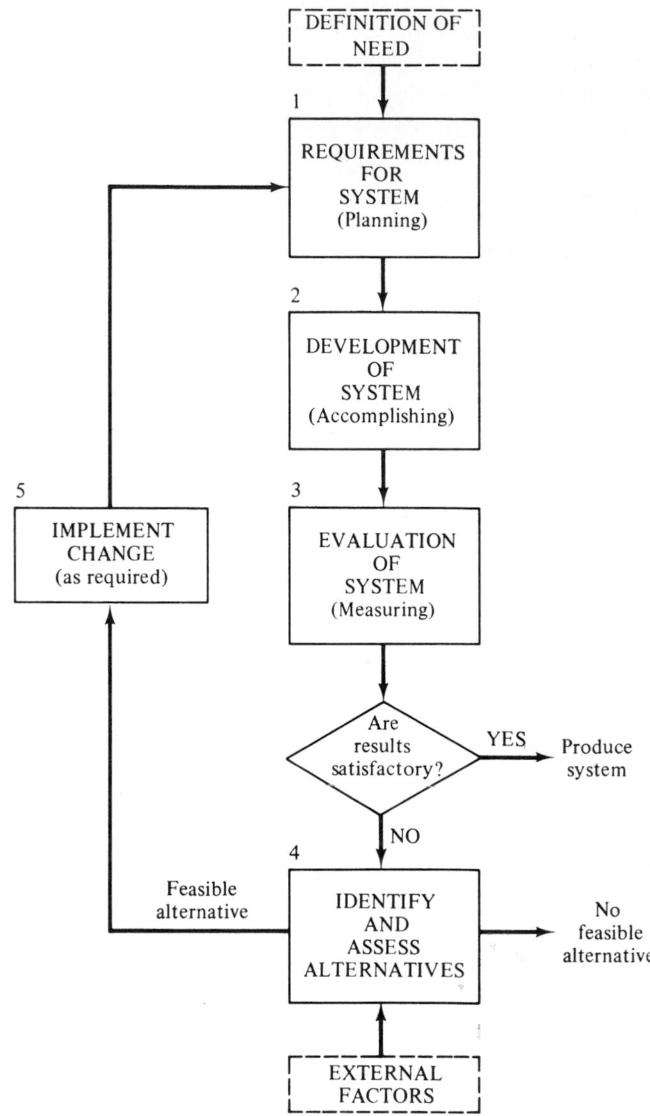

Figure 6.1 System requirements and evaluation relationships.

a new requirement is established, the question is how will we be able to determine if the requirement will be met and what test and evaluation method should be implemented to verify that this is the case?

Given an initial definition of system-level requirements, the ongoing iterative process of evaluation may then begin. Early in system design, analytical techniques may be used to predict and evaluate the anticipated characteristics of a specific system configuration. CAD methods are often used to present an early three-dimensional view of a design, showing potential interferences, accesses, size of components, and so on. The definition of operation requirements and maintenance concept, accomplishing functional analyses, performing trade-offs and optimization, and the establishment of a preferred configuration are elements of a continuing design and evaluation effort. Various aspects of the design are reviewed on the basis of specified requirements, and corrective action is initiated in areas where non-responsiveness to these requirements is indicated. Changes at this stage in the life cycle are relatively easy to incorporate, and the cost of change is generally low.

As the design progresses to the point where mock-ups and engineering service test models are developed, the evaluation process becomes more meaningful because actual hardware and software are now available. Further, when prototypes and production models evolve, as the system approaches its intended operational configuration, the effectiveness of the evaluation assumes even more significance.

When viewing overall test requirements, it should be realized that a *true* test (that which is relevant from the standpoint of assessing total system performance and effectiveness) constitutes the evaluation of a system deployed in an operational user environment and subjected to actual operating conditions. For example, an aircraft or a power plant should be tested while it is performing its intended mission in an actual operational situation. User personnel should accomplish operator and maintenance functions with the designated operating or maintenance procedures, support equipment, and so on. In these situations, actual experience in a realistic environment can be recorded and subsequently evaluated to reflect a true representation of the system design. A demonstration of this type can be accomplished best by the user during standard operations supported through the employment of normal resources (i.e., the requirements specified for operational use and system support functions in the life cycle).

Idealistically it is best to wait until the system is fully operational before accomplishing an evaluation of system performance, effectiveness, and supportability. But waiting is usually not practical from the standpoint of allowing for possible corrective action. In the event that evaluation indicates noncompliance (i.e., the system as presently designed will not meet the operational requirements and fulfill its purpose), corrective action should be initiated as early as possible in the system life cycle.

Taking corrective action after equipment is produced and fully deployed in the field can result in extensive and costly modification programs. It is best to establish an overall test program which allows for the evaluation of hardware, software, and its support elements in an evolutionary manner. As is the case for the earlier design evaluation using analytical techniques, an evaluation of hardware, software, facility work spaces, or equivalent, is accomplished, beginning with the development of the first engineering model and extending through the test of the system deployed in the field. This evaluation process may include various types of demonstrations.

6.2 CATEGORIES OF SYSTEM/COMPONENT TESTING

The evolution of system evaluation is illustrated in Figure 6.2. Various categories of testing are identified by program phase, and the effectiveness of the evaluation effort increases when progressing through types 1, 2, 3, and 4 testing. These steps are neces-

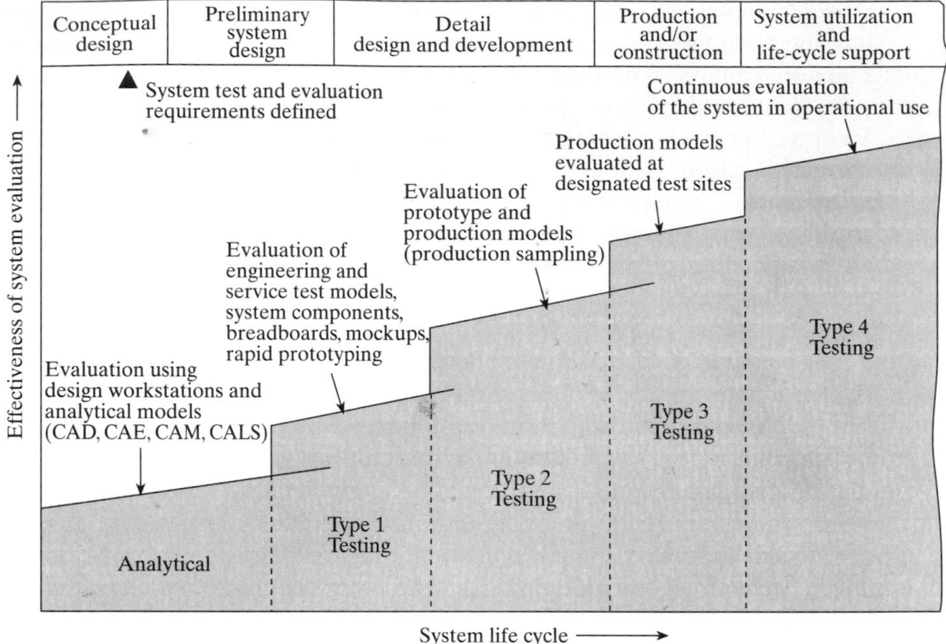

Figure 6.2 Stages of system evaluation during the life cycle.

sary to ensure, on a continuing basis, that system requirements are being met. Incorporating changes after the system is in operational use can be extremely costly to both the producer and consumer; thus, it is preferable to eliminate (or avoid) potential problems as early as possible in the life cycle through the best means available.[1]

Type 1 Testing

During the early phases of detail design, breadboards, bench-test models, engineering models, engineering software, and service test models are built with the intent of verifying certain performance and physical design characteristics. These models, representing either an entire system or a designated system component, usually operate functionally (electrically and mechanically) but do not by any means represent production equipment. In the development of software, the application of rapid prototyping is sometimes used to verify design adequacy.[2]

[1] The scope and depth of testing will depend on the type of the system, whether new design and development are required, and the risks involved. Comprehensive testing may be required where design is new and many unknowns exist, whereas commercial off-the-shelf, proven entities will not require much testing except to assure the proper integration with other elements of the system. Not too much or too little testing should be accomplished. Thus, good planning is essential.

[2] Individual components are frequently tested in lieu of testing the entire system. For instance, it may not be feasible to test an entire petroleum distribution system, whereas testing a segment of pipeline may be sufficient to verify design adequacy. The objective at this point is to verify design adequacy to the maximum extent.

Tests may involve operating and logistic support actions that are directly comparable to tasks performed in a real operational situation (e.g., measuring a performance parameter, accomplishing a remove-replace action, meeting a servicing requirement). Although these tests are not formal demonstrations in a fully operational sense, information pertinent to actual system characteristics can be derived and used as an input in overall system evaluation and assessment. Such testing is usually performed in the producer/supplier's facility by engineering technicians using "jury-rigged" test equipment and engineering notes for test procedures. It is during this initial phase of testing that changes to the design can be incorporated at minimum cost.

Type 2 Testing

Formal tests and demonstrations are accomplished during the latter part of detail design when preproduction prototype equipment, software, and the like are available. Prototype equipment is similar to production equipment (that which will be deployed for operational use), but it is not necessarily fully qualified at this point in time.[3] A test program may constitute a series of individual tests tailored to the need. Such a program might include the following:

1. *Performance tests*—Tests are accomplished to verify individual system performance characteristics. For instance, tests are designed to determine whether the electric motor will provide the necessary output, whether the pipeline will withstand certain fluid pressures, whether the airplane will perform its intended mission successfully, whether the process will provide x widgets per given period, and so on. Also, it is necessary to verify form, fit, interchangeability, product safety, and other comparable features.
2. *Environmental qualification*—Temperature cycling, shock and vibration, humidity, wind, salt spray, dust and sand, fungus, acoustic noise, pollution emission, explosion proofing, and electromagnetic interference tests are conducted. These factors are oriented to what the various system elements will be subjected to during operation, maintenance, and during transportation and handling functions. In addition, the effects of the item being tested on the overall environment will be noted.
3. *Structural tests*—Tests are conducted to determine material characteristics relative to stress, strain, fatigue, bending, torsion, and general decomposition.
4. *Reliability qualification*—Tests are accomplished on one or more system elements to determine the MTBF and MTBM. Also, special tests are often designed to measure component life, to evaluate degradation, and to determine modes of failure.[4]

[3] Qualified equipment refers to a production configuration that has been verified through the successful completion of performance tests, environmental qualification tests (e.g., temperature cycling, shock, vibration), reliability qualification, maintainability demonstration, and compatibility tests. Type 2 testing primarily refers to that activity associated with the qualification of a system for operational use.

[4] Reliability testing is covered further in Chapter 12.

5. *Maintainability demonstration*—Tests are conducted on one or more system elements to assess the values for mean active maintenance time ($\overline{M}$), mean corrective maintenance time ($\overline{M}ct$), mean preventive maintenance time ($\overline{M}pt$), maintenance labor-hours per operating hour (MLH/OH), and so on. In addition, maintenance tasks, task times and sequences, prime equipment-test equipment interfaces, maintenance personnel quantities and skills, maintenance procedures, and maintenance facilities are verified to varying degrees. The elements of logistic support are initially evaluated on an individual basis.[5]
6. *Support equipment compatibility tests*—Tests are often accomplished to verify compatibility among the prime equipment, test and support equipment, and transportation and handling equipment.
7. *Personnel test and evaluation*—Tests are often accomplished to verify the relationships between people and equipment, the personnel quantities and skill levels required, and training needs. Both operator and maintenance tasks are evaluated.
8. *Technical data verification*—The verification of operational and maintenance procedures is accomplished.
9. *Software verification*—The verification of operational and maintenance software is accomplished. This includes computer software units (CSUs), computer software configuration items (CSCIs), hardware-software compatibility, software reliability and maintainability, and related testing.

The specific requirements for testing evolve from the technical performance measures (TPMs) described in Section 3.5. The depth of testing will be influenced by the established levels of importance reflected by the priorities indicated.

The ideal situation is to plan and schedule individual tests such that they can be accomplished on an integrated basis as one overall test. Data output from one test may be beneficial as an input to another test. The intent is to provide the proper emphasis, consistent with the need, and to eliminate redundancy and excessive cost. Proper test planning is essential.

Another aspect of testing in this category involves production sampling tests when multiple quantities of an item are produced. The tests defined earlier basically "qualify" the item; that is, the equipment hardware configuration meets the requirements for production and operational use. However, once an item is initially qualified, some assurance must be provided that all subsequent replicas of that item are equally qualified. Thus, in a multiple-quantity situation, samples are selected from the production line and tested.[6]

Production sampling tests may cover certain critical performance characteristics, reliability, or any other designated parameter that may significantly vary from one serial-numbered item to the next, or may vary as a result of a production process. Samples may be selected on the basis of a percentage of the total equipment produced or may tie in with x number of pieces of equipment in a given calendar time interval. This

[5] Maintainability demonstration is covered further in Chapter 13.

[6] These tests are in addition to the normal performance tests that are accomplished on every individual system element after fabrication and assembly and before delivery to the customer.

depends on the peculiarities of the system and the complexities of the production process. From production sampling tests, one can measure system growth (or degradation) throughout the production/construction phase.

Type 2 tests are generally performed in the producer or supplier's facility by people at that facility. Test and support equipment, designated for operational use, and preliminary technical manual procedures are employed where possible. User personnel often observe and/or participate in the testing activities. Design changes as a result of corrective action are handled through a formalized engineering change procedure.

Type 3 Testing

Formal tests and demonstrations, started after initial system qualification and prior to the completion of production, are accomplished at a designated field test site by user personnel.[7] Operational test and support equipment, operational spares, and formal operator and maintenance procedures are used. Testing is generally continuous, accomplished over an extended period of time, and covers the evaluation of a number of system elements scheduled through a series of simulated operational exercises.

This is the first time that all elements of the system are operated and evaluated on an integrated basis. The compatibility of the prime equipment with software and the elements of maintenance and logistic support is verified as well as the compatibility of the various elements of support with each other. Turnaround times and transportation times, stock levels, personnel effectiveness factors, and other related operational and support parameters are measured. In essence, system performance (based on certain use conditions) and operational readiness characteristics (i.e., operational availability, dependability, system effectiveness, etc.) can be determined to a certain extent.[8]

Type 4 Testing

During the operational-use phase, formal tests are sometimes conducted to gain further insight in a specific area. It may be desirable to vary the mission profile or the system utilization rate to determine the impact on total system effectiveness, or it might be feasible to evaluate several alternative support policies to see whether system operational availability can be improved. Even though the system is designed and operational in the field, this is actually the first time that we really know its true capability. Hopefully, the system will accomplish its objective in an efficient manner; however, there is still the possibility that improvements can be realized by varying basic operational and maintenance support policies.

Type 4 testing is accomplished at one or more operational sites (in a realistic environment) by user operating and maintenance personnel, and supported through the normal logistics capability. The elements of maintenance and logistic support as

[7] The test site may constitute a ship at sea, an aircraft or space vehicle in flight, a facility in the arctic or located in the middle of the desert, or a mobile land vehicle traveling between two points.

[8] Type 3 testing does not represent a complete operational situation; however, tests can be designed to provide a close approximation.

designated through predictions and analyses, earlier testing, and so on, are evaluated in the context of the total system.

6.3 PLANNING FOR TEST AND EVALUATION

Test planning should actually begin as part of the advance system planning activity during conceptual design. If a system requirement is to be specified, then there must be a way to evaluate the system later to ensure that the requirement has been met; hence, testing considerations are intuitive at an early point in time.

Throughout the various stages of system development, a number of individual subsystem or equipment tests may be specified. Often there is a tendency to develop a test to measure one system characteristic, develop another test to measure a different parameter, and so on. However, the amount of testing specified may be overwhelming and prove to be quite costly. Test requirements must be considered on an integrated basis. Where possible, individual tests are reviewed in terms of resource requirements and output results and are scheduled in such a manner as to gain the maximum benefit possible. For instance, maintainability data can be obtained from reliability tests resulting in a possible reduction in the amount of maintainability testing required. Support equipment compatibility data and personnel data can be obtained from both reliability and maintainability testing; thus, it might be feasible to schedule reliability qualification testing first and maintainability demonstration second. In some instances, the combining of tests may be feasible as long as the proper characteristics are measured and the data output is compatible with the initial testing objectives.

For each program, an integrated test and evaluation plan is prepared, usually for implementation beginning in the preliminary design phase.[9] Although the specific content may vary somewhat depending on system requirements, the plan will generally include the following:

1. The definition and schedule of all test requirements, including anticipated test output (in terms of what the test is to accomplish) for each individual test and integrated where possible. In determining test requirements, some components of the system may go through each category of the test, whereas other components may undergo only a limited amount of testing. This is a function of the degree of design definition and the risks associated with each item in question. An example of this approach is illustrated in Figure 6.3.
2. The definition of organization, administration, and control responsibilities (organization functions, organizational interfaces, monitoring of test activities, cost control, and reporting).

[9] In the defense sector, the Test and Evaluation Master Plan (TEMP) is prepared during conceptual design and usually includes coverage of test objectives, technical and operating characteristics of the system, critical issues and interfaces, developmental test and evaluation requirements, operational test and evaluation requirements, test resource requirements, test procedures, and test reporting (see Figure 2.4).

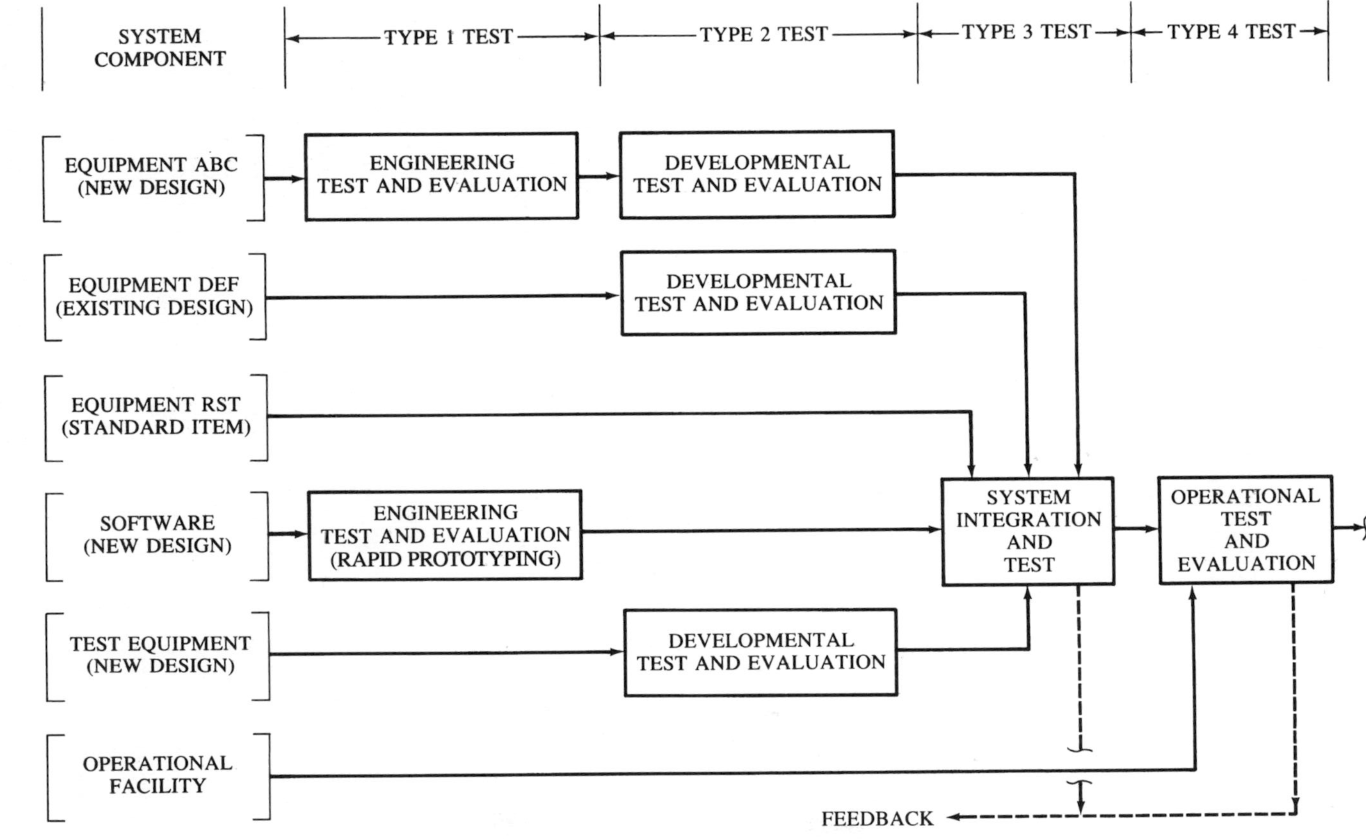

Figure 6.3 Evolution of test requirements.

3. The definition of test conditions and maintenance and logistic resource requirements (test environment, facilities, test and support equipment, spares and repair parts, test personnel, software, and test procedures).
4. A description of the test preparation phase for each type of testing (selection of specific test method, training of test personnel, acquisition of logistic resource requirements, and preparation of facilities).
5. A description of the formal test phase (test procedures and test data collection, reduction, and analysis methods).
6. A description of conditions and provisions for a retest phase (methods for conducting additional testing as required due to a reject situation).
7. The identification of test documentation (test reporting requirements).

The basic test plan serves as a valuable reference throughout system design, development, and production. It indicates what is to be accomplished, the requirements for testing, a schedule for the processing of equipment and material for test support, data collection and reporting methods, and so on. It is an integrating device for a number of individual tests in various categories, and a change in any single test requirement will likely affect the total plan.

6.4 PREPARATION FOR SYSTEM TEST AND EVALUATION

After initial planning, and prior to the start of formal evaluation, a period of time is set aside for test preparation. During this period, the proper conditions must be established to ensure effective results. Although there is some variance, depending on the type of evaluation, these conditions or prerequisites include the selection of the item(s) to be tested, establishment of test procedures, test site selection, selection and training of test personnel, preparation of test facilities and resources, the acquisition of support equipment and test supply support.

Selection of Test Item(s)

The equipment/software configuration used in the test must be representative of the operational item to the maximum extent possible. For Type I tests, engineering models are used that are not often directly comparable with operational equipment; however, most subsequent testing will be accomplished at such a time that equipment (or software) representing the final configuration is available. A prerequisite task involves selecting the test model by serial number, defining the configuration in terms of incorporated versus unincorporated engineering changes (if any), and ensuring that it is available at the time needed.

Test and Evaluation Procedures

Fulfillment of test objectives involves the accomplishment of both operator and maintenance tasks. Completion of these tasks should follow formal approved procedures, which are generally in the form of technical manuals developed during the latter phases of detail design. Following approved procedures is necessary to ensure that the system is operated and maintained in a proper manner. Deviation from approved procedures may result in the introduction of personnel-induced failures and will distort maintenance frequencies and task times as recorded in the test data. The identification of the procedures to be used in testing should be included in the evaluation plan.

Test-Site Selection

A single test site may be selected to evaluate the product in a designated environment; several test sites may be selected to evaluate the product under different environmental conditions (e.g., arctic versus tropical areas, mountainous versus flat terrain); or a variety of geographical locations may be selected to evaluate the product in terms of different markets and user conditions. Any one or combination of these methods may be appropriate, depending on the product type and program requirements.

Test Personnel and Training

Test personnel will include (1) individuals who actually operate and maintain the system and equipment during the test and (2) the supporting engineers, technicians, data recorders, analysts, and test administrators, as appropriate. Individuals assigned to the operation and maintenance of the system should possess the backgrounds and skill levels similar to consumer personnel who will normally operate and support the system throughout its life cycle. The required proficiency level(s) is attained through a combination of formal and on-the-job training.

Test Facilities and Resources

The necessary facilities, test chambers, capital equipment, environmental controls, special instrumentation, and associated resources (e.g., heat, water, air conditioning, gas, telephone, power, and lighting) must be identified and scheduled. In many instances, new design and construction are required, which directly affects the scheduling and duration of the test preparation period. A detailed description of the test facility and the facility layout should be included in test planning documentation and in subsequent test reports.

Test and Support Equipment

Test and support equipment requirements are initially considered in the maintenance concept and later defined through the maintenance and supportability analysis. During the latter phases of detail design, the necessary support items are procured and should be available for types 2, 3, and 4 testing. In the event that the proper type of support equipment is not available and alternative items are required, such items must be identified in the test and evaluation plan. The use of alternative items generally results in distorted test data (e.g., maintenance times), tends to cause personnel induced failures in the system, and often forces a change in test procedures and facility requirements.

Test Supply Support

Supply support constitutes all materials, data, personnel, and related activities associated with the requirements, provisioning, and procurement of spare and repair parts and the sustaining maintenance of inventories for support of the system throughout its life cycle. Specifically, this includes the following:

1. Initial and sustaining requirements for spares, repair parts, and consumables for the prime elements of the system. Spares are major replacement items and are repairable, while repair parts are nonrepairable, smaller components. Consumables refer to fuel, oil, lubricants, liquid oxygen, nitrogen, and so on.
2. Initial and sustaining requirements for spares, repair parts, material, and consumables for the various elements of logistic support (i.e., test and support equipment, transportation and handling equipment, training equipment, and facilities).
3. Facilities and warehousing required for the storage of spares, repair parts, and consumables. This involves consideration of space requirements, location, and environmental needs.
4. Personnel requirements for the accomplishment of supply support activities, such as provisioning, cataloging, receipt and issue, inventory management and control, shipment, and disposal of material.
5. Technical data requirements for supply support, which include initial and sustaining provisioning data, catalogs, material stock lists, receipt and issue reports, and material disposition reports.

The types of spare and repair parts needed at each level of maintenance are dependent on the maintenance concept and the supportability analysis. For type 3 and 4 testing, spare and repair parts will generally be required for all levels since these tests primarily involve an evaluation of the system as an entity and its total maintenance and logistic support capability. The complete maintenance cycle, supply support provisions (the type and quantity of spares specified at each level), transportation times, turnaround times, and related factors are evaluated as appropriate. In certain instances, the producer's facility may provide depot-level support. Thus, it is important to establish a realistic supply system (or a close approximation) for the item being tested.

The type and quantity of spare and repair parts and the procedures for spare parts inventory control throughout the test program should be specified in the test and evaluation plan. Usage rates, reorder requirements, procurement lead times, and out-of-stock conditions during the test are recorded and included in the test report.

6.5 TEST PERFORMANCE AND REPORTING

With the necessary prerequisites established, the next step is to commence with the formal test and demonstration of the system. This requires operating and supporting the system in a prescribed manner as defined in the system test and evaluation plan.

Throughout this process, data are collected and analyzed, which leads to assessment of system performance and effectiveness characteristics.

The assessment of performance and effectiveness of a system requires the availability of operational and maintenance histories of the various system elements. Performance and effectiveness parameters are established early in the life cycle with the development of operational requirements, the maintenance concept, and the establishment of technical performance measures (TPMs). These parameters describe the characteristics of the system that are considered paramount in fulfilling the need objectives. With the system in an operational status, the following questions arise:

1. What is the *true* performance and effectiveness of the system?
2. What is the *true* effectiveness of the maintenance and logistic support capability?
3. Are the initially specified requirements being met?

Providing answers to these questions requires a formalized data-information and feedback capability with the proper output. A data subsystem must be designed, developed, and implemented to achieve a specific set of objectives, and these objectives must relate directly to the preceding questions and whatever other questions the engineer or manager needs to answer. The establishment of the data subsystem capability is basically a two-step process: (1) the identification of requirements and the applications for such; and (2) the design, development, and implementation of a capability that will satisfy the identified requirements.

Test Data Requirements

The purpose of a test data and information feedback subsystem is twofold.

1. It provides ongoing data that are analyzed to evaluate and assess the performance, effectiveness, operation, maintenance, logistic support capability, and so on, for the system in the field. The systems engineer or manager needs to know exactly how the system is doing, and needs the answer relatively quickly. Thus, certain types of information must be provided at designated times.
2. It provides historical data (covering existing systems in the field) that are applicable in the design and development of new systems having a similar function and nature. Engineering potential in the future certainly depends on the ability to capture experiences of the past, and subsequently be able to apply the results in terms of what to do and what not to do in new design situations.

Supporting the above requires a capability that is both responsive to a repetitive need in an expeditious and timely manner (i.e., the manager's need for assessment information), and one that incorporates the provisions for data storage and retrieval. It is necessary to determine the specific elements of data reporting. These factors are combined to identify total volume requirements for the subsystem and the type, quantity, and frequency of data reports.

The elements of data identified are related to the operational and support requirements for the system. An analysis of these elements will provide certain evaluative- and verification-type functions covering the characteristics of the system that are to be assessed. A listing of sample applications appropriate in the assessment of system characteristics is presented in Figure 6.4. When determining the reliability of an item (i e., the probability that an item will operate satisfactorily in a given environment for a specified period), the data required will include the system operating time to failure and the time-to-failure distribution, which can be generated from a history of the particular item or a set of identical and independent items. When verifying spare/repair part demand rates, the data required should include a history of all item replacements, system/equipment operating time at replacement, and disposition of the replaced items (whether the item is condemned or repaired and returned to stock). An evaluation of organizational effectiveness will require identification of the number of assigned personnel and skill level, the tasks accomplished by the organization and the labor-hours and elapsed time expended in task accomplishment. Assessment objectives such as these (comparable to those listed in the figure) serve as the basis for the identification of the specific data elements needed.

Development of a Data Subsystem

With the overall subsystem objectives defined, the next step is to identify the specific data factors that must be acquired and the method for acquisition. A format for data collection must be developed and should include both success data and maintenance data. *Success data* constitute information covering system operation and use on a day-to-day basis, and the information should be comparable with the factors listed in Figure 6.5. *Maintenance data* cover each event involving scheduled and unscheduled maintenance. The events are recorded and referenced in system operational information reports, and the factors recorded in each instancc are illustrated in Figure 6.6.

The format for data collection may vary considerably, and the information desired may be different for each system. However, most of the factors in Figures 6.5 and 6.6 are common for all systems and must be addressed in the design of a new data subsystem. In any event, the following provisions should apply:

1. The data collection forms should be simple to understand and complete (preferably on single sheets), as the task of recording the data may be accomplished under adverse environmental conditions by a variety of personnel skill levels. If the forms are difficult to understand, they will not be completed properly (if at all), and the needed data will not be available.
2. The factors specified on each form must be clear and concise and not require considerable interpretation and manipulation to obtain.
3. The factors specified must have a meaning in terms of application. The usefulness of each factor must be verified.

These considerations are important and cannot be overemphasized. All the analytical methods, prediction techniques, models, and so on, discussed earlier have little meaning without the proper input data. Our ability to evaluate alternatives and predict in the future depends on the availability of good historical data, and the source of such stems from the type of data-information feedback subsystem developed at this stage. This subsystem must not only incorporate the forms for recording the right type

1. *General Operational and Support Factors*
   (a) Evaluation of mission requirements (operational scenarios).
   (b) Evaluation of performance factors (capacity, output, range, accuracy, size, weight, volume, mobility, etc.) (i.e., TPMs).
   (c) Verification of system use (modes of operation and operating hours).
   (d) Verification of cost and system effectiveness, operational availability, dependability, reliability, maintainability, human factors, safety (i.e., TPMs).
   (e) Evaluation of levels and location of maintenance.
   (f) Evaluation of operation and maintenance function/tasks by level and location.
   (g) Verification of repair-level policies.
   (h) Verification of frequency distributions for maintenance actions and repair times.
   (i) Evaluation of software and software interfaces with other system elements.
2. *Test and Support Equipment*
   (a) Verification of support equipment type and quantity by maintenance level and location.
   (b) Verification of support equipment availability.
   (c) Verification of support equipment use (frequency of use, location, percentage of time used, flexibility of use).
   (d) Evaluation of maintenance requirements for support equipment (scheduled and unscheduled maintenance, downtime, logistic resource requirements).
3. *Supply Support (Spare and Repair Parts)*
   (a) Verification of spare and repair part types and quantities by maintenance level and location.
   (b) Evaluation of supply responsiveness. (Is a spare available when needed?)
   (c) Verification of item replacement rates, condemnation rates, attrition rates.
   (d) Verification of inventory turnaround and supply pipeline times.
   (e) Evaluation of maintenance requirements for shelf items.
   (f) Evaluation of spare and repair part replacement and inventory policies.
   (g) Identification of shortage risks.
4. *Personnel and Training*
   (a) Verification of personnel quantities and skills by maintenance level and location.
   (b) Verification of elapsed times and labor-hour expenditures by personnel skill level.
   (c) Evaluation of personnel skill mixes.
   (d) Evaluation of personnel training policies.
   (e) Verification of training equipment and data requirements.
5. *Transportation and Handling*
   (a) Verification of transportation and handling equipment type and quantity by maintenance level and location.
   (b) Verification of availability and utilization of transportation and handling equipment.
   (c) Evaluation of delivery response times.

Figure 6.4 Data information subsystem applications (requirements).

6. *Facilities*
   (a) Verification of facility adequacy and utilization (operation, maintenance, and training facilities).
   (b) Evaluation of logistic resource requirements for support and operation, maintenance, and training facilities.
7. *Technical Data*
   (a) Verification of adequacy of data coverage (level, accuracy, and method of information presentation) in operating and maintenance manuals.
   (b) Verification of adequacy of field data, collection, analysis, and corrective action subsystem.
8. *Operational and Maintenance Software*
   (a) Verification of the compatibility of software with other elements of the system.
   (b) Verification of software reliability and maintainability.
9. *Consumer Response*
   (a) Evaluation of degree of consumer satisfaction.
   (b) Verification that consumer needs are met.

Figure 6.4 *(cont.)* Data information subsystem applications (requirements).

System Operational Information Report

1. Report number, report date, and individual preparing report.
2. System nomenclature, part number, manufacturer, serial number.
3. Description of system operation by date (mission type, profile, and duration).
4. Equipment use by date (operating time, cycles of operation, etc.).
5. Description of personnel, transportation and handling equipment, software, and facilities required for system operation.
6. Recording of maintenance events by date and time (reference maintenance event reports).

Figure 6.5 System success data.

of data, but must also consider the personnel factors (skill levels, motivation, etc.) involved in the data recording process. The individual who must complete the appropriate form(s) must understand the system and the purposes for which the data are being collected. If this individual is not properly motivated to do a thorough job in recording events, the resulting data will, of course, be highly suspect.

Once the appropriate data forms are distributed and completed by the responsible line organizations, a means must be provided for the retrieval, formatting, sorting, and processing of the data for reporting purposes. Field data are collected and sent to a designated centralized facility for analysis and processing. The results are disseminated to engineering and management for decision making and entered into a database for retention and possible future use.

Maintenance Event Report

1. *Administrative data*
   (a) Event report number, report date, and individual preparing report.
   (b) Work order number.
   (c) Work area and time of work (month, day, hour).
   (d) Activity (organization) identification.
2. *System factors*
   (a) System element part number and manufacturer.
   (b) System element serial number.
   (c) System operating time when event occurred (when discovered).
   (d) Segment of mission when event occurred.
   (e) Description of event (describe symptom of failure for unscheduled actions).
3. *Maintenance factors*
   (a) Maintenance requirement (repair, calibration, servicing, etc.).
   (b) Description of maintenance tasks.
   (c) Maintenance downtime.
   (d) Active maintenance times.
   (e) Maintenance delays (time awaiting spare part, delay for test equipment, work stoppage, awaiting personnel assistance, delay for weather, etc.).
4. *Logistics factors*
   (a) Start and stop times for each maintenance technician by skill level.
   (b) Technical manual or maintenance procedure used (procedure number, paragraph, date, comments on procedure adequacy).
   (c) Test and support equipment used (item nomenclature, part number, manufacturer, serial number, time of item usage, operating time on test equipment when used).
   (d) Description of facilities used.
   (e) Description of replacement parts (type and quantity).
      i. Nomenclature, part number, manufacturer, serial number, and operating time on replaced item. Describe disposition.
      ii. Nomenclature, part number, manufacturer, serial number, and operating time on installed item.
5. *Other information*
   Include any additional data considered appropriate and related to the maintenance event.

Figure 6.6 System maintenance data.

System Evaluation and Corrective Action

Figure 6.7 illustrates a system evaluation and corrective-action loop. The evaluation aspect responds to the type of subjects listed in Figure 6.4 and can address both the system as an entity or individual segments of the system on an independent basis. Figure

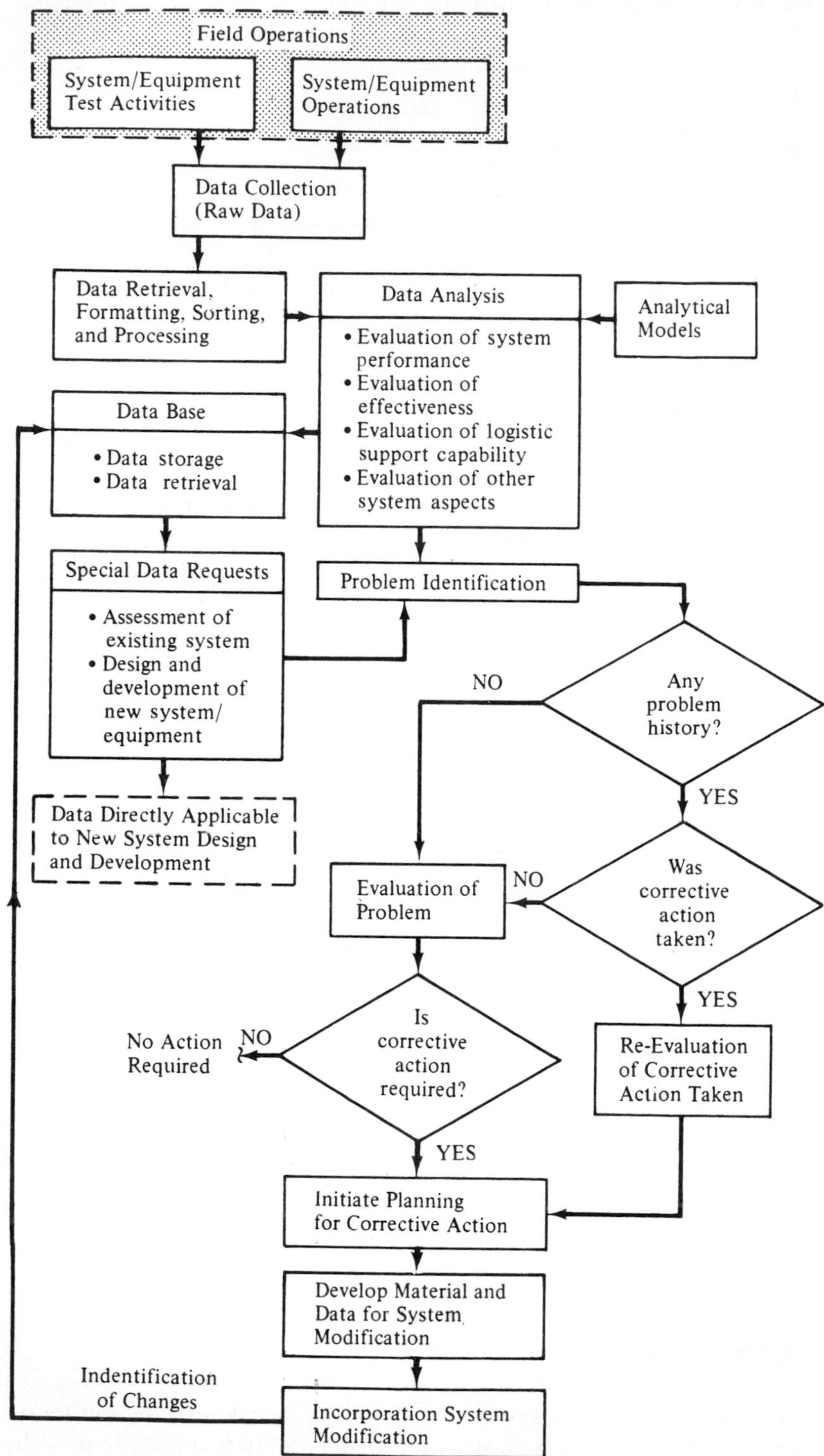

Figure 6.7 System evaluation and corrective action loop.

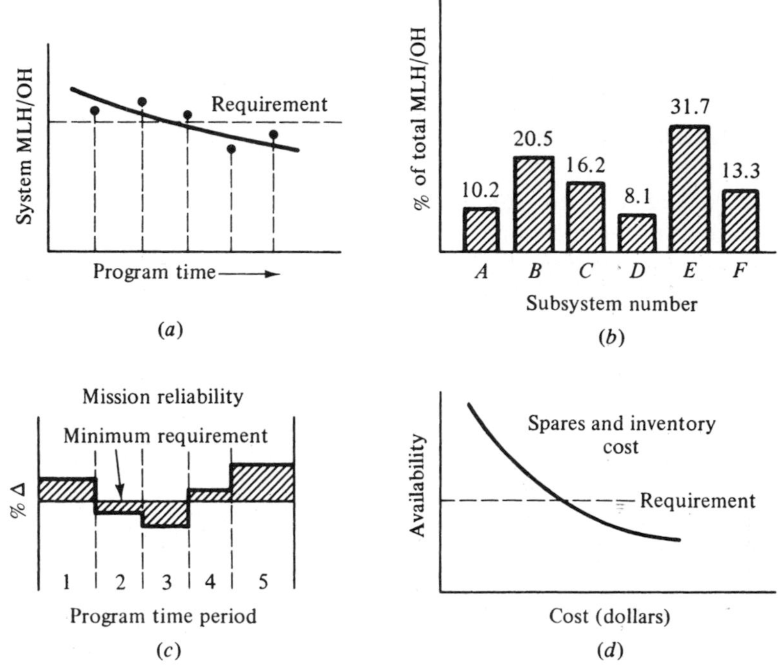

Figure 6.8 Evaluation factors (example).

6.8 presents some typical examples of evaluation factors. The evaluation approach and the analytical techniques (i.e., tools) used are basically the same as described earlier for design decision making, the only difference being the data input. The evaluation effort can be applied on a continuing basis to provide certain system measures at designated points thoughout the life cycle (see [a] and [c] of Figure 6.8), or it may constitute a one-time investigation.

Problem areas are identified at various stages in the evaluation and are reviewed in terms of the feasibility for corrective action. Referring to Figure 6.8(b), subsystem E is a likely candidate for investigation since the MLH/OH for that item is 31.7. In Figure 6.8(c), one may wish to investigate program time period 3 to determine why the mission reliability was so poor at that time. Corrective action may be accomplished in response to a system deficiency (i.e., the system fails to meet the specified requirements), or may be accomplished to improve system performance, effectiveness, or logistic support. If corrective action is to be accomplished, the necessary planning and implementation steps are a prerequisite to ensure the complete compatibility of all elements of the system throughout the change process.

Test Reporting

The final effort in the evaluation process constitutes the preparation of an appropriate test report. A test report should reference the initial system evaluation planning document and should describe all test conditions, incorporated system modifications during the test (if any), test data, and the results of data analysis. These results may include appropriate recommendations for operation and support of the system as applicable to the utilization phase.

6.6 SYSTEM MODIFICATION

When a change occurs in a procedure, the prime equipment, an element of software, or an item of logistic support, the change in most instances will affect many different elements of the system. Procedural changes will affect personnel and training requirements and necessitate a change in the technical data (equipment operating and/or maintenance instructions). Hardware changes will affect spare and repair parts, test and support equipment, technical data, and training requirements. Software changes may affect the hardware and the technical data. Each change must be thoroughly evaluated in terms of its impact on other elements of the system prior to a decision on whether or not to incorporate the change. The feasibility of incorporating a change will depend on the extensiveness of the change, its impact on the system's ability to accomplish its mission, the time in the life cycle when the change can be incorporated, and the cost of change implementation. A minimum amount of evaluation and planning are required to make a rational decision as to whether the change is feasible.

If a change is to be incorporated, the necessary change control procedures must be implemented. A change control board is established to review and evaluate the potential impact of the change, the effectivity of the change (the serial numbered item on which the change is to be incorporated), retrofit provisions, and so on. The various components of the system (including all the elements of logistic support) must track. Otherwise, the probability of inadequate logistic support and unnecessary waste is high. The concepts and concerns of configuration change control in the production/construction stage are equally as appropriate in the operation and system support phase. In fact, a change involving the operational system is more critical, since we are dealing with a later stage in the life cycle. Changes will usually be more costly as the life cycle progresses. An abbreviated version of a change control process is illustrated in Figure 6.9.[10]

QUESTIONS AND PROBLEMS

1. Describe the system evaluation process. What is the objective?
2. How is system evaluation accomplished in the following phases: Conceptual design? Preliminary system design? Detail design? Production?
3. Define the basic categories of test and their applications. How do they fit into the total system evaluation process?
4. When does the planning for system evaluation commence? What information is included?
5. Select a system of your choice and develop a detailed system test and evaluation plan.

[10] Systems engineering must play a major role relative to configuration management and change control if a reasonable baseline is to be maintained throughout system design and development. Otherwise, there is no assurance of an unified approach in the system definition process.

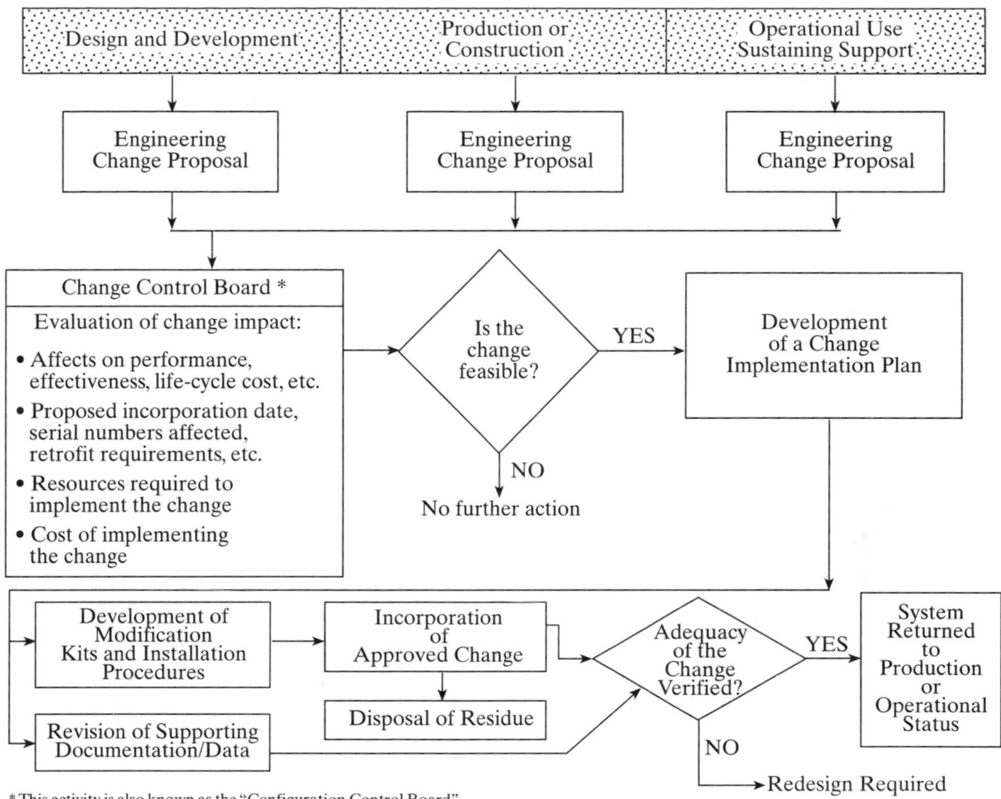

Figure 6.9 Change control process.

6. How are specific system test requirements determined?
7. Why is it important to establish the proper level of logistic support for system evaluation? What would probably happen in the absence of adequate test procedures? What would happen in the event of inadequate training of test personnel? What would probably happen in the absence of the proper test and support equipment?
8. What purpose(s) does a good data subsystem serve?
9. Why are system success data important? What type of information is required?
10. What data are required to measure the following: cost-effectiveness, system effectiveness, operational availability, and life-cycle cost?
11. In the event that system evaluation indicates noncompliance with a specified system requirement, what should be done?
12. If during evaluation the customer makes a recommendation for product improvement, what should be done?
13. Why is change control so important? How do system changes affect logistic support?
14. What role does systems engineering fulfill in the system evaluation function?
15. Briefly describe the producer and customer functions in system evaluation.
16. What benefits are derived from test reporting?
17. How does evaluation fit into the systems engineering process?

SYSTEMS ANALYSIS AND DESIGN EVALUATION

PART III

ALTERNATIVES AND MODELS IN DECISION MAKING

Engineers, whether engaged in research, design, development, construction, production, operations, or a synthesis of these activities, are concerned with the efficient use of limited resources. When known opportunities fail to hold sufficient promise for the employment of resources, more promising opportunities are sought. This view accompanied by initiative leads to exploratory activities aimed at finding the better opportunities. In such activities, steps are taken into the unknown to find new possibilities that may then be evaluated to determine if they could be superior to those now known.

An understanding of the decision-making process usually requires simplification of the complexity facing the decision maker. Conceptual simplifications of reality, or models, are a means to this end. In this chapter, a structure for alternatives and a general decision evaluation approach using models is presented to facilitate decision making in design and operations.

7.1 ALTERNATIVES IN DECISION MAKING

A complete and all-inclusive alternative rarely emerges in its final state. It begins as a hazy but interesting idea. The attention of the individual or group is then directed to analysis and synthesis, and the result is a definite proposal. In its final form, an alternative should consist of a complete description of its objectives and its requirements in terms of benefit and cost.

Both different ends and different methods are embraced by the term *alternative*. All proposed alternatives are not necessarily attainable. Some are proposed for analysis even though there seems to be little likelihood that they will prove feasible. The idea is that it is better to consider many alternatives than to overlook one that might be preferred. Alternatives that are not considered cannot be adopted, no matter how desirable they may actually be.

Limiting and Strategic Factors

Those factors that stand in the way of attaining objectives are known as *limiting factors*. An important element of the systems engineering process is the identification of the limiting factors restricting accomplishment of a desired objective. Once the limiting factors have been identified, they are examined to locate *strategic factors*, those factors that can be altered to make progress possible.

The identification of strategic factors is important, for it allows the decision maker to concentrate effort on those areas in which success is obtainable. This may require inventive ability, or the ability to put known things together in new combinations, and is distinctly creative in character. The means that will achieve the desired objective may consist of a procedure, a technical process, or a mechanical, organizational, or managerial change. Strategic factors limiting success may be circumvented by operating on engineering, human, and economic factors individually and jointly.

An important element of the process of defining alternatives is the identification of the limiting factors restricting the accomplishment of a desired objective. Once the limiting factors have been identified, they are examined to locate those strategic factors that can be altered in a cost-efficient way so that a selection from among the alternatives may be made.

Comparing Alternatives Equivalently

To compare alternatives equivalently, it is important that they be converted to a common measure. This conversion to a common measure permits comparison on the basis of equivalence. Models and optimization are essential in the conversion process and permit alternatives to be compared equivalently.

On completion of the conversion step, quantitative and qualitative outputs and inputs for each alternative form the basis for comparison and decision. Quantitative measures should be obtained with the use of suitable models, and decisions between alternatives should be made on the basis of their differences. Thus, all identical factors can be canceled out for the comparison of any two or more alternatives at any step of a decision-making process.

After a situation has been carefully analyzed and the possible outcomes have been evaluated as accurately as possible, a decision must be made. Several outcomes are evaluated for each alternative that is examined. We know that all the outcomes cannot occur, and that at any given time any outcome can occur. Thus, decision making or selection from among alternatives is often done under risk or uncertainty.

In addition to the alternatives formally set up for evaluation, another alternative is almost always present—that of making no decision. The decision not to decide may be a result of either active consideration or passive failure to act; it is usually motivated by the thought that there will be opportunities in the future that may prove more desirable than any known at present.

7.2 MODELS IN DECISION MAKING

Models and their manipulation (the process of simulation) are useful tools in systems analysis. A *model* may be used as a representation of a system to be brought into being, or to analyze a system already in being. Experimental investigation using a model yields design or operational decisions in less time and at less cost than direct manipulation of the system itself. This is particularly true when it is not possible to manipulate reality because the system is not yet in existence, or when manipulation is costly and disruptive as with complex industrial systems.

There is a fundamental difference between models used in science and engineering. Science is concerned with the natural world, whereas engineering is concerned primarily with the human-made world. Science uses models to gain an understanding of the way things are in the natural world. Engineering uses models of the human-made world in an attempt to achieve what ought to be. The validated models of science are used in engineering to establish bounds for engineering creations and to improve the products of such creations.

Classification of Models

When used as a noun the word "model" implies representation. An aeronautical engineer may construct a wooden model of a possible configuration for a proposed aircraft type. An architect might represent a proposed building with a scale model of the building. An industrial engineer may use templates to represent a proposed layout of equipment in a factory. The word "model" may also be used as an adjective, carrying with it the implication of ideal. Thus, a man may be referred to as a model husband or a child praised as a model student. Finally, the word "model" may be used as a verb, as is the case where a person models clothes.

Models are designed to represent a system under study, by an idealized example of reality, to explain the essential relationships involved. They can be classified by distinguishing physical, analog, schematic, and mathematical types. Physical models look like what they represent, analogue models behave like the original, schematic models graphically describe a situation or process, and mathematical models symbolically represent the principles of a situation being studied. These model types are used successfully in systems engineering and analysis.

Physical models. Physical models are geometric equivalents, either as miniatures, enlargements, or duplicates made to the same scale. Globes are one example. They are used to demonstrate the shape and orientation of continents, water bodies, and other geographic features of the earth. A model of the solar system is used to demonstrate the orientation of the sun and planets in space. A model of an atomic structure would be similar in appearance but at the other extreme in dimensional reproduction. Each of these models represents reality and is used for demonstration.

Some physical models are used in the simulation process. An aeronautical engineer may test a specific tail assembly design with a model airplane in a wind tunnel. A pilot plant might be built by a chemical engineer to test a new chemical process for the purpose of locating operational difficulties before full-scale production. An environmental chamber is often used to create conditions anticipated for a component under test.

The use of templates in plant layout is an example of experimentation with a physical model. Templates are either two- or three-dimensional replicas of machinery and equipment that are moved about on a scale-model area. The relationship of distance is important, and the templates are manipulated until a desirable layout is obtained. Such factors as noise generation, vibration, and lighting are also important but are not a part of the experimentation and must be considered separately.

Analogue models. Analogue comes from the Greek word *analogia*, which means proportion. This explains the concept of an *analogue model*; the focus is on similarity in relations. Analogues are usually meaningless from the visual standpoint.

Analogue models can be physical in nature, such as where electric circuits are used to represent mechanical systems, hydraulic systems, or even economic systems. Analogue computers use electronic components to model power distribution systems, chemical processes, and the dynamic loading of structures. The analogue is represented by physical elements. When a digital computer is used as a model for a system, the analogue is more abstract. It is represented by symbols in the computer program and not by the physical structure of the computer components.

The analogue may be a partial subsystem, or it may be an almost complete representation of the system under study. For example, the tail assembly design being tested in a wind tunnel may be complete in detail but incomplete in the properties being studied. The wind tunnel test may examine only the aerodynamic properties and not the structural, weight, or cost characteristics of the assembly. From this it is evident that only those features of an analogue model that serve to describe reality should be considered. These models, like other types, suffer from certain inadequacies.

Schematic models. A schematic model is developed by reducing a state or event to a chart or diagram. The schematic model may or may not look like the realworld situation it represents. It is usually possible to achieve a much better understanding of the real-world system described by the model through use of an explicit coding process employed in the construction of the model. The execution of a football play may be diagrammed on a blackboard with a simple code. It is the idealized aspect of this schematic model that permits this insight into the football play.

An organization chart is a common schematic model. It is a representation of the state of formal relationships existing between various members of the organization. A human-machine chart is another example of a schematic model. It is a model of an event; that is, the time-varying interaction of one or more people and one or more machines over a complete work cycle. A flow process chart is a schematic model that describes the order or occurrence of several events that constitute an objective, such as the assembly of an automobile from a multitude of component parts.

In each case, the value of the schematic model lies in its ability to describe the essential aspects of the existing situation. It does not include all extraneous actions and relationships but rather concentrates on a single facet. Thus, the schematic model is not in itself a solution but only facilitates a solution. After the model has been carefully analyzed, a proposed solution can be defined, tested, and implemented.

Mathematical models. A mathematical model employs the language of mathematics and, like other models, may be a description and then an explanation of the system it represents. Although its symbols may be more difficult to comprehend than verbal symbols, they do provide a much higher degree of abstraction and precision in their application. Because of the logic it incorporates, a mathematical model may be manipulated in accordance with established mathematical procedures.

Almost all mathematical models are used either to predict or to control. Such laws as Boyle's law, Ohm's law, and Newton's laws of motion are formulated mathematically and may be used to predict certain outcomes when dealing with physical phenomena. Outcomes of alternative courses of action may also be predicted if a measure of evaluation is adopted. For example, a linear programming model may predict the profit associated with various production quantities of a multiproduct process. Mathematical models may be used to control an inventory. In quality control, a mathematical model may be employed to monitor the proportion of defects that will be accepted from a supplier. Such models maintain control over a state of reality.

Mathematical models directed to the study of systems differ from those traditionally used in the physical sciences in two important ways. First, because the system being studied usually involves social and economic factors, these models must often incorporate probabilistic elements to explain their random behavior. Second, mathematical models formulated to explain existing or planned operations incorporate two classes of variables: those under the control of a decision maker and those not directly under control. The objective is to select values for controllable variables so that some measure of effectiveness is optimized. Thus, these models are of great benefit in systems engineering and systems analysis.

Models and Indirect Experimentation

Models and the process of simulation provide a convenient means of obtaining factual information about a system being designed or a system in being. In component design it is customary and feasible to build several prototypes, test them, and then modify the design based on the test results. This is often not possible in systems engineering because of the cost involved and the length of time required over the system life cycle. A major part of the design process requires decisions based on a model of the system rather than decisions derived from the system itself.

Direct and indirect experimentation. In direct experimentation, the object, state, or event, and/or the environment is subject to manipulation, and the results are observed. For example, a couple might rearrange the furniture in their living room by this method. Essentially, they move the furniture and observe the results. This process may then be repeated with a second move and perhaps a third, until all logical alternatives have been exhausted. Eventually, one such move is subjectively judged best; the furniture is returned to this position, and the experiment is completed. Direct experimentation, such as this, may be applied to the rearrangement of equipment in a factory. Such a procedure is time-consuming, disruptive, and costly. Hence, simulation or indirect experimentation is employed with templates representing the equipment to be moved.

Direct experimentation in aircraft design would involve constructing a full-scale prototype that would be flight-tested under real conditions. Although this is an essential step in the evolution of a new design, it would be costly as the first step. The usual procedure is evaluating several proposed configurations by building a model of each and then testing in a wind tunnel. This is the process of indirect experimentation, or *simulation*. It is extensively used in situations where direct experimentation is not economically feasible.

In systems analysis indirect experimentation is effected through the formulation and manipulation of decision models. This makes it possible to determine how changes in those aspects of the system under control of the decision maker affect the modeled system. Indirect experimentation enables the systems analyst to evaluate the probable outcome of a given decision without changing the operational system itself. In effect, indirect experimentation in the study of operations provides a means for making quantitative information available to the decision maker without disturbing the operations under his or her control.

Simulation through indirect experimentation. In most design and operational situations, the objective sought is the optimization of a performance measure economically. Rarely, if ever, can this be done by direct experimentation with a system under development or a system in being.

The primary use of simulation in systems engineering is to explore the effects of alternative system characteristics on system performance without actually producing and testing each candidate system. Most models used will fit the classification given earlier, and many will be mathematical. The type used will depend on the questions to be answered. In some instances, simple schematic diagrams will suffice. In others, mathematical or probabilistic representations will be needed. In many cases, simulation with the aid of an analogue or digital computer will be required.

In most systems engineering undertakings, several models must be formulated. These models form a hierarchy ranging from considerable aggregation to extreme detail. At the start of a systems project, knowledge of the system is sketchy and general. As the design progresses this knowledge becomes more detailed, and consequently, the models used for simulation should be detailed.

Models in Design and Operations

The delay in the development of models to explain and describe complex systems may be attributed to an early preoccupation with the physical and biological sciences. This is understandable because during much of history the limiting factors in the satisfaction of human wants were predominantly physical and biological. But with the accumulation of scientific knowledge, human beings have been able to design complex systems to meet their needs. Engineers and managers are becoming increasingly aware that experience, intuition, and judgment are insufficient for the effective pursuit of design and operational objectives. Models are useful in design and operations because they take the decision maker part way to the point of decision.

In formulating a mathematical decision model, one attempts to consider all components of the system that are relevant to the system's effectiveness and cost. Because of the impossibility of including all factors in constructing the evaluation function, it is common practice to consider only those on which the outcome is believed to depend significantly. This necessary viewpoint sometimes leads to the erroneous conclusion that certain segments of the environment are actually isolated from each other. Although it may be feasible to consider only those relationships that are significantly pertinent, one should remember that all system elements are interdependent.

Models for design decisions and for operational decision making are abstractions of the system under study. Like all abstractions, models involve several assumptions: assumptions about the operating characteristics of components, about the behavior of people, and about the nature of the environment. These assumptions must be fully understood and evaluated when models are used for decision making in design and operations.

Manipulation of the model can lead to model modification to reduce the misfit between the model and the real world. This *validation* process has a recurrent pattern analogous to the scientific method. It consists of three steps: (1) postulate a model, (2) test the model prediction or explanation against measurements or observations, and (3) modify the model to reduce the misfit.

The validation process should continue until the model is supported reasonably by evidence from measurement and observation. As a model evolves to a state of validity, it provides postulates of reality that can be depended on. A model may have achieved a state of validity even though it gives biased results. Here the model is valid because it is consistent; that is, it gives results that do not vary.

A decision model cannot be classified as accurate or inaccurate in any absolute sense. It may be considered to be accurate if it is an idealized substitute of the actual system under study. If manipulation of the model would yield the identical result that manipulation of reality would have yielded, the model would be true. If, however, one knew what the manipulation of reality would have yielded, the process of simulation would be unnecessary. Hence, it is difficult to test a decision model except by an intuitive check for reasonableness.

There is no available theory by which thc best model for a given system simulation can be selected. The choice of an appropriate model is determined as much by the background of the systems analyst as the system itself.

7.3 DECISION EVALUATION THEORY

Decision evaluation is an important part of systems engineering and analysis. Evaluation is needed as a basis for choice among alternatives that arise from design activities, as well as for optimizing systems already in operation. In either case, equivalence provides the common evaluation measure on which choice can be based.

This section presents two general categories of decision evaluation models. The first is based on money flows over time and designated *money flow modeling*. The second often embraces money flows but always incorporates optimization. It is designated *economic optimization modeling*.

Evaluation by Money Flow Modeling

Economic equivalence is expressed as the present equivalent, annual equivalent, or future equivalent amount, as well as the internal rate-of-return and the payback period. Each of these will be explained and developed in Chapter 8.

A general equivalence function subsuming each equivalence expression may be stated as

$$\text{PE, AE, or FE} = f(F_t, i, n) \tag{7.1}$$

where $t = 0, 1, 2, \ldots, n$ and where

F_t = positive or negative money flow at the end of year t
i = annual rate of interest
n = number of years

The product life cycle is the underlying money flow generator over its acquisition and utilization phases as shown for a hypothetical situation in Figure 7.1 (also refer to Figure 2.1). For acquisition by purchase (Figure 7.1a), the mapping of the general

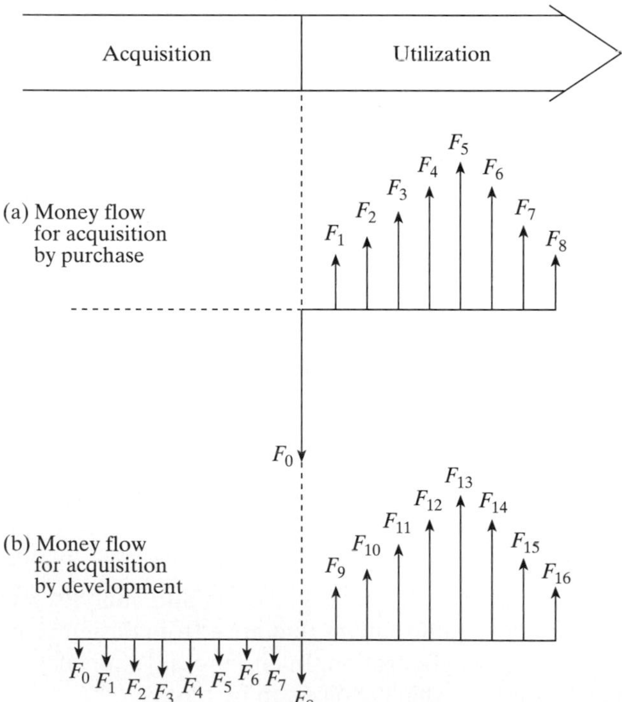

Figure 7.1 Money flows for two modes of acquisition/utilization.

equivalence function over the life cycle is simple. Here acquisition occurs instantaneously, with F_0 as the first cost or initial investment. Net benefits or revenues in this example occur at the end of each year for 8 years; F_1 through F_8. Salvage value or cost, if any, is included in F_8.

When acquisition requires expenditures for design and development (Figure 7.1b), mapping of the general equivalence function over the life cycle is more complex. Here acquisition spans several years, with expenditures composed of F_0 through F_8 (F_1 through F_3 for conceptual and preliminary design, F_4 through F_6 for detail design and development, and F_7 through F_8 for production and/or construction). Net benefits or revenues occur at the end of each year for 8 years; F_9 through F_{16}. Salvage value or cost, if any, is included in F_{16}.

The present equivalent, annual equivalent, and future equivalent amounts will be shown to be consistent bases for the evaluation of a single alternative, or for the comparison of mutually exclusive alternatives. Any one of these may be used in accordance with the general economic equivalence function of Equation 7.1. Additionally, Equation 7.1 provides a general structure for evaluation by the internal rate of return or payout period approaches. Chapter 8 provides the needed analytical foundation for decision evaluation in accordance with Equation 7.1.

Evaluation by Economic Optimization Modeling

Decision evaluation often requires a combination of both money flow modeling and economic optimization. When investment cost, periodic costs, or project life is a function of one or more decision variables, it is important to optimize over these variables as a prerequisite to the determination of economic equivalence. This optimization is linked to decision evaluation through one or more money flows, which, in turn, are used in calculating a measure of economic equivalence. Optimization requires that an evaluation measure be derived from an economic optimization model.

An economic optimization function is a mathematical model formally linking an evaluation measure, E, with controllable decision variables, X, and system parameters, Y, which cannot be directly controlled by the decision maker. It provides a means for testing decision variables in the presence of system parameters. This test is an indirect experiment performed mathematically, which results in an optimized value for E. The functional relationship, in its unconstrained form, may be expressed as

$$E = f(X, Y) \tag{7.2}$$

As an example of the optimization function structure for an unconstrained decision situation, consider the determination of an optimal procurement quantity for inventory operations. Here the evaluation measure is cost, and the objective is to choose a procurement quantity in the face of demand, procurement cost, and holding cost, so that total cost is minimized. The procurement quantity is the variable directly under the control of the decision maker. Demand, procurement cost, and holding cost are not directly under his or her control. Use of the optimization function allows the decision maker to arrive at a value for the variable under his or her control that trades-off conflicting cost elements.

The optimization function may be extended to both operational and design decisions involving alternatives. This extension involves the identification and isolation of design or decision-dependent system parameters, Y_d, from design or decision-independent system parameters, Y_i. Accordingly, Equation 7.2 can be restated in unconstrained form as

$$E = f(X, Y_d, Y_i) \tag{7.3}$$

As an example of the application of this version of the decision evaluation function, consider the establishment of a procurement and inventory system to meet the demand for an item that is available from one of several sources. Decision variables are the procurement level and the procurement quantity. For each source under consideration there exists a set of source-dependent parameters. These are the item cost per unit, the procurement cost per procurement, the replenishment rate, and the procurement lead time. Uncontrollable (source independent) system parameters include the demand rate, the holding cost per unit per period, and the shortage penalty cost. The objective is to determine the procurement level, and procurement quantity, and the procurement source so that total system cost will be minimized.

In design decision making, consider the deployment of a population of repairable equipment units to meet a demand. Three decision variables may be identified: the number of units to deploy, the number of maintenance channels to provide, and the age at which units should be retired. Controllable (design dependent) system parameters include the reliability, maintainability, energy efficiency, and design life. Uncontrollable system parameters include the cost of energy, the time value of money, and the penalty cost incurred when there are insufficient units operational to meet demand. Because the controllable parameters are design dependent, the objective is to develop design alternatives in the face of decision variables so that the design alternative that will minimize total system cost can be identified.

The place and role of Equation 7.2 and Equation 7.3 differs in systems analysis and design evaluation. When considered in terms of the life cycle, each applies to a major domain as exhibited in Figure 7.2. Equation 7.2 is applicable to the optimization of operations already in being, whereas Equation 7.3 is useful for choosing from among mutually exclusive design alternatives based on design-dependent parameters.

The design-dependent parameter approach facilitates the life-cycle complete evaluation of alternative system designs. This approach involves separation of the system design space (represented by design-dependent parameters) from the optimization space (represented by design variables). Terms are defined as follows:

1. *Design-dependent parameters* (Y_d). These are factors with values under the control of the designer(s) and that are impacted by the specialty disciplines during the development process. Every instance of the design-dependent parameter set represents a distinct candidate system or design alternative, because these parameters represent the design space. Examples of design dependent parameters include reliability, producibility, maintainability, time to altitude, and throughput.
2. *Design-independent parameters* (Y_i). These are factors beyond control of the designer(s) but that impact the effectiveness of all candidate systems or design

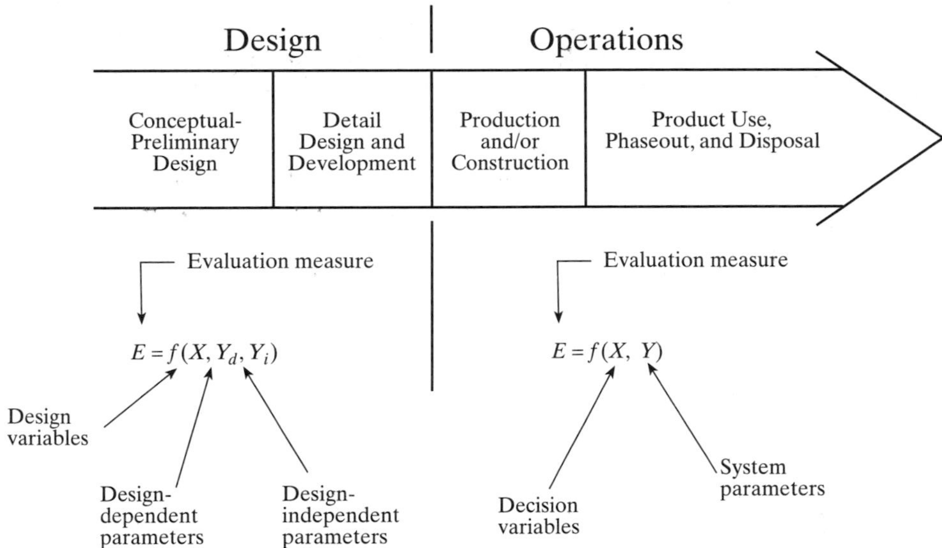

Figure 7.2 Economic optimization in design and operations.

alternatives, and can significantly alter their relative "goodness" or desirability. Examples of design-independent parameters include labor rates, material cost, energy cost, inflation and interest factors, and so on.

3. *Design variables* (X). These are factors that define the design optimization space. Each candidate system is optimized over the set of design variables before being compared with the other alternatives. In this way, equivalence is assured.

A summary of the evolution and development of the decision evaluation function is given in Table 7.1. Cited references in the table are given in complete form in Appendix F. Models in subsequent chapters are developed in accordance with the decision evaluation functions presented in this section and summarized in Table 7.1.

7.4 THE DECISION EVALUATION MATRIX

A particular decision can result in one of several outcomes, depending on which of several future events occurs. For example, a decision to go sailing can result in a high degree of satisfaction if the day turns out to be sunny, or in a low degree of satisfaction if it rains. These levels of satisfaction would be reversed if the decision were made to stay home. Thus, for the two states of nature, sun and rain, there are different payoffs depending on the alternative chosen.

A decision evaluation matrix is a formal way of exhibiting the interaction of a finite set of alternatives and a finite set of possible futures (or states of nature). In this usage, alternatives have the meaning presented in Section 7.1; that is, they are courses of action among which a decision maker expects to choose. The states of nature are

TABLE 7.1 Forms of the Decision Evaluation Function (References in Appendix F)

REFERENCES	FUNCTIONAL FORM	APPLICATION
Churchman, Ackoff, & Arnoff (1957)	$E = f(x_i, y_j)$ E = system effectiveness x_i = variables under direct control y_j = variables not subject to direct control	Operations
Fabrycky, Ghare, & Torgersen (1984)	$E = f(X, Y); g(X) \lesseqgtr B$ E = evaluation measure X = controllable variables Y = uncontrollable variables	Operations
Banks & Fabrycky (1987)	$E = f(X, Y_d, Y_i); g(X, Y_d) \lesseqgtr C$ E = evaluation measure X = procurement level and procurement quantity Y_d = source-dependent parameters Y_i = source-independent parameters	Procurement Operations
Fabrycky & Blanchard (1991)	$E = f(X, Y_d, Y_i); g(X, Y_d) \lesseqgtr C$ E = evaluation measure X = design variables Y_d = design-dependent parameters Y_i = design-independent parameters	Design Optimization

normally not natural events, such as rain, sleet, or snow, but are a wide variety of future outcomes over which the decision maker has no direct control.

The general decision evaluation matrix is a model depicting the positive and negative results that will occur for each alternative under each possible future. In abstract form, this model is structured as shown in Figure 7.3. Its symbols are defined as follows:

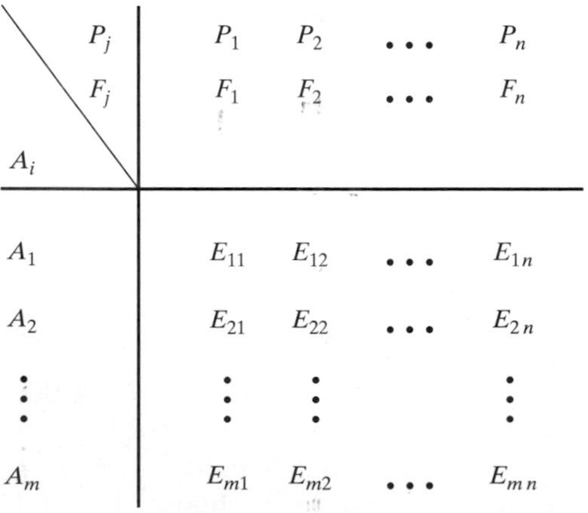

Figure 7.3 The decision evaluation matrix.

A_i = an alternative available for selection by the decision maker, where $i = 1, 2, \ldots, m$

F_j = a future not under control of the decision maker, where $j = 1, 2, \ldots, n$

P_j = the probability that the jth future will occur, where $j = 1, 2, \ldots, n$

E_{ij} = evaluation measure (positive or negative) associated with the ith alternative and the jth future determined from Equation 7.1 or 7.3

Several assumptions underlie the application of this decision evaluation matrix model to decision making under assumed certainty, risk, and uncertainty. Foremost among these is the presumption that all viable alternatives have been considered and all possible futures have been identified. Possible futures not identified can significantly affect the actual outcome relative to the planned outcome.

Evaluation measures in the matrix model are associated with outcomes that may be either objective or subjective. The most common case is one in which the outcome values are objective and, therefore, subject to quantitative expression in cardinal form. For example, the payoffs may be profits expressed in dollars, yield expressed in pounds, costs (negative payoffs) expressed in dollars, or other desirable or undesirable measures. Subjective outcomes, conversely, are those that are valued on an ordinal or ranking scale. Examples are expressions of preference, such as a good corporate image being preferred to a poor image, higher-quality outputs being preferred to those of lower quality, and so forth.

Other assumptions of importance in the evaluation matrix representation of decisions are

1. The occurrence of one future precludes the occurrence of any other future (futures are mutually exclusive).
2. The occurrence of a specific future is not influenced by the alternative selected.
3. The occurrence of a specific future is not known with certainty, even though certainty is often assumed for analysis purposes.

7.5 DECISIONS UNDER ASSUMED CERTAINTY

In dealing with physical aspects of the environment, physical scientists and engineers have a body of systematic knowledge and physical laws on which to base their reasoning. Such laws as Boyle's law, Ohm's law, and Newton's laws of motion were developed primarily by collecting and comparing many comparable instances and by the use of an inductive process. These laws may then be applied with a high degree of certainty to specific instances. They are supplemented by many models for physical phenomena that enable conclusions to be reached about the physical environment that match the facts with narrow limits. Much is known with certainty about the physical environment.

Much less, particularly of a quantitative nature, is known about the environment within which operational decisions are made. Nonetheless, the primary aim of operations research and management science is to bring the scientific approach to bear to a maximum feasible extent. This is done with the aid of conceptual simplifications and models of reality, the most common being the assumption of a single known future. It is not claimed that knowledge about the future is in hand. Rather, the suppression of risk and uncertainty is one of the ways in which the scientific approach simplifies reality to gain insight. Such insight can assist greatly in decision making, provided that its shortcomings are recognized and accommodated.

The evaluation matrix for decision making under assumed certainty is not a matrix at all. It is a vector with as many evaluations as there are alternatives, with the outcomes constituting a single column. This decision vector is a special case of the matrix of Figure 7.3. It appears as in Figure 7.4, with the payoffs represented by E_i, where $i = 1, 2, \ldots, m$. The single future, which is assumed to occur with certainty, actually carries a probability of unity ($P = 1.0$) in the matrix. All other futures are suppressed by carrying probabilities of zero ($P = 0.0$).

When the outcomes, E_i, are stated in monetary terms (cost or profit), the decision rule or principle of choice is simple. If the alternatives are equal in all other respects, one would choose the alternative that minimizes cost or maximizes profit. In the case of cost, one would choose

$$\min_i \{E_i\} \qquad \text{for } i = 1, 2, \ldots, m$$

For profit, one would choose

$$\max_i \{E_i\} \qquad \text{for } i = 1, 2, \ldots, m$$

It is often not possible to accept the premise that only the cost or the profit differences are important, with intangibles and irreducibles having little or no effect. Unquantifiable nonmonetary factors may be significant enough to outweigh calculated costs or profit differences among alternatives. In other cases, the outcome is not easily expressed in monetary terms, or even in quantitative terms of some other evaluation measure, such as time, percentage of market, and so forth. Valid qualitative comparisons may be made when the quantitative outcomes cannot stand alone and when the outcomes are nonquantitative.

The use of outcome scales often makes possible a somewhat rational choice from among several nonquantifiable alternatives, each with an outcome rating determined by expert opinion, estimation, history, or other means. Foremost among these are ordi-

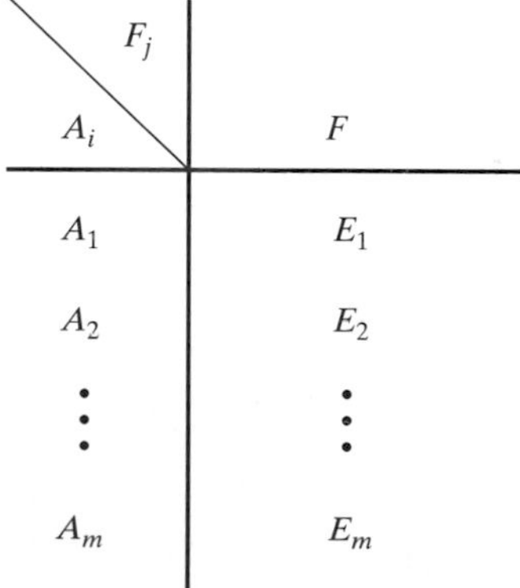

Figure 7.4 Decision evaluation vector.

nal comparisons. Where intangibles and irreducibles are significant, ranking each outcome above or below every other outcome leads to a preferred choice. To do this, each outcome can be compared to a common standard, or the outcome can be paired and compared. As an example of the paired approach, suppose that four alternatives are assumed to lead (with certainty) to four outcomes E_1, E_2, E_3, and E_4. Suppose that the six possible pairs are arranged according to preference as follows.

$$E_1 > E_3 \quad E_2 > E_3 \quad E_2 > E_1$$

$$E_2 > E_4 \quad E_3 > E_4 \quad E_1 > E_4$$

where the symbol $>$ is used to indicate that the outcome identified first is preferred to its counterpart. In these comparisons, E_2 is preferred three times; E_1 twice; E_3 once; and E_4 not at all. Accordingly, the preference ranking is

$$E_2 > E_1 > E_3 > E_4$$

7.6 DECISIONS UNDER RISK

There is usually little assurance that predicted futures will coincide with actual futures. The physical and economic elements on which a course of action depends may vary from their estimated values because of chance causes. Not only are the estimates of future costs problematical but, in addition, the anticipated future worth of most ventures is known only with a degree of assurance. This lack of certainty about the future makes decision making one of the most challenging tasks faced by individuals, industry, and government.

Decision making under risk occurs when the decision maker does not suppress acknowledged ignorance about the future but makes it explicit through the assignment of probabilities. Such probabilities may be based on experimental evidence, expert opinion, subjective judgment, or a combination of these.

Consider the following example. *A* computer systems firm has the opportunity to bid on two related contracts being advertised by a municipality. The first pertains to the selection and installation of hardware for a central computing facility together with required software. The second involves the development of a distributed computing network involving the selection and installation of hardware and software. The firm may be awarded either contract C_1 or contract C_2, or both contract C_1 and C_2. Thus, there are three possible futures.

Careful consideration of the possible approaches leads to the identification of five alternatives. The first is for the firm to subcontract the hardware selection and installation, but to develop the software itself. The second is for the firm to subcontract the software development, but to select and install the hardware itself. The third is for the firm to handle both the hardware and software tasks itself. The fourth is for the firm to bid jointly with a partner firm on both the hardware and software projects. The fifth alternative is for the firm to serve only as project manager, subcontracting all hardware and software tasks.

With the possible futures and various alternatives identified, the next step is to determine payoff values. Also to be determined are the probabilities for each of the three futures, where the sum of these probabilities must be unity. Suppose that these determinations lead to the profits and probabilities given in Table 7.2.

Table 7.2 is structured in accordance with the format of the decision evaluation matrix model exhibited in Figure 7.1. It is observed that the firm anticipates a profit of $100,000 if alternative A_1 is chosen, and contract C_1 is secured. If contract C_2 is secured, the profit would also be $100,000. However, if both contract C_1 and C_2 are secured, the profit anticipated is $400,000. Similar information is exhibited for the other alternatives, with each row of the matrix representing the outcome expected for each future (column) for a particular alternative.

Before proceeding to the application of criteria for the choice from among alternatives, the decision evaluation matrix should be examined for dominance. Any alternatives that are clearly not preferred, regardless of the future that occurs, may be dropped from consideration. If the outcomes for alternative x are better than the outcomes for alternative y for all possible futures, alternative x is said to *dominate* alternative y, and y can be eliminated as a possible choice.

The computer systems firm, facing the evaluation matrix of Table 7.2, may eliminate A_5 from consideration because it is dominated by all other alternatives. This means that the possible choice of serving only as project manager is inferior to each and every one of the other alternatives, regardless of the way in which the projects are awarded. Therefore, the matrix can be reduced to that given in Table 7.3. The decision criteria in the sections that follow may be used to assist in the selection from among alternatives A_1 through A_4.

TABLE 7.2 Decision Evaluation Matrix (Profit in Thousands of Dollars)

	PROBABILITY:	(0.3)	(0.2)	(0.5)
	FUTURE:	C_1	C_2	$C_1 + C_2$
	A_1	100	100	400
	A_2	−200	150	600
Alternative	A_3	0	200	500
	A_4	100	300	200
	A_5	−400	100	200

Aspiration-Level Criterion

Some form of aspiration level exists in most personal and professional decision making. An aspiration level is some desired level of achievement such as profit, or some undesirable result level to be avoided, such as loss. In decision making under risk, the aspiration level criterion involves selecting some level of achievement that is to be met, followed by a selection of that alternative which maximizes the probability of achieving the stated aspiration level.

The computer systems firm is now at the point of selecting from among alternatives A_1 through A_4, as presented in the reduced matrix of Table 7.3. Under the aspiration level criterion, management must set a minimum aspiration level for profit and possibly a maximum aspiration level for loss. Suppose that the profit level is set to be at least $400,000, and the loss level is set to be no more than $100,000. Under these aspiration level choices, alternatives A_1, A_2, and A_3 qualify as to profit potential, but

TABLE 7.3 Reduced Decision Evaluation Matrix (Profit in Thousands of Dollars)

PROBABILITY:		(0.3)	(0.2)	(0.5)
FUTURE:		C_1	C_2	$C_1 + C_2$
Alternative	A_1	100	100	400
	A_2	−200	150	600
	A_3	0	200	500
	A_4	100	300	200

alternative A_2 fails the loss test and must be eliminated. The choice could now be made between A_1 and A_3 by some other criterion, even though both satisfy the aspiration level criterion.

Most-Probable-Future Criterion

A basic human tendency is to focus on the most probable outcome from among several that could occur. This approach to decision making suggests that all except the most probable future be disregarded. Although somewhat equivalent to decision making under certainty, this criterion works well when the most probable future has a significantly high probability so as to partially dominate.

Under the most probable future criterion, the computer systems firm would focus its selection process from among the four alternatives on the profits associated with the future designated $C_1 + C_2$ (both contracts awarded). This is because the probability of this future occuring is 0.5, the most probable possibility. Alternative A_2 is preferred by this approach.

The most probable future criterion could be applied to select between A_1 and A_3, as identified under the aspiration level criterion. If this is done, the firm would choose alternative A_3.

Expected-Value Criterion

Many decision makers strive to make choices that will maximize expected profit or minimize expected loss. This is ordinarily justified in repetitive situations where the choice is to be made over and over again with increasing confidence that the calculated expected outcome will be achieved. This criterion is viewed with caution only when the payoff consequences of possible outcomes are disproportionately large, making a result that deviates from the expected outcome a distinct possibility.

The calculation of the expected value requires weighing all payoffs by their probabilities of occurrence. These weighed payoffs are then summed across all futures for each alternative. For the computer systems firm, alternatives A_1 through A_4 yield the following expected profits (in thousands):

$$A_1: \quad \$100(0.3) + \$100(0.2) + \$400(0.5) = \$250$$
$$A_2: \quad -\$200(0.3) + \$150(0.2) + \$600(0.5) = \$270$$
$$A_3: \quad \$0(0.3) + \$200(0.2) + \$500(0.5) = \$290$$
$$A_4: \quad \$100(0.3) + \$300(0.2) + \$200(0.5) = \$190$$

From this analysis it is clear that alternative A_3 would be selected. Further, if this criterion were to be used to resolve the choice of either A_1 or A_3 under the aspiration level approach, the choice would be alternative A_3.

Comparison of Decisions

It is evident that there is no one best selection when these criteria are used for decision making under risk. The decision made is dependent on the decision criterion adopted by the decision maker. For the example of this section, the alternatives selected under each criterion were

Aspiration level criterion: A_1 or A_3
Most probable future criterion: A_2
Expected value criterion: A_3

If the application of the latter two criteria to the resolution of A_1 or A_3 chosen under the aspiration level criterion is accepted as valid, then A_3 is preferred twice and A_2 once. From this it might be appropriate to suggest that A_3 is the best alternative arising from the use of these three criteria.

7.7 DECISIONS UNDER UNCERTAINTY

It may be inappropriate or impossible to assign probabilities to the several futures identified for a given decision situation. Often no meaningful data are available from which probabilities may be developed. In other instances, the decision maker may be unwilling to assign a subjective probability, as is often the case when the future could prove to be unpleasant. When probabilities are not available for assignment to future events, the situation is classified as decision making under uncertainty.

As compared with decision making under certainty and under risk, decisions under uncertainty are made in a more abstract environment. In this section several decision criteria will be applied to the example of Section 7.6 to illustrate the formal approaches that are available.

Laplace Criterion

Suppose that the computer systems firm is unwilling to assess the futures in terms of probabilities. Specifically, the firm is unwilling to differentiate between the likelihood at acquiring contract C_1, contract C_2, and contract C_1 and contract C_2. In the absence of these probabilities one might reason that each possible state of nature is as likely to occur as any other. The rationale of this assumption is that there is no stated basis for one state of nature to be more likely than any other. This is called the *Laplace principle* or the *principle of insufficient reason* based on the philosophy that nature is assumed to be indifferent.

Under the Laplace principle, the probability of the occurrence of each future state of nature is assumed to be $1/n$, where n is the number of possible future states.

To select the best alternative one would compute the arithmetic average for each. For the decision matrix of Table 7.3, this is accomplished as shown in Table 7.4. Alternative A_3 results in a maximum profit of $233,000 and would be selected.

TABLE 7.4 Computation of Average Profit (Thousands of Dollars)

ALTERNATIVE	AVERAGE PAYOFF
A_1	($100 + $100 + $400) ÷ 3 = $200
A_2	(−$200 + $150 + $600) ÷ 3 = $183
A_3	($0 + $200 + $500) ÷ 3 = $233
A_4	($100 + $300 + $200) ÷ 3 = $200

Maximin and Maximax Criteria

Two simple decision rules are available for dealing with decisions under uncertainty. The first is the *maximin* rule, based on an extremely pessimistic view of the outcome of nature. The use of this rule would be justified if it is judged that nature will do its worst. The second is the *maximax* rule, based on an extremely optimistic view of the future. Use of this rule is justified if it is judged that nature will do its best.

Because of the pessimism embraced by the maximin rule, its application will lead to the alternative that assures the best of the worst possible outcomes. If E_{ij} is used to represent the payoff for the ith alternative and the jth state of nature, the required computation is

$$\max_i \{\min_j E_{ij}\}.$$

Consider the decision situation described by the decision matrix of Table 7.3. The application of the maximin rule requires that the minimum value in each row be selected. Then the maximum value is identified from these and associated with the alternative that would produce it. This procedure is illustrated in Table 7.5. Selection of either alternative A_1 or A_4 assures the firm of a profit of at least $100,000 regardless of the future.

TABLE 7.5 Profit by the Maximin Rule (Thousands of Dollars)

ALTERNATIVE	$\min_j E_{ij}$
A_1	$ 100
A_2	−200
A_3	0
A_4	100

The optimism of the maximax rule is in sharp contrast to the pessimism of the maximin rule. Its application will choose the alternative that assures the best of the best possible outcomes. As before, if E_{ij} represents the payoff for the ith alternative and the jth state of nature, the required computation is

$$\max_i \{\max_j E_{ij}\}.$$

Consider the decision situation of Table 7.3 again. The application of the maximax rule requires that the maximum value in each row be selected. Then the maximum value is identified from these and associated with the alternative that would produce it. This procedure is illustrated in Table 7.6. Selection of alternative A_2 is indicated. Thus, the decision maker may receive a profit of $600,000 if the future is benevolent.

TABLE 7.6 Profit by the Maximax Rule (Thousands of Dollars)

ALTERNATIVE	$\max_j E_{ij}$
A_1	$400
A_2	600
A_3	500
A_4	300

A decision maker who chooses the maximin rule considers only the worst possible occurrence for each alternative and selects the alternative that promises the best of the worst possible outcomes. In the example where A_1 was chosen, the firm would be assured of a profit of at least $100,000, but it could not receive a profit any greater than $400,000. Or, if A_4 were chosen, the firm could not receive a profit any greater than $300,000. Conversely, the firm that chooses the maximax rule is optimistic and decides solely on the basis of the highest profit offered for each alternative. Accordingly, in the example in which A_2 was chosen, the firm faces the possibility of a loss of $200,000 while seeking a profit of $600,000.

Hurwicz Criterion

Because the decision rules presented previously are extreme, they are shunned by many decision makers. Most people have a degree of optimism or pessimism somewhere between the extremes. A third approach to decision making under uncertainty involves an index of relative optimism and pessimism. It is called the *Hurwicz rule.*

A compromise between optimism and pessimism is embraced in the Hurwicz rule by allowing the decision maker to select an index of optimism, α, such that $0 \leq \alpha \leq 1$. When $\alpha = 0$ the decision maker is pessimistic about nature, while an $\alpha = 1$ indicates optimism about nature. Once α is selected, the Hurwicz rule requires the computation of

$$\max_i \{\alpha[\max_j E_{ij}] + (1 - \alpha)[\min_j E_{ij}]\}$$

where E_{ij} is the payoff for the ith alternative and the jth state of nature.

As an example of the Hurwicz rule, consider the payoff matrix of Table 7.3 with $\alpha = 0.2$. The required computations are shown in Table 7.7 and alternative A_1 would be chosen by the firm.

Additional insight into the Hurwicz rule can be obtained by graphing each alternative for all values of α between zero and one. This makes it possible to identify the value of α for which each alternative would be favored. Such a graph is shown in Figure 7.5. It may be observed that alternative A_1 yields a maximum expected profit for

TABLE 7.7 Profit by the Hurwicz Rule with $\alpha = 0.2$ (Thousands of Dollars)

ALTERNATIVE	$\alpha[\max E_{ij}] + (1 - \alpha)[\min E_{ij}]$
A_1	$0.2(\$400) + 0.8(\$100) = \$160$
A_2	$0.2(\$600) + 0.8(\$-200) = \$-40$
A_3	$0.2(\$500) + 0.8(0) = \$100$
A_4	$0.2(\$300) + 0.8(\$100) = \$140$

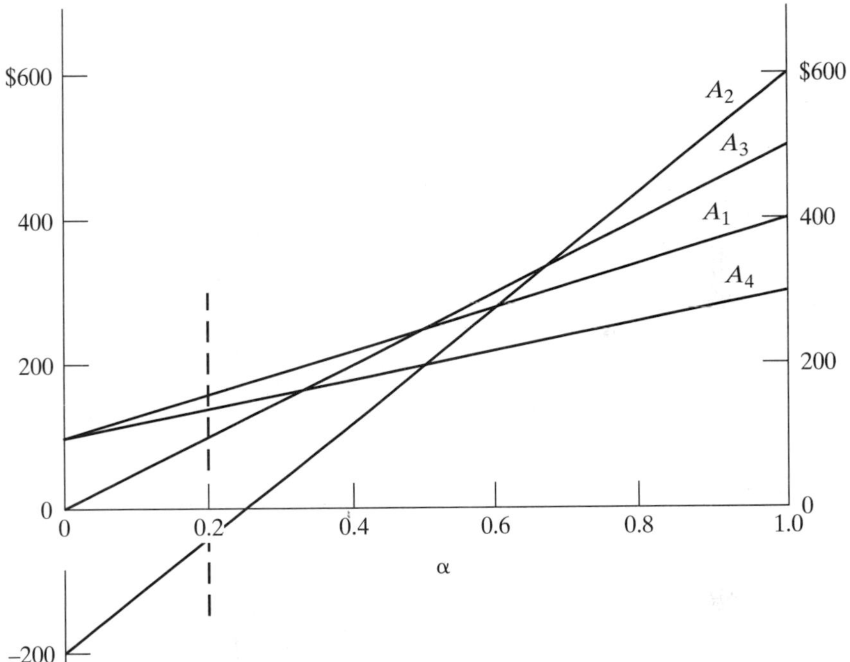

Figure 7.5 Values for the Hurwicz rule for four alternatives.

all values of $\alpha \le \frac{1}{2}$. Alternative A_3 exhibits a maximum for $\frac{1}{2} \le \alpha \le \frac{2}{3}$, and alternative A_2 gives a maximum for $\frac{2}{3} \le \alpha \le 1$. There is no value of α for which alternative A_4 would be best except at $\alpha = 0$, where it is as good an alternative as A_1.

When $\alpha = 0$, the Hurwicz rule gives the same result as the maximin rule, and when $\alpha = 1$, it is the same as the maximax rule. This may be shown for the case where $\alpha = 0$ as

$$\max_i \{0[\max_j E_{ij}] + (1 - 0)[\min_j E_{ij}]\} = \max_i [\min_j E_{ij}]$$

For the case where $\alpha = 1$,

$$\max_i \{1[\max_j E_{ij}] + (1 - 1)[\min_j E_{ij}]\} = \max_i [\max_j E_{ij}]$$

Thus, the maximin rule and the maximax rule are special cases of the Hurwicz rule.

The philosophy behind the Hurwicz rule is that focus on the most extreme outcomes or consequences bounds or brackets the decision. By use of this rule, the decision maker may weight the extremes in such a manner as to reflect their relative importance.

Comparison of Decisions

As was the case for the decision criteria applied for decision making under risk, it is evident that there is no one best criterion for decision making under uncertainty. The decision made is dependent on the decision criterion adopted by the decision maker. For the examples of this section, the alternatives selected were

Laplace criterion: A_3
Maximin criterion: A_1 or A_4
Maximax criterion: A_2
Hurwicz criterion ($\alpha = 0.2$): A_1

Examination of the selections recommended by the five decision rules indicates that each has its own merit. Several factors may influence a decision maker's choice of a rule in a given decision situation. The decision maker's attitude toward uncertainty (pessimistic or optimistic) and his or her personal utility function are important influences. Thus, the choice of a particular decision rule for a given decision situation must be based on subjective judgment.

7.8 THE DECISION EVALUATION DISPLAY

The decision evaluation concepts and theory presented thus far were based on single criterion situations. This single criterion domain is the exception rather than the rule. Multiple criteria considerations arise when both economic and noneconomic elements are present in the evaluation. In these situations, decision evaluation is facilitated by the use of a decision evaluation display exhibiting both cost and effectiveness measures. In this section, the decision evaluation display is presented as one means for dealing with multiple criteria in design and operations.

Effectiveness is a measure of mission fulfillment for a product or system in terms of a stated need. Mission fulfillment may be expressed by one or more figures of merit, depending on the type of product or system and the objectives to be achieved. Some common effectiveness measures are shown on the right side of Figure 7.6 (also see Figure 2.9).

Life-cycle cost, shown on the left side of Figure 7.6, and one or more effectiveness measures may be displayed simultaneously as an aid in decision making. A decision evaluation display, as shown in Figure 7.7 is one way of doing this. Note that effectiveness requirements (or thresholds) are shown on the display. These are useful to the decision maker in assessing, subjectively, the degree to which each alternative satisfies effectiveness criteria. Equivalent cost or profit, shown on the horizontal axis,

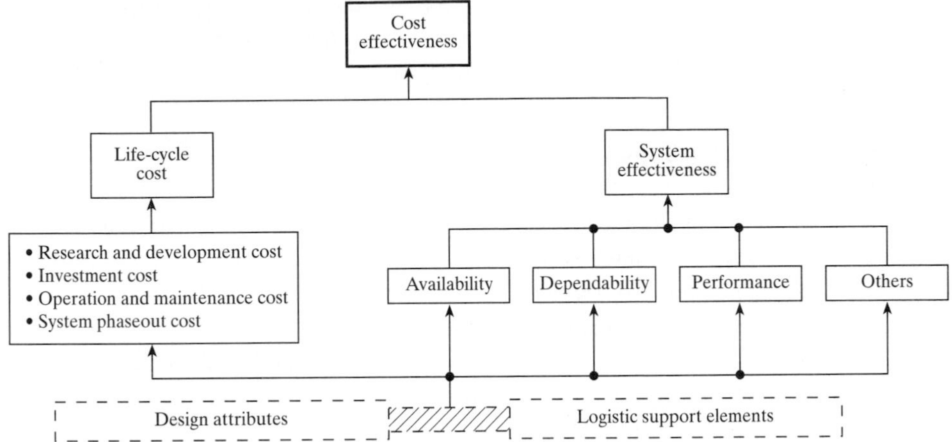

Figure 7.6 Elements for consideration in cost-effectiveness evaluation.

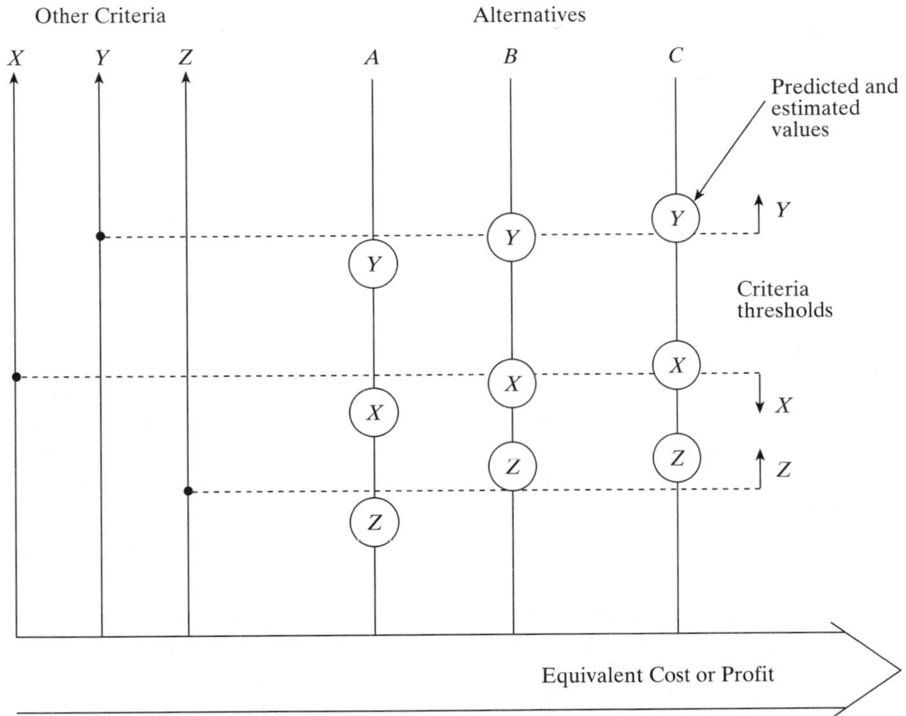

Figure 7.7 General decision evaluation display.

is an objective measure. The goal is to select the alternative with the lowest cost or greatest profit that satisfies most of the effectiveness measures.

Equivalent cost or profit entered on the horizontal axis of Figure 7.7 is obtained, for each alternative, from Equation 7.1. Alternatively, if equivalent cost or profit involves economic optimization, the specific values entered on the horizontal axis are obtained from Equation 7.3. Each of these applications of the decision evaluation display will be illustrated in subsequent chapters.

QUESTIONS AND PROBLEMS

1. A complete and all-inclusive alternative rarely emerges in its final state. Explain.
2. Should decision making be classified as an art or as a science?
3. Contrast limiting and strategic factors.
4. Discuss the various meanings of the word model.
5. Describe briefly physical models, schematic models, and mathematical models.
6. How do mathematical models directed to decision situations differ from those traditionally used in the physical sciences?
7. Contrast direct and indirect experimentation.
8. Write the general form of the evaluation function for money flow modeling and define its symbols.
9. Identify a decision situation and indicate the variables under the control of the decision maker and those not directly under his or her control.
10. Contrast the similarities and differences in the economic optimization functions given by Equations 7.2 and 7.3 (see Figure 7.2).
11. Why is it not possible to formulate a model that accurately represents reality.
12. What means may be used to simplify model construction?
13. Under what conditions may a properly formulated model become useless as an aid in decision making?
14. Explain the nature of the cost components that should be considered in deciding how frequently to review a dynamic environment.
15. What caution must be exercised in the use of models?
16. Discuss several specific reasons why models are of value in decision making.
17. What should be done with those facets of a decision situation that cannot be explained by the model?
18. Formulate an evaluation matrix for a hypothetical decision situation of your choice.
19. Formulate an evaluation vector for a hypothetical decision situation under assumed certainty.
20. Develop an example to illustrate the application of paired outcomes in decision making among a number of nonquantifiable alternatives.
21. What approaches may be used to assign probabilities to future outcomes?
22. What is the role of dominance in decision making among alternatives?
23. Give an example of an aspiration level in decision making.
24. When would one follow the most probable future criterion in decision making?
25. What drawback exists in using the most probable future criterion?
26. How does the Laplace criterion for decision making under uncertainty actually convert the situation to decision making under risk?
27. Discuss the maximin and the minimax rules as special cases of the Hurwicz rule.
28. The cost of developing an internal training program for office automation is unknown but described by the following probability distribution:

COST	PROBABILITY OF OCCURRENCE
\$ 80,000	0.20
95,000	0.30
105,000	0.25
115,000	0.20
130,000	0.05

What is the expected cost of the course? What is the most probable cost? What is the maximum cost that will occur with a 95% assurance?

29. Net profit has been calculated for five investment opportunities under three possible futures. Which alternative should be selected under the most probable future criterion; the expected value criterion?

	(0.3) F_1	(0.2) F_2	(0.5) F_3
A_1	\$ 100,000	\$100,000	\$380,000
A_2	−200,000	160,000	590,000
A_3	0	180,000	500,000
A_4	110,000	280,000	200,000
A_5	400,000	90,000	180,000

30. Daily positive and negative payoffs are given for five alternatives and five futures in the following matrix. Which alternative should be chosen to maximize the probability of receiving a payoff of at least 9? What choice would be made by using the most probable future criterion?

	(0.15) F_1	(0.20) F_2	(0.30) F_3	(0.20) F_4	(0.15) F_5
A_1	12	8	−4	0	9
A_2	10	0	5	10	16
A_3	6	5	10	15	−4
A_4	4	14	20	6	12
A_5	−8	22	12	4	9

31. The following matrix gives the payoffs in utiles (a measure of utility) for three alternatives and three possible states of nature:

	STATE OF NATURE		
	S_1	S_2	S_3
A_1	50	80	80
A_2	60	70	20
A_3	90	30	60

Which alternative would be chosen under the Laplace principle? The maximin rule? The maximax rule? The Hurwicz rule with $\alpha = 0.75$?

32. The following payoff matrix indicates the costs associated with three decision options and four states of nature:

	STATE OF NATURE			
	S_1	S_2	S_3	S_4
Option T_1	20	25	30	35
Option T_2	40	30	40	20
Option T_3	10	60	30	25

Select the decision option that should be selected for the maximin rule; the maximax rule; the Laplace rule; the minimax regret rule; and the Hurwicz rule with $\alpha = 0.2$. How do the rules applied to the cost matrix differ from those that are applied to a payoff matrix of profits?

33. The following matrix gives the expected profit in thousands of dollars for five marketing strategies and five potential levels of sales.

	LEVELS OF SALES				
	L_1	L_2	L_3	L_4	L_5
Strategy M_1	10	20	30	40	50
Strategy M_2	20	25	25	30	35
Strategy M_3	50	40	5	15	20
Strategy M_4	40	35	30	25	25
Strategy M_5	10	20	25	30	20

Which marketing strategy would be chosen under the maximin rule? The maximax rule? The Hurwicz rule with $\alpha = 0.4$?

34. Graph the Hurwicz rule for all values of α using the payoff matrix of Problem 33.

35. The following decision evaluation matrix gives the expected savings in maintenance costs (in thousands of dollars) for three policies of preventive maintenance and three levels of operation of equipment. Given the probabilities of each level of operation, $P_1 = 0.3$, $P_2 = 0.25$, and $P_3 = 0.45$, determine the best policy based on the most probable future criterion.

POLICY	LEVEL OF OPERATION		
	L_1	L_2	L_3
M_1	10	20	30
M_2	22	26	26
M_3	40	30	15

Also, determine the best policy under uncertainty, using the Laplace rule, the Maximax rule, and the Hurwicz rule with $\alpha = 0.2$.

36. Use the decision evaluation display of Figure 7.7 to make visible a design decision situation of your choice from Part II of this text.

8 MODELS FOR ECONOMIC EVALUATION

Economic considerations are important in systems engineering, the process of bringing systems into being. For systems already in existence, economic considerations often provide a basis for the analysis of system operation and retirement. There are numerous examples of structures, processes, and systems that exhibit excellent physical design but have little economic merit. The essential prerequisite of successful engineering application is economic feasibility.

Many design and operational alternatives may be described in terms of their receipts and disbursements over time. When this is the case, reducing these monetary values to a common base is essential in decision making. This chapter presents interest factor applications and related techniques of economic analysis useful in comparing alternatives on an equivalent economic basis.

8.1 INTEREST AND INTEREST FORMULAS

The time value of money in the form of an interest rate is an important element in most decision situations involving money flow over time. Because money can earn at a certain interest rate, it is recognized that a dollar in hand at present is worth more than a dollar to be received at some future date. A lender may consider interest received as a gain or profit, whereas a borrower usually considers interest to be a charge or cost.

The interest rate is the ratio of the borrowed money to the fee charged for its use over a period, usually a year. This ratio is expressed as a percentage. For example, if $100 is paid for the use of $1,000 for one year, the interest rate is 10%. In compound interest, the interest earned at the end of the interest period is either paid at that time or earns interest upon itself. This compounding assumption is usually used in the economic evaluation of alternatives.

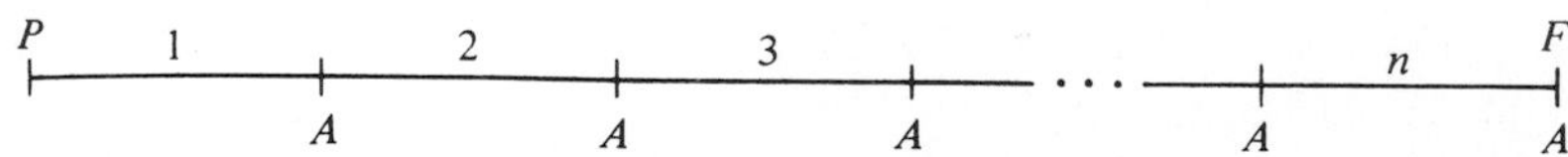

Figure 8.1 The money-flow model.

A schematic model for money flow over time is shown in Figure 8.1. It is the basis for the derivation of interest factors and may be applied to any phase of the system life cycle for the purpose of life-cycle cost or income analysis (refer to Figure 7.1). Let

i = nominal annual rate of interest
n = number of interest periods, usually annual
P = principal amount at a time assumed to be the present
A = single amount in a series of n equal amounts at the end of each interest period
F = amount, n interest periods hence, equal to the compound amount of P, or the sum of the compound amounts of A, at the interest rate i

Single-Payment Compound-Amount Formula

When interest is permitted to compound, the interest earned during each interest period is added to the principal amount at the beginning of the next interest period. Using the terms defined, the relationship among F, P, n, and i can be developed as shown in Table 8.1. The resulting factor, $(1 + i)^n$, is known as the *single-payment compound-amount factor*[1] and is designated as

$$\left(\overset{F/P,\,i,\,n}{\quad}\right)$$

This factor may be used to express the equivalence between a present amount, P, and a future amount, F, at an interest rate i for n years. The formula is

$$F = P(1 + i)^n$$

or

$$F = P\left(\overset{F/P,\,i,\,n}{\quad}\right) \tag{8.1}$$

TABLE 8.1 Single-Payment Compound-Amount Formula

YEAR	AMOUNT AT BEGINNING OF YEAR	INTEREST EARNED DURING YEAR	COMPOUND AMOUNT AT END OF YEAR
1	P	Pi	$P + Pi = P(1 + i)$
2	$P(1 + i)$	$P(1 + i)i$	$P(1 + i) + P(1 + i)i = P(1 + i)^2$
3	$P(1 + i)^2$	$P(1 + i)^2 i$	$P(1 + i)^2 + P(1 + i)^2 i = P(1 + i)^3$
n	$P(1 + i)^{n-1}$	$P(1 + i)^{n-1} i$	$P(1 + i)^{n-1} + P(1 + i)^{n-1} i = P(1 + i)^n = F$

[1] Values for this factor and those that follow are tabulated in Appendix D.

The compound amount of \$1,000 in 6 years at 10% interest compounded annually may be found from Equation 8.1 as

$$F = \$1{,}000(1 + 0.10)^6$$
$$= \$1{,}000(1.772) = \$1{,}772.00$$

or by use of the factor designation and its tabular value from Appendix D, Table D.3,

$$F = \$1{,}000\overset{F/P,\,10,\,6}{(\;1.772\;)} = \$1{,}772.00$$

Single-Payment Present-Amount Formula

The single-payment compound-amount formula may be solved for P and expressed as

$$P = F\left[\frac{1}{(1 + i)^n}\right]$$

The resulting factor, $1/(1 + i)^n$, is known as the *single-payment present-amount factor* and is designated

$$\overset{P/F,\,i,\,n}{(\qquad)}$$

This factor may be used to express the equivalence between a future amount, F, and a present amount, P, at an interest rate i for n years. The formula is

$$P = F\overset{P/F,\,i,\,n}{(\qquad)} \tag{8.2}$$

As an example, assume that it is desired to accumulate \$1,000 in 4 years. If the interest rate is 8%, the amount of money that must be deposited may be found from Equation 8.2 as

$$P = \$1{,}000\overset{P/F,\,8,\,4}{(0.7350)} = \$735.00$$

Equal-Payment Series Compound-Amount Formula

In some situations, a series of receipts or disbursements occuring uniformly at the end of each year may be encountered. The sum of the compound amounts of this series may be determined by reference to Figure 8.1.

The A dollars deposited at the end of the nth year will earn no interest and will contribute only A dollars to F. The A dollars deposited at the end of period $n - 1$ will earn interest in the amount of Ai, and $A(1 + i)$ will be contributed to the sum. The amount at the end of period $n - 2$ will contribute $A(1 + i)^2$. The sum of this series will be

$$F = A(1) + A(1 + i) + A(1 + i)^2 + \cdots + A(1 + i)^{n-2} + A(1 + i)^{n-1}$$

Multiplying this series by $(1 + i)$ gives

$$F(1 + i) = A[(1 + i) + (1 + i)^2 + (1 + i)^3 + \cdots + (1 + i)^{n-1} + (1 + i)^n]$$

Subtracting the first expression from the second gives

$$F(1 + i) - F = A[(1 + i)^n - 1]$$

$$Fi = A[(1 + i)^n - 1]$$

$$F = A\left[\frac{(1 + i)^n - 1}{i}\right]$$

The resulting factor, $[(1 + i)^n - 1]/i$, is known as the *equal-payment series compound-amount factor* and is designated

$$\overset{F/A,\,i,\,n}{(\quad)}$$

This factor may be used to express the equivalence between an equal-payment series, A, and a future amount F, at an interest rate i for n years. The formula is

$$F = A\overset{F/A,\,i,\,n}{(\quad)} \tag{8.3}$$

Consider an example where \$500 amounts are invested at the end of each year for four years at 9% interest. The total that will be accumulated at the end of the 4-year period is

$$F = \$500\overset{F/A,\,9,\,4}{(4.5730)} = \$2{,}286.50$$

Equal-Payment Series Sinking-Fund Formula

The equal-payment series compound-amount formula may be solved for A and expressed as

$$A = F\left[\frac{i}{(1 + i)^n - 1}\right]$$

The resulting factor, $i/[(1 + i)^n - 1]$, is known as the *equal-payment series sinking-fund factor* and is designated

$$\overset{A/F,\,i,\,n}{(\quad)}$$

This factor may be used to express the equivalence between a future amount, F, and an equal-payment series, A, at an interest rate i for n years. The formula is

$$A = F\overset{A/F,\,i,\,n}{(\quad)} \tag{8.4}$$

As an example, suppose that it is desired to deposit a series of uniform, year-end amounts over 10 years to provide a total of \$5,000. The amount that should be deposited each year at an interest rate of 8% is

$$A = \$5{,}000\overset{A/F,\,8,\,10}{(0.0690)} = \$345.00$$

Equal-Payment Series Capital-Recovery Formula

The substitution of $P(1 + i)^n$ for F in the equal-payment series sinking-fund formula results in

$$A = P(1 + i)^n \left[\frac{i}{(1 + i)^n - 1} \right]$$

$$= P\left[\frac{i(1 + i)^n}{(1 + i)^n - 1} \right]$$

The resulting factor, $i(1 + i)/[(1 + i)^n - 1]$, is known as the *equal-payment series capital-recovery factor* and is designated

$$\overset{P/A,\, i,\, n}{(\quad)}$$

This factor may be used to express the equivalence between future equal-payment series, A, and a present amount, P, at an interest rate i for n years. The formula is

$$A = P\overset{P/A,\, i,\, n}{(\quad)} \tag{8.5}$$

As an example of the application of this formula, assume that a deposit of $100,000 at 8% for 8 years is made in a savings account. This will enable a uniform amount to be withdrawn every year, for the next 8 years. The magnitude of this amount is

$$A = \$100{,}000\overset{A/P,\, 8,\, 8}{(0.1740)} = \$17{,}400.00$$

Equal-Payment Series Present-Amount Formula

The equal-payment series capital-recovery formula can be solved for P and expressed as

$$P = A\left[\frac{(1 + i)^n - 1}{i(1 + i)^n} \right]$$

The resulting factor $[(1 + i)^n - 1]/(1 + i)^n$, is known as the *equal-payment series present-amount factor* and is designated

$$\overset{P/A,\, i,\, n}{(\quad)}$$

This factor may be used to express the equivalence between future equal-payment series, A, and a present amount, P, at an interest rate i for n years. The formula is

$$P = \overset{P/A,\, i,\, n}{(\quad)} \tag{8.6}$$

As an example of the application of this formula, assume that an expenditure now will save $4,000 per year in operating costs over the next 10 years. If an interest rate of 15% is used, the present-equivalent amount of these savings is

$$P = \$4{,}000\overset{P/A,\, 15,\, 10}{(\ 5.0188\ )} = \$20{,}075.20$$

Geometric-Gradient-Series Formula

In many situations, annual money flows increase or decrease over time by a constant percentage. If g is used to designate the percentage change in the magnitude of the money flows from one period to the next, the magnitude of the tth flow is related to flow F_1 as

$$F_t = F_1(1 + g)^{t-1} \qquad t = 1, 2, \ldots, n$$

When g is positive, the series will increase as illustrated in Figure 8.2. When g is negative, the series will decrease.

To derive an expression for the present amount, P, the relationship between F_1 and F_t can be used, together with the single-payment present-amount factor, as

$$P = F_1\left[\frac{(1+g)^0}{(1+i)^1}\right] + F_1\left[\frac{(1+g)^1}{(1+i)^2}\right] + F_1\left[\frac{(1+g)^2}{(1+i)^3}\right] + \cdots + F_1\left[\frac{(1+g)^{n-1}}{(1+i)^n}\right]$$

Multiply each term by $(1 + g)/(1 + g)$ and simplify

$$P = \frac{F_1}{1+g}\left[\frac{(1+g)^1}{(1+i)^1} + \frac{(1+g)^2}{(1+i)^2} + \frac{(1+g)^3}{(1+i)^3} + \cdots + \frac{(1+g)^n}{(1+i)^n}\right]$$

Let

$$\frac{1}{1+g'} = \frac{1+g}{1+i}$$

where g' is the *growth-free rate*. Substituting for each item gives

$$P = \frac{F_1}{1+g}\left[\frac{1}{(1+g')^1} + \frac{1}{(1+g')^2} + \frac{1}{(1+g')^3} + \cdots + \frac{1}{(1+g')^n}\right]$$

The terms within the brackets constitute the equal-payment series present-amount factor for n years. Therefore,

$$P = \frac{F_1}{1+g}\left[\frac{(1+g')^n - 1}{g'(1+g')^n}\right]$$

or

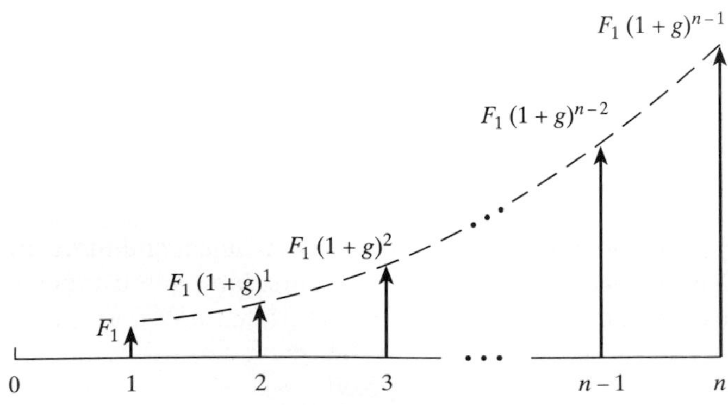

Figure 8.2 A geometric-gradient series with $g > 0$.

$$P = F_1\left[\frac{\overset{P/A,\,g',\,n}{(\qquad)}}{1+g}\right] \tag{8.7}$$

The factor within brackets is the *geometric-gradient-series factor*. Its use requires finding g' from

$$g' = \frac{1+i}{1+g} - 1 \tag{8.8}$$

The geometric-gradient-series factor may also be used for decreasing-gradient evaluations. In this case g will be negative and will result in a positive value for g' for all positive values of i. As an example, suppose that a revenue flow of \$250,000 during the first year is expected to decrease by 10% per year. The present equivalent of the anticipated gross revenue at an interest rate of 17% over 8 years may be found as follows:

$$g' = \frac{1+0.17}{1-0.10} - 1 = 0.30 \quad \text{or} \quad 30\%$$

$$P = \$250{,}000\,\frac{\overset{P/A,\,30,\,8}{(2.9247)}}{1-0.10} = \$812{,}417$$

In this example, the value for the P/A factor was available directly from the 30% interest table in Appendix D.

Summary of Interest Formulas

The interest factors derived in the previous paragraphs express relationships among P, A, F, i, and n. In any given application, an equivalence between F and P, P and F, F and A, A and F, P and A, or A and P is expressed for an interest rate i and a number of years n. Table 8.2 may be used to select the interest formula needed in a given situation. The factor designations summarized in the last column make it possible to set up a problem symbolically before determining the values of the factors involved.

Interest tables based on the designations summarized in Table 8.2 make it unnecessary to remember the name of each formula. Such tables are given in Appendix D for a range of values for i and n. Each column in the table is headed by a designation that identifies the function of the tabular value as established by the money-flow diagram. Each table is based on \$1 for a specific value of i, a range of n, and all six factors.

8.2 DETERMINING ECONOMIC EQUIVALENCE

If two or more situations are to be compared, their characteristics must be placed on an equivalent basis. Two things are said to be equivalent when they have the same effect. For instance, the torques produced by applying forces of 100 and 200 lb, 2 and 1 ft, respectively, from the fulcrum of the lever are equivalent because each produces a torque of 200 ft-lb.

TABLE 8.2 Summary of Interest Formulas

FORMULA NAME	FUNCTION	FORMULA	DESIGNATION
Single-payment compound amount	Given P Find F	$F = P(1+i)^n$	$F = P\left(\overset{F/P,i,n}{\quad}\right)$
Single-payment present amount	Given F Find P	$P = F\left[\frac{1}{(1+i)^n}\right]$	$P = F\left(\overset{P/F,i,n}{\quad}\right)$
Equal-payment-series compound amount	Given A Find F	$F = A\left[\frac{(1+i)^n - 1}{i}\right]$	$F = A\left(\overset{F/A,i,n}{\quad}\right)$
Equal-payment-series sinking fund	Given F Find A	$A = F\left[\frac{i}{(1+i)^n - 1}\right]$	$A = F\left(\overset{A/F,i,n}{\quad}\right)$
Equal-payment-series present amount	Given A Find P	$P = A\left[\frac{(1+i)^n - 1}{i(1+i)^n}\right]$	$P = A\left(\overset{P/A,i,n}{\quad}\right)$
Equal-payment-series capital recovery	Given P Find A	$A = P\left[\frac{i(1+i)^n}{(1+i)^n - 1}\right]$	$A = P\left(\overset{A/P,i,n}{\quad}\right)$

Two monetary amounts are equivalent when they have the same value in exchange. Three factors are involved in the equivalence of sums of money: (1) the amount of the sums, (2) the time of occurrence of the sums, and (3) the interest rate. In this connection, the general equivalence function of Equation 7.1 applies. It is directly applicable to alternatives described by money flows over time according to the convention established in Figure 8.1 and illustrated conceptually in Figure 7.1. Simple equivalence calculations and equivalence function diagrams are presented in this section.

Interest Formula Equivalence Calculations

Equivalence calculations based on simple applications of the interest formulas derived in Section 8.1 are illustrated in this section. More comprehensive applications are presented later. In each case, the determination of economic equivalence is in accordance with Equation 8.7.

At an interest rate of 10% with $n = 8$ years, a P of \$1 is equivalent to an F of \$2.144. This may be stated as

$$F = \$1\overset{F/P,10,8}{(2.1440)} = \$2.1440$$

A practical application of this statement of equivalence is that \$1 spent today must result in a revenue receipt, or avoid a cost of \$2.144, 8 years hence, if the interest rate is 10%.

A reciprocal situation is where $i = 12\%$, $n = 10$ years, and $F = \$1$. The P that is equivalent to \$1 is

$$P = \$1\overset{P/F,12,10}{(0.3220)} = \$0.3220$$

A practical application of this statement of equivalence is that no more than \$0.322 can be spent today to secure a revenue receipt or to avoid a cost of \$1, 10 years hence if the interest rate is 12%.

If $i = 8\%$ and $n = 20$ years, the F that is equivalent to an A of \$1 is

$$F = \$1\overset{F/A,8,20}{(45.7620)} = \$45.7620$$

A practical application of this equivalence statement is that \$1 spent each year for 20 years must result in a revenue receipt, or it must avoid a cost of \$45.762, 20 years hence if the interest rate is 8%.

A reciprocal situation is where $i = 12\%$, $n = 6$ years, and $F = \$1$. The A that is equivalent to \$1 is

$$A = \$1\overset{A/F,12,6}{(0.1232)} = \$0.1232$$

A practical application of this equivalence statement is that \$0.1232 must be received each year for 6 years to be equivalent to the receipt of \$1, 6 years hence.

If $i = 9\%$, $n = 10$ years, and A is \$1, the P that is equivalent to \$1 is

$$P = \$1\overset{P/A,9,10}{(6.4177)} = \$6.4177$$

A practical application of this statement of equivalence is that an investment of \$6.4177 today must yield an annual benefit of \$1 each year for 10 years if the interest rate is 9%.

A reciprocal situation is where $i = 14\%$, $n = 7$ years, and $P = \$1$. The A that is equivalent to \$1 is

$$A = \$1\overset{A/P,14,7}{(0.2332)} = \$0.2332$$

A practical application of this statement of equivalence is that \$1 can be spent today to capture an annual saving of \$0.2332 per year over seven years if the interest rate is 14%.

Equivalence Function Diagrams

A useful technique in the determination of equivalence is to plot present value as a function of the interest rate. For example, what value of i will make a P of \$1,500 equivalent to an F of \$5,000 if $n = 9$ years? Symbolically,

$$\$5{,}000 = \$1{,}500\overset{F/P,i,9}{(\qquad)}$$

The solution is illustrated graphically in Figure 8.3, showing that i is slightly less than 13%.

Present value may also be plotted as a function of n as a useful technique for determining equivalence. For example, what value of n will make a P of \$4,000 equivalent to an F of \$8,000 if $i = 8\%$? Symbolically,

$$\$8{,}000 = \$4{,}000\overset{F/P,8,n}{(\qquad)}$$

The solution is illustrated graphically in Figure 8.4, showing that n is between 9 and 10 years.

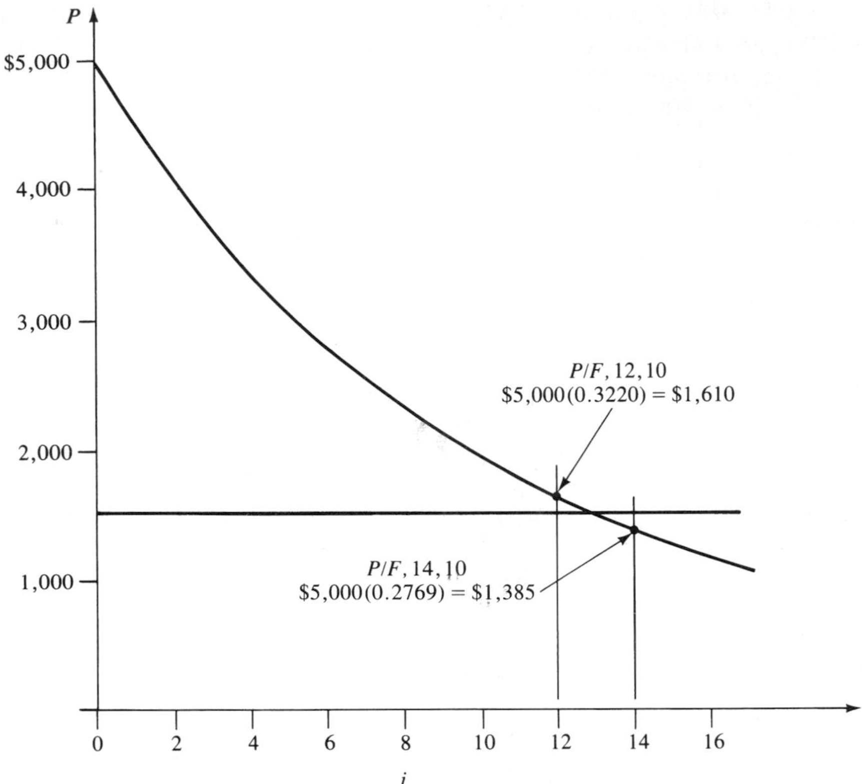

Figure 8.3 Equivalence-function diagram for i.

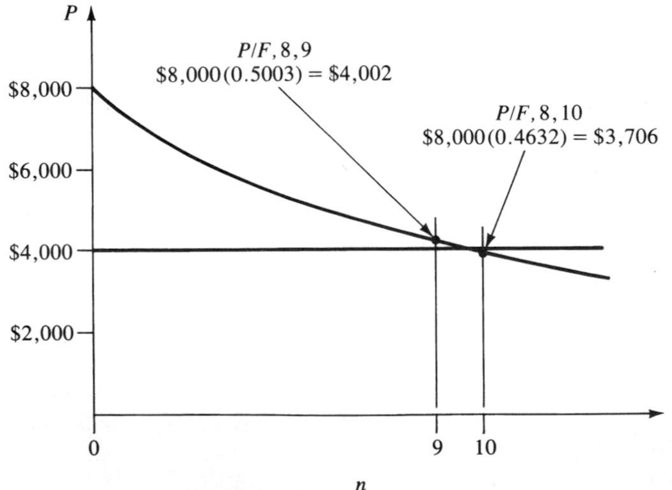

Figure 8.4 Equivalence-function diagram for n.

Equivalence functions can be plotted for A or F as a function of i or n if desired. In these cases, equivalence would be found between the two quantities A or F as a function of i or n.

8.3 EVALUATING A SINGLE ALTERNATIVE UNDER CERTAINTY

In most cases, a decision is reached by selecting one option from among two or more available alternatives. Sometimes, however, the decision is limited to acceptance or rejection of a single alternative. In such a case, the decision will be based on the relative merit of the alternative and other opportunities believed to exist, even though none of the latter has been crystallized into definite proposals.

When only one specified possibility exists, it should be evaluated within a framework that will permit its desirability to be compared with other opportunities that may exist but are unspecified. The most common bases for comparison are the present-equivalent evaluation, the annual-equivalent evaluation, and future-equivalent evaluation, the rate-of-return evaluation, and the service-life evaluation.

The following example illustrates the several bases of evaluation of a single alternative under certainty about future outcomes. Assume that a waste heat recirculation system is contemplated for a small building. The anticipated costs and savings of this project are given in Table 8.3. Consider an interest rate of 12% as the cost of money and take January 1, 19 × 0, to be the present.

For convenience, costs and savings that may occur during the year are assumed to occur at the end of that year or the start of the next year, considered to be the same point. There is some small error in the practice of considering money flows to be year-end amounts. This error is insignificant, however, in comparison with the usual errors in estimates, except under high interest rates. The costs and savings can be represented by the money flow diagram shown in Figure 8.5

TABLE 8.3 Disbursements and Savings for a Single Alternative

ITEM	DATE	DISBURSEMENTS	SAVINGS
Initial cost	1-1-19 × 0	$28,000	—
Saving, first year	1-1-19 × 1	—	$9,500
Saving, second year	1-1-19 × 2	—	9,500
Overhaul cost	1-1-19 × 2	2,500	—
Saving, third year	1-1-19 × 3	—	9,500
Saving, fourth year	1-1-19 × 4	—	9,500
Salvage value	1-1-19 × 4	—	8,000

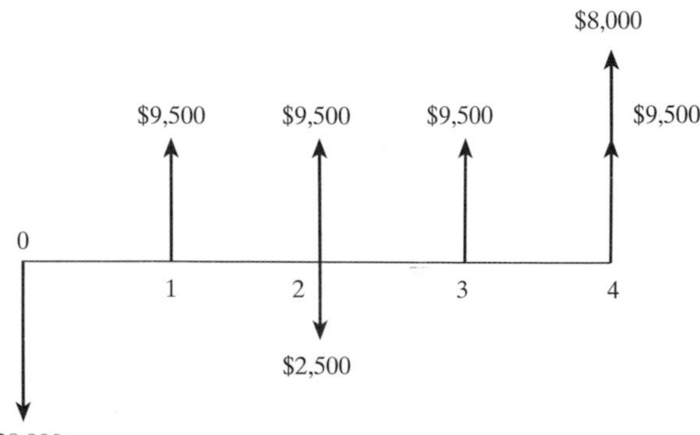

Figure 8.5 Money flow diagram from Table 8.3.

Present Equivalent Evaluation

The present equivalent evaluation is based on finding a present equivalent amount, *PE*, that represents the difference between present equivalent savings and present equivalent costs for an alternative at a given interest rate. Thus, the present equivalent amount at an interest rate i over n years is found by applying Equation 8.2.

$$PE(i) = F_0(\overset{P/F,i,0}{\quad}) + F_1(\overset{P/F,i,1}{\quad}) + F_2(\overset{P/F,i,2}{\quad}) + \cdots + F_n(\overset{P/F,i,n}{\quad})$$

$$= \sum_{t=0}^{n} F_t(\overset{P/F,i,t}{\quad}) \tag{8.9}$$

But because $(\overset{P/F,i,t}{\quad}) = (1 + i)^{-t}$,

$$PE(i) = \sum_{t=0}^{n} F_t(1 + i)^{-t} \tag{8.10}$$

For the waste heat recirculation system example, the present equivalent amount at 12% using Equation 8.9 is

$$PE(12) = -\$28{,}000(\overset{P/F,12,0}{1.0000}) + \$9{,}500(\overset{P/F,12,1}{0.8929}) + \$7{,}000(\overset{P/F,12,2}{0.7972})$$

$$+ \$9{,}500(\overset{P/F,12,3}{0.7118}) + \$17{,}500(\overset{P/F,12,4}{0.6355})$$

$$= \$3{,}946$$

Or by using Equation 8.10

$$PE(12) = -\$28{,}000(1.12)^0 + \$9{,}500(1.12)^{-1} + \$7{,}000(1.12)^{-2}$$

$$+ \$9{,}500(1.12)^{-3} + \$17{,}500(1.12)^{-4}$$

$$= \$3{,}946$$

Because the present equivalent amount is greater than zero, this is a desirable undertaking at 12%.

Annual Equivalent Evaluation

The annual equivalent evaluation is similar to the present equivalent evaluation except that the difference between savings and costs is now expressed as an annual equivalent amount, *AE*, at a given interest rate as follows:

$$AE(i) = PE(i)(\overset{A/P,i,n}{\quad}) \tag{8.11}$$

But because

$$PE(i) = \sum_{t=0}^{n} F_t(1 + i)^{-t}$$

and

$$\overset{A/P,i,n}{(\quad)} = \left[\frac{i(1+i)^n}{(1+i)^n - 1}\right]$$

$$AE(i) = \left[\sum_{t=0}^{n} F_t(1+i)^{-t}\right]\left[\frac{i(1+i)^n}{(1+i)^n - 1}\right] \tag{8.12}$$

For the waste-heat recirculation example, the annual equivalent evaluation using Equation 8.11 gives

$$AE(12) = \$3{,}946\overset{A/P,12,4}{(0.3292)} = \$1{,}299$$

Or, by using Equation 8.12

$$\begin{aligned} AE(12) = {} & [-\$28{,}000(1.12)^0 + \$9{,}500(1.12)^{-1} + \\ & + \$7{,}000(1.12)^{-2} + \$9{,}500(1.12)^{-3} + \\ & + \$17{,}500(1.12)^{-4}]\left[\frac{0.12(1.12)^4}{(1.12)^4 - 1}\right] \\ = {} & \$1{,}299 \end{aligned}$$

These results mean that if \$28,000 is invested on January 1, 19 × 0, a 12% return will be received plus an equivalent of \$1,299 on January 1, 19 × 1, 19 × 2, 19 × 3, and 19 × 4.

Future Equivalent Evaluation

The future equivalent basis for comparison is based on finding an equivalent amount, *FE*, that represents the difference between the future equivalent savings and future equivalent costs for an alternative at a given interest rate. Thus, the future equivalent amount at an interest rate i over n years is formed by applying Equation 8.1

$$\begin{aligned} FE(i) &= F_0\overset{F/P,i,n}{(\quad)} + F_1\overset{F/P,i,n-1}{(\quad)} + \cdots + F_{n-1}\overset{F/P,i,1}{(\quad)} + F_n\overset{F/P,i,0}{(\quad)} \\ &= \sum_{t=0}^{n} F_t\overset{F/P,i,n-t}{(\quad)} \end{aligned} \tag{8.13}$$

But because

$$\overset{F/P,i,n-t}{(\quad)} = (1+i)^{n-t}$$

$$FE(i) = \sum_{t=0}^{n} F_t(1+i)^{n-t} \tag{8.14}$$

or,

$$FE(i) = PE(i)\overset{F/P,i,n}{(\quad)} \tag{8.15}$$

For the waste-heat recirculation example, the future equivalent evaluation using Equation 8.13 gives

$$FE(12) = -\$28{,}000\overset{F/P,12,4}{(1.5740)} + \$9{,}500\overset{F/P,12,3}{(1.4050)} + \$7{,}000\overset{F/P,12,2}{(1.2542)}$$
$$+ \$9{,}500\overset{F/P,12,1}{(1.1208)} + \$17{,}500\overset{F/P,12,0}{(1.0000)}$$
$$= \$6{,}211$$

Or, by the use of Equation 8.15

$$FE(12) = \$3{,}946\overset{F/P,12,4}{(1.5740)} = \$6{,}211$$

Because the future equivalent amount is greater than zero, this is a desirable venture at 12%.

Rate-of-Return Evaluation

The rate-of-return evaluation is probably the best method for comparing a specific proposal with other opportunities believed to exist but not delineated. The rate of return is a universal measure of economic success because the return from different classes of opportunities are usually well established and generally known. This permits comparison of an alternative against accepted norms. This characteristic makes the rate-of-return comparison well adapted to the situation where the choice is to accept or reject a single alternative.

Rate of return is a widely accepted index of profitability. It is defined as the interest rate that causes the equivalent receipts of a money flow to be equal to the equivalent disbursements of that money flow. The interest rate that reduces the $PE(i)$, $AE(i)$, or $FE(i)$ of a series of receipts and disbursements to zero is another way of defining the rate of return. Mathematically, the rate of return for an investment proposal is the interest rate i^* that satisfies the equation

$$0 = PE(i^*) = \sum_{t=0}^{n} F_t(1 + i^*)^{-t} \qquad (8.16)$$

where the proposal has a life of n years.

As an example of the rate-of-return evaluation, the energy saving system can be evaluated on the basis of the rate of return that would be secured from the invested funds. In effect, a rate of interest will be specified that makes the receipts and disbursements equivalent. This can be done either by equating present equivalent, annual equivalent, or future equivalent amounts.

Equating the present-equivalent amount of receipts and disbursements at $i = 15\%$ gives

$$[\$9{,}500\overset{P/A,15,4}{(2.8850)} + \$8{,}000\overset{P/F,15,4}{(0.5718)}] = [\$28{,}000 + \$2{,}500\overset{P/F,15,2}{(0.7562)}]$$
$$[\$27{,}408 + \$4{,}574] = [\$28{,}000 + \$1{,}890]$$
$$\$31{,}982 = \$29{,}800$$

At $i = 20\%$

$$[\$9{,}500\overset{P/A,20,4}{(2.5887)} + \$8{,}000\overset{P/F,20,4}{(0.4823)}] = [\$28{,}000 + \$2{,}500\overset{P/F,20,2}{(0.6945)}]$$

$$[\$24{,}592 + \$3{,}858] = [\$28{,}000 + \$1{,}736]$$

$$\$28{,}450 = \$29{,}736$$

Interpolating,

$$i = 15\% + (5)\left[\frac{[31{,}982 - 29{,}800] - 0}{[31{,}982 - 29{,}800] - [28{,}450 - 29{,}736]}\right]$$

$$i = 18.15\%$$

From these results, a present equivalent graph can be developed as shown in Figure 8.6. From the graph it is evident that the rate of return on the venture is just over 18%. This means that the investment of $28,000 in the system should yield a 18.15% rate of return over the 4-year period.

Payout Evaluation

Often a proposed system can be evaluated in terms of how long it will take the system to pay for itself from benefits, revenues, or savings. Systems that tend to pay for themselves quickly are desirable because there is less uncertainty with estimates of short duration.

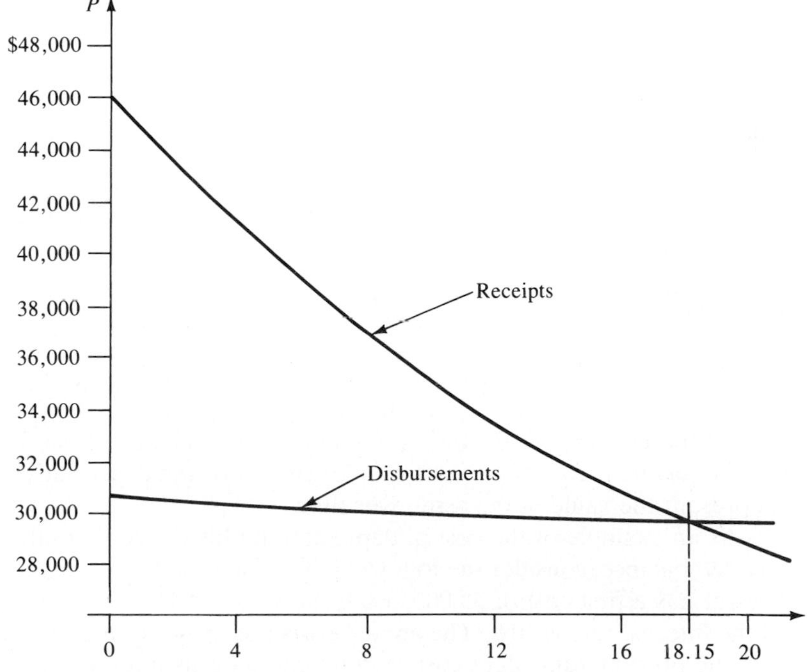

Figure 8.6 Solution for the rate-of-return.

The payout period is the amount of time required for the difference in the present value of receipts (savings) to equal the present value of the disbursements (costs); the annual equivalent receipts and disbursements may also be equated. For the present equivalent approach

$$0 \leq \sum_{t=0}^{n^*} F_t(1 + i)^{-t} \tag{8.17}$$

The smallest value of n^* that satisfied the preceding expression is the payout duration.

For a duration of 3 years the equation for present equivalent of savings and disbursements is

$$\begin{aligned}
[\$9{,}500\overset{P/A,12,3}{(2.4018)} + \$8{,}000\overset{P/F,12,3}{(0.7118)}] &= [\$28{,}000 + \$2{,}500\overset{P/F,12,2}{(0.7972)}] \\
[\$22{,}817 + \$5{,}694] &= [\$28{,}000 + \$1{,}993] \\
\$28{,}511 &= \$29{,}993
\end{aligned}$$

For a duration of 4 years, the equation for present equivalent of savings and disbursements is

$$\begin{aligned}
[\$9{,}500\overset{P/A,12,4}{(3.0374)} + \$8{,}000\overset{P/F,12,4}{(0.6355)}] &= [\$28{,}000 + \$2{,}500\overset{P/F,12,2}{(0.7972)}] \\
[\$28{,}855 + \$5{,}084] &= [\$28{,}000 + \$1{,}993] \\
\$33{,}939 &= \$29{,}993
\end{aligned}$$

Interpolating gives

$$n = 2 + \frac{[22{,}433 - 26{,}007] - 0}{[22{,}433 - 26{,}007] - [28{,}511 - 26{,}220]}$$

$$n = 3.3 \text{ years}$$

From these results, a present-equivalent graph can be developed, as shown in Figure 8.7. From this graph it is evident that the payout period on the energy-saving system is just over three years.

Annual Equivalent Cost of an Asset

A useful application of the annual equivalent evaluation approach pertains to the cost of an asset. The cost of any asset is made up of two components, the cost of depreciation and the cost of interest on the undepreciated balance. It can be shown that the annual equivalent cost of an asset is independent of the depreciation function chosen to represent the value of the asset over time.

As an example of the cost of depreciation plus the cost of interest on the undepreciated balance, consider the following example based on straight line depreciation. An asset has a first cost of \$5,000, a salvage value of \$1,000, a service life of 5 years, and the interest rate is 10%. The annual costs are shown in Figure 8.8.

The present equivalent cost of depreciation plus interest on the the undepreciated balance from Equation 8.9 is

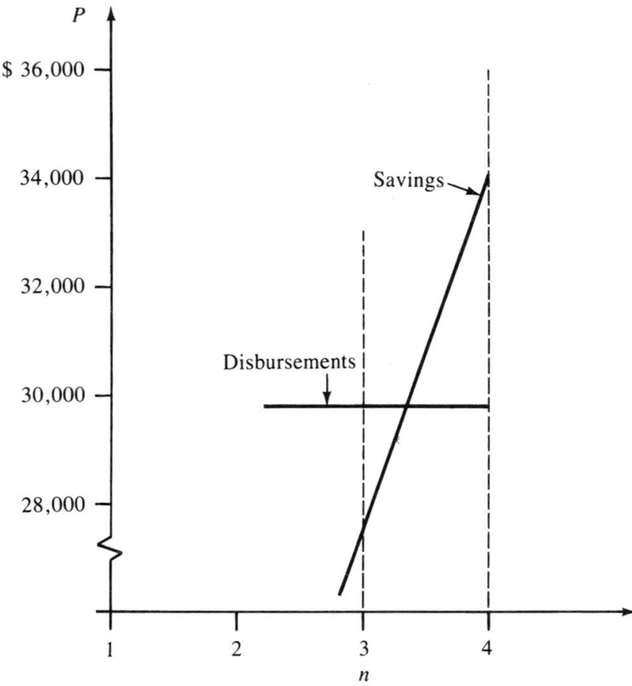

Figure 8.7 Solution for the payout period.

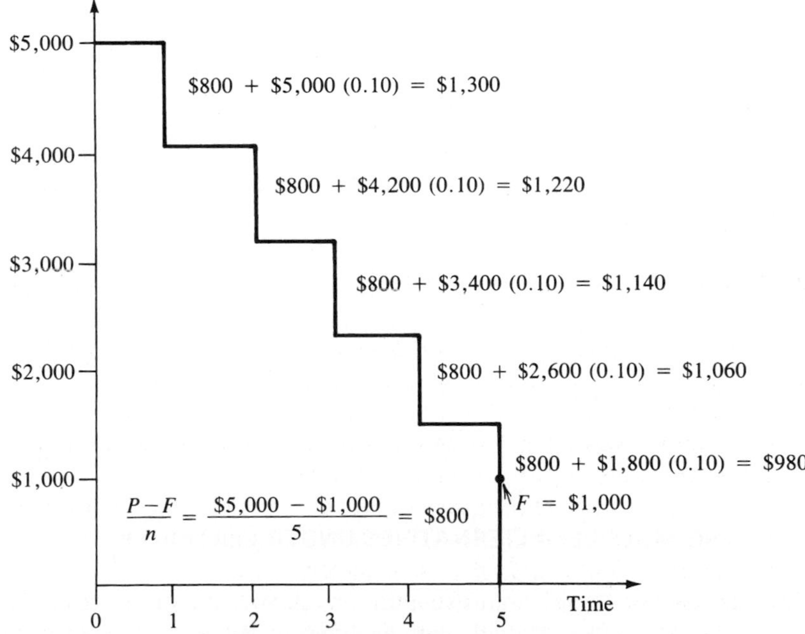

Figure 8.8 Value-time function.

$$PE(10) = \$1{,}300\overset{P/F,10,1}{(0.9091)} + \$1{,}220\overset{P/F,10,2}{(0.8265)} + \$1{,}140\overset{P/F,10,3}{(0.7513)}$$
$$+ \$1{,}060\overset{P/F,10,4}{(0.6830)} + \$980\overset{P/F,10,5}{(0.6209)}$$
$$= \$4{,}379$$

and the annual equivalent cost of the asset from Equation 8.11 is

$$AE(10) = \$4{,}379\overset{A/P,10,5}{(0.2638)} = \$1{,}155$$

Regardless of the depreciation function that describes the reduction in value of a physical asset over time, the annual equivalent cost of an asset may be expressed as the annual equivalent first cost minus the annual equivalent salvage value. This annual equivalent cost is the amount an asset must earn each year if the invested capital is to be recovered along with a return on the investment. The annual equivalent cost is derived as follows:

$$A = P\overset{A/P,i,n}{(\quad)} - F\overset{A/F,i,n}{(\quad)}$$
$$\overset{A/F,i,n}{(\quad)} = \overset{A/P,i,n}{(\quad)} - i$$

But because

$$A = P\overset{A/P,i,n}{(\quad)} - F[\overset{A/P,i,n}{(\quad)} - i]$$
$$A = (P - F)\overset{A/P,i,n}{(\quad)} + F(i) \tag{8.18}$$

This means that the annual equivalent cost of any asset may be found from knowledge of its first cost, P, its anticipated service life, n, its estimated salvage value, F, and the interest rate, i. For example, an asset with a first cost of \$5,000, a service life of 5 years, and a salvage value of \$1,000 will lead to an annual equivalent cost of

$$(\$5{,}000 - \$1{,}000)\overset{A/P,10,5}{(0.2638)} + \$1{,}000(0.10) = \$1{,}155$$

if the interest rate is 10%.

The annual equivalent cost formula is useful in a "should cost" study. For example, suppose that a contractor is to use equipment, tooling, and so on, for a certain defense contract that will cost \$3,250,000 and will depreciate to \$750,000 at the end of the 4-year production contract. If money costs the contractor 10%, the annual equivalent cost of the contractor's equipment should be

$$(\$3{,}250{,}000 - \$750{,}000)\overset{A/P,10,4}{(0.3155)} + \$750{,}000(0.10) = \$863{,}750$$

8.4 EVALUATING MULTIPLE ALTERNATIVES UNDER CERTAINTY

When several mutually exclusive alternatives provide service of equal value, it is desirable to compare them directly with each other. Where service of unequal value is provided by multiple alternatives, each alternative must be evaluated as a single alternative and is accepted or rejected on the basis of one or more of the comparisons suggested in Section 8.3. In many cases, however, the available alternatives do provide outputs that are identical or equal in value. Under this condition, the objective is to select the alternative that provides the desired service at least cost. This will be developed and presented in Section 8.5.

Assume that a defense contractor is considering the purchase of new test equipment. Semiautomatic test equipment will cost \$110,000 and can be expected to last 6 years, with a salvage value of \$10,000. Operating costs will be \$28,000 per year. Fully automatic equipment will cost \$160,000, should last 6 years, and have a salvage value of \$14,000. The operating costs will be \$12,000 per year. The services provided by the equipment is assumed here to be identical. With a desired interest rate of 14%, the alternative that meets the criterion of least cost should be selected.

This is an example of acquisition by purchase as in Figure 7.1a. The general equivalence function of Equation 7.1 will be used as a basis for comparing the alternatives.

Present Equivalent Evaluation

Under this method, the two alternatives may be compared on the basis of equivalent cost at a time taken to be the present. The present equivalent cost of the semiautomatic test equipment at 14% is

$$PE(14) = \$110{,}000 + \$28{,}000\overset{P/A,14,6}{(3.8887)} - \$10{,}000\overset{P/F,14,6}{(0.4556)} = \$214{,}328$$

The present equivalent cost of the fully automatic equipment is

$$PE(14) = \$160{,}000 + \$12{,}000\overset{P/A,14,6}{(3.8887)} - \$14{,}000\overset{P/F,14,6}{(0.4556)} = \$200{,}286$$

This comparison shows the present equivalent cost of the fully automatic test equipment to be less than the present equivalent cost of the semiautomatic test equipment by \$14,042 (\$214,328 − \$200,286).

Annual Equivalent Evaluation

The annual equivalent costs are taken as an equal-cost series over the life of the assets. The annual equivalent cost of the semiautomatic test equipment is

$$AE(14) = \$110{,}000\overset{A/P,14,6}{(0.2572)} + \$28{,}000 - \$10{,}000\overset{A/F,14,6}{(0.1175)} = \$55{,}118$$

and the annual equivalent cost of the fully automatic test equipment is

$$AE(14) = \$160{,}000\overset{A/P,14,6}{(0.2572)} + \$12{,}000 - \$14{,}000\overset{A/F,14,6}{(0.1175)} = \$51{,}507$$

The annual equivalent difference of \$55,118 minus \$51,507 or \$3,611 is the annual equivalent cost superiority of the fully automatic test equipment. As a verification, the annual equivalent amount of the present equivalent difference is

$$\$14{,}042\overset{A/P,14,6}{(0.2572)} = \$3{,}611$$

Rate-of-Return Evaluation

The previous cost comparisons indicated that the fully automatic test equipment was more desirable at an interest rate of 14%. At some higher interest rate, the two alternatives will be identical in equivalent cost, and beyond that interest rate, the semiautomatic test equipment will be less expensive because of its lower initial cost.

The interest rate at which the costs of the two alternatives are identical can be determined by setting the present equivalent amounts for the alternatives equal to each other and solving for the interest rate, i. Thus,

$$\$110{,}000 + \$28{,}000(\overset{P/A,i,6}{\quad}) - \$10{,}000(\overset{P/F,i,6}{\quad})$$

$$= \$160{,}000 + \$12{,}000(\overset{P/A,i,6}{\quad}) - \$14{,}000(\overset{P/F,i,6}{\quad})$$

$$\$16{,}000(\overset{P/A,i,6}{\quad}) + \$4{,}000(\overset{P/F,i,6}{\quad}) = \$50{,}000$$

For $i = 20\%$

$$\$16{,}000(\overset{P/A,20,6}{3.3255}) + \$4{,}000(\overset{P/F,20,6}{0.3349}) = \$54{,}548$$

For $i = 25\%$

$$\$16{,}000(\overset{P/A,25,6}{2.9514}) + \$4{,}000(\overset{P/F,25,6}{0.2622}) = \$48{,}271$$

Then, by interpolation,

$$i = 20 + (5)\frac{\$[54{,}548 - \$50{,}000]}{\$[54{,}548 - \$48{,}271]}$$

$$i = 23.6\%$$

When funds earn less than 23.6%, the fully automatic test equipment will be most desirable. When funds earn more than 23.6%, the semiautomatic equipment would be preferred.

Payout Evaluation

The service life of 6 years for each of the two test equipment alternatives is only the result of estimates and may be in error. If the services are needed for shorter or longer periods and if the assets are capable of providing the service for a longer period of time, the advantage may pass from one alternative to the other. Just as there is an interest rate at which two alternatives may be equal, there may be a service life at which the equivalent costs may be identical. This service life may be obtained by setting the alternatives equal to each other and solving for the life, n. Thus, for an interest rate of 14%,

$$\$110{,}000 + \$28{,}000(\overset{P/A,14,n}{\quad}) - \$10{,}000(\overset{P/F,14,n}{\quad})$$

$$= \$160{,}000 + \$12{,}000(\overset{P/A,14,n}{\quad}) - \$14{,}000(\overset{P/F,14,n}{\quad})$$

$$\$16{,}000(\overset{P/A,14,n}{\quad}) + \$4{,}000(\overset{P/F,14,n}{\quad}) = \$50{,}000$$

For $n = 4$ years

$$\$16{,}000\overset{P/A,14,4}{(2.9137)} + \$4{,}000\overset{P/F,14,4}{(0.5921)} = \$48{,}988$$

For $n = 5$ years

$$\$16{,}000\overset{P/A,14,5}{(3.4331)} + \$4{,}000\overset{P/F,14,5}{(0.5194)} = \$57{,}007$$

Then, by interpolation,

$$n = 4 + (1)\frac{[\$50{,}000 - \$48{,}988]}{[\$57{,}007 - \$48{,}988]}$$

$$n = 4.12 \text{ years}$$

Thus, if the desired service were to be used for less than 4.12 years, the semiautomatic equipment would be the most desirable.

Life-Cycle Economic Evaluation

An example of the evaluation of multiple alternatives under certainty where the equivalence measure is not determined at a traditional point in time is considered next. Suppose that there are two design concepts being considered for the acquisition of a complex system for which the life-cycle costs (in millions of dollars) are estimated to be as shown in Figure 8.9.

Although other choices are available the design concepts may be compared by finding their equivalent costs at the end of the production phase (the point when the system goes into operation). For Design A this equivalent cost at 10% is

$$\$10 + \$5\overset{F/P,10,1}{(1.100)} + \$1\overset{F/P,10,2}{(1.210)} + \$0.8\overset{F/P,10,3}{(1.331)} + \$0.6\overset{F/P,10,4}{(1.464)}$$
$$+ \$1\overset{P/A,10,7}{(4.8684)} = \$23.52 \text{ million}$$

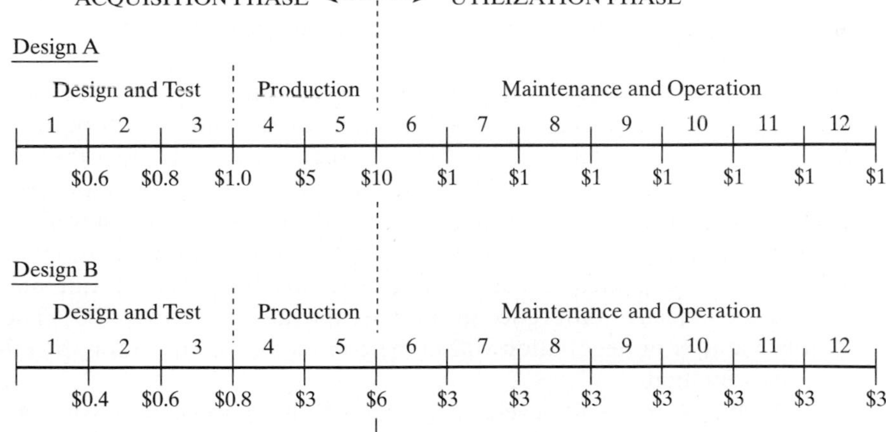

Figure 8.9 Two design alternatives for system acquisition.

And for Design B, the equivalent life-cycle cost is

$$\$6 + \$3(\overset{F/P,10,1}{1.100}) + \$0.8(\overset{F/P,10,2}{1.210}) + \$0.6(\overset{F/P,10,3}{1.331}) + \$0.4(\overset{F/P,10,4}{1.464})$$

$$+ \$3(\overset{P/A,10,7}{4.8684}) = \$26.26 \text{ million}$$

From the preceding analysis, it is evident that Design A would be preferred since \$23.52 is less than \$26.26. These are decision values and are not to be considered the costs of Designs A and B.

8.5 EVALUATIONS INVOLVING MULTIPLE CRITERIA

Design evaluation and multiple criteria were first introduced in Section 2.5. Multiple criteria must be considered when both economic and noneconomic factors are involved in decision making. This difficult situation can be simplified and clarified by use of the decision evaluation display as presented and explained in Section 7.8. Two examples of evaluations involving multiple criteria are presented in this section. They are the same examples presented in Section 8.4 but expanded to recognize noneconomic factors.

Suppose that the defense contractor is inclined to procure the fully automatic test equipment in accordance with the economic evaluations. On the basis of economic factors alone, the fully automatic acquisition and use plan would be best. However, if the contractor is facing a budget constraint because of start up costs, or if there is some concern for the dependability of the test equipment, another alternative may have to be considered.

To illustrate, assume that no more than \$150,000 can be allocated to the test equipment portion of the defense contract. Also assume that the test equipment must be highly dependable, as would be the case with the semiautomatic choice. This is due to concern for the performance of the software component of the automatic equipment in contrast to the performance of the semiautomatic equipment, where a human provides oversight.

The multiple criteria situation described is illustrated with the aid of a decision evaluation display in Figure 8.10. Note that the automatic alternative exhibits the lowest present equivalent cost, but it violates each of the other criteria (First cost = \$160,000 > \$150,000 and Dependability ⩾ MH). The semiautomatic alternative promises high dependability as required, and it also has a first cost well within the budget constraint (\$110,000 < \$150,000). Now the choice is no longer strictly objective. Preference for dependability and conformance to a budget constraint must be weighed against the objective difference in present equivalent cost (\$14,000). The final choice is now a matter of negotiation within the contractors organization, together with subjective evaluation.

Turn next to the life-cycle economic evaluation in Section 8.4 (Figure 8.9). Suppose, in the example, that the acquisition of the complex system is budget constrained as follows:

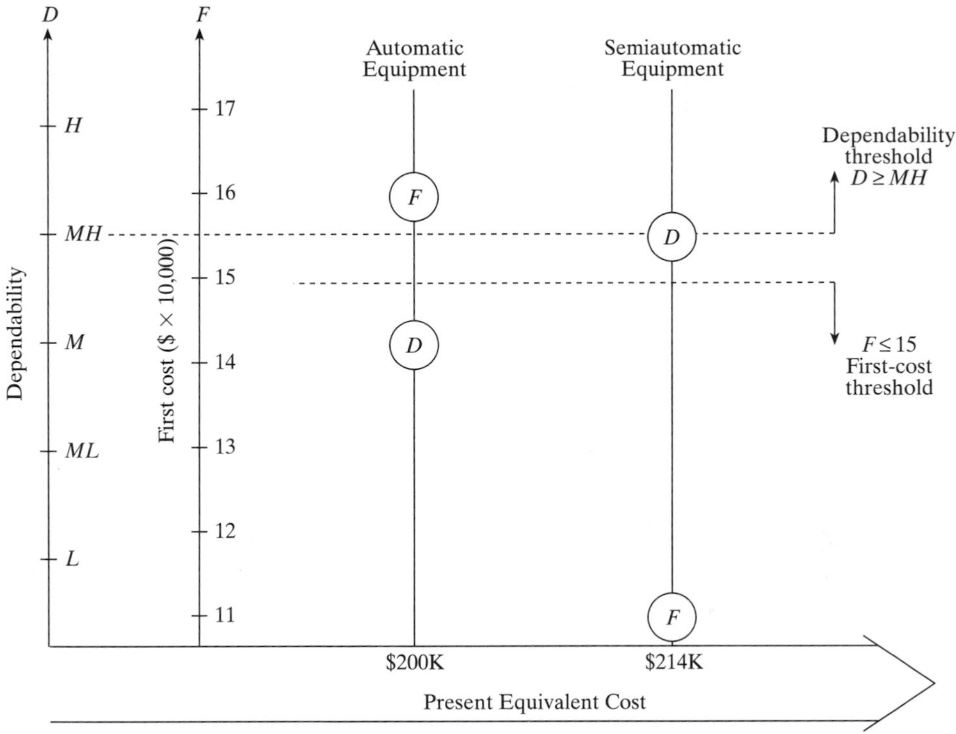

Figure 8.10 Decision evaluation display for test equipment purchase.

1. The capital requirement in any one year of the acquisition phase should not exceed $7,500,000.
2. The maintenance and operating cost per year should not exceed $2,000,000.

Another requirement to be met is that the MTBF must be greater than 2,500 hours. For this example, the MTBF under Design A is predicted to be 3,000 hours whereas for Design B it is predicted to be only about 1,500 hours. The higher maintenance and operating cost for Design B ($3,000,000) reflects the more frequent maintenance situation when compared with Design A, which is only $1,000,000. All other effectiveness and performance factors are assumed to be identical for Designs A and B.

The multiple criteria situation described above is illustrated with the aid of a decision evaluation display in Figure 8.11. Although Design A exhibits the lowest equivalent cost ($23.52 million versus $26.26 million), it violates the acquisition budget. Design B violates both the MTBF requirement and the annual budget for maintenance and operations. The alternatives are shown along with the economic and noneconomic factors and can be studied by the decision maker. As before, the choice has to do with preference and subjective evaluation along with negotiation.

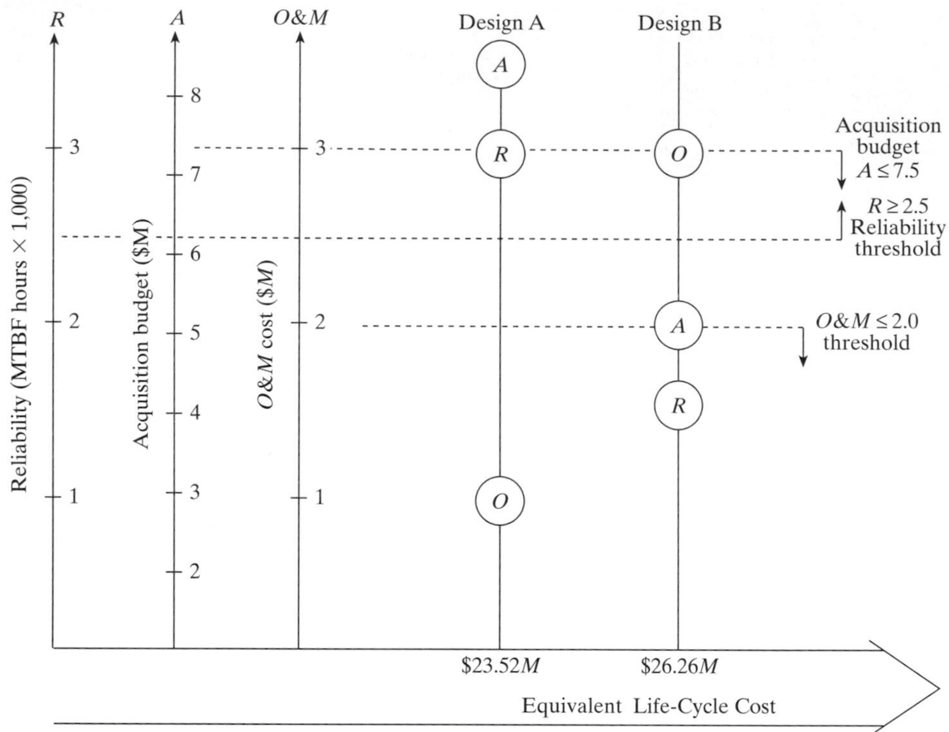

Figure 8.11 Decision evaluation display for two system designs.

8.6 EVALUATING MULTIPLE ALTERNATIVES WITH MULTIPLE FUTURES

An engineering firm is engaged in the design of a unique machine tool. Because the new machine tool is technologically advanced, the firm is considering setting up a new production facility separate from its other operations.

Because the firm's own finances are not adequate to set up this new facility, there is a need to borrow capital from financial sources. The firm has identified three sources, namely, A, B, and C, which would lend money for this venture.

Each of these financial sources will lend money based on the anticipated sales of the product and will use a varying interest rate based on the anticipated demand for the product. If the product demand is anticipated to be low over the next few years, the financial backers will charge a high interest rate. This is because the backers think that the firm may not borrow money in the future if the product fails because of low demand.

Conversely, if the financial sources feel that the demand will be high, they will be willing to lend the money at a lower interest rate. This is because of the possibility of future business with the firm owing to its present product being popular.

The engineering firm has fairly accurate knowledge of the receipts and disbursements associated with each demand level and the financial sources under consideration. The firm has knowledge about the initial outlay and anticipated receipts over 3 years for each demand level as is summarized in Table 8.4. The interest rates quoted by each financial source are summarized in Table 8.5.

TABLE 8.4 Anticipated Disbursements and Receipts for Machine Tool Production

DEMAND LEVEL	INITIAL OUTLAY	ANNUAL RECEIPTS
Low	$ 500	$400
Medium	1,300	700
High	2,000	900

TABLE 8.5 Anticipated Interest Rates for Production Financing

FINANCING SOURCE	DEMAND LEVEL	INTEREST RATE
	Low	15
A	Medium	13
	High	7
	Low	14
B	Medium	12
	High	8
	Low	15
C	Medium	11
	High	6

TABLE 8.6 Present Equivalent Payoff for Three Financing Sources

	DEMAND LEVEL		
Source	Low	Medium	High
A	$413	$353	$362
B	429	382	320
C	343	411	406

The engineering firm wishes to select a financing source in the face of three demand futures, not knowing which demand future will occur. Accordingly, a decision evaluation matrix is developed as a first step. Using the present equivalent approach, nine evaluation values are derived and displayed in Table 8.6, one for each financing source for each demand level. For example, if financing comes from Source A, and the demand level is low, the present equivalent payoff is

$$PE(15) = -\$500 + \$400\overset{P/A,15,3}{(2.2832)} = \$413$$

If the firm applied probabilities to each of the demand futures, the financing decision becomes one under risk. If the probability of a low demand is 0.3, the probability of a medium demand is 0.2, and the probability of a high demand is 0.5, the expected present equivalent amount for each financing source is calculated as follows:

$$\text{Source A: } \$413(0.3) + \$353(0.2) + \$362(0.5) = \$376$$

$$\text{Source B: } \$429(0.3) + \$382(0.2) + \$320(0.5) = \$365$$

$$\text{Source C: } \$343(0.3) + \$411(0.2) + \$406(0.5) = \$388$$

Accordingly, financing Source C is selected by this decision rule.

In the event the firm has no basis for assigning probabilities to demand futures, the financing decision must be made under uncertainty. Under the Laplace criterion, each future is assumed to be equally likely and the present equivalent payoffs are calculated as follows:

$$\text{Source A: } \$(413 + 353 + 362) \div 3 = \$376$$

$$\text{Source B: } \$(429 + 382 + 320) \div 3 = \$377$$

$$\text{Source C: } \$(343 + 411 + 406) \div 3 = \$386$$

Financing Source C would be selected by this decision rule.

By the Maximin rule, the firm calculates the minimum present equivalent amount which could occur for each financing source and then selects the source which provides a maximum. The minimums for each source are

Source A: $353

Source B: $320

Source C: $343

Financing Source A would be selected by this decision rule as the one that will maximize the minimum present equivalent amount.

By the Maximax rule, the firm calculates the maximum present equivalent amount which could occur for each financing source and then selects the source that provides a maximum. The maximums for each source are

Source A: $413

Source B: $429

Source C: $411

Financing Source B would be selected by this decision rule as the one that will maximize the present equivalent amount.

A summary of the sources selected by each decision rule is given in Table 8.7.

TABLE 8.7 Summary of Financing Source Selections

	FINANCING SOURCE		
Decision rule	A	B	C
Expected value			X
Laplace			X
Maximin	X		
Maximax		X	

8.7 BREAK-EVEN ECONOMIC EVALUATIONS

Break-even analysis may be graphical or mathematical in nature. This economic evaluation technique is useful in relating fixed and variable costs to the number of hours of operation, the number of units produced, or other measures of operational activity. In each case, the break-even point is of primary interest in that it identifies the range of the decision variable within which the most desirable economic outcome may occur.

When the cost of two or more alternatives is a function of the same variable, it is usually useful to find the value of the variable for which the alternatives incur equal cost. Several examples will be presented in this section.

Make-or-Buy Evaluation

Often a manufacturing firm has the choice of making or buying a certain component for use in the product being produced. When this is the case, the firm faces a make-or-buy decision.

Suppose, for example, that a firm finds that it can buy from a vendor the electric power supply for the system it produces for \$8 per unit. Alternatively, suppose that it can manufacture an equivalent unit for a variable cost of \$4 per unit. It is estimated that the additional fixed cost in the plant would be \$12,000 per year if the unit is manufactured. The number of units per year for which the cost of the two alternatives breaks even would help in making the decision.

First, the total annual cost is formulated as a function of the number of units for the make alternative. It is

$$TC_M = \$12{,}000 + \$4N$$

And the total annual cost for the buy alternative is

$$TC_B = \$8N$$

Break-even occurs when $TC_M = TC_B$ or

$$\$12{,}000 + \$4N = \$8N$$

$$\$4N = \$12{,}000$$

$$N = 3{,}000 \text{ units}$$

These cost functions and the break-even point are shown in Figure 8.12. For requirements in excess of 3,000 units per year, the make alternative would be more economical. If the rate of use is likely to be less than 3,000 units per year, the buy alternative should be chosen.

If the production requirement changes during the course of the production program, the break-even choice in Figure 8.12 can be used to guide the decision of whether to make or to buy the power unit. Small deviations below and above 3,000 units per year make little difference. However, the difference can be significant when the production requirement is well above or below 3,000 units per year.

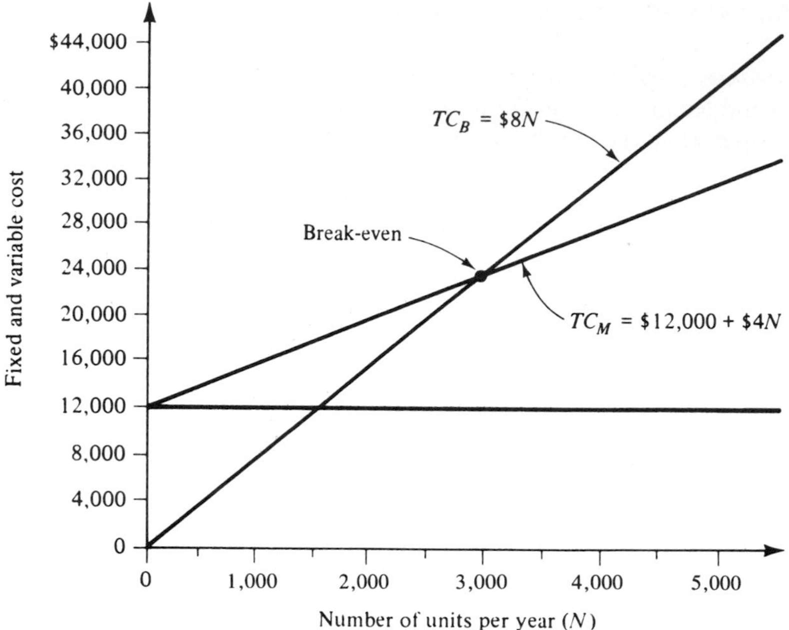

Figure 8.12 Break-even for make or buy.

Lease-or-Buy Evaluation

As another example of break-even analysis, consider the decision to lease or buy a piece of equipment. Assume that a small electronic computer is needed for data processing in an engineering office. Suppose that the computer can be leased for $50 per day, which includes the cost of maintenance. Alternatively, the computer can be purchased for $25,000.

The computer is estimated to have a useful life of 15 years with a salvage value of $4,000 at the end of that time. It is estimated that annual maintenance costs will be $2,800. If the interest rate is 9% and it costs $50 per day to operate the computer, how many days of use per year are required for the two alternatives to break even?

First, the annual cost if the computer is leased is

$$\begin{aligned} TC_L &= (\$50 + \$50)N \\ &= \$100N \end{aligned}$$

And the annual equivalent total cost if the computer is bought is

$$\begin{aligned} TC_B &= (\$25{,}000 - \$4{,}000)\overset{A/P,9,15}{(0.1241)} + \$4{,}000(0.09) + \$2{,}800 + \$50N \\ &= \$2{,}606 + \$360 + \$2{,}800 + \$50N \\ &= \$5{,}766 + \$50N \end{aligned}$$

The first three terms represent the fixed cost and the last term is the variable cost. Break-even occurs when $TC_L = TC_B$ or

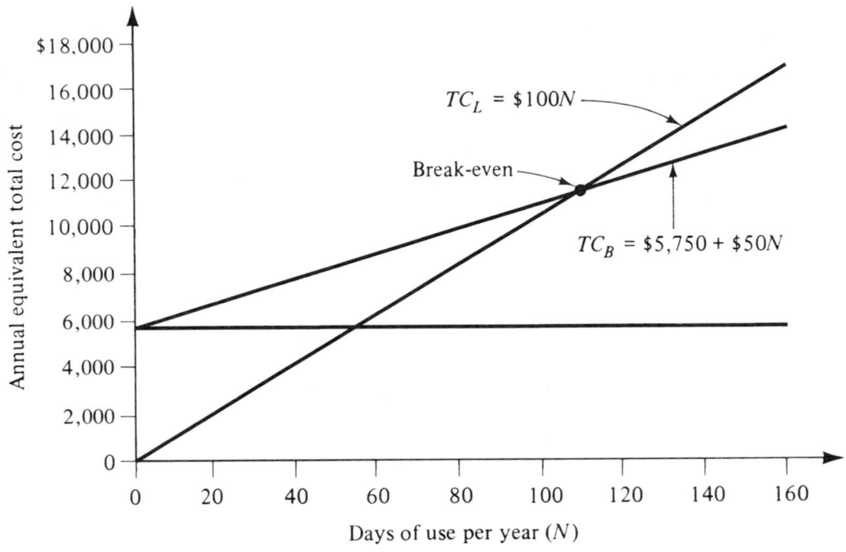

Figure 8.13 Break-even for lease or buy.

$$\$100N = \$5{,}766 + \$50N$$
$$\$50N = \$5{,}766$$
$$N = 115 \text{ days}$$

A graphical representation of this decision situation is shown in Figure 8.13. For all levels of use exceeding 115 days per year, it would be more economical to purchase the computer. If the level of use is anticipated to be below 115 days per year, the computer should be leased.

Equipment Selection Evaluation

Suppose that a fully automatic controller for a machine center can be fabricated for \$140,000 and that it will have an estimated salvage value of \$20,000 at the end of four years. Maintenance cost will be \$12,000 per year, and the cost of operation will be \$85 per hour.

As an alternative, a semiautomatic controller can be fabricated for \$55,000. This device will have no salvage value at the end of a 4-year service life. The cost of operation and maintenance is estimated to be \$140 per hour.

With an interest rate of 10%, the annual equivalent total cost for the automatic attachment as a function of the number of hours of use per year is

$$\begin{aligned} TC_A &= (\$140{,}000 - \$20{,}000)\overset{A/P,10,4}{(0.3155)} + \$20{,}000(0.10) + \$12{,}000 + \$85N \\ &= \$51{,}800 + \$85N \end{aligned}$$

And the annual equivalent total cost for the semiautomatic attachment as a function of the number of hours of use per year is

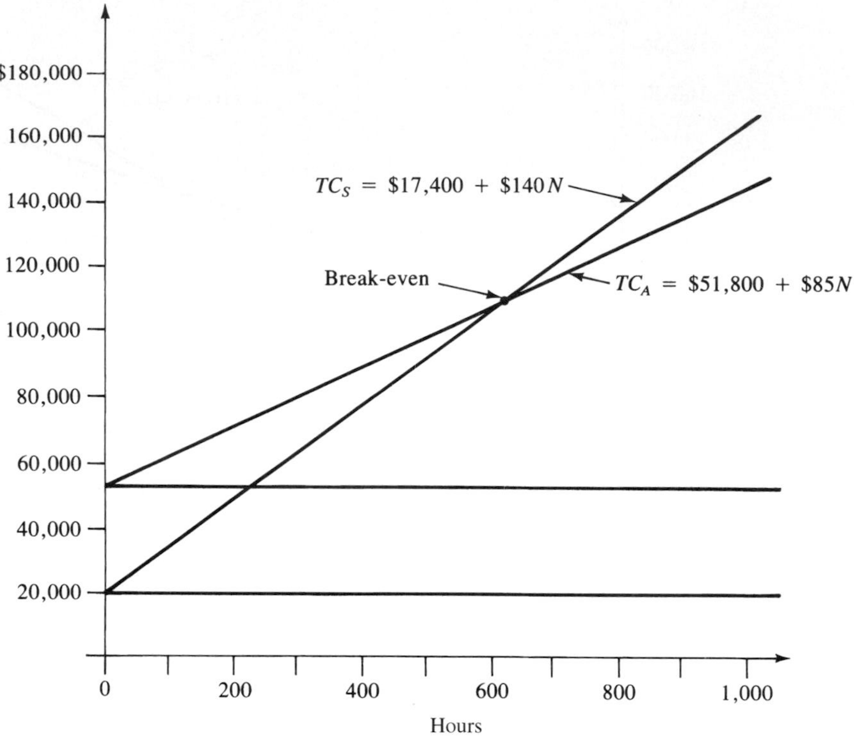

Figure 8.14 Break-even for equipment selection.

$$TC_s = \$55{,}000(\overset{A/P,\,10,\,4}{0.3155}) + \$140N$$

$$= \$17{,}400 + \$140N$$

Break-even occurs when $TC_A = TC_S$, or

$$\$51{,}800 + \$85N = \$17{,}400 + \$140N$$

$$\$55N = \$34{,}400$$

$$N = 625 \text{ hr}$$

Figure 8.14 shows the two cost functions and the break-even point. For rates of use exceeding 625 hours per year, the automatic controller would be more economical. However, if it is anticipated that the rate of use will be less than 625 hours per year, the semiautomatic controller should be fabricated.

Profitability Evaluation

There are two aspects of production operations. One consists of assembling the production system of facilities, material, and people, and the other consists of the sale of the goods produced. The economic success of an enterprise depends upon its ability to

TABLE 8.8 Fixed and Variable Manufacturing Costs

	FIXED COSTS	VARIABLE COSTS
Capital recovery with return		
$\$140{,}000\overset{A/P,12,8}{(0.1877)}$	\$26,278	
Insurance and taxes	2,000	
Repairs and maintenance	1,722	\$0.50/unit
Materials		9.50/unit
Labor, electricity, space		11.00/unit
Totals	\$30,000	\$21.00/unit

carry on these activities to the end that there may be a net difference between receipts and the cost of production.

Linear break-even analysis is useful in evaluating the effect on profit of proposals for new operations not yet implemented and for which no data exist. Consider a proposed activity consisting of the manufacturing and marketing of a certain plastic item for which the sale price per unit is estimated to be \$30,000. The machine required in the operation will cost \$140,000 and will have an estimated life of 8 years. It is estimated that the cost of production, including power, labor, space, and selling expense, will be \$21 per unit sold. Material will cost \$9.50 per unit. An interest rate of 8% is considered necessary to justify the required investment. The estimated costs associated with this activity are as given in Table 8.8.

The difficulty of making a clear-cut separation between fixed and variable costs becomes apparent when attention is focused on the item for repair and maintenance. In practice it is very difficult to distinguish between repairs that are a result of deterioration that takes place with the passage of time and those that result from the wear and tear of use. However, in theory, the separation can be made as shown in this example and is in accord with fact, with the exception, perhaps, of the assumption that repairs from wear and tear will be in direct proportion to the number of units manufactured. To be in accord with actualities, depreciation also undoubtedly should have been separated so that a part would appear as variable cost.

In this example, $F = \$30{,}000$, $TC = \$30{,}000 + \$21N$, and $I = \$30N$. If F, TC, I are plotted as N varies from 0 to 6,000 units, the results will be as shown in Figure 8.15.

The cost of producing the plastic item will vary with the number made per year. The production cost per unit is given by $F/N + V$. If production cost per unit, variable cost per unit, and income per unit are plotted as N varies from 0 to 6,000, the results will be as shown in Figure 8.16. It will be noted that the fixed cost per unit may be infinite. Thus, in determining unit costs, fixed cost has little meaning unless the number of units to which it applies are known.

Income for most enterprises is directly proportional to the number of units sold. However, the income per unit may easily be exceeded by the sum of the fixed and the variable cost per unit for low production volumes. This is shown by comparing the total cost per unit curve and the income per unit curve shown in Figure 8.16.

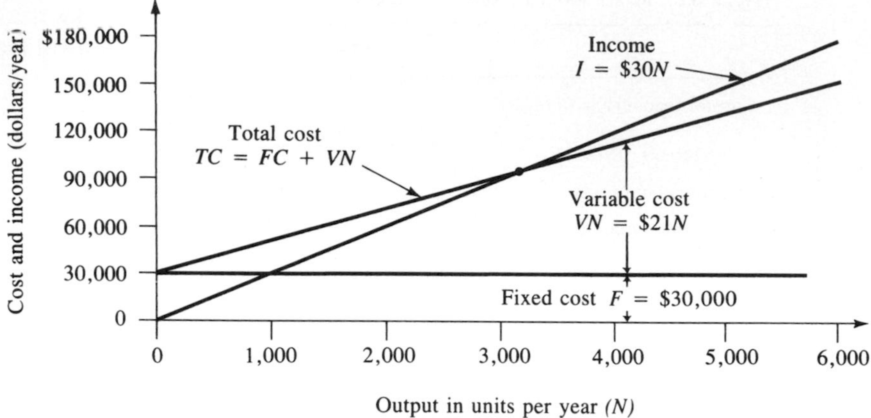

Figure 8.15 Fixed cost, variable cost, and income per year.

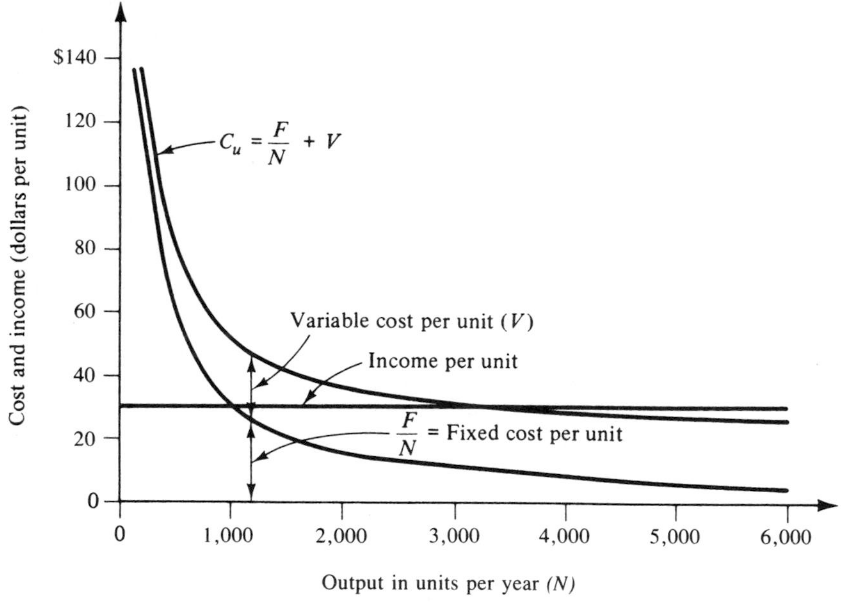

Figure 8.16 Fixed cost, variable cost, and income per unit.

8.8 BREAK-EVEN ANALYSIS UNDER RISK

In Section 8.7, the make-or-buy analysis was aided by Figure 8.15 showing the break-even point on which a make-or-buy decision is based. The analysis was based on certainty about the demand for the electric power supply. Ordinarily, the decision maker does not know with certainty what the demand level will be.

Consider demand ranging from 1,500 units to 4,500 units with probabilities as given in Table 8.9. Entries for the cost to make and the cost to buy are calculated from the equations in Section 8.7 as was exhibited in Figure 8.12.

The expected demand may be calculated from the probabilities associated with each demand level as given in Table 8.9.

$$1{,}500(0.05) + 2{,}000(0.10) + 2{,}500(0.15) + 3{,}000(0.20) + 3{,}500(0.25) + 4{,}000(0.15) + 4{,}500(0.10) = 3{,}175$$

From the equations for this decision situation, the expected total cost to make is

$$TC_M = \$12{,}000 + \$4(3{,}175) = \$24{,}700$$

And the expected total cost to buy is

$$TC_B = \$8(3{,}175) = \$25{,}400$$

Accordingly, the power supply should be made. Alternatively, this decision might be based on the most probable future criterion. From Table 8.9 it is evident that a demand level of 3,500 is most probable. Therefore, the cost to make of \$26,000 is compared with cost to buy of \$28,000 and the decision would be to make the item. This is the same decision as for the expected demand approach.

TABLE 8.9 Demand Range for Make or Buy

	DEMAND IN UNITS						
	1500	2000	2500	3000	3500	4000	4500
Probability	0.05	0.10	0.15	0.20	0.25	0.15	0.10
Cost to make	\$18,000	20,000	22,000	24,000	26,000	28,000	30,000
Cost to buy	\$12,000	16,000	20,000	24,000	28,000	32,000	36,000

QUESTIONS AND PROBLEMS

1. How much money must be invested to accumulate \$1,000 in 8 years at 9% compounded annually?
2. What amount will be accumulated by each of the following investments?
   (a) \$8,000 at 12% compounded annually over 10 years.
   (b) \$52,500 at 8% compounded annually over 5 years.
3. What is the present equivalent amount of a year-end series of receipts of \$600 over 5 years at 8% compounded annually?
4. What interest rate compounded annually is involved if \$4,000 results in \$10,000 in 6 years?
5. How many years will it take for \$4,000 to grow to \$7,000 at an interest rate of 10% compounded annually?
6. What interest rate is necessary for a sum of money to double itself in 8 years? What is the approximate product of i and n (i as an integer) that establishes the doubling period?

7. An asset was purchased for $5,200 with the anticipation that it would serve for 12 years and be worth $600 as scrap. After 5 years of operation the asset was sold for $1,800. The interest rate is 14%.

(a) What was the anticipated annual equivalent cost of the asset?

(b) What was the actual annual equivalent cost of the asset?

8. An epoxy mixer purchased for $3,300 has an estimated salvage value of $500 and an expected life of 3 years. An average of 25 pounds per month will be processed by the mixer.

(a) Calculate the annual equivalent cost of the mixer with an interest rate of 8%.

(b) Calculate the annual equivalent cost per pound mixed with an interest rate of 12%.

9. The table below shows the receipts and disbursements for a given venture. Determine the desirability of the venture for a 14% interest rate based on the present equivalent comparison and the annual equivalent comparison.

END OF THE YEAR	RECEIPTS	DISBURSEMENTS
0	$ 0	$25,000
1	600	0
2	500	400
3	500	0
4	1,200	100

10. A new automatic controller can be installed for $30,000 and will have a $3,000 salvage value after ten years. This controller is expected to decrease operating cost by $4,000 per year.

(a) What rate of return is expected if the controller is used for 10 years?

(b) For what life will the controller give a return of 15%?

11. Transco plans on purchasing a bus for $75,000 that will have a capacity of 80 passengers. As an alternative, a larger bus can be purchased for $95,000 which will have a capacity of 100 passengers. The salvage value of either bus is estimated to be $8,000 after a ten year life. If an annual net profit of $200 can be realized per passenger, which alternative should be recommended using an interest rate of 15%?

12. An office building and its equipment are insured to $710,000. The present annual insurance premium is $0.85 per $100 of coverage. A sprinkler system with an estimated life of 20 years and no salvage value can be installed for $18,000. Annual maintenance and operating cost is estimated to be $360. The premium will be reduced to $0.40 per $100 coverage if the sprinkler system is installed.

(a) Find the rate of return if the sprinkler system is installed.

(b) With interest at 12%, find the payout period for the sprinkler system.

13. The design of a system is to be pursued from one of two available alternatives. Each alternative has a life-cycle cost associated with an expected future. The costs for the corresponding futures are given in the table below (in millions of dollars). If the probabilities of occurrence of the futures are 30%, 50%, and 20%, respectively, which alternative is most desirable from an expected cost viewpoint using an interest rate of 10%.

DESIGN 1						YEARS						
Future	1	2	3	4	5	6	7	8	9	10	11	12
Optimistic	0.4	0.6	5.0	7.0	0.8	0.8	0.8	0.8	0.8	0.8	0.8	0.8
Expected	0.6	0.8	1.0	5.0	10.0	1.0	1.0	1.0	1.0	1.0	1.0	1.0
Pessimistic	0.8	0.9	1.0	7.0	10.0	1.2	1.2	1.2	1.2	1.2	1.2	1.2

DESIGN 2						YEARS						
Future	1	2	3	4	5	6	7	8	9	10	11	12
Optimistic	0.4	0.4	0.4	1.0	3.0	2.5	2.5	2.5	2.5	2.5	2.5	2.5
Expected	0.6	0.8	1.0	3.0	6.0	3.0	3.0	3.0	3.0	3.0	3.0	3.0
Pessimistic	0.6	0.8	1.0	5.0	6.0	3.1	3.1	3.1	3.1	3.1	3.1	3.1

14. Prepare a decision evaluation matrix for the design alternatives in Problem 13, and then choose the alternative that is best under the following decision rules: Laplace, Maximax, Maximin, and Hurwicz with $a = 0.6$. Assume that the choice is under uncertainty.

15. A campus building can be air conditioned by piping chilled water from a central refrigeration plant. Two competing proposals are being considered for the piping system, as outlined in the table. On the basis of a 10-year life, find the number of hours of operation per year for which the cost of the two systems will be equal if the interest rate is 9%.

	6-IN. SYSTEM	8-IN. SYSTEM
Motor size in horsepower	30	15
Installed cost of pump and pipe	$3,200	$4,400
Installed cost of motor	450	300
Salvage value of system	500	600
Energy cost per hour of operation	$ 0.32	$ 0.18

16. Fence posts for a large cattle ranch are currently purchased for $2.19 each. It is estimated that equivalent posts can be cut from timber on the ranch for a variable cost of $0.77 each, which is made up of the value of the timber plus labor cost. Annual fixed cost for required equipment is estimated to be $800. Two thousand posts will be required each year. What will be the annual saving if posts are cut?

17. A manufacturer can buy a required component from a supplier for $96 per unit delivered. Alternatively, the firm can manufacture the component for a variable cost of $46 per unit. It is estimated that the additional fixed cost would be $8,000 per year if the part is manufactured. Find the number of units per year for which the cost of the two alternatives will break even.

18. A marketing company can lease a fleet of automobiles for its sales personnel for $35 per day plus $0.09 per mile for each vehicle. As an alternative, the company can pay each salesperson $0.30 per mile to use his or her own automobile. If these are the only costs to the company, how many miles per day must a salesperson drive for the two alternatives to break even?

19. An electronics manufacturer is considering the purchase of one of two types of laser trimming machines. The sales forecast indicated that at least 8,000 units will be sold per year. Machine A will increase the annual fixed cost of the plant by $20,000 and will reduce variable cost by $5.60 per unit. Machine B will increase the annual fixed cost by $5,000 and will reduce variable cost by $3.60 per unit. If variable costs are now $20 per unit produced, which machine should be purchased?

20. Machine A costs $20,000, has zero salvage value at any time, and has an associated labor cost of $1.14 for each piece produced on it. Machine B costs $36,000, has zero salvage value at any time, and has an associated labor cost of $0.85. Neither machine can be used except to produce the product described. If the interest rate is 10% and the annual rate of production is 20,000 units, how many years will it take for the cost of the two machines to break even?

21. An electronics manufacturer is considering two methods for producing a circuit board. The board can be hand-wired at an estimated cost of $0.98 per unit and an annual fixed equipment cost of $1,000. A printed equivalent can be produced using equipment costing $18,000 with a service life of 9 years and salvage value of $1,000. It is estimated that the labor cost will be $0.32 per unit and that the processing equipment will cost $450 per year to maintain. If the interest rate is 13%, how many circuit boards must be produced each year for the two methods to break even?

22. It is estimated that the annual sales of a new product will be 2,000 the first year and increased by 1,000 per year until 5,000 units are sold during the fourth year. Proposal A is to purchase equipment costing $12,000 with an estimated salvage value of $2,000 at the end of four years. Proposal B is to purchase equipment costing $28,000 with an estimated salvage value of $5,000 at the end of 4 years. The variable cost per unit under proposal A is estimated to be $0.80, but is estimated to be only $0.25 under proposal B. If the interest rate is 9%, which proposal should be accepted for a four-year production period?

23. The fixed cost of a machine (capital recovery, interest, maintenance, space charges, supervision, insurance, and taxes) is F dollars per year. The variable cost of operating the machine (power, supplies, and other items, but excluding direct labor) is V dollars per hour of operation. If N is the number of hours the machine is operated per year, TC the annual total cost of operating the machine, TC_h the hourly cost of operating the machine, t the time in hours to process 1 unit of product, and M the machine cost of processing 1 unit of product per year, write expressions for (a) TC, (b) TC_h, and (c) M.

24. In problem 23, $F = \$600$ per year, $t = 0.2$ hr, $V = \$0.50$ per hr, and N varies from 0 to 10,000 in increments of 1,000.

(a) Plot values of M as a function of N.

(b) Write an expression for the total cost of direct labor and machine cost per unit, TC_u, using the symbols in Problem 23 and letting W equal the hourly cost of direct labor.

25. A certain firm has the capacity to produce 800,000 units per year. At present it is operating at 75% of capacity. The income per unit is $0.10 regardless of output. Annual fixed costs are $28,000, and the variable cost is $0.06 per unit. Find the annual profit or loss at this capacity and the capacity for which the firm will break even.

26. An arc welding machine that is used for a certain joining process costs $10,000. The machine has a life of 5 years and a salvage value of $1,000. Maintenance, taxes, insurance, and other fixed costs amount to $500 per year. The cost of power and supplies is $3.20 per hour of operation and the operator receives $15.80 per hour. If the cycle time per unit of product is 60 min and the interest rate is 8%, calculate the cost per unit if (a) 200, (b) 600, (c) 1,200, and (d) 2,500 units of product are made per year.

27. A certain firm has the capacity to produce 650,000 units of product per year. At present, it is operating at 65% of capacity. The firm's annual income is $416,000. Annual fixed costs are $192,000 and the variable costs are $0.38 per unit of product.

(a) What is the firm's annual profit or loss?

(b) At what volume of sales does the firm break even?

(c) What will be the profit or loss at 70%, 80%, and 90% of capacity on the basis of constant income per unit and constant variable cost per unit?

28. A chemical company owns two plants, A and B, that produce an identical product. The capacity of plant A is 60,000 gallons while that of B is 80,000 gallons. The annual fixed cost of plant A is $260,000 per year and the variable cost is $3.20 per gallon. The corresponding values for plant B are $280,000 and $3.90 per gallon. At present, plant A is being operated at 35% of capacity and plant B is being operated at 40% of capacity.

(a) What would be the total cost of production of plants A and B?

(b) What is the total cost and the average cost of the total output of both plants?

(c) What would be the total cost to the company and cost per gallon if all production were transferred to plant A?

(d) What would be the total cost to the company and cost per gallon if all production were transferred to plant B?

9 Optimization in Design and Operations

The design of complex systems that are optimum in some sense is a most important challenge facing the systems engineer. This challenge is paralleled by the frequent task of optimizing the operation of systems already in being. In the former case, optimum values of design variables are sought. In the latter, optimum values for policy variables are desired. This distinction is important only as a means for contrasting engineering design and systems analysis. The modeling approaches are essentially the same.

In this chapter, the general approach to the formulation and manipulation of mathematical models is presented. The specific mathematical methods used vary in degree of complexity and depend on the system under study. Examples will be used to illustrate design situations for static systems and for dynamic systems. Both unconstrained and constrained examples will be presented. Classical optimization techniques will be illustrated first, followed by some nonclassical approaches, including techniques of linear and dynamic programming.

9.1 CLASSICAL OPTIMIZATION THEORY

Optimization is the process of seeking the best. In systems engineering and analysis, this process is applied to each alternative in accordance with the decision evaluation theory presented in Section 7.3. In so doing, the general decision evaluation function given by Equation 7.3 is used along with the decision evaluation display of Figure 7.7. Before presenting specific applications of optimization in design and operations, this section provides a review of calculus-based methods for optimizing functions with one or more decision variables. Optmization in design and operations, this section provides a review of calculus-based methods for optimizing functions with one or more decision variables.

The Slope of a Function

The slope of a function $y = f(x)$ is defined as the rate of change of the dependent variable, y, divided by the rate of change of the independent variable, x. If a positive change in x results in a positive change in y, the slope is positive. Conversely, a positive change in x resulting in a negative change in y indicates a negative slope.

If $y = f(x)$ defines a straight line, the difference between any two points x_1 and x_2 represents the change in x, and the difference between any two points y_1 and y_2 represents the change in y. Thus, the rate of change of y with respect to x is $(y_1 - y_2)/(x_1 - x_2)$; the slope of the straight line. This slope is constant for all points on the straight line; $\Delta y/\Delta x$ = constant.

For a nonlinear function, the rate of change of y with respect to changes in x is not constant but changes with changes in x. The slope must be evaluated at each point on the curve. This can be done by assuming an arbitrary point on the function p, for which the x and y values are x_0 and $f(x_0)$, as shown in Figure 9.1. The rate of change of y with respect to x at point p is equal to the slope of a line tangent to the function at that point. It is observed that the rate of change of y with respect to x differs from that at p at other points on the curve.

Differential Calculus Fundamentals

Differential calculus is a mathematical tool for finding successively better and better approximations for the slope of the tangent line shown in Figure 9.1. Consider another point on $f(x)$ designated q, situated at an x distance from point p equal to Δx and situated at a y distance from point p equal to Δy. Then the slope of a line segment through points p and q would have a slope $\frac{\Delta y}{\Delta x}$ But this is the average slope of $f(x)$ between the designated points. In classical optimization, it is the instantaneous rate of change at a given point that is sought.

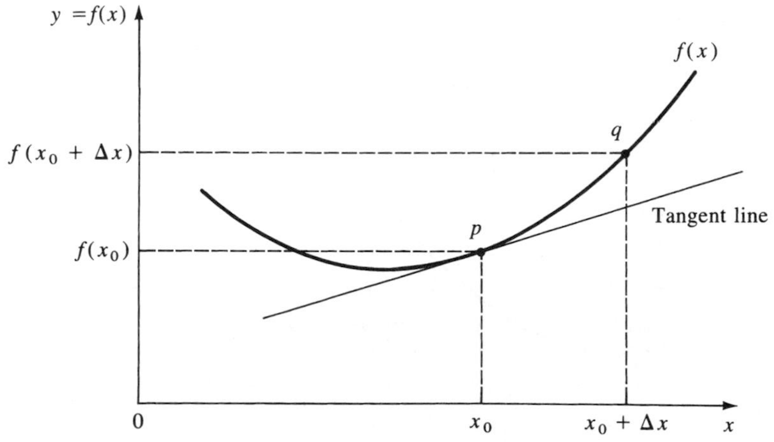

Figure 9.1 Slope of a function.

The instantaneous rate of change in $f(x)$ at $x = x_0$ can be found by letting $\Delta x \to 0$. Referring to Figure 9.1, this can be stated as

$$\lim_{\Delta x \to 0} \frac{f(x_0 + \Delta x) - f(x_0)}{\Delta x} \tag{9.1}$$

It is noted that as Δx becomes smaller and smaller, the line segment passing through points p and q approaches the tangent line to point p. Thus, the slope of the function at x_0 is given by Equation 9.1 when Δx is infinitesimally small.

Equation 9.1 is an expression for slope of general applicability. It is called a *derivative*, and its application is a process known as *differentiation*. For the function $y = f(x)$, the symbolism dy/dx or $f'(x)$ is most often used to denote the derivative.

As an example of the process of differentiation utilizing Equation 9.1, consider the function $y = f(x) = 8x - x^2$. Substituting into Equation 9.1 gives

$$\begin{aligned}
\lim_{\Delta x \to 0} &= \frac{[8(x + \Delta x) - (x + \Delta x)^2] - [8x - x^2]}{\Delta x} \\
&= \frac{8x + 8\Delta x - x^2 - 2x\,\Delta x - \Delta x^2 - 8x + x^2}{\Delta x} \\
&= \frac{8\Delta x - 2x\,\Delta x - \Delta x^2}{\Delta x} \\
&= 8 - 2x - \Delta x
\end{aligned}$$

As $\Delta x \to 0$ the derivative, $dy/dx = 8 - 2x$, the slope at any point on the function. For example, at $x = 4$, $dy/dx = 0$, indicating that the slope of the function is zero at $x = 4$.

An alternative example of the process of differentiation, not tied to Equation 9.1, might be useful to consider. Begin with a function of some form such as $y = f(x) = 3x^2 + x + 2$. Let x increase by an amount Δx. Then y will increase by an amount Δy as shown in Figure 9.2, giving

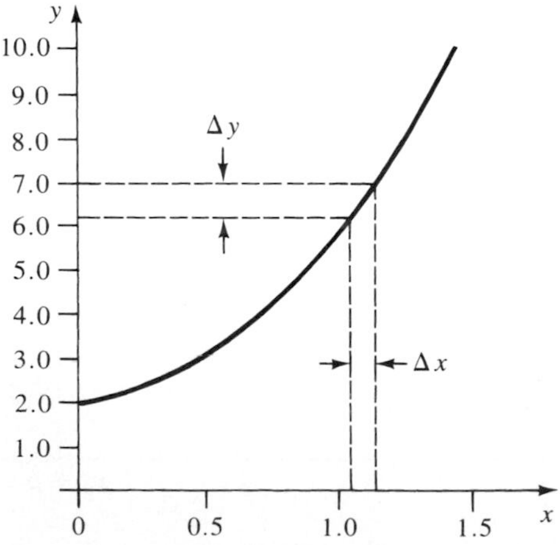

Figure 9.2 Function $3x^2 + x + 2$.

$$y + \Delta y = 3(x + \Delta x)^2 + (x + \Delta x) + 2$$
$$= 3[x^2 + 2x(\Delta x) + (\Delta x)^2] + (x + \Delta x) + 2$$
$$\Delta y = 3[x^2 + 2x(\Delta x) + (\Delta x)^2] + (x + \Delta x) + 2 - y$$

but $y = 3x^2 + x + 2$, giving

$$\Delta y = 3x^2 + 6x(\Delta x) + 3(\Delta x)^2 + x + \Delta x + 2 - 3x^2 - x - 2$$
$$= 6x(\Delta x) + 3(\Delta x)^2 + \Delta x$$

The average rate of change of Δy with respect to Δx is

$$\frac{\Delta y}{\Delta x} = \frac{6x(\Delta x)}{\Delta x} + \frac{3(\Delta x)^2}{\Delta x} + \frac{\Delta x}{\Delta x}$$
$$= 6x + 3(\Delta x) + 1$$

But because

$$\lim_{\Delta x \to 0} = \frac{dy}{dx} = f'(x)$$

the instantaneous rate of change is $6x + 1$.

Instead of proceeding as earlier for each case encountered, certain rules of differentiation have been developed for a range of common functional forms. Some of these are

1. Derivative of a constant: If $f(x) = k$, a constant, then $dy/dx = 0$.
2. Derivative of a variable: If $f(x) = x$, a variable, then $dy/dx = 1$.
3. Derivative of a straight line: If $f(x) = ax + b$, a straight line, then $dy/dx = a$.
4. Derivative of a variable raised to a power: If $f(x) = x^n$, a variable raised to a power, then $dy/dx = nx^{n-1}$.
5. Derivative of a constant times a function: If $f(x) = k[g(x)]$, a constant times a function, then

$$\frac{dy}{dx} = k\left[\frac{dg(x)}{dx}\right]$$

6. Derivative of the sum or difference of two functions: If $f(x) = g(x) \pm h(x)$, a sum or difference of two functions, then

$$\frac{dy}{dx} = \frac{dg(x)}{dx} \pm \frac{dh(x)}{dx}$$

7. Derivative of the product of two functions: If $f(x) = g(x)h(x)$, a product of two functions then

$$\frac{dy}{dx} = g(x)\frac{dh(x)}{dx} + h(x)\frac{dg(x)}{dx}$$

8. Derivative of an exponential function: If $f(x) = e^x$, an exponent of e, then

$$\frac{dy}{dx} = e^x$$

9. Derivative of the natural logarithm: If $f(x) = \ln x$, natural log of x, then

$$\frac{dy}{dx} = \frac{1}{x}$$

The methods presented previously may be used to find *higher-order* derivatives. When a function is differentiated, another function results that may also be differentiated if desired. For example, the derivative of $8x - x^2$ was $8 - 2x$. This was the slope of the function. If the slope is differentiated, the higher-order derivative is found to be -2. This is the rate of change in dy/dx and is designated d^2y/dx^2, of $f''(x)$. The process of differentiation can be extended to even higher orders as long as there remains some function to differentiate.

Partial Differentiation

Partial differentiation is the process of finding the rate of change whenever the dependent variable is a function of more than one independent variable. In the process of partial differentiation, all variables except the dependent variable, y, and the chosen independent variable are treated as constants and differentiated as such. The symbolism $\partial y/\partial x_1$ is used to represent the partial derivative of y with respect to x_1.

As an example, consider the following function of two independent variables, x_1 and x_2:

$$4x_1^2 + 6x_1x_2 - 3x_2$$

The partial derivative of y with respect to x_1 is

$$\frac{\partial y}{\partial x_1} = 8x_1 + 6x_2 - 0$$

The partial derivative of y with respect to x_2 is

$$\frac{\partial y}{\partial x_2} = 0 + 6x_1 - 3$$

Partial differentiation permits a view of the slope along one dimension at a time of an n-dimensional function. By combining this information from such a function, the slope at any point on the function can be found. Of particular interest is the point on the function where the slope along all dimensions is zero. For the previous function, this point is found by setting the partial derivatives equal to zero and solving for x_1 and x_2 as follows:

$$8x_1 + 6x_2 = 0$$

$$6x_1 - 3 = 0$$

from which

$$x_1 = 1/2 \text{ and } x_2 = -2/3$$

The meaning of this point is explained in the paragraphs that follow.

Unconstrained Optimization

Unconstrained optimization simply means that no constraints are placed on the function under consideration. Under this assumption a *necessary* condition for x^* to be an optimum point on $f(x)$ is that the first derivative be equal to zero. This is stated as

$$\left.\frac{df(x)}{dx}\right|_{x=x^*} = 0 \tag{9.2}$$

The first derivative being zero at the stationary point x^* is not *sufficient*, because it is possible for the derivative to be zero at a point of inflection on $f(x)$. It is sufficient for x^* to be an optimum point if the second derivative at x^* is positive or negative; x^* being a minimum if the second derivative is positive and a maximum if the second derivative is negative.

If the second derivative is also zero, a higher-order derivative is sought until the first nonzero one is found at the nth derivative as

$$\left.\frac{d^n f(x)}{dx^n}\right|_{x=x^*} = 0 \tag{9.3}$$

If n is odd, then x^* is a point of inflection. If n is even and if

$$\left.\frac{d^n f(x)}{dx^n}\right|_{x=x^*} < 0 \tag{9.4}$$

then x^* is a local maximum. But if

$$\left.\frac{d^n f(x)}{dx^n}\right|_{x=x^*} > 0 \tag{9.5}$$

then x^* is a local minimum.

As an example, suppose that the minimum value of x is sought for the function $f(x) = 2x^3 - 3x^2 + 1$ exhibited in Figure 9.3. The necessary condition for a minimum to exist is

$$\frac{df(x)}{dx} = 6x^2 - 6x = 0$$

Both $x = 0$ and $x = 1$ satisfy the previous condition. Accordingly, the second derivative is taken as

$$\frac{d^2 f(x)}{dx^2} = 12x - 6$$

and evaluated at $x = 0$ and $x = 1$. At $x = 0$ the second derivative is -6, indicating that this is a maximum point. At $x = 1$ the second derivative is 6, indicating that this is the minimum sought. These points are shown on Figure 9.3.

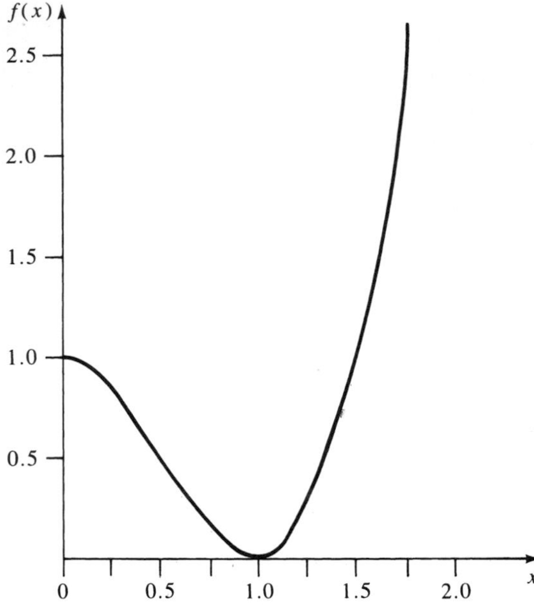

Figure 9.3 Function $2x^3 - 3x^2 + 1$.

For functions of more than one independent variable, the vector $x^* = [x_1, x_2, \ldots, x_n]$ will have a stationary point on a function $f(x)$ if all elements of the vector of first partial derivatives evaluated at x^* are zero. Symbolically,

$$F(x) = \left[\frac{\partial F(x)}{\partial x_1}, \frac{\partial F(x)}{\partial x_2}, \ldots, \frac{\partial F(x)}{\partial x_n}\right] = 0 \tag{9.6}$$

Consider a function in two variables, x_1 and x_2, given as

$$f(x_1, x_2) = x_1^2 - 8x_2 + 2x_2^2 - 6x_1 + 30$$

The necessary condition requires that

$$\frac{\partial f(x_1, x_2)}{\partial x_1} = 2x_1 - 6 = 0$$

and

$$\frac{\partial f(x_1, x_2)}{\partial x_2} = 4x_2 - 8 = 0$$

The optimal value x_1^* is found to be 3 and the optimal value x_2^* is found to be 2. These may be substituted into $f(x_1, x_2)$ to give the optimal value.

$$f(x_1, x_2) = (3)^2 - 8(2) + 2(2)^2 - 6(3) + 30 = 13$$

To determine whether this value for $f(x_1, x_2)$ is a maximum or a minimum, the second partials are required. These are

$$\frac{\partial(2x_1 - 6)}{\partial x_1} = 2 \qquad \text{and} \qquad \frac{\partial(4x_2 - 8)}{\partial x_2} = 4$$

Each is greater than zero, indicating that a minimum may have been found. To verify this, the following relationship must be satisfied:

$$\left(\frac{\partial^2 y}{\partial x_1^2}\right)\left(\frac{\partial^2 y}{\partial x_2^2}\right) - \left(\frac{\partial^2 y}{\partial x_1 \partial x_2}\right)^2 > 0 \tag{9.7}$$

where

$$\frac{\partial^2 y}{\partial x_1 \partial y_2} = \frac{\partial(\partial y/\partial x_1)}{\partial x_2}$$

Because

$$\frac{\partial y}{\partial x_1} = 2x_1 - 6$$

$$\frac{\partial^2 y}{\partial x_1 \partial x_2} = \frac{\partial(2x_1 - 6)}{\partial x_2} = 0$$

Substituting back into Equation 9.7 gives $(2)(4) - 0 = 8$, verifying that the value found for $f(x_1, x_2)$ is a minimum.

9.2 UNCONSTRAINED CLASSICAL OPTIMIZATION

There are several economic decision situations in design and operations to which optimization approaches must be applied before alternatives can be compared equivalently. Many economic decisions are characterized by two or more cost factors that are affected differently by common design or policy variables. Certain costs may vary directly with an increase in the value of a variable, whereas others may vary inversely. When the total cost of an alternative is a function of increasing and decreasing cost components, a value may exist for the common variable, or variables, which will result in a minimum cost for the alternative. This section presents selected examples from classical situations which illustrate the application of unconstrained optimization theory in design and operations.

Optimizing Investment or First Cost

Investment (or first cost) has a significant effect on the equivalent life-cycle cost of an undertaking. Accordingly, it is important that investment cost be minimized before a life-cycle cost analysis is performed. In this section the classical situation of bridge design will be used to illustrate the appropriate approach.

Before engaging in detail design, it is important to optimally allocate the anticipated capital investment to superstructure and to piers. This can be accomplished by recognizing that there exists an inverse relationship between the cost of superstructure and the number of piers. As the number of piers increases, the cost of superstructure

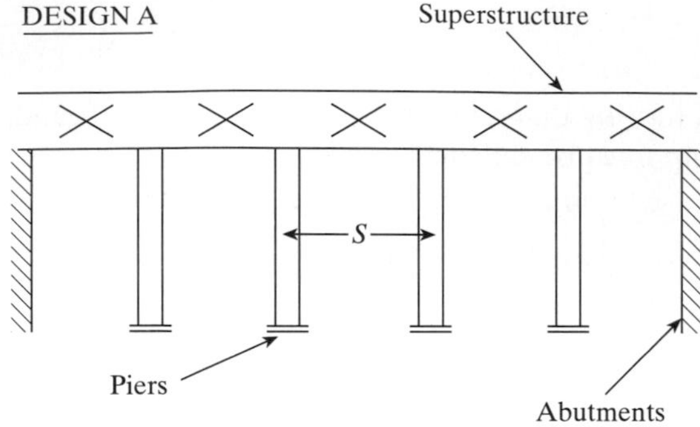

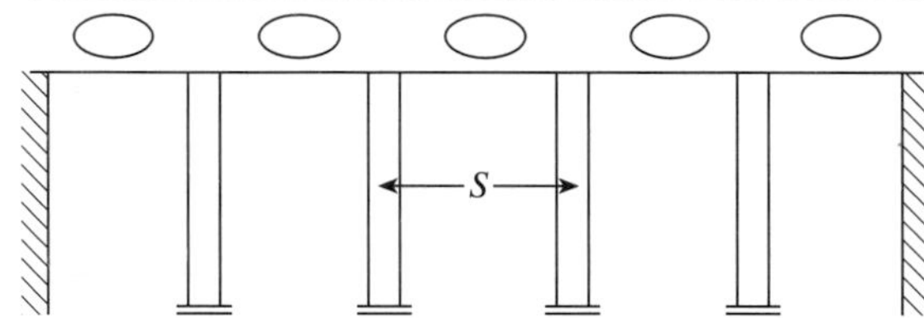

Figure 9.4 Two bridge superstructure design alternatives.

decreases. Conversely, the cost of superstructure increases as the number of piers decreases. Pier cost for the bridge is directly related to the number specified. This is a classical design situation involving increasing and decreasing cost components, the sum of which will be a minimum for a certain number of piers. Figure 9.4 illustrates two bridge superstructure designs with the span between piers indicated by S.

A general mathematical model may be derived for the evaluation of investment in superstructure and piers. The decision evaluation function of Equation 7.2 applies as

$$E = f(X, Y)$$

where

E = evaluation measure (total first cost)

X = design variable of the span between piers

Y = system parameters of the bridge length, the superstructure weight, the erected cost of superstructure per pound, and the cost of piers per pier.

Equation 7.3 is the design evaluation function applicable to the evaluation of alternative design configurations, where design-dependent parameters arise in selecting the best alternative. Let

L = bridge length (feet)

W = superstructure weight (pounds per foot)

S = span between piers (feet)

C_s = erected cost of superstructure (dollars per pound)
C_p = installed cost of piers (dollars per pier)

Assume that the weight of the superstructure is linear over a certain span range, $W = AS + B$, with the Parameters A and B having been established by a sound statistical procedure. Accordingly, the superstructure cost (SC) will be

$$\text{SC} = (AS + B)(L)(C_S) \tag{9.8}$$

The total cost of piers, PC, will be

$$\text{PC} = \left(\frac{L}{S} + 1\right)(C_P) \tag{9.9}$$

where two abutments are included as though they were piers.

The total first cost of the bridge is expressed as

$$\begin{aligned} \text{TFC} &= \text{SC} + \text{PC} \\ &= (AS + B)(L)(C_S) + \left(\frac{L}{S} + 1\right)(C_P) \\ &= ASLC_S + BLC_S + \frac{LC_P}{S} + C_P \end{aligned} \tag{9.10}$$

To find the optimum span between piers, differentiate Equation 9.10 with respect to S and equate the result to zero as follows:

$$\frac{d(\text{TFC})}{dS} = ALC_S - \frac{LC_P}{S^2} = 0$$

$$S^* = \sqrt{\frac{C_P}{AC_S}} \tag{9.11}$$

The minimum total first cost for the bridge is found by substituting Equation 9.11 for S in Equation 9.12 to obtain

$$\text{TFC}^* = 2\sqrt{AC_PL^2C_S} + BLC_S + C_P \tag{9.12}$$

Equation 9.11 can be used to find the optimal pier spacing for a given bridge design. However, to evaluate alternative bridge designs, Equation 9.12 would be used first followed by a single application of Equation 9.11 to find the best alternative. This procedure is illustrated by examples in the following sections.

Single design alternative. Assume that a bridge to serve a 1,000-foot crossing is to be fabricated from steel with a certain girder design configuration (see Figure 9.4). For this design alternative, the weight of the superstructure in pounds per foot is estimated to be linear (over a limited span range) and is expressed as

$$W = 16S + 600$$

Also, assume that the superstructure is expected to cost \$0.65 per pound erected. Piers are anticipated to cost \$80,000 each in place, and this amount will also be used as the estimated cost for each abutment.

From Equation 9.11, the optimum span between piers is found to be

$$S^* = \sqrt{\frac{\$80{,}000}{16 \times \$0.65}} = 87.7 \text{ feet}$$

This result is a theoretical spacing and must be adjusted to obtain an integer number of piers. The required adjustment gives 12 piers (11 spans) for a total cost from Equation 9.10 of $2,295,454. The span will be 90.9 feet, slightly greater than the theoretical minimum. A check can be made by considering 13 piers (12 spans) with each span being 83.3 feet. In this case, the total cost is higher ($2,296,667), so the 12-pier design would be adopted.

Total cost as a function of the pier spacing (and the number of piers) is summarized in Table 9.1. Note that a minimum occurs when 12 piers are specified; actually, 10 piers and two abutments. The optimal pier spacing is 90.9 feet.

TABLE 9.1 TFC as a Function of the Span between Piers

Span (feet)	Number of Piers	Pier Cost	Superstructure Cost	Total First Cost
142.8	8	$640,000	$1,875,714	$2,515,714
111.1	10	800,000	1,545,555	2,345,555
100.0	11	880,000	1,430,000	2,310,000
90.9	12	960,000	1,335,454	2,295,454
83.3	13	1,040,000	1,256,666	2,296,667
76.9	14	1,120,000	1,190,000	2,310,000
71.4	15	1,200,000	1,132,857	2,332,857
66.6	16	1,280,000	1,083,333	2,363,333
58.8	18	1,440,000	1,001,765	2,441,765

Multiple design alternatives. To introduce optimal design for multiple alternatives, assume that there is another superstructure configuration under consideration for the bridge design described in the preceding section (see Figure 9.4). The weight in pounds per foot for the alternative configuration is estimated from paramedic methods to be

$$W = 22S + 0$$

Assume that all other factors are the same as for the previous design.

A choice between the design alternatives is made by finding the optimum total cost for each alternative utilizing Equation 10.5. This gives $2,294,280 for Design A and $2,219,158 for Design B. Thus Design B would be chosen.

For Design B, the optimum span between piers is found from Equation 9.11 to be

$$S^* = \sqrt{\frac{\$80{,}000}{22 \times \$0.65}} = 74.8 \text{ feet}$$

and the lowest-cost integer number of piers is found to be 14 (13 spans) with a span of 76.9 feet.

Table 9.2 summarizes this design decision by giving the total cost as a function of the number of piers for each alternative. The theoretical total cost function for each alternative is exhibited in Figure 9.5 along with an indication of the number of piers.

TABLE 9.2 TFC for Two Bridge Design Alternatives

Number of Piers	Pier Cost	Superstructure Cost		Total First Cost	
		Design A	Design B	Design A	Design B
8	$640,000	$1,875,714	$2,042,857	$2,515,714	$2,682,857
10	800,000	1,545,555	1,588,889	2,345,555	2,388,889
11	880,000	1,430,000	1,430,000	2,310,000	2,310,000
12	960,000	1,335,454	1,300,000	2,295,454	2,260,000
13	1,040,000	1,256,666	1,191,666	2,296,667	2,231,666
14	1,120,000	1,190,000	1,100,000	2,310,000	2,220,000
15	1,200,000	1,132,857	1,021,438	2,332,857	2,221,438
16	1,280,000	1,083,333	953,333	2,363,333	2,233,333
18	1,440,000	1,001,765	841,176	2,441,765	2,281,176

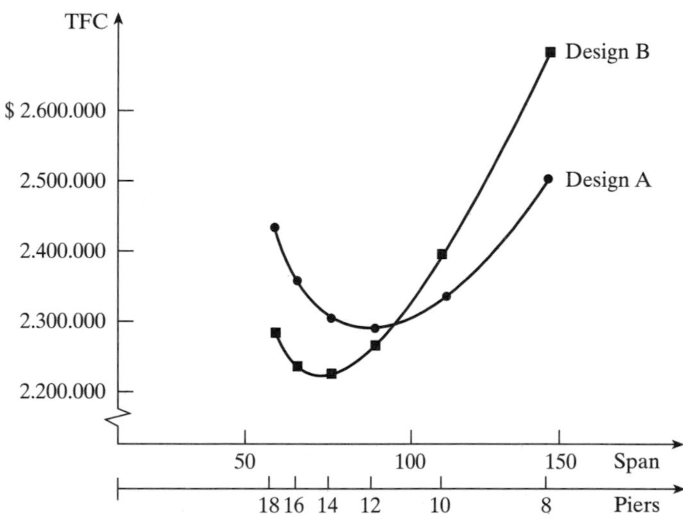

Figure 9.5 Total first cost for two bridge design alternatives.

Optimizing Life-Cycle Cost

Investment cost was optimized in the derivations and example applications just presented. When the annual cost of maintenance or operations differs between design alternatives, it is important to formulate a life-cycle cost function including both first cost and operating cost as a basis for choosing the best alternative.

Bridge evaluation including maintenance. Equation 9.12 gave the total first cost of the bridge. It can be augmented by Equation 8.7. To give the present equivalent life-cycle cost (PELCC) as follows:

$$\text{PELCC} = [2\sqrt{AC_PL^2C_S} + BLC_S + C_P] + M\left[\frac{\overset{P/A,g',n}{(\quad)}}{1+g}\right] \qquad (9.13)$$

where terms not previously defined are

n = anticipated service life of the bridge (years)
M = maintenance cost of the bridge in the first year
g = percentage increase in maintenance cost in each subsequent year

As an example of the application of Equation 9.13, assume that the two bridge design alternatives in Figure 9.4 differ in the maintenance cost to be experienced. Design A is estimated to cost \$100,000 the first year with a 5% increase in each subsequent year. Design B will cost \$90,000 to maintain in the first year with a 7% increase each subsequent year. If the interest rate is 10%, the PELCC for Design A as a function of its life, n, is

$$\text{PELCC}_A = [2\sqrt{16(\$80{,}000)(1{,}000)^2(\$0.65)} + 600(1{,}000)(\$0.65) + \$80{,}000] + \$100{,}000\left[\frac{\overset{P/A,\,g',\,n}{(\qquad)}}{1 + 0.05}\right]$$

where

$$g' = \frac{1+i}{1+g} - 1 = \frac{1+0.10}{1+0.05} - 1 = 4.76\%$$

The PELCC for Design B as a function of n is

$$\text{PELCC}_B = [2\sqrt{22(\$80{,}000)(1{,}000)^2(\$0.65)} + \$80{,}000] + \$90{,}000\left[\frac{\overset{P/A,\,g',\,n}{(\qquad)}}{1 + 0.07}\right]$$

$$g' = \frac{1+i}{1+g} - 1 = \frac{1+0.10}{1+0.07} - 1 = 2.80\%$$

Assuming a service life of 40 years ($n = 40$), the present equivalent cost for Design A is

$$\text{PELCC}_A = \$2{,}294{,}281 + \$100{,}000\left[\frac{\overset{P/A,\,4.76,\,40}{(17.7382)}}{1.05}\right] = \$3{,}983{,}633$$

And, for Design B it is

$$\text{PELCC}_B = \$2{,}219{,}159 + \$90{,}000\left[\frac{\overset{P/A,\,2.80,\,40}{(23.8807)}}{1.07}\right] = \$4{,}227{,}816$$

From this, it appears that Design A is best.

A break-even illustration of the present equivalent life-cycle cost of Designs A and B is exhibited in Figure 9.6. From this illustration, it is evident that Design A becomes better than Design B after 21 years. Thus the choice of Design A is confirmed.[1]

[1] Note that Design B was preferred when only investment cost is considered (see Figure 9.5 and Table 9.2). The inclusion of maintenance cost leads to a clear preference for Design A, illustrating the importance of life-cycle thinking.

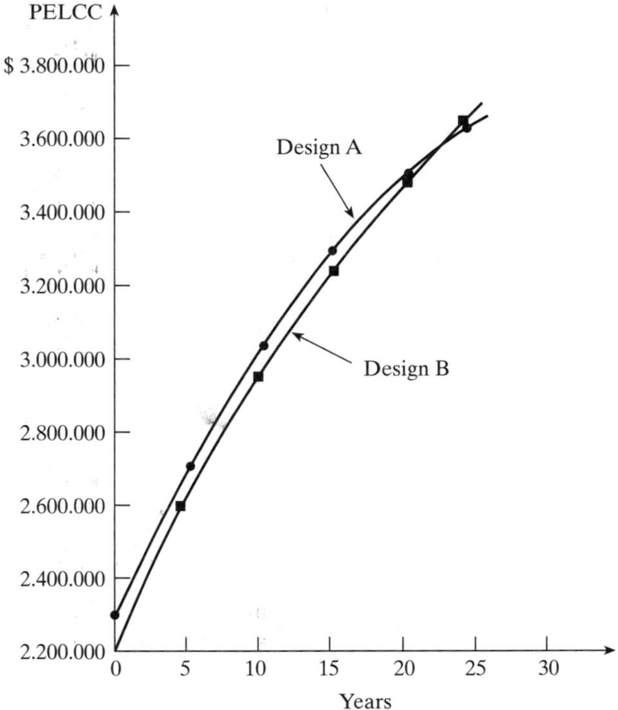

Figure 9.6 Cumulative present equivalent life-cycle cost for two bridge design alternatives.

Optimum Equipment Life

The optimum service life of an asset may be projected from knowledge of its first cost and by estimating maintenance and operation cost. Generally, the objective is to minimize the annual equivalent cost over the service life. This optimum life is also known as the minimum cost service life or the optimum replacement interval.

In this section, a general model is developed for finding the optimum life of an asset. This model is then applied to a determination of the economic life of an existing asset and the selection of an asset from among competing design alternatives. In each case, the life-cycle cost provides the basis for decision.

Evaluation by mathematical approach. Operating and maintenance costs generally increase with an increase in the age of equipment. This rising trend is offset by a decrease in the annual equivalent cost of capital recovery with return. Thus, for some age, there will be a minimum life-cycle cost or a minimum-cost life.

If the time value of money is neglected, the average annual life-cycle cost of an asset with constantly increasing maintenance cost is

$$\text{ALCC} = \frac{P}{n} + O + (n - 1)\frac{M}{2} \tag{9.14}$$

where

ALCC = average annual life-cycle cost
P = first cost of the asset
O = annual constant portion of operating cost (equal to first-year operation cost, of which maintenance is a part)
M = amount by which maintenance costs increase each year
n = life of the asset (years)

The minimum-cost life can be found mathematically using differential calculus as

$$\frac{d\text{ALCC}}{dn} = \frac{P}{n^2} + \frac{M}{2} = 0$$

$$n^* = \sqrt{\frac{2P}{M}} \tag{9.15}$$

Therefore, minimum average life-cycle cost is given by

$$\begin{aligned}\text{ALCC}^* &= P\sqrt{\frac{M}{2P}} + O + \left(\sqrt{\frac{2P}{M}} - 1\right)\frac{M}{2} \\ &= \sqrt{\frac{PM}{2}} + O + \sqrt{\frac{PM}{2}} - \frac{M}{2} \\ &= 2\sqrt{\frac{PM}{2}} + O - \frac{M}{2} \\ &= \sqrt{2PM} + O - M/2 \end{aligned} \tag{9.16}$$

As an example of the application of Equation 9.16, consider an asset with a first cost of \$18,000, a salvage value of zero at any age, a first year operating and maintenance cost of \$4,000, and with maintenance costs increasing by \$1,000 in each subsequent year. The minimum-cost life from Equation 9.15 is

$$n^* = \sqrt{\frac{2(\$18{,}000)}{\$1{,}000}} = 6 \text{ years}$$

The minimum total cost at this life is found from Equation 9.16 as

$$\text{ALCC}^* = \sqrt{2(\$18{,}000)(\$1{,}000)} + \$4{,}000 - \$500 = \$9{,}500$$

This situation can also be seen by tabulating the cost components and the average life-cycle cost as shown in Table 9.3. The minimum life-cycle cost occurs at a life of 6 years. Figure 9.7 illustrates the nature of the increasing and decreasing cost components and the optimum life of 6 years.

Consider an alternative equipment configuration that is more reliable than the previous configuration, leading to operating and maintenance costs in the first year of \$3,200, and with an increase each subsequent year equal to \$800. This configuration has a first cost of \$25,600. The minimum cost life from Equation 9.15 is

$$n^* = \sqrt{\frac{2(\$25{,}600)}{\$800}} = 8 \text{ years.}$$

TABLE 9.3 Average Annual Life-Cycle Cost of an Asset (First Configuration)

End of Year	Maintenance Cost at End of Year	Cumulative Maintenance Cost	Average Maintenance Cost	Average Capital Cost	Average Life-Cycle Cost
A	B	$C = \Sigma B$	$D = C/A$	$E = \$18{,}000/A$	$F = D + E$
1	\$4,000	\$4,000	\$4,000	\$18,000	\$22,000
2	5,000	9,000	4,500	9,000	13,500
3	6,000	15,000	5,000	6,000	11,000
4	7,000	22,000	5,500	4,500	10,000
5	8,000	30,000	6,000	3,600	9,600
6	9,000	39,000	6,500	3,000	9,500
7	10,000	49,000	7,000	2,571	9,571
8	11,000	60,000	7,500	2,250	9,750
9	12,000	72,000	8,000	2,000	10,000

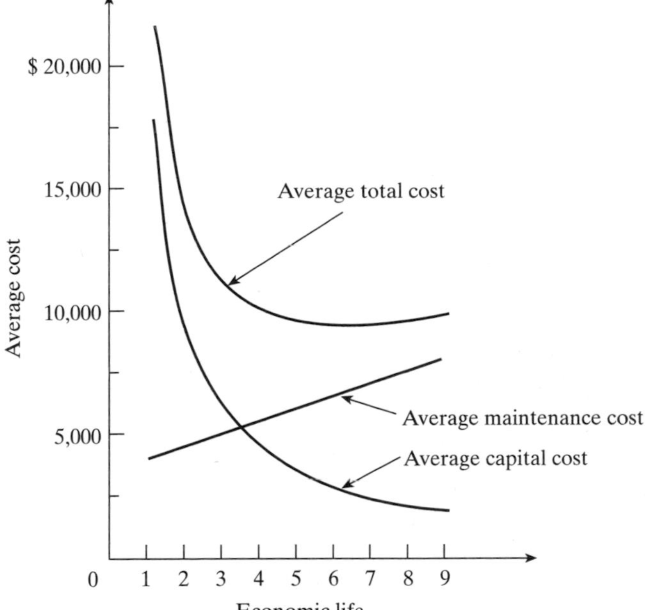

Figure 9.7 Component and LCC for an asset (first configuration).

The minimum life-cycle cost at this age is found from Equation 9.16 as

$$\text{ALCC}^* = \sqrt{2(\$25{,}600)(\$800)} + \$3{,}200 - \$400 = \$9{,}200$$

This situation can be tabulated as before to illustrate the cost components and the average life-cycle cost. This is shown in Table 9.4. The minimum total cost occurs at a life of eight years and is \$9,200. Figure 9.8 illustrates the nature of the increasing and decreasing cost components and the optimum life of 8 years. This is the preferred configuration from an average life-cycle cost standpoint and from the standpoint of service life.

TABLE 9.4 Average Annual Life-Cycle Cost of an Asset (Alternative Configuration)

End of Year	Maintenance Cost at End of Year	Cumulative Maintenance Cost	Average Maintenance Cost	Average Capital Cost	Average Life-Cycle Cost
A	B	$C = \Sigma B$	$D = C/A$	$E = \$25{,}600/A$	$F = D - E$
1	$3,200	$3,200	$3,200	$25,600	$28,800
2	4,000	7,200	3,600	12,800	16,400
3	4,800	12,000	4,000	8,533	12,533
4	5,600	17,600	4,400	6,400	10,800
5	6,400	24,000	4,800	5,120	9,920
6	7,200	31,200	5,200	4,267	9,467
7	8,000	39,200	5,600	3,657	9,257
8	8,800	48,000	6,000	3,200	9,200
9	9,600	57,600	6,400	2,844	9,244
10	10,400	68,000	6,800	2,560	9,360
11	11,200	79,200	7,200	2,327	9,527
12	12,000	91,200	7,600	2,133	9,733

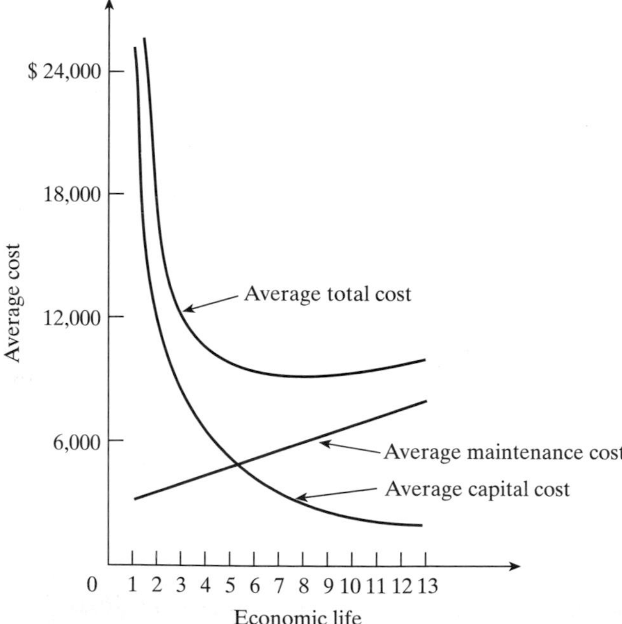

Figure 9.8 Component and LCC for an asset (alternative configuration).

Evaluation by tabular approach. Often it is not possible to find the minimum-cost life mathematically. When nonlinearities enter and the time value of money is taken into consideration, a tabular approach must be used. Consider the following example. The economic future of an asset with a first cost of $15,000, with linearly decreasing salvage values and with operating costs beginning at $1,000, increasing by $300 in the second year and thereafter (the increase each year is $55 more than the increase in the preceding year), and with an interest at a rate of 12%, is shown in Table 9.5. To find the

TABLE 9.5 Tabular Calculations for Optimum Life

End of Year	Salvage Value When Asset Retired	Operating Costs	Annual Equivalent Cost of Asset	Annual Equivalent Operating Cost	Annual Equivalent LCC
1	$12,000	$1,000	$4,800	$1,000	$5,800
2	9,000	1,300	4,630	1,142	5,772
3	6,000	1,655	4,468	1,294	5,762
4	3,000	2,065	4,310	1,455	5,765
5	0	2,530	4,161	1,624	5,785

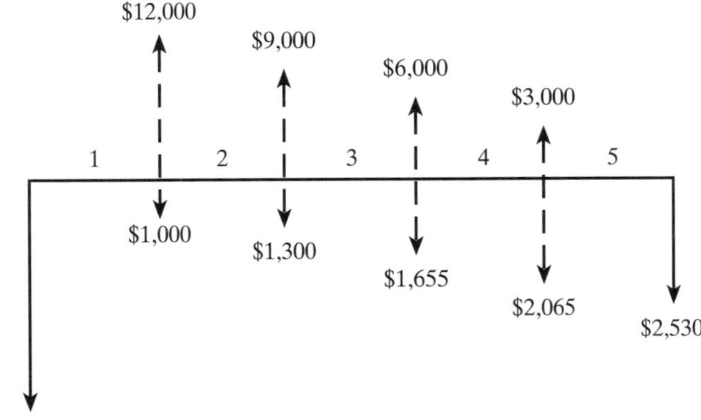

Figure 9.9 Money flows for an asset (first configuration).

asset's optimum life, it is necessary to identify the relevant money flows associated with retaining the asset 1, 2, 3, 4, or 5 years. These flows are depicted in Figure 9.9 and are the basis for the annual equivalent life-cycle cost calculations shown in Table 9.5.

Table 9.5 illustrates the tabular method for determining the economic life of an asset. If the asset were retired after 3 years, it would have a minimum annual equivalent life-cycle cost of $5.762, and this is the life most favorable for comparison purposes. Suppose that the need for the asset discussed and tabulated in Table 9.5 can be met by an alternative physical configuration. The initial investment needed for the alternative asset is $12,500, again with linearly decreasing salvage values and with operating costs beginning at $1,600, increasing by $200 in the second year and thereafter the increase each year is $100 more than the increase in the preceding year, and with an interest rate of 12%. To find this asset's optimum life, it is necessary to identify as before, the relevant money flows. These money flows are depicted in Figure 9.10 and are the basis for the annual equivalent life-cycle cost calculations shown in Table 9.6.

In this example, both the first case and the alternative have an optimum life of 3 years. However, the alternative is to be preferred in that the annual equivalent life-cycle cost is lower by $5,762 − $5,538 = $224.

TABLE 9.6 Tabular Calculations for Optimum Life (Alternative)

End of Year	Salvage Value When Asset Retired	Operating Costs	Annual Equivalent Cost of Asset	Annual Equivalent Operating Cost	Annual Equivalent LCC
1	$10,000	$1,600	$4,000	$1,600	$5,600
2	7,500	1,800	3,859	1,694	5,553
3	5,000	2,100	3,723	1,815	5,538
4	2,500	2,500	3,592	1,958	5,550
5	0	3,000	3,468	2,122	5,590

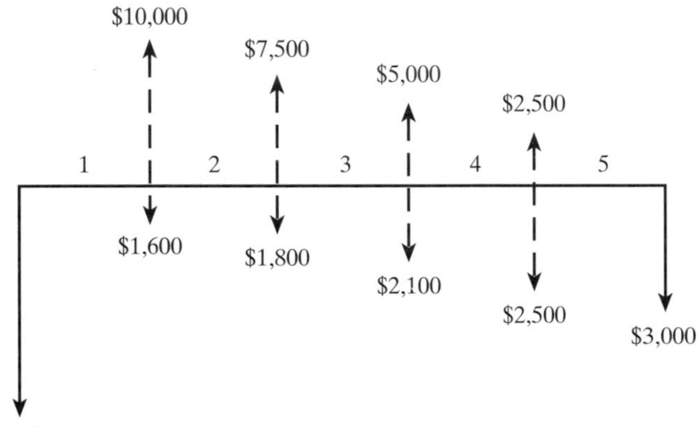

Figure 9.10 Money flows for an asset (alternative configuration).

Optimizing Procurement and Inventory Operations

The bridge design and equipment selection examples illustrated the application of classical optimization theory to systems having structure without activity. In this section, classical optimization theory is applied to inventory systems with single-item, single-source, and multisource characteristics. Here the objective is to determine the optimum procurement and inventory policy so that the system will operate at minimum cost over time.

When to procure and how much to procure are policy variables of importance in the single-source system. The decision evaluation function of Equation 7.2 applies as

$$E = f(X, Y)$$

where

E = evaluation measure (total system cost)
X = policy variables of when to procure and how much to procure
Y = system parameters of demand, procurement lead time, replenishment rate, item cost, procurement cost, holding cost, and shortage cost

Equation 7.3 is the decision evaluation function applicable to the multisource system, where source dependent parameters are the key to determining the best procurement source. Both versions of the function are used in their unconstrained forms in this section.

Inventory decision evaluation model. A mathematical model can be formulated for the general deterministic case (demand and lead time constant) from the inventory flow geometry shown in Figure 9.11. The following symbolism will be adopted:

TC = total system cost per period
L = procurement level
Q = procurement quantity
D = demand rate in units per period
T = lead time in periods
N = number of periods per cycle
R = replenishment rate in units per period
C_i = item cost per unit
C_p = procurement cost per procurement
C_h = holding cost per unit per period
C_s = shortage cost per unit short per period

It is assumed that the replenishment rate is greater than the demand rate and that unsatisfied demand is not lost. From Figure 9.11 the number of periods per cycle can be expressed as

$$N = \frac{Q}{D} \tag{9.17}$$

Also, the following relationships are evident:

$$(n_1 + n_2)(R - D) = (n_3 + n_4)D \tag{9.18}$$

$$n_1 + n_2 = \frac{Q}{R} \tag{9.19}$$

and

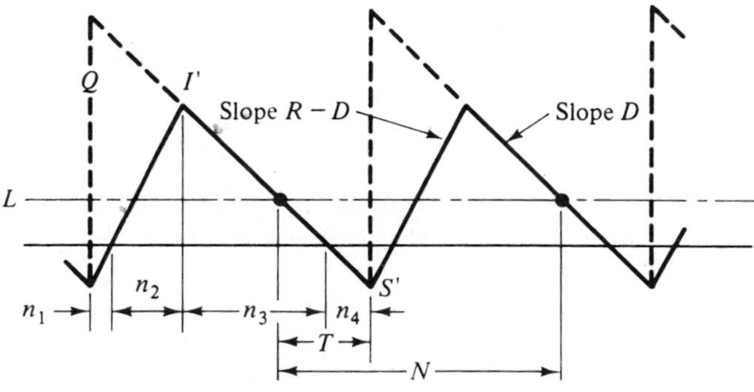

Figure 9.11 Deterministic inventory system geometry.

$$n_3 + n_4 = \frac{I' + DT - L}{D} \tag{9.20}$$

From Equations 9.18, 9.19, and 9.20,

$$I' = Q\left(1 - \frac{D}{R}\right) + L - DT \tag{9.21}$$

The total number of unit periods of stock on hand during the inventory cycle, I, is

$$I = \frac{I'}{2}(n_2 + n_3)$$

$$= \frac{I'^2}{2(R - D)} + \frac{I'^2}{2D} \tag{9.22}$$

Substituting Equation 9.21 for I' gives

$$I = \frac{[Q(1 - D/R) + L - DT]^2}{2}\left(\frac{1}{R - D} + \frac{1}{D}\right) \tag{9.23}$$

The total number of unit periods of shortage during the cycle, S, is

$$S = \frac{S'}{2}(n_1 + n_4)$$

$$= \frac{S'^2}{2(R - D)} + \frac{S'^2}{2D}$$

But because $S' = DT - L$,

$$S = \frac{(DT - L)^2}{2}\left(\frac{1}{R - D} + \frac{1}{D}\right) \tag{9.24}$$

The total system cost per period will be a summation of the item cost for the period, the procurement cost for the period, the holding cost for the period, and the shortage cost for the period, or

$$TC = IC + PC + HC + SC \tag{9.25}$$

Item cost for the period will be the product of the item cost per unit and the demand rate in units per period, or

$$IC = C_i D \tag{9.26}$$

The procurement cost for the period will be the procurement cost per procurement divided by the number of periods per inventory cycle, or

$$PC = \frac{C_p}{N}$$

$$= \frac{C_p D}{Q} \tag{9.27}$$

Holding cost for the period will be the product of the holding cost per unit per period and the average number of units on hand during the period, or

$$HC = \frac{C_h I}{N}$$

$$= \frac{C_h D}{Q}\left\{\frac{[Q(1 - D/R) + L - DT]^2}{2}\left(\frac{1}{R - D} + \frac{1}{D}\right)\right\}$$

Note the existence of the following relationship:

$$\frac{D}{Q}\left(\frac{1}{R - D} + \frac{1}{D}\right) = \frac{1}{Q(1 - D/R)} \tag{9.28}$$

Using Equation 9.34, the holding-cost component becomes

$$HC = \frac{C_h}{2Q(1 - D/R)}\left[Q\left(1 - \frac{D}{R}\right) + L - DT\right]^2 \tag{9.29}$$

Shortage cost for the period will be the product of the shortage cost per unit short per period and the average number of units short during the period, or

$$SC = \frac{C_s S}{N}$$

$$= \frac{C_s D}{Q}\left[\frac{(DT - L)^2}{2}\left(\frac{1}{R - D} + \frac{1}{D}\right)\right] \tag{9.30}$$

Equation 9.30 may be written as

$$SC = \frac{C_s(DT - L)^2}{2Q(1 - D/R)} \tag{9.31}$$

The total system cost per period will be a summation of the four cost components developed above and may be expressed as

$$TC = C_i D + \frac{C_p D}{Q} + \frac{C_h}{2Q(1 - D/R)}\left[Q\left(1 - \frac{D}{R}\right) + L - DT\right]^2 + \frac{C_s(DT - L)^2}{2Q(1 - D/R)} \tag{9.32}$$

The minimum-cost procurement level and procurement quantity may be found by setting the partial derivatives equal to zero and solving the resulting equations. Modifying Equation 9.32 gives

$$TC = C_i D + \frac{C_p D}{Q} + \frac{C_h Q(1 - D/R)}{2} - C_h(DT - L) + \frac{C_h(DT - L)^2}{2Q(1 - D/R)} + \frac{C_s(DT - L)^2}{2Q(1 - D/R)} \tag{9.33}$$

By taking the partial derivative of Equation 9.43 with respect to Q and then with respect to $DT - L$, and setting both equal to zero, one obtains

$$\frac{\partial TC}{\partial Q} = -\frac{C_p D}{Q^2} + \frac{C_h(1 - D/R)}{2} - \frac{C_h(DT - L)^2}{2Q^2(1 - D/R)} - \frac{C_s(DT - L)^2}{2Q^2(1 - D/R)} = 0 \tag{9.34}$$

$$\frac{\partial TC}{\partial(DT - L)} = -C_h + \frac{C_h(DT - L)}{Q(1 - D/R)} + \frac{C_s(DT - L)}{Q(1 - D/R)} = 0 \tag{9.35}$$

Equation 9.35 may be expressed as

$$\frac{DT - L}{Q} = \frac{C_h(1 - D/R)}{C_h + C_s} \tag{9.36}$$

Substituting Equation 9.36 into Equation 9.34 gives

$$-\frac{C_p D}{Q^2} + \frac{C_h(1 - D/R)}{1} - \frac{C_h^2(1 - D/R)}{2(C_h + C_s)^2} - \frac{C_s C_h^2(1 - D/R)}{2(C_h + C_s)^2} = 0$$

$$\frac{C_p D}{Q^2} = \frac{C_h C_s(1 - D/R)}{2(C_h + C_s)}$$

Solving for Q gives

$$Q^* = \sqrt{\frac{1}{1 - D/R}}\sqrt{\frac{2C_p D}{C_h} + \frac{2C_p D}{C_s}} \tag{9.37}$$

Substituting Equation 9.37 into Equation 9.36 gives

$$L^* = DT - \frac{C_h(1 - D/R)}{C_h + C_s}\sqrt{\frac{1}{1 - D/R}}\sqrt{\frac{2C_p D}{C_h} + \frac{2C_p D}{C_s}}$$

$$L^* = DT - \sqrt{1 - \frac{D}{R}}\sqrt{\frac{2C_p D}{C_s(1 + C_s/C_h)}} \tag{9.38}$$

Equations 9.37 and 9.38 may now be substituted back into Equation 9.33 to give an expression for the minimum total system cost.[2] After several steps, the result is

$$TC^* = C_i D + \sqrt{1 - \frac{D}{R}}\sqrt{\frac{2C_p C_h C_s D}{C_h + C_s}} \tag{9.39}$$

Single source alternative. As an example of an inventory system with only one procurement source alternative, consider a situation in which an item having the following system parameters is to be manufactured:

[2] The policy variables, Q^* and L^*, supposedly occur at the minimum of the total cost function. Actually, there are three possibilities that may occur when the first partials of a function are set equal to zero and solved: a maximum, a minimum, or a point of inflection may be found. Although the nature of the total cost function derived here is such that a minimum is always found, a formal test could be applied using Equation 9.7 to verify that the minimum has been found.

TABLE 9.7 Total System Cost as a Function of L and Q

L \ Q	242	243	244	245	246	247	248
−6	56.399	56.394	56.390	56.386	56.383	56.380	56.377
−5	56.387	56.383	56.380	56.377	56.374	56.372	56.371
−4	56.378	56.375	56.372	56.370	56.368	56.367	56.366
−3	56.371	56.369	56.367	56.366	56.365	56.364	56.364
−2	56.366	56.365	56.364	56.363	56.363	56.364	56.364
−1	56.364	56.364	56.364	56.364	56.365	56.366	56.367
0	56.364	56.365	56.366	56.367	56.368	56.370	56.373
1	56.367	56.369	56.370	56.372	56.374	56.377	56.380
2	56.373	56.375	56.377	56.380	56.383	56.386	56.390
3	56.381	56.383	56.386	56.390	56.394	56.398	56.403
4	56.391	56.394	56.398	56.403	56.407	56.442	56.418

D = 10 units per period
R = 20 units per period
T = 8 periods
C_i = \$4.82 per unit
C_p = \$100.00 per procurement
C_h = \$0.20 per unit per period
C_s = \$0.10 per unit per period

Total system cost may be tabulated as a function of Q and L by use of Equation 9.32. The resulting values in the region of the minimum-cost point are given in Table 9.7. The minimum-cost Q and L may be found by inspection. As in the previous examples, the surface generated by the TC values is seen to be rather flat. Also, the selection of integral values of Q and L does not affect the total cost appreciably.

The minimum-cost procurement quantity and minimum-cost procurement level may be found directly by substituting into Equations 9.37 and 9.38 as follows.

$$Q^* = \sqrt{\frac{1}{1 - 10/20}}\sqrt{\frac{2(\$100.00)(10)}{\$0.20} + \frac{2(\$100.00)(10)}{\$0.10}} = 244.95$$

$$L^* = 10(8) - \sqrt{1 - \frac{10}{20}}\sqrt{\frac{2(\$100.00)(10)}{\$0.10(1 + \$0.10/\$0.20)}} = -1.65$$

The resulting total system cost at the minimum-cost procurement quantity and procurement level may be found from Equation 9.39 as

$$TC^* = \$4.82(10) + \sqrt{1 - \frac{10}{20}}\sqrt{\frac{2(\$100.00)(\$0.20)(\$0.10)(10)}{\$0.20 + \$0.10}} = \$56.363$$

Accordingly, the item should be manufactured in lots of 245 units. Procurement action should be initiated when the stock on hand is negative; that is, when the backlog is two units. The total system cost will be \$56.36 per period under these optimum operating conditions.

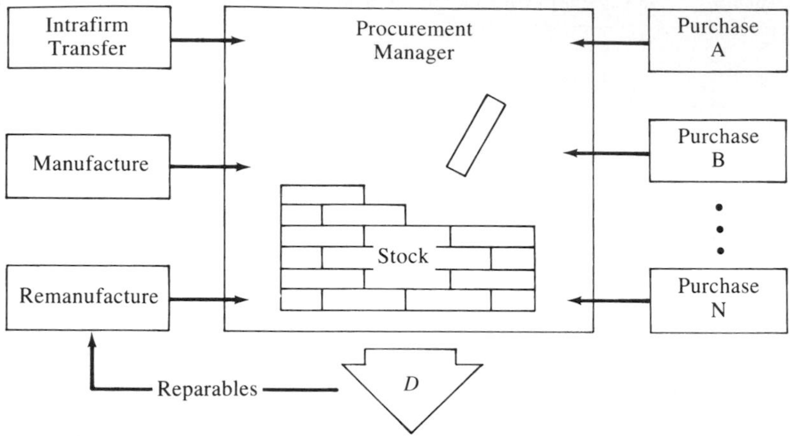

Figure 9.12 Multiple replenishment sources.

Multiple source alternatives. Consider the following situation. A single-item inventory is maintained to meet demand. When the number of units on hand falls to a predetermined level, action is initiated to procure a replenishment quantity from one of several possible sources as is illustrated in Figure 9.12. The object is to determine the procurement level, the procurement quantity, and the procurement source in the light of system and cost parameters, so that the sum of all costs associated with the system will be minimized.

The procurement lead time, rate of replenishment, item cost, and procurement cost are source dependent. Thus, the decision evaluation function of Equation 7.3 applies with the named source dependent parameters. All other system and cost parameters are source independent. Explicit recognition of source-dependent parameters permits the use of Equation 9.39 to find the minimum cost procurement source. The procedure is to compare TC^* for all sources and to choose that source that results in a minimum TC^*. Equations 9.38 and 9.37 may then be applied to find the minimum cost procurement level and minimum cost procurement quantity.

An example of the procedure is the situation in which a system manager is experiencing a demand of four units per period for an item that may be either manufactured or purchased from one of two vendors. Holding cost per unit per period is $0.24 and shortage cost per unit short per period is $0.17. Values for source-dependent parameters are given in Table 9.8.

TABLE 9.8 Source-Dependent Parameters for Multi-Source System

PARAMETER	MANUFACTURE	VENDOR A	VENDOR B
R	12.00	∞	∞
T	6.00	3.00	4.00
C_i	$19.85	$17.94	$18.33
C_p	$17.32	$18.70	$17.50

The procurement source resulting in a minimum total system cost can be found from Equation 9.39. For the Manufacturing alternative, it is

$$TC^* = \$19.85(4) + \sqrt{1 - \frac{4}{12}}\sqrt{\frac{2(\$17.32)(\$0.24)(\$0.17)(4)}{\$0.24 + \$0.17}}$$

$$= \$82.43$$

For the alternative designated Vendor A, it is

$$TC^* = \$17.94(4) + \sqrt{1 - \frac{4}{\infty}}\sqrt{\frac{2(\$18.70)(\$0.24)(\$0.17)(4)}{\$0.24 + \$0.17}}$$

$$= \$75.62$$

For the alternative designated Vendor B, it is

$$TC^* = \$18.33(4) + \sqrt{1 - \frac{4}{\infty}}\sqrt{\frac{2(\$17.50)(\$0.24)(\$0.17)(4)}{\$0.24 + \$0.17}}$$

$$= \$77.05$$

On the basis of this analysis, the alternative designated Vendor A would be chosen as the minimum cost procurement source.

The minimum cost procurement quantity for this source may be found from Equation 9.37 to be

$$Q^* = \sqrt{1 - \frac{4}{\infty}}\sqrt{\frac{2(\$18.70)(4)}{\$0.24} + \frac{2(\$18.70)(4)}{\$0.17}}$$

$$= 38.78$$

The minimum cost procurement level for this source, from Equation 9.38 is

$$L^* = 4(3) - \sqrt{1 - \frac{4}{\infty}}\sqrt{\frac{2(\$18.70)(4)}{\$0.17(1 + \$0.17/\$0.24)}}$$

$$= -10.69$$

Thus, the system manager would initiate procurement action when the stock level falls to -10.69 units, for a procurement quantity of 38.78 units, from the procurement source designated Vendor A.

9.3 CONSTRAINED CLASSICAL OPTIMIZATION

In design and operations alike, physical and economic limitations often exist which act to limit system optimization globally. These limitations arise for a variety of reasons and generally cannot be removed by the decision maker. Accordingly, there may be no choice except to find the best or optimum solution subject to the constraints.

In this section, the example applications already presented will be revisited to illustrate two physical constraint situations: span constrained bridge design and space constrained inventory system operations. Constrained optimization by linear and dynamic programming, for both physical and economic constraints is treated in Sections 9.5 and 9.6.

Span Constrained Bridge Design

Table 9.2 gave total costs as a function of the number of piers for two bridge design alternatives. Suppose that there exists a constraint on the pier spacing, expressed as a minimum spacing of 110 feet to permit the safe passage of barge traffic. With this requirement, it is evident from Table 9.2 and Figure 9.13 that Design A now replaces Design B as the minimum cost choice.

This particular constraint leads to a design decision reversal and to a penalty in total system cost. Although a reversal will not always occur, a penalty in total cost usually will. In this example, the penalty is found from Table 9.2 to be \$2,388,889 − \$2,345,555 or \$43,334.

Space Constrained Inventory Operations

Warehouse space is often a scarce resource. It may be expressed in cubic units designated W. Each unit in stock consumes a certain amount of space designated w. The units on hand require a total amount of space that must not exceed the amount avail-

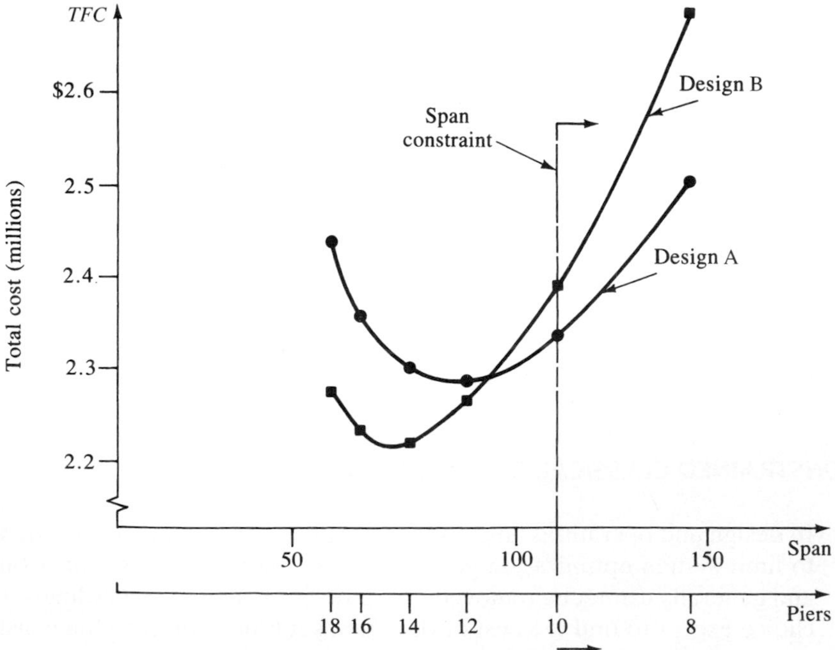

Figure 9.13 Span constrained bridge design.

able. In general, this may be formally stated as $I'w \leq W$, where I' is the maximum accumulation of stock.

The space constrained inventory model. Consider the inventory operations of Figure 9.12. The total cost for this model was given by Equation 9.32. For this model, the maximum inventory, I', was given by Equation 9.21. If $I'w \leq W$, then $I' \leq W/w$, or

$$I' = Q\left(1 - \frac{D}{R}\right) + L - DT \leq \frac{W}{w}$$

If the warehouse restriction is active, then $I' = W/w$, or

$$I' = Q\left(1 - \frac{D}{R}\right) + L - DT - \frac{W}{w} = 0 \tag{9.40}$$

For an active warehouse restriction, the minimum total cost will occur along the intersection of the surface defined by Equation 9.32 and the plane defined by Equation 9.40. The new TC equation is determined by substituting Equation 9.40 into Equation 9.32 giving

$$TC = C_iD + \frac{C_pD}{Q} + \frac{C_h(W/w)^2}{2Q(1 - D/R)} + \frac{C_s[Q(1 - D/R) - W/w]^2}{2Q(1 - D/R)} \tag{9.41}$$

After some manipulation,

$$TC = a_1 + \frac{a_2}{Q} = a_3Q \tag{9.42}$$

where

$$a_1 = C_iD - C_s\left(\frac{W}{w}\right)$$

$$a_2 = C_pD + \frac{(C_h + C_s)(W/w)^2}{2(1 - D/R)}$$

and

$$a_3 = \frac{C_s(1 - D/R)}{2}$$

The optimal value of Q can be obtained by differentiating Equation 9.42 with respect to Q, setting the result equal to zero, and solving for Q^* as

$$\frac{\partial TC}{\partial Q} = -\frac{a_2}{Q^2} + a_3 = 0$$

and

$$Q^* = \sqrt{\frac{a_2}{a_3}}$$

which can be written in terms of the original parameters as

$$Q^* = \sqrt{\frac{2C_pD}{C_s(1 - D/R)} + \frac{(C_h + C_s)(W/w)^2}{C_s(1 - D/R)^2}} \tag{9.43}$$

The optimal L^* can be determined from Equation 9.40 with $Q = Q^*$ as

$$L^* = DT + \frac{W}{w} - Q^*\left(1 - \frac{D}{R}\right) \tag{9.44}$$

Finally, the minimum total cost is given by substituting the optimal values of Q into Equation 9.41, or using Equation 9.42 with

$$TC^* = a_1 + \frac{a_2}{Q^*} + a_3Q^* \tag{9.45}$$

Space constrained single-source alternative. As an example of the application of Equations 9.40 through 9.45, suppose that a procurement manager will purchase an item having the following parameters:

D = 6 units per period
R = ∞ units per period
T = 7 periods
C_i = \$34.75 per unit
C_p = \$23.16 per procurement
C_h = \$0.30 per unit per period
C_s = \$0.30 per unit per period
W = 100 cubic units of space
w = 24 cubic units per item

With no restrictions, the optimal $Q^* = 43.04$ using Equation 9.37, $L^* = 20.48$ using Equation 9.38, and $TC^* = \$214.96$ using Equation 9.39. If $R = \infty$, Equation 9.40 can be written as

$$I'^* = Q^* + L^* - DT \tag{9.46}$$

When the optimal values are used, Equation 9.46 yields

$$I'^* = 43.04 + 20.48 - 6(7) = 21.52$$

Because $W/w = 100/24 < 21.52 = I'^*$, the warehouse restriction is active.

Next, determine the optimal policy values for the constrained warehouse where $I'^* = W/w = 100/24 = 4.17$. From Equation 9.43 with $1 - D/R \rightarrow$ unity,

$$Q^* = \sqrt{\frac{2(\$23.16)(6)}{\$0.30} + \frac{(\$0.30 + \$0.30)(4.17)^2}{\$0.30}} = 30.57$$

Next, using Equation 9.44 with $1 - D/R \rightarrow$ unity, we determine L^* as

$$L^* = 6(7) + 4.17 - 30.57 = 15.60$$

Finally, determine the optimal total cost, using Equation 9.41 with $1 - D/R \rightarrow$ unity, as

$$TC^* = \$34.75(6) + \frac{\$23.16(6)}{30.57} + \frac{\$0.30(4.17)^2}{2(30.57)} + \frac{\$0.30[6(7) - 15.17]^2}{2(30.57)} = \$216.55$$

The difference in TC at the optimum is approximately \$1.60 per period. The total cost as a function of the procurement quantity both without and with the warehouse restriction are shown in Figure 9.14.

Space constrained multiple-source alternatives. The explicit recognition of source dependent parameters permits the use of models for the constrained system to be used for the single-item, multisource system. The procedure is to evaluate each source and to find that source which will result in a minimum total cost subject to the restriction.

As an example, suppose that the system is constrained by a total warehouse space of 1000 cubic units. Also suppose that the item requires 120 cubic units of space. Determine if the restriction is active for Vendor A using Equation 9.46 as

$$I'^* = 38.7806 - 10.6917 - 4(3) = 16.0889$$

Because $W/w = 1000/120 < 16.0889 = I'^*$, the warehouse restriction is active.

Next, determine the optimal policy values for the constrained situation with $I'^* = 1000/120 = 8.3333$. From Equation 9.43 with $1 - D/R \rightarrow$ unity, Q^* is found as

$$Q^* = \sqrt{\frac{2(\$18.70)4}{\$0.17} + \frac{(\$0.24 + \$0.17)(8.3333)^2}{\$0.17}} = 32.3648$$

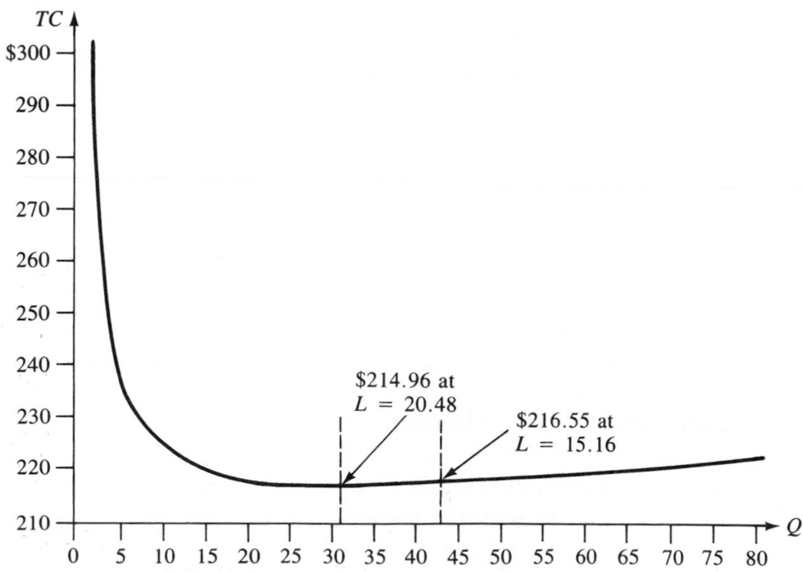

Figure 9.14 Total cost with restriction.

Using Equation 9.44 with $1 - D/R \rightarrow$ unity, L^* is found to be

$$L^* = 4(3) + 8.3333 - 32.3648 = -12.0315$$

As the last step, determine the optimal total cost using Equation 9.41 with $1 - D/R \rightarrow$ unity, as

$$TC^* = \$17.94(4) + \frac{\$18.70(4)}{32.3648} + \frac{\$0.24(8.3333)^2}{2(32.3648)} + \frac{\$0.17[(32.3648) - 8.3333]^2}{2(32.3648)} = \$75.85$$

A penalty of \$75.85 less \$75.62 = \$0.23 occurs due to the warehouse constraint.

It is possible for the restricted source choice to differ from the source choice when no restriction exists. Accordingly, the other source possibilities in the example should be evaluated in the face of the warehouse restriction. The results of this evaluation are given in Table 9.9. Inspection reveals that Vendor A remains the best choice for the case where the warehouse capacity is 1000 cubic units.

TABLE 9.9 s Optimal Policies for Restricted Multisource System

SOURCE	L^*	Q^*	TC^*
Manufacture	5.6710	40.0000	\$82.5158
Vendor A	−12.0315	32.3648	75.8454
Vendor B	−7.1470	31.4820	77.2550

9.4 EVALUATION INVOLVING OPTIMIZATION AND MULTIPLE CRITERIA

Optimization is not an end in itself. It is only a means for bringing mutually exclusive alternatives into comparable (or equivalent) states. Like money-flow modeling, it is an essential step in comparing alternatives equivalently. There is no need to go further if only economic criteria are involved.

When multiple criteria are present in a decision situation, neither economic optimization nor money flow analysis are sufficient. Although necessary, these steps must be augmented with information about the degree to which each alternative meets (or exceeds) specific criteria. One means for consolidating and displaying this information is through the approach introduced in Section 7.3. In this section, two examples of the approach are presented based on situations presented in Sections 9.2 and 9.3.

Superstructure Design Preference

Consider the superstructure design alternatives illustrated in Figure 9.4. If the public (customer) is indifferent to the superstructure configuration for the bridge, then a choice can be made on the basis of economic factors alone. Recall that Design B is preferred for the unconstrained case and Design A for the constrained. This was based only on first cost considerations.

Now suppose that the Design A configuration has some undesirable aesthetic and safety characteristics in comparison with Design B. Specifically, suppose that the aesthetics committee rates Design A at only 5 on a 10-point scale versus a 9^+ for Design B. Further, the safety officer is concerned that Design A will be more risky to youth who venture to climb within the superstructure. On a L/ML/M/MH/H scale, this design is rated M and Design B is rated MH.

The previous ratings for aesthetics and safety can now be exhibited along with total first cost as in Figure 9.15. Here the decision evaluation display shows the economic advantages identified and rated. The total first cost difference of $43,334 is no longer the only basis for the decision. Further, if aesthetics must rate at least 8 and safety greater than M, then Design A fails on the asthetics measure. It is now left to negotiation in the face of subjective evaluation as the basis for choice.

Procurement Source Preference

Manufacturing, or procurement form one of two vendors, are identified as possible procurement sources by the source-dependent parameters in Table 9.8. The total cost for each source was found from Equation 9.39 and it differs. However, suppose there are other considerations.

Manufacturing is judged to provide a source of supply without interruption and is rated 9^+ on a 10-point scale, Vendors A and B are not likely to be as dependable and a rating of 6 is applied to A and a 7 to B. As to quality, Vendor A is judged to be able

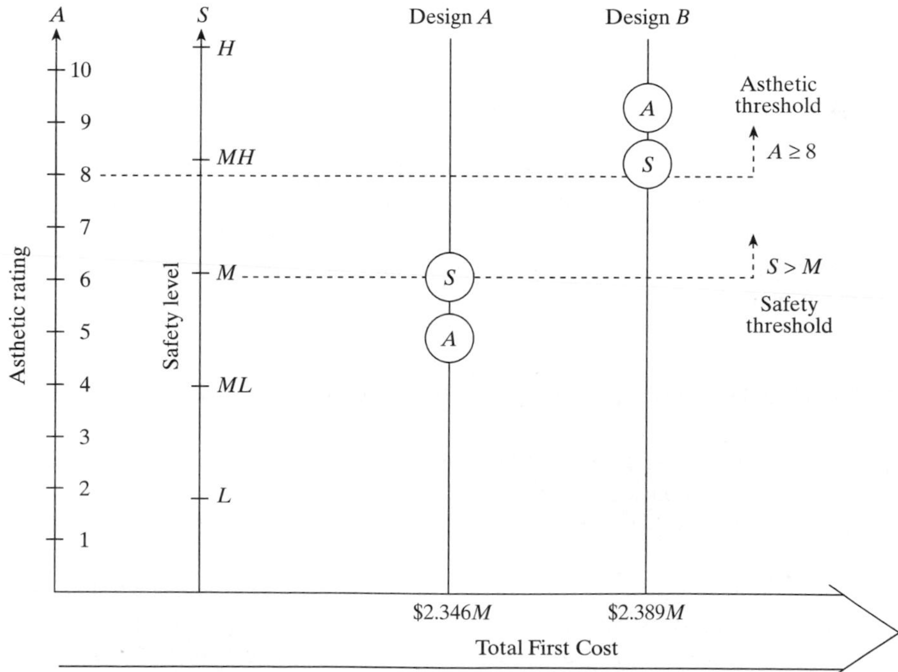

Figure 9.15 Decision evaluation display for superstructure alternatives.

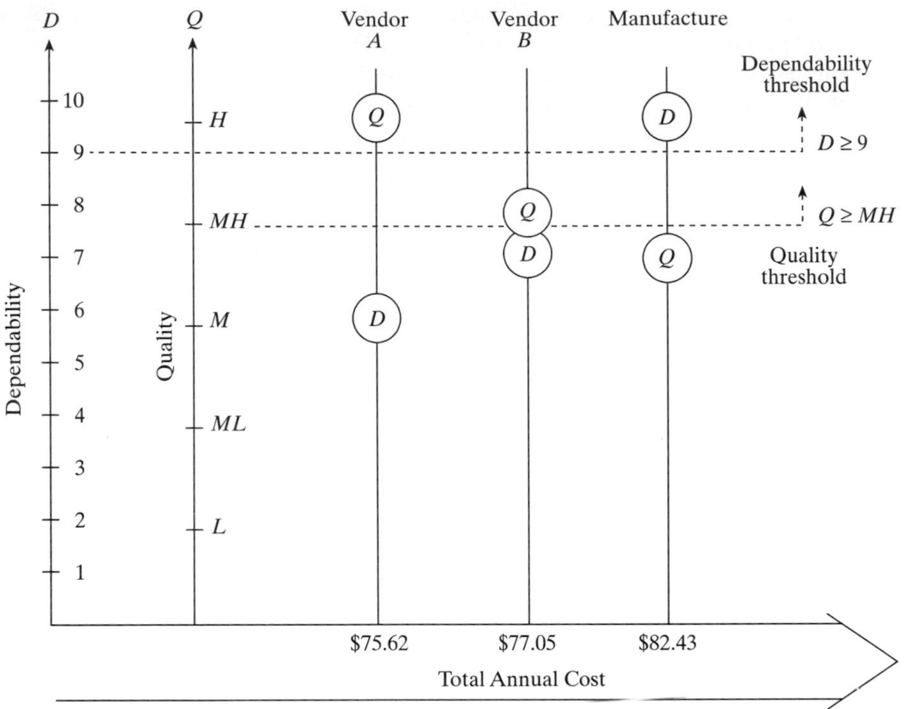

Figure 9.16 Decision evaluation display for procurement alternatives.

to provide a quality level just above H, Vendor B a level of MH^+, and manufacturing a level of just below MH. These values and the total costs are displayed in Figure 9.16.

Suppose that a dependability rating of at least 9 is desired, with a quality rating of at least MH. The decision is now reduced to a trade-off of two noneconomic factors against total cost. Choice can only be made by a knowledgeable person or group by the application of subjective evaluation and negotiation, aided by the decision evaluation display.

9.5 CONSTRAINED OPTIMIZATION BY LINEAR PROGRAMMING[3]

The general linear programming model may be stated mathematically as that of optimizing the evaluation function,

$$E = \sum_{j=1}^{n} e_j x_j \tag{9.47}$$

subject to the constraints

[3] The mathematical formulation of the general linear programming model, together with the simplex optimization algorithm, was developed by George Dantzig in 1947.

$$\sum_{j=1}^{n} a_{ij}x_j = c_i \qquad i = 1, 2, \ldots, m$$

$$x_j \geq 0 \qquad j = 1, 2, \ldots, n$$

Optimization requires either maximization or minimization, depending on the measure of evaluation involved. The decision maker has control of the vector of variables, x_j. Not directly under his or her control is the vector of evaluation coefficients, e_j, the matrix of constraints, a_{ij}, and the vector of constants, c_i. This structure is in accordance with the constrained version of the decision evaluation function in Equation 7.2.

Graphical Optimization Methods

The mathematical statement of the general linear programming model may be explained in graphical terms. There exist n variables that define an n-dimensional space. Each constraint corresponds to a hyperplane in this space. These constraints surround the region of feasible solution by hypersurfaces so that the region is the interior of a convex polyhedron. Because the evaluation function is linear in the n variables, the requirement that this function have some constant value gives a hyperplane that may or may not cut through the polyhedron. If it does, one or more feasible solutions exist. By chancing the value of this constant, a family of hyperplanes parallel to each other is generated. The distance from the origin to a member of this family is proportional to the value of the evaluation function.

Two limiting hyperplanes may be identified: one corresponds to the largest value of the evaluation function for which the hyperplane just touches the polyhedron; the other corresponds to the smallest value. In most cases, the limiting hyperplanes just touch a single vertex of the limiting polyhedron. This outermost limiting point is the solution that optimizes the evaluation function.

Although this is a graphical description, a graphical solution is not convenient when more than three variables are involved. The following examples illustrate the general linear programming model and its graphical optimization method for operations having two and three activities.

Graphical maximization for two activities. As an example of the case where two activities compete for scarce resources, consider the following production system. Two products are to be manufactured. A single unit of product A requires 2.4 min of punch press time and 5.0 min of assembly time. The profit for product A is \$0.60 per unit. A single unit of product B requires 3.0 min of punch press time and 2.5 min of welding time. The profit for product B is \$0.70 per unit. The capacity of the punch press department available for these products is 1,200 min per week. The welding department has idle capacity of 600 min per week and the assembly department can supply 1.500 min of capacity per week. The manufacturing and marketing data for this production system are summarized in Table 9.10.

In this example, two products compete for the available production time. The objective is to determine the quantity of product A and the quantity of product B to produce so that total profit will be maximized. This will require maximizing

TABLE 9.10 Manufacturing and Marketing Data for Two Products

DEPARTMENT	PRODUCT A	PRODUCT B	CAPACITY
Punch press	2.4	3.0	1,200
Welding	0.0	2.5	600
Assembly	5.0	0.0	1,500
Profit	\$0.60	\$0.70	

$$TP = \$0.60A + \$0.70B$$

subject to

$$2.4A + 3.0B \leq 1{,}200$$
$$0.0A + 2.5B \leq 600$$
$$5.0A + 0.0B \leq 1{,}500$$
$$A \geq 0 \quad \text{and} \quad B \geq 0$$

The graphical equivalent of the algebraic statement of this two-product system is shown in Figure 9.17. The set of linear restrictions define a region of feasible solutions. This region lies below $2.4A + 3.0B = 1{,}200$ and is restricted further by the requirements that $B \leq 240$, $A \leq 300$, and that both A and B be non-negative. Thus, the scarce resources determine which combinations of the activities are feasible and which are not feasible.

The production quantity combinations of A and B that fall within the region of feasible solutions constitute feasible production programs. That combination or com-

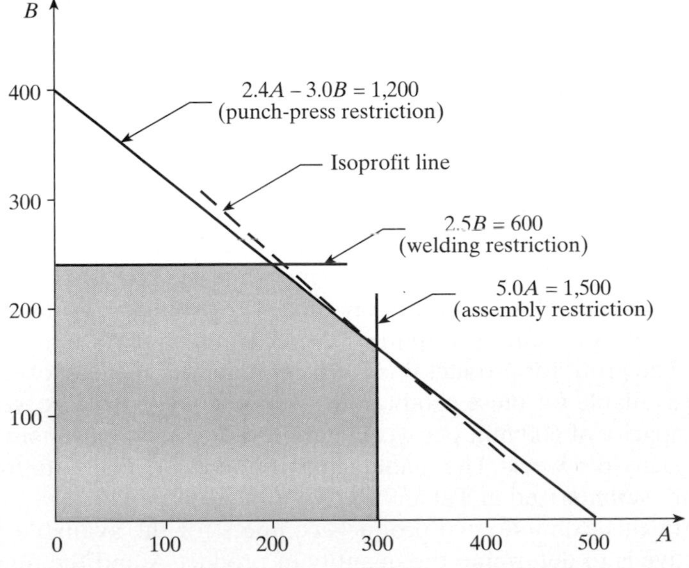

Figure 9.17 Graphical optimization for a two-product production operation.

binations of A and B which maximize profit are sought. The relationship between A and B is $1.167A = B$. This relationship is based on the relative profit of each product. The total profit realized will depend on the production quantity combination chosen. Thus, there is a family of isoprofit lines, one of which will have at least one point in the region of the feasible production quantity combinations and be a maximum distance from the origin. The member that satisfies this condition intersects the region of feasible solutions at the extreme point $A = 300$, $B = 160$. This is shown as a dashed line in Figure 9.17 and represents a total profit of $\$0.60(300) + \$0.70(160) = \$292$. No other production quantity combination would result in a higher profit.

Alternative production programs with the same profit might exist in some cases. This would occur when the isoprofit line lies parallel to one of the limiting restrictions. For example, if the relative profits of product A and product B were $A = 1.25B$, the isoprofit line in Figure 9.17 would coincide with the restriction $2.4A + 3.0B = 1{,}200$. In this case, the isoprofit line would touch the region of feasible solutions along a line instead of at a point. All production quantity combinations along the line would maximize profit.

Graphical maximization for three activities. When three activities compete for scarce resources, a three-dimensional space is involved. Each constraint is a plane in this space, and all constraints taken together identify a volume of feasible solutions. The evaluation function is also a plane, its distance from the origin being proportional to its value. The optimum value for the evaluation function occurs when this plane is located so that it is at the extreme point of the volume of feasible solutions.

As an example, suppose that the production operations for the previous system are to be expanded to include a third product, designated C. A single unit of product C will require 2.0 min of punch press time, 1.5 min of welding time, and 2.5 min of assembly time. The profit associated with product C is \$0.50 per unit. Manufacturing and marketing data for this revised production situation are summarized in Table 9.11.

In this example, three products compete for the available production time. The objective is to determine the quantity of product A, the quantity of product B, and the quantity of product C to produce so that total profit will be maximized. This will require maximizing

$$TP = \$0.60A + \$0.70B + \$0.50C$$

subject to

TABLE 9.11 Manufacturing and Marketing Data for Three Products

DEPARTMENT	PRODUCT A	PRODUCT B	PRODUCT C	CAPACITY
Punch press	2.4	3.0	2.0	1,200
Welding	0.0	2.5	1.5	600
Assembly	5.0	0.0	2.5	1,500
Profit	\$0.60	\$0.70	\$0.50	

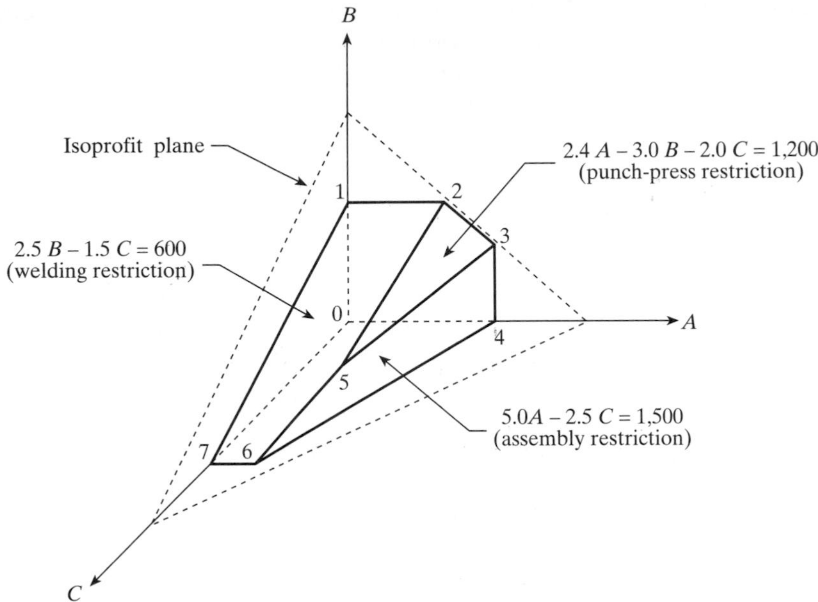

Figure 9.18 Graphical optimization for a three-product production operation.

$$2.4A + 3.0B + 2.0C \leq 1{,}200$$
$$0.0A + 2.5B + 1.5C \leq 600$$
$$5.0A + 0.0B + 2.5C \leq 1{,}500$$
$$A \geq 0, \quad B \geq 0, \quad \text{and} \quad C \geq 0$$

The graphical equivalent of the algebraic statement of this three-product production situation is shown in Figure 9.18. The set of restricting planes defines a volume of feasible solutions. This region lies below $2.4A + 3.0B + 2.0C = 1{,}200$ and is restricted further by the requirement that $2.5B + 1.5C \leq 600$, $5.0A + 1.5C \leq 1{,}500$, and that A, B, and C be nonnegative. Thus, the scarce resources determine which combinations of the activities are feasible and which are not feasible.

The production quantity combinations of A, B, and C that fall within the volume of feasible solutions constitute feasible production programs. That combination or combinations of A, B, and C which maximizes total profit is sought. The expression $0.60A + 0.70B + 0.50C$ gives the relationship among A, B, and C based on the relative profit of each product. The total profit realized will depend upon the production quantity combination chosen. Thus, there exists a family of isoprofit planes, one for each value of total profit. One of these planes will have at least one point in the volume of feasible solutions and will be a maximum distance from the origin. The plane that maximizes profit will intersect the volume at an extreme point. This calls for the computation of total profit at each extreme point as given in Table 9.12. The coordinates of each extreme point were found from the restricting planes, and the associated profit was calculated from the total profit equation.

TABLE 9.12 Total Profit Computations at Extreme Points of Figure 9.14

	COORDINATE			
Point	A	B	C	Profit
0	0	0	0	$ 0
1	0	240	0	$168.00
2	200	240	0	$288.00
3	300	160	0	$292.00
4	300	0	0	$180.00
5	180	96	240	$295.20
6	100	0	400	$260.00
7	0	0	400	$200.00

Inspection of the total profit values in Table 9.12 indicates that profit is maximized at point 5, which has the coordinates 180, 96, and 240. This means that if 180 units of product A, 96 units of product B, and 240 units of product C are produced, profit will be maximized. No other production quantity combination will result in a higher profit. Also, no alternative optimum solutions exist, because the total profit plane intersects the volume of feasible solutions at only a single point.

Addition of a third product to the two-product production system increased total profit from $292.00 per week to $295.20 per week. This increase in profit results from reallocations of the available production time and from greater utilization of the idle capacity of the welding department. The number of units of product A was reduced from 200 to 180 and the number of units of B was reduced from 240 to 96. This made it possible to add 240 units of product C to the production program, with the resulting increase in profit.

The Simplex Optimization Algorithm

Many problems to which the general linear programming model may be applied are n-dimensional in that n activities compete for scarce resources. In these cases graphical optimization methods cannot be applied and a numerical optimization technique known as the simplex algorithm can be used.

The *simplex optimization algorithm* is an iterative technique that begins with a feasible solution, tests for optimality, and proceeds toward an improved solution. It can be shown that the algorithm will finally lead to an optimal solution if such a solution exists. In this section, the simplex method will be applied to the three-product production problem presented above. Reference to the graphical solution will explain certain facets of the computational procedure.

The simplex matrix. The three-product production problem required the maximization of a total profit equation subject to certain constraints. These constraints must be converted to equalities of the form specified by the general linear programming model. This requires the addition of three "slack" variables to remove the inequalities. Thus, the constraints become

TABLE 9.13 Initial Matrix for a Three-Product Production Problem

e_j			0	0	0	0.60	0.70	0.50		
e_i	*Sol*	b	S_1	S_2	S_3	A	B	C	θ	
0	S_1	1,200	1	0	0	2.4	3.0	2.0	400	
0	S_2	600	0	1	0	0	2.5	1.5	240	r
0	S_3	1,500	0	0	1	5.0	0	2.5	∞	
E_j		0	0	0	0	0	0	0		
$E_j - e_j$			0	0	0	−0.60	−0.70	−0.50		
							k			

$$2.4A + 3.0B + 2.0C + S_1 = 1{,}200$$
$$0.0A + 2.5B + 1.5C + S_2 = 600$$
$$5.0A + 0.0B + 2.5C + S_3 = 1{,}500$$

The amount of departmental time not used in the production program is represented by a slack variable. Thus, each slack variable takes on whatever value is necessary for the equality to exist. If nothing is produced, the slack variables assume values equal to the total production time available in each department. This gives an initial feasible solution expressed as $S_1 = 1{,}200$, $S_2 = 600$, and $S_3 = 1{,}500$. Each slack variable is one of the x_j variables in the model. Because, however, no profit can be derived from idle capacity, total profit may be expressed as

$$TP = \$0.60A + \$0.70B + \$0.50C + \$0S_1 + \$0S_2 + \$0S_3$$

The initial matrix required by the simplex algorithm may now be set up as shown in Table 9.13. The first column is designated e_i and gives the coefficients applicable to the initial feasible solution. These are all zero, since the initial solution involves the allocation of all production time to the slack variables. The second column is designated *Sol* and gives the variables in the initial solution. These are the slack variables that were introduced. The third column is designated b and gives the number of minutes of production time associated with the solution variables of the previous column. These reflect the total production capacity in the initial solution. Each of the next three columns is headed by slack variables with elements of zero or unity, depending upon which equation is served by which slack variable. The e_j heading for these columns carries an entry of zero corresponding to a zero profit. The last three columns are headed by the activity variables, with elements entered from the restricting equations. The e_j heading for these columns is the profit associated with each activity variable. The last column, designated θ, is used during the computational process.

Testing for optimality. After an initial solution has been obtained, it must be tested to see if a program with a higher profit can be found. The optimality test is accomplished with the aid of the last two rows in Table 9.13. The required steps are

1. Enter values in the row designated E_j from the expression $E_j = \Sigma\, e_i a_{ij}$, where a_{ij} are the matrix elements in the ith row and the jth column.
2. Calculate $E_j - e_j$, for all positions in the row designated $E_j - e_j$.
3. If $E_j - e_j$ is negative for at least one j, a better program is possible.

Application of the optimality test to the initial feasible solution is shown in the last two rows of Table 9.13. The first element in the E_j row is calculated as $0(1{,}200) + 0(600) + 0(1{,}500) = 0$. The second is $0(1) + 0(0) = 0$. All values in this row will be zero because all e_i values are zero in the initial feasible solution. The first element in the $E_j - e_j$ row is $0 - 0 = 0$, the second is $0 - 0 = 0$, the third is $0 - 0 = 0$, the fourth is $0 - 0.60 = -0.60$, and so forth. Because $E_j - e_j$ is negative for at least one j, this initial solution is not optimal.

Iteration toward an optimal program. If the optimality test indicates that an optimal program has not been found, the following iterative procedure may be employed.

1. Find the minimum value of $E_j - e_j$ and designate this column k. The variable at the head of this column will be the incoming variable.
2. Calculate entries for the column designated θ from $\theta_i = b_i/a_{ik}$.
3. Find the minimum positive value of θ_i and designate this row r. The variable to the left of this row will be the outgoing variable.
4. Set up a new matrix with the incoming variable substituted for the outgoing variable. Calculate new elements, a'_{ij}, as $a'_{rj} = a_{rj}/a_{rk}$ for $i = r$ and $a'_{ij} = a_{ij} - a_{ik}a'_{rj}$ for $i \neq r$.
5. Perform the optimality test.

Apply steps 1, 2, and 3 of the foregoing procedure to the initial matrix. In Table 9.13 step 1 designates B as the incoming variable. Values for θ_i are calculated from step 2. Step 3 designates S_2 as the outgoing variable. The affected column and row are marked with a k and an r, respectively, in Table 9.13.

Steps 4 and 5 require a new matrix, as shown in Table 9.14. The incoming variable B, together with its associated profit, replaces the outgoing variable S_2 with its

TABLE 9.14 First Iteration for a Three-Product Production Problem

e_j			0	0	0	0.60	0.70	0.50	
e_j	*Sol*	b	S_1	S_2	S_3	A	B	C	θ
0	S_1	480	1	−1.20	0	2.40	0	0.20	200
0.70	B	240	0	0.40	0	0	1	0.60	∞
0	S_3	1,500	0	0	1	5.00	0	2.50	300
E_j		168	0	0.28	0	0	0.70	0.42	
$E_j - e_j$			0	0.28	0	−0.60	0	−0.08	
						k			

TABLE 9.15 Second Iteration for a Three-Product Production Problem

e_j			0	0	0	0.60	0.70	0.50		
e_j	*Sol*	b	S_1	S_2	S_3	A	B	C	θ	
0.60	A	200	0.416	−0.50	0	1	0	0.084	2,381	
0.70	B	240	0	0.40	0	0	1	0.60	400	
0	S_3	500	−2.08	2.50	1	0	0	2.08	240	r
E_j		288	0.25	−0.02	0	0.60	0.70	0.47		
$E_j - e_j$			0.25	−0.02	0	0	0	−0.03		
								k		

profit. All other elements in this row are calculated from the first formula of step 4. Elements in the remaining two rows are calculated from the second formula of step 4. The optimality test indicates that an optimal solution has not yet been reached. Note that after this iteration the profit at point 1 of Figure 9.18 appears. Comparison of the results in Table 9.13 and Table 9.14 with the total profit computations in Table 9.12 indicates that the isoprofit plane, which began at the origin, has now moved away from this initial position to point 1. The gain from this iteration was \$168 − \$0 = \$168.

Because the first iteration did not yield an optimal solution, it is necessary to repeat steps 1 through 5. Steps 1, 2, and 3 are applied to Table 9.14, designating A as the incoming variable and S_1 as the outgoing variable. The incoming variable, together with its associated profit, replaces the outgoing variable as shown in Table 9.15. All other elements in this new matrix are calculated from the formulas in step 4. Applications of the optimality test indicate that the solution indicated is still not optimal. Table 9.12 and Figure 9.15 show that the isoprofit plane is now at point 2. The gain from this iteration was \$288 − \$168 = \$120.

Table 9.15 did not yield an optimal solution, requiring the reapplication of steps 1 through 5. Steps 1, 2, and 3 designate C as the incoming variable and S_3 as the outgoing variable. This incoming variable, together with its associated profit, replaces the outgoing variable as shown in Table 9.16. All other elements in the matrix are calculated from the formulas of step 4. Application of the optimality test indicates that the solution exhibited by Table 9.16 is optimal. Table 9.12 and Figure 9.18 indicate that the isoprofit plane is now at point 5. The gain from this iteration was \$295.20 − \$288 = \$7.20.

Minimizing by the simplex algorithm. The computational algorithm just presented may be used without modification for problems requiring minimization if the signs of the cost coefficients are changed from positive to negative. The principle that maximizing the negative of a function is the same as minimizing the function then applies. If these coefficients are entered in the simplex matrix with their negative signs, the value of the solution will decrease as the computations proceed.

TABLE 9.16 Third Iteration for a Three-Product Production Problem

e_j			0	0	0	0.60	0.70	0.50	
e_i	*Sol*	b	S_1	S_2	S_3	A	B	C	θ
0.60	*A*	180	0.50	−0.60	−0.04	1	0	0	
0.70	*B*	96	0.60	−0.32	−0.29	0	1	0	
0.50	*C*	240	−1	1.20	0.48	0	0	1	
E_j		295.20	0.22	0.016	0.016	0.60	0.70	0.50	
$E_j - e_j$			0.22	0.016	0.016	0	0	0	

9.6 CONSTRAINED OPTIMIZATION BY DYNAMIC PROGRAMMING[4]

The dynamic programming model is based on the principle of optimality, which states the following: *An optimal policy has the property that, whatever the initial state and initial decisions are, the remaining decisions must constitute an optimal policy regarding the state resulting from the first decision.* This principle implies a sequential decision process in which a problem involving several variables is broken down in a sequence of simpler problems each involving a single variable, and prior decisions are not affected by subsequent decisions. Then, each component problem can be solved by the best available procedure, usually enumeration when the evaluation functions are not simple functions. Thus, by using the principle of optimality, a sequence of univariate enumerations is substituted for multivariate enumeration. This almost always results in a significant increase in computational efficiency.

Consider a problem involving three decision variables, each capable of assuming 10 different values. A multivariate enumeration involves $10 \times 10 \times 10$, or 1,000 evaluations. The sequence of three univariate enumerations involves only $10 + 10 + 10$, or 30 evaluations. Best results are achieved when the effectiveness functions are additive, that is, when it can be assumed that the returns from different activities can be measured in a common unit. In addition, the return from any activity must be independent of the level of other activities, and the total return must be the sum of returns from individual activities. The discussion in this section and examples in subsequent sections are based on the assumption of separate evaluation functions, the sequential decision process, and the application of the principle of optimality.

The Dynamic Programming Model

A general dynamic programming model may be developed quite easily from the principle of optimality for single-dimensional processes. The programming situation involves a certain quantity of an economic resource, such as machines, space, money,

[4] Dynamic programming owes much of its development to the pioneering work of Richard Bellman and his associates at the Rand Corporation during the 1950s.

or personnel, which can be allocated to a number of different activities. A conflict arises from the numerous ways in which the allocation can be made. A certain return is derived from the allocation of all or part of the resource to a given activity. The magnitude of the return depends jointly on the specific activity and the quantity of the resource allocated. The objective is to allocate the available resource among the various activities so as to maximize the total return.

Let the number of activities be designated by N, enumerated in fixed order, 1, 2, ..., N. Associated with each activity is a return function which gives the dependence of the return from the activity upon the quantity of the resource allocated. The quantity of the resource allocated to the ith activity may be designated x_i, and $g_i(x_i)$ denotes the return function. Because it was assumed that the activities are independent and that the returns from all activities are additive, the total return may be expressed as

$$R(x_1, x_2, \ldots, x_N) = g_1(x_1) + g_2(x_2) + \cdots + g_N(x)_N \tag{9.48}$$

The limited quantity of the resource, Q, leads to the constraint

$$Q = x_1 + x_2 + \cdots + x_N \qquad x_i \geq 0$$

The objective is to maximize the total return over all x_i, subject to the foregoing constraint.

The problem of maximizing the total return from the allocation process is viewed as one member of a sequence of allocation processes. The quantity of the resource, Q, and the number of activities, N, will not be assumed to be fixed, but may take on any value subject to the restriction that N be an integer. It is artificially required that the allocations be made one at a time; that is, a quantity of the resource is allocated to the Nth activity, then to the $(N-1)$th activity, and so on.

Because the maximization of R depends on Q and N, the dependence must be explicitly stated by a sequence of functions

$$f_N(Q) = \max_{x_i} \{R(x_1, x_2, \ldots, x_N)\} \tag{9.49}$$

The function $f_N(Q)$ represents the maximum return from the allocation of the resource Q to the N activities. If it is assumed that $g_i(0) = 0$ for each activity, it is evident that

$$f_N(0) = 0 \qquad N = 1, 2, \ldots, \text{and so forth}$$

and that

$$f_1(Q) = g_1(Q) \tag{9.50}$$

These two statements express the return to be expected from the Nth activity if no resources are allocated and the return to be expected from the first activity if all the resources are allocated to it.

A functional relation connecting $f_N(Q)$ and $f_{N-1}(Q)$ for arbitrary values of N and Q may be developed by the following procedure. If x_N, $0 \leq x_N \leq Q$, is the allocation of the resource to the Nth activity, then, regardless of the value of x_N, a quantity of resource $Q - x_N$ will remain. The return from the $N - 1$ activities may be expressed as $f_{N-1}(Q - x_N)$. Therefore, the total return from the N activity process may be expressed as

$$g_N(x_N) + f_{N-1}(Q - x_N)$$

An optimal choice of x_N would be that choice which maximizes the foregoing function. Thus the fundamental dynamic programming model may be written as

$$f_N(Q) = \max_{0 \le x_N \le Q} \{g_N(x_N) + f_{N-1}(Q - x_N)\} \qquad N = 2, 3, \ldots \tag{9.51}$$

where $f_1(Q)$ is determined from Equation 9.50. This model is based on the concept stated as the principle of optimality.

Dynamic Programming Applications

Equation 9.58 provides a method for obtaining the sequence $f_N(Q)$ once $f_1(Q)$ is known. Because $f_1(Q)$ determines $f_2(Q)$, it is also true that $f_2(Q)$ leads to an evaluation of $f_3(Q)$. This recursive relationship progresses in this manner until $f_{N-1}(Q)$ determines $f_N(Q)$, at which time the process stops.

As a general application of this procedure, consider the network of Figure 9.19. The maximum path through the network is desired. There are three possible starting points (A, B, and C) and three possible ending points (a, b, and c). The problem might be solved by identifying all possible paths and calculating the value of each. It is evident, however, that this method might result in the omission of a path. In addition, this process requires much calculation.

By using the concept of dynamic programming, the problem is viewed as a stagewise process. The maximum path from stage 1 to stage 2 is entered for each terminal point of stage 2. This occurs because whatever the starting point chosen, the path to the terminal point must be a maximum for an optimal two-stage policy to exist. Thus, the optimal two-stage policy is easily identified with little calculation. The optimal three-stage policy is found by again calculating the maximum of each terminal point from a knowledge of the results of the optimal two-stage policies. Continuing to the fourth stage with the same philosophy results in an optimal fourstage policy. This maximum path will have a value of 20 and is indicated by the dashed line.

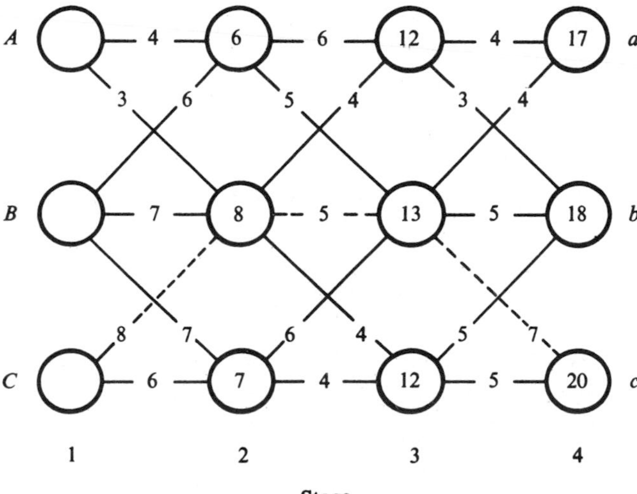

Figure 9.19 A network illustration of dynamic programming.

Consider the resource allocation situation next. The entire set of values $f_N(Q)$ in the range 0 to Q may be found by assuming a finite grid of points. Each element of the sequence $f_N(Q)$ may be evaluated and tabulated at each of these grid points. Maximization of Equation 9.51 may be performed when $N = 1$ because $f_1(Q) = g_1(Q)$. A set of values $f_1(Q)$ may be computed and tabulated. Equation 9.51 may then be used to compute $f_2(Q)$ for $N = 2$; that is,

$$f_2(Q) = \max_{0 \leq x_2 \leq Q} \{g_2(x_2) + f_1(Q - x_2)\}$$

The maximization process begins by evaluating $g_2(0) + f_1(Q)$ and $g_2(\Delta) + f_1(Q - \Delta)$. The larger of these values is retained and compared with the values $g_2(2\Delta) + f_1(Q - 2\Delta)$. As before, the larger of these values is retained and compared with $g_2(3\Delta) + f_1(Q - 3\Delta)$. This process is continued until all values of Δ are considered. The result is a table of values for $f_2(Q)$ at the Q values of 0, Δ, 2Δ, ..., as given by Table 9.17. For any given value of Q, subject to the grid established, the table gives the corresponding allocation, x_2, for the second activity, $N - 2$ because only a two-stage process was considered. Once the allocation x_2 to be the second activity is determined, the allocation of the remaining resources, $Q - x_2$, to the first activity may be determined. The computational procedure can be extended for N stages, giving an expanded table and a means for determining all allocations.

Capital investment application. As a simple application of the concept developed to this point consider the following situation. A conflict arises from the numerous ways in which a fixed amount of capital can be allocated to several activities. A certain return is derived as a result of allocating all or part of the capital to a given activity. The total return depends upon the manner in which the allocation is made. Therefore, the objective is to find that allocation which maximizes the total return.

Suppose that 8 units of capital are available and that three different activities are under consideration. The return functions for each venture are given in Table 9.17, with each return being measured in a common unit. The return from each activity is independent of the allocations to the other activities. In addition, the total return is the sum of the individual returns. Each activity exhibits a return function that increases for a portion of its range and then levels off. This is due to the law of diminishing returns and is typical of many activities.

A finite grid of points is established from this problem by the discrete nature of the return functions. Direct enumeration of all ways in which the 8 units of capital can

TABLE 9.17 Tabular Representation of Computational Scheme

Q	x_1	$f_1(Q)$	x_2	$f_2(Q)$	...
0					
Δ					
2Δ					
⋮					

be allocated to the three activities would be possible. The dynamic programming model given by Equation 9.51 may, however, be used with a saving in computational effort.

The return expected from the first activity if all the available capital is allocated to it is determined from Equation 9.57 as

$$
\begin{aligned}
f_1(0) &= g_1(0) = 0 \\
f_1(1) &= g_1(1) = 5 \\
f_1(2) &= g_1(2) = 15 \\
f_1(3) &= g_1(3) = 40 \\
f_1(4) &= g_1(4) = 80 \\
f_1(5) &= g_1(5) = 90 \\
f_1(6) &= g_1(6) = 95 \\
f_1(7) &= g_1(7) = 98 \\
f_1(8) &= g_1(8) = 100
\end{aligned}
$$

This completes the computation of $f_1(Q)$. Each value is entered in the first stage of Table 9.18.

From the results of $f_1(Q)$, $f_2(Q)$ may be computed by use of Equation 9.51. It is evident that when $Q = 0$, $f_2(Q) = 0$. When $Q = 1$,

$$f_2(1) = \max_{0 \le x_2 \le 1} g_2(x_2) + f_1(1 - x_2)$$

For values of x_2 ranging from 0 to 1, this gives

$$f_2(1) = \max \begin{Bmatrix} g_2(0) + f_1(1) = 5 \\ g_2(1) + f_1(0) = 5 \end{Bmatrix}$$

When $Q = 2$,

$$f_2(2) = \max_{0 \le x_2 \le 2} \{g_2(x_2) + f_1(2 - x_2)\}$$

For values of x_2 ranging from 0 to 2, this gives

TABLE 9.18 Return Functions for Capital Investment

Q	$g_1(Q)$	$g_2(Q)$	$g_3(Q)$
0	0	0	0
1	5	5	4
2	15	15	26
3	40	40	40
4	80	60	45
5	90	70	50
6	95	73	51
7	98	74	52
8	100	75	53

$$f_2(2) = \max \begin{Bmatrix} g_2(0) + f_1(2) = 15 \\ g_2(1) + f_1(1) = 10 \\ g_2(2) + f_1(0) = 15 \end{Bmatrix}$$

When $Q = 3$,

$$f_2(3) = \max_{0 \le x_2 \le 3} \{g_2(x_2) + f_1(3 - x_2)\}$$

For values of x_2 ranging from 0 to 3, this gives

$$f_2(3) = \max \begin{Bmatrix} g_2(0) + f_1(3) = 40 \\ g_2(1) + f_1(2) = 20 \\ g_2(2) + f_1(1) = 20 \\ g_2(3) + f_1(0) = 40 \end{Bmatrix}$$

This process is continued until $f_2(8)$ is evaluated. The maximum value of $f_2(Q)$ is identified for each value of Q and entered in the second stage of Table 9.18 together with its associated value of x_2. An arbitrary choice may be made when a tie is involved.

The third stage is considered next. Using the results of $f_2(Q)$, $f_3(Q)$ may be computed by use of Equation 9.51. As before, when $Q = 0$, $f_3(Q) = 0$. When $Q = 1$,

$$f_3(1) = \max_{0 \le x_3 \le 1} \{g_3(x_3) + f_2(1 - x_3)\}$$

For values of x_3 ranging from 0 to 1, this gives

$$f_3(1) = \max \begin{Bmatrix} g_3(0) + f_2(1) = 5 \\ g_3(1) + f_2(0) = 4 \end{Bmatrix}$$

When $Q = 2$,

$$f_3(2) = \max_{0 \le x_3 \le 2} \{g_3(x_3) + f_2(2 - x_3)\}$$

For values of x_3 ranging from 0 to 2, this gives

$$f_3(2) = \max \begin{Bmatrix} g_3(0) + f_2(2) = 15 \\ g_3(1) + f_2(1) = 9 \\ g_3(2) + f_2(0) = 26 \end{Bmatrix}$$

When $Q = 3$,

$$f_3(3) = \max_{0 \le x_3 \le 3} \{g_3(x_3) + f_2(3 - x_3)\}$$

For values of x_3 ranging from 0 to 3, this gives

$$f_3(3) = \max \begin{Bmatrix} g_3(0) + f_2(3) = 40 \\ g_3(1) + f_2(2) = 19 \\ g_3(2) + f_2(1) = 31 \\ g_3(3) + f_2(0) = 40 \end{Bmatrix}$$

TABLE 9.19 Tabular Solution for a Capital Investment Problem

Q	x_1	$f_1(Q)$	x_2	$f_2(Q)$	x_3	$f_3(Q)$
0	0	0	0	0	0	0
1	1	5	1	5	0	5
2	2	15	2	15	2	26
3	3	40	3	40	3	40
4	4*	80	0	80	0	80
5	5	90	0	90	0	90
6	6	95	2	95	2	106
7	7	98	3	120	3	120
8	8	100	4*	140	0*	140

Again this process is continued until $f_3(8)$ is evaluated. The maximum value of $f_3(Q)$ is identified for each value of Q and entered in the third stage of Table 9.19 together with its associated value of x_3.

Table 9.19 may now be used to find the maximum return allocation of capital. This maximum is found in the third stage of the table to be 140 units. The allocation of capital associated with this return may be found by noting that $x_3 = 0$ at $f_3(Q) = 140$. Therefore, 8 units of capital remain for the two-stage process giving a value for x_2 of 4 units. This leaves 4 units for the first stage: $x_1 = 4$. Each allocation is indicated with an asterisk in Table 9.19.

Note that Table 9.19 can be used to find the maximum return investment policy for investments ranging from 1 to 8 units of capital. It may also be used to find the optimal policy for a reduced number of activities. For example, if 6 units of capital are to be invested in activities 1 and 2, the solution will be found by noting that $x_2 = 2$ at $f_2(Q) = 95$. This means that 4 units will remain for the first activity; $x_1 = 4$. Thus, the maximization of R depends on Q and N, as was expressed in Equation 9.49.

Shipping application. A given number of items, each with a different weight and value, are to make up a shipment the total weight of which must not exceed a certain maximum. The objective is to select the number of each item to include in the shipment so that its value will be a maximum. Problems such as this arise in shipping operations where the value of the shipment may be measured in terms of its worth to the receiver or where its value is a function of the shipping revenue. The latter case will be considered in the example of this section.

Suppose that four items with different weights and values are to form a shipment with a total weight of not more than 9 tons. Table 9.20 gives the weight and net profit

TABLE 9.20 Weights and Values of Items

ITEM	WEIGHT (TONS)	NET PROFIT
1	2	$ 50
2	4	120
3	5	170
4	3	80

TABLE 9.21 Return Functions for Shipping Problem

Q	$g_1(Q)$	$g_2(Q)$	$g_3(Q)$	$g_4(Q)$
0	0	0	0	0
1	0	0	0	0
2	50	0	0	0
3	50	0	0	80
4	100	120	0	80
5	100	120	170	80
6	150	120	170	160
7	150	120	170	160
8	200	240	170	160
9	200	240	170	240

to be derived from each item. In this problem, the weight of the shipment is a restriction and constitutes a resource to be distributed or allocated to the four items. Thus, the concept outlined previously is applicable.

The first step in the solution is to determine the return functions for each individual item. This is accomplished by considering different weights to be allocated to each item and by identifying the whole number of items that can be accommodated within these weights. By reference to Table 9.20, the return functions of Table 9.21 are developed. Each function gives the return to be expected from allocating designated amounts of the scarce resource (shipping weight) to each activity (item).

First, the return to be expected from the first item if all the available weight is allocated to it must be determined from Equation 9.50 as

$$f_1(0) = g_1(0) = 0$$

$$f_1(1) = g_1(1) = 0$$

$$f_1(2) = g_1(2) = 50$$

$$f_1(3) = g_1(3) = 50$$

$$f_1(4) = g_1(4) = 100$$

$$f_1(5) = g_1(5) = 100$$

$$f_1(6) = g_1(6) = 150$$

$$f_1(7) = g_1(7) = 150$$

$$f_1(8) = g_1(8) = 200$$

$$f_1(9) = g_1(9) = 200$$

This completes the computation of $f_1(Q)$. Each value is entered in the first stage of Table 9.22.

From the results of $f_1(Q)$, $f_2(Q)$ may be computed by the use of Equation 9.51. When $Q = 0$, $F_2(0) = 0$. When $Q = 1$,

$$f_2(1) = \max_{0 \le x_2 \le 1} \{g_2(x_2) + f_1(1 - x_2)\}$$

TABLE 9.22 Tabular Solution for a Shipping Problem

Q	x_1	$f_1(Q)$	x_2	$f_2(Q)$	x_3	$f_3(Q)$	x_4	$f_4(Q)$
0	0*	0	0	0	0	0	0	0
1	1	0	0	0	0	0	0	0
2	2	50	0	50	0	50	0	50
3	3	50	1	50	1	50	3	80
4	4	100	4*	120	0	120	0	120
5	5	100	5	120	5	170	0	170
6	6	150	4	170	6	170	1	170
7	7	150	5	170	5	220	0	220
8	8	200	8	240	0	240	3	250
9	9	200	9	240	5*	290	0*	290

For values of x_2 ranging from 0 to 1, this gives

$$f_2(1) = \max \begin{Bmatrix} g_2(0) + f_1(1) = 0 \\ g_2(1) + f_1(0) = 0 \end{Bmatrix}$$

When $Q = 2$,

$$f_2(2) = \max_{0 \le x_2 \le 2} \{g_2(x_2) + f_1(2 - x_2)\}$$

For values of x_2 ranging from 0 to 2, this gives

$$f_2(2) = \max \begin{Bmatrix} g_2(0) + f_1(2) = 50 \\ g_2(1) + f_1(1) = 0 \\ g_2(2) + f_1(0) = 0 \end{Bmatrix}$$

When $Q = 3$,

$$f_2(3) = \max_{0 \le x_2 \le 3} \{g_2(x_2) + f_1(3 - x_2)\}$$

For values of x_2 ranging from 0 to 3, this gives

$$f_2(3) = \max \begin{Bmatrix} g_2(0) + f_1(3) = 50 \\ g_2(1) + f_1(2) = 50 \\ g_2(2) + f_1(1) = 0 \\ g_2(3) + f_1(0) = 0 \end{Bmatrix}$$

This process is continued until $f_2(9)$ is evaluated. The maximum values of $f_2(Q)$ are identified for each value of Q and entered in the second stage of Table 9.22 together with their associated values of x_2.

The third stage is considered next. Using the results of $f_2(Q)$, $f_3(Q)$ may be computed by use of Equation 9.51. As before, when $Q = 0$, $f_3(Q) = 0$. When $Q = 1$,

$$f_3(1) = \max_{0 \le x_3 \le 1} \{g_3(x_3) + f_2(1 - x_3)\}$$

For values of x_3 ranging from 0 to 1, this gives

$$f_3(1) = \max \begin{Bmatrix} g_3(0) + f_2(1) = 0 \\ g_3(1) + f_2(0) = 0 \end{Bmatrix}$$

When $Q = 2$,

$$f_3(2) = \max_{0 \le x_3 \le 2} \{g_3(x_3) + f_2(2 - x_3)\}$$

For values of x_3 ranging from 0 to 2, this gives

$$f_3(2) = \max \begin{Bmatrix} g_3(0) + f_2(2) = 50 \\ g_3(1) + f_2(1) = 0 \\ g_3(2) + f_2(0) = 0 \end{Bmatrix}$$

When $Q = 3$,

$$f_3(3) = \max_{0 \le x_3 \le 3} \{g_3(x_3) + f_2(3 - x_3)\}$$

For values of x_3 ranging from 0 to 3, this gives

$$f_3(3) = \max \begin{Bmatrix} g_3(0) + f_2(3) = 50 \\ g_3(1) + f_2(2) = 50 \\ g_3(2) + f_2(1) = 0 \\ g_3(3) + f_2(0) = 0 \end{Bmatrix}$$

Again, this process is continued until $f_3(9)$ is evaluated. The maximum values of $f_3(Q)$ are identified for each value of Q and entered in the third stage of Table 9.22 together with their associated values of x_3.

By continuing this pattern for the fourth stage, Table 9.21 is completed. The maximum profit is found in stage 4 to be \$290. The number of tons to be allocated to item 4 is found by noting that $x_4 = 0$ at $f_4(Q) = \$290$. Because 9 tons is still available, x_3 will be 5. This leaves 4 tons for item 2 and none for item 1. Thus, the shipment resulting in a maximum profit will contain one each of items 2 and 3.

QUESTIONS AND PROBLEMS

1. Specify the dimensions of the sides of a rectangle of perimeter p so that the area it encloses will be maximum.
2. The cost per unit produced at a certain facility is represented by the function

$$UC = 2x^2 - 10x + 50$$

where x is in thousands of units produced. For what value of x would unit cost be minimized (other than zero)? What is the minimum cost at this volume? Show that the value found is truly a minimum.

3. Advertising expenditures have been found to relate to profit approximately in accordance with the function

$$P = x^3 - 100x^2 + 3{,}125x$$

where x is the expenditure in thousands of dollars. What advertising expenditure would produce the maximum profit? What profit is expected at this expenditure? Show that the derived result is truly a maximum.

4. The cost of producing and selling a certain item is \$220x + \$15,000 for the first 1,000 units, \$120x + \$115,000 for a production range between 1,000 and 2,500 units, and \$205x – \$97,500 for more than 2,500 units, where x is the number of units produced. If the selling price is \$200 per unit, and all units produced are sold, find the level of production that will maximize profit.

5. Ethyl acetate is made from acetic acid and ethyl alcohol. Let x = pounds of acetic acid input, y = pounds of ethyl alcohol input, and z = pounds of ethyl acetate output. The relationship of output to input is

$$\frac{z^2}{(1.47x - z)(1.91y - z)} = 3.9$$

(a) Determine the output of ethyl acetate per pound of acetic acid, where the ratio of acetic acid of ethyl alcohol is 2.0, 1.0, and 0.67, and graph the result.

(b) Graph the cost of material per pound of ethyl acetate for each of the ratios given and determine the ratio for which the material cost per pound of ethyl acetate is a minimum if acetic acid costs \$0.80 per pound and ethyl alcohol costs \$0.92 per pound.

6. It has been found that the heat loss through the ceiling of a building is 0.13 Btu per hour per square foot of area per degree Fahrenheit. If the 2,200-ft^2 ceiling is insulated, the heat loss in Btu per hour per degree temperature difference per square foot of area is taken as

$$\frac{1}{(1/0.13) + (t/0.27)}$$

where t is the thickness in inches. The in-place cost of insulation 2, 4, and 6 in, thick is \$0.18, \$0.30, and \$0.44 per square foot, respectively. The building is heated to 75°F 3,000 hrs per year by a gas furnace with an efficiency of 50%. The mean outside temperature is 45°F and the natural gas used in the furnace costs \$4.40 per 1,000 ft^3 and has a heating value of 2,000 Btu per 1,000 ft^3. What thickness of insulation, if any, should be used if the interest rate is 10% and the resale value of the building 6 years hence is enhanced \$850 if insulation is added, regardless of the thickness?

7. An overpass is being considered for a certain crossing The superstructure design under consideration will be made of steel and will have a weight per foot depending on the span between piers in accordance with $W = 32(S) + 1{,}850$. Piers will be made of concrete and will cost \$185,000 each. The superstructure will be erected at a cost of \$0.70 per pound. If the number of piers required is to be one less than the number of spans, find the number of piers that will result in a minimum total cost for piers and superstructure if $L = 1{,}250$ ft.

8. Two girder designs are under consideration for a bridge for a 1,200-foot crossing. The first is expected to result in a superstructure weight per foot of $22(S) + 800$, where S is the span between piers The second should result in superstructure weight per foot of $20(S) + 1{,}000$. Piers and two required abutments are estimated to cost \$220,000 each. The superstructure will be erected at a cost of \$0.55 per pound. Choose the girder design that will result in a minimum cost and specify the optimum number of piers.

9. What is the cost advantage of choosing the best girder design for the bridge described in Problem 8? If the number of piers is determined from the best girder design alternative, but the other design alternative is adopted, what cost penalty is incurred?

10. A used automobile can be purchased by a student to provide transportation to and from school for $5,500 as is (i.e., the auto will have no warranty). First-year maintenance cost is expected to be $350 and the maintenance costs will increase by $100 per year thereafter. Operation costs for the automobile will be $1,200 for every year the auto is used and its salvage value decreases by 15% per year.
   (a) What is the economic life without considering the time value of money?
   (b) With interest at 16%, what is the economic life?

11. As an alternative to the used automobile in Problem 13, the student can purchase a new "utility" model for $6,800 with a three-year warranty. First-year maintenance cost is expected to be $50 and the maintenance cost will increase by $50 per year thereafter. Operation costs for this new automobile are expected to be $850 for each year of use and its salvage value decreases by 20% per year. What is the economic advantage of the new automobile without interest; with interest at 16%?

12. Special equipment can be designed and built for $80,000. This equipment will have a salvage value of $70,000 in year 1, and decrease by $10,000 through year 8. Operation and maintenance costs will start at $18,000 in year 1, and increase by 5% per year through year 8. The organization proposing this design can borrow money with a 10% annual interest rate. Using the tabular approach, find the economic life of this equipment.

13. Company X is considering using an impact wrench with a torque-sensitive clutch to fasten bolts on one of their assembly lines. The impact wrench, which can be purchased for $950, has a maximum life of 4 years. The impact wrench will have a salvage value of $600 in year 1, and will decrease in value by $200 through year 4. The impact wrench will have operation and maintenance costs of $100 in year one, $250 in year 2, $325 in year 3, and $400 in year 4. Company X has an MARR of 25%. Using the tabular approach, find the impact wrench's economic value. Show the impact wrench's economic value graphically.

14. Plot an inventory flow diagram similar to Figure 9.11 if $R = \infty$. Derive optimum values for Q, L, and TC under this assumption and verify the result by substituting into Equations 9.38, 9.39, and 9.40.

15. Plot an inventory flow diagram similar to Figure 9.11 if $R = \infty$ and $C_s = \infty$. Derive optimum values of Q, L, and TC under these assumptions and verify the result by substituting into Equations 9.38, 9.39, and 9.40.

16. An engine manufacturer requires 82 pistons per day in its assembly operations. No shortages are to be allowed. The machine shop can produce 500 pistons per day. The cost associated with initiating manufacturing action is $400, and the holding cost is $0.45 per piston per day. The manufacturing cost is $105 per piston.
   (a) Find the minimum-cost production quantity.
   (b) Find the minimum-cost procurement level if production lead time is 8 days.
   (c) Calculate the total system cost per day.

17. A subcontractor has been found who can supply pistons to the manufacturer described in Problem 16. Procurement cost will be $90 per purchase order. The cost per unit is $108.
   (a) Calculate the minimum total system cost per day for purchasing from the subcontractor.
   (b) What is the economic advantage of adopting the minimum-cost source?

18. The demand for a certain item is 12 units per period. No shortages are to be allowed. Holding cost is $0.02 per unit per period. Demand can be met either by purchasing or manufacturing, with each source described by the data given in the table.

	PURCHASE	MANUFACTURE
Procurement lead time	18 periods	13 periods
Item cost	\$11.00	\$9.60
Procurement cost	\$20.00	\$90.00
Replenishment rate	∞	25 units/period

(a) Find the minimum-cost procurement source and calculate its economic advantage over its alternative source.

(b) Find the minimum-cost procurement quantity.

(c) Find the minimum-cost procurement level.

19. An electronic equipment manufacturing firm has a demand of 250 units per period. It costs \$400 to initiate manufacturing action to produce at the rate of 600 units per period. The unit production cost is \$90. The holding cost is \$0.15 per unit, and the shortage cost is \$3.25 per unit short per period for unsatisfied demand. Determine (a) the minimum-cost manufacturing quantity, (b) the minimum-cost procurement level if production lead time is 12 periods, and (c) the total system cost per period.

20. Derive expressions for the minimum-cost procurement quantity and minimum-cost procurement level when C_h is assumed to be infinite. Name a real-world situation where such a model would apply.

21. If the span between piers must be at least 200 feet in Problem 7, what cost penalty is incurred for this constraint?

22. Suppose that no more than 6 piers can be utilized in the bridge design of Problem 7. What is the cost penalty incurred for this constraint if these piers cost \$210,000 each?

23. An item with a demand of 300 units per period is to be purchased and no shortages are allowed. The item costs \$1.30, holding cost is \$0.02 per unit per period, and it costs \$28.00 to process a purchase order. Each item consumes 2 cubic feet of warehouse space. The warehouse contains 1,500 cubic feet of space. Find the minimum cost procurement quantity and procurement level under this restriction if the lead time is 2 periods. What is the cost penalty due to the warehouse restriction?

24. The demand for a certain item is 20 units per period. Unsatisfied demand causes a shortage cost of \$0.60 per unit per period. The cost of initiating manufacturing action is \$48.00 and the holding cost is \$0.04 per unit per period. Production cost is \$7.90 per unit and the item may be produced at the rate of 60 units per period. Each item consumes 3 cubic units of warehouse space. The warehouse space reserved for this item is limited to 300 cubic units. Find the minimum cost procurement quantity and procurement level under this restriction if the lead time is 3 periods. What is the cost penalty per period due to the warehouse restriction?

25. Solve graphically for the values of x and y that maximize the function

$$Z = 2.2x + 3.8y$$

subject to the constraints

$$2.4x + 3.2y \leq 140$$

$$0.0x + 2.6y \leq 80$$

$$4.1x + 0.0y \leq 120$$

$$x \geq 0 \quad \text{and} \quad y \geq 0$$

26. A small machine shop has capability in turning, milling, drilling, and welding. The machine capacity is 16 hr per day in turning, 8 hr per day in milling, 16 hr per day in drilling, and 8 hr per day in welding. Two products, designated A and B, are under consideration. Each will yield a net profit of \$0.35 per unit and will require the amount of machine time shown in the table. Solve graphically for number of units of each product that should be scheduled to maximize profit.

	PRODUCT A	PRODUCT B
Turning	0.046	0.124
Milling	0.112	0.048
Drilling	0.040	0.000
Welding	0.000	0.120

27. Solve graphically for the values of x, y, and z that maximize the function

$$P = 7.8x + 9.4y + 2.6z$$

subject to the constraints

$$4.2x + 11.7y + 3.5z \leq 1{,}800$$
$$0.8x + 4.3y + 1.9z \leq 2{,}700$$
$$12.7x + 3.8y + 2.5z \leq 950$$
$$x \geq 0, \quad y \geq 0, \quad \text{and} \quad z \geq 0$$

28. Eight units of capital can be invested in three activities with the return from each activity given in the accompanying table. Determine the capital allocation to each activity that will maximize the total return. What will be the total return if the available capital is reduced by 2 units?

Q	$g_1(Q)$	$g_2(Q)$	$g_2(Q)$
0	0	0	0
1	2	2	2
2	3	4	4
3	4	5	6
4	6	7	6
5	8	7	8
6	9	8	9
7	10	9	9
8	10	11	12

29. Determine the number of each of four items to include in a shipment of repair items so that the value of the shipment will be a maximum. The total weight of the shipment must not exceed 15 kilos.

Item	Weight (kilos)	Value (utils)
1	2	5
2	4	7
3	3	7
4	5	9

10 QUEUING THEORY AND ANALYSIS

The *queuing* or *waiting-line system* under study within this chapter may be described as follows. A facility or group of facilities is maintained to meet the demand for service created by a population of individuals or units. These individuals form a queue or waiting line and receive service in accordance with a predetermined waiting-line discipline. In most cases, the serviced units rejoin the population and again become candidates for service. In other cases, the individuals form a waiting line at the next stage in the system.

Systems having these characteristics are common in many operations in which people, materials, equipment, or vehicles form waiting lines. The public forms waiting lines at cafeterias, doctor's offices, and theaters. In production, the flow of items in process produces a waiting line at each machine center. In maintenance, equipment to be repaired waits for service at maintenance facilities. In transportation, waiting lines form at toll gates, traffic signals, docks, and loading ramps. In each case, the objective is to optimize across a set of decision variables in the face of system parameters.

10.1 THE QUEUING SYSTEM

A multiple-channel queuing system is illustrated schematically in Figure 10.1. It exists because the population shown requires service. In satisfying the demand on the system, the decision maker must establish the level of service capacity to provide. This will involve increasing or decreasing the service capacity by altering the service rate at existing channels or by adding or deleting channels. The following paragraphs describe the components of the waiting-line system and indicate their importance in the decision evaluation process.

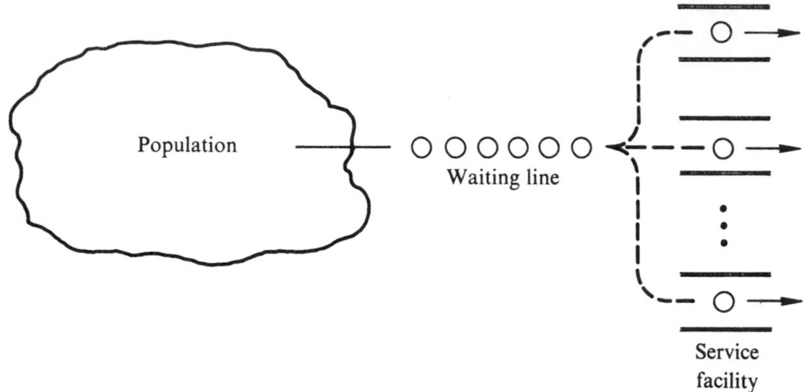

Figure 10.1 A multiple-channel queuing system.

The Arrival Mechanism

The demand for service is the primary stimulus on the waiting-line system and the justification for its existence. As previously indicated, the waiting-line system may exist to meet the service demand created by people, materials, equipment, or vehicles. The characteristics of the arrival pattern depend on the nature of the population giving rise to the demand for service.

Waiting-line systems come into being because there exists a population of individuals or units requiring service from time to time. Usually, the arrival population is best thought of as a group of items, some of which depart and join the waiting line. For example, if the population is composed of all airborne aircraft, then flight schedules and random occurrences will determine the number of aircraft that will join the landing pattern of a given airport during a given time interval. If the population is composed of telephone subscribers, then the time of day and the day of the week, as well as many other factors, will determine the number of calls placed on an exchange. If the population consists of production machines, the deterioriation, wear-out, and use rates will determine the departure mechanism causing the machines to join a waiting line of machines requiring repair service.

The arrival population, although always finite in size, may be considered infinite under certain conditions. If the departure rate is small relative to the size of the population, the number of units that potentially may require service will not be seriously depleted. Under this condition, the population may be considered infinite. Models used to explain the behavior of such systems are much easier to formulate than are models for a finite population. Examples of populations that may usually be treated as infinite are automobiles that may require passage over a bridge, customers who may potentially patronize a theater, telephone subscribers who may place a call, and production orders that may require processing at a specific machine center.

In some cases, the proportion of the population requiring service may be fairly large when compared with the population itself. In these cases, the population is seriously depleted by the departure of individuals to the extent that the departure rate will not remain stable. Because models used to explain waiting-line systems depend on the stability of the arrival rate, finite cases must be given special treatment.

Examples of waiting-line operations that might be classified as finite are equipment items that may require repair, semiautomatic production facilities that require operator attention, and company cafeterias that serve a captive population. Queuing systems for the case where the infinite population assumption does not hold will be treated in Section 10.6. Then, a general application of finite queuing, called the repairable equipment population system, is presented in Section 10.7.

The Waiting Line

In any queuing system, a departure mechanism exists that governs the rate at which individuals leave the population and join the queue. This departure mechanism is responsible for the formation of the waiting line and the need to provide service. Formation of the queue is a discrete process. Individuals or items joining the waiting line do so as integer values. The number of units in the waiting line at any point is an integer value. Rarely, if ever, is the queuing process continuous.

Individuals or items becoming a part of the waiting line take a position in the queue in accordance with a certain waiting-line discipline. The most common discipline is that of first come, first served. Other priority rules that may exist are the random selection process; the relative urgency rule; first come, last served; and disciplines involving a combination of these. In addition individuals may remain in the queue for a period and then rejoin the population. This behavior is called *reneging.*

When a unit joins the waiting line, or is being serviced, a waiting cost is incurred. Waiting cost per unit per period will depend on the units in question. If expensive equipment waits for operator attention, or requires maintenance, the loss of profit may be sizable. Vehicles waiting in queue at a toll gate incur a waiting cost due to interruption of trip progress. Customers waiting at a checkout counter become irritated and the proprietor suffers a loss of good will.

Increasing the level of service capacity will cause a decrease in both the length of the waiting line and the time for each service. As a result, the waiting time will be decreased. Because waiting cost to the system is a product of the number of units waiting and the time duration involved, this action will decrease this cost component. But since increasing the service facility capacity increases the service cost, it is appropriate to seek a reduction in waiting cost for the system only up to the point where the saving justifies the added facility cost.

The Service Mechanism

The rate at which units requiring service are serviced is assumed to be a variable directly under the control of the decision maker. This variable can be assigned a specific value to create a minimum-cost waiting-line system.

Service is the process of providing the activities required by the units in the waiting line. It may consist of collecting a toll, filling an order, providing a necessary repair, or completing a manufacturing operation. In each case, the act of providing the service causes a unit decrease in the waiting line. The service mechanism, like the arrival mechanism, is discrete, because items are processed on a unit basis.

The service facility may consist of a single channel, or it may consist of several channels in parallel, as in Figure 10.1. If it consists of only a single channel, all arrivals must eventually pass through it. If several channels are provided, items may move from the waiting line into the first channel that becomes empty. The rate at which individuals are processed depends on the service capacity provided at the individual channels and the number of channels in the system.

The service may be provided by human beings only, by human beings aided by tools and equipment, or by equipment alone. For example, collecting a fee is essentially a clerk's task, which requires no tools or equipment. Repairing a vehicle, on the other hand, requires a mechanic aided by tools and equipment. Processing a phone call dialed by the subscriber seldom requires human intervention and is usually paced by the automatic equipment. These examples indicate that service facilities can vary widely with respect to the person-machine mix used to provide the required service.

Each channel of the service facility represents a capital investment plus operating and maintenance costs. In addition, wages for personnel may be involved, together with associated overhead rates. The capability of the channel to process units requiring service is a function of the resources expended at the channel. For example, the channel may consist of a single repairperson with modest tools, or it may be a crew of technicians with complex tools and equipment. The cost of providing such a facility will depend on the characteristics of the personnel and equipment employed.

Because increasing the service capacity will result in a reduction in the waiting line, it is appropriate to adjust service capacity so that the sum of waiting cost and service cost is a minimum. The general structure of evaluation models directed to this objective will be presented next.

Queuing System Evaluation

For a queuing system already in being, an evaluation function of the form in Equation 7.2 is applicable. Specifically,

$$E = f(X, Y)$$

where

E = measure of evaluation

X = policy variable concerning the level of service capacity to provide

Y = system parameters of the arrival rate, the service rate, the waiting cost, and the service facility cost

If the system is being designed, an evaluation function of the form in Equation 7.3 is applicable. Specifically,

$$E = f(X, Y_d, Y_i)$$

where

E = measure of evaluation

X = design variable concerning the level of service capacity to specify

Y_d = design-dependent parameters (usually reliability and/or maintainability of the units requiring service)

Y_i = design-independent parameters of waiting cost and service facility cost

The following sections are devoted to developing decision evaluation models with the preceding characteristics. Let

TC = total system cost per period

A = number of periods between arrivals

S = number of periods to complete one unit of service

C_w = cost of waiting per unit per period

C_f = service facility cost for servicing one unit

Additional notation will be adopted and defined as required for deriving specific decision models.

10.2 MONTE CARLO ANALYSIS OF QUEUING

Decision models for probabilistic waiting-line systems are usually based on certain assumptions regarding the mathematical forms of the arrival and service-time distributions. Monte Carlo analysis, however, does not require that these distributions obey certain theoretical forms. Waiting-line data are produced as the system is simulated over time. Conclusions can be reached from the output statistics, whatever the form of the distributions assumed. In addition, the detailed numerical description that results from Monte Carlo analysis assists greatly in understanding the probabilistic queuing process. This section illustrates the application of Monte Carlo analysis to an infinite population, single-channel waiting-line system (see Appendix B.3).

Arrival- and Service-Time Distributions

The probabilistic waiting-line system usually involves both an arrival time and a service-time distribution. Monte Carlo analysis requires that the form and parameters of these distributions be specified. The cumulative distributions may then be developed and used as a means for generating arrival- and service-time data.

For the example under consideration, assume that the time between arrivals, A_x, has an empirical distribution with a mean of 6.325 periods ($A_m = 6.325$). Service time, S_x, will be assumed to have a normal distribution with a mean of 5.000 periods ($S_m = 5.000$) and a standard deviation of 1 period. These distributions are exhibited in Figure 10.2. The probabilities associated with each value of A_x and S_x are indicated. By summing these individual probabilities from left to right, the cumulative distributions of Figure 10.3 result. These distributions may be used with a table of random rectangular variates to generate arrival- and service-time random variables.

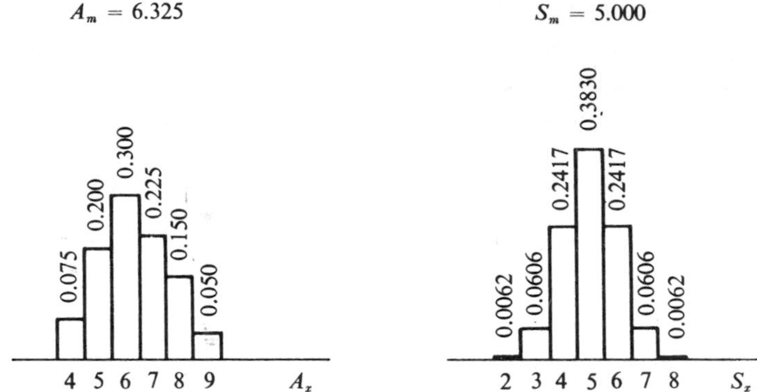

Figure 10.2 Arrival and service time distributions.

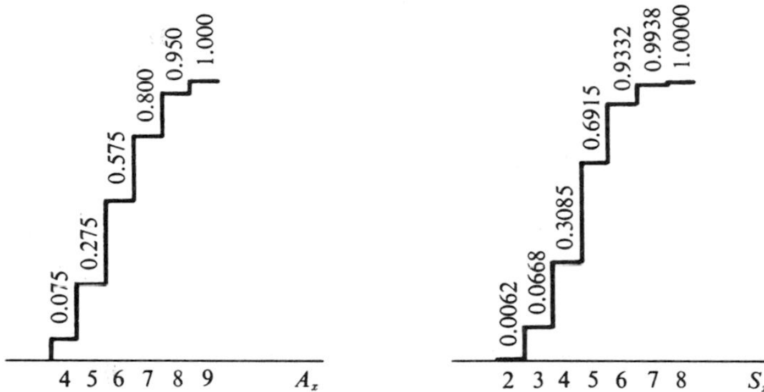

Figure 10.3 Cumulative arrival and service time distributions.

The Monte Carlo Analysis

The queuing process under study is assumed to begin when the first arrival occurs. A unit will move immediately into the service facility if it is empty. If the service facility is not empty, the unit will wait in the queue. Units in the waiting line will enter the service facility on a first-come, first-served basis. The objective of the Monte Carlo analysis is to simulate this process over time. The number of unit periods of waiting in the queue and in service may be observed for each of several service rates. The service rate resulting in a minimum-cost system may then be adopted.

The waiting-line process resulting from the arrival and service time distributions of Figure 10.2 is shown in Figure 10.4. The illustration reads from left to right, with the second line being a continuation of the first, and so forth. The interval between vertical lines represents two periods, the heavy dots represent arrivals, the slanting path a unit in service, and the arrows a service completion. When a unit cannot move directly into the service channel it waits in the queue, which is represented by a horizontal path.

The probabilistic waiting-line process illustrated in Figure 10.4 involves 400 periods and was developed in the following manner. First, the sequence of arrivals was established by the use of random rectangular variates and the cumulative arrival dis-

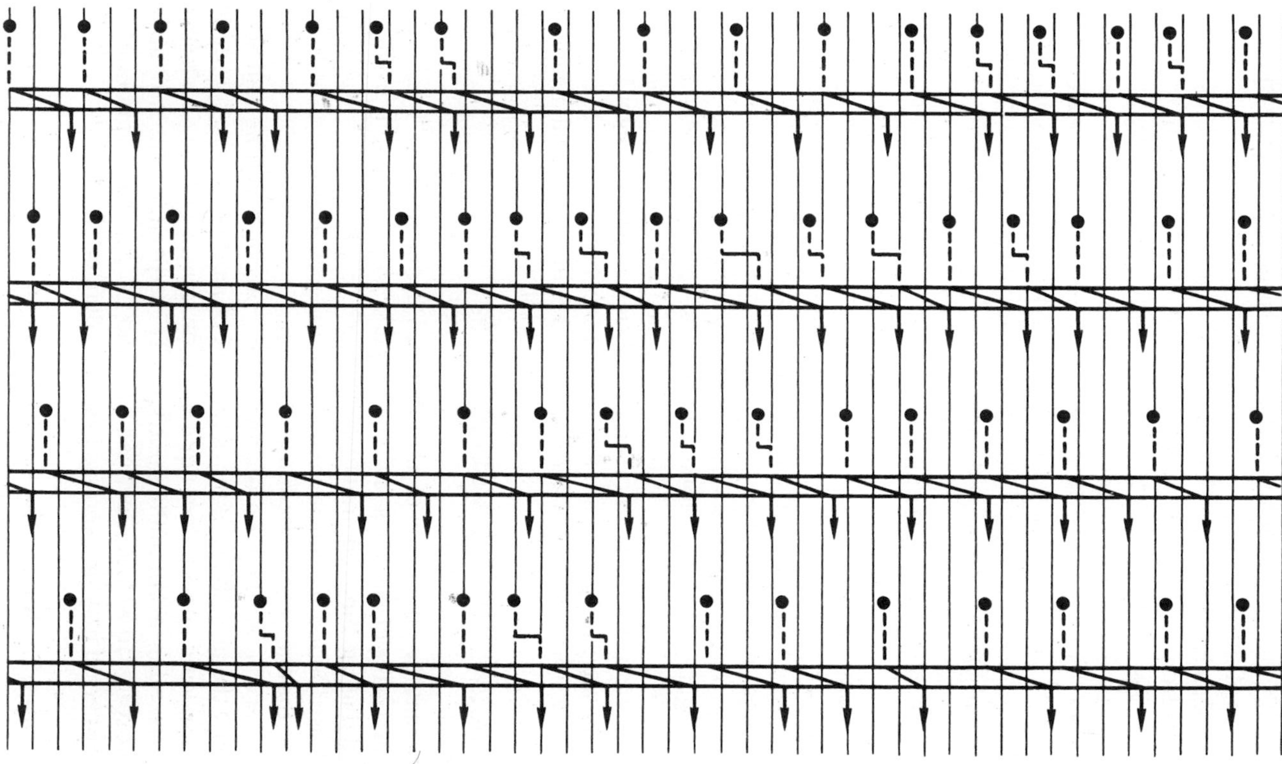

Figure 10.4 Single-channel queuing analysis by Monte Carlo.

tribution of Figure 10.3. Next, each arrival was moved into the service channel if it was available. This availability is a function of the arrival pattern and the service durations selected with the aid of random rectangular variates and the cumulative service-time distribution of Figure 10.3.

Specifically, the Monte Carlo analysis proceeded as follows. Random rectangular variates from Appendix C, were chosen as 5668, 3513, 2060, 7804, 0815, 2639, 9845, 6549, 6353, 7941, and so on. These correspond to arrival intervals of 6, 6, 5, 7, 5, 5, 9, 7, 7, 7, and so forth. Next, random rectangular variates were chosen as 323, 249, 404, 275, 879, 404, 740, 779, 441, 384, and so on. These correspond to service durations of 5, 4, 5, 4, 6, 5, 6, 6, 5, 5, and so forth. These service times determine the time an arrival enters the service channel and the time it is discharged. By proceeding in this manner, the results of Figure 10.4 are obtained.

Economic Evaluation of Results

The 400 periods simulated produced a waiting pattern involving 337 unit periods of waiting in service and 23 unit periods of waiting in the queue. The total number of unit periods of waiting for the 400-period sample was 360.

Suppose that waiting cost per unit per period is \$9.60 and that it costs \$16.10 per period to provide the service capability indicated by the service-time distribution of Figure 10.4. The total system cost for the study period is, therefore, \$9.60(360) + \$16.10(400) = \$9,896. This total system cost may be compared with the total system cost for alternative service policies by performing a Monte Carlo simulation for the alternative policies. Although this process is time-consuming, it is applicable to many situations that cannot be treated by mathematical means.

10.3 SINGLE-CHANNEL QUEUING MODELS

Assume that the population of units that may demand service is infinite with the number of arrivals per period a random variable with a Poisson distribution. It is also assumed that the time required to service each unit is a random variable with an exponential distribution. Events are recognized to have occurred at the time of arrival of a unit or at the time of completion of a service.

Under the assumption of Poisson arrivals and exponential service, it can be shown that the probability of the occurrence of an event (arrival or service completion) during a specific interval does not depend on the time of the occurrence of the immediately preceding event of the same kind. The expected number of arrivals per period may be expressed as $1/A_m$, defined as λ, and the expected number of service completions per period may be expressed as $1/S_m$, defined as μ.

Probability of n Units in the System

Under the foregoing assumptions, the probability that an arrival occurs between the time t and time $t + \Delta t$ is $\lambda \Delta t$. Similarly, the probability that a service completion occurs between time t and time $t + \Delta t$, given that a unit is being serviced at time t, is $\mu \Delta t$. Let

$$n = \text{number of units in the system at time } t, \text{ including the unit being served, if any}$$

$$P_n(t) = \text{probability of } n \text{ units in the system at time } t.$$

Because the time interval Δt is small, it can be assumed that the probability of more than one arrival or service completion during the interval is negligible. Consider the event that there are n units in the system at time $t + \Delta t$ with $n \geq 1$ expressed as

Event $\{n$ units in the system at time $t + \Delta t\}$

= Event $\{n$ units in the system at time t, no arrivals during interval Δt, and no service completions during interval $\Delta t\}$, or
Event $\{n + 1$ units in the system at time t, no arrivals during interval Δt, and one service completion during interval $\Delta t\}$, or
Event $\{n - 1$ units in the system at time t, one arrival during interval Δt, and no service completion during interval $\Delta t\}$

The probability of the event n units in the system at time $t + \Delta t$ can be written as the sum of the probabilities of these three mutually exclusive events as

$$\begin{aligned} P_n(t + \Delta t) &= \{P_n(t)[1 - \lambda\Delta t][1 - \mu\Delta t]\} \\ &\quad + \{P_{n+1}(t)[1 - \lambda\Delta t]\mu\Delta t\} + \{P_{n-1}(t)\lambda\Delta t[1 - \mu\Delta t]\} \\ &= P_n(t) - (\lambda + \mu)P_n(t)\Delta t + \lambda\mu P_n(t)(\Delta t)^2 + \mu P_{n+1}(t)\Delta t \\ &\quad - \lambda\mu P_{n+1}(t)(\Delta t)^2 + \lambda P_{n-1}(t)\Delta t - \lambda\mu P_{n-1}(t)(\Delta t)^2 \end{aligned} \tag{10.1}$$

Terms involving $(\Delta t)^2$ can be neglected. Subtracting $P_n(t)$ from both sides and dividing by Δt yields

$$\frac{P_n(t + \Delta t) - P_n(t)}{\Delta t} = -(\lambda + \mu)P_n(t) + \mu P_{n+1}(t) + \lambda P_{n-1}(t)$$

In the limit,

$$\lim_{\Delta t \to 0} \frac{P_n(t + \Delta t) - P_n(t)}{\Delta t} = \frac{d}{dt}P_n(t) = -(\lambda + \mu)P_n(t) + \mu P_{n+1}(t) + \lambda P_{n-1}(t) \tag{10.2}$$

For the special case $n = 0$,

Event $\{0$ units in the system at time $t + \Delta t\}$

= Event $\{0$ units in the system at time t and no arrivals during the interval $\Delta t\}$, or
Event $\{1$ unit in the system at time t, no arrivals during the interval Δt, and one service completion during the interval $\Delta t\}$

The probability of no units in the system at time $t + \Delta t$ can be written as the sum of the probabilities of these two mutually exclusive events as

$$\begin{aligned} P_0(t + \Delta t) &= \{P_0(t)[1 - \lambda\Delta t]\} + \{P_1(t)[1 - \lambda\Delta t]\mu\Delta t\} \\ &= P_0(t) - \lambda P_0(t)\Delta t + \mu P_1(t)\Delta t - \lambda\mu P_1(t)(\Delta t)^2 \end{aligned} \tag{10.3}$$

Again neglecting terms involving $(\Delta t)^2$, subtracting $P_0(t)$ from both sides, and dividing by Δt, we obtain

$$\frac{P_0(t + \Delta t) - P_0(t)}{\Delta t} = -\lambda P_0(t) + \mu P_1(t)$$

In the limit,

$$\frac{d}{dt} P_n(t) = \lim_{\Delta t \to 0} \frac{P_0(t + \Delta t) - P_0(t)}{\Delta t} = -\lambda P_0(t) + \mu P_1(t) \tag{10.4}$$

Equations 10.2 and 10.4 are called the *governing equations* of a Poisson arrival and exponential service single-channel queue. The differential equations constitute an infinite system for which the general solution is rather difficult to obtain. Figure 10.5 shows an example of the nature of the solution for a particular case.

As long as the probabilities $P_n(t)$ are changing with time, the queue is considered to be in a transient state. From Figure 10.5 it should be noted that this change in $P_n(t)$ becomes smaller and smaller as the time increases. Eventually, there will be little change in $P_n(t)$, and the queue will have reached a steady state. In the steady state the rate of change $dP_n(t)/dt$ can be considered to be zero and the probabilities considered to be independent of time. The steady-state governing equations can be written as

$$(\lambda + \mu)P_n = \mu P_{n+1} + \lambda P_{n-1} \tag{10.5}$$

and

$$\lambda P_0 = \mu P_1 \tag{10.6}$$

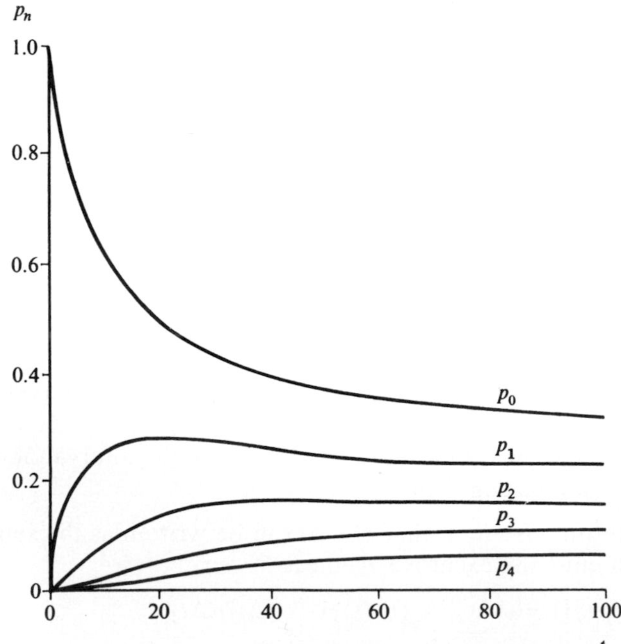

Figure 10.5 Transient solution to governing equations.

Equations 10.5 and 10.6 constitute an infinite system of algebraic equations that can be solved by substituting $P_{n+1} = P_n \cdot \rho$ into Equation 10.5 giving

$$(\lambda + \mu)\rho P_{n-1} = (\rho^2\mu + \lambda)P_{n-1}$$

or

$$(\lambda + \mu)\rho = \rho^2\mu + \lambda$$

where

$$\rho = \frac{\lambda}{\mu}$$

Substitution in Equation 10.6 gives the same result. Using this substitution, the general solution can be written as

$$P_1 = \rho P_0$$

and

$$P_n = \rho P_{n-1}$$
$$= \rho^n P_0 \tag{10.7}$$

However, because $\Sigma_{n=0}^{\infty} P_n = 1$,

$$1 = \sum_{n=0}^{\infty} P_0\rho^n = P_0 \sum_{n=0}^{\infty} \rho^n$$
$$= P_0\left(\frac{1}{1-\rho}\right) \tag{10.8}$$

Hence,

$$P_0 = 1 - \frac{\lambda}{\mu}$$

and

$$P_n = \left(1 - \frac{\lambda}{\mu}\right)\left(\frac{\lambda}{\mu}\right)^n \tag{10.9}$$

The requirement for the convergence of the sum $\Sigma_{n=0}^{\infty} (\lambda/\mu)^n$ is that λ/μ be less than 1. This means that the arrival rate λ must be less than the service rate μ for the queue to reach steady state.

As an example of the significance of Equation 10.9 in waiting-line operations, suppose that a queue is experiencing Poisson arrivals with a mean rate of 1/10 unit per period and that the service duration is distributed exponentially with a mean of 4 periods. The service rate is, therefore, 1/4, or 0.25 unit per period. Probabilities associated with each value of n may be calculated as follows:

$$P_0 = (0.6)(0.4)^0 = 0.600$$
$$P_1 = (0.6)(0.4)^1 = 0.240$$

$$P_2 = (0.6)(0.4)^2 = 0.096$$
$$P_3 = (0.6)(0.4)^3 = 0.039$$
$$P_4 = (0.6)(0.4)^4 = 0.015$$
$$P_5 = (0.6)(0.4)^5 = 0.006$$
$$P_6 = (0.6)(0.4)^6 = 0.003$$
$$P_7 = (0.6)(0.4)^7 = 0.001$$

Figure 10.6 exhibits the probability distribution of n units in the system. Certain important characteristics of the waiting-line system can be extracted from this distribution. For example, the probability of 1 or more units in the system is 0.4, the probability of no units in the system is 0.6, the probability of more than 4 units in the system is 0.01, and so forth. Such information as this is useful when there is a restriction on the number of units in the system. By altering the arrival population or the service rate or both, the probability of the number of units in the system exceeding a specified value may be controlled.

Mean Number of Units in the System

The mean number of units in the system may be expressed as

$$n_m = \sum_{n=0}^{\infty} nP_n = \sum_{n=0}^{\infty} n(1 - \rho)\rho^n$$
$$= (1 - \rho) \sum_{n=0}^{\infty} n\rho^n$$

Let $g = \sum_{n=0}^{\infty} n\rho^n$, then

$$\rho g = \sum_{n=0}^{\infty} n\rho^{n+1} = \sum_{n=1}^{\infty} (n - 1)\rho^n$$

Subtracting ρg from g,

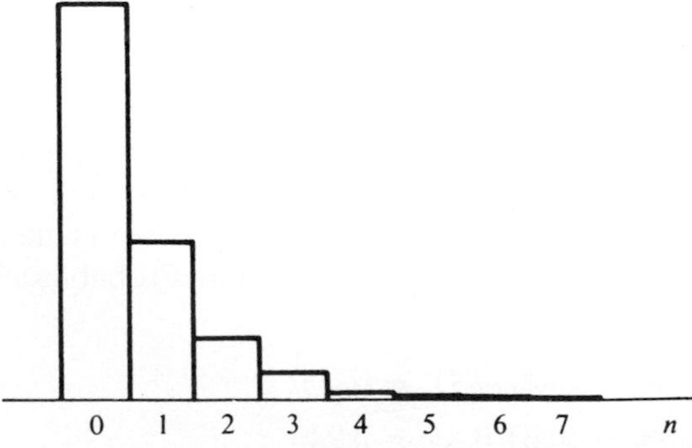

Figure 10.6 Probability distribution of n units in the system.

$$(1 - \rho)g = \sum_{n=0}^{\infty} n\rho^n - \sum_{n=1}^{\infty} (n - 1)\rho^n$$

$$= \sum_{n=1}^{\infty} n\rho^n - \sum_{n=1}^{\infty} n\rho^n + \sum_{n=1}^{\infty} \rho^n$$

$$= \sum_{n=1}^{\infty} \rho^n = \rho \sum_{n=0}^{\infty} \rho^n = \frac{\rho}{1 - \rho}$$

Hence,

$$n_m = (1 - \rho)g = \frac{\rho}{1 - \rho}$$

Substituting $\rho = \lambda/\mu$ yields

$$n_m = \frac{\lambda}{\mu - \lambda} \tag{10.10}$$

For the example given previously, the mean number of units in the system is

$$n_m = \frac{0.10}{0.25 - 0.10} = 0.667$$

Average Length of the Queue

The average length of the queue m_m can be expressed as the average number of units in the system less the average number of units being serviced

$$m_m = \frac{\lambda}{\mu - \lambda} - \frac{\lambda}{\mu}$$

$$= \frac{\lambda^2}{\mu(\mu - \lambda)} \tag{10.11}$$

For the previous example, the average length of queue is

$$m_m = \frac{(0.10)^2}{0.25(0.25 - 0.10)} = 0.267$$

that is, the average length of a nonempty waiting line. The probability that the queue is nonempty is given by

$$P(m > 0) = 1 - P_0 - P_1$$

$$= 1 - (1 - \rho) - (1 - \rho)\rho = \rho^2 \tag{10.12}$$

And the average length of the nonempty queue is

$$
\begin{aligned}
(m|m > 0)_m &= \frac{m_m}{P(m > 0)} \\
&= \frac{\lambda^2/\mu(\mu - \lambda)}{\rho^2} \\
&= \frac{\mu}{\mu - \lambda}
\end{aligned} \tag{10.13}
$$

For the previous numerical example, the probability that the queue is nonempty is

$$P(m > 0) = \rho^2 = \frac{(0.10)^2}{(0.25)^2} = 0.16$$

and the average length of the nonempty queue is

$$\frac{0.25}{0.25 - 0.10} = 1.667$$

Distribution of Waiting Time

In a probabilistic queuing system, waiting time spent by a unit before it goes into service is a random variable that depends on the status of the system at the time of arrival and also on the times required to service the units already waiting for service. In the case of a single-channel system, an arriving unit can go immediately into service only if there are no other units in the system. In all other cases the arriving unit will have to wait.

Two different events can be identified under the Poisson arrival and exponential service assumption for the waiting time, w:

1. Event $\{w = 0\}$ is identical to Event {0 units in the system}.
2. Event {waiting time is in the interval w and $w + \Delta w$} is a composite event of there being n units in the system at the time of the arrival, $n - 1$ services being completed during the time w, and the last service being completed within the interval w and $w + \Delta w$.

The probability of the first event occuring is given by

$$
\begin{aligned}
P(w = 0) &= P_0 \\
&= 1 - \frac{\lambda}{\mu}
\end{aligned} \tag{10.14}
$$

The second event is illustrated in Figure 10.7. There will be one such event for every n. Furthermore,

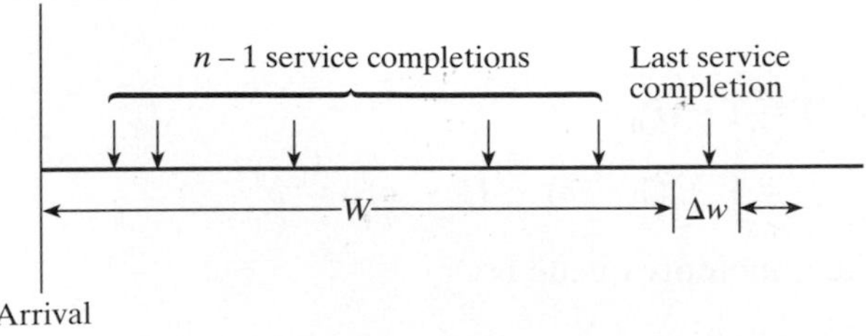

Figure 10.7 Waiting time when there are n units in the system.

$$P(w \le \text{ waiting time} \le w + \Delta w) = f(w)\Delta w$$

$$= \sum_{n=1}^{\infty} \{P_n \cdot P[(n-1)\text{services in time } w]\}$$

$$\cdot P(\text{one service completion in time } \Delta w)$$

$$= \sum_{n=1}^{\infty} \left(1 - \frac{\lambda}{\mu}\right)\left(\frac{\lambda}{\mu}\right)^n \left[\frac{(\mu w)^{n-1} e^{-\mu w}}{(n-1)!}\right] \mu \Delta w$$

$$= \sum_{n=1}^{\infty} \left(\frac{\lambda}{\mu}\right)^n \left[\frac{(\mu w)^{n-1}}{(n-1)!}\right] e^{-\mu w}\left(1 - \frac{\lambda}{\mu}\right) \mu \Delta w.$$

Let $k = n - 1$. Then

$$f(w)\Delta w = \sum_{k=0}^{\infty} \left(\frac{\lambda}{\mu}\right)^k \frac{\mu^k w^k}{k!} \left(\frac{\lambda}{\mu}\right) e^{-\mu w}\left(1 - \frac{\lambda}{\mu}\right) \mu \Delta w \tag{10.15}$$

However,

$$\sum_{k=0}^{\infty} \left(\frac{\lambda}{\mu}\right)^k \frac{\mu^k w^k}{k!} = \sum_{k=0}^{\infty} \left(\frac{\lambda w}{k!}\right)^k = e^{\lambda w} \tag{10.16}$$

Substituting Equation 10.16 into Equation 10.15 gives

$$f(w)\Delta w = e^{\lambda w}\left(\frac{\lambda}{\mu} e^{-\mu w}\right) \mu \Delta w \left(1 - \frac{\lambda}{\mu}\right)$$

$$= \lambda\left(1 - \frac{\lambda}{\mu}\right) e^{-(\mu-\lambda)w} \Delta w$$

or

$$f(w) = \lambda\left(1 - \frac{\lambda}{\mu}\right) e^{-(\mu-\lambda)w} \tag{10.17}$$

Equations 10.14 and 10.17 describe the waiting-time distribution for a single-channel queue with Poisson arrivals and exponential service. This distribution is partly discrete and partly continuous.

The mean time an arrival spends waiting for service can be obtained from the waiting-time distribution as

$$w_m = 0 \cdot P(w = 0) + \int_{w>0} w\lambda\left(1 - \frac{\lambda}{\mu}\right) e^{-(\mu-\lambda)w} dw$$

After several steps,

$$w_m = \frac{\lambda}{\mu(\mu - \lambda)} \tag{10.18}$$

For the numerical example presented previously,

$$w_m = \frac{0.10}{0.25(0.25 - 0.10)}$$
$$= 2.667 \text{ periods}$$

The average time that a unit spends in the waiting-line system is composed of the average waiting time and the average time required for service, or

$$t_m = \frac{\lambda}{\mu(\mu - \lambda)} + \frac{1}{\mu}$$
$$= \frac{\lambda + \mu - \lambda}{\mu(\mu - \lambda)} = \frac{1}{\mu - \lambda} \tag{10.19}$$

For the preceding example, the average time spent in the system will be

$$t_m = \frac{1}{0.25 - 0.10} = 6.667 \text{ periods}$$

Minimum-Cost Service Rate

The expected total system cost per period is the sum of the expected waiting cost per period and the expected facility cost per period, that is,

$$TC_m = WC_m + FC_m$$

The expected waiting cost per period is obtained as the product of cost of waiting per period, the expected number of units arriving per period, and the average time each unit spends in the system, or

$$WC_m = C_w(\lambda)\frac{1}{\mu - \lambda}$$
$$= \frac{C_w\lambda}{\mu - \lambda} \tag{10.20}$$

Alternativelv, the expected waiting cost per period can be obtained as the product of cost of waiting per period and the mean number of units in the system during the period, or

$$WC_m = C_w(n_m)$$
$$= \frac{C_w\lambda}{\mu - \lambda} \tag{10.21}$$

The expected service cost per period is the product of the cost of servicing one unit and the service rate in units per period, or

$$FC_m = C_f(\mu) \tag{10.22}$$

The expected total system cost per period is the sum of these cost components and may be expressed as

$$TC_m = \frac{C_w \lambda}{\mu - \lambda} + C_f(\mu) \qquad (10.23)$$

A minimum-cost service rate may be found by differentiating with respect to μ, setting the result equal to zero, and solving for μ as follows:

$$\frac{dTC_m}{d\mu} = -C_w\lambda(\mu - \lambda)^{-2} + C_f = 0$$

$$(\mu - \lambda)^2 C_f = \lambda C_w$$

$$\mu = \lambda + \sqrt{\frac{\lambda C_w}{C_f}} \qquad (10.24)$$

As an application of the preceding model, consider the following Poisson arrival- and exponential service-time situation. The mean time between arrivals is eight periods, the cost of waiting is \$0.10 per unit per period, and the facility cost for serving 1 unit is \$0.165. The expected waiting cost per period, the expected facility cost per period, and the expected total system cost per period are exhibited as a function of μ in Table 10.1. The expected waiting cost per period is infinite when $\mu = \lambda$ and decreases as μ increases. The expected facility cost per period increases with increasing values of μ. The minimum expected total system cost occurs when μ is 0.4 unit per period.

The minimum-cost service rate may be found directly by substituting into Equation 10.24 as follows:

$$\mu = 0.125 + \sqrt{\frac{(0.125)(\$0.10)}{0.165}}$$

$$= 0.125 + 0.275 = 0.400 \text{ unit per period}$$

TABLE 10.1 Cost Components for Exponential Service Duration

μ	WC_m	FC_m	TC_m
0.125	\$ ∞	\$0.0206	\$ ∞
0.150	0.5000	0.0248	0.5248
0.200	0.1667	0.0330	0.1997
0.250	0.1000	0.0413	0.1413
0.300	0.0714	0.0495	0.1209
0.400	0.0455	0.0660	0.1115
0.500	0.0333	0.0825	0.1158
0.600	0.0263	0.0990	0.1253
0.800	0.0185	0.1320	0.1505
1.000	0.0143	0.1650	0.1793

10.4 MULTIPLE-CHANNEL QUEUING MODELS

In the previous sections, arriving units were assumed to have been serviced through a single service facility. In many practical situations, however, there are several alternative service facilities. One example is the toll plaza on the turnpike, where several toll booths may serve the arriving traffic.

In the multiple-channel case the service facility will have c service channels, each capable of serving one unit at a time. An arriving unit will go to the first available service channel that is not busy. If all channels are busy, additional arrivals will form a single queue. As soon as any busy channel completes service and becomes available, it accepts the first unit in the queue for service. The steady-state probabilities in such a system are defined as

$$P_{m,n}(t) = \text{probability that there are } n \text{ units waiting in queue and } m \text{ channels are busy at time } t$$

It must be noted that m can only be an integer between 0 and c, and that n is zero unless $m = c$.

The determination of the steady-state probabilities follows the logic used in the single-channel case, although the solution becomes quite involved. The results are given below for the Poisson arrival and exponential service situation. Define

$$\rho = \frac{\lambda}{c\mu}$$

then

$$P_{c,n} = P_{0,0}\left(\frac{\lambda}{\mu}\right)^c \frac{1}{c!}\rho^n$$

$$P_{m,0} = P_{0,0}\left(\frac{\lambda}{\mu}\right)^m \frac{1}{m!}$$

$$P_{0,0} = \frac{1}{(\lambda/\mu)^c(1/c!)[1/(1-\rho)] + \sum_{r=0}^{r=c-1}(\lambda/\mu)^r(1/r!)} \tag{10.25}$$

As an example, consider a three-channel system with Poisson arrivals at a mean rate of 0.50 units per period and exponential service at each channel with a mean service rate of 0.25 units per period. Under these conditions ρ is $[0.50/(3 \times 0.25)] = 2/3$ and

$$P_{0,0} = \frac{1}{(0.50/0.25)^3(1/3!)[1/(1-2/3)] + \sum_{0}^{2}(0.50/0.25)^r(1/r!)}$$

$$= \frac{1}{4+1+2+2} = \frac{1}{9}$$

Average Length of the Queue

The average queue length is obtained from the expression

$$m_m = P_{0,0} \frac{(\lambda/\mu)^{c+1}}{(c-1)!(c-\lambda/\mu)^2} \tag{10.26}$$

For the example considered,

$$m_m = \frac{1}{9}\left[\frac{(0.50/0.25)^4}{2!(3-0.50/0.25)^2}\right]$$

$$= \frac{1}{9}\left[\frac{(2)^4}{2}\right] = \frac{8}{9} = 0.89 \text{ units}$$

Mean Number of Units in the System

The mean number of units in the system is

$$n_m = m_m + \frac{\lambda}{\mu} \tag{10.27}$$

For the example considered,

$$n_m = 0.89 + \frac{0.50}{0.25} = 2.89 \text{ units}$$

Mean Waiting Time

The mean waiting time can be obtained from the expression for m_m as

$$w_m = \frac{m_m}{\lambda} \tag{10.28}$$

In this example, $w_m = 0.89/0.50$, or 1.78 periods.

Average Delay or Holding Time

The average delay is obtained as the sum of waiting and service times as

$$d_m = w_m + \frac{1}{\mu}. \tag{10.29}$$

In the example, average delay is 1.78 + 4, or 5.78 periods.

Probability That an Arriving Unit Must Wait

The probability of a delay is the same as the probability that all channels are occupied. This is

$$Pr(w > 0) = \sum_{0}^{\infty} P_{c,n}$$

$$= P_{0,0}\left(\frac{\lambda}{\mu}\right)^{c} \frac{1}{c!(1 - \rho)} \tag{10.30}$$

In the example, the probability that an arriving unit has to wait is

$$P(w > 0) = \frac{1}{9}\left(\frac{0.50}{0.25}\right)^{3} \frac{1}{3!(1 - 2/3)}$$

$$= \frac{1}{9}(8)\frac{1}{2} = \frac{4}{9} = 0.444$$

10.5 QUEUING WITH NONEXPONENTIAL SERVICE

The assumption that the number of arrivals per period obeys a Poisson distribution has a sound practical basis. Although it cannot be said that the Poisson distribution always adequately describes the distribution of the number of arrivals per period, much evidence exists to indicate that this is often the case. Intuitive considerations add support to this assumption because arrival rates are usually independent of time, queue length, or any other property of the waiting-line system. Evidence in support of the exponential distribution of service durations is not as strong. Often this distribution is assumed for mathematical convenience, as in previous sections. When the service-time distribution is nonexponential, the development of decision models is more difficult. Therefore, this section will present models with nonexponential service without proof.

Poisson Arrivals with Constant Service Times

When service is provided automatically by mechanical means, or when the service operation is mechanically paced, the service duration might be constant. Under these conditions the service-time distribution has a variance of zero. The mean number of units in the system is given by

$$n_m = \frac{(\lambda/\mu)^2}{2[1 - (\lambda/\mu)]} + \frac{\lambda}{\mu} \tag{10.31}$$

and the mean waiting time is

$$w_m = \frac{\lambda/\mu}{2\mu[1 - (\lambda/\mu)]} + \frac{1}{\mu} \tag{10.32}$$

The expected total system cost per period is the sum of the expected waiting cost per period and the expected facility cost per period

$$TC_m = WC_m + FC_m$$

The expected waiting cost per period is the product of the cost of waiting per unit per period and the mean number of units in the system during the period, or

$$\begin{aligned} WC_m &= C_w(n_m) \\ &= C_w\left\{\frac{(\lambda/\mu)^2}{2[1 - (\lambda/\mu)]} + \frac{\lambda}{\mu}\right\} \end{aligned}$$

The expected facility cost per period is the product of the cost of servicing one unit and the service rate in units per period, or

$$FC_m = C_f(\mu)$$

The expected total system cost per period is the sum of these cost components and may be expressed as

$$TC_m = C_w\left\{\frac{(\lambda/\mu)^2}{2[1 - (\lambda/\mu)]} + \frac{\lambda}{\mu}\right\} + C_f(\mu) \tag{10.33}$$

As an application of the foregoing model, consider the example of the previous section. Instead of the parameter μ being the expected value from an exponential distribution, however, assume that it is a constant. The expected waiting cost per period, the expected service cost per period, and the expected total system cost per period is exhibited as a function of μ in Table 10.2. Although the expected waiting-cost function differs from the previous example, the minimum-cost service interval is still 0.4 units per period.

The example may be more easily compared by graphing the expected total cost functions as shown in Figure 10.8. The upper curve is the expected total system cost function when μ is an expected value from an exponential distribution. The lower curve is the expected total system cost when μ is a constant. No significant difference in the minimum-cost policy is evident for the example considered.

TABLE 10.2 Cost Components for Constant Service Duration

μ	WC_m	FC_m	TC_m
0.1250	\$ ∞	\$0.0206	\$ ∞
0.1500	0.2913	0.0248	0.3161
0.2000	0.1145	0.0330	0.1475
0.2500	0.0750	0.0413	0.1163
0.3000	0.0566	0.0495	0.1061
0 4000	0.0383	0.0660	0.1043
0.5000	0.0292	0.0825	0.1117
0.6000	0.0236	0.0990	0.1226
0.8000	0.0170	0.1320	0.1490
1.0000	0.0134	0.1650	0.1785

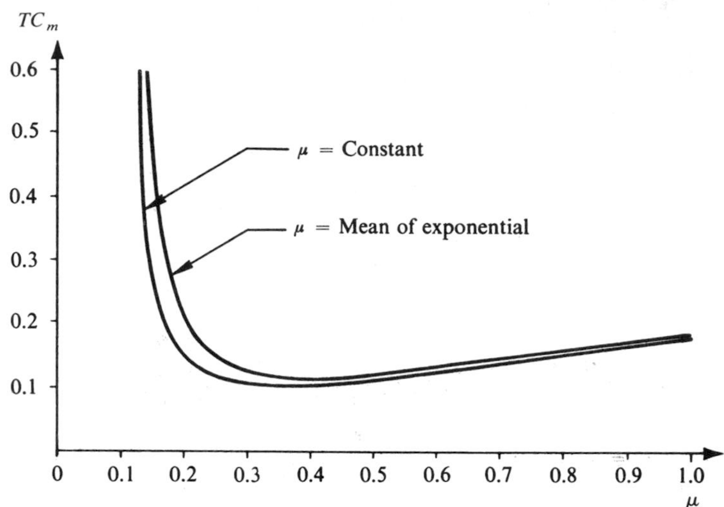

Figure 10.8 TC_m as a function of the service rate.

Poisson Arrivals with Any Service-Time Distribution

For further generality, it is desirable to have expressions for pertinent system characteristics regardless of the form of the service-time distribution. If σ^2 is the variance of the service-time distribution, the mean number of units in the system is given by

$$n_m = \frac{(\lambda/\mu)^2 + \lambda^2\sigma^2}{2[1 - (\lambda/\mu)]} + \frac{\lambda}{\mu} \tag{10.34}$$

and the mean waiting time is

$$w_m = \frac{(\lambda/\mu^2) + \lambda\sigma^2}{2[1 - (\lambda/\mu)]} + \frac{1}{\mu} \tag{10.35}$$

Equation 10.34 reduces to Equation 10.31 and Equation 10.35 reduces to Equation 10.32 when $\sigma^2 = 0$. In addition, because the variance of an exponential distribution is $(1/\mu)^2$, Equation 10.34 reduces to Equation 10.10 and Equation 10.35 reduces to Equation 10.23 when this substitution is made.

The expected total system cost per period is the sum of the expected waiting cost per period and the expected facility cost per period; that is,

$$TC_m = WC_m + FC_m$$

The expected waiting cost per period is the product of the cost of waiting per unit per period and the mean number of units in the system during the period. The expected facility cost per period may be taken as the product of the cost of serving one unit and the service rate in units per period. Therefore, the expected total system cost per period is

$$TC_m = C_w\left\{\frac{(\lambda/\mu)^2 + \lambda^2\sigma^2}{2[1 - (\lambda/\mu)]} + \frac{\lambda}{\mu}\right\} + C_f(\mu) \tag{10.36}$$

As an example of the application of this model, consider the following situation: The number of arrivals per hour has a Poisson distribution with a mean of 0.2 unit. The cost of waiting per unit per hour is $2.10 and the cost of servicing one unit is $4.05. The decision maker may choose one of two service policies. The first will result in a service rate of 0.4 units per hour with a service time variance of 3 hours. The second will result in a service rate of 0.5 units per hour with a service time variance of 4 hours. The first policy will result in an expected total system cost of

$$TC_m = \$2.10\left\{\frac{(0.2/0.4)^2 + (0.2)^2(3)}{2[1 - (0.2/0.4)]} + \frac{0.2}{0.4}\right\} + \$4.05(0.4)$$

$$= \$1.83 + \$1.62 = \$3.45$$

The second policy will result in an expected total system cost of

$$TC_m = \$2.10\left\{\frac{(0.2/0.5)^2 + (0.2)^2(4)}{2[1 - (0.2/0.5)]} + \frac{0.2}{0.5}\right\} + \$4.05(0.5)$$

$$= \$1.40 + \$2.03 = \$3.43$$

From these results, it is evident that it makes little difference which policy is adopted.

10.6 FINITE POPULATION QUEUING MODELS

Finite waiting-line models must be applied to those waiting-line systems where the population is small relative to the arrival rate. In these systems, units leaving the population significantly affect the characteristics of the population and the arrival probabilities. It is assumed that both the time between calls for service for a unit of the population and the service times are distributed exponentially.

Finite Queuing Theory

Units leave the population when they fail, and the number of these failed units can be determined by an approach similar to that used in Section 10.3. Let

N = number of units in the population
M = number of service channels in the repair facility
λ = failure rate of an item, 1/MTBF
μ = repair rate of a repair channel, 1/MTTR
n = number of failed items
P_n = steady-state probability of n failed items
P_0 = probability that no items failed
$M\mu$ = maximum possible repair rate
λ_n = failure rate when n items already failed
μ_n = repair rate when n items already failed

The failure rate of an item is expressed as $\lambda = 1/\text{MTBF}$ and the failure rate of the entire population when n items already failed can be expressed as $\lambda_n = (N - n)\lambda$, where $N - n$ is the number of operational units, each of which fails at a rate of λ. Similarly, the repair rate of a repair channel is expressed as $\mu = 1/\text{MTTR}$, and the repair rate of the entire repair facility when n items have already failed can be expressed as

$$\mu_n = \begin{Bmatrix} n\mu & \text{if } n \in 1, 2, \ldots, M - 1 \\ M\mu & \text{if } n \in M, M + 1, \ldots, N \end{Bmatrix} \tag{10.37}$$

An analysis using birth-death processes is employed to determine the probability distribution, P_n, for the number of failed items. In the birth-death process, the state of the system is the number of failed items (state = 0, 1, 2, ..., N). The rates of change between the states are the breakdown rate, λ_n, and the repair rate, μ_n. This gives

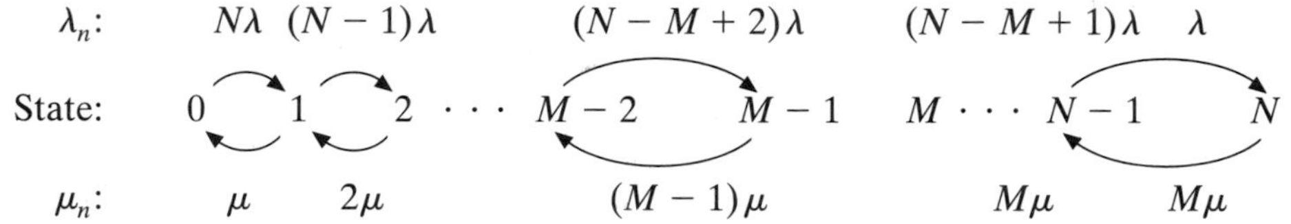

λ_n: $N\lambda$ $(N-1)\lambda$ $(N-M+2)\lambda$ $(N-M+1)\lambda$ λ

State: 0 1 2 · · · $M-2$ $M-1$ M · · · $N-1$ N

μ_n: μ 2μ $(M-1)\mu$ $M\mu$ $M\mu$

If steady-state operation of the system is assumed, this yields

$$\begin{aligned}
N\lambda P_0 &= \mu P_1 \\
N\lambda P_0 + 2\mu P_2 &= [\mu + (N-1)\lambda]P_1 \\
(N-1)\lambda P_1 + 3\mu P_3 &= [2\mu + (N-2)\lambda]P_2 \\
&\vdots \\
(N-M+2)\lambda P_{M-2} + M\mu P_M &= [(M-1)\mu + (N-M+1)\lambda]P_{M-1} \\
&\vdots \\
2\lambda P_{N-2} + M\mu P_N &= (M\mu + \lambda)P_{N-1} \\
\lambda P_{N-1} &= M\mu P_N
\end{aligned}$$

Additionally,

$$\sum_{n=0}^{N} P_n = 1$$

Solving these balance equations gives

$$P_0 = \left(\sum_{n=0}^{N} C_n\right)^{-1} \tag{10.38}$$

where

$$C_n = \begin{Bmatrix} \dfrac{N!}{(N-n)!\,n!}\left(\dfrac{\lambda}{\mu}\right)^n & \text{if } n = 0, 1, 2, \ldots, M \\ \dfrac{N!}{(N-n)!\,M!\,M^{n-M}}\left(\dfrac{\lambda}{\mu}\right)^n & \text{if } n = M+1, M+2, \ldots, N \end{Bmatrix} \tag{10.39}$$

Equations 10.38 and 10.39 can now be used to find the steady-state probability of n failed units as $P_n = P_0 C_n$ for $n = 0, 1, 2, \ldots, N$.

For example, assume that a finite population of 10 units exists, with each unit having a mean time between failure of 32 hours. Also assume that the mean time to service an item in a single service channel is 8 hours. From these data, the arrival rate and the service rate is found to be

$$\lambda = 1/32 \quad \text{and} \quad \mu = 1/8$$

from which

$$\frac{\lambda}{\mu} = 0.25$$

First, compute C_n for $n = 0$ and 1 using the first expression in Equation 10.39 as

$$C_0 = \frac{10!\,(0.25)^0}{10!\,0!} = 1$$

$$C_1 = \frac{10!\,(0.25)^1}{9!\,1!} = 2.5$$

Continue computing C_n for $n = 2, 3, \ldots, 10$ using the second expression in Equation 10.39 as

$$C_2 = \frac{10!\,(0.25)^2}{8!\,1!\,1^1} = 5.625$$

$$C_3 = \frac{10!\,(0.25)^3}{7!\,1!\,1^2} = 11.250$$

$$C_4 = \frac{10!\,(0.25)^4}{6!\,1!\,1^3} = 19.6875$$

$$C_5 = \frac{10!\,(0.25)^5}{5!\,1!\,1^4} = 29.53125$$

$$C_6 = \frac{10!\,(0.25)^6}{4!\,1!\,1^5} = 36.9140625$$

$$C_7 = \frac{10!\,(0.25)^7}{3!\,1!\,1^6} = 36.9140625$$

$$C_8 = \frac{10!\,(0.25)^8}{2!\,1!\,1^7} = 27.685547$$

$$C_9 = \frac{10!\,(0.25)^9}{1!\,1!\,1^8} = 13.842773$$

$$C_{10} = \frac{10!\,(0.25)^{10}}{0!\,1!\,1^9} = 3.460693$$

$$\sum_{n=0}^{10} C_n = 188.410888$$

From Equation 10.38

$$P_0 = \frac{1}{\sum_{n=0}^{10} C_n} = \frac{1}{188.410888} = 0.0053076$$

P_n for $n = 0, 1, 2, \ldots, N$ can now be computed from $P_n = P_0 C_n = 0.0053076\,(C_n)$ as follows:

$$P_0 = 0.0053076 \times 1 = 0.0053076$$

$$P_1 = 0.0053076 \times 2.5 = 0.0132690$$

$$P_2 = 0.0053076 \times 5.625 = 0.0298553$$

$$P_3 = 0.0053076 \times 11.250 = 0.0597105$$

$$P_4 = 0.0053076 \times 19.6875 = 0.1044934$$

$$P_5 = 0.0053076 \times 29.53125 = 0.1567400$$

$$P_6 = 0.0053076 \times 36.9140625 = 0.1959251$$

$$P_7 = 0.0053076 \times 36.9140625 = 0.1959251$$

$$P_8 = 0.0053076 \times 27.685547 = 0.1469438$$

$$P_9 = 0.0053076 \times 13.842773 = 0.0734719$$

$$P_{10} = 0.0053076 \times 3.460693 = 0.0183680$$

The steady-state probabilities of n failed units calculated previously can now be used to find the mean number of failed units from

$$\sum_{n=0}^{10} n \times P_n$$

as

$$0 \times 0.0053076 + 1 \times 0.0132690 + \cdots + 10 \times 0.0183680 = 6.023 \text{ units}$$

Finite Queuing Tables[1]

The finite queuing derivations in the previous section and the appropriate numerical applications provide complete probabilities for n units having failed, P_n. From these probabilities, the expected number of failed units was obtained.

In this section an approach based on the Finite Queuing Tables is presented. For convenience, the notation used in these tables will be adopted. Let

T = mean service time

U = mean time between calls for service

[1] L. G. Peck and R. N. Hazelwood, *Finite Queuing Tables* (New York: John Wiley & Sons, Inc., 1958).

H = mean number of units being serviced
L = mean number of units waiting for service
J = mean number of units running or productive

Appendix E gives a portion of the Finite Queuing Tables (for populations of 10, 20, and 30 units). Each set of values is indexed by N, the number of units in the population. Within each set, data are classified by X, the service factor, and M, the number of service channels. Two values are listed for each value of N, X, and M. The first is D, the probability of a delay, expressing the probability that an arrival will have to wait. The second is F, an efficiency factor needed in the calculation of H, L, and J.

The service factor is a function of the mean service time and the mean time between calls for service,

$$X = \frac{T}{T + U} \tag{10.40}$$

The mean number of units being serviced is a function of the efficiency factor, the number of units in the population, and the service factor,

$$H = FNX \tag{10.41}$$

The mean number of units waiting for service is a function of the number of units in the population and the efficiency factor,

$$L = N(1 - F) \tag{10.42}$$

Finally, the mean number of units running or productive is a function of the number of units in the population, the efficiency factor, and the service factor,

$$J = NF(1 - X) \tag{10.43}$$

A knowledge of N, T, and U for the waiting-line system under study, the expressions given previously, and a set of Finite Queuing Tables makes it easy to find mean values for important queuing parameters. For example, the mean number of failed units calculated in the previous section to be 6.023 can be found from Equations 10.40, 10.41, and 10.42 as

$$X = \frac{8}{8 + 32} = 0.20$$

$$H = 0.497(10)(0.20) = 0.994$$

$$L = 10(1 - 0.497) = 5.03$$

This gives $H + L = 6.024$ as the mean number failed. Other examples are given in the sections that follow.

Number of Service Channels Under Control

Assume that a population of 20 units exists, with each unit having a mean time between required service of 32 minutes. Each service channel provided will have a mean service time of 8 minutes. Both the time between arrivals and the service interval are distributed exponentially. The number of channels to be provided is under management

TABLE 10.3 Cost as a Function of the Number of Service Channels

M (A)	F (B)	H (C)	L (D)	$H + L$ (E)	WAITING COST (F)	SERVICE COST (G)	TOTAL COST (H)
8	0.999	4.00	0.02	4.02	$20.10	$80	$100.00
7	0.997	3.99	0.06	4.05	20.25	70	90.25
6	0.988	3.95	0.24	4.19	20.95	60	80.95
5	0.963	3.85	0.74	4.59	22.95	50	72.95
4	0.895	3.58	2.10	5.68	28.40	40	68.40
3	0.736	2.94	5.28	8.22	41.10	30	71.10
2	0.500	2.00	10.00	12.00	60.00	20	80.00

control. The cost of providing one channel with a mean service-time capacity of 8 minutes is $10 per hour. The cost of waiting is $5 per unit per hour. The service factor for this system is

$$X = \frac{T}{T + U} = \frac{8}{8 + 32} = 0.20$$

Table 10.3 provides a systematic means for finding the minimum-cost number of service channels. The values in columns A and B are entered from Appendix E, Table E.2, with $N = 20$ and $X = 0.20$. The mean number of units being serviced is found from Equation 10.41 and entered in column C. The mean number of units waiting for service is found from Equation 10.42 and is entered in column D. The mean number of units waiting in queue and in service is given in column E. The data of columns A and E may be multiplied by their respective costs to give the total system cost.

First, multiplying $5 per unit per hour by the mean number of units waiting gives the waiting cost per hour in column F. Second, multiplying $10 per channel per hour by the number of channels gives the service cost per hour in column G. Finally, adding the expected waiting cost and the service cost gives the expected total system cost in column H. The minimum cost number of channels is found to be four.

In this example, the cost of waiting was taken to be $5 per unit per hour. If this is due to lost profit, resulting from unproductive units, the same solution may be obtained by maximizing profit. As before, the values in columns A and B of Table 10.4 are

TABLE 10.4 Profit as a Function of the Number of Service Channels

M (A)	F (B)	J (C)	GROSS PROFIT (D)	SERVICE COST (E)	NET PROFIT (F)
8	0.999	15.98	$79.90	$80	$−0.10
7	0.997	15.95	79.75	70	9.75
6	0.988	15.81	79.05	60	19.05
5	0.963	15.41	77.05	50	27.05
4	0.895	14.32	71.60	40	31.60
3	0.736	11.78	58.90	30	28.90
2	0.500	8.00	40.00	20	20.00

entered from Appendix E, Table E.2, with $N = 20$ and $X = 0.200$. The mean number of units running or productive is found from Equation 10.43 and is entered in column C.

The profit per hour in column D is found by multiplying the mean number of productive units by $5 profit per productive unit per hour. The cost of service per hour in column E is obtained by multiplying the number of channels by $10 per channel per hour. Finally, the net profit in column F is found by subtracting the service cost per hour from the profit per hour. As before, the number of channels that should be used is four. This example illustrates that either the minimum-cost or the maximum-profit approach may be used with the same results.

Mean Service Time under Control

Assume that a population of 10 units is to be served by a single service channel. The mean time between calls for service is 30 mins. If the mean service rate is 60 units per hour, the service cost will be $100 per hour. The service cost per hour is inversely proportional to the time in minutes to service one unit, expressed as $\$100/T$. Both the time between calls for service and the service duration are distributed exponentially. Lost profit because of units waiting in the system is $15 per hour.

Column A in Table 10.5 gives the capacity of the channel expressed as the mean service time in minutes per unit processed. The service factor for each service time is found from Equation 10.38 and entered in column B. The efficiency factors in column C are found by interpolation in Appendix E, Table E.1, for $N = 10$ and the respective service factors of column B. The mean number of units running is found from Equation 10.41 and entered in column D. The data given in column A and column D may now be used to find the service capacity that results in a maximum net profit.

The expected profit per hour in column E is found by multiplying the mean number of units running by $15 per hour. The cost of service per hour is found by dividing $100 by the value for T in column A. These costs are entered in column F. By subtracting the cost of service per hour from the expected gross profit per hour, the expected net profit per hour in column G is found. The mean service time resulting in an expected maximum profit is three periods.

TABLE 10.5 Profit as a Function of the Mean Service Time

T (A)	X (B)	F (C)	J (D)	GROSS PROFIT (E)	SERVICE COST (F)	NET PROFIT (G)
1	0.032	0.988	9.56	$143.20	$100.00	$43.20
2	0.062	0.945	8.86	132.90	50.00	82.90
3	0.091	0.864	7.85	117.75	33.33	84.42
4	0.118	0.763	6.73	101.00	25.00	76.00
5	0.143	0.674	5.77	86.51	20.00	66.51

Service Factor under Control

Suppose that a population of production equipment is under study with the objective of deriving minimum cost maintenance policy. It is assumed that both the time between calls for maintenance for a unit of the population and the service times are distributed exponentially. Two parameters of the system are subject to management control: (1) by increasing the repair capability (reducing T) the average machine downtime will be reduced; and (2) alternative policies of preventive maintenance will alter the mean time between breakdowns, U. Therefore, the problem of machine maintenance reduces to one of determining the service factor, X, that will result in a minimum-cost operation. This section will present methods for establishing and controlling the service factor in maintenance operations.

As machines break down, they become unproductive with a resulting economic loss. This loss may be reduced by reducing the service factor. But a decrease in the service factor requires either a reduction in the repair time or a more expensive policy of preventive maintenance, or both. Therefore, the objective is to find an economic balance between the cost of unproductive machines and the cost of establishing a specific service factor.

The analysis of this situation is facilitated by developing curves giving the percentage of machines not running as a function of the service factor. Figure 10.9 gives curves for selected populations when one service channel is provided. Each cure is developed from Equation 10.41 and the Finite Queuing Tables. As was expected, the percentage of machines not running increases as the service factor increases.

As an example of the determination of the minimum-cost service factor, suppose that eight machines are maintained by a mechanic and one helper. Each machine pro-

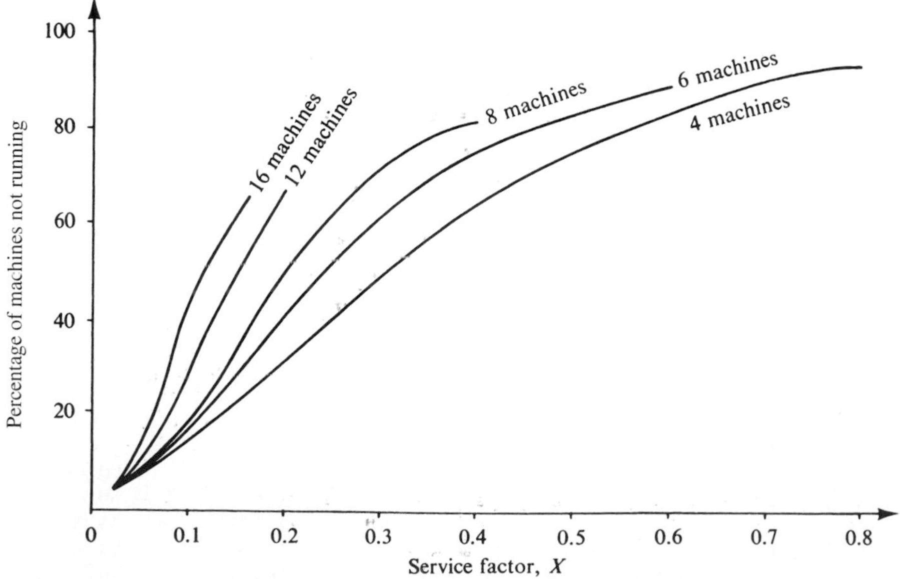

Figure 10.9 Percentage of machines not running as a function of the service factor.

duces a profit of $22.00 per hour while it is running. The mechanic and the assistant cost the company $28.80 per hour. Three policies of preventive maintenance are under consideration. The first will cost $80 per hour. After considering the increase in the mean time between breakdowns and the effect on service time, it is estimated that the resulting service factor will be 0.04. The second policy of preventive maintenance will cost only $42 per hour but will result in a service factor of 0.10. The third alternative involves no preventive maintenance at all; hence, it will cost nothing, but a service factor in excess of 0.2 will result. The time between calls for service and the service times are distributed exponentially. By reference to Figure 10.9, the results of Table 10.6 are developed. From the last column, it is evident that the second preventive maintenance alternative should be adopted.

TABLE 10.6 Three Policies of Preventive Maintenance

Maintenance Policy	Service Factor, X	Machines Not Running	Cost of Lost Profit	Cost of Maintenance	Cost of Mechanic	Total Cost
1	0.04	0.5	$11.00	$80	$28.80	$119.80
2	0.10	1.6	35.20	42	28.80	106.00
3	0.20	4.1	90.20	0	28.80	119.00

10.7 REPAIRABLE EQUIPMENT POPULATION SYSTEMS

The finite population queuing system presented in this section is an extension of the situations described in Section 10.6. It uses the design evaluation function of Equation 7.3 and the decision evaluation display for selecting the best repairable equipment population system (REPS) design.

A finite population of repairable equipment is procured and maintained in operation to meet a demand. As repairable equipment units fail or become unserviceable, they are repaired and returned to service. As they age, the older units are removed from the system and replaced with new units. The general equipment design problem is to determine the population size, the replacement age of units, the number of repair channels, the design mean time between failures, and the design mean time to repair, so that the life-cycle cost of the system will be minimized with consideration for other criteria.

Introduction to REPS

Two repairable equipment population system problems are treated in this system design example. The first is to determine the population size, the replacement age of units, and the number of repair channels so that the sum of all costs associated with the system will be minimized. The second problem is to generate and evaluate candidate systems by specifying the unit mean time between failures, MTBF, and the unit mean time to repair, MTTR, as a function of unit cost, as well as the population size, the replacement age of units, and the number of repair channels.

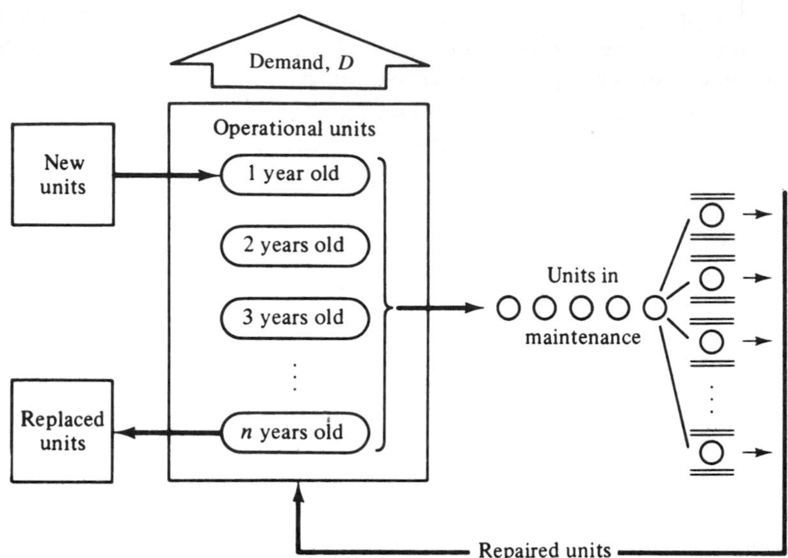

Figure 10.10 The repairable equipment population system.

The operational system. The repairable equipment population system illustrated in Figure 10.10 is designed and deployed to meet a demand, D. Units within the system can be separated into two groups; those in operation and available to meet demand, and those out of operation and hence unavailable to meet demand. It is assumed that units are not discarded upon failure, but are repaired and returned to operation.

As units age, they become less reliable and their maintenance costs increase. Accordingly, it is important to determine the optimum replacement age. It is assumed that the number of new units procured each year is constant and that the number of units in each age group is equal to the ratio of the total number required in the population and the desired number of age groups. Although the analysis deals with the life cycle of the units, the objective is to optimize the total system of which the units are a part.

In REPS Problem I, the decision is to specify a population of units, a number of maintenance channels, and a replacement schedule for bringing new units into the system. For REPS Problem II, the decision is extended to include the unit's reliability and maintainability characteristics. In either case the system is to be employed to meet the demand for equipment optimally.

The foregoing paragraphs have described, in general terms, the finite population queuing system under investigation. This model describes the operation of numerous systems. For example, both the airlines and the military operate and maintain aircraft with these system characteristics. In ground transit, vehicles, such as rental automobiles, taxis, and commercial trucks, constitute repairable equipment populations. Production equipment types, such as weaving looms, drill presses, and autoclaves, are populations of equipment which fit the repairable classification. But the repairable unit may also be an inventory of components for the larger entities mentioned. For example, aircraft hydraulic pumps, automobile starters and alternators, truck engines, electric motors, and electronic controllers for equipment also constitute repairable equipment population systems.

Scope and assumptions. Repairable equipment population systems normally come into being over a nonsteady-state buildup phase. They then operate over a steady-state interval of years, after which a phase-out period is entered. Only the steady-state mode of operation will be considered here.

The following assumptions are adopted in the development of the mathematical model and algorithm for REPS:

1. The interarrival times are exponentially distributed.
2. The repair times are exponentially distributed.
3. The number of units in the population is small such that finite population queuing formulations must be used.
4. The interarrival times are statistically independent of the repair times.
5. The repair channels are parallel and each is capable of similar performance.
6. The population size will always be larger than or at least equal to the number of service channels.
7. Each channel performs service on one unit at a time.
8. MTBF and MTTR values vary for each age group and represent the expected value for these variables for that age group.
9. Units completing repair return to operation with the same operational characteristics as their age group.

System design evaluation. REPS Problem I may be evaluated by using the decision evaluation function of Equation 7.2. Here the objective is to find optimal values for controllable design variables in the face of uncontrollable system parameters. In REPS Problem II, the decision evaluation function of Equation 7.3 is used to seek the best candidate system. This is accomplished by establishing values for controllable design-dependent parameters in the face of uncontrollable design-independent parameters and optimal values for design variables.

Three design variables are identified in the repairable equipment population system. These controllables are the number of units to deploy, the replacement age of units, and the number of repair channels. Optimal values are sought for these variables so that the sum of all costs associated with the repairable equipment population system will be minimized.

In Design Problem I the focus is entirely on optimizing design variables as the only controllable factors. This situation arises when the system is in existence and the objective is to optimize its operation in the face of uncontrollable system parameters. The focus shifts to seeking the best candidate system in Design Problem II. In this activity, optimal values for design variables are secondary. They are needed as a means for comparing candidate systems equivalently. Then, when the best system design is identified, they are used to specify its optimal operation.

Demand is the primary stimulus on the repairable equipment population system and the justification for its existence. This uncontrollable system parameter is assumed to be constant over time. Other uncontrollable system parameters are economic in nature. They include the shortage penalty cost which arises when there are insufficient units operational to meet demand, the cost of providing repair capability, and the time value of money on invested capital.

Some system parameters are uncontrollable in REPS Problem I, but controllable in REPS Problem II. These are the design MTBF and MTTR, the energy efficiency of equipment units, the design life of units, and the first cost and salvage value of these units. It is through these design-dependent system parameters that the best candidate system may be identified.

Evaluation Function Formulation

A mathematical model for system design evaluation can be formulated using the structure outlined in Chapter 7. The model uses annual equivalent life-cycle cost as the evaluation measure expressed as

$$AELCC = PC + OC + RC + SC \tag{10.44}$$

where

$AELCC$ = annual equivalent life-cycle cost
PC = annual equivalent population cost
OC = annual operating cost
RC = annual repair facility cost
SC = annual shortage penalty cost

Annual equivalent population cost. The annual equivalent cost of a deployed population of N equipment units is

$$PC = C_i N$$

with

$$C_i = P(\overset{A/P,i,n}{\quad}) - B(\overset{A/F,i,n}{\quad}) \tag{10.45}$$

Alternatively, $(P - B)(\overset{A/P,i,n}{\quad}) + Bi$ could be used as given by Equation 8.18.

Book value, B, in Equation 10.45 is used to represent the original value of a unit minus its accumulated depreciation at any point in time. The depreciation of a unit over its lifetime by the straight-line method gives an expression for book value as

$$B = P - n\frac{P - F}{L} \tag{10.46}$$

where

C_i = annual equivalent cost per unit
P = first or acquisition cost of a unit

F = estimated salvage value of a unit
B = book value of a unit at the end of year n
L = estimated design life of the unit
N = number of units in the population
n = retirement age of units $n > 1$
i = annual interest rate

Annual operating cost. The annual cost of operating a population of N deployed equipment units is

$$OC = (EC + LC + PMC + \text{Other})N \tag{10.47}$$

where

EC = annual cost of energy consumed
LC = annual cost of operating labor
PMC = annual cost of preventive maintenance

Other annual operating costs may be incurred. These include all recurring annual costs of keeping the population of equipment units in service such as storage cost, insurance premiums, and taxes.

Annual repair facility cost. The annual cost of providing a repair facility to repair failed equipment units is

$$RC = C_r M \tag{10.48}$$

where

C_r = annual fixed and variable repair cost per repair channel
M = number of repair channels

If there are a number of repair channel components with different estimated lives, then C_r is the sum of their annual costs. Some of the repair facility cost items that could be included are the cost of the building, maintenance supplies, test equipment, and so on, expressed on a per channel basis. The administrative, maintenance manpower, and other overhead costs would also be computed on a yearly basis and on a per channel basis.

Annual shortage penalty cost. When failed equipment units cause the number in an operational state to fall below the demand, an out-of-operation or shortage cost is incurred. The annual shortage cost is the product of the shortage cost per unit short per year and the expected number of units short expressed as

$$SC = C_s[E(S)] \tag{10.49}$$

The expected number of units short can be found from the probability distribution of n units short, P_n, as developed from Equations 10.38 and 10.39 in Section 10.6.

Define the quantity $N - D$ as the number of extra units to be held in the population. For $n = 0, 1, 2, \ldots, N - D$ there is no shortage of units. However, when

$$n = N - D + 1, \text{ a shortage of 1 units exists}$$

$$n = N - D + 2, \text{ a shortage of 2 units exists}$$

$$\vdots$$

$$n = N, \text{ a shortage of } D \text{ units exists}$$

The expected number of units short, $E(S)$, can be found by multiplying the number of units short by the probability of that occurrence as

$$E(S) = \sum_{j=1}^{D} jP_{(N-D+j)} \tag{10.50}$$

REPS Problem I

In this REPS example, the decision maker has no control over system parameters but can only choose the number of equipment units to procure and deploy, the age at which units should be replaced, and the number of channels in the repair facility.

Assume that the demand, D, is for 15 identical equipment units. Table 10.7 lists system parameters for this design example.

TABLE 10.7 System Parameters For REPS Problem 1

PARAMETER	VALUE	
Unit acquisition cost	$52,000	
Unit design life	6 years	
Unit salvage value at end of design life	$7,000	
Unit operating cost		
Energy and fuel	$500	
Operating labor	450	
Preventive maintenance	400	
Other operating costs	400	
Annual repair channel cost	$45,000	
Annual shortage cost	$73,000	
Annual interest rate	10%	
Age cohorts	MTBF	MTTR
0–1	0.20	0.03
1–2	0.24	0.04
2–3	0.29	0.05
3–4	0.29	0.05
4–5	0.26	0.06
5–6	0.22	0.07

TABLE 10.8 Design Variables for REPS Problem 2

Population, N	Repair channels, M	Retirement age, n
19	3	4

Table 10.8 exhibits a set of design variables for the example under consideration. The example computations that follow are based on these variables and the system parameters in Table 10.7.

Annual equivalent population cost. First compute the book value, B, of the units at retirement age after four years using Equation 10.46 as

$$\$52{,}000 - 4\left(\frac{\$52{,}000 - \$7{,}000}{6}\right) = \$22{,}000$$

The annual equivalent unit cost per unit from Equation 10.45 is

$$C_i = \$52{,}000\left[\frac{0.10(1.10)^4}{(1.10)^4 - 1}\right] - \$22{,}000\left[\frac{0.10}{(1.10)^4 - 1}\right]$$
$$= \$16{,}404 - \$4{,}740 = \$11{,}664$$

from which the annual equivalent population cost is

$$PC = \$11{,}664(19) = \$221{,}616$$

Annual operating cost. Annual operating cost for the deployed population is found from Equation 10.47 to be

$$OC = (\$500 + \$450 + \$400 + \$400)(19) = \$33{,}250$$

Annual repair facility cost. The annual equivalent repair channel cost for three channels from Equation 10.48 is

$$RC = \$45{,}000(3) = \$135{,}000.$$

Annual shortage cost. Calculation of the shortage cost is based on the MTBF and MTTR values from Table 10.7 for years 1 to 4. From these values, the average MTBF and MTTR for the population can be computed as[2]

[2] Because the units are homogeneous, the aggregate MTBF and MTTR for the population can be found as

$$\text{MTBF} = \frac{1}{n}\sum_{j=1}^{n} \text{MTBF}_j$$
$$\text{MTTR} = \frac{1}{n}\sum_{j=1}^{n} \text{MTTR}_j$$

where the subscript j represent age groups and n is the number of these age groups. This follows from the superposition of Poisson processes.

$$\text{MTBF} = (1/4)(0.20 + 0.24 + 0.29 + 0.29) = 0.2550$$

$$\text{MTTR} = (1/4)(0.03 + 0.04 + 0.05 + 0.05) = 0.0425$$

The failure rate of an item and the repair rate at a repair channel are given by

$$\lambda = (1/\text{MTBF}) = (1/0.2550) = 3.9215$$

$$\mu = (1/\text{MTTR}) = (1/0.0425) = 23.5294$$

from which $\left(\frac{\lambda}{\mu}\right) = 1/6$. Next compute C_n for $n = 0, 1, \ldots, 3$ from Equation 10.39 as

$$C_0 = \frac{19!(1/6)^0}{19!0!} = 1$$

$$C_1 = \frac{19!(1/6)^1}{18!1!} = 3.1665$$

$$C_2 = \frac{19!(1/6)^2}{17!2!} = 4.7496$$

$$C_3 = \frac{19!(1/6)^3}{16!3!} = 4.4856$$

Computing C_n for $n = 4, \ldots, 19$, also from Equation 10.39

$$C_4 = \frac{19!(1/6)^4}{15!3!3^1} = 3.9813$$

$$C_5 = \frac{19!(1/6)^5}{14!3!3^2} = 3.3224$$

$$C_6 = \frac{19!(1/6)^6}{13!3!3^3} = 2.5840$$

$$\vdots$$

$$C_{18} = \frac{19!(1/6)^{18}}{1!3!5^{15}} = 0.0000$$

$$C_{19} = \frac{19!(1/6)^{19}}{0!3!3^{16}} = 0.0000$$

Now,

$$\sum_{n=0}^{19} C_n = 27.9390$$

and from Equation 10.38

$$P_0 = \frac{1}{\sum_{n=0}^{19} C_n} = \frac{1}{27.9390} = 0.0358$$

P_n for $n = 1, 2, \ldots, N$ can now be computed from $P_n = P_o C_n = (0.0358)\, C_n$ as follows:

$$P_0 = 0.0358 \times 1 \qquad = 0.0358$$
$$P_1 = 0.0358 \times 3.1665 = 0.1134$$
$$\vdots$$
$$P_5 = 0.0358 \times 3.3224 = 0.1189$$
$$P_6 = 0.0358 \times 2.5840 = 0.0925$$
$$\vdots$$
$$P_{18} = 0.0358 \times 0.0000 = 0.0000$$
$$P_{19} = 0.0358 \times 0.0000 = 0.0000$$

Now the expected number of units short can be calculated from Equation 10.50 as

$$\begin{aligned} E(S) &= \sum_{j=1}^{D} jP_{(N-D+j)} = \sum_{j=1}^{15} jP_{(4+j)} \\ &= 1(0.1189) + 2(0.0925) + \cdots + 15(0.0000) \\ &= 1.00663 \end{aligned}$$

From which the annual shortage cost is

$$\begin{aligned} SC &= C_S[E(S)] \\ &= \$73{,}000(1.00663) = \$73{,}484 \end{aligned}$$

The total system annual equivalent cost may now be summarized as

$$\begin{aligned} TC &= PC + OC + RC + SC \\ &= \$221{,}616 + \$33{,}250 + \$135{,}000 + \$73{,}484 = \$463{,}350 \end{aligned}$$

As can be seen from Table 10.9, this solution is actually the optimum with neighbouring points given for N, M, and n.

TABLE 10.9 Points in the Optimum Region

Retirement Age, n	Number of Units, N	Number of repair channels, M: 2	3	4
3	19	\$598,395	\$465,985	\$469,130
	18	592,920	464,770	465,755
4	19	600,720	463,350*	464,295
	20	610,775	466,610	468,755
5	19	643,050	480,375	467,735

*Optimum.

The shortage distribution can be calculated from the P_n values and plotted as a histogram of $Pr(S = s) = N - D + s$. In this example, $Pr(S = s) = P_{4+s}$

$$Pr(S = 0) = 0.622$$

$$Pr(S = 1) = 0.119$$

$$Pr(S = 2) = 0.093$$

$$Pr(S = 3) = 0.067$$

$$Pr(S = 4) = 0.046$$

$$Pr(S = 5) = 0.027$$

$$Pr(S = 6) = 0.015$$

$$Pr(S = 7) = 0.008$$

$$Pr(S = 8) = 0.003$$

The shortage probability histogram for this example is exhibited in Figure 10.11.

In the next section the adequacy of this "baseline" design will be considered in the face of design requirements. Additionally, challengers in the form of alternative candidate systems will be presented and evaluated.

REPS Problem II

In REPS Problem II, the decision maker has control over a set of design-dependent parameters in addition to the set of design variables. Accordingly, the decision evaluation function of Equation 7.3 applies. In this section, design-dependent parameters will be altered to seek a REPS candidate more acceptable than the baseline design.

Considering design requirements. In the absence of design requirements, the baseline design might be acceptable. However, assume that it is not acceptable due to the existence of the following design requirements:

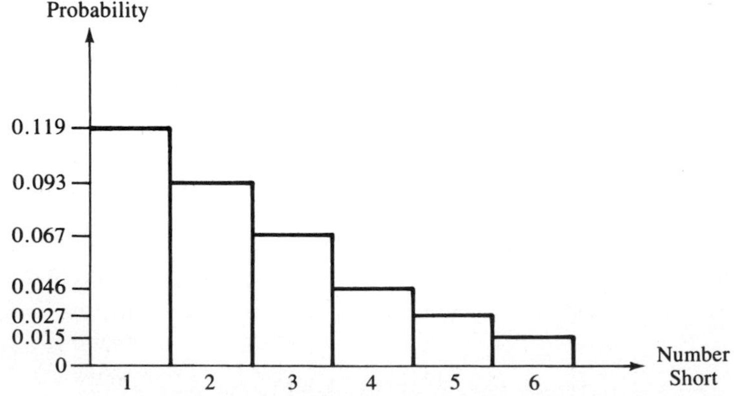

Figure 10.11 Shortage probability histogram.

1. The first cost of the deployed population must not exceed $900,000.
2. The probability of no units short of demand must be at least 0.70.
3. The mean MTBF for the unit over its life must be at least 0.20 year.
4. A unit must not be kept in service more than 4 years.

Note that the baseline design does not meet requirement 1 and 2, but it does meet requirements 3 and 4. Also its annual equivalent life-cycle cost is noted to be $463,350. Figure 10.12 shows a design evaluation display for this situation with the baseline design indicated. However, because it does not meet all design requirements one or more alternative candidates must be designed that will meet all requirements at an acceptable life-cycle cost.

Generating candidate systems. Suppose that two candidate systems are generated in the face of the demand for 15 equipment units and the design independent parameters of Table 10.7. Design-dependent parameters for these candidates are given in Table 10.10.

Optimization over the design variables for each set of design-dependent parameters gives the results summarized in Table 10.11 for the candidate systems. Also included are the optimized results for the baseline design. The following observations are made:

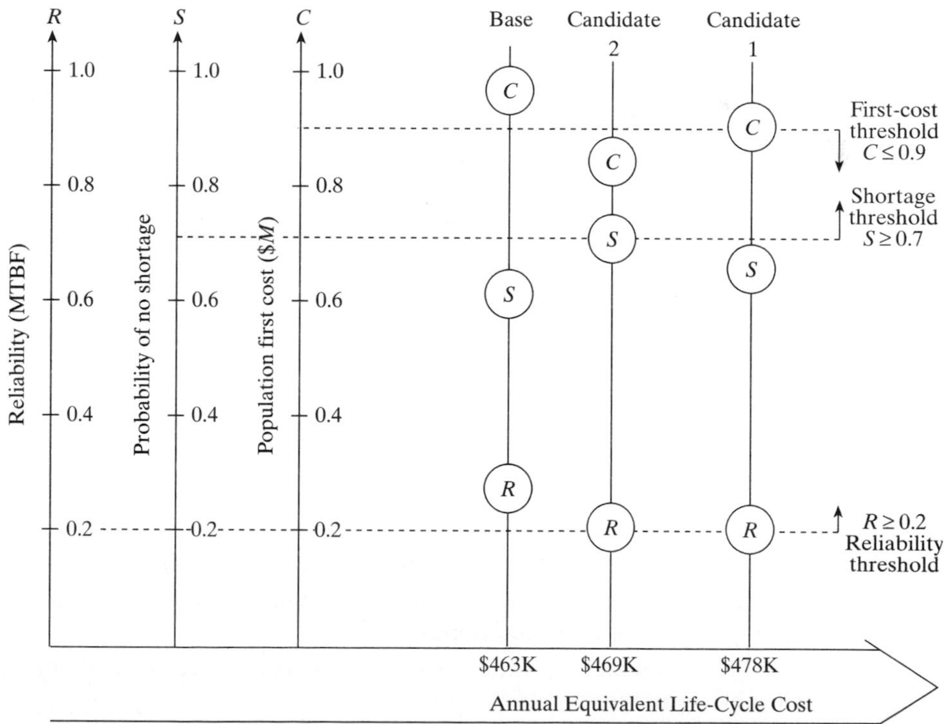

Figure 10.12 Design decision evaluation display for REPS.

TABLE 10.10 Design-Dependent Parameters for Candidate Systems

Parameter	Candidate System 1		Candidate System 2	
Unit acquisition cost	$45,000		$43,000	
Unit design life	6 years		6 years	
Unit salvage value at end of design life	$6,000		$5,000	
Unit operating cost				
Energy and fuel	$600		$800	
Operating labor	500		700	
Preventative maintenance	400		400	
Other operating costs	400		400	
Age cohorts	MTBF	MTTR	MTBF	MTTR
0–1	0.16	0.04	0.18	0.04
1–2	0.21	0.04	0.21	0.04
2–3	0.26	0.05	0.25	0.05
3–4	0.26	0.06	0.25	0.05
4–5	0.26	0.06	0.23	0.06
5–6	0.24	0.06	0.20	0.06

TABLE 10.11 Evaluation Summary for REPS Candidates

	Baseline Design	Candidate System 1	Candidate System 2
Number deployed, N	19	20	20
Repair channels, M	3	4	4
Retirement age, n	4	5	4
Unit cost, P	$52,000	$45,000	$43,000
Mean MTBF	0.26	0.23	0.22
Probability of no units short	0.622	0.663	0.730
Annual equivalent life-cycle cost	$463,350	$478,470	$468,825

1. Only Candidate System 2 meets all design requirements.
2. Candidate System 2 has the lowest cost per unit and, therefore, the lowest investment cost for the deployed population.
3. Candidate System 2 has the highest probability of meeting the demand for 15 serviceable equipment units.
4. Candidate System 2 consumes the most energy and requires the highest expenditure for operating labor.

5. Candidate System 2 has an annual equivalent life-cycle cost penalty of $468,825 less $463,350 or $5,475 over the baseline design but has a lower investment cost than Candidate System 1.

There may be other candidate systems that meet all design requirements and which have life-cycle costs equal to or less than the LCC for Candidate System 2. The objective of REPS Problem II is to formalize the evaluation of candidates, with each candidate being characterized by its design-dependent parameter values, comparison of the candidates is made only after the optimized (minimized) values for life-cycle cost are found. This is in keeping with the concept of equivalence set forth in Equations 7.1 and 7.3.

Summary

This queuing system design application models and optimizes a repairable equipment population system composed of operational units and repair facilities used to keep the units operational. In REPS Problem I it is assumed that the units are procured with a predetermined MTBF and MTTR, with these values varying as a function of the age of units. The annual equivalent life-cycle cost is minimized by finding the population size, the replacement age, and the number of repair channels.

In REPS Problem II, a decision is made by considering units with MTBF and MTTR characteristics which are a function of the first cost of the units. Here it is assumed that improvements in MTBF and MTTR can be obtained at a price. This leads to the multiple application of the optimization routine for REPS Problem I, an application to each possible MTBF and MTTR case. The result is the specification of design MTBF and MTTR values for the units to be deployed.

QUESTIONS AND PROBLEMS

1. Under what conditions is it necessary to use Monte Carlo analysis in the study of a queuing system?
2. Without changing the standard deviation of S_s in Figure 10.2, change its mean from 5 to 6 and then redo 100 periods of the Monte Carlo analysis. Discuss the observed difference.
3. Why is it essential that μ be greater than λ in a probabilistic waiting-line process?
4. Suppose that the time between arrivals is distributed exponentially with a mean of 6 mins and that service time is constant and equal to 5 mins. Service is provided on a first-come, first served basis. No units are in the system at 8:00 A.M. Use Monte Carlo analysis to estimate the total number of unit minutes of waiting between 8:00 A.M. and 12:00 noon.
5. Use Monte Carlo analysis to verify Equation 10.9 if the number of arrivals per period has a Poisson distribution with a mean of 0.10 and if the service duration is distributed exponentially with a mean of four periods. Plot the histogram of the number of units in the system and compare with Figure 10.6.
6. Suppose that arrivals are distributed according to the Poisson distribution with a mean of 0.125 unit per period and that the service duration is distributed exponentially with a mean

of five periods. Develop the probability distribution of n units in the system. What is the probability of there being more than 4 units in the system?

7. The arrival rate for a certain waiting-line system obeys a Poisson distribution with a mean of 0.5 unit per period. It is required that the probability of one or more units in the system not exceed 0.20. What is the minimum service rate that must be provided if the service duration will be distributed exponentially?
8. What is the expected number of units in the system and the expected waiting time for the conditions of Problem 6?
9. The expected number of units in a waiting-line system experiencing Poisson arrivals with a rate of 0.4 unit per period must not exceed 8. What is the minimum service rate that must be provided if the service duration will be distributed exponentially? What will be the expected waiting time?
10. If the cost of waiting in Problem 7 is $5.00 per unit per period, and the service facility costs $2.50 per unit served, what is the total system cost? If the service rate is doubled at a total cost of $3.50 per unit served, what is the total system cost?
11. Plot the mean number of units in the system, and the mean waiting time, as a function of λ/μ, if the number of arrivals per period obey the Poisson distribution and if the service duration is distributed exponentially. What is the significance of this illustration?
12. The number of arrivals per period is distributed according to the Poisson with an expected value of 0.75 unit per period. The cost of waiting per unit per period is $3.20. The facility cost for serving one unit per period is $5.15. What expected service rate should be established if the service duration will be distributed exponentially? What is the expected total system cost?
13. The expected waiting time in a waiting-line system with Poisson arrivals at a rate of 1.5 units per day must not exceed 6 days. What is the minimum constant service rate that must be provided? What is the expected number in the system?
14. Trucks arrive at a loading dock in a Poisson manner at the rate of 3.5 per day. The cost of waiting per truck per day is $168. A three-person crew that can load at a constant rate of four trucks per day costs $224 per day. Compute the total system cost for the operation.
15. Customers arrive at a bank at the rate of 0.2 per period and the service rate is 0.4 customers per period. If there is one teller, what is the average length of the line? If the arrival rate increases to 0.6 per period and the number of tellers is increased to two, how does the average length of the line change. Assume Poisson arrivals and an exponential service-time distribution.
16. Cars arrive at a toll plaza at the rate of 60 cars per hour. There are four booths each capable of servicing 30 cars per hour. Assuming a Poisson distribution for arrivals and an exponential distribution for the service time, determine (a) the probability of no cars in the toll plaza, (b) the average length of the line, and (c) the average time spent by a car in the plaza.
17. In Problem 16, if it costs $17.20 per hour to operate a toll booth and the waiting cost per car is $0.25 per minute spent at the plaza, how many booths should be operated to minimize total cost?
18. In a three-channel queuing system the arrival rate is I unit per time period and the service rate is 0.6 unit per time period. Determine the probability that an arriving unit does not have to wait.
19. Use Equation 10.39 to extend the example in Section 10.6 to the case where there are two service channels. Compare the mean number waiting with the single channel assumption.
20. Each truck in a fleet of 30 delivery trucks will return to a warehouse for reloading at an average interval of 160 minutes. An average of 20 minutes is required by the driver and one ware-

houseman to load the next shipment. If the warehouse personnel are busy loading previous arrivals, the driver and his unit must wait in line. Both the time between arrivals and the loading time are distributed exponentially. The cost of waiting in the system is $18.20 per hour per truck and the total cost per warehouseman is $10.65 per hour. Find the minimum cost number of warehousemen to employ.

21. A population of 10 cargo aircraft each produces a profit of $1,360 per 24-hour period when not waiting to be unloaded. The time between arrivals is distributed exponentially with a mean of 144 hours. The unloading time at the ramp is distributed exponentially with a mean of 18 hours. It costs $42 per hour to lease a ramp with this unloading capacity. How many ramps should be leased?

22. Each unit in a 10-unit population returns for service at an average interval of 20 minutes. If the cost of waiting is $10 per hour per unit and the cost of service is $\$15/T$ per unit per hour, find the optimum service time if there are two service channels. Assume Poisson arrivals and exponential service times.

23. A population of 30 chemical processing units is to be unloaded and loaded by a single crew. The mean time between calls for this operation is 68 mins. If the mean service rate is one unit per minute, the cost of the crew and equipment will be $16 per minute. This cost will decrease to $12, $9, and $7 for service intervals of two, three, and four minutes, respectively. The time between calls-for service and the service duration are distributed exponentially. If it costs $28 per hour for each unit that is idle, find the minimum cost service interval.

24. Plot the percentage of machines not running as a function of the service factor for a population of 20 machines with exponential arrivals and services (a) if a single channel is employed and (b) if two channels are employed.

25. A population of 10 production cells is to be served by an unknown number of service channels. The mean time between service calls is 15 minutes. The cost per service channel is $60 per hour at a mean service rate of 60 cells per hour and the service cost per channel per hour is inversely proportional to the service rate. Both the time between calls for service and the service times are distributed exponentially. If the profit from each production cell is $10 per hour, determine the optimum mix of service channels and service times.

26. A group of 12 machines are repaired by a single repairperson when they break down. Each machine yields a profit of $5.20 per hour while running. The mechanic costs the firm $20.40 per hour, including overhead. At the present time no preventive maintenance is used. It is proposed that one technician be employed to perform certain routine maintenance and adjustment tasks. This will cost $16.60 per hour but will reduce the service factor from 0.13 to 0.08. What is the economic advantage of implementing preventive maintenance?

27. Use Equation 7.2 to discuss and explain the essential aspects of the evaluation of REPS Problem I.

28. Use Equation 7.3 to discuss and explain the essential aspects of the evaluation of REPS Problem II.

29. Perform the calculations to verify that the annual equivalent annual life-cycle cost for Candidate System 2 in Table 10.11 is $468,825.

30. Show that the probability of one or more units short for Candidate System 2 in Table 10.11 is 0.27.

11 Control Concepts and Techniques

Most systems are deployed and then operate in an environment that changes over time. Except for static systems, a changing environment can lead to system instability unless control action is applied. The value of dynamic systems may be enhanced through control action applied during operations. Control of portions or all of a system can help maintain system performance within specified tolerances, or to increase the worth of system output.

The concern for control began with physical systems and servo theory. Engineers who turned their attention to large-scale systems in the 1950s were trained in servomechanisms and other control devices. A large-scale, human-made system is one in which there are many states, control variables, and constraints. There may be several control loops and different measures of performance for each subsystem. Classical control theory cannot be used directly to select measures of effectiveness and to optimize outputs. Nevertheless, the concepts can be used to provide a basis for structuring system description and internal relationships so that feedback and adaptive phenomena are incorporated in system design.

11.1 SOME CONTROL CONCEPTS

A control problem may arise from the need to regulate speed, temperature, quality, quantity, or to determine the trajectory of an aircraft or space vehicle. In each case, the allocation of scarce resources over time is involved. Control variables must be related in some way to state variables describing the system characteristic or condition being controlled.

The classical example of control is the determination of a missile trajectory. Control variables are the amount, timing, and direction of the thrust applied to the missile, where the thrust available is constrained by fuel availability. State variables describe the trajectory incorporating mass, position, and velocity. The thrusts and the state variables are related by differential equations in the light of the effectiveness function to be optimized. The objective is usually to maximize the payload that can be delivered to a given destination.

Control problems also arise in large-scale person-equipment systems of activity. In air traffic control, state variables are the number of aircraft enroute, the number of aircraft waiting to land at available airports, the number of aircraft waiting on the ground, the available number of communication channels, and so on. Control variables in air traffic control are the instructions given by the traffic controllers.

The Elements of a Control System

Every control system has four basic elements. These elements always occur in the same sequence and have the same relationship to each other. They are

1. A controlled characteristic or condition.
2. A sensory device or method for measuring the characteristic or condition.
3. A control device that will compare measured performance with planned performance.
4. An activating device that will alter the system to bring about a change in the output characteristic or condition being controlled.

Figure 11.1 illustrates the relationships among the four elements of a control system. The output of the system (Figure 11.1, Block 1) is the characteristic or condition to be measured. This output may be speed, temperature, quality, or any other characteristic or condition of the system under consideration.

The sensory device (Figure 11.1, Block 2) measures output performance. System design must incorporate such devices as tachometers, thermocouples, thermometers, transducers, inspectors, and other physical or human sensors. Information gathered from a sensory device is essential to the operation of a control system.

The third element in a control system is the control device (Figure 11.1, Block 3). This element exists to determine the need for control action based on the information provided by the sensory device. It may be a minicomputer or microcomputer, but need be no more than a visual or hand-calculated comparison depending upon the situation.

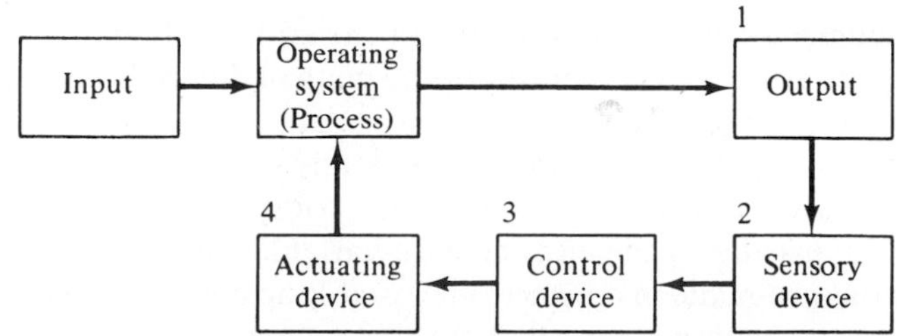

Figure 11.1 Control system elements and relationships.

It is important that this element in the control system be able to detect significant differences between planned output and actual output.

Any out-of-control signal received from the control device must be acted upon by the actuating device (Figure 11.1, Block 4). This device is the fourth element in a control system. Its role is that of implementation. Actuating devices may be mechanical, electromechanical, hydraulic, pneumatic, or they may be human, as in the case where a machine operator changes the setting on a machine to alter the dimension of a part being produced.

Information as the Medium of Control

It is the flow of measurement information, which, when converted into corrective information, makes it possible for a characteristic or condition of a system to be controlled. To control every characteristic or condition of a system's output is usually not possible. Accordingly, it is important to make a wise choice of the characteristic to be measured. It is important that the successful control of the characteristic chosen lead to successful control of the system. This requirement is usually met if the controlled characteristic is stated in the language to be used in the corrective loop. Before information is gathered, one must determine what is to be transmitted.

Information to be compared with the control standard must be expressed in the same terms as the standard to facilitate a decision regarding the control status of the system. This information can often be secured by sampling the output. Of course, a sample is only an estimate of a group or population. Statistical sampling based on probability theory can, however, provide valid control information subject to some error.

The sensor element in Figure 11.1 provides information that becomes the basis for control action. Output information is compared with the characteristic or condition to be controlled. Significant deviations are noted and corrective information is provided to the activating device.

Information received from the control device causes the actuating device to respond. However, every control system has a finite delay in its action time which can cause problems of over correction. This difficulty can be overcome by reducing the time lag between output measurement and input correction or by introducing a lead time in the control system.

Types of Control Systems

There are two basic types of control systems, open-loop and closed-loop. Each has its own characteristic. The basic difference is whether or not the control device is an integral part of the system it controls.

In *open-loop control*, the optimal control action is completely specified at the initial time in the system cycle. In *closed-loop control*, the optimal control action is determined both by the initial conditions and by the current state-variable values. Whereas in open-loop control all decisions are-made in advance, in closed-loop control decisions are revised in the light of new information received about the system state.

Common examples of open-loop and closed-loop control are the laundry dryer and the home thermostat. In the laundry dryer the length of the drying cycle is set once by human action. A home thermostat, on the other hand, provides control signals to the furnace whenever temperature falls below a predetermined level. When a system is regulated by a person, as in the case of a laundry dryer, attention must be paid to see that the desired output is obtained. In this open-loop example, the drying time must be determined by experience and be properly set into the timer.

If control is exercised in terms of system operation, as with a thermostat, closed-loop control is occurring. The significant difference in this type of control system is that the control device is an element of the heating system it serves. In closed-loop control the four elements of a control system all act together.

An essential part of a closed-loop system is feedback. *Feedback* is a process in which the output of a system is measured continuously or at predetermined intervals as in sample-data control. Generally, costs are incurred in measuring a characteristic or condition. These costs may be in terms of added workload, energy, weight, and so on. Open-loop control requires only initial measurements and is preferred unless there are random elements either in measurement or in system operation.

11.2 STATISTICAL PROCESS CONTROL

Variation is inherent in the output characteristic or condition of most processes. Patterns of statistical variation exist in the number of units processed through maintenance per time period, the dimension of a machined part, the procurement lead time for an item, and the number of defects per unit quantity of product produced. Statistical process control is a methodology for testing to see if an operating system is in control.

A statistical control chart is a control device (Figure 11.1, Block 3) that uses measurement information from a sensory device (Figure 11.1, Block 2) and provides system change information to an actuating device (Figure 11.1, Block 4). Control charts use samples taken from system output and control limits to determine if an operation is in control. When a sample value falls outside the control limits, a stable system of statistical variation probably no longer exists. Action may then be taken to find the cause of the "out-of-control" condition so that control may again be established, or compensation made for the change that has occurred.

Patterns of Statistical Variation

A stable or probabilistic steady-state pattern of variation exists when the parameters of the statistical distribution describing an operation remain constant over time. Steady-state variation of this type is normally an exception rather than the rule. Many operations will produce a probabilistic non–steady-state pattern of variation over time. When this is the case, the mean or the variance of the distribution describing the pattern change with time. Sometimes, the form of the distribution will also undergo chance.

Control limits may often be placed about an initial stable pattern of variation to detect subsequent changes from that pattern, a statistical inference is made when a

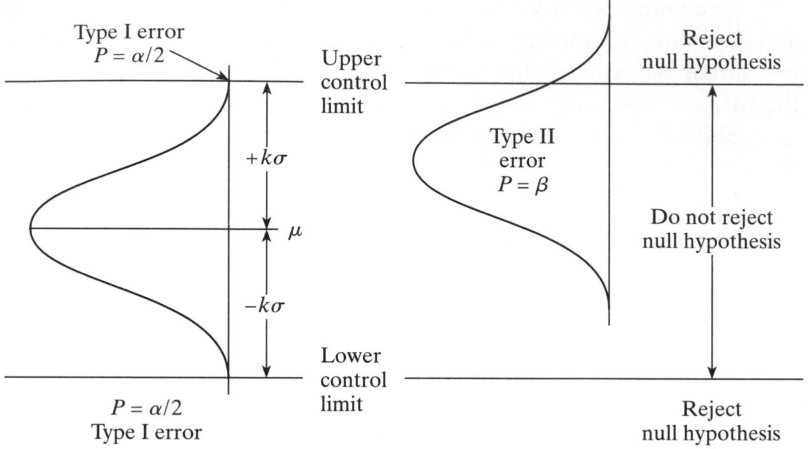

Figure 11.2 A stable and a changed pattern of variation.

sample value is considered. If the sample falls within the control limits, the process under study is said to be *in control.* If the sample falls outside the limits, the process is deemed to have changed and is said to be *out of control.*

Figure 11.2 illustrates control limits placed at a distance $k\sigma$ from the mean of the initial stable pattern of variation. If a sample value falls outside the control limits, and the process has not changed, a Type I error of probability α has been made. Conversely, if a sample value falls within these limits and the process has changed, a Type II error of probability β has been made. In the first case, the null hypothesis has been rejected in error. In the second, the null hypothesis has not been rejected, with a resulting error.

If the control limits are set relatively far apart, it is unlikely that a Type I error will be made. The control model, however, is then not likely to detect small shifts in the parameter. In this case the probability, β, of a Type II error is large. If the other approach is taken, and the limits are placed relatively close to the initial stable pattern of variation, the value of α will increase. The advantage gained is greater sensitivity of the model for the shifts that may occur. The ultimate criterion will be the costs associated with the making of Type I and Type II errors. The limits should be established to minimize the sum of these two costs.

Control Charts for Variables

A *control chart* is a graphical representation of a mathematical model used to monitor a random variable process to detect changes in a parameter of that process. Charting statistical data is a test of the null hypothesis that the process from which the sample came has not changed. A control chart is employed to distinguish between the existence of a stable pattern of variation and the occurrence of an unstable pattern. If an unstable pattern of variation is detected, action may be initiated to discover the cause of the instability. Removal of the assignable cause should permit the process to return to the stable state.

Control charts for variables are used for continuous operations. Two charts are available for operations of this type. The $\overline{X}$ *chart* is a plot over time of sample means taken from a process. It is primarily employed to detect changes in the mean of the process from which the samples came. The *R chart* is a plot over time of the ranges of these same samples. It is employed to detect changes in the dispersion of the process. These charts are often employed together in control operations.

Constructing the $\overline{X}$ chart. The $\overline{X}$ chart receives its input as the mean of a sample taken from the process under study. Usually, the sample will contain four or five observations, a number sufficient to make the Central Limit Theorem applicable.[1] Accepting an approximately normal distribution of the sample means allows the establishment of control limits with a predetermined knowledge of the probability of making a Type I error. It is not necessary to know the form of the distribution of the process.

The first step in constructing an $\overline{X}$ chart is to estimate the process mean, μ, and the process variance, σ^2. This requires taking m samples each of size n and calculating the mean, $\overline{X}$, and the range, R, for each sample. Table 11.1 illustrates the format that may be used in the calculations. The mean of the sample means, $\overline{\overline{X}}$, is used as an estimate of μ and is calculated as

$$\overline{\overline{X}} = \frac{\sum_{i=1}^{m} \overline{X}}{m} \tag{11.1}$$

and the mean of the sample ranges, $\overline{R}$, is calculated as

$$\overline{R} = \frac{\sum_{i=1}^{m} R}{m} \tag{11.2}$$

The expected ratio between the average range, $\overline{R}$, and the standard deviation of the process has been computed for various sample sizes, n. This ratio is designated d_2 and is expressed as

$$d_2 = \frac{\overline{R}}{\sigma}$$

TABLE 11.1 Computational Format for Determining $\overline{X}$ and R

Sample Number	Sample values	Mean $\overline{X}$	Range R
1	$x_{11}, x_{12}, \cdots, x_{1n}$	$\overline{X}_1$	R_1
2	$x_{21}, x_{22}, \cdots, x_{2n}$	$\overline{X}_2$	R_2
$\vdots$	$\vdots \quad \vdots \quad \vdots$	$\vdots$	$\vdots$
m	$x_{m1}, x_{m2}, \cdots, x_{mn}$	$\overline{X}_m$	R_m

[1] The *Central Limit Theorem* states that the distribution of sample means approaches normality as the sample size approaches infinity, provided that the sample values are independent.

TABLE 11.2 Factors for the Construction of $\overline{X}$ and R Charts

Sample size, n	$\overline{X}$ Chart		R Chart	
	d_2	A_2	D_3	D_4
2	1.128	1.880	0	3.267
3	1.693	1.023	0	2.575
4	2.059	0.729	0	2.282
5	2.326	0.577	0	2.115
6	2.534	0.482	0	2.004
7	2.704	0.419	0.076	1.924
8	2.847	0.373	0.136	1.864
9	2.970	0.337	0.184	1.816
10	3.078	3.308	0.223	1.777

Therefore, σ can be estimated from the sample statistic $\overline{R}$ as

$$\sigma = \frac{\overline{R}}{d_2} \tag{11.3}$$

Values of d_2 as a function of the sample size may be found in Table 11.2.

The mean of the $\overline{X}$ chart is set at $\overline{\overline{X}}$. The control limits are normally set at $\pm 3\sigma_{\overline{X}}$, which results in the probability of making a Type I error of 0.0027. Because

$$\sigma_{\overline{X}} = \frac{\sigma}{\sqrt{n}}$$

substitution into Equation 11.3 gives

$$\sigma_{\overline{X}} = \frac{\overline{R}}{d_2\sqrt{n}}$$

and

$$3\sigma_{\overline{X}} = \frac{3\overline{R}}{d_2\sqrt{n}} \tag{11.4}$$

The factor $3/d_2\sqrt{n}$ has been tabulated as A_2 in Table 11.2. Therefore, the upper and lower control limits for the $\overline{X}$ chart may be specified as

$$UCL_{\overline{X}} = \overline{\overline{X}} + A_2\overline{R} \tag{11.5}$$

$$LCL_{\overline{X}} = \overline{\overline{X}} - A_2\overline{R} \tag{11.6}$$

Constructing the *R* chart. The R chart is constructed in a manner similar to the $\overline{X}$ chart. If the $\overline{X}$ chart has already been completed, $\overline{R}$ has been calculated from Equation 11.2. Tabular values of 3σ control limits for the range have been compiled for varying sample sizes and are included in Table 11.2. The upper and lower control limits for the R chart are then specified as

$$UCL_R = D_4\overline{R} \tag{11.7}$$

$$LCL_R = D_3\overline{R} \tag{11.8}$$

Because $D_3 = 0$ for sample size of $n \leq 6$ in Table 11.2, the $LCL_R = 0$. Actually, 3σ limits yield a negative lower control limit which is recorded as zero. This means that with samples of six or fewer, it will be impossible for a value on the R chart to fall outside the lower limit. Thus the R chart will not be capable of detecting reductions in the dispersion of the process output.

Application of the $\overline{X}$ and R charts. Once the control limits have been specified for each chart, the data used in constructing the limits are plotted. Should all values fall within both sets of limits, the charts are ready for use. Should one or more values fall outside one set of limits, however, further inquiry is needed. A value outside the limits on the $\overline{X}$ chart indicates that the process may have undergone some change in regard to its central tendency. A value outside the limits on the R chart is evidence that the process variability may now be out of control. In either case, one should search for the source of the change in process behavior. If one or two values fall outside the limits and an assignable cause can be found, then these one or two values may be discarded and revised control limits calculated. If the revised limits contain all the remaining values, the control chart is ready for implementation. If they do not, the procedure may be repeated before using the control chart.

Assume that control charts are to be established to monitor the weight in ounces of the contents of containers being filled on an assembly line. The containers should hold at least 10 oz, and to guarantee this weight, the process must be set to deliver slightly more. Samples of five have been taken every 30 minutes. The sample data, together with the sample means and sample ranges, are given in Table 11.3.

An R chart is first constructed from these data. Using Equation 11.2, $\overline{R}$ is calculated as 2.05. The control limits are then determined from Equations 11.7 and 11.8 as

$$UCL_R = 2.115(2.05) = 4.34$$

$$LCL_R = 0(2.05) = 0$$

These limits are used to construct the R chart of Figure 11.3. Since all values fall within the control limits, the R chart is accepted as a means of assessing subsequent process variation. Had a point fallen outside the calculated limits, that point would have had to be discarded and limits recalculated.

Attention should next be directed to the $\overline{X}$ chart. The mean of the sample means, $\overline{\overline{X}}$, is found from Equation 11.1 to be 11.19. The mean of the sample ranges, $\overline{R}$, has already been calculated as 2.05. Preliminary control limits for the $\overline{X}$ chart can now be calculated from Equations 11.5 and 11.6 as

$$UCL_{\overline{X}} = 12.19 + 0.577(2.05) = 13.37$$

$$LCL_{\overline{X}} = 12.19 - 0.577(2.05) = 11.01$$

These limits are used to construct the $\overline{X}$ chart shown in Figure 11.4.

TABLE 11.3 Weight of the Contents of Containers (oz.)

Sample number	Sample values					Mean $\overline{X}$	Range R
1	11.3	10.5	12.4	12.2	12.0	11.7	1.9
2	9.6	11.7	13.0	11.4	12.8	11.7	3.4
3	11.4	12.4	11.7	11.4	12.4	11.9	1.0
4	12.0	11.9	13.2	11.9	12.2	12.2	1.3
5	12.4	11.9	11.7	11.6	10.5	11.6	1.9
6	13.8	12.5	13.9	11.9	11.4	12.7	2.5
7	13.3	11.6	13.2	10.7	11.4	12.0	2.6
8	11.1	11.3	13.2	12.8	12.0	12.1	2.1
9	12.5	11.9	13.8	11.6	13.0	12.6	2.2
10	12.1	11.7	12.0	11.7	12.9	12.1	1.2
11	11.7	12.6	12.3	11.2	10.8	11.7	1.8
12	13.8	12.3	12.4	14.1	11.3	12.8	2.8
13	10.6	11.8	13.1	12.8	11.7	12.0	2.5
14	12.0	11.2	12.1	11.7	12.1	11.8	0.9
15	11.5	13.1	13.9	11.9	10.7	12.2	3.2
16	13.4	12.6	12.4	11.9	11.8	12.4	1.6
17	12.1	13.1	14.1	11.4	12.3	12.6	2.7
18	11.5	13.2	12.4	12.6	12.2	12.4	1.7
19	13.8	14.2	13.5	13.2	12.8	13.5	1.4
20	11.5	11.4	13.1	11.6	10.8	11.7	2.3

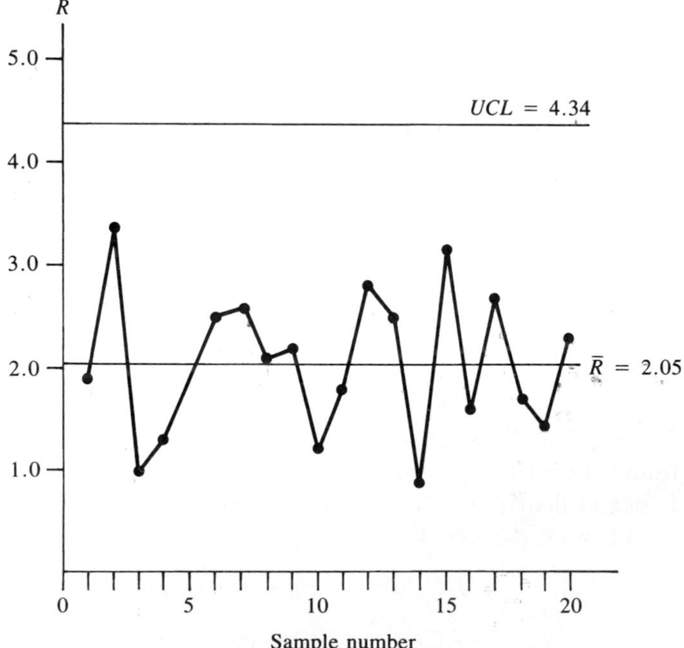

Figure 11.3 An R chart for the data of Table 11.3.

The 20 sample means may now be plotted. It is noted that the mean of sample 19 exceeds the upper control limit. This would indicate that at this point in time, the universe from which this sample was selected was not exhibiting a stable pattern of variation. Some change occurred between the time of selecting sample 18 and sample 19. It is further noted that after sample 19, the process returned to its original state. One may base action on the assumption that these statements are true, particularly if one thinks that some recognized assignable cause effected the change. Actually, the mean of sample 19 might have exceeded the control limit by chance, and a Type I error might have been made. Alternatively, the pattern of variation might have shifted some time before sample 18 or not returned after this time. Then a Type II error would have been made at these other points in time.

The data of sample 19 should now be discarded and the control limits recalculated for the remaining pattern of variation. The mean of the remaining sample means is 12.12. The range is not recalculated because it is assumed that the process variation did not change when sample 19 was selected. The control limits are now revised to

$$UCL_{\overline{X}} = 12.12 + 0.577(2.05) = 13.30$$

$$LCL_{\overline{X}} = 12.12 - 0.577(2.05) = 10.94$$

These are indicated on the control chart of Figure 11.4 and again checked to determine that no sample means exceed these limits. It is noted that no further changes are necessary, and the control chart of Figure 11.4, with the revised limits, and the chart of Figure 11.3 may now be employed to monitor the process. This control model may be implemented to test future process variation to see that it does not change from that used to construct the chart.

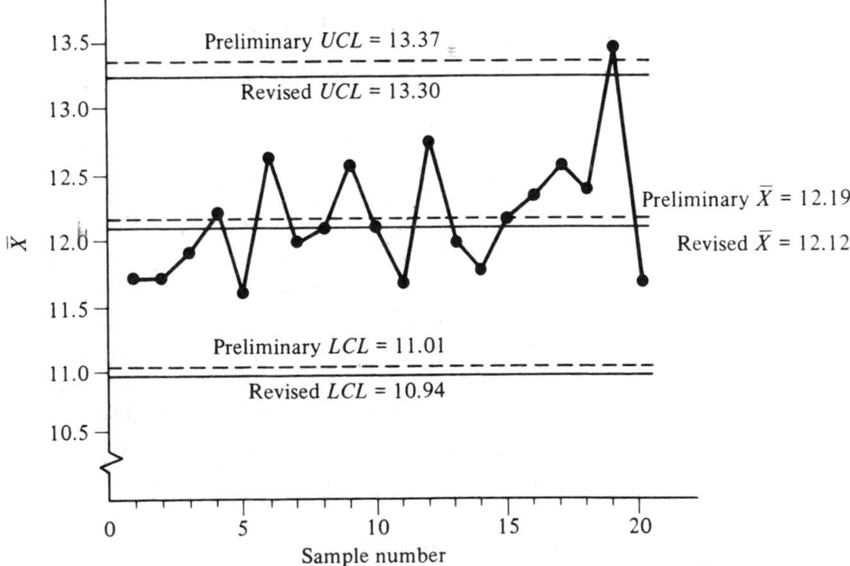

Figure 11.4 An $\overline{X}$ chart for the data of Table 11.3.

Control Charts for Attributes

A system characteristic or condition is sometimes expressible in a two-value manner. A defense system may be operationally ready or it may be down for maintenance. The surface finish of a workpiece may be acceptable or not. A maintenance crew may be working or it may be idle. When a two-value classification is made, the proportion of observations falling in one class during a predetermined time period may be monitored over time with a *p* chart.

Often observations yield a multivalue but still discrete classification. An employee may suffer none, one, two, or more lost-time accidents during a given time period. The number of demands on a supply system will be a discrete number during a specified period. A maintenance mechanic may perform an assigned task perfectly, or may have made one or more errors. When a discrete classification is made, a system characteristic or condition can be monitored over time with a *c* chart.

The *p* chart. When a characteristic or condition is sampled and then placed into one of two defined classes, the proportion of units falling into one class may be controlled over time or from one sample to another with a *p* chart. The applicable probability distribution is the binomial. The mean of this distribution and its standard deviation may be expressed as given in Equation 13.4.

$$\mu = np$$
$$\sigma = \sqrt{np(1 - p)}$$

These parameters may be expressed as proportions by dividing by the sample size, n. If $\bar{p}$ is then defined as an estimate of the proportion parameter μ/n, and s_p as an estimate of σ/n, then these statistics can be expressed as

$$\bar{p} = \frac{\text{total number in the class}}{\text{total number of observations}}$$

$$s_p = \sqrt{\frac{\bar{p}(1 - \bar{p})}{n}} \tag{11.9}$$

The application of the *p* chart will be illustrated with an example of an activity-sampling study. This activity measurement technique is used to obtain information about the activities of workers or machines, usually in less time and at lower cost than by conventional means. Random and instantaneous observations are taken by classifying the activity at a point in time into one and only one category. In the most simple form, the categories *idle* and *busy* are used. The control chart is useful in work-sampling studies in that the observed proportions can be verified as in control and following a stable pattern of variation or out of control with an unstable pattern. In the latter case, a search can be undertaken for an assignable cause.

Consider the case of an activity-sampling study involving computer terminals in an office. The objective was to determine the proportion of time the terminals were in use as opposed to the time they were idle. One hundred observations were taken each day over all working days in a month. The number of times the terminals were in use,

TABLE 11.4 Number of Times a Day the Terminals are in Use

Working day	Times in use	Proportion	Working day	Times in use	Proportion
1	22	0.22	12	46	0.46
2	33	0.33	13	31	0.31
3	24	0.24	14	24	0.24
4	20	0.20	15	22	0.22
5	18	0.18	16	22	0.22
6	24	0.24	17	29	0.29
7	24	0.24	18	31	0.31
8	29	0.29	19	21	0.21
9	18	0.18	20	26	0.26
10	27	0.27	21	24	0.24
11	31	0.31	Total	546	

and the proportion for each day are given in Table 11.4. From this table and Equation 11.9, $\bar{p}$ is established as

$$\bar{p} = \frac{546}{(21)(100)} = 0.260$$

The standard deviation of the data is calculated with Equation 11.11 as

$$s_p = \sqrt{\frac{(0.26)(0.74)}{100}} = 0.044$$

With $\bar{p} = 0.26$ as the best estimate of the population proportion, a control chart may now be constructed. Control limits in activity sampling are usually established at $\bar{p} \pm 2s_p$, and these limits will not vary from one day to another in this example, since a constant sample size has been maintained for each day. With these control limits, the probability of making a Type I error may be defined as $\alpha = 0.0456$ if the normal distribution is used as an approximation. Such an approximation is realistic for this example since the binomial will approximate the normal distribution if $n > 50$ and $0.20 < p < 0.80$. If these requirements were not met or a more accurate estimate were needed, the binomial distribution would have had to be employed. The control limits for the p chart are defined and calculated as

$$\begin{aligned} UCL_p &= \bar{p} + 2s_p \\ &= 0.260 + 2(0.044) = 0.348 \qquad (11.11) \end{aligned}$$

$$\begin{aligned} LCL_p &= \bar{p} - 2s_p \\ &= 0.260 - 2(0.044) = 0.172 \qquad (11.12) \end{aligned}$$

The p chart is constructed as shown in Figure 11.5. A plot of the data indicates that day 12 was not typical of the pattern of use established by the rest of the month. Subsequent investigation reveals that personnel from another office were also using the terminals that one day because their equipment had preceded them in a move to

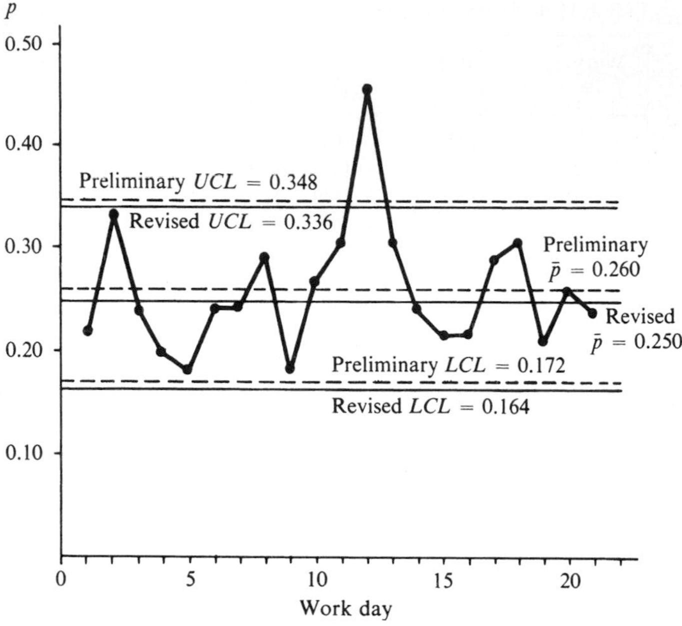

Figure 11.5 A p chart of daily computer terminal usage.

another building. As this is an atypical situation, the sample is discarded and a revised mean and standard deviation are calculated as

$$\bar{p} = \frac{500}{(20)(100)} = 0.250$$

$$s_p = \sqrt{\frac{(0.25)(0.75)}{100}} = 0.043$$

Revised control limits are now calculated and placed on the same control chart as

$$UCL_p = \bar{p} + 2s_p$$
$$= 0.250 + 2(0.043) = 0.336$$
$$LCL_p = \bar{p} - 2s_p$$
$$= 0.250 - 2(0.043) = 0.164$$

It is noted that no further days fall outside these limits and that the chart and data, without day 12, can be taken as a stable pattern of variation of terminal usage. Deviations from this in-control condition can be noted and appropriate control action taken.

The *c* chart. Some systems exhibit characteristics expressible numerically. For example, the number of arrivals per hour seeking service at a toll booth is of interest when deciding on the level of service capability to provide. If the number of arrivals per hour deviates from the stable pattern of variation, it may be necessary to compensate by either opening or closing certain toll booths.

The Poisson distribution is usually used to describe the number of arrivals per time period. Here, the opportunity for the occurrence of an event, n, is large but the probability of each occurrence, p, is small. The mean and the variance of the Poisson distribution are equal as $\mu = \sigma^2 = np$. These parameters can be estimated from the statistics with $\bar{c}$ and s_c^2 defined as these estimates. In many applications, values for n and p cannot be determined, but their product np can be established. Then the mean and variance can be estimated as

$$\bar{c} = s_c^2 = \frac{\Sigma(np)}{n} \tag{11.13}$$

Consider the application of a c chart to the arrival process previously described. Data for the past 20 hours have been collected and are presented in Table 11.5. The mean of the arrival population may be estimated from $\bar{c}$ with Equation 11.13 as

$$\bar{c} = \frac{84}{20} = 4.20$$

and the standard deviation may be estimated as

$$\begin{aligned} s_c &= \sqrt{\bar{c}} \\ &= \sqrt{4.20} = 2.05 \end{aligned}$$

With these estimates, the control chart may now be constructed. Control limits may be established as $\bar{c} \pm 3s_c$. The probability of making a Type I error can be determined from the cumulative values of Appendix C, Table C.1. For the example under consideration,

$$\begin{aligned} UCL_c &= \bar{c} + 3s_c \\ &= 4.2 + 3(2.05) = 10.35 \end{aligned} \tag{11.14}$$

TABLE 11.5 Number of Arrivals per Hour Demanding Service at a Toll Booth

Hour number	*np* Number of arrivals	Hour number	*np* Number of arrivals
1	6	11	6
2	4	12	4
3	3	13	2
4	5	14	2
5	4	15	4
6	6	16	8
7	5	17	2
8	4	18	3
9	2	19	5
10	5	20	4
		Total = 84	

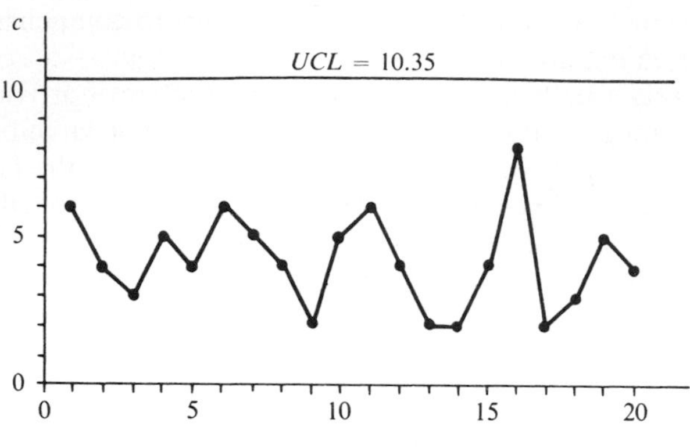

Figure 11.6 A c chart of the number of arrivals per hour.

$$LCL_c = \bar{c} - 3s_c$$
$$= 4.2 - 3(2.05) < 0 \tag{11.15}$$

There is no lower control limit. The probability of making a Type I error is the probability of 11 or more arrivals in a given hour from a population with $\bar{c} = 4.2$. This is $1 - P(10 \text{ or less})$ or $1 - 0.994 = 0.006$. Alternatively, the control limit could have been defined as a probability limit. If it were thought desirable to define $\alpha \leq 0.01$, the control limit would have been specified as 9.5. Under this control policy the probability of detecting 10 or more arrivals would satisfy $\alpha \leq 0.01$.

In the *c* chart of Figure 11.6, all values fell within the control limits and the chart can be utilized as originally formulated. Action will be initiated to alter the service capability to meet the new demand pattern. At the same time, a new chart based upon the recent data will be constructed and implemented. In this way the decision maker is informed of conditions in the system that require modification of operating policy.

11.3 OPTIMUM POLICY CONTROL

A statistical control chart may be used to monitor the optimum solution obtained from a decision model. Most decision models require monitoring because they do not automatically adapt themselves to the instability of the decision environment. These models are usually formulated at a point in time for a specific set of input parameters. A solution derived from a decision model will be optimal only as long as the input parameters retain the values initially established.[2]

Optimum policy control will be illustrated for the case where an optimum service factor has been determined for a maintenance situation.[3] Once the service factor resulting in a minimum-cost operation has been established, it might be desirable to implement a control model for detecting the effects of a shift in the arrival rate or the service

[2] In terms of $E = f(X, Y)$, this means that E will remain optimal for specific values of X only as long as Y values retain their value initially established.

[3] This example follows the computational procedure of Section 10.6.

rate or both. This may be accomplished by constructing a control chart for the number of machines not running. The statistical control charts presented cannot be applied directly to this variable, since its distribution is badly skewed. The expected form of the distribution is known in advance; however, a factor usually missing in other control applications.

As an example of the application of control charts in waiting-line operations, consider a one worker/N machine situation.[4] Suppose that 20 automatic machines are to be run by one worker, and that the minimum-cost service factor is found to be 0.03. It is assumed that the machines require service at randomly distributed times, that the service times are distributed exponentially, and that machines are serviced on a first-come, first-served basis. The probability of n machines in the queue and in service (not running) is given by

$$P_n = \frac{N!}{(N-n)!}\rho^n P_0 \tag{11.6}$$

where

P_n = probability that n machines are not running at any point in time
P_0 = probability that all machines are running at any point in time
ρ = ratio of λ/μ
N = number of machines assigned
n = number of machines not running at any point in time

Values for P_n are given in column C of Table 11.6 for $0 \le n \le N$. The first entry is found by dividing the first entry in column B by 2.30005387. The second results from dividing the second entry in column B by the same value, and so on. The results for P_n are plotted in Figure 11.7. The distribution of n is seen to be extremely skewed, being strictly convex. Unlike some control chart applications, the distribution of the variable to be controlled is known. It may be used to determine the control limits.

Because the probability that all machines are running is approximately 0.435, the lower control limit is obviously zero. The upper control limit is all that needs to be determined, because the variable n can go out of control only at the top. Thus, the entire critical range will be at the upper end of the distribution. It will be desirable to set this limit so that it will not be violated too frequently. Because n is an integer, the magnitude of the critical range can be observed as a function of n from column D in Table 11.6. If six machines not running is chosen as a point in control and seven is a point out of control, the probability of designating the system out of control when it is really in control is

$$1 - \sum_{n=0}^{6} P_n = 0.0059$$

Therefore, a control limit set at 6.5 should be satisfactory. The control chart applied to a period of operation might appear as in Figure 11.8. It would be concluded that the service factor has not changed during this period of observation.

[4] Adapted from R. W. Llewellyn, "Control Charts for Queuing Applications," *Journal of Industrial Engineering*, Vol. 11 No. 4 (July-August 1960).

TABLE 11.6 Calculation of P_n and ΣP_n

n (A)	$\frac{P_n}{P_0}$ (B)	P_n (C)	ΣP_n (D)
0	1.000000000	0.434772422	0.434772422
1	0.600000000	0.260863453	0.695635875
2	0.342000000	0.148692168	0.844328043
3	0.184680000	0.080293771	0.924621814
4	0.094186800	0.040949823	0.965571637
5	0.045209664	0.019655915	0.985227552
6	0.020344349	0.008845162	0.994072714
7	0.008544627	0.003714968	0.997787682
8	0.003332405	0.001448838	0.999236520
9	0.001199666	0.000521582	0.999758102
10	0.000395890	0.000172122	0.999930224
11	0.000118767	0.000051637	0.999981861
12	0.000032067	0.000013942	0.999995803
13	0.000007696	0.000003346	0.999999149
14	0.000001616	0.000000703	0.999999852
15	0.000000291	0.000000127	0.999999979
16	0.000000044	0.000000019	0.999999998
17	0.000000005	0.000000002	1.000000000
18	0.000000000	0.000000000	1.000000000
19	0.000000000	0.000000000	1.000000000
20	0.000000000	0.000000000	1.000000000
	2.300053887		

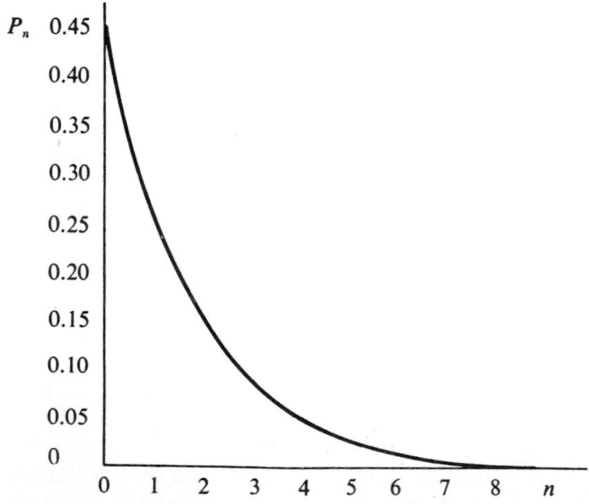

Figure 11.7 Probability distribution of the number of machines not running.

For the application of this control model to be valid, two conditions must be met. First, the observation must be made at random times. Second, the observations must be spaced far enough apart, so that the results are independent. For example, if the ser-

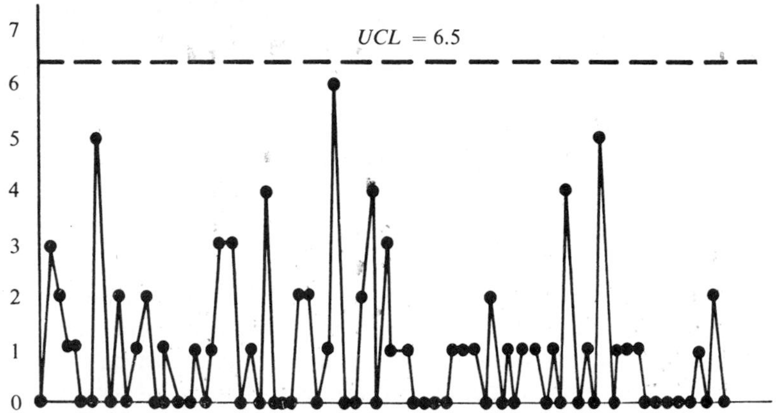

Figure 11.8 Control chart for the number of machines not running.

vice time averages 5 minutes and if six machines are idle in an observation, then an observation taken 10 minutes later would probably indicate at least four idle machines. The readings should be far enough apart for a waiting line to be dissipated. To ensure both randomness and spacing, all numbers from 45 to 90 could be taken from a random number table with their order preserved: numbers from 00 to 44 and from 91 to 99 would be dropped. These numbers can then be used to space the observations. This would yield an average of eight random observations per day with a minimum spacing of 45 minutes and a maximum spacing of 90 minutes.

If the control chart indicates that n is no longer in control, corrective action must be taken. This will require investigation to determine whether the service factor has changed because of a change in the mean time between calls or a change in the mean service time, or both. Specific items that might be studied are the policy of preventive maintenance, the age of the machines, the capability of the operator, or material characteristics. Once the assignable cause for the out-of-control condition is located, it may be corrected so that the optimal operating condition may be restored.

11.4 PROJECT CONTROL WITH CPM AND PERT

A project is composed of a series of activities directed to the accomplishment of a desired objective. Most projects are nonrepetitive and confront the decision maker with a situation in which prior experience and information for control is nonexistent. Therefore, techniques for project control should meet two conditions: (1) during the planning phase of a project they should make possible the determination of a logical, preferably optimal, project plan; and (2) during the execution of a project they should make possible the evaluation of a project progress against the plan.

The planning and control of a large-scale project may be accomplished by utilizing an activity network as the planning and control medium. Project planning and control techniques may be classified as deterministic and designated CPM (critical path method) or probabilistic and designated PERT (program evaluation and review technique).

CPM and PERT are similar in their logical structure. Projects are represented by sets of required activities, with the principal difference being the treatment of estimated time for the completion of each activity. In PERT, activity time is considered to be a random variable with an assumed probability distribution.

Critical Path Methods (CPMs)

Many engineering projects involving design, development, and construction operations are nonrepetitive in nature. Although a particular project is to be executed only once, many interdependent activities are required. Some must be executed simultaneously. Each will require personnel and equipment, the cost of which will depend upon the resource commitment. The total project duration and its overall cost depend upon the activities on the *critical path* and the resources allocated to them. Critical path methods (CPM) for dealing with this situation are presented in this section.

Activity and event networks. A network is the basic structural entity behind all critical path methods. It is used to portray the interrelationships among the activities associated with a project. Figure 11.9 illustrates a network that represents the activities and events in connection with the preparation of a foundation. Four events and three activities are shown.

An event in CPM is represented by a circle. It indicates the completion of an activity. In the case of an initial event it indicates project start or the initiation of the associated activities. Events *A* and *C* in Figure 11.9 indicate the start of excavation and the start of equipment move activities. Event *D* represents the completion of foundation preparation.

Activities in CPM are represented by arrows that interconnect events. These arrows symbolize the effort required to complete an event in units of time. Their direction specifies the order in which the events must occur. For example, activities *BD* (set forms) and *CD* (equipment move-in) are required in order for event *D* (complete

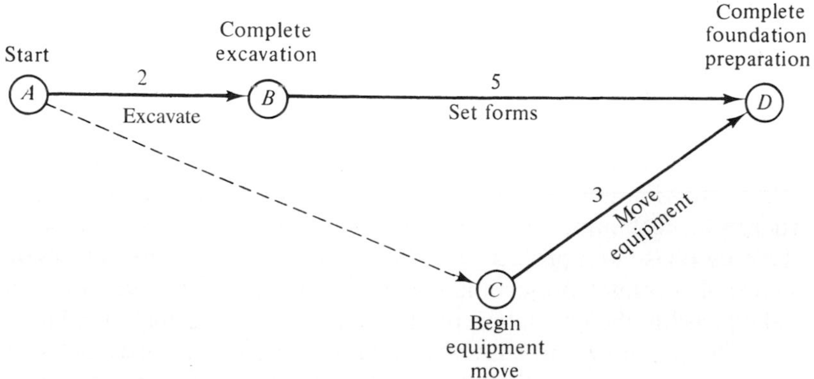

Figure 11.9 Activity-event network for preparing a foundation.

preparation) to be achieved. The precedence relationships between event D and event B and between event D and event C are established by the arrows. Events B and C must be realized before the foundation is prepared (event D).

In some network representations it is necessary or useful to create dummy activities to clarify precedence relationships. A dummy activity is illustrated in Figure 11.9 by a broken arrow that connects activity A with activity C, establishing that event C is preceded by event A. Even though dummy activities do not require time and do not consume resources, they may alter the completion time of a project by establishing the order in which events are performed.

Associated with each activity is an elapsed time estimate. This estimate is usually shown above the arrow in the network. The length of the arrows need not be proportional to the duration of the activity. In Figure 11.9 the time estimates are weeks.

Two important absolute times for each event must be identified. The first is the earliest time, T_E, which indicates the calender time at which an event can occur, provided all previous activities were completed at their earliest possible times. The second is the latest time, T_L, for each event, defined as the latest calendar time at which an event can be completed without delaying initiating the following activities and completion of the project.

The difference between the latest and the earliest times for each event is called the *slack time* for that event. Positive slack for an event indicates the maximum amount of time that it can be delayed without-delaying subsequent events and the overall project. In the example, event B has no slack since the latest time for completion of excavation and the earliest time for setting the forms is the same point in time. Any delay in realizing event B would cause a delay in the project. From this it is evident that the activities of excavation and setting the forms are critical. The activity of equipment move-in is noncritical to the realization of event D because the start of equipment move-in can occur at any time on or before 3 weeks before the completion of the project. Thus, 4 weeks is the slack time for event C.

Finding the critical path. In CPM all critical activities must be identified and given special attention. This is necessary because any delay in performing these activities may lead to a delay in subsequent activities and the project as a whole. Often it is possible to identify the critical path by tracing the activities along a sequence of events which have zero slack time. The sum of the activity times along this critical path gives the shortest possible elapsed time for project completion. This is 7 weeks in the foundation sample.

As an example of a systematic procedure for finding the critical path, consider a project requiring the construction of a certain structure. Five major events must be realized through seven activities as illustrated in Figure 11.10. The number appearing above each arrow is the estimated duration of the activity in months, with the arrow indicating the precedence relationship.

The procedure for finding the critical path starts by determining the earliest time and the latest time for each event. By subtracting T_E and T_L for each event, the slack time is found. To facilitate this process each event can be labeled (T_E, T_L, S), where S represents the slack time for the event. First, T_Es are determined for each event and

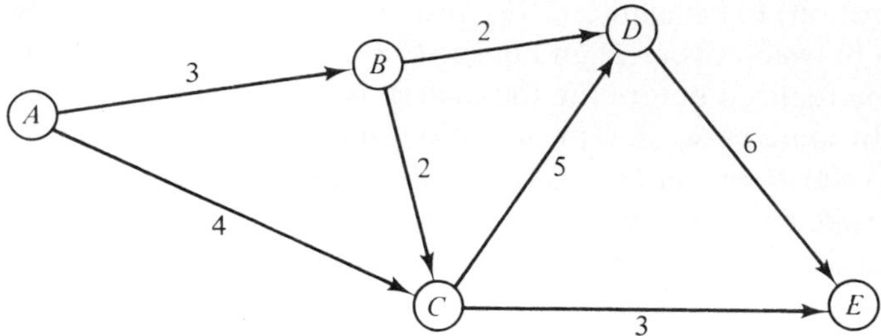

Figure 11.10 A network for the construction of a certain structure.

are labeled (T_E, ,). Next, T_Ls are found and the labels become (T_E, T_L,). Finally, S is found by subtracting T_E from T_L and completing the label (T_E, T_L, S) for each event.

The earliest time for each event is determined by adding the earliest time for the event immediately preceding to the activity duration connecting the two events. When there is more than one preceding event, the largest value of the sum is chosen as the earliest time for the event under consideration. The process for determining these earliest times proceeds forward from the initial event, which is assumed to have an earliest time of zero. This process is illustrated in Figure 11.11 and Table 11.7 for the example under consideration.

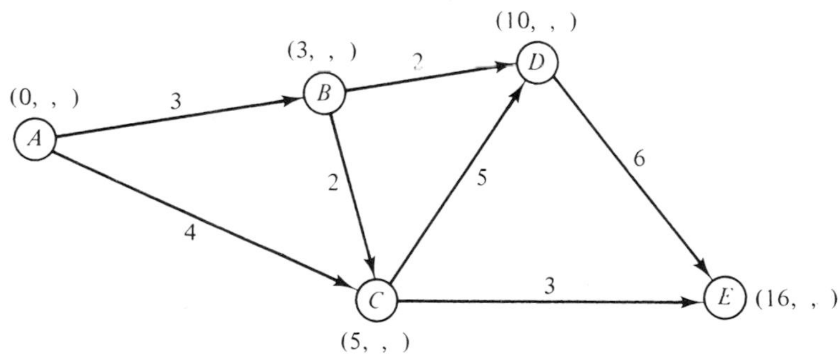

Figure 11.11 The earliest time for each event.

TABLE 11.7 Computation of Earliest Times

Event	Immediate predecessor	T_E for immediate predecessor plus activity time	Largest sum = the earliest time
A	None	$0 + 0 = 0$	0
B	A	$0 + 3 = 3$	3
C	A	$0 + 4 = 4$	
	B	$3 + 2 = 5 \longleftarrow$	5
D	B	$3 + 2 = 5$	
	C	$5 + 5 = 10 \longleftarrow$	10
E	C	$5 + 3 = 8$	
	D	$10 + 6 = 16 \longleftarrow$	16

TABLE 11.8 Computation of Latest Times

Event	Immediate successor	T_L for immediate successor minus activity time	Smallest difference = the latest time
E	None	$16 - 0 = 16$	16
D	E	$16 - 6 = 10$	10
C	D	$10 - 5 = 5 \longleftarrow$	5
	E	$16 - 3 = 13$	
B	C	$5 - 2 = 3 \longleftarrow$	3
	D	$10 - 2 = 8$	
A	B	$3 - 3 = 0 \longleftarrow$	0
	C	$5 - 4 = 1$	

The latest time for each event can be determined in a manner similar to that used in finding the earliest time except that the process starts with the final event and proceeds backward. The latest time for this final event is the same as its earliest time, for, if it were not, the project would be delayed. For each event the latest time is determined by subtracting the duration time of an activity connecting the event and an event immediately succeeding from the latest time of the event immediately succeeding. If there is more than one event immediately succeeding, the smallest value of the difference is chosen to be the latest time for the event under consideration. A slack time for each event is then found by computing the difference between T_L and T_E as illustrated in Figure 11.12 and Table 11.8.

The critical path can now be identified with the T_L, T_E, and S values entered in Figure 11.12. This is done by tracing all zero slack events and viewing them as a path. In this example the critical path is $A \rightarrow B \rightarrow C \rightarrow D \rightarrow E$ with a minimum project duration of 16 months.

Economic aspects of CPM. The CPM example above was presented under the assumption that the activities are performed in normal time with a normal allocation of resources. This is called a *normal schedule*. When one or more activities are performed with additional resources to shorten their time durations, a *crash schedule* is said to exist.

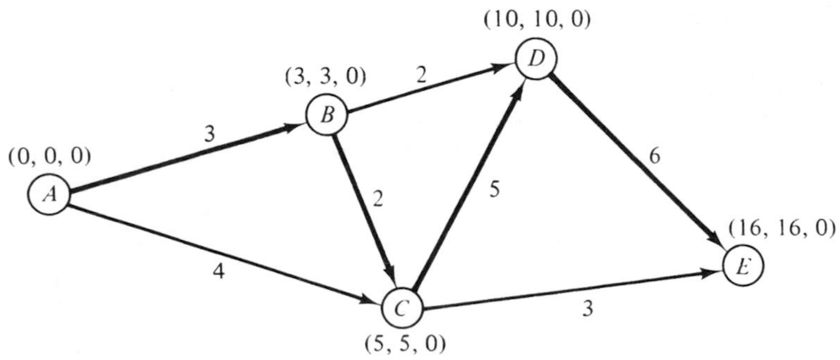

Figure 11.12 Labeling of events and identification of critical path.

Two costs are associated with a project. Direct costs exist for each activity. These costs increase with an increase in the resources allocated for the purpose of decreasing the activity time. Indirect costs exist which are associated with the overall project duration. These increase in direct proportion to an increase in the total time required for project completion. This relationship should be self-evident, for overhead costs continue independent of activity levels and they depend on the passage of time.

In most projects it is of interest to investigate the economic aspects of increasing direct costs as activities are expedited in the light of decreasing overhead costs. An optimum allocation of resources to each activity is sought so that the sum of direct and indirect costs will be a minimum.

The first step in finding the optimum project schedule is to find the critical path under normal resource conditions. This is the starting point. Next, the activity on the critical path that has the least impact on direct cost is shortened. When shortening this activity, other critical paths may appear. If they do, it is necessary to reduce the duration of activities on these paths an equal number of time units for each path. No further activity time reduction should be attempted either when a limit has been reached or when the indirect costs saved are not greater than the extra expenditure of direct resources. This process is repeated for all activities on the critical path or paths.

Consider the example project shown in Figure 11.13. Indirect costs of $1,000 can be saved for each day removed from the total project duration. Event times shown are those that would occur under normal conditions. The label for each event represents T_E, T_L, and S as defined earlier. The critical path is $A \to B \to D \to E$.

Data on activity durations under normal and crash conditions are given in Table 11.9. These data are daily costs computed from the assumption that each additional worker adds a direct cost of $50 per day. Additional equipment costs are those estimated to arise for each day reduced from the normal activity duration.

The project depicted in Figure 11.13 will take 11 days to complete under the normal activity times shown. No crash cost is incurred and therefore there is no saving in indirect cost.

A 10-day schedule can be achieved by reducing BD by one day. From Table 11.9 a crash cost of $700 is found. A saving of $1,000 in indirect cost will result. No further reduction in activity BD should be attempted because the critical path will change. This 10-day schedule is shown in Figure 11.14.

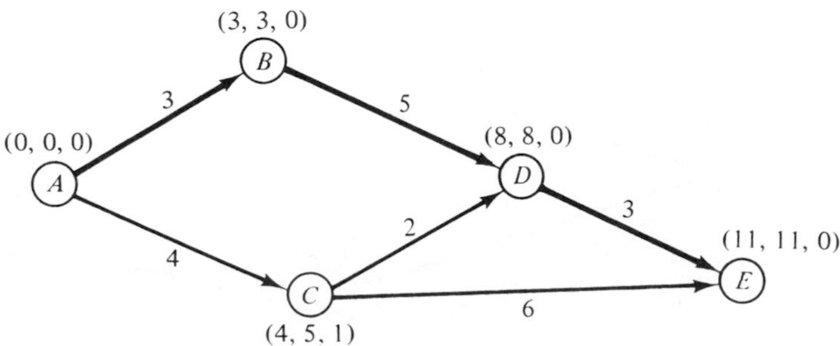

Figure 11.13 A CPM network with normal activities.

TABLE 11.9 Computation of Crash Costs

Activity	Number of days reduction	Additional labor cost: Additional workers	Additional labor cost: Working days	Additional labor cost: Additional costs	Additional equipment cost	Total crash cost
AB	1	2	2	$200	$ 600	$ 800
AC	1	3	3	450	300	750
BD	1	1	4	200	500	700
	2	3	3	450	450	900
	3	3	2	300	1,700	2,000
CD	0	0	2	0	0	0
CE	1	0	5	0	950	950
	2	4	4	800	1,110	1,900
DE	0	0	3	0	0	0

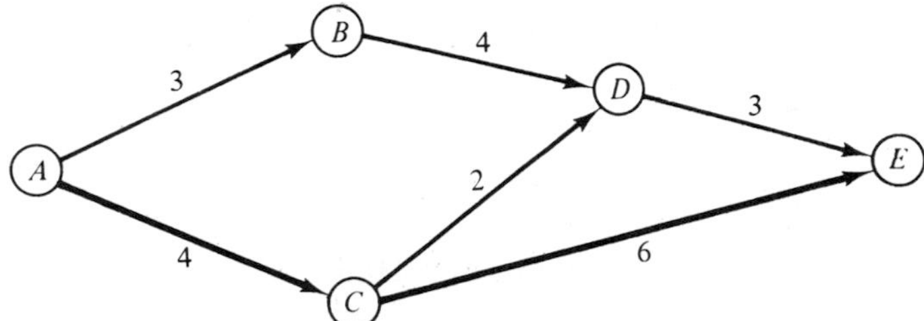

Figure 11.14 A ten-day schedule.

Two critical paths now exist: the original path and a new path $A \to C \to E$. Reducing activity BD one more day yields a 9-day schedule. This leads to a crash cost of \$900, which is less than the crash cost for shortening BD one day and AB one day. The duration of one activity in the critical path $A \to C \to E$ should also be shortened by one day at the same time. A one-day reduction of AC costs \$750 which is less than shortening CE by one day. A saving of \$2,000 in indirect cost will result. This 9-day schedule is shown in Figure 11.15.

In the 9-day schedule, path $A \to B \to D \to E$ and path $A \to C \to E$ remain critical as in the 10-day schedule. Further equal time reduction should be sought in these critical paths.

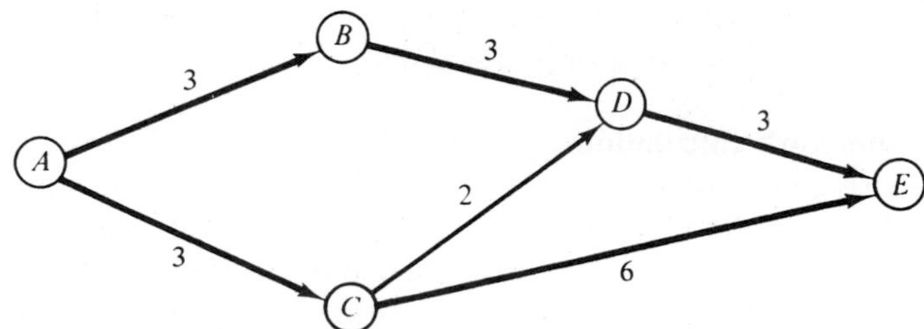

Figure 11.15 A nine-day schedule.

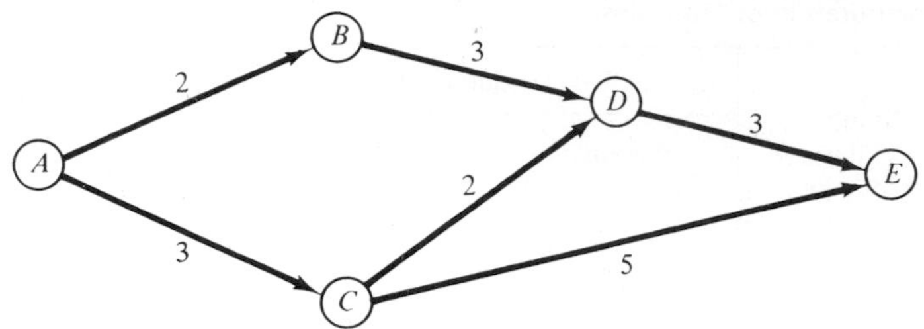

Figure 11.16 An eight-day schedule.

TABLE 11.10 Net Savings for Crash Schedules

Crash schedule	Crash cost	Indirect Cost saving	Net saving
10-day	$ 700	$1,000	$ 300
9-day	1,650	2,000	350
8-day	3,400	3,000	−400

An 8-day schedule is obtained by reducing AB by 1 day at a crash cost of $800 and BD by 2 days at a crash cost of $900 for a total of $1,700. This is better than a 3-day reduction in activity BD. It is necessary to reduce CE and AC by 1 day each at a crash cost of $950 plus $750 or $1,700. This is better than a 2-day reduction in CE. A saving of $3,000 is possible by this schedule. Three critical paths result: $A \to B \to D \to E$, $A \to C \to D \to E$, and $A \to C \to E$, each with a duration of 8 days. No further reductions are possible. The resulting schedule is shown in Figure 11.16.

Finally, it is necessary to find the optimum schedule. This is accomplished by summarizing the crash costs and the savings in Table 11.10. A maximum net saving occurs for a schedule of 9 days.

Program Evaluation and Review Technique (PERT)

PERT is ideally suited to early project planning where precise time estimates are not readily available. The advantages of PERT is that it offers a way of dealing with random variation, making it possible to allow for chance in the scheduling of activities. PERT may be used as a basis for computing the probability that the project will be completed on or before its scheduled date.

PERT network calculations. A PERT network is illustrated in Figure 11.17. One starts with an end objective (e.g., event H in Figure 11.17) and works backward until event A is identified. Each event is labeled, coded, and checked in terms of program time frame. Activities are then identified and checked to ensure that they are properly sequenced. Some activities can be performed on a concurrent basis, and others must be accomplished in series. For each completed network, there is one beginning event and one ending event, and all activities must lead to the ending event.

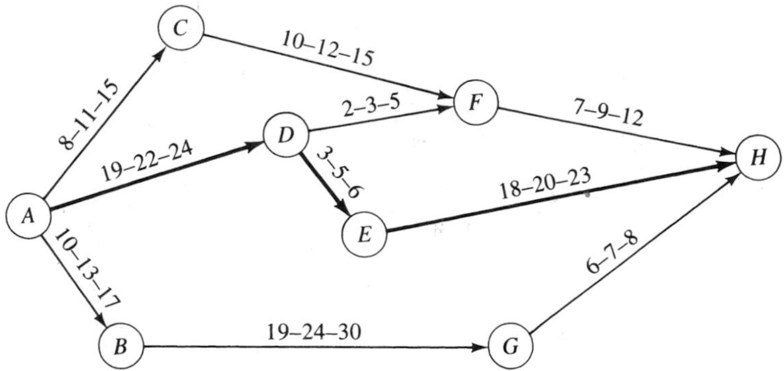

Figure 11.17 A PERT network with events and activity times.

TABLE 11.11 Example of PERT Calculations

Event number (1)	Previous events (2)	t_a (3)	t_b (4)	t_c (5)	t_e (6)	σ^2 (7)	T_E (8)	T_L (9)	S (10)	PT (11)	Prob. (12)
H	*G*	6	7	8	7.0	0.108				46.0	0.266
	F	7	9	12	9.2	0.693					
	E	18	20	23	20.2	0.693	46.8	46.8	0.0		
G	*B*	19	24	30	24.2	3.349	37.7	39.8	2.1		
F	*D*	2	3	5	3.2	0.250					
	C	10	12	15	12.2	0.693	23.4	37.6	14.2		
E	*D*	3	5	6	4.8	0.250	26.6	26.6	0.0		
D	*A*	19	22	24	21.8	0.693	21.8	21.8	0.0		
C	*A*	8	11	15	11.2	1.369	11.2	25.4	14.2		
B	*A*	10	13	17	13.5	1.369	13.5	15.6	2.1		

The next step in developing a PERT network is to estimate activity times and to relate these times to the probability of their occurrences. An example of calculations for a typical PERT network is presented in Table 11.11. The following steps are required:

1. Column 1 lists each event, starting from the last event and working downward to the start.
2. Column 2 lists all previous events that have been indicated prior to the event listed in column 1.
3. Columns 3 to 5 give the optimistic (t_a), the most likely (t_b), and the pessimistic (t_c) times, in weeks or months, for each activity. Optimistic time means that there is little chance that the activity can be completed before this time, while pessimistic time means that there is little likelihood that the activity will take longer. Time value (t) is defined as a random variable with a beta distribution and with range from a to b and mode m as shown in Figure 11.18.
4. Column 6 gives the expected or mean time (t_e) from

$$t_e = \frac{t_a + 4t_b + t_c}{6} \tag{11.7}$$

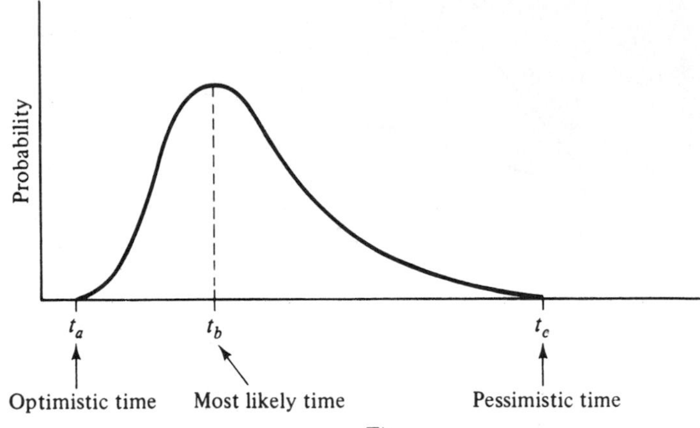

Figure 11.18 β-distribution of activity times.

5. Column 7 gives the variance, σ^2, from

$$\sigma^2 = \left(\frac{t_c - t_a}{6}\right)^2 \tag{11.8}$$

6. Column 8 gives the earliest expected time (T_E) as the sum of all times (t_e) for each activity and the cumulative total of the expected times through the preceding event, remaining on the same path throughout the network. When several activities lead to an event, the highest time value (t_e) is used. For example, path $A \rightarrow D \rightarrow E \rightarrow H$ in Figure 11.17 totals 46.8.
7. Column 9 gives the latest allowable time (T_L) calculated by starting with the latest time for the last event, which equals T_E (where T_E equals 46.8). Then one works backward, subtracting the value in column 6 for each activity, staying in the same path.
8. Column 10 gives the slack time (S) as the difference between the latest allowable time (T_L) and the earliest expected time (T_E) from

$$S = T_L - T_E$$

9. Columns 11 and 12 give the required project time (PT) for the network. Assuming that PT is 46.0, it is necessary to determine the probability of meeting this requirement. This probability is determined from

$$Z = \frac{PT - T_L}{\sqrt{\Sigma \text{ variances}}}$$

Z is the area under the normal distribution curve and the variance is the sum of the individual variances applicable to the critical path activities in Figure 11.17 (path $A \rightarrow D \rightarrow E \rightarrow H$). For this example,

$$Z = \frac{46.0 - 46.8}{\sqrt{0.693 + 0.250 + 0.693}} = -0.625$$

From Appendix C, Table C.3, the calculated value of −0.625 represents an area of approximately 0.266; that is, the probability of meeting the scheduled time of 46 weeks is 0.266.

When evaluating the resultant probability value (column 12 of Table 11.11), management must decide on the range of factors allowable in terms of risk. If the probability factor is too low, additional resources may be applied to the project to reduce the activity schedule times and improve the probability of success. Conversely, if the probability factor is too high, this may indicate that excess resources are being applied, some of which should be diverted to other projects.

PERT-cost. An extension of PERT to include economic considerations brings cost factors into project control decisions. A cost network can be superimposed on a PERT network by estimating the total cost and cost slope for each activity time.

When implementing PERT cost, there is always the time-cost option that enables the decision maker to evaluate alternatives relative to the allocation of resources for activity accomplishment. In many instances time can be saved by applying more resources or, conversely, cost may be reduced by extending the time to complete an activity. The time-cost option can be attained by applying the following steps:

1. Determine alternative time and cost estimates for all activities and select the lowest-cost alternative in each instance.
2. Calculate the critical path time for the network. If the calculated value is equal to or less than the total time permitted, select the lowest-cost option for each activity and check to ensure that the total of the incremental activity times does not exceed the allowable overall program completion time. If the calculated value exceeds the program time permitted, review the activities along the critical path, select the alternative with the lowest cost slope, and then reduce the time value to be compatible with the program requirement.
3. After the critical path has been established in terms of the lowest-cost option, review all the network paths with slack time and shift activities to extend the times and reduce costs wherever possible. Activities with the steepest cost-time slopes should be addressed first.

The time-cost option can be illustrated by referring to two paths of the network presented in Figure 11.17. Data associated with each path are given in Table 11.12. In this example, path $A \rightarrow D \rightarrow E \rightarrow H$ is critical, and the alternative path $(A \rightarrow C \rightarrow F \rightarrow H)$ has a slack of 14.2. If one is concerned with the critical path (since it is the governing factor in the overall network relative to program success or failure), the shifting of resources from the alternative path to critical-path activities should be considered. For instance, assume that \$200 is reallocated from activity *HF* to activity *HE*. As a result, activity *HF* will now require 12.2 units of time, and activity *HE* will require 18.7 units of time. The total time along the critical path is reduced, and the project is completed in 45.3 units of time. Conversely, the time along the alternative path is increased to 35.6 units, and the slack is reduced to 9.7 units. The entire network must

TABLE 11.12 Time-Cost Options

Path $A \to D \to E \to H$				Path $A \to C \to F \to H$			
Event number	Previous event	t_e	Estimated activity cost ($)	Event number	Previous event	t_e	Estimated activity cost ($)
H	*E*	20.2	3,500	*H*	*F*	9.2	1,050
E	*D*	4.8	940	*F*	*C*	12.2	2,350
D	*A*	21.8	4,800	*C*	*A*	11.2	1,870
Total		46.8	$9,240	Total		32.6	$5,370

now be reevaluated to determine the effects of the changes on the critical path. This process may be continued by trading off resources against time until the most acceptable result is obtained.

11.5 TOTAL QUALITY CONTROL

Total quality control (TQC) is a management concept being stimulated by the need to compete in global markets where higher quality, lower cost, and more rapid development are essential to market leadership. Management for total quality is rapidly emerging as a means for improving the quality-cost characteristics of products, processes, and services.

Approaches now being promulgated emphasize that conformance to design specifications is necessary, but not sufficient. Quality losses begin to accumulate whenever a product parameter deviates from its nominal or optimal value. Accordingly, it is being recognized that quality must be designed into products and processes. Furthermore, optimal values for product and process parameters must be established and controlled throughout the product life cycle.

Statistical Process Control

Statistical process control (SPC) was introduced in Section 11.2 and related to the process or operating system being controlled in Figure 11.1. Measurement is the key to evaluation in SPC. This measurement is applied to system output for the purpose of detecting a difference between actual values and ideal or target values.

By applying statistical control to a process properly centered on product specifications, considerable faith can be placed in the quality of the output product. Statistical control permits producers to identify and eliminate assignable causes of variation, while reducing random or unassignable causes. Furthermore, products produced under statistical control generally need not be subjected to sampling inspection. Sampling inspection seeks to find and eliminate defective items, a practice that is proving to be too disruptive and costly.

Statistical process control is based on the assumption of a stable process. If stable, one can measure a product characteristic that reflects the behavior of the process. The measured characteristic will have a steady-state distribution (in a probabilistic sense); that is, each sample of the product from the process will have the identical statistical distribution.

A stable process that yields product which satisfies specifications may later become unstable. Instability may arise from special causes. Continued monitoring of the process using statistical process control can detect the transition to an unstable state and the presence of special causes. A process that passes the statistical test is not always stable. However, the probability is very small that an unstable process will continue to produce output that passes the statistical tests (refer to Figure 11.2).

Charts are used in statistical process control for conducting stability tests. When the value of a characteristic can be measured, $\overline{X}$ and R charts are used; when the fraction of a two-valued characteristic is being measured, a p chart is appropriate; when the overall number of defects is being measured, the c chart is applicable. These were presented in Section 11.2 and are attributed to the work of Walter Shewhart in the 1920s.

The Shewhart approach to variation focuses more on process stability than on conformance to specifications. However, process control must not ignore specifications. Engineers should define specifications and statistical tolerances as part of the design process with the process capability in mind. A production process is said to have a process capability index, I, defined as

$$I = \frac{z}{3\sigma} \tag{11.9}$$

where $z = \min\,[(\overline{\overline{X}} - \text{lower specification limit})(\text{upper tolerance limit} - \overline{\overline{X}})]$ and $\overline{\overline{X}}$ is the grand average (from Equation 11.1). The process capability index is a measure of the ability of a process to produce quality product.

Although process control methods were first applied to manufacturing, their use is not restricted to production. Statistical process control is also used to evaluate and improve engineering processes. It is a means to obtain useful information to feed back into product and production system design. In addition to control charts, Pareto curves, cause and effect diagrams, CPNI networks, and PERT are used to evaluate and improve engineering and production processes.

As real-time information becomes more readily available in design, production, and support, practitioners are beginning to measure not only product parameters but also changes in process control parameters. Such applications are closely related to control engineering and use many of its basic principles.

Experimental Design

Experimental design (ED) was first introduced in the 1920s by Sir Ronald Fisher. Its subsequent development led to gains in the efficiency of experimentation by changing factors, not one at a time, but together in a factorial design. Fisher introduced the concept of randomization (so that trends resulting from unknown disturbing factors would not bias results), the idea that a valid estimate of experimental error could be obtained from the design, and blocking to eliminate systematic differences introduced by using different lots.

Experimental designs were introduced into industry in the 1930s. At that time, the Industrial and Agricultural Section of the Royal Statistical Society was inaugurated in London and papers on applications to glass manufacturing, light bulb production, textile spinning, and the like were presented and discussed. This led to new statistical methods such as variance component analysis to reduce variation in textile production.

During World War II, the need for designs that could screen large numbers of factors led to the introduction of fractional factorial designs and other orthogonal arrays. These designs have been widely applied in industry and many successful industrial examples are described in papers and books dating from the 1950s. Also in the early 1950s, George Box developed response surface methods for the improvement and optimization of industrial processes by experimentation.

Modern engineering research and development involves considerable experimentation and analysis. Statistically designed experiments enable the engineer to achieve development objectives in less time with less resources. It is not unusual for development cycle time and cost to be reduced by one-half when experimental design methods are integrated into a concurrent engineering development process. When product quality and process design are addressed early in the life cycle, many decisions that affect product life, reliability, and producibility can be evaluated before the product is produced.

Parameter Design

Parameter design (PD) is based on the idea of controllables and uncontrollables in design and operations as introduced in Chapter 7 (refer to Section 7.3). Parameter optimization involves the selection of controllable values in the product and the processes of manufacture and maintenance, so that some measure of merit will be improved (or optimized if possible).

Parameter optimization is traditionally used when the goal is to improve some performance measure. It is now being applied to variability reduction for increased quality and lower cost. Although parameter design to reduce variability may be accomplished by a variety of methods, there is one technique which is gaining recognition. This technique is attributable to Genichi Taguchi.

Taguchi uses the term *quality engineering* (QE) to describe an approach to achieving both improved quality and low cost. He identifies two methods of dealing with variability: parameter optimization and tolerance design. The first attempts to minimize the affects of variation and the second seeks to remove its causes.

There have been many examples of the successful application of parameter optimization. The most widely reported is the case of tile manufacturing in Japan. Similar improvements have been demonstrated in automobile manufacturing, electronic component production, computer operations, IC chip bonding processes, ultrasonic welding, and the design of disc brakes and engines.

The concept of a quality loss function is used in parameter optimization to capture the effect of variability. This function represents the loss to society resulting from variability of function as well as from harmful side effects during the production and use of a product. Loss is calculated from the view of society with an emphasis on the viewpoint of the customer. For example, the loss to an automobile owner when a part fails includes the cost of repair, the cost of lost wages, and the inconvenience of not having use of the car.

In many applications the quality loss can be approximated by a quadratic function of the important design parameter. The function is assumed to be analytic, and so it can be expanded in a Taylor series about some known optimal value T. Then the first two terms of the expansion can be assumed to be zero and the principal term becomes

$$L(y) = (y - T)^2 \frac{L''(T)}{2!}$$

or

$$L(y) = k(y - T)^2$$

where

$L(y)$ = the loss as a function of y
T = the optimum or target value for parameter y
$L''(T)$ = the second derivative of L evaluated at T
k = the loss coefficient for the particular application

Taguchi used the loss function to measure the effect of variation whether or not the specification is satisfied. He also used the orthogonal experimental design for efficiently evaluating the effect of individual parameter settings in the face of noise.

Quality Engineering

Quality engineering as promulgated by Taguchi involves the following activities:

1. *Plan an experiment.* Identify the main function, side effects, and failure modes. Identify noise factors and testing conditions for evaluating quality loss, the quality characteristic to be observed, the quality function to be optimized, and the controllable parameters and their most likely settings.
2. *Perform the experiment.* Conduct a limited number of experiments that simultaneously vary all of the parameters according to the pattern of an appropriate orthogonal matrix. Collect data on the results of the experiments.
3. *Analyze and verify the results.* Analyze the data, determine the optimum parameter settings and predict the performance under these settings. Conduct a verification experiment to confirm the results obtained at the optimum settings.

As with statistical process control, the Taguchi method relies on a hypotheses. Part of the evaluation includes a series of tests to verify that the hypothesis should not be rejected based on experimental evidence. To compare the effects of changing the values of different parameters, Taguchi introduces the signal-to-noise ratio measure

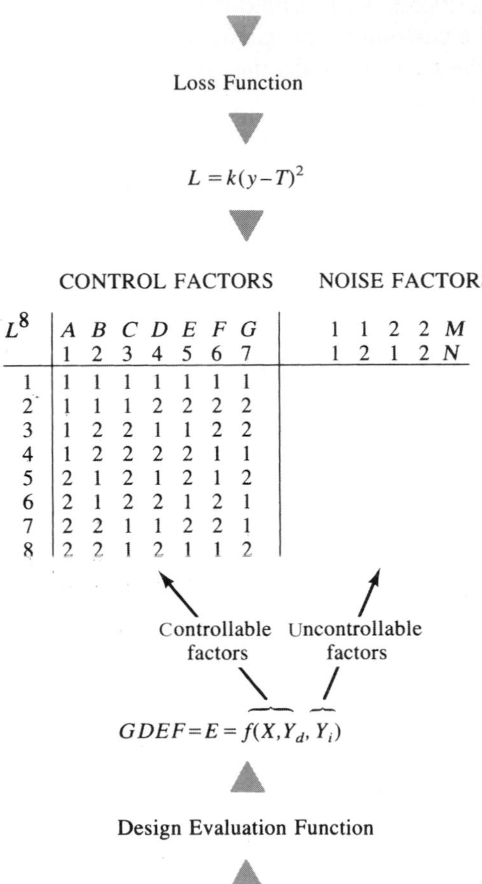

L^8	A 1	B 2	C 3	D 4	E 5	F 6	G 7	1 1	1 2	2 1	2 2	M N
1	1	1	1	1	1	1	1					
2	1	1	1	2	2	2	2					
3	1	2	2	1	1	2	2					
4	1	2	2	2	2	1	1					
5	2	1	2	1	2	1	2					
6	2	1	2	2	1	2	1					
7	2	2	1	1	2	2	1					
8	2	2	1	2	1	1	2					

Figure 11.19 Parameter design and design-dependent parameter relationships.

embracing a simple addition of the contribution of individual parameter settings. This additive property is the hypothesis to be tested.

The relationship of Taguchi parameter design and the design-dependent parameter approach introduced in Equation 7.3 is illustrated in Figure 11.19. A summary of the relationship follows:

1. Parameter Design Approach (Taguchi).
   a. *Loss function.* Quadratic in the deviation of actual parameter value from target value and related to the specific design by k.
   b. *Experimental design.* A factorial experiment relating control factors at two levels to noise factors at two or more levels.

2. Design-Dependent Parameter Approach (Equation 7.3).
   a. *Design evaluation function.* A life-cycle cost function optimized on design variables for each set of design-dependent parameters.
   b. *Indirect experimentation.* The use of modeling, Monte Carlo sampling, and sensitivity analysis to evaluate the LCC penalty for various combinations of design-dependent parameters in the face of variation in design-independent parameters.

Design-dependent parameter approaches, as introduced in Chapter 7 and subsequently illustrated, and parameter design (Taguchi style) have much in common. Taguchi would call the concurrent design process and the design evaluation function approach *system design*. System design involves innovation and knowledge from the science and engineering fields. It includes the selection of materials, parts, and product parameter values in the product design stage (Figure 2.3, top life cycle) and the selection of production equipment and values for process factors in the process design stage (Figure 2.3, middle life cycle).

Tentative nominal values for design-dependent parameters are then tested over specified ranges. This step is Taguchi's parameter design. Its purpose it to determine the best combination of parameter levels or values. Parameter design determines product parameter values that are least sensitive to joint changes in design independent parameter noise and environmental noise factors.

Indirect experimentation through modeling and simulated sampling is a means of determining the best combination of levels for design-dependent parameters. The loss function best used is the design evaluation function of Equation 7.3. Taguchi parameter design is an approach yet to be adopted to conceptual and preliminary design. Because the product and the production process does not yet exist, *ED* approaches are not feasible except on surrogate systems.

Tolerance design, as defined by Taguchi, is expensive and is, therefore, to be minimized. It is in this area that SPC can be most helpful. The objective, however, is to define design-dependent parameters in such a way that production, operations, and support capabilities are not adversely affected by design-dependent parameter drift. The objective of the Taguchi and the design-dependent parameter approaches is to achieve robustness against noise factors, so that life-cycle cost may be minimized.

QUESTIONS AND PROBLEMS

1. Speed is one characteristic or condition of an automobile which must be controlled. Discuss the role of each element of a control system for speed control.
2. Sketch a diagram such as Figure 11.1 which would apply to a thermostatically controlled heating system. Discuss each block.
3. Give an example of open-loop control; closed-loop control.
4. What is the relationship among an unstable pattern of variation, control limits, and a Type I error? a Type II error?

5. Samples of $n = 10$ were taken from a process for a period of time. The process average was estimated to be $\overline{X} = 0.0250$ in, and the process range was estimated as $\overline{R} = 0.0020$ in. Specify the control limits for an X chart and for an R chart.

6. Control charts by variables are to be established on the tensile strength in pounds of a yarn. Samples of five have been taken each hour for the past 20 hours. These were recorded as shown in the table.

Hour																			
1	2	3	4	5	6	7	8	9	10	11	12	13	14	15	16	17	18	19	20
50	44	44	48	47	47	44	52	44	13	47	49	47	43	44	45	45	50	46	48
51	46	44	52	46	44	46	46	46	44	44	48	51	46	43	47	45	49	47	44
49	50	44	49	46	43	46	45	46	49	44	41	50	46	40	51	47	45	48	49
42	47	47	49	48	40	48	42	46	47	42	46	48	48	40	48	47	49	46	50
43	48	48	46	50	45	46	55	43	45	50	46	42	46	46	46	46	48	45	46

(a) Construct an $\overline{X}$ chart based on these data.
(b) Construct an R chart based on these data.

7. A lower specification limit of 42 lb is required for the condition of Problem 6. Sketch the relationship between the specification limit and the control limits. What proportion, if any, of the yarn will be defective?

8. The total number of accidents during the long weekends and the number of fatalities for a 10-year period are given in the table. Assuming that the process is in control with regard to the proportion of fatalities, construct a p chart and record the data of the last eight weekends.

Weekend	Number of accidents	Number of fatalities	Weekend	Number of accidents	Number of fatalities
1	2,378	426	16	3,943	523
2	3,375	511	17	3,950	557
3	3,108	498	18	4,358	536
4	3,756	525	19	4,217	533
5	3,947	564	20	3,959	547
6	2,953	475	21	4,108	554
7	3,075	490	22	4,379	579
8	3,173	504	23	4,455	598
9	3,479	528	24	4,753	585
10	3,545	555	25	4,276	543
11	3,865	537	26	3,868	507
12	3,747	529	27	3,947	523
13	4,011	569	28	3,665	575
14	3,108	470	29	4,078	569
15	3,207	510	30	4,025	578

9. During a 4-week inspection period, the number of defects listed in the table were found in a sample of 400 electronic components. Construct a *c* chart for these data. Does it appear as though there existed an assignable cause of variation during the inspection period?

Date	Number of defects	Date	Number of defects
1	7	11	6
2	8	12	8
3	9	13	16
4	8	14	2
5	3	15	4
6	9	16	2
7	5	17	6
8	6	18	5
9	15	19	3
10	9	20	7

10. A survey during a safety month showed the number of defective cars on a highway as given in the table. A sample of 100 cars was taken during each day. Construct a *c* chart for the data. Is there any assignable cause of variation during this period?

Date	Defective cars	Date	Defective cars	Date	Defective cars
1	12	11	9	21	9
2	15	12	7	22	8
3	13	13	13	23	17
4	9	14	12	24	16
5	14	15	6	25	13
6	17	16	18	26	18
7	8	17	15	27	12
8	21	18	5	28	9
9	12	19	14	29	11
10	14	20	12	30	19

11. One operator is assigned to run 14 automatic machines. The minimum-cost service factor is 0.2. The machines require attention at random, and the service duration is distributed exponentially. Machines receive service on a first-come, first-served basis. Specify the upper control limit if the probability of designating the system out of control when it is really in control must not exceed 0.05.

12. Find the critical path in the network shown.

13. Eleven activities and eight events constitute a certain research and development project. The activities and their expected completion times are as follows:

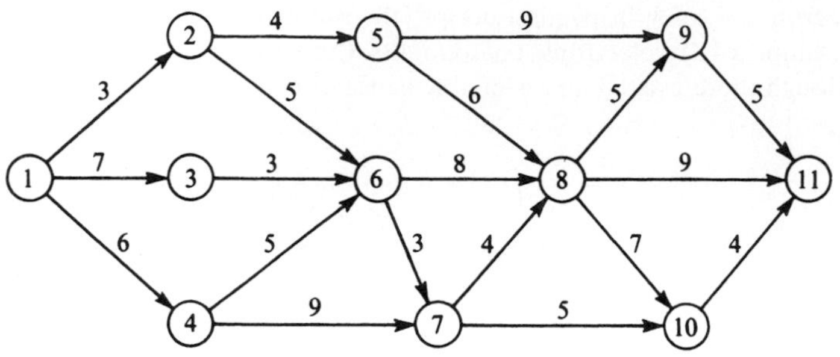

Activity	Completion time in weeks
$A \to B$	5
$A \to C$	6
$A \to D$	3
$B \to E$	10
$B \to F$	7
$C \to E$	8
$D \to E$	2
$E \to F$	1
$E \to G$	2
$F \to H$	5
$G \to H$	6

(a) Represent the project in the form of an activity-event network.
(b) Calculate T_E and T_L for each event and label each with T_E, T_L, and S.
(c) Identify the critical path and calculate the shortest possible time for project completion.

14. In Problem 13, assume that event C must occur before event D occurs. Modify the activity-event network for this assumption and show how the critical path would be changed.

15. Nine activities and six events are required to execute and complete a certain construction program. The activities and their completion times in weeks under normal and under expedited conditions are as follows:

	Normal		Expedited	
Activity	Duration	Cost	Duration	Cost
AB	10	$3,000	8	$ 5,000
AC	5	2,500	4	3,600
AD	2	1,100	1	1,200
BC	6	3,000	3	12,000
BE	4	8.500	4	8,500
CE	7	9,800	5	10,200
CF	3	2,700	1	3,500
DF	2	9,200	2	9,200
EF	4	300	1	4,600

A linear crash cost-time relationship exists between the normal and expedited conditions. Overhead of $11,000 per week can be saved if the program is completed earlier than the normal schedule.

(a) Represent the program in the form of an activity-event network.
(b) Find the earliest time within which the program can be completed under normal conditions.
(c) Find the minimum cost schedule for the program.

16. Eight maintenance activities, their normal durations in weeks, and the crew sizes for the normal condition are as follows:

Activity	Duration in weeks	Crew size
AC	6	8
BC	6	7
BE	2	6
CD	3	2
CE	4	1
DF	7	4
EF	4	6
FG	5	10

(a) Represent the maintenance project in the form of an activity-event network.
(b) Identify the critical path and find the minimum time for project completion.
(c) Extra crew members can be used to expedite activities *BC*, *CE*, *DF*, and *EF* at a cost given below:

Extra crew	Weeks saved	Crash cost per week
1	1	$100
2	2	120
3	2	200
4	3	250

If there is a penalty cost of $1,250 per week beyond the minimum maintenance time of 17 weeks, recommend the minimum cost schedule.

17. Determine the critical path of the PERT network shown. What is the second most critical path?

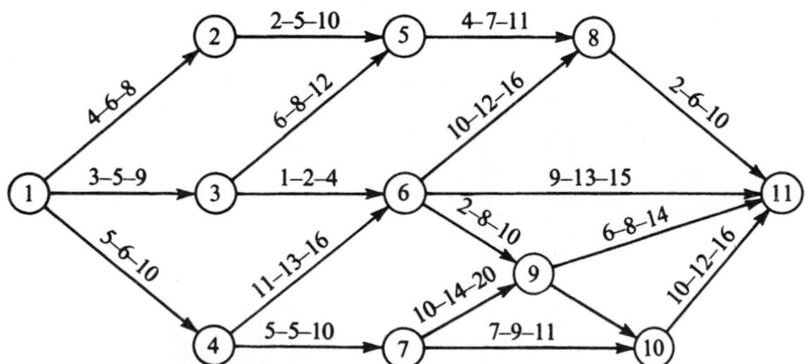

18. Calculate the probability of meeting a scheduled time of 50 units for the PERT network shown.

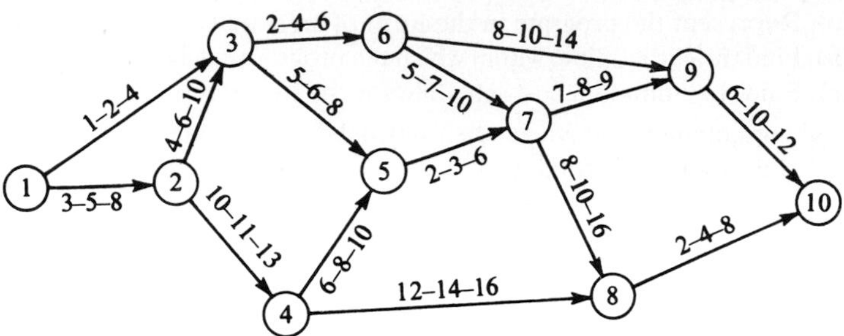

DESIGN FOR OPERATIONAL FEASIBILITY

PART IV

12 DESIGN FOR RELIABILITY

System design and development is accomplished through the systems engineering process, which requires the appropriate application of scientific and engineering efforts to ensure that the ultimate product is operationally feasible. Operational feasibility implies that the system will perform as intended in an effective and efficient manner for as long as necessary. The accomplishment of such requires the proper integration of design-related specialties such as reliability, maintainability, usability (human factors), and so on, into the total engineering design effort (refer to Figure 2.14). Some of these design-related specialties are presented in Part IV.

One of the most significant design parameters requiring attention is *reliability*. Many systems (or products) in use today are highly sophisticated and will fulfill most expectations when operating. However, experience has indicated that these systems are inoperative much of the time, requiring extensive maintenance and the expenditure of scarce support resources. Unreliable systems are unable to fulfill the mission for which they were designed. In an environment of scarce resources, it is essential that reliability be considered as a major system parameter during the design process. Reliability is a characteristic inherent in design.

12.1 DEFINITION AND EXPLANATION OF RELIABILITY

Reliability may be defined simply as the probability that a system or product will accomplish its designated mission in a satisfactory manner or, in more specific terms, the probability that it will perform in a satisfactory manner for a given period when used under specified operating conditions. Inherent within this definition are the elements of *probability, satisfactory performance, time or mission-related cycle,* and *specified operating conditions.*[1]

Probability, the first element in the reliability definition, is usually stated in quantitative terms representing a fraction or a percent specifying the number of times that one can expect an event to occur in a total number of trials. For instance, a statement that the probability of survival of an item operating 80 hours is 0.75 (or 75%) indicates that one can expect the item to function properly for at least 80 hours 75 times out of 100. When there are several supposedly identical items operating under similar conditions, one may wish to specify that 75% of the items will operate in a satisfactory manner when required and for the duration of the designated mission. It can be expected that failures will occur at different points of time; thus, failures are described in probabilistic terms. The fundamental definition of reliability is heavily dependent on the concepts derived from probability theory as described in Appendix B.[2]

The second element in the reliability definition is *satisfactory performance*, indicating that specific criteria must be established that describe what is considered to be satisfactory. This relates back to the definition of system operational requirements, the functions to be accomplished, and the TPMs described in Sections 3.4 and 3.5. What must the system do in order to satisfy the needs of the customer?

The third element, *time*, is one of the most important because it represents a measure against which the degree of system performance can be related. One must know the time parameter to assess the probability of completing a mission or a designated function as scheduled. Of particular interest is the ability to predict the probability of an item surviving (without failure) for a designated period. Reliability is often expressed in terms of *mean time between failure* (MTBF) or *mean time to failure* (MTTF), making time critical in reliability measurement.[3]

The *specified operating conditions* under which a system or product is expected to function constitute the fourth significant element of the reliability definition. These conditions include environmental factors, such as the geographical location where the system is expected to operate and the anticipated period of time, the operational profile, and the potential impacts resulting from temperature cycling, humidity, vibration and shock and so on. Such factors must not only address the conditions during the period when the system is actually operating, but during the period when the system (or a portion thereof) is in a storage mode or being transportated from one location to the next. Experience has indicated that the transportation, handling, and storage modes are sometimes more critical from a reliability standpoint than are the conditions experienced during actual system operational use.[4]

[1] Reliability must be directly related to one or more specific mission scenarios, or the system operational requirements described in Section 3.4.

[2] Refer to the literature in Appendix F for additional coverage of reliability definitions, concepts, and principles.

[3] Reliability may also be specified in terms of the number of cycles of operation or the number of successful functions completed. In the world of software, failures may be related to calendar time, processor time, the number of transactions per period, the number of faults per line item or module of code, and so on. The specific TPMs, or measures of reliability, must be related to the type of system and its make-up.

[4] The functional analysis described in Section 3.6 should aid in the identification and definition of operating functions, transportation functions, and so on; i.e., specifed operating conditions.

The four elements discussed here are critical in determining the reliability of a system or product. As reliability is an inherent characteristic of design, it is essential that these elements be adequately considered at program inception. Reliability requirements must be addressed in the definition of operational requirements for the system (Section 3.4), the appropriate reliability TPMs must be identified and prioritized (Section 3.5), and the resultant design dependent parameters (DDPs) must be *designed in* from the beginning (Section 2.5).

12.2 MEASURES OF RELIABILITY

The evaluation of any system or product in terms of reliability is based on precisely defined reliability concepts and measures. This section is concerned with the development of selected reliability measures and terms. A basic understanding of these is required before discussing reliability program functions as related to the system/product design. Terms such as the reliability function, failure rate, probability density function, reliability model, and so on, are briefly defined herein.

The Reliability Function

The *reliability function*, also known as the *survival function*, is determined from the probability that a system (or product) will be successful at least for some specified time t. The reliability function, $R(t)$, is defined as

$$R(t) = 1 - F(t) \tag{12.1}$$

where $F(t)$ is the probability that the system will fail by time t. $F(t)$ is basically the *failure distribution function* or *unreliability function*. If the random variable t has a density function of $f(t)$, the expression for reliability is

$$R(t) = 1 - F(t) = \int_t^{\infty} f(t)\, dt \tag{12.2}$$

If the time to failure is described by an exponential density function, then, from Appendix B.2

$$f(t) = \frac{1}{\theta} e^{-t/\theta} \tag{12.3}$$

where θ is the mean life and t the time period of interest. The reliability at time t is

$$R(t) = \int_t^{\infty} \frac{1}{\theta} e^{-t/\theta}\, dt = e^{-t/\theta} \tag{12.4}$$

Mean life (θ) is the arithmetic average of the lifetimes of all items considered, which for the exponential function is MTBF. Thus,

$$R(t) = e^{-t/M} = e^{-\lambda t} \tag{12.5}$$

where λ is the instantaneous failure rate and M the MTBF.

If an item has a constant failure rate, the reliability of that item at its mean life is approximately 0.37. Thus, there is a 37% probability that a system will survive its mean life without failure. Mean life and failure rate are related as

$$\lambda = \frac{1}{\theta} \tag{12.6}$$

Figure 12.1 illustrates the exponential reliability function, where time is given in units of t/M. The illustration focuses on the reliability function for the exponential distribution, which is commonly used in many applications.

The failure characteristics of different items are not necessarily the same. There are several well-known probability distribution functions that have been found in practice to describe the failure characteristics of different equipments. These include the binomial, exponential, normal, Poisson, γ, and Weibull distributions. Thus, one should not assume that any one distribution is applicable in all instances (refer to Appendix B.2).

The Failure Rate

The rate at which failures occur in a specified time interval is called the *failure rate* for that interval. The failure rate per hour is expressed as

$$\lambda = \frac{\text{number of failures}}{\text{total operating hours}} \tag{12.7}$$

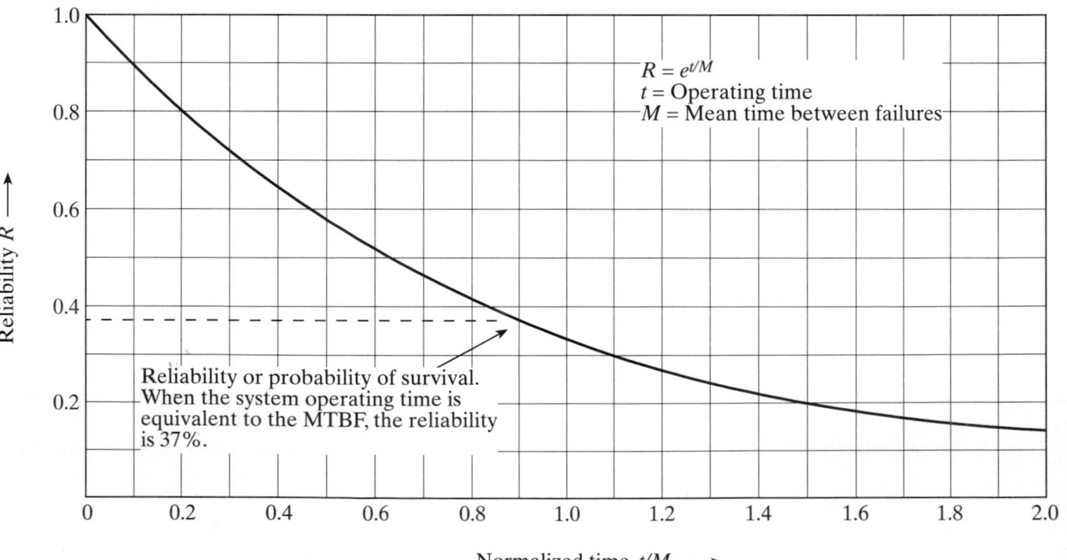

Figure 12.1 Reliability curve for the exponential distribution.

The failure rate may be expressed in terms of failures per hour, percentage of failures per 1,000 hours, or failures per million hours. As an example, suppose that 10 components were tested for 600 hours under specified operating conditions. The components (which are not repairable) failed as follows: component 1 failed after 75 hours, component 2 failed after 125 hours, component 3 failed after 130 hours, component 4 failed after 325 hours, and component 5 failed after 525 hours. Thus, there were 5 failures and the total operating time was 4,180 hours. Using Equation 12.7, the calculated failure rate per hour is

$$\lambda = \frac{5}{4{,}180} = 0.001196$$

As another example, suppose that the operating cycle for a given system is 169 hours as illustrated in Figure 12.2. During that time, 6 failures occur at the points indicated. A failure is defined as an instance when the system is not operating within a specified set of parameters. The failure rate, or corrective maintenance frequency, per hour is

$$\lambda = \frac{\text{number of failures}}{\text{total mission time}} = \frac{6}{142} = 0.04225$$

Assuming an exponential distribution, the system mean life or the mean time between failure (MTBF) is

$$\text{MTBF} = \frac{1}{\lambda} = \frac{1}{0.04225} = 23.6686 \text{ hours}$$

Figure 12.3 presents a reliability nomograph (for the exponential failure distribution) which facilitates calculations of MTBF, λ, $R(t)$, and operating time. If the MTBF is 200 hours ($\lambda = 0.005$), and the operating time is 2 hours, the nomograph gives a reliability value of 0.99.

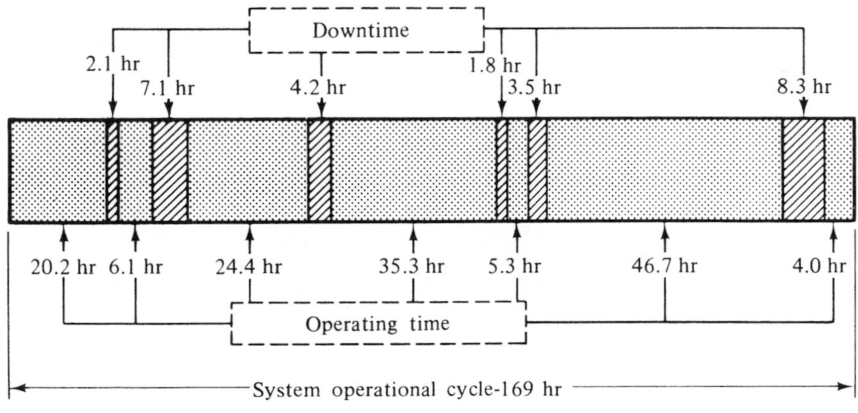

Figure 12.2 A system operational cycle.

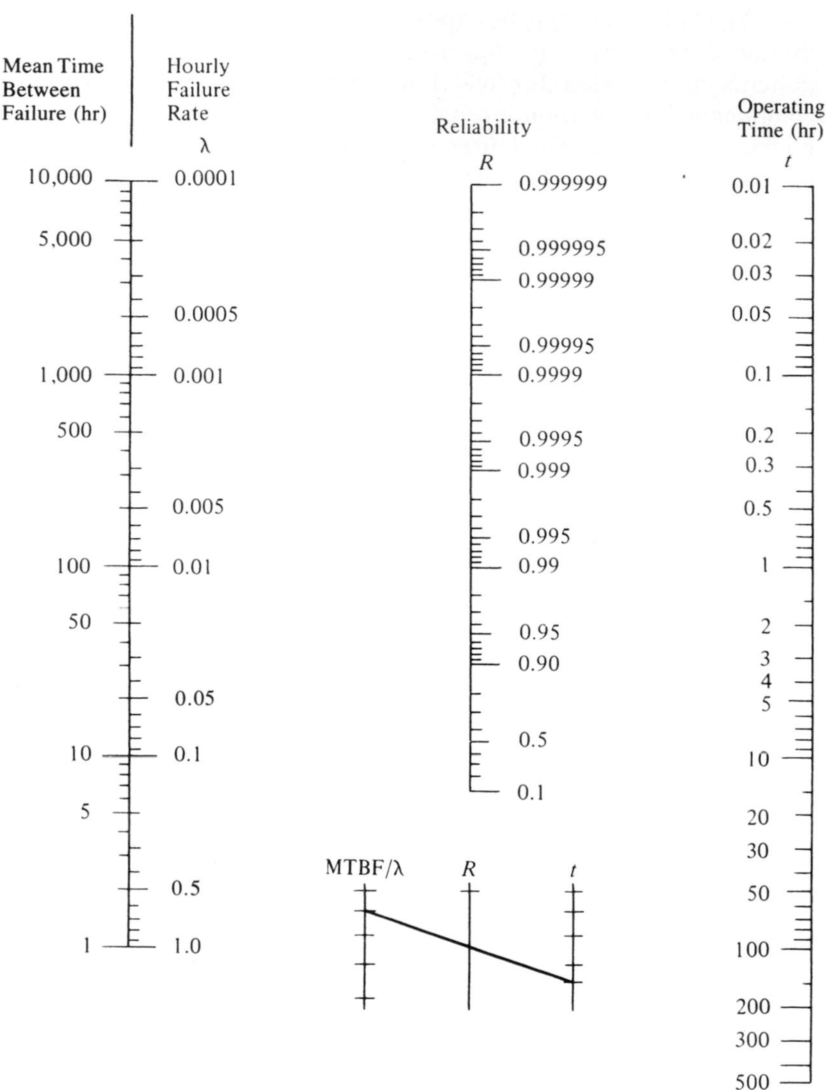

Figure 12.3 Reliability nomograph for the exponential failure distribution. Given equipment mean time to failure of hourly failure rate and operating time, solve for reliability. Connect MTBF and t values with straight line. Read R. *Source:* NAVAIR 00-65-502/NAVORD OD 41146, *Reliability Engineering Handbook*, Naval Air Systems Command and Naval Ordnance Systems Command, revised March 1968.

When assuming the negative exponential distribution, the failure rate is considered to be relatively constant during normal system operation if the system design is mature. That is, when equipment is produced and the system is initially distributed for operational use, there are usually a higher number of failures resulting from component variations and mismatches, manufacturing processes, and so on. The initial fail-

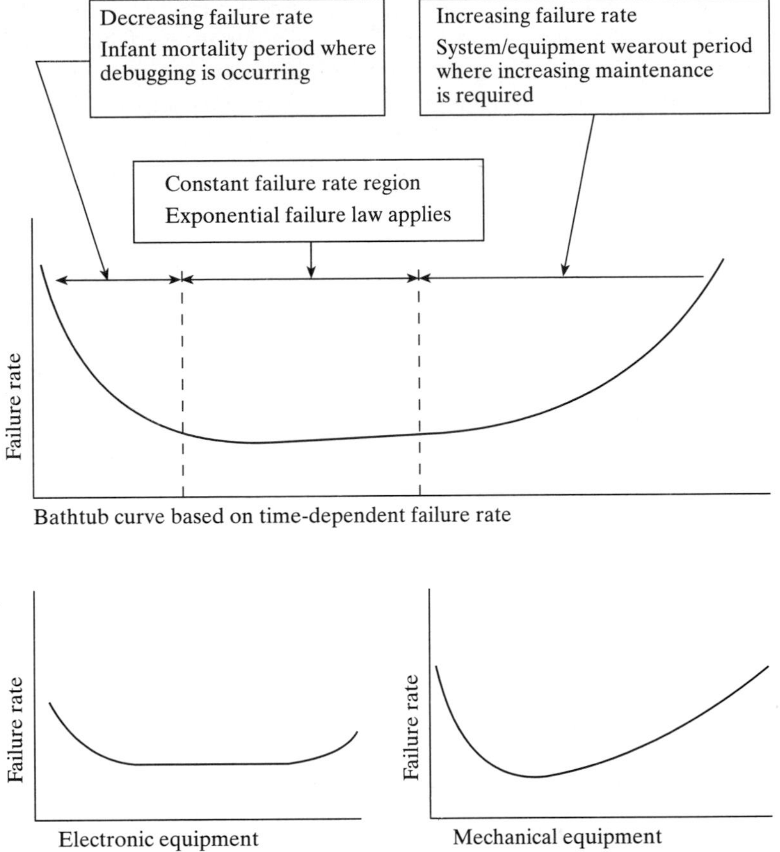

Figure 12.4 Typical failure-rate curve relationships.

ure rate is higher than anticipated, but gradually decreases and levels off, as illustrated in Figure 12.4, during the *debugging* period. Similarly, when the system reaches a certain age, there is a *wearout* period when the failure rate increases. The relatively level portion of the curve in Figure 12.4 is the constant failure-rate region, where the exponential failure law applies.

Figure 12.4 illustrates certain relative relationships. Actually, the curve may vary considerably depending on the type of system and its operational profile. Further, if the system is continually being modified for one reason or another, the failure rate may not be constant. In any event, the illustration does provide a good basis for discussion when considering relative failure-rate trends.

In the domain of software, failures may be related to calendar time, processor time, the number of transactions per period, the number of faults per module of code, and so on. Expectations are usually based on an operational profile and criticality to the mission. Thus, an accurate description of the mission scenario (s) is required (refer to Section 3.4). As the system evolves from the design and development stage to the operational use phase, the ongoing maintenance of software often becomes a major issue. Although the failure rate of equipment generally assumes the profiles in Figure 12.4, the maintenance of software often has a negative effect on the overall system reliability. The performance of software maintenance on a continuing basis, along with the incorporation of system changes in general, usually impacts the overall failure rate as shown in Figure 12.5. When a change or modification is incorporated, "bugs" are usually introduced, and it takes a while for these to be "worked out" of the system.

When defining failures (and failure rates) from a *systems* perspective, any time that the system is not functioning properly a failure has occurred. Failures may be classified as inherent "primary" or "catastrophic" component failures, secondary or dependent failures, manufacturing defects, operator-induced failures, maintenance-induced failures, failures resulting from wearout, material damage because of handling, and so on. Table 12.1 shows an overall composite failure rate including these various factors,

TABLE 12.1 Composite Failure Rate

Contributor	Assumed factor (Instances/Hr)
Inherent "primary" failure rate	0.000392
Dependent or "secondary" failure rate	0.000072
Manufacturing defects	0.000002
Operator-induced failure rate	0.000003
Maintenance-induced failure rate	0.000012
Equipment handling damage rate	0.000005
Wearout rate	0.000001
Total composite factor	0.000487

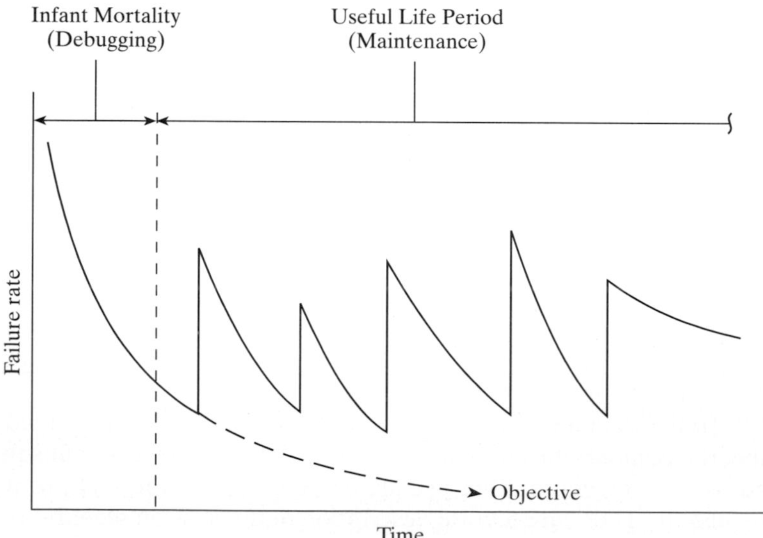

Figure 12.5 Failure-rate curve with maintenance (software application).

which, in turn, leads to determining the frequency of corrective (unscheduled) maintenance. When defining failures in a *pure* reliability sense, it is not uncommon to deal only with primary and secondary failures (i.e., the first two items in the table), assuming that other contributors are insignificant. This may lead to an incorrect reliability MTBF prediction. Thus, from a systems engineering view, all of the factors that may contribute to the overall reliability of the system, and its elements must be considered.[5]

Component Relationships

After defining the basic reliability function and some of the measures associated with system failures, it is appropriate to consider their application in design. Components are selected and combined in a serial manner, in parallel, or combinations thereof. Block diagrams are developed to show component relationships and for accomplishing reliability analysis.[6]

Series networks. The series relationship is probably the most commonly used and is the simplest to analyze. It is illustrated in Figure 12.6. In a *series network* all components must operate in a satisfactory manner if the system is to function properly. Assuming that a system includes subsystem A, subsystem B, and subsystem C, the reliability of the system is the product of the reliabilities for the individual subsystems expressed as

$$R = (R_A)(R_B)(R_C). \tag{12.8}$$

As an example, suppose that an electronic system includes a transmitter, a receiver, and a power supply. The transmitter reliability is 0.8521, the receiver reliability is 0.9712, and the power supply reliability is 0.9357. The overall reliability for the electronic system is

$$R_s = (0.8521)(0.9712)(0.9357) = 0.7743$$

If a series system is expected to operate for a specified time period, its required overall reliability can be derived. Substituting Equation 12.5 into Equation 12.8 gives

$$R_s = (e^{-\lambda_1 t})(e^{-\lambda_2 t}) \cdots (e^{-\lambda_n t})$$

for a series with n components. This may be expressed as

$$R_s = e^{-(\lambda_1 + \lambda_2 + \cdots + \lambda_n)t} \tag{12.9}$$

Suppose that a series system consists of four subsystems and is expected to operate for 1,000 hours. The four subsystems have the following MTBFs: subsystem A,

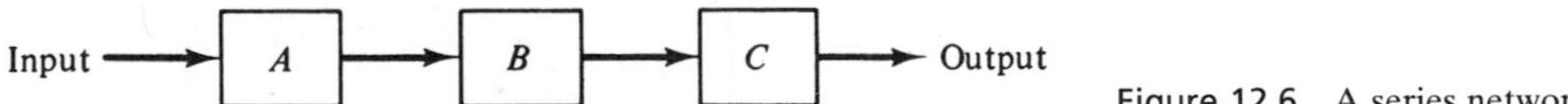

Figure 12.6 A series network.

[5] The issues pertaining to operator-induced and maintenence-induced failures are discussed in Chapter 14, and manufacturing defects are mentioned in Chapter 16.

[6] The development of reliability block diagrams should evolve directly from the functional analysis, and the development of functional block diagrams described in Sections 3.6 and 4.1.

MTBF = 6,000 hours; subsystem B, MTBF = 4,500 hours; subsystem C, MTBF = 10,500 hours; subsystem D, MTBF = 3,200 hours. The objective is to determine the overall reliability of the series network where

$$\lambda_A = \frac{1}{6{,}000} = 0.000167 \text{ failure/hour}$$

$$\lambda_B = \frac{1}{4{,}500} = 0.000222 \text{ failure/hour}$$

$$\lambda_C = \frac{1}{10{,}500} = 0.000095 \text{ failure/hour}$$

$$\lambda_D = \frac{1}{3{,}200} = 0.000313 \text{ failure/hour}$$

The reliability of the series network is found from Equation 12.9 as

$$R = e^{-(0.000797)(1{,}000)} = 0.4507$$

This means that the probability of the system surviving (i.e., reliability) for 1,000 hours is 45.1%. If the requirement were reduced to 500 hours, the reliability would increase to about 67%.

Parallel networks. A pure *parallel network* is one where several of the same components are in parallel and where all the components must fail to cause total system failure. A parallel network with two components is illustrated in Figure 12.7. Assuming that components A and B are identical, the system will function if either A or B, or both, are working. The reliability is expressed as

$$R = R_A + R_B - (R_A)(R_B) \tag{12.10}$$

Consider next a network with three components in parallel, as shown in Figure 12.8. The network reliability is expressed as

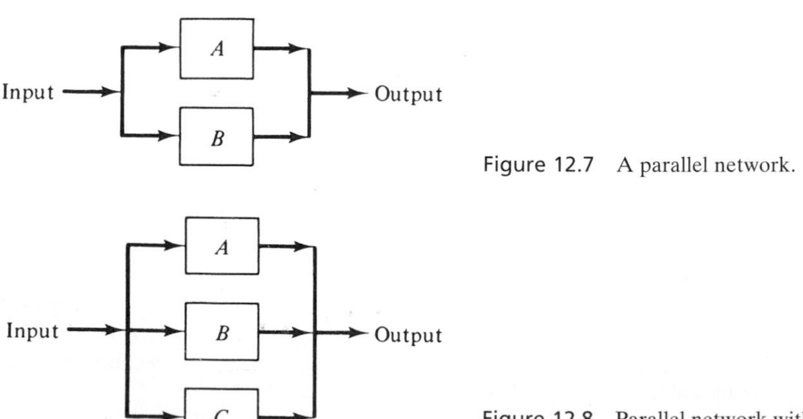

Figure 12.7 A parallel network.

Figure 12.8 Parallel network with three components.

$$R = 1 - (1 - R_A)(1 - R_B)(1 - R_C) \tag{12.11}$$

If components A, B, and C are identical, the reliability expression can be simplified to

$$R = 1 - (1 - R)^3$$

for a system with three parallel components. For a system with n identical components, the reliability is

$$R = 1 - (1 - R)^n \tag{12.12}$$

Parallel redundant networks are used primarily to improve system reliability as Equations 12.11 and 12.12 indicate mathematically. Consider the following example. A system includes two identical subsystems in parallel and the reliability of each subsystem is 0.95. The reliability of the system is found from Equation 12.10 as

$$R = 0.95 + 0.95 - (0.95)(0.95) = 0.9975$$

Suppose that the reliability of the preceding system needs improvement beyond 0.9975. By adding a third identical subsystem in parallel, the system reliability is found from Equation 12.11 to be

$$R = 1 - (1 - 0.95)^3 = 0.999875$$

Note that there is a reliability improvement of 0.002375 over the previous configuration.

If the subsystems are not identical, Equation 12.10 can be used. For example, a parallel redundant network with two subsystems, $R_A = 0.75$ and $R_B = 0.82$, gives a system reliability of

$$R = 0.75 + 0.82 - (0.75)(0.82) = 0.955$$

The effects of redundancy on reliability in design, presented in a simple generic sense, is illustrated in Figure 12.9. Redundancy can be applied in design at different hierarchical indenture levels of the system (refer to Figure 3.13). At the subsystem level, it

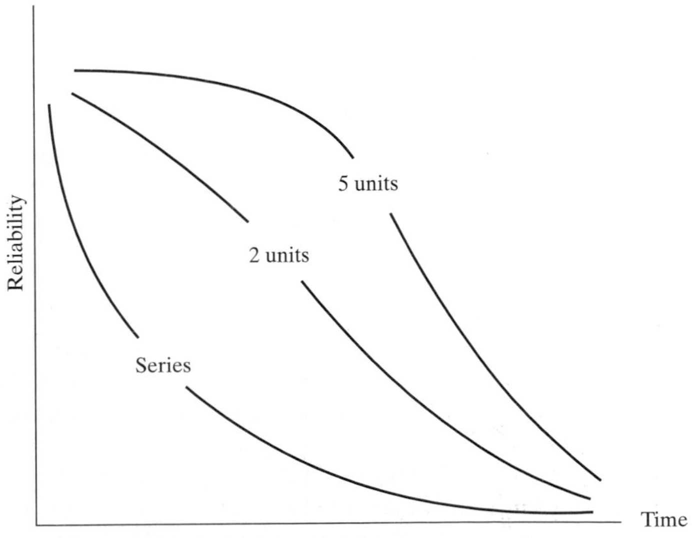

Figure 12.9 Effects of redundancy on reliability in design.

may be appropriate to incorporate parallel functional capabilities, where the system will continue to operate if one path fails to function properly. The flight control capability (incorporating electronic, digital, and mechanical alternatives) in an aircraft is an example where there are alternate paths in case of a failure in any one. At the detailed piece-part level, redundancy may be incorporated to improve the reliability of critical functions, particularly in areas where the accomplishment of maintenance is not feasible. For example, in the design of many electronic circuit boards, redundancy is often built in for the purposes of improving reliability, while at the same time the accomplishment of maintenance is not practical.

The application of redundancy in design is a key area for evaluation. Although redundancy per se does improve reliability, the incorporation of extra components in the design requires additional space, and the costs are higher. This leads to several questions: Is redundancy really required in terms of criticality relative to system operation and the accomplishment of the mission? At what level should redundancy be incorporated? What type of redundancy should be considered ("active" or "standby")? Should maintainability provisions be considered? Are there any alternative methods for improving reliability (e.g., improved part selection, part derating)? In essence, there are many interesting and related concerns that require further investigation.

Combined series-parallel networks. Various levels of reliability can be achieved through the application of a combination of series and parallel networks. Consider the three examples illustrated in Figure 12.10.

The reliability of the first network in Figure 12.10 is given by

$$R = R_A(R_B + R_C - R_B R_C) \tag{12.13}$$

For the second network the reliability is given by

$$R = [1 - (1 - R_A)(1 - R_B)][1 - (1 - R_C)(1 - R_D)] \tag{12.14}$$

And for the third network the reliability is given by

$$R = [1 - (1 - R_A)(1 - R_B)(1 - R_C)][R_D][R_E + R_F - (R_E)(R_F)] \tag{12.15}$$

Combined series-parallel networks such as those in Figure 12.10 require that the analyst first evaluate the redundant elements to obtain unit reliability. Overall system reliability is then determined by finding the product of all unit reliabilities.

Related Factors

There are several factors that may not be directly categorized as reliability figures-of-merit but that are closely related. Some of these are availability, effectiveness, mean time between maintenance, and others, as described in the paragraphs that follow.

Mean time between maintenance (MTBM). The degree to which reliability characteristics are incorporated in design leads to the requirements for maintenance, where a prime metric is *mean time between maintenance* (MTBM)—refer to Section 3.4 and Figure 3.7. MTBM may be broken down to include the *mean time between maintenance*

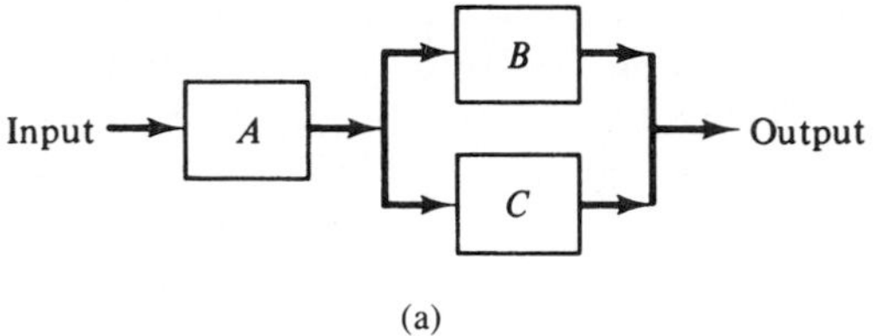

(a)

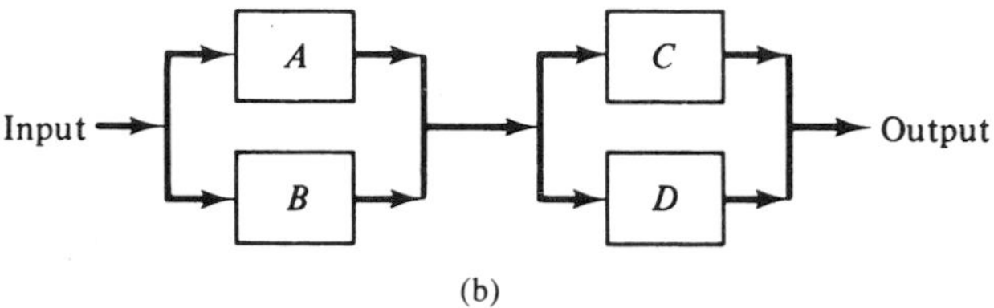

(b)

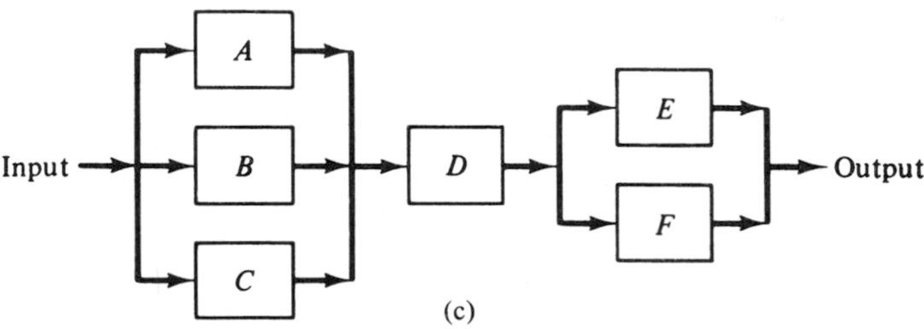

(c)

Figure 12.10 Some combined series-parallel networks.

unscheduled ($MTBM_u$) and *mean time between maintenance scheduled* ($MTBM_s$). The $MTBM_u$ covers all corrective or unscheduled maintenance required as a result of system failures. Referring to Table 12.1, if *all* of the failures (independent of cause) are addressed, then the MTBF should be equivalent to the reliability $MTBM_u$. However, in many instances the MTBF factor includes only primary and secondary failures, whereas the other factors in Table 12.1 are not considered. Further, there are instances when a system failure is based on an operator's perception; reported; and, when checked out, cannot be verified. In these cases, although an actual failure may not have been reported, an unscheduled maintenance action did result in response to the operator's report. From a system's perspective, this should also be counted as a failure because the operator (an element of the system) reported an "out-of-spec" condition, perceived or otherwise.

Referring to Figure 12.11, it is assumed that a system has completed a number of missions and that there were 100 failures reported. On investigation, 75 of the 100 maintenance actions resulted in the removal and replacement of an item at the organizational level of maintenance. The item was then checked out at the intermediate level, with a fault being verified on 48 occasions. By inspection, one can see that there

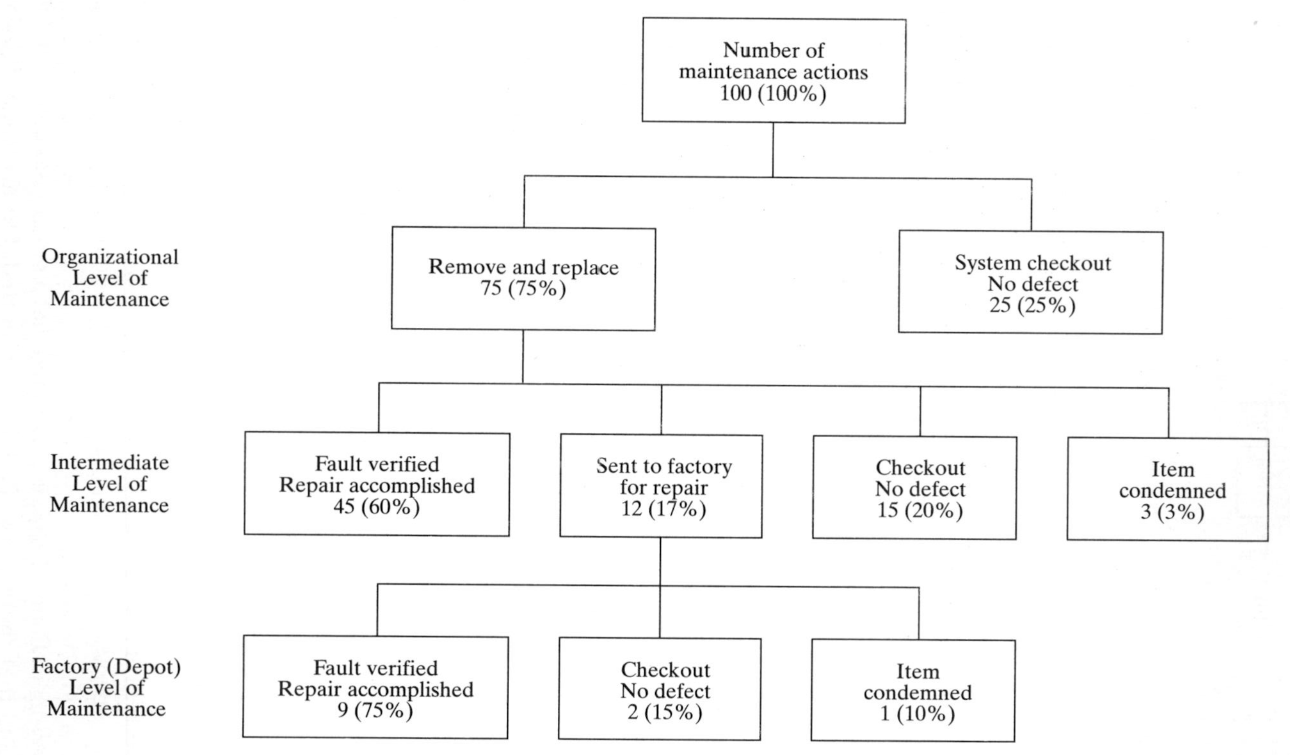

Total Verified Failures—58 of 100 Maintenance Actions

Figure 12.11 System XYZ unscheduled maintenance actions.

were 100 maintenance actions overall, 75 removals and replacements at the system level, and 58 verified failures (at all levels). Sometimes, when calculating MTBF, only the confirmed equipment failures (i.e., 58) are considered. Conversely, the calculated MTBM_u value should consider the 100 maintenance actions, which is of particular interests from a systems engineering perspective.[7]

Availability (A). Availability may be expressed and defined in three ways.

1. *Inherent availability* (A_i). Inherent availability is "the probability that a system or equipment, when used under stated conditions in an *ideal* support environment (i.e., readily available tools, spares, maintenance personnel, etc.), will operate satisfactorily at any point in time as required." It excludes preventive or scheduled maintenance actions, logistics delay time, and administrative delay time, and is expressed as

$$A_i = \frac{\text{MTBF}}{\text{MTBF} + \overline{\text{M}}\text{ct}} \tag{12.16}$$

   where MTBF is the mean time between failure and $\overline{\text{M}}$ct the mean corrective maintenance time. The term $\overline{\text{M}}$ct is equivalent to the mean time to repair (MTTR).

2. *Achieved availability* (A_a). Achieved availability is "the probability that a system or equipment, when used under stated conditions in an *ideal* support environment (i.e., readily available tools, spares, personnel, etc., will operate satisfactorily at any point in time," This definition is similar to the definition for A_i except that preventive (i.e., scheduled) maintenance is included. It excludes logistics delay time and administrative delay time, and is expressed as

$$A_a = \frac{\text{MTBM}}{\text{MTBM} + \overline{\text{M}}} \tag{12.17}$$

   where MTBM is the mean time between maintenance and $\overline{\text{M}}$ the mean active maintenance time. MTBM and $\overline{\text{M}}$ are functions of corrective (unscheduled) and preventive (scheduled) maintenance actions and times, respectively.

3. *Operational availability* (A_0). Operational availability is the "probability that a system or equipment, when used under stated conditions in an *actual* operational environment, will operate satisfactorily when called upon." It is expressed as

$$A_0 = \frac{\text{MTBM}}{\text{MTBM} + \text{MDT}} \tag{12.18}$$

   where MDT is the mean maintenance down time. The reciprocal of MTBM is the frequency of maintenance which, in turn, is a significant factor in determining logistic support requirements. MDT includes active maintenance time ($\overline{\text{M}}$), logistics delay time, and administrative delay time.[8]

[7] The MTBM factor, which is derived through the development of the maintenance concept described in Section 3.4, is also used as a metric for maintainability (Chapter 13) and supportability (Chapter 15).

[8] MTBM, MDT, $\overline{\text{M}}$ and $\overline{\text{M}}$ct are discussed further in Chapter 13.

The term *availability* is used differently in different situations. If one is to impose an availability figure-of-merit as a *design requirement* for a given equipment supplier, and the supplier has no control over the operational environment in which that equipment is to function, then A_a or A_i might be appropriate figures-of-merit against which the supplier's equipment can be properly assessed. Conversely, if one is to assess a system in a realistic operational environment, then A_0 is a preferred figure-of-merit to employ for assessment purposes. Further, availability may be applied at any time in the overall mission profile representing a point estimate or may be more appropriately related to a specific segment of the mission where the requirements are different from other segments. Thus, one must define precisely what is meant by "availability" and how it is to be applied. In any event, reliability is a major factor in the determination of system availability.

System effectiveness (SE). System effectiveness may be defined as "the probability that a system can successfully meet an overall operational demand within a given time when operated under specified conditions" or "the ability of a system to do the job for which it was intended." *System effectiveness*, like operational availability, is a term used in a broad context to reflect the technical characteristics of a system (i.e., performance, availability, supportability, dependability, etc.) and may be expressed differently depending on the specific mission application. Sometimes a single figure-of-merit is used to express system effectiveness and sometimes multiple figures-of-merit are employed. The objective is to reflect system design attributes and logistic support elements as is illustrated in Figure 12.12.

Cost effectiveness (CE). *Cost-effectiveness* relates to the measure of a system in terms of mission fulfillment (system effectiveness) and total life-cycle cost, and can be expressed in various ways, depending on the specific mission or system parameters that one wishes to evaluate. In essence, cost-effectiveness includes the elements illustrated in Figure 12.12 and can be expressed as indicated in Equations 12.19 through 12.23. Reliability is a major factor in determining the cost effectiveness of a system.[9]

$$\text{Figure of merit (FOM)} = \frac{\text{system benefits}}{\text{life-cycle cost}} \tag{12.19}$$

$$\text{FOM} = \frac{\text{system capacity}}{\text{life-cycle cost}} \tag{12.20}$$

$$\text{FOM} = \frac{\text{system effectiveness}}{\text{life-cycle cost}} \tag{12.21}$$

[9] The terms *system effectiveness* (SE) and *cost-effectiveness* (CE) are defined and applied differently depending on the system, its mission, and the metrics (TPMs) used to define the requirements as described in Section 3.5. The discussion herein is *generic* and must be tailored to the specific system in question. Further, Figure 12.12, which should evolve from Figure 2.9, is presented to illustrate some overall relationships in conceptual terms.

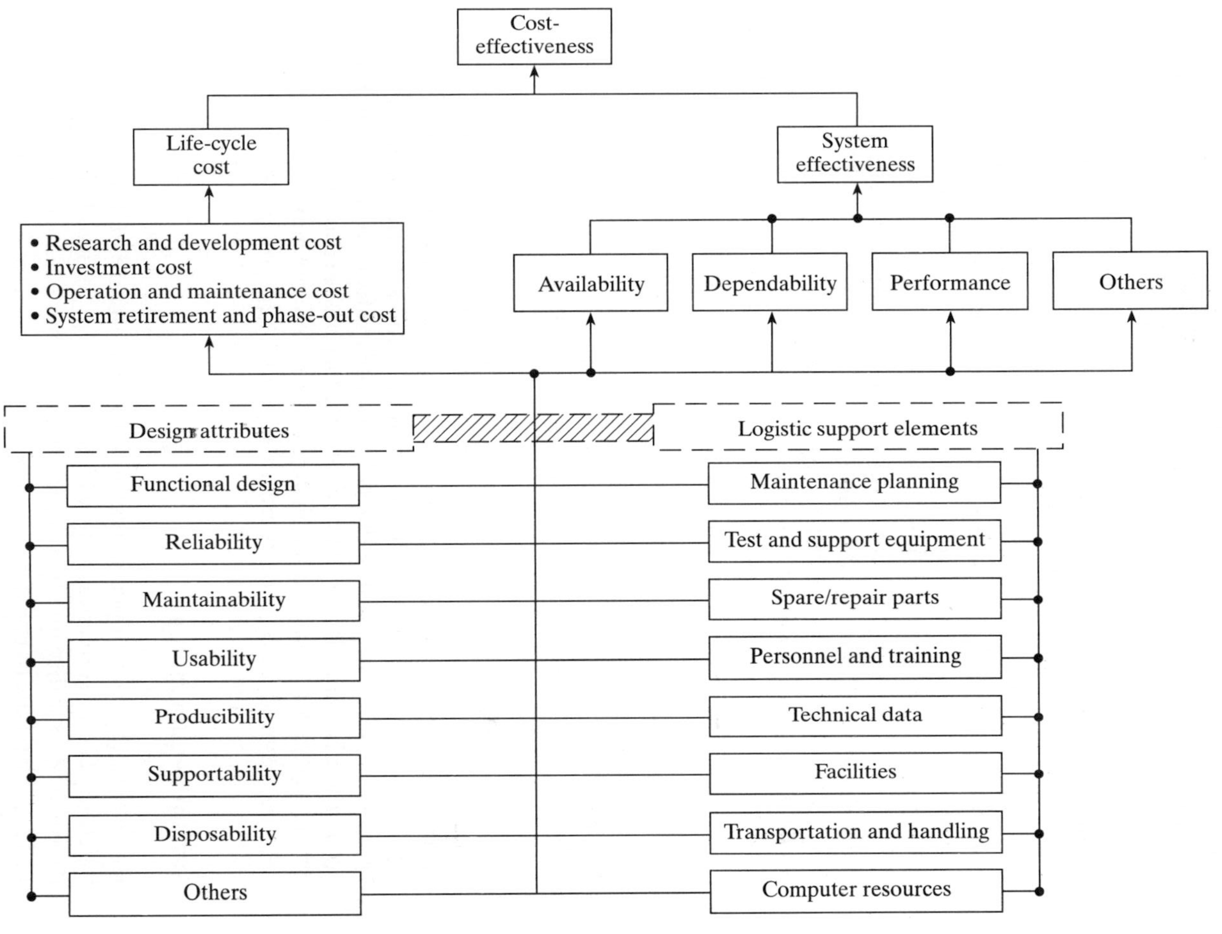

Figure 12.12 The elements of cost-effectiveness.

$$\text{FOM} = \frac{\text{availability}}{\text{life-cycle cost}} \tag{12.22}$$

$$\text{FOM} = \frac{\text{supportability}}{\text{life-cycle cost}} \tag{12.23}$$

12.3 RELIABILITY IN THE SYSTEM LIFE CYCLE

Reliability, as an inherent characteristic of design, must be considered in the overall systems engineering process beginning with the conceptual design phase. Referring to Figure 12.13 (which is an extension of Figure 2.2), qualitative and quantitative reliability requirements are developed through the accomplishment of feasibility analysis (Section 3.2), the development of operational requirements and the maintenance concept (Section 3.4), and the identification and prioritization of TPMs (Section 3.5). The applicable measures of reliability must be established, their importance with respect to other system metrics must be delineated, and the requirements for design must be identified through the development of the appropriate DDPs (refer to Figure 2.9).

Given the specification of reliability requirements in Block 1 of the figure (also included in the type A system specification—refer to Section 3.8), the next step is to develop a reliability model for the purposes of allocating top-level requirements down to the subsystem level and below as part of the system allocation process described in Section 4.2. Having defined the basic requirements, one evolves through the iterative process of system synthesis, analysis, and evaluation (refer to Figure 2.10). To facilitate this process, there are several different analysis methods/tools that can be effectively applied in evolving from conceptual design to preliminary and detail design (i.e., Blocks 2 and 3 of Figure 12.13). As the design progresses and models of the system are developed (both analytical and physical), the ongoing evaluation process occurs and reliability testing is included as part of the overall system test and evaluation activity described in Chapter 6. From this point on, the evaluation effort continues through the production/construction and system utilization phases (Blocks 4 and 5), with the appropriate feedback and system modification for corrective action or for product improvement as required. In essence, there are reliability requirements in each phase of the life cycle as illustrated in Figure 12.13.

System Requirements

Every system is developed in response to a customer need to fulfill some anticipated function. The effectiveness with which the system fulfills this function is the ultimate measure of its utility and its value to the customer (i.e., user). As conveyed in Figure 12.12, system effectiveness is a composite of many factors, with reliability being a major contributor in determining the usefulness of a system.

Reliability requirements, specified both in quantitative and qualitative terms, are defined within the context of the system operational requirements and the maintenance concept described in Section 3.4. This includes

1. Definition of system performance factors, the mission profile, and system requirements (use conditions, duty cycles, and how the system is to be operated).

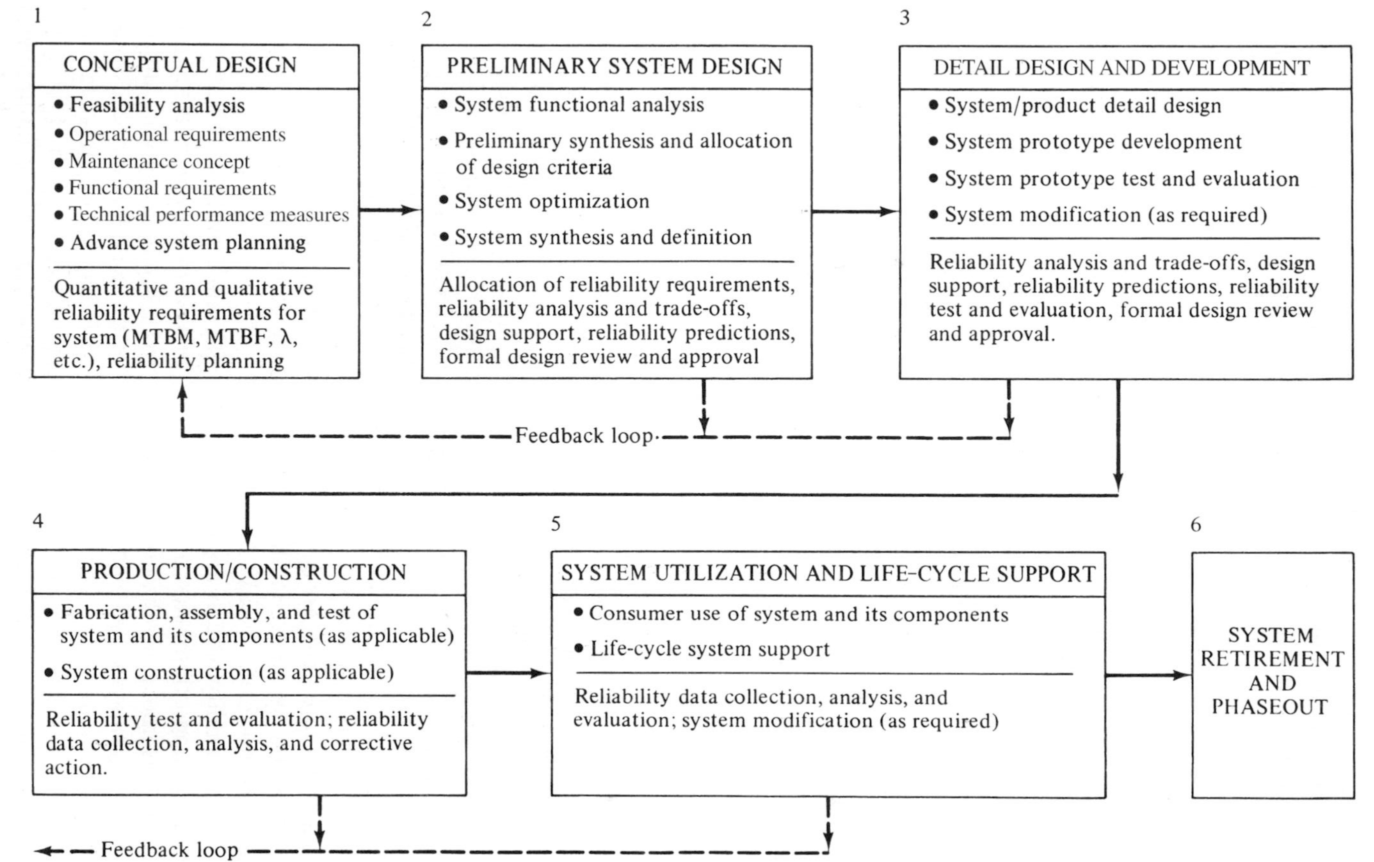

1
CONCEPTUAL DESIGN
• Feasibility analysis
• Operational requirements
• Maintenance concept
• Functional requirements
• Technical performance measures
• Advance system planning
Quantitative and qualitative reliability requirements for system (MTBM, MTBF, λ, etc.), reliability planning
2
PRELIMINARY SYSTEM DESIGN
• System functional analysis
• Preliminary synthesis and allocation of design criteria
• System optimization
• System synthesis and definition
Allocation of reliability requirements, reliability analysis and trade-offs, design support, reliability predictions, formal design review and approval
3
DETAIL DESIGN AND DEVELOPMENT
• System/product detail design
• System prototype development
• System prototype test and evaluation
• System modification (as required)
Reliability analysis and trade-offs, design support, reliability predictions, reliability test and evaluation, formal design review and approval.
Feedback loop
4
PRODUCTION/CONSTRUCTION
• Fabrication, assembly, and test of system and its components (as applicable)
• System construction (as applicable)
Reliability test and evaluation; reliability data collection, analysis, and corrective action.
5
SYSTEM UTILIZATION AND LIFE-CYCLE SUPPORT
• Consumer use of system and its components
• Life-cycle system support
Reliability data collection, analysis, and evaluation; system modification (as required)
6
SYSTEM RETIREMENT AND PHASEOUT
Feedback loop

Figure 12.13 Reliability requirements in the system life cycle (refer to Figure 2.2).

2. Definition of the operational life cycle (the anticipated time that the system will be in the inventory and in operational use).
3. Definition of the environment in which the system is expected to operate and be maintained (temperature, humidity, vibration, etc.). This should include a range of values as applicable and should cover all transportation, handling, maintenance, and storage modes.

The basic question at this point is: What reliability should the system have to accomplish its intended mission successfully, throughout the specified life cycle, and in the environment defined? If the operational requirements specify that the system must function 24 hours a day for 360 days a year without failure, the system reliability requirements may be rather stringent. Conversely, if the system is only required to operate 2 hours per day for 260 days/year, the specified requirements may be different. In any event, quantitative and qualitative reliability requirements must be defined for the system based on the foregoing considerations. As presented in Section 12.2, quantitative requirements are usually expressed in terms of $R(t)$, MTBM, MTBF, λ, successful operational cycles per period, or combinations thereof. For software, reliability may be specified as the number of errors per module of software or lines of code, number of processor errors, the time to first failure, or some equivalent measure impacting system failure.

Reliability Models

With the top-level requirements defined, the system is then described in functional terms as presented in Section 3.6. Referring to Figure 3.11, functions are identified at the system level and then broken down to the next lower level, and so on. Ultimately, one evolves from the identification of functions to the grouping and packaging of system elements into hardware, software, and so on, as shown in Figure 3.13, which serves as the framework for the allocation of reliability requirements.

Figure 12.14, an extension of Figure 3.11, shows a simple description of the system at Level I. The reliability requirement for the system (e.g., R, MTBF) is specified for the entire network identified at this level. For example, the reliability of Block 3 may be expressed as a probability of survival of 0.95 for a 4-hour period of operation. Similar requirements are specified for Blocks 1, 2, 4, and 5. These, when combined, will indicate the system reliability. Given this, one can progress down to the component level as shown at Level V.

Reliability models are developed for the initial purpose of requirements allocation (or apportionment) during the early phases of design, and for the subsequent accomplishment of reliability prediction, stress-strength analysis, and related design synthesis, analysis, and evaluation efforts. For the completion of these tasks, it is often necessary to develop an overall reliability block diagram as shown in Figure 12.15. As the design evolves and components are being selected for each of the elements shown, input-output values are identified for each block, predicted failure rates are determined, and the results are combined to provide a reliability prediction of the overall configuration as it exists at the time.

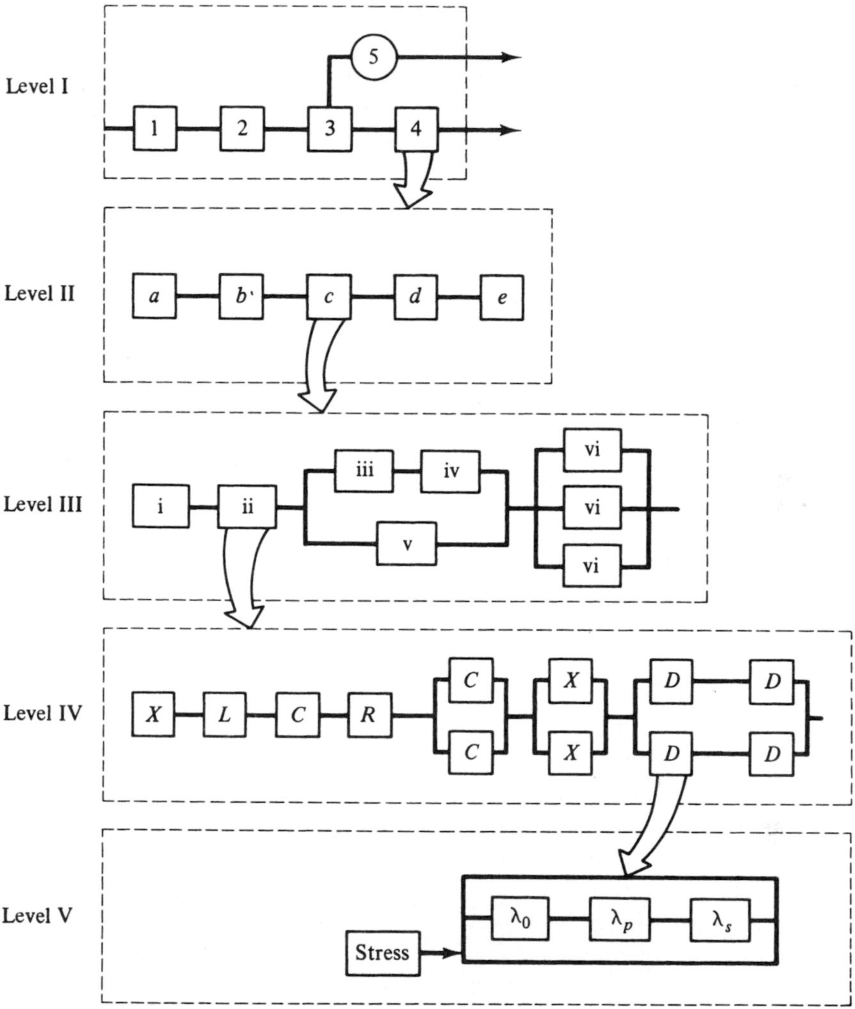

Figure 12.14 Progressive expansion of reliability block diagram. *Source:* NAVAIR 00-65-502/NAVORD OD 41146, *Reliability Engineering Handbook,* Naval Air Systems Command and Naval Ordnance Systems Command, revised March 1968.

The reliability model (i.e., block diagram) can be used not only for allocation and prediction, but for accomplishing a stress analysis, as an input for the FMECA, as a basis for determining maintenance requirements, and so on.

Reliability Allocation

The allocation of reliability requirements is accomplished as part of the process described in Section 4.2. Top-level requirements are specified for the system, and these requirements are then allocated to subsystem level, unit level, and down to the level

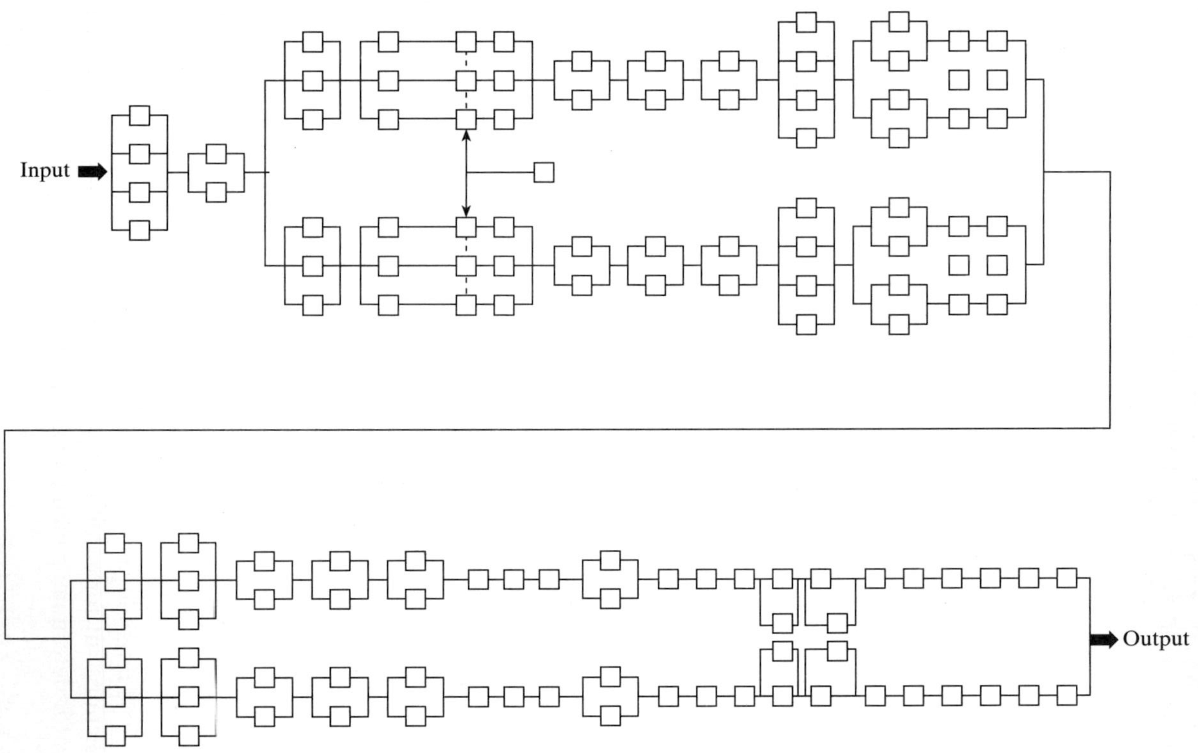

Figure 12.15 Expanded reliability block diagram of system.

needed to provide a meaningful input to the design (i.e., establishment of the right input design criteria at the proper level).

The approach used in the allocation of reliability requirements is not "fixed," may be somewhat subjective at times, and will vary depending on the system type and complexity. Referring to Figure 12.16, it is assumed that at the system level there is a MTBF requirement of 450 hours. The question is: What should be specified as a reliability requirement for Units A, B, and C? In response, the following steps are appropriate:

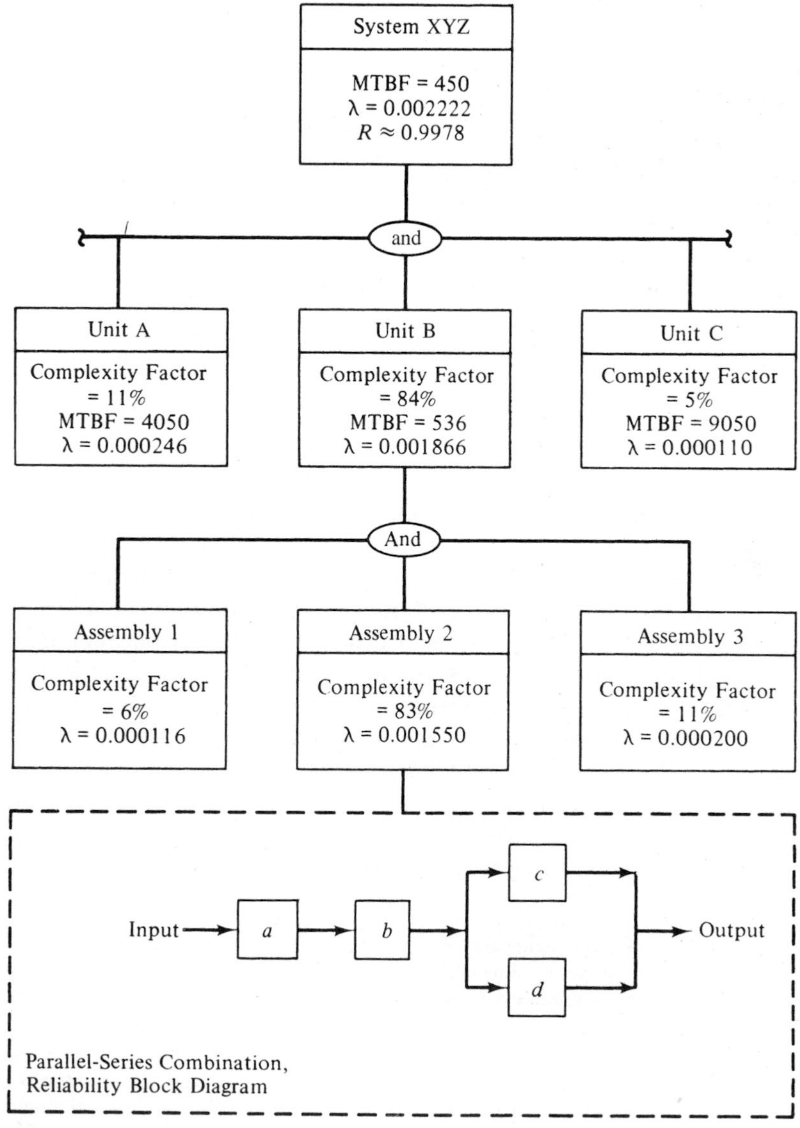

Figure 12.16 Allocation of reliability requirements.

1. Identify the elements of the system where the design is known and where reliabilty data are available or can be readily assessed. For example, through the selection of commercial off-the-shelf (COTS) items, it is expected that field data may be available from the manufacturer and that one can acquire a relatively close assessment of item reliability based on past experience. Such experience factors should be assigned to the appropriate elements of the system, and their respective contribution to the 450-hour requirement should be determined.[10]
2. Identify the areas that are new and where design information is not available. Assign complexity or weighting factors to each applicable block. Complexity factors may be based on an estimate of the number and relationship of component parts, the duty cycle, whether it is anticipated that an item will be subjected to temperature extremes, and so on. Experience of the designer and his/her knowledge of similar applications in the past will help in this area. The portion of the system reliability requirement that is not already allocated to the areas of known design is allocated using the assigned weighting factors.

The end result should constitute a series of lower-level values that, when combined, will represent the system reliability requirement initially specified. The reliability factors established for the three units in Figure 12.16 serve as design criteria and are included in the appropriate development or product specifications used in the acquisition of the units (refer to Figure 3.2).[11]

Component Selection and Application

The reliability of a system depends largely on the reliability of its component parts, and the selection of parts must be compatible with the requirements of the particular application of those parts. The process of procuring what is advertised as being a reliable part is not adequate in itself and does not guarantee a reliable system. The specific application of the component is of prime importance, particularly when considering factors such as part tolerances and drift characteristics, electrical and environmental stresses, and so on. In electrical and electronic systems, part tolerances, drift characteristics, electrical stresses, and environmental stresses can have a major impact on system reliability and on the individual failure modes of various system components. For mechanical items the same concern prevails when considering part tolerances, component stress-strength factors, and material fatigue characteristics. Consequently, a fundamental approach in attaining a high level of reliability is to select and apply those

[10] When evaluating field data from past applications of COTS items, or equivalent, care must be taken to ensure that the use factors, duty cycles, stresses, and so forth are similar. A given COTS item can demonstrate a high degree of reliability in one application but may not in another application where the mission scenarios and stresses are different.

[11] As the design progresses, it may become apparent that a given reliability requirement can not be met (e.g., the MTBF of 536 hours for Unit B). In such a case, it may be necessary to accomplish a *reallocation* where the requirement for Unit B is "relaxed." However, if this occurs, the requirements for Units A and C will have to be "tightened" if the overall system requirements are to be met. There may be some trade-offs necessary at this point in arriving at a balanced set of requirements.

components and materials of known reliabilities and capable of meeting system requirements. Major emphasis in the design for reliability should consider

1. The selection of *standardized* components and materials to the extent feasible.
2. The evaluation of all components and materials prior to design acceptance. This includes studying the operating features, tolerances, stresses, and other characteristics of the component(s) in its intended application.
3. The use of only those component parts capable of meeting reliability objectives, and not substitute parts that appear to meet specification requirements but whose quality can be degraded through manufacture.

When considering component application(s) in system design, there are trade-offs involving the use of components in series and in parallel (or combinations thereof) that will allow for the achievement of the required system reliability. Figures 12.17 and 12.18 illustrate this fact. Figure 12.17 shows the relationship among system reliability, individual component reliability, and the quantity of *series* components included in the system design. As indicated, the reliability of a series network can be improved by decreasing the number of components or increasing the individual component reliabilities. It is evident that the system reliability deteriorates rather rapidly as the number of components increases.

Figure 12.18 illustrates the relationship among system reliability, individual component reliability, and the quantity of components in parallel. The paralleling of components is often considered as a means to improve system reliability. However, the

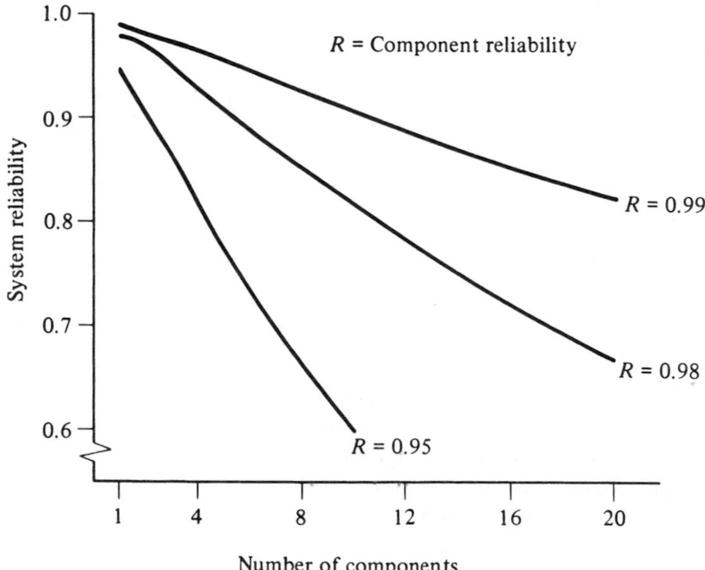

Figure 12.17 Series system reliability. *Source:* K. C. Kapur and L. R. Lamberson, *Reliability in Engineering Design*. New York: John Wiley & Sons, 1977.

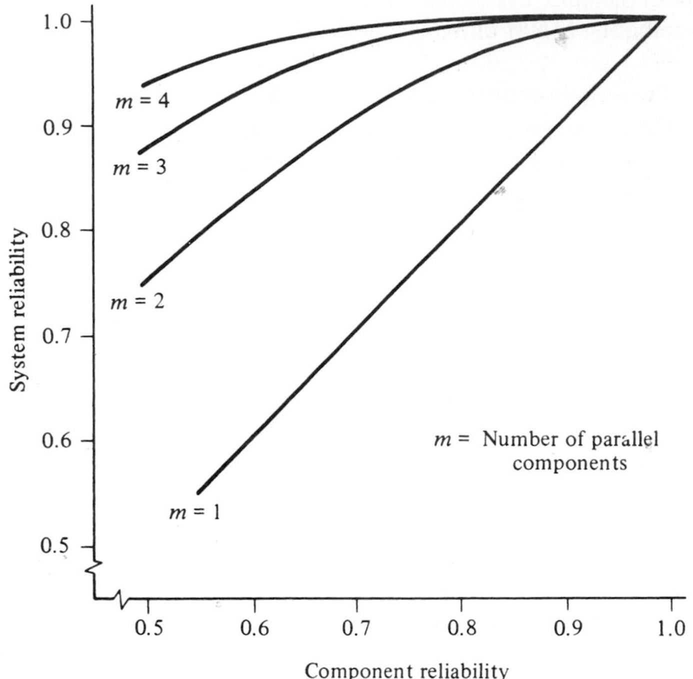

Figure 12.18 Reliability improvement by parallel components. *Source:* K. C. Kapur and L. R. Lamberson, *Reliability in Engineering Design*. New York: John Wiley & Sons, 1977.

overall gains may be marginal due to the added costs involved. It may be difficult to design parallel networks, particularly when dealing with mechanical devices. Also, it can be seen that the gain in system reliability by adding more than three parallel components is relatively small. However, the consideration of parallel components, in conjunction with series networks, should be accomplished in arriving at alternative ways to meet system reliability requirements.

Component Part Derating

Derating refers to the techniques of using a component part under stress conditions considerably below the rated value(s) to achieve a reliability margin in design. For example, if a 200-V capacitor is used in a 200-V application, then the part is operating at its full rating and it should function satisfactorily. However, if the capacitor is used in a 100-V application, it will in all probability last longer. Many electronic parts, such as resistors, transformers, transistors, and so on, tend to last longer when operated at lower power or voltage levels than at the rated values. Derating constitutes the application of parts in design operating at less than the rated values so as to improve system reliability.

Figure 12.19 illustrates a sample derating curve for a transistor. Similar curves are available for other components. Referring to the figure, the full rating is indicated

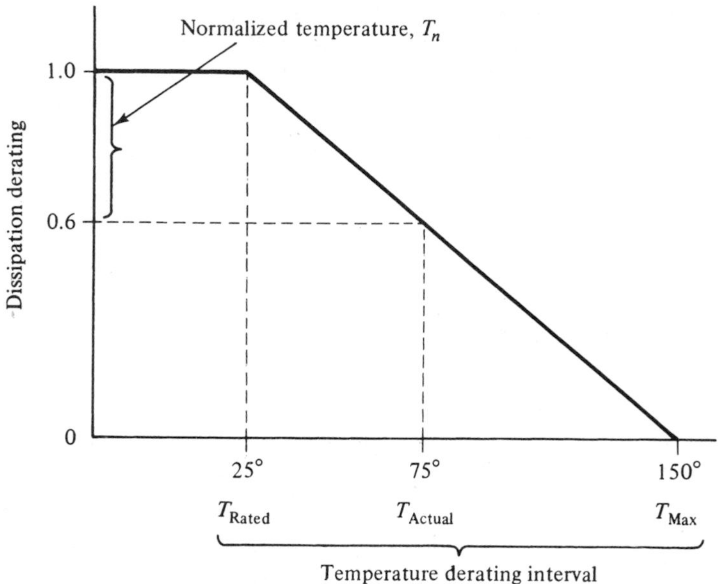

Figure 12.19 Dissipation derating curve.

along with the derated values presented as a function of ambient temperature. Given that the actual ambient temperature is expected to be 75°F, one can determine the derated value as

$$T_{\text{normalized}} = \frac{T_{\text{actual}} - T_{\text{rated}}}{T_{\max} - T_{\text{rated}}} \tag{12.24}$$

$$= \frac{75 - 25}{150 - 25} = \frac{50}{125} = 0.4$$

$$\text{derated value} = 1.0 - 0.4 = 0.6$$

In other words, the transistor should be incorporated into the system design and used at 0.6 of its fully rated value. This, in turn, can be related to a delta reduction in the component failure rate.

For many component parts, particularly in the case of electronic parts, data based on testing combined with actual experience have been converted into convenient tables that provide failure rates for various parts at derated values. These data are often used for design and in reliability prediction.[12]

[12] Sources for reliability data include (a) MIL-HDBK-217, *Reliability Prediction of Electronic Equipment.* Washington, D.C.: Department of Defense, (latest edition); (b) EPRD-97, "Electronic Parts Reliability Data," Reliability Analysis Center, Rome, NY, 1997; and (c) NPRD-95, "Nonelectronic Parts Reliability Data," Reliability Analysis Center, Rome, NY, 1995.

Redundancy in Design

Under certain conditions in system design it may be necessary to consider the use of redundancy to enhance system reliability by providing two or more functional paths (or channels of operation) in areas that are critical for successful mission accomplishment. But the application of redundancy per se will not necessarily solve all problems, because it usually implies increased weight and space, increased power consumption, greater complexity, and higher cost. Conversely, the use of redundancy may be the only solution for reliability improvement in specific situations.

Redundancy can be applied at several levels, as illustrated in Figure 12.20. At the system level Block G is redundant with the other blocks in the network and is at a different level than Block C, which is redundant with Block D. From the block diagram, the paths that will result in successful system operation are *A*, *B*, *C*, *E*; *A*, *B*, *C*, *F*; *A*, *B*, *D*, *E*; *A*, *B*, *D*, *F*; and *G*.

The probability of success, or the reliability, of each path may be calculated using the multiplication rule expressed in Equation 12.8. The calculation of system reliability (considering all paths combined) requires knowledge of the type of redundancy used and the individual block reliabilities. For operating (or active) redundancy, where all blocks are fully energized during an operational cycle, the appropriate equations in Section 12.2 may be used to calculate system reliability. Equations 12.10 through 12.12 and the related examples illustrate the gain in reliability that can be obtained through redundancy.

When design problems become somewhat more complex than the illustrations presented in Section 12.2, the number of possible events becomes greater. For example, in Figure 12.8, the following possibilities exist:

1. Subsystem *A*, *B*, and *C* are all operating.
2. Subsystems *A* and *B* are operating while *C* is failed.

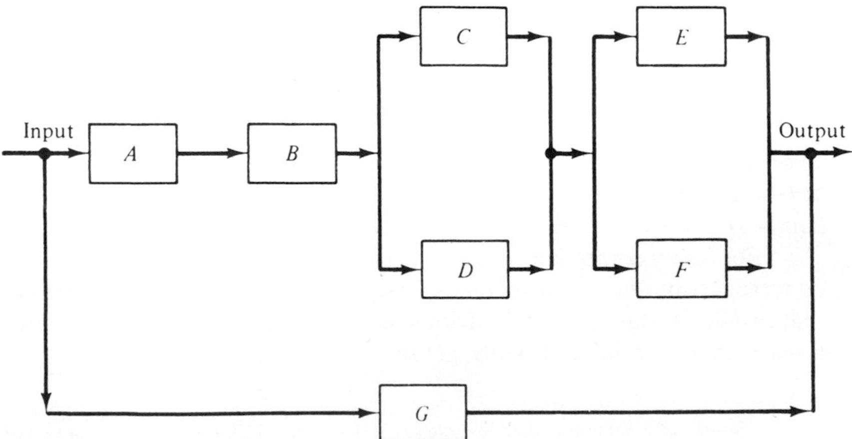

Figure 12.20 Reliability block diagram illustrating redundancy at system and subsystem levels.

3. Subsystems A and C are operating while B is failed.
4. Subsystems B and C are operating while A is failed.
5. Subsystem A is operating while B and C are failed.
6. Subsystem B is operating while A and C are failed.
7. Subsystem C is operating while A and B are failed.
8. Subsystems A, B, and C are all failed.

In the interest of simplicity, let R represent the reliability and $1 - R = Q$, the unreliability. Then

$$R^3 \quad \text{represents } A, B, \text{ and } C \text{ operating}$$

$$3R^2Q \begin{cases} \text{represents } A \text{ and } B \text{ operating, } C \text{ failed} \\ \text{represents } A \text{ and } C \text{ operating, } B \text{ failed} \\ \text{represents } B \text{ and } C \text{ operating, } A \text{ failed} \end{cases}$$

$$3RQ^2 \begin{cases} \text{represents } A \text{ operating, } B \text{ and } C \text{ failed} \\ \text{represents } B \text{ operating, } A \text{ and } C \text{ failed} \\ \text{represents } C \text{ operating, } A \text{ and } B \text{ failed} \end{cases}$$

$$Q^3 \quad \text{represents } A, B, \text{ and } C \text{ failed}$$

Because the sum of R^3, $3R^2Q$, $3RQ^2$, and Q^3 represents all possible events, then

$$R^3 + 3R^2Q + 3RQ^2 + Q^3 = 1 \tag{12.25}$$

Referring to Figure 12.8, it is assumed that the reliability of each block (i.e., blocks A, B, and C) is 0.95. Then the reliability of the network is determined as

$$\begin{aligned} R &= R^3 + 3R^2Q + 3RQ^2 \\ &= (0.95)^3 + 3(0.95)^2(0.05) + 3(0.95)(0.05)^2 = 0.999875 \end{aligned}$$

Note that this value agrees with the results of Section 12.2 and Equation 12.11.

Referring to Figure 12.20, suppose that it is desired to calculate the reliability for the network as illustrated. The approach is to first calculate the reliability for the redundant subsystems C, D, E, and F; apply the product rule for A, B, and the resulting networks in this path (i.e., C, D, E, and F); and then determine the combined reliability of the two overall redundant paths (to include subsystem G). It is assumed that the reliability of each of the individual subsystems is as follows:

$$\text{Subsystem } A = 0.97$$

$$\text{Subsystem } B = 0.98$$

$$\text{Subsystem } C = 0.92$$

$$\text{Subsystem } D = 0.92$$

$$\text{Subsystem } E = 0.93$$

$$\text{Subsystem } F = 0.90$$

$$\text{Subsystem } G = 0.99$$

The reliability of the redundant network including subsystems C and D is

$$R_{CD} = R_C + R_D - (R_C)(R_D) = 0.9936$$

The reliability of the redundant network including subsystems E and F is

$$R_{EF} = R_E + R_F - (R_E)(R_F) = 0.9930$$

The reliability of the path is

$$R_{ABCDEF} = (R_A)(R_B)(R_{CD})(R_{EF}) = 0.9379$$

The reliability of the combined network in Figure 12.20 is

$$R_{\text{system}} = R_{ABCDEF} + R_G - (R_{ABCDEF})(R_G) = 0.999379$$

Thus far, the discussion has addressed only one form of redundancy, operating redundancy, where all subsystems are fully energized throughout the system operating cycle. In actual practice, however, it is often preferable to use standby redundancy. Figure 12.21 provides a simple illustration of standby redundancy where subsystem A is operating full time, and subsystem B is standing by to take over operation if subsystem A fails. The standby unit (i.e., subsystem B) is not operative until a failure-sensing device senses a failure in subsystem A and switches operation to subsystem B, either automatically or through manual selection. Because of the fact that subsystem B is not operating unless a failure of subsystem A occurs, the system reliability for such a configuration is higher than for a comparable system where both subsystems A and B are operating continuously.

When determining the reliability of standby systems, the Poisson distribution may be used because standby systems display the constant λt characteristic of this distribution (refer to Appendix B.2 for discussion of the Poisson process). In essence, the probability of no failure is represented by the first term, $e^{-\lambda t}$; the probability of one failure is $(\lambda t)(e^{-\lambda t})$; and so on.

Referring to Figure 12.21, where one operating subsystem and one standby subsystem are grouped together, one must consider the probability that no failure or one failure will occur (with one subsystem remaining in satisfactory condition). This combined probability is expressed as

$$P(\text{one standby}) = e^{-\lambda t} + (\lambda t)e^{-\lambda t} \tag{12.27}$$

where λt is the expected number of failures. If one operating subsystem and two standby subsystems are grouped together, then the combined probability is

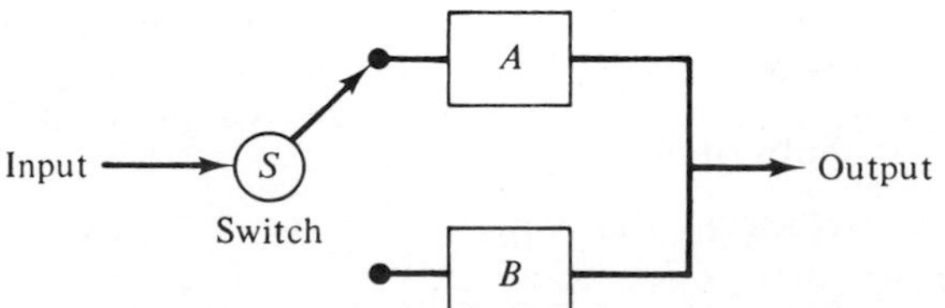

Figure 12.21 Standby redundant network.

$$P(\text{two standbys}) = e^{-\lambda t} + (\lambda t)e^{-\lambda t} + \frac{(\lambda t)^2 e^{-\lambda t}}{2!}$$

An additional term in the Poisson distribution is added for each subsystem in standby.

As an example, suppose that one must determine the system reliability for a configuration consisting of one operating subsystem and one identical standby operating for a period of 200 hours. This configuration is illustrated in Figure 12.21, and it is assumed that the reliability of the switch is 100%. The failure rate (λ) for each subsystem is 0.002 failure per hour. Using Equation 12.27, the reliability is

$$\begin{aligned} R &= e^{-\lambda t}(1 + \lambda t) \text{ and } \lambda t = (0.002)(200) = 0.4 \\ &= e^{-0.4}(1 + 0.4) \\ &= (0.67032)(1.4) = 0.9384 \end{aligned}$$

To illustrate the difference between operating redundancy and standby redundancy, assume that both of the subsystems in Figure 12.21 are operating throughout the mission. The reliability of the configuration is determined as

$$R = 1 - (1 - 0.67032)^2 = 0.8913$$

As anticipated, the reliability of the standby system is higher (0.9384) than the reliability of the system using operating redundancy (0.8913).

12.4 RELIABILITY ANALYSIS METHODS

Throughout design (as part of the iterative process of synthesis, analysis, and evaluation), there are several reliability tools that can be effectively used in support of the objectives described in this text. Of particular interest are the failure mode, effects, and criticality analysis (FMECA); the fault tree analysis (FTA); the critical useful life analysis; the stress-strength analysis; and reliability growth analysis. These are discussed briefly.

Failure Mode, Effects, and Criticality Analysis (FMECA)

The FMECA is a design technique that can be applied to identify and investigate potential system (product or process) weaknesses. It includes the necessary steps for examining all ways in which a system failure can occur, the potential effects of failure on system performance and safety, and the seriousness of these effects. The FMECA can be used initially during the conceptual and preliminary design stages, and can be subsequently applied as system definition evolves through detail design and development. Although the analysis is best used to impact "before-the-fact" enhancements to system design, it can also be used as an "after-the-fact" tool to evaluate and improve existing systems on a continuing basis.

FMECA can be applied to either a functional entity or a physical entity. For instance, Figure 12.22 conveys a partial illustration of the major functions of a package handling plant, or a series of activities for the processing of packages for distribution.

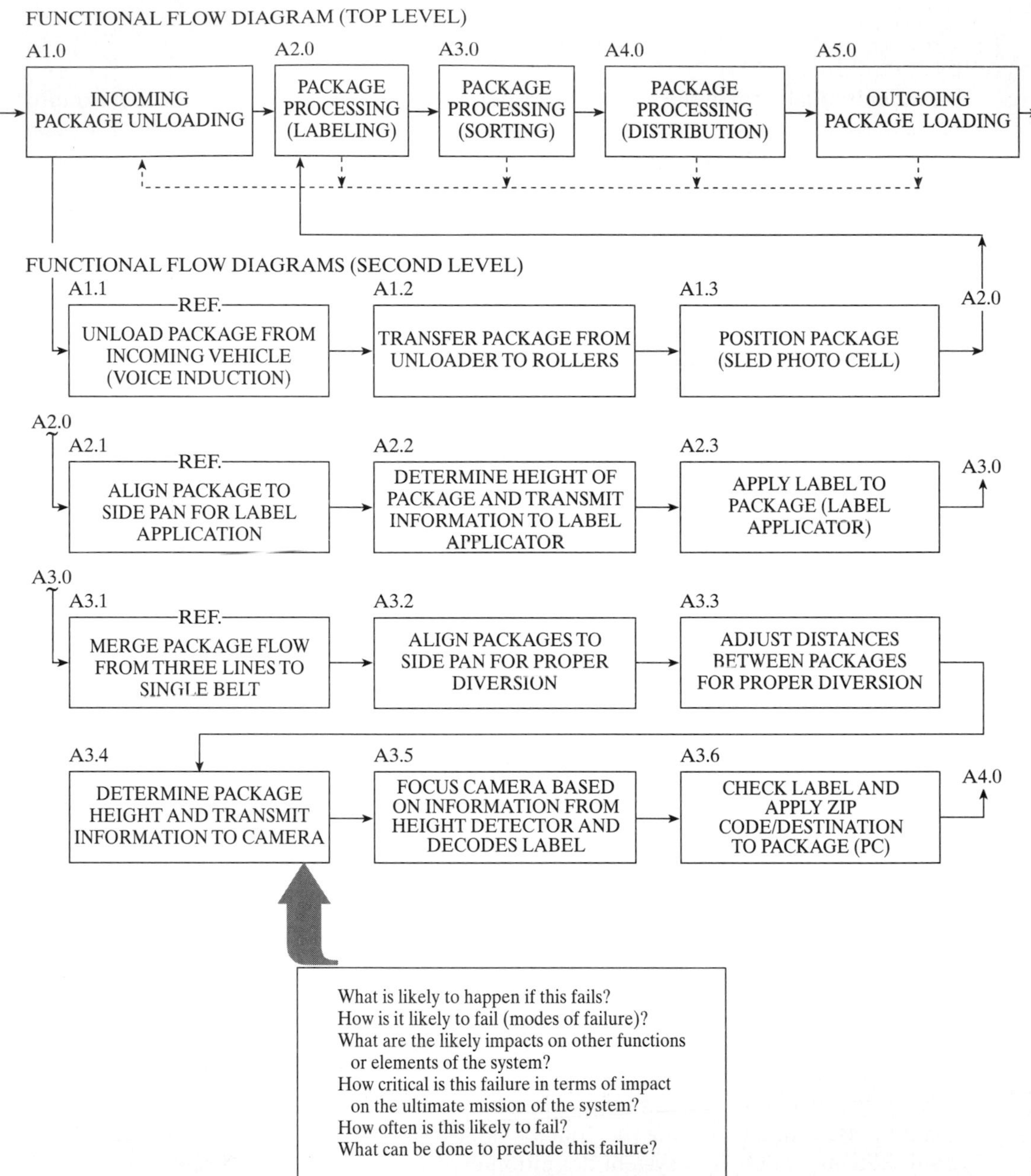

Figure 12.22 Package-handling plant functional flow diagram.

The question is: What is likely to happen if Function A3.4 were to fail, and what are the possible consequences of such a failure? Conversely, one might ask to same question when referring to a physical entity such as Assembly 3 in Figure 3.13. In other words, the FMECA needs to address both the *product* and the *process*.

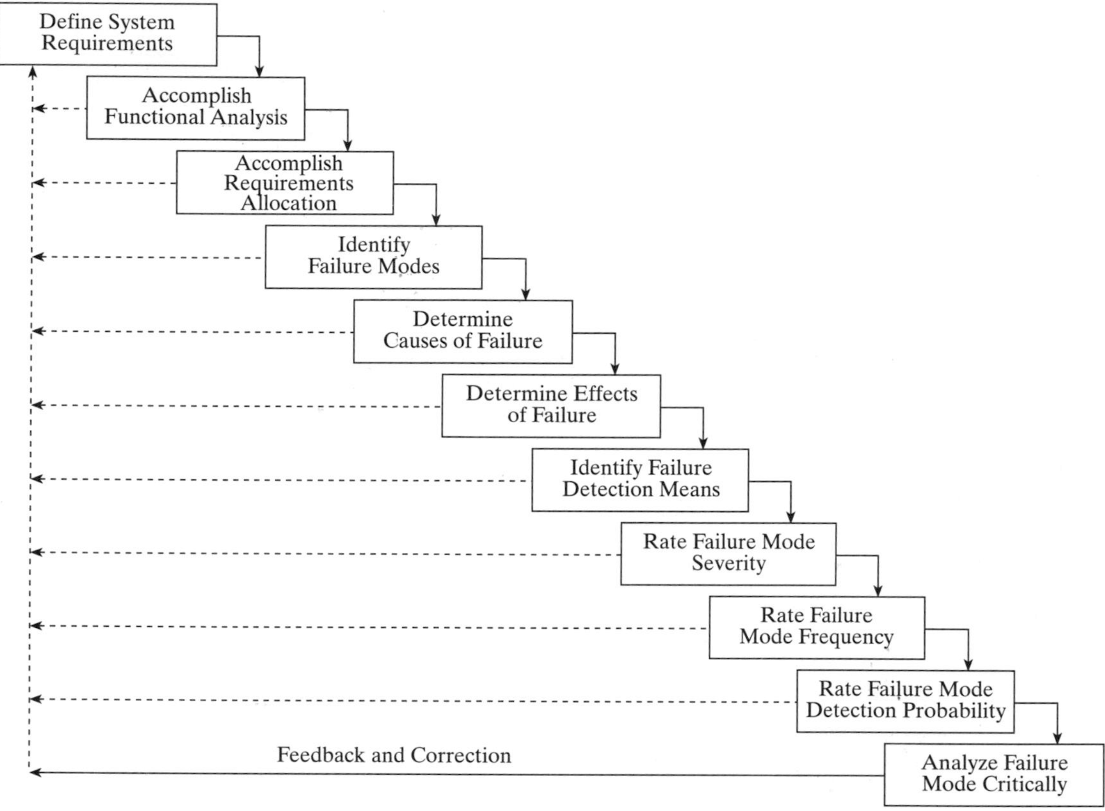

Figure 12.23 Failure mode, effects, and criticality analysis process.

The approach used in conducting a FMECA is shown in Figure 12.23. A brief description of the steps follows[13]:

1. *Define system (product or process) requirements.* Describe the system in question, the expected outcomes, and the relevant technical performance measures (TPMs). Figure 12.24 shows an example where the FMECA can be applied to the manufacturing process for a gasket, as well as addressing the gasket within the automobile. One can first conduct a FMECA on the "stamping operation," determine the modes and effects of failure on the other operations within the process, and assess the impacts on the gasket. Then, a product-oriented FMECA can be accomplished assessing the affects of gasket failure on the engine block and on the automobile as an entity.

[13] Three good references include Instruction Manual, *Potential Failure Mode and Effects Analysis*, Ford Motor Company, 1988; Instruction Manual, *Failure Mode and Effects Analysis*, Saturn Quality System, Saturn Corporation, 1990; and *Potential Failure Mode and Effects Analysis (FMEA)*, Reference Manual FMEA-1, developed by FMEA teams at Ford Motor Company, General Motors, Chrysler, Goodyear, Bosch, and Kelsey-Hayes, under the auspices of ASQC and AIAG.

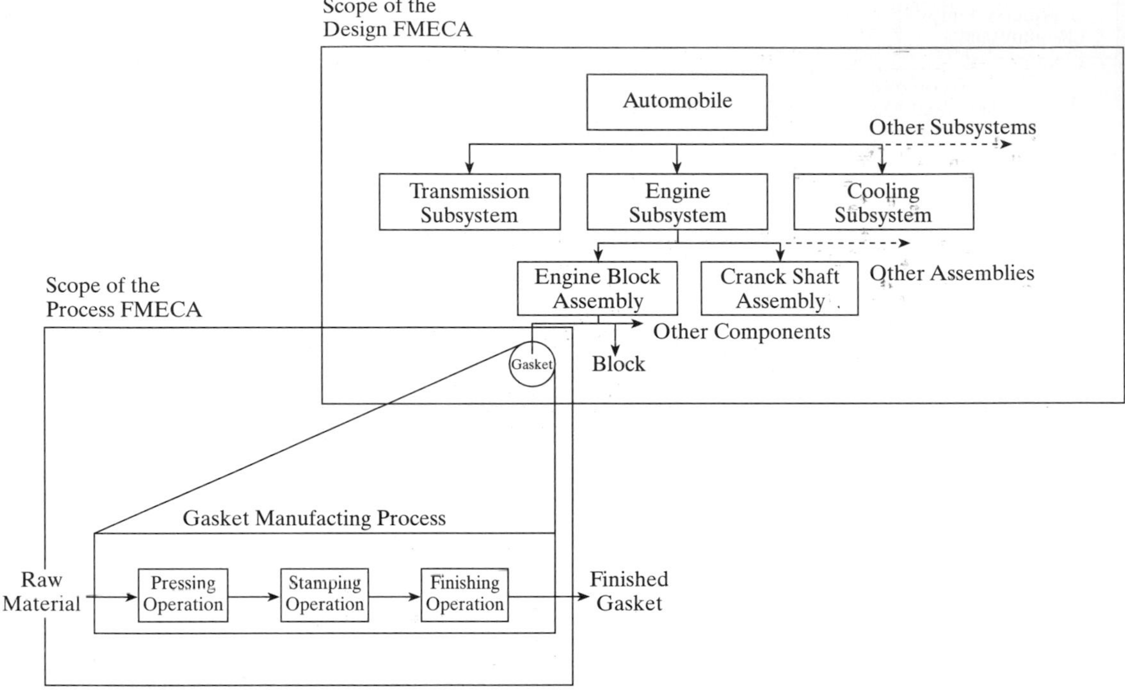

Figure 12.24 Sample design and process Failure Mode, Effects, and Criticality Analysis (FMECA).

2. *Accomplish functional analysis.* This involves defining the system in functional terms as described in Sections 3.6 and 4.1. A system may be broken down into functional entities early in the life cycle and subsequently into a physical packaging scheme (refer to Figures 4.6 and 12.24).
3. *Accomplish requirements allocation.* This is a top-down breakout of system-level requirements as described in Section 4.2.
4. *Identify failure modes.* A "failure mode" is the manner in which a system element fails to accomplish its function. For example, a switch may fail in an "open" position; a pipe may "rupture"; a given material may "shear" because of stress; a document may fail to be delivered on time; and so on.
5. *Determine causes of failure.* This involves analyzing the process or product to determine the actual cause(s) responsible for the occurrence of a failure. Typical causes might include abnormal equipment stresses during operation, aging and wearout, a software coding error, poor workmanship, defective materials damage because of transportation and handling, or operator- and maintenance-induced faults. Although experience with similar systems, or the availability of good data from the field, is preferred, using an Ishikawa "cause-and-effect" dia-

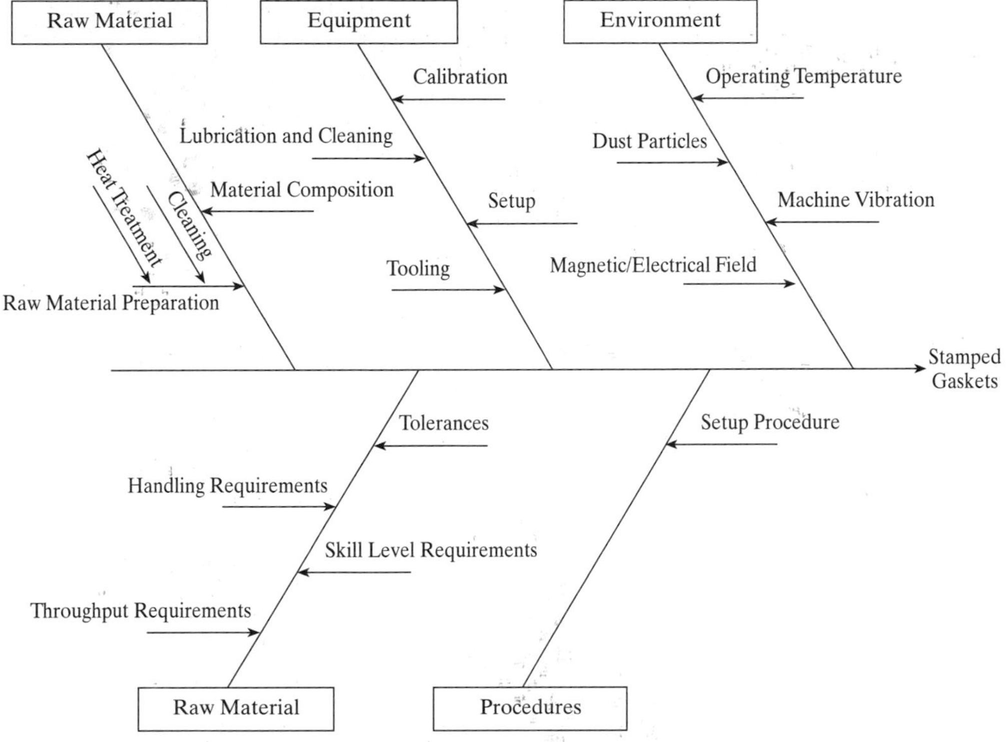

Figure 12.25 Ishikawa cause-and-effect "fishbone" diagram.

gram can prove to be highly effective in delineating potential failure causes. Figure 12.25 illustrates such a diagram.[14]

6. *Determine the effects of failure.* Failures impact, often in multiple ways, the performance and effectiveness of not only the associated functional element but the overall system. It is important to consider the effects of failure on other elements at the same level in the system hierarchical structure, at the next higher level, and on the overall system. Referring to Figure 12.24, if the stamping operation fails, what are the effects on the finishing operation, on the gasket itself, on the automobile, and ultimately on the customer?
7. *Identify failure detection means.* For a process-oriented FMECA, this refers to the current process controls which may detect the occurrence of failures or defects. However, when the FMECA has a design focus, this refers to the existence of any design features, aids, gauges, readout devices, condition monitoring provisions, or evaluation procedures that will result in the detection of potential failures.
8. *Rate failure mode severity.* This refers to the *seriousness* of the effect or impact of a particular failure. If a failure occurs, will this cause the death of the operator

[14] K. Ishikawa, *Guide to Quality Control*, 2nd rev., English ed. (Tokyo, Japan: Asian Productivity Organization, 1982); and K. Ishikawa, *Introduction to Quality Control*, (London, England: Chapman and Hall, 1991).

and the system to be destroyed, or will it cause only a slight degradation in performance? For the purpose of illustration, the degree of severity may be expressed quantitatively (using a checklist approach) on a scale of 1 to 10 with *minor* effects being 1, *low* effects being 2 to 3, *moderate* effects being 4 to 6, *high* effects being 7 to 8, and *very high* effects being 9 to 10. The level of severity can be related to issues pertaining to safety or the degree of customer dissatisfaction.

9. *Rate failure mode frequency*. Given that a function or physical component within the system may fail in a variety of ways, this step addresses the frequency of occurrence of each individual failure mode. The sum of all modal failure frequencies for a system element must equal its failure rate. For the purposes of quantification, a scale of 1 to 10 may be used with *remote* (failure is unlikely) being 1, *low* (relatively few failures) being 2 to 3, *moderate* (occasional failures) being 4 to 6, *high* (repeated failures) being 7 to 8, and *very high* (failure is almost inevitable) being 9 to 10. These ratings can be based on the expected number of failures per unit of time, or equivalent.

10. *Rate failure mode detection probability*. This pertains to the probability that process controls, design features/aids, verification procedures, and so on, will detect potential failures in time to prevent a major system catastrophy. For a process application, this refers to the probability that a set of process controls currently in place will be in a position to detect and verify a failure before its effects are transferred to a subsequent process, or to the end consumer/customer. For the purposes of quantification, a scale of 1 to 10 may be used with *very high* being 1 to 2, *high* being 3 to 4, *moderate* being 5 to 6, *low* being 7 to 8, *very low* being 9, and *absolute certainty of nondetection* being 10.

11. *Analyze failure mode criticality*. The objective is to consolidate the preceding information in an effort to delineate the more critical aspects of system design. Criticality, in this context, is a function of *severity, frequency,* and *probability of detection,* and may be expressed in terms of a *risk priority number* (RPN). RPN can be determined from

$$\text{RPN} = (\text{severity rating})(\text{frequency rating})(\text{probability of detection rating}) \tag{12.28}$$

The RPN reflects failure mode criticality. On inspection, one can see that a failure mode with a high frequency of occurrence, with significant impact on system performance, and that is difficult to detect is likely to have a very high RPN.

12. *Initiate recommendations for product/process improvement.* This pertains to the iterative process of identifying areas with high RPNs, evaluating the causes, and the subsequent initiation of recommendations for product/process improvement (refer to the feedback loop in Figure 12.23). A Pareto analysis, such as illustrated in Figure 12.26, can be accomplished to make visible the high-priority items that need to be addressed.

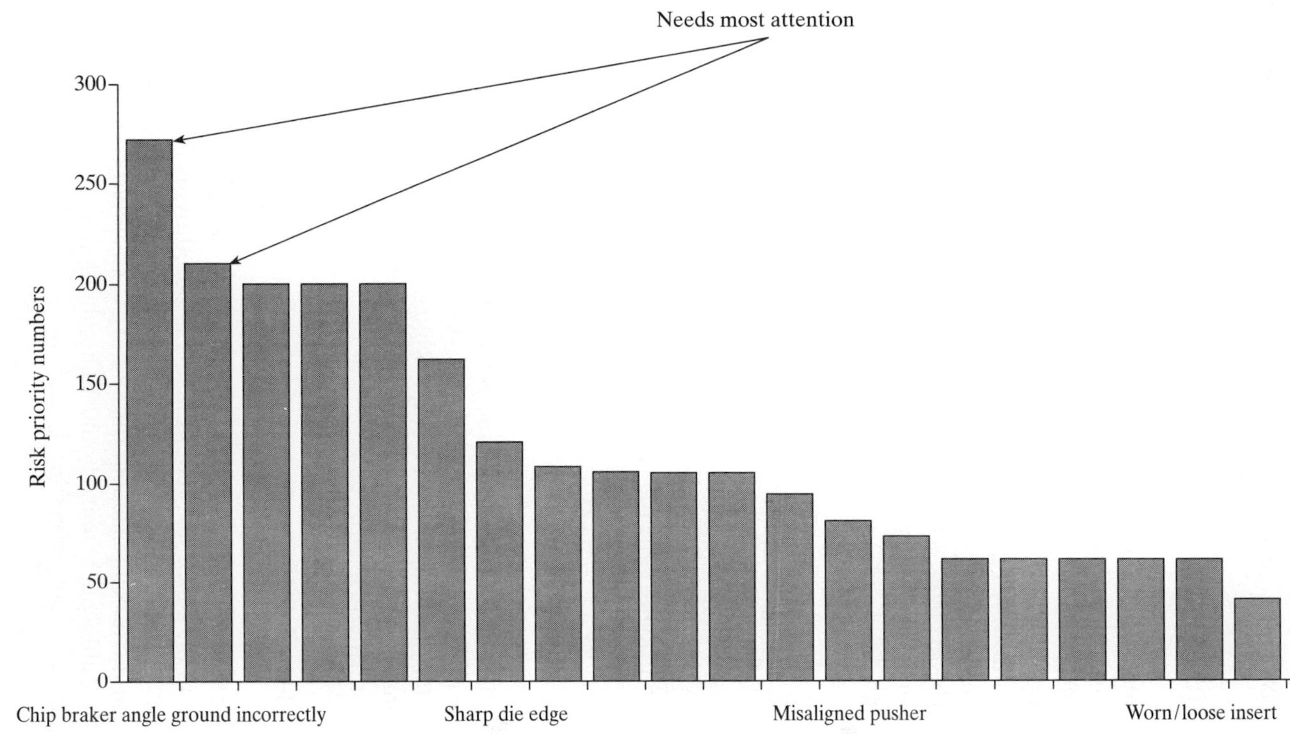

Figure 12.26 Pareto analysis–potential causes of failure (partial).

Figure 12.27 shows a partial example of a popular format that is used for recording the results of the FMECA. The information was derived from the functional flow analysis and expanded to include the results from the steps presented in Figure 12.23.

Fault-Tree Analysis (FTA)

The fault-tree analysis (FTA) is a deductive approach involving the graphical enumeration and analysis of different ways in which a particular failure can occur and the probability of its occurrence. It may be applied during the early stages of design, is oriented to specific failure modes, and is developed using a top-down fault-tree structure, such as illustrated in Figure 12.28. A separate fault tree is developed for every critical failure mode.

The first step is to identify a top-level event. One needs to be specific in defining this event, and it must be clearly observable and measurable. For example, it could be delineated as the "system catches fire" rather than the "system fails." Once the top-level event has been clearly defined, it is necessary to construct a *causal* hierarchy in the form of a fault tree. The causes for the top event are determined, using a technique such as Ishikawa's *cause-and-effect* diagram (refer to Figure 12.25); each of these causes is next investigated for its causes; and so on. The next step is to determine the reliability of the top-level event by determining the probabilities of all relevant input events, and the subsequent consolidation of these probabilities in accordance with the underlying logic of the tree. If the reliability of the top-level event turns out to be unacceptable, then corrective action is required.[15]

The FTA can be effectively applied in the early phases of design when potential problems are suspected. It is narrow in focus and often easier to accomplish than the FMECA, requiring less input data to complete. For large and complex systems, which are highly software intensive and where there are many interfaces, the use of the FTA may be preferred in lieu of the FMECA. Conversely, accomplishing the FMECA provides more information for design evaluation, the establishing of "cause-and-effect" relationships across the spectrum of the system, and for pinpointing high-cost (high-risk) contributors.

Critical-Useful-Life Analysis

A critical-useful-life item is one that, because of its relatively short life, is incapable of satisfying the functional requirements imposed by its application unless corrective or preventive maintenance is performed. These "short-life" items, often identified through the accomplishment of the FMECA, should be eliminated from the design if at all possible. Such items should be highlighted along with their expected life in terms of calendar time, operating cycles, or system operating hours Alternative design approaches should be considered to enhance overall system effectiveness and minimize

[15] The objective here is to convey only the basic concepts and principles of the FTA. Review of the reliability literature in Appendix F is recommended for a more indepth coverage.

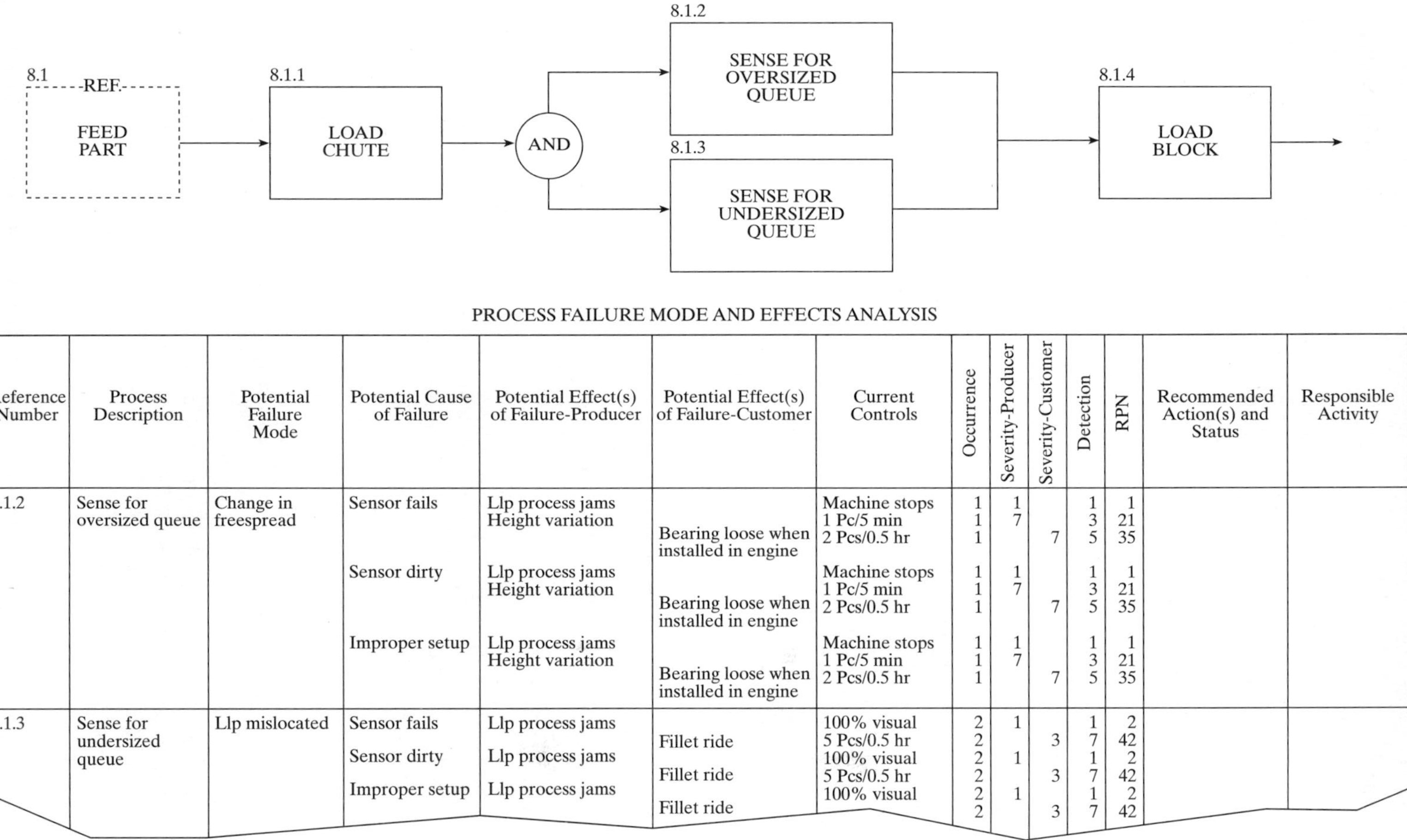

PROCESS FAILURE MODE AND EFFECTS ANALYSIS

Reference Number	Process Description	Potential Failure Mode	Potential Cause of Failure	Potential Effect(s) of Failure-Producer	Potential Effect(s) of Failure-Customer	Current Controls	Occurrence	Severity-Producer	Severity-Customer	Detection	RPN	Recommended Action(s) and Status	Responsible Activity
8.1.2	Sense for oversized queue	Change in freespread	Sensor fails	Llp process jams		Machine stops	1	1		1	1		
				Height variation		1 Pc/5 min	1	7		3	21		
					Bearing loose when installed in engine	2 Pcs/0.5 hr	1		7	5	35		
			Sensor dirty	Llp process jams		Machine stops	1	1		1	1		
				Height variation		1 Pc/5 min	1	7		3	21		
					Bearing loose when installed in engine	2 Pcs/0.5 hr	1		7	5	35		
			Improper setup	Llp process jams		Machine stops	1	1		1	1		
				Height variation		1 Pc/5 min	1	7		3	21		
					Bearing loose when installed in engine	2 Pcs/0.5 hr	1		7	5	35		
8.1.3	Sense for undersized queue	Llp mislocated	Sensor fails	Llp process jams		100% visual	2	1		1	2		
					Fillet ride	5 Pcs/0.5 hr	2		3	7	42		
			Sensor dirty	Llp process jams		100% visual	2	1		1	2		
					Fillet ride	5 Pcs/0.5 hr	2		3	7	42		
			Improper setup	Llp process jams		100% visual	2	1		1	2		
					Fillet ride		2		3	7	42		

Figure 12.27 FMECA process results (partial).

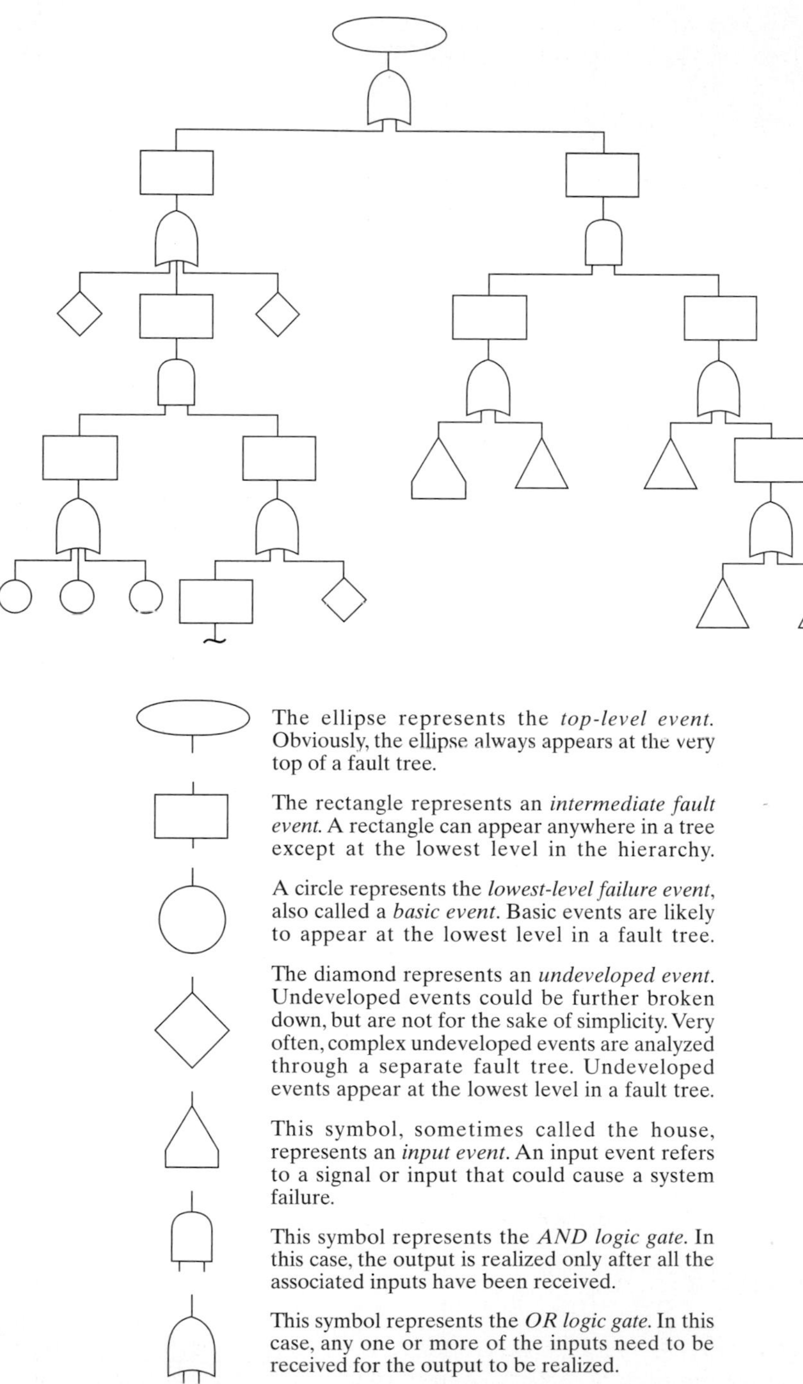

The ellipse represents the *top-level event.* Obviously, the ellipse always appears at the very top of a fault tree.
The rectangle represents an *intermediate fault event.* A rectangle can appear anywhere in a tree except at the lowest level in the hierarchy.
A circle represents the *lowest-level failure event,* also called a *basic event.* Basic events are likely to appear at the lowest level in a fault tree.
The diamond represents an *undeveloped event.* Undeveloped events could be further broken down, but are not for the sake of simplicity. Very often, complex undeveloped events are analyzed through a separate fault tree. Undeveloped events appear at the lowest level in a fault tree.
This symbol, sometimes called the house, represents an *input event.* An input event refers to a signal or input that could cause a system failure.
This symbol represents the *AND logic gate.* In this case, the output is realized only after all the associated inputs have been received.
This symbol represents the *OR logic gate.* In this case, any one or more of the inputs need to be received for the output to be realized.

Figure 12.28 Sample fault-tree analysis format and symbology.

cost. Otherwise, these short-life items will lead to the requirement for maintenance, personnel, spares/repair parts, and possibly other elements of support.

Stress-Strength Analysis

Of major concern in the design for system reliability are the stress-strength characteristics of its components. Component parts are designed and manufactured to operate in a specified manner when utilized under nominal conditions. If additional stresses are imposed because of electrical loads, temperature, vibration, shock, humidity, and so on, then unexpected failures will occur and the reliability of the system will be less than anticipated. Also, if materials are used in a manner where nominal strength characteristics are exceeded, fatigue occurs and the materials may fail much earlier than expected. In any event, overstress conditions will result in reliability degradation and understress conditions may be costly as a result of overdesign; that is, incorporating more than what is actually necessary to do the job.

A stress-strength analysis is often undertaken to evaluate the probability of identifying a situation(s) where the value of stress is much larger than (or the strength much less than) the nominal value. Such an analysis may be accomplished through the following steps:

1. For selected components, determine nominal stresses as a function of loads, temperature, vibration, shock, physical properties, and time.
2. Identify factors affecting maximum stress, such as stress concentration factors, static and dynamic load factors, stresses as a result of manufacturing and heat treating, environmental stress factors, and so on.
3. Identify critical stress components and calculate critical mean stresses (e.g., maximum tensile stress, shear stress).
4. Determine critical stress distributions for the specified useful life. Analyze the distribution parameters and identify component safety margins. Applicable distributions may include normal, Poisson, Gamma, Weibull, log-normal, or variations thereof.
5. For those components that are critical and where the design safety margins are inadequate, corrective action must be initiated. This may constitute component-part substitution or a complete redesign of the system element in question.

Reliability models are used to facilitate the stress-strength analysis process (refer to Figure 12.15). On the basis of such an analysis, component-part failure rates (λ) are adjusted as appropriate to reflect the effects of the stresses of the parts involved.

Reliability Prediction

As engineering data become available, reliability prediction is accomplished as a check on design in terms of the system requirement and the factors specified through allocation. The predicted values of MTBM and/or MTBF are compared against the requirement, and areas of incompatibility are evaluated for possible design improvement.

Predictions are accomplished at different times in the system development process and will vary somewhat depending on the type of data available. Basic prediction techniques are summarized as follows:

1. *Prediction may be based on the analysis of similar equipment.* This technique should only be used when the lack of data prohibits the use of more sophisticated techniques. The prediction uses MTBF values for similar equipments of similar degrees of complexity performing similar functions and having similar reliability characteristics. The reliability of the new equipment is assumed to be equal to that of the equipment which is most comparable in terms of performance and complexity. Part quantity and type, stresses, and environmental factors are not considered. This technique is easy to perform but not very accurate.
2. *Prediction may be based on an estimate of active element groups* (AEG). The AEG is the smallest functional building block that controls or converts energy. An AEG includes one active element (relay, transistor, pump, machine) and a number of passive elements. By estimating the number of AEGs and using complexity factors, one can predict MTBF.
3. *Prediction may be accomplished from an equipment parts count.* There are a variety of methods used that differ somewhat due to data source, the number of part type categories, and assumed stress levels. Basically, a design parts list is used and parts are classified in certain designated categories. Failure rates are assigned and combined to provide a predicted MTBF at the system level. A representative approach is illustrated in Table 12.2.
4. *Prediction may be based on a stress analysis* (discussed earlier). When detailed equipment design is relatively firm, the reliability prediction becomes more sophisticated. Part types and quantities are determined, failure rates are applied, and stress ratios and environmental factors are considered. The interaction effects between components are addressed. This approach is peculiar and varies

TABLE 12.2 Reliability Prediction Data Summary

COMPONENT PART	λ/PART (%/1000 HOURS)	QUANTITY OF PARTS	(λ/PART) (QUANTITY)
Part *A*	0.161	10	1.610
Part *B*	0.102	130	13.260
Part *C*	0.021	72	1.512
Part *D*	0.084	91	7.644
Part *E*	0.452	53	23.956
Part *F*	0.191	3	0.573
Part *G*	0.022	20	0.440

Failure Rate (λ) = 48.995%/1000 hours $\qquad \Sigma = 48.995\%$

$$\text{MTBF} = \frac{1000}{0.48995} = 2041 \text{ hours}$$

Source: MIL-HDBK-217, Military handbook, *Reliability and Failure Rate Data for Electronic Equipment* (Washington, D.C.: Department of Defense).

somewhat with each particular system/product design. Computer methods are often used to facilitate the prediction process.

The figures derived through reliability prediction constitute a direct input to maintainability prediction data, supportability analysis, and the determination of specific support requirements (test and support equipment, spare and repair parts, etc.). Reliability basically determines the frequency of corrective maintenance and the quantity of maintenance actions anticipated throughout the life cycle; thus, it is imperative that reliability prediction results be as accurate as possible.

Reliability Growth Modeling

New systems or products often display a lower reliability during the early phases of design. Referring to Figure 5.12, the design MTBF requirement for the system being developed is 485 hours; yet, the predicted MTBF at the conceptual design review was 445 hours. The question is: What needs to be accomplished in order to "grow" the reliability to the required 485 hours before the delivery of the system to the customer?

In response, a plan needs to be established that involves measurement, the identification of potential problem areas, the initiation of changes for improvement, test and evaluation (i.e., assessment), etc., accomplished iteratively throughout the entire period of development. This concept of "test, analyze, and fix" (resulting in reliability "growth") was formally addressed in the mid-1960s by James Duane. Duane derived an empirical relationship based on reliability improvement observed with respect to a range of aircraft components. Through this experience, a set of curves was developed and applied to new system acquisitions.[16]

Referring to Figure 12.29, the objective is to develop a reliability growth plan based on past experience. Through the iterative application of various reliability analysis methods (i.e., FMECA, FTA, and reliability prediction), potential problem areas in design can be identified, causes-and-effects can be determined, and changes can be incorporated with the goal of improving the reliability of the system overall. This approach can be implemented as part of an overall *continuous product/process improvement* effort. The success of such is dependent on being able to evolve through each cycle of activity in a timely manner.[17]

12.5 DESIGN REVIEW AND EVALUATION

The review of reliability in system design is accomplished as an inherent part of the process described in Sections 3.9, 4.6, and 5.7. The characteristics of the system (and its elements) are evaluated in terms of compliance with the initially specified reliability requirements for the system. If the requirements appear to be fulfilled, the design is approved as is. If not, the appropriate changes are initiated for corrective action.

[16] J. T. Duane, "Learning Curve Approach to Reliability Monitoring," *IEEE Transactions: Aerospace*, 2(2), 1964.

[17] Implementation of the "test, analysis, and fix" concept requires top-management interest and support, and being able to review and incorporate the appropriate changes in an expeditious manner. A lengthy cycle will not allow for the growth desired.

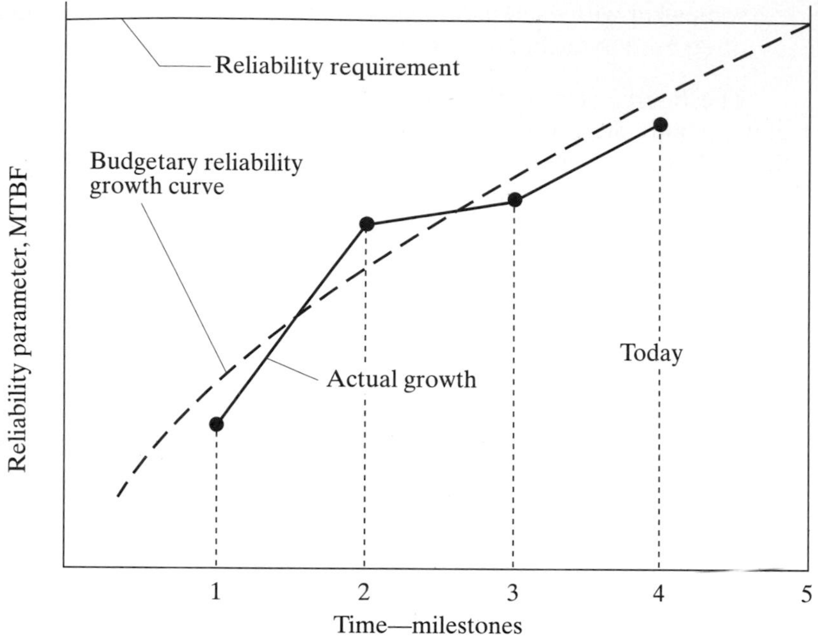

Figure 12.29 A sample reliability growth plan.

In support of conducting reliability reviews, one may wish to develop a design review checklist, including questions such as those listed below.

1. Have reliability quantitative and qualitative requirements for the system been adequately defined and specified?
2. Are the reliability requirements compatible with other system requirements? Are they realistic?
3. Has system or product design complexity been minimized?
4. Have system failure modes and effects been identified?
5. Has the system or product wearout period been defined?
6. Are system, subsystem, unit, and component-part failure rates known?
7. Have component parts with excessive failure rates been identified?
8. Has the system mean life been determined?
9. Have adequate derating factors been established and adhered to where appropriate?
10. Is protection against secondary failures (resulting from primary failures) incorporated where possible?
11. Have all critical-useful-life items been eliminated from system design?
12. Has the use of adjustable components been minimized?
13. Have cooling provisions been incorporated in design "hot-spot" areas?
14. Have all hazardous conditions been eliminated?
15. Have all system reliability requirements been met?

The items covered certainly are not intended to be all-inclusive, but merely represent a sample of possible interest areas (it should be noted that the answer to those questions that are applicable should be *yes*). The response to these questions is primarily dependent on the results of the reliability analyses and predictions discussed earlier.

12.6 RELIABILITY TEST AND EVALUATION

Reliability testing is conducted as part of the system test and evaluation effort discussed in Chapter 6. Specifically, reliability testing is accomplished under the categories of Type II and III testing described in Section 6.2.

As is required for any category of testing, there is a planning phase, a test preparation phase, the actual test and evaluation itself, the data collection and analysis phase, and test reporting. The general requirements for each are discussed in Chapter 6. Only those procedures applicable to reliability test methods are covered in this chapter.

The ultimate objective of reliability testing is to determine whether the system (or product) under test meets the specified MTBF requirements. To accomplish this, the system is operated in a prescribed manner for a designated period and failures are recorded and evaluated as the test progresses. System acceptance is based on demonstrating a minimum acceptable life.

There are a number of different test methods and statistical procedures in practice today which are designed to measure system reliability. Most of these assume that the exponential distribution is applicable. The criteria for system acceptance (or rejection) are based on statistical assumptions involving test sample size, consumer-producer decision risk factors, test data confidence limits, and so on. These assumptions often vary from one application to the next, depending on the type of system, the mission that the system is expected to perform, and whether new design techniques are used in system development reflecting a potential high-risk area. In view of this, it is impossible to cover all facets of reliability testing within the confines of this chapter. However, in order to provide some understanding of the approaches used, it is intended to briefly cover reliability sequential qualification testing, reliability acceptance testing, and life testing.

Reliability Sequential Qualification Testing

Reliability qualification testing is conducted to provide an evaluation of system development progress, as well as the assurance that specified requirements have been met prior to proceeding to the next phase (i.e., the production or construction phase of life cycle). Initially, a reliability MTBF is established for the system, followed by allocation and the definition of design criteria. System design is accomplished and reliability analyses and predictions are made to evaluate the design configuration (on an analytical basis) relative to compliance with system requirements. If the predictions indicate compliance, the design progresses to the construction of a prototype, or preproduction model, of the system and qualification testing commences.

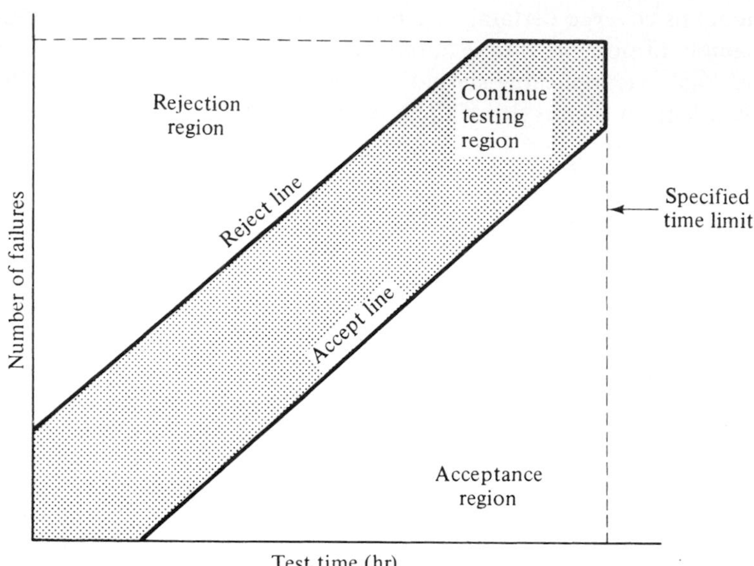

Figure 12.30 Sequential test plan.

When a reliability sequential test is conducted, there are three possible decisions: (1) accept the system, (2) reject the system, or (3) continue to test. Figure 12.30 illustrates a typical sequential test plan. The system under test is operated in a manner reflecting actual customer utilization in a realistic environment. The objective is to simulate (to the extent possible) a mission profile, or at least to subject the system under test to conditions similar to those that will be present when the system is in use by the customer. The accomplishment of this objective generally involves an environmental testing facility and the operation of the system or equipment through different duty cycles and environmental conditions. Figure 12.31 illustrates a sample duty cycle. The system is operated through a series of these duty cycles for a designated period of time.

Referring to the sequential test plan in Figure 12.30, system operating hours are accumulated along the abscissa and failures are plotted against the ordinate. If enough operating time is acquired without too many failures, a decision is made to accept the system and testing is discontinued. Conversely, if there are a significant quantity of failures occurring early in test operations, then a reject decision may prevail and the system is unacceptable as is. A marginal condition results in a decision to continue testing until a designated point in time. In essence, the sequential test plan allows for an early decision. Highly reliable systems will be accepted with a minimum amount of required testing. If the system is unreliable, this will also be readily evident at an early point in time. In this respect, sequential testing is extremely beneficial. Conversely, if the system inherent reliability is marginal, the amount of test time involved can be rather extensive and costly. An actual example of sequential testing is presented later.

Sequential test plans are highly influenced by the risks that both the producer and consumer are willing to accept in connection with decisions made as a result of testing. These risks are defined as

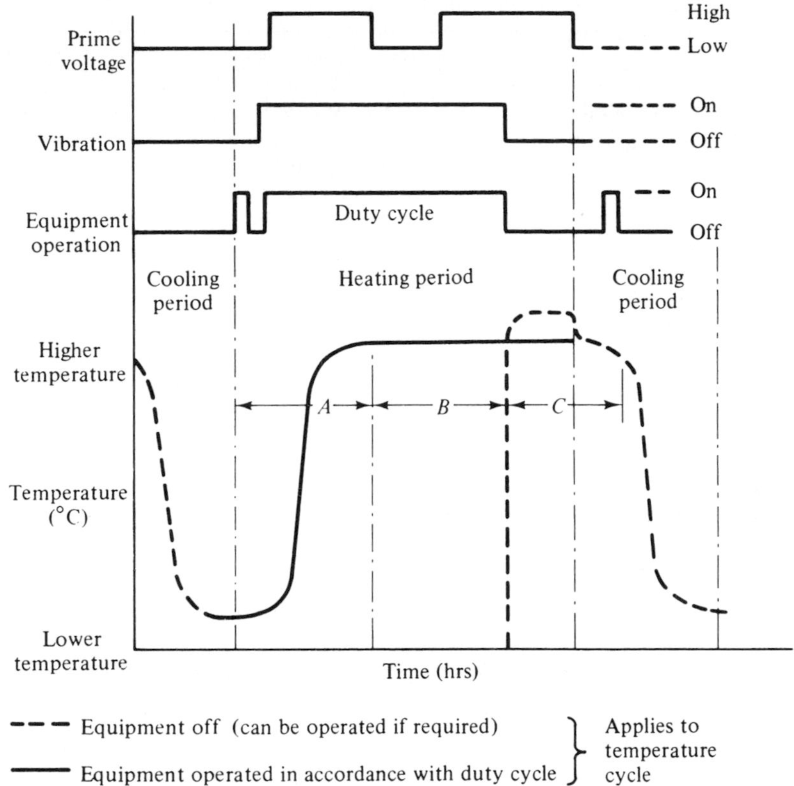

Figure 12.31 Sample environmental test cycle. *Source:* MIL-STD-781, *Reliability Design Qualification and Production Acceptance Tests: Exponential Distribution.* Washington, D.C.: Department of Defense.

1. *Producer's risk* (α). The probability of rejecting a system when the measured MTBF is equal to or better than the specified MTBF. In other words, this refers to the probability of rejecting a system when it really should be accepted, which constitutes a risk to the system manufacturer or producer (also known as a Type I error).
2. *User's or consumer's risk* (β). The probability of accepting a system when the measured MTBF is less than the specified MTBF. This refers to the probability of accepting a system that actually should be rejected, which constitutes a risk to the user (also known as a Type II error).

The probability of making an incorrect decision on the basis of test results must be addressed in a manner similar to hypothesis testing. An assumption is made and a

TABLE 12.3 Risks in Sample Testing

True state of affairs	Accept H_0 and Reject H_1 (i.e., MTBF = 100)	Reject H_0 and Accept H_1 (i.e., MTBF $\neq$ 100)
H_0 is true (i.e., MTBF = 100)	High probability $1 - \alpha$ (i.e., 0.90)	Low probability Error, α (i.e., 0.10)
H_0 is false and H_1 is true (i.e., MTBF $\neq$ 100)	Low probability Error, β (i.e., 0.10)	High probability $1 - \beta$ (i.e., 0.90)

test is accomplished to support (or disprove) that assumption. A null hypothesis (H_0) is established; that is, a statement or conjecture about a parameter such as the true MTBF is equal to 100. The alternative hypothesis (H_1) is "the MTBF is not equal to 100." When testing an item (representing a sample of a total population), the question arises as to whether to accept H_0. The desired result is to accept when the null hypothesis is true and reject when false, or to minimize the chances of making an incorrect decision. The relationship of risks in sample testing illustrated in Table 12.3. In selecting a sequential test plan, it is necessary to first identify two values of MTBF:

1. Specified MTBF representing the system requirement (assume that θ_0 represents this value).
2. Minimum MTBF that is considered to be acceptable based on the results of testing (assume that θ_1 represents this value).

Given θ_0 and θ_1, one must next decide on the values of producer's risk (α) and consumer's risk (β). Most test plans in use today accept risk values ranging between 5 and 25%. For example, a test where $\alpha = 0.10$ means that in 10 of 100 instances the test plan will reject items that should have been accepted. Risk values are generally negotiated in the test planning phase.

Referring to Figure 12.30, the design of the sequential test plan is based on the values of θ_0, θ_1, α and β. For instance, the *accept line* slope is based on the following expression:

$$t_1 = \frac{\ln(\theta_0/\theta_1)}{1/\theta_1 - 1/\theta_0}(r) - \frac{\ln[\beta/(1-\alpha)]}{1/\theta_1 - 1/\theta_0} \tag{12.29}$$

where r is the number of failures (the accept line is plotted for assumed values of r). The ratio of θ_0 to θ_1 is known as the *discrimination ratio*, a standard test parameter. The *reject line* slope is based on Equation 12.30.

$$t_2 = \frac{\ln(\theta_0/\theta_1)}{1/\theta_1 - 1/\theta_0}(r) - \frac{\ln[(1-\beta)/\alpha]}{1/\theta_1 - 1/\theta_0} \tag{12.30}$$

Determination of the specified time limit in Figure 12.30 is generally based on a multiple of θ_1.

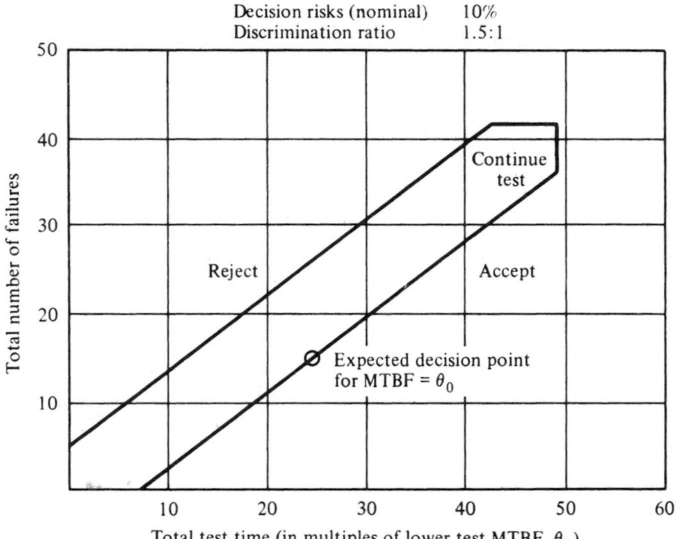

	Total test time*			Total test time*	
Number of failures	Reject (equal or less)	Accept (equal or more)	Number of failures	Reject (equal or less)	Accept (equal or more)
0	N/A	6.60	21	18.92	32.15
1	N/A	7.82	22	20.13	33.36
2	N/A	9.03	23	21.35	34.58
3	N/A	10.25	24	22.56	35.79
4	N/A	11.46	25	23.78	37.01
5	N/A	12.68	26	24.99	38.22
6	0.68	13.91	27	26.21	39.44
7	1.89	15.12	28	27.44	40.67
8	3.11	16.34	29	28.65	41.88
9	4.32	17.55	30	29.85	43.10
10	5.54	18.77	31	31.08	44.31
11	6.75	19.98	32	32.30	45.53
12	7.97	21.20	33	33.51	46.74
13	9.18	22.41	34	34.73	47.96
14	10.40	23.63	35	35.94	49.17
15	11.61	24.84	36	37.16	49.50
16	12.83	26.06	37	38.37	49.50
17	14.06	27.29	38	39.59	49.50
18	15.27	28.50	39	40.82	49.50
19	16.49	29.72	40	42.03	49.50
20	17.70	30.93	41	49.50	N/A

*Total test time is total unit hours of equipment on time and is expressed in multiples of the lower test MTBF.

Figure 12.32 Accept-reject criteria for a sample test plan. *Source:* MIL-STD-781, *Reliability Design Qualification and Production Acceptance Tests: Exponential Distribution.* Washington, D.C.: Department of Defense.

As indicated, there are a number of different sequential test plans that have been used for reliability testing. One example is illustrated in Figure 12.32. The established α and β decision risks are 10% (each), the discrimination ratio agreed on is 1.5:1, and the accept-reject criteria are as specified.

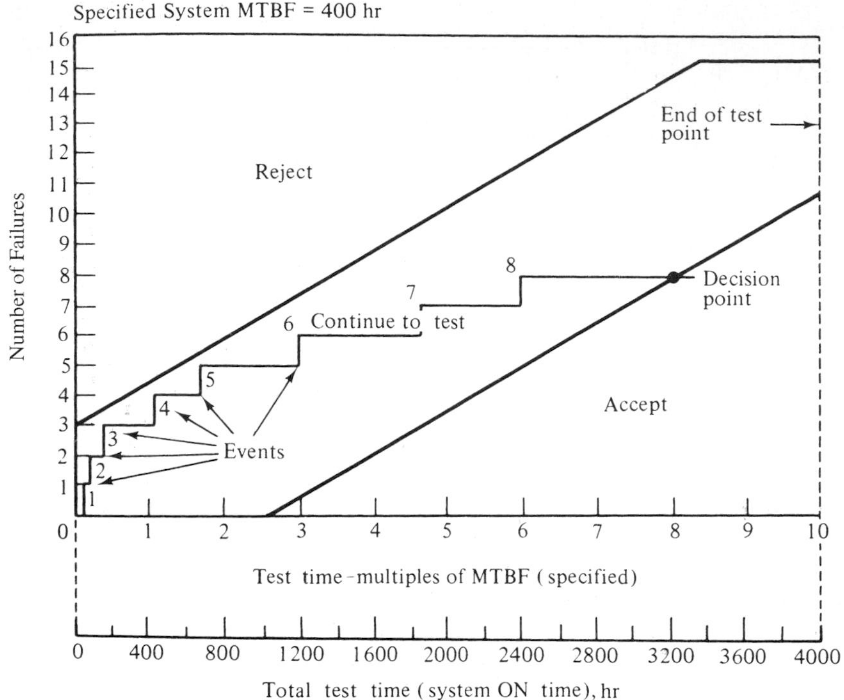

Figure 12.33 System XYZ reliability test plan.

Another approach is illustrated in Figure 12.33, representing actual experience in testing a specific system (to be designated as system *XYZ*). Referring to the figure, the specified MTBF for the system is 400 hours, and the maximum designated testing time for the sequential test plan used is 4,000 hours, or 10 times the specified MTBF. The test approach involves the selection of a designated quantity of equipments, operating the equipment under certain performance conditions over an extended period of time, and monitoring the equipment for failure. Failures are noted as events, corrected through the appropriate maintenance actions, and the applicable equipment is returned to full operational status for continued testing. An analysis of each event should determine the cause of failure, and trends may be established if more than one failure is traceable to the same cause. This may be referred to as a pattern failure, and in such cases, a change should be initiated to eliminate the occurrence of future failures of the same type.

Referring to Figure 12.33, the system is accepted after 3,200 hours of testing, and eight events are recorded. Thus, the specified MTBF requirement is fulfilled. In this instance, the test sample is six pieces of equipment. If a single model is designated for test with a minimum of 4,000 test hours required, the length of the test program (assuming continuous testing) will obviously extend beyond 6 months. This may not be feasible in terms of the overall program schedule and the delivery of qualified equipment. Conversely, the use of two or more models will result in a shorter test program. Sometimes the implementation of accelerated test conditions will result in an additional number of failures. These effects should be evaluated, and an optimum balance should be established between a smaller sample size and longer test times and a larger sample size and shorter test times.

Reliability Acceptance Testing

As indicated previously, reliability testing may be accomplished as part of qualification testing prior to commencing with full-scale production (which has been discussed) and during full-scale production on a sampling basis. To determine the effects of the production process on system reliability, it may be feasible to select a sample number of equipments from each production lot and test them in the same manner as described above. The sample may be based on a percentage of the total equipments spread over the entire production period, or a set number of equipment(s) selected during a given calendar time period (e.g., the two items of equipment per month throughout the production phase.) In any event, the selected equipment is tested and an assessed MTBF is derived from the test data. This value is compared against the specified MTBF and the measured value determined from earlier qualification testing. Growth MTBF trends (or negative trends) may be determined by plotting the resultant values as testing progresses.

Reliability Life Testing

There are two basic forms of life testing in use: (1) life tests based on a fixed-test time, and (2) life tests based on the occurrence of a predetermined number of failures. The first approach to life testing (based on time) assumes that a computed fixed test time will be planned and a specified quantity of failures will be predetermined. System acceptance occurs if the actual number of failures during test is equal to (or less than) the predetermined quantity of failures at the end of the scheduled test time.

The second approach to life testing is (based on failures) assumes that a test plan is developed specifying a predetermined quantity of failures and a computed test time dependent on an expected system failure rate. Testing continues until the specified quantity of failures occurs. System acceptance occurs if the test time is equal to (or is greater than) the computed time at the point of last failure.

Operational Reliability Assessment

The activities discussed previously have included requirements definition, allocation, analyses and predictions, and qualification and acceptance testing. The measure of system reliability has been based on a combination of analytical studies and the demonstration of prototype system elements or equipment. The opportunity to observe the system being used by the consumer in a realistic environment has not been possible until this point in time. Now, it is essential that a "true" assessment of the system reliability be made.

This assessment of system reliability in an operational environment is best accomplished through the establishment of an effective data collection, analysis, and system evaluation capability, as described in Section 6.5. The purpose is twofold.

1. To provide ongoing data that can be analyzed to determine the true reliability of the system while performing its intended mission.
2. To provide historical data that can be beneficially used in the design and development of new systems and equipment having a similar function and nature. Such data will facilitate the accomplishment and accuracy of analyses and predictions in the future.

Figures 6.5 and 6.6 illustrated both the *success* and *maintenance* data elements that are considered important for the system as a whole. Of specific interest in the reliability assessment task are (1) the operational status and condition of the system at the time of failure, (2) the maintenance requirements necessary to restore the system to full operational status, and (3) the details associated with the actual cause of the failure and the effects on other elements of the system.

The first two items of concern are adequately covered through the data elements included in the *system operational information report* (Figure 6.5) and the maintenance event report (Figure 6.6). However, the details associated with the cause and effects of failures are not sufficiently covered. Thus, these two reports should be supplemented with a *failure analysis report.*

A failure analysis report should be prepared for each system failure and should provide information such as the time of failure, symptom of failure, effects of failure on system operation, effects of the failure on other elements of the system, the actual cause of system failure, and the specific failure mode. Enough data are required to determine *what* happened, *how* the failure occurred, and *why* did it occur.

These data should be collected throughout the system operational life cycle (to the maximum extent possible) and analyzed to determine trends and inherent weaknesses in the system. Major areas of deficiency should be corrected through the initiation of changes and the appropriate system modification(s).

QUESTIONS AND PROBLEMS

1. Identify and discuss the four major elements in the reliability definition. Include an illustration of application through the selection of a system of your choice and the specification of reliability requirements for that system.
2. Why is reliability important in system design? When in the life-cycle process should it be considered? To what extent should reliability be emphasized in system design, and what are some of the factors that govern this?
3. What are the quantitative measures of reliability (discuss measures for hardware, software, personnel, facilities, and data)?
4. Define operational availability, achieved availability, inherent availability, system effectiveness, and cost effectiveness.

5. Refer to Figure 12.4. Reliability predictions/estimations are usually based on what portion of the "bathtub" curve? What is likely to happen if the system is delivered to the customer during the early "infant mortality" portion of the curve? What would you do to extend the flat portion of the curve?
6. Refer to Figure 12.5. Identify some of the possible "causes" for the jagged portion of the curve?
7. A system consists of four subassemblies connected in series. The individual subassembly reliabilities are as follows:

$$\text{Subassembly } A = 0.98$$
$$\text{Subassembly } B = 0.85$$
$$\text{Subassembly } C = 0.90$$
$$\text{Subassembly } D = 0.88$$

Determine the overall system reliability.

8. A system consists of three subsystems in parallel (assume operating redundancy). The individual subsystem, reliabilities are as follows:

$$\text{Subsystem } A = 0.98$$
$$\text{Subsystem } B = 0.85$$
$$\text{Subsystem } C = 0.88$$

Determine the overall system reliability.

9. Refer to Figure 12.20. Determine the overall network reliability if the individual reliabilities of the subsystems are as follows:

$$\text{Subsystem } A = 0.95$$
$$\text{Subsystem } B = 0.97$$
$$\text{Subsystem } C = 0.92$$
$$\text{Subsystem } D = 0.94$$
$$\text{Subsystem } E = 0.90$$
$$\text{Subsystem } F = 0.88$$
$$\text{Subsystem } G = 0.98$$

10. The failure rate (λ) of a device is 22 failures per million hours. Two standby units are added to the system. Determine the MTBF of the system.
11. Calculate the reliability of a system consisting of one operating unit and one identical standby unit operating for a period of 200 hours. The failure rate (λ) of each unit is 0.003 failure/hour and the failure sensing switch reliability is 1.0.
12. A system consists of five subsystems with the following MTBFs:

$$\text{Subsystem } A\text{: MTBF} = 10{,}540 \text{ hours}$$
$$\text{Subsystem } B\text{: MTBF} = 16{,}220 \text{ hours}$$
$$\text{Subsystem } C\text{: MTBF} = 9{,}500 \text{ hours}$$
$$\text{Subsystem } D\text{: MTBF} = 12{,}100 \text{ hours}$$
$$\text{Subsystem } E\text{: MTBF} = 3{,}600 \text{ hours}$$

The five subsystems are connected in series. Determine the probability of survival for an operating period of 1,000 hrs.

13. Ten components were tested for 500 hours, each within prescribed operating conditions. Component 1 failed after 30 hours; component 2 failed after 85 hours; component 3 failed after 220 hours; and component 4 failed after 435 hours. Determine the overall composite failure rate (λ) for the system.

14. Suppose that you have the following design data on an equipment item and that, based on this data, you may wish to accomplish a reliability prediction. What is the predicted MTBF?

Component	Quantity of parts used	Failure rate (% 1,000 hrs)
A	16	0.135
B	75	0.121
C	32	0.225
D	44	0.323
E	60	0.120
F	15	0.118
G	28	0.092

15. Assume that there is a requirement for a new system with a specified performance capability and a reliability of 70%. In response to an "invitation to bid," three supplier configurations have been proposed and are reflected in Figure 12.34. The component reliability factors are noted in the following table:

Component	Reliability	Component	Reliability	Component	Reliability
A	0.84	G	0.87	M	0.83
B	0.86	H	0.88	N	0.85
C	0.89	I	0.89	O	0.84
D	0.86	J	0.86	P	0.89
E	0.87	K	0.85	Q	0.89
F	0.82	L	0.86		

The overall cost associated with each of the supplier proposals is $57,000 for Configuration A, $39,000 for Configuration B, $42,000 for Configuration C.

(a) Determine the system reliability for each of the three configurations.

(b) In evaluating the three alternatives, employing cost-effectiveness criteria, which configuration would you select as being preferred?

16. When determining overall system failure rate, what factors must be considered? Why?

17. Describe the steps that you would take in developing a *reliability model.*

18. Describe the process that you would follow in identifying the appropriate reliability requirements at the *system-level,* and on down to the *subsystem level, unit level,* and below.

19. What is the purpose of component *derating*?

20. What is the purpose of reliability allocation? The FMECA? The FTA? Stress-strength analysis? Reliability growth analysis? Reliability test and evaluation? Describe how these method-

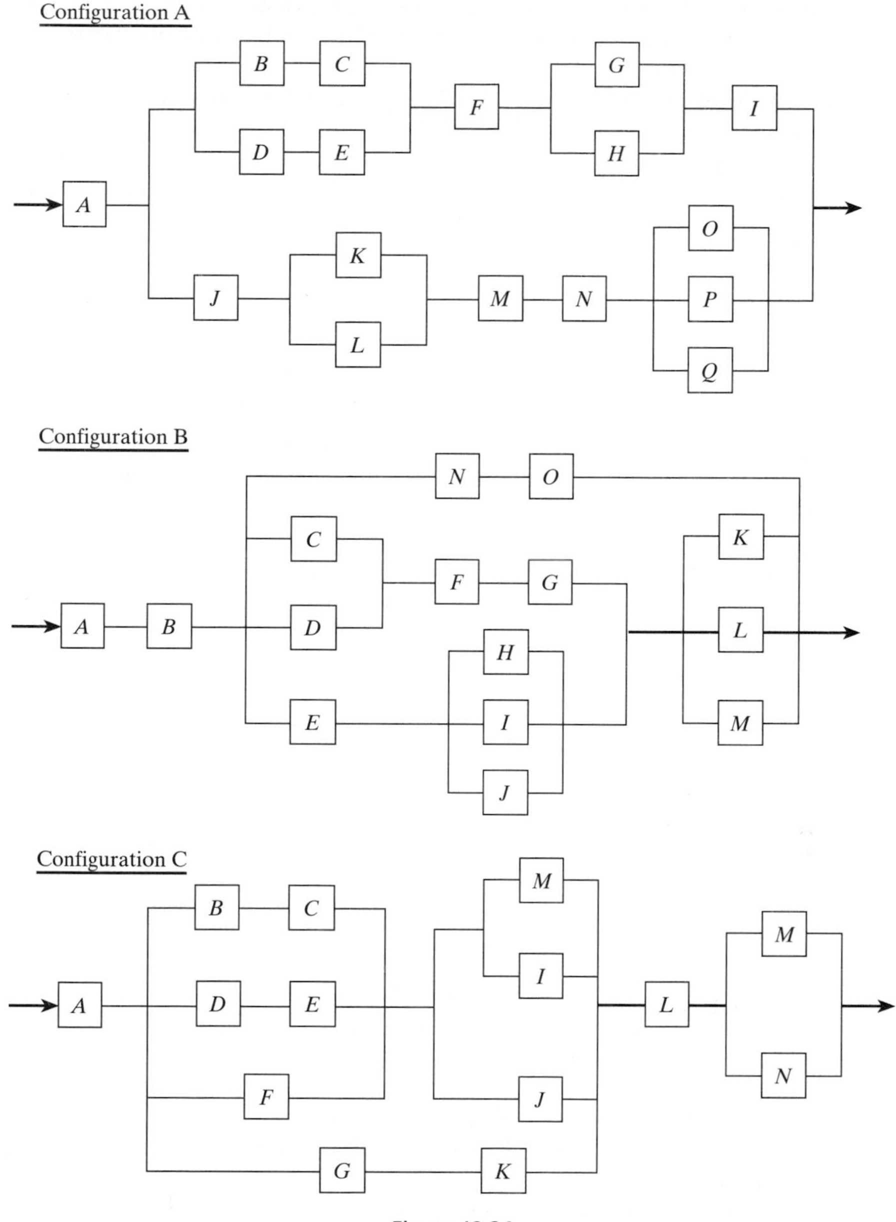

Figure 12.34

ologies/tools might be related (include a flow chart showing their application and the possible interfaces that might exist).

21. What information does the FMECA provide? When in the system life cycle can it be accomplished? What benefits are provided? Problems?

22. Select a system of your choice, and describe the system in functional terms (construct a functional block diagram), Select an element of that system (i.e., product or process) and conduct a FMECA.
23. What is the purpose of the FTA? How does it differ from the FMECA? When would you use the FTA in lieu of the FMECA?
24. Assume that there is a 500-hour MTBF requirement for a new system being developed, At the last design review, the predicted MTBF was 400 hours, What steps would you take to improve the reliability and ensure that the requirement will be met? What techniques/tools would you employ to facilitate your objective?
25. At what stage in the system life cycle are the requirements for reliability test and evaluation determined? What steps would you take to ensure that these requirements will be met (i.e., "validation" steps)?
26. Describe some of the advantages of reliability sequential testing. Identify some of the disadvantages. What is meant by *life* testing? *Accelerated* testing?
27. Assume that you are planning to produce a multiple quantity of "like" products, What steps would you take to ensure that all of the products delivered will reflect the same reliability characteristics?
28. Define what is meant by *producer's* risk and *consumer's* risk?
29. Describe in your own words how reliability (and the degree to which reliability characteristics are incorporated in the design of a system) can impact maintainability, human factors, supportability, producibility, disposability, and cost.

13 Design for Maintainability

The systems engineering process requires the application of scientific and engineering efforts to ensure that the final product is operationally feasible. One of the more important design parameters in this process is *maintainability*. Maintainability is a design characteristic dealing with the ease, accuracy, safety, and economy in the performance of maintenance functions. It can be specified, controlled and measured, and maintainability methods and techniques are available to facilitate this process. Maintainability must be considered along with performance, reliability, human factors, producibility, supportability, disposability, life-cycle cost, and other factors in systems design.

Systems (or products) in use today are highly sophisticated and will fulfill most expectations when operating. However, experience has indicated that the reliability of many of these systems is marginal and that systems are inoperative much of the time. Unreliable systems are unable to fulfill the mission for which they were designed and they require extensive maintenance. In an environment where resources are becoming scarcer, it is essential that system maintenance requirements be minimized and that maintenance costs be reduced. Accordingly, it is essential that maintainability be considered as a major system parameter in the design process.

13.1 DEFINITION AND EXPLANATION OF MAINTAINABILITY

An objective in systems engineering is to design and develop a system or product that can be maintained effectively, safely, in the least amount of time, at the least cost, and with a minimum expenditure of support resources (e.g., people, materials, test equipment, and facilities) without adversely affecting the mission of that system. Maintainability is the *ability* of a product to be maintained, whereas maintenance constitutes a series of actions to be taken to restore or retain a product in an effective operational state. Maintainability is a design dependent parameter. Maintenance is a result of design.

Maintainability, as a characteristic of design, can be expressed in terms of maintenance times, maintenance frequency factors, and maintenance cost. These terms may be presented as different figures-of-merit; therefore, maintainability may be defined on the basis of a combination of factors, such as the following:

1. A characteristic of design and installation which is expressed as the probability that an item will be retained in or restored to a specified condition within a given period of time, when maintenance is performed in accordance with prescribed procedures and resources.
2. A characteristic of design and installation that is expressed as the probability that maintenance will not be required more than x times in a given period, when the system is operated in accordance with prescribed procedures.
3. A characteristic of design and installation that is expressed as the probability that the maintenance cost for a system will not exceed y dollars per designated period, when the system is operated and maintained in accordance with prescribed procedures.

Specific maintainability requirements are derived from the definition of system operational requirements and the maintenance concept discussed in Section 3.4. They must be tailored to the applicable mission scenario(s). The appropriate metrics are developed through the identification and prioritization of the technical performance measures (TPMs) covered in Section 3.5.

13.2 MEASURES OF MAINTAINABILITY

Maintainability, defined in the broadest sense, can be measured in terms of a combination of different maintenance factors. From a system perspective, it is assumed that maintenance can be broken down into the following general categories:

1. *Corrective maintenance.* Unscheduled maintenance accomplished, as a result of failure, to *restore* a system or product to a specified level of performance. This includes the possible ongoing modification of software to bring it to the proper operational state in the event that it has not achieved the desired level of maturity when the system is delivered to the customer.
2. *Preventive maintenance.* Scheduled maintenance accomplished to *retain* a system at a specified level of performance by providing systematic inspection, detection, servicing, or the prevention of impending failures through periodic item replacements.

When dealing with software, there are several additional categories that are often used. *Adaptive maintenance* refers to the continuing process of modifying software to be responsive to changing requirements in data or in the processing environment (but within the original functional structure). *Perfective maintenance* is a term often used to describe the modification of software for the purposes of enhancing performance, packaging, and other evolving needs of the consumer.[1]

Of particular interest herein are the measures of maintainability pertaining to "downtime" (i.e., the elapsed time when the system is not operating because of maintenance), personnel labor hours (i.e., the total number of people, skill levels, and the labor hours expended in the accomplishment of maintenance), maintenance frequencies (i.e., the MTBM factors discussed in Section 12.2), maintenance cost, and related factors dealing with the various elements of support (i.e., spares/repair parts, test equipment, transportation and handling factors). The measures most commonly used are described in this section.

Maintenance Elapsed-Time Factors

The elapsed-time category includes corrective and preventive active maintenance times, administrative and logistics delay times, and total maintenance downtime.

Mean corrective maintenance time ($\overline{\mathrm{Mct}}$). When a system fails, a series of steps is required to repair or restore the system to its full operational status. These steps include failure detection, fault isolation, disassembly to gain access to the faulty item, repair, and so on, as illustrated in Figure 13.1. Completion of these steps for a given failure constitutes a corrective maintenance cycle.

Throughout the system utilization phase, there will be a number of individual maintenance actions involving the series of steps illustrated in the figure. The mean corrective maintenance time ($\overline{\mathrm{Mct}}$), or the mean time to repair (MTTR) that is equivalent, is a composite value representing the arithmetic average of these individual maintenance cycle times (Mct_i).

For the purposes of illustration, Table 13.1 includes data covering a sample of 50 corrective maintenance repair actions on a typical system. Each of the times indicated represents the completion of one corrective maintenance cycle. Based on the set of raw data presented, which constitutes a random sample, a frequency distribution table and frequency histogram may be prepared, as illustrated in Table 13.2 and Figure 13.2, respectively.

TABLE 13.1 Corrective Maintenance Times (Mct_i in min)

40	58	43	45	63	83	75	66	93	92
71	52	55	64	37	62	72	97	76	75
75	64	48	39	69	71	46	59	68	64
67	41	54	30	53	48	83	33	50	63
86	74	51	72	87	37	57	59	65	63

[1] A. P. Sage, and J. D. Palmer, *Software Systems Engineering* (New York: John Wiley & Sons, Inc., 1990); and B. W. Boehm, *Software Engineering Economics* (Upper Saddle River, N. J.: Prentice Hall, Inc., 1981).

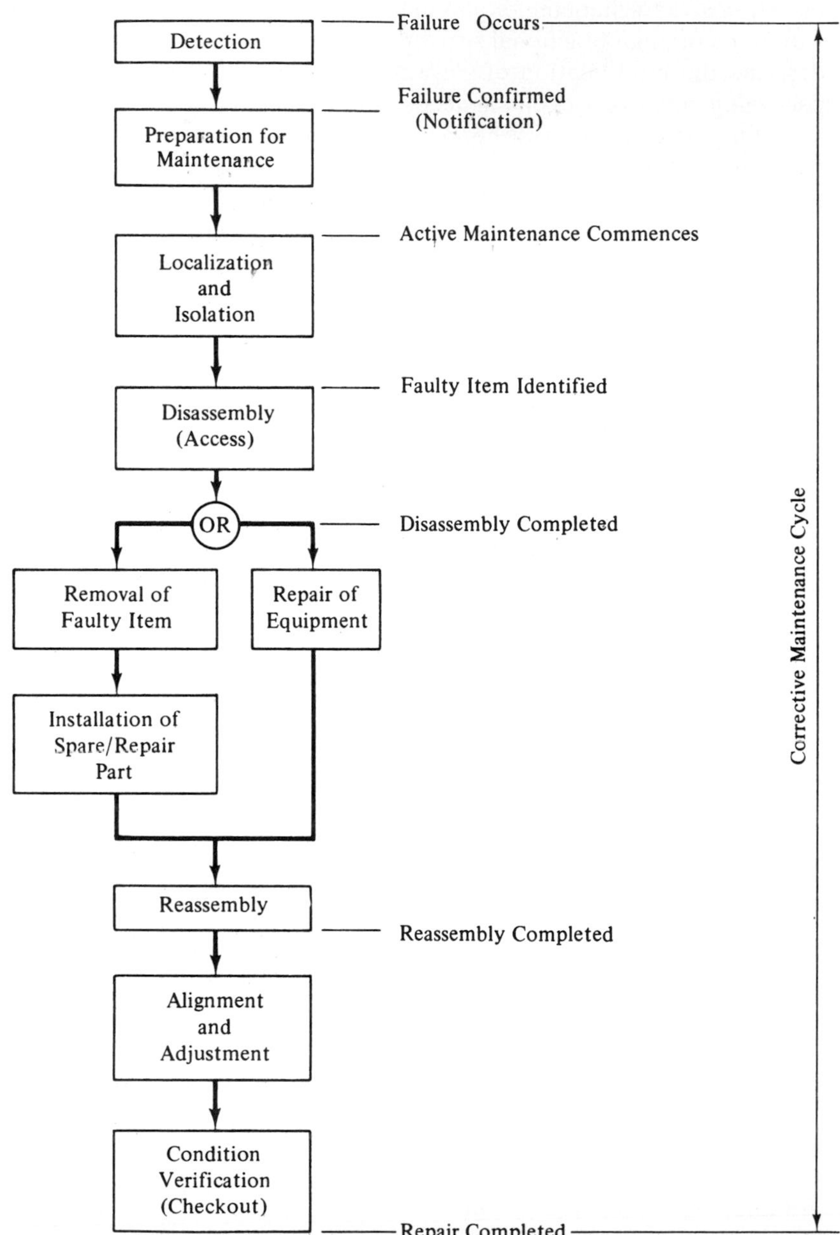

Figure 13.1 Corrective maintenance cycle.

Referring to Table 13.2, the range of observations is between 97 and 30 minutes, or a total of 67 minutes. This range can be divided into class intervals, with a class interval width of 10 assumed for convenience. A logical starting point is to select class intervals of 20 to 29, 30 to 39, and so on. In such instances, it is necessary to establish the dividing point between two adjacent intervals, such as 29.5, 39.5, and so on.

TABLE 13.2 Frequency Distribution

Class interval	Frequency	Cumulative frequency
29.5–39.5	5	5
39.5–49.5	7	12
49.5–59.5	10	22
59.5–69.5	12	34
69.5–79.5	9	43
79.5–89.5	4	47
89.5–99.5	3	50

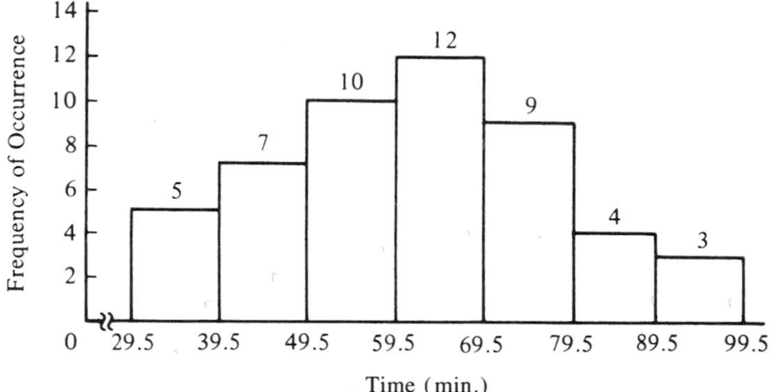

Figure 13.2 Histogram of maintenance times.

Given the distribution of repair times, one can plot a histogram showing time values in minutes and the frequency of occurrence as in Figure 13.2. By determining the midpoint of each class interval, a frequency polygon can be developed as illustrated in Figure 13.3. This provides an indication of the form of the probability distribution applicable to repair times for this particular system.

As additional corrective maintenance actions occur and data points are plotted for the system in question, the curve illustrated in Figure 13.3 may take the shape of

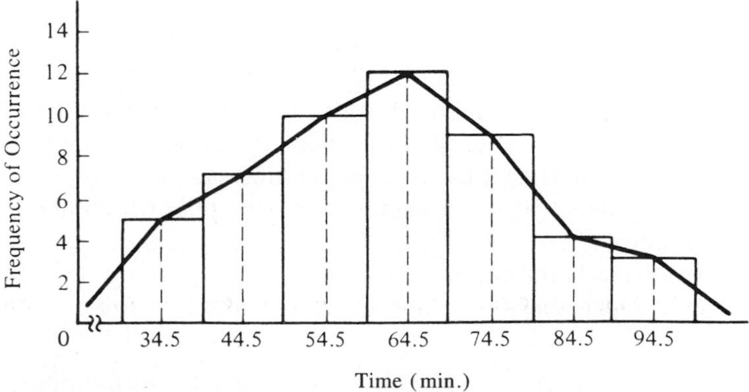

Figure 13.3 Frequency polygon.

TABLE 13.3 Variance Data

TOTAL	$Mct_i - \overline{Mct}$	$(Mct_i - \overline{Mct})^2$
40	−22	484
71	+ 9	81
75	+13	169
67	+ 5	25
etc.	etc.	etc.
Total		13,013

the normal distribution. The normal curve is defined by the arithmetic mean ($\overline{Mct}$) and the standard deviation (σ). From the maintenance repair times presented in Table 13.1, the arithmetic mean is determined as[2]:

$$\overline{Mct} = \frac{\sum_{i=1}^{n} Mct_i}{n} = \frac{3{,}095}{50} = 61.9 \quad \text{(assume 62)} \tag{13.1}$$

where Mct_i is the total active corrective maintenance cycle time for each maintenance action and n is the sample size. Thus, the average value for the sample of 50 maintenance actions is 62 minutes.

The standard deviation (σ) measures the dispersion of maintenance time values. When a standard deviation is calculated, it is convenient to generate a table giving the deviation of each task time from the mean of 62. Table 13.3 illustrates this for only four individual task times, although all 50 tasks should be treated. The total value of 13,013 does cover all 50 tasks.

The standard deviation of the sample normal distribution curve can now be determined as

$$\sigma = \sqrt{\frac{\sum_{i=1}^{n} (Mct_i - \overline{Mct})^2}{n - 1}} = \sqrt{\frac{13{,}013}{49}} = 16.3 \text{ min} \quad \text{(assume 16)} \tag{13.2}$$

One may wish to determine what percentage of the total repair actions falls within the range 46 to 78 minutes. This can be calculated by converting the range to multiples of standard deviation. In this case, the interval of 46 to 78 equals $\overline{Mct} \pm 1\sigma$, or 62 minutes $\pm$ the standard deviation of 16 minutes. When normality is assumed, it can be stated that 68% of the total population sampled falls within the range 46 to 78 minutes and that 99.7% of the sample population lies within the range $\overline{Mct} \pm 3\sigma$, or 14 to 110 minutes.

As a typical application, it may be desirable to determine the percentage of total population repair times that lies between 40 and 50 minutes. Graphically, this is rep-

[2] Refer to Appendix B.2 for discussion on probability distribution models. Both the normal and log-normal distributions are assumed when dealing with the distribution of repair times.

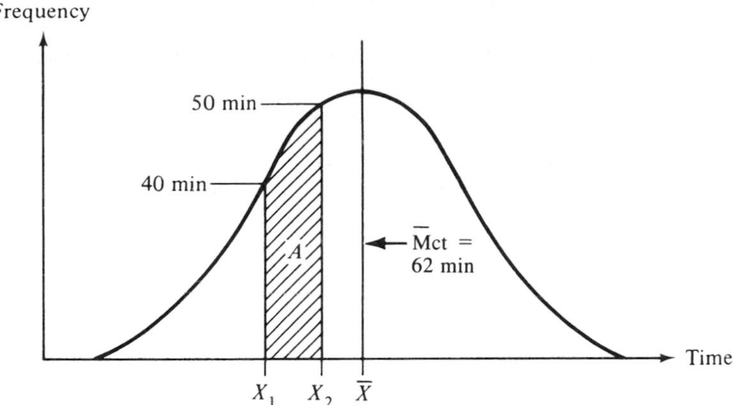

Figure 13.4 Normal distribution of repair times.

resented in Figure 13.4. The problem is to find the percent represented by the shaded area. This can be calculated as follows:

1. Convert maintenance times of 40 and 50 minutes, into standard values (Z) or the number of standard deviations above and below the mean of 62 minutes.

$$Z \text{ for } 40 \text{ min} = \frac{X_1 - \overline{X}}{\sigma} = \frac{40 - 62}{16} = -1.37$$

$$Z \text{ for } 50 \text{ min} = \frac{X_2 - \overline{X}}{\sigma} = \frac{50 - 62}{16} = -0.75$$

   The maintenance times of 40 and 50 minutes represent -1.37 and -0.75 standard deviations below the mean (because the values are negative).
2. Point X_1 ($Z = -1.37$) represents an area of 0.0853, and point X_2 ($Z = -0.75$) represents an area of 0.2266, as given in Appendix C.3.
3. The shaded area A in Figure 13.4 represents the difference in area or, area $X_2 - X_1 = 0.2266 - 0.0853 = 0.1413$. Thus, 14.13% of the population of maintenance times are estimated to lie between 40 and 50 minutes.

Next, confidence limits should be determined. Because the 50 maintenance tasks represent only a sample of all maintenance actions on the equipment being evaluated, it is possible that another sample of 50 maintenance actions on the same equipment could have a mean value either greater or less than 62 minutes. The original 50 tasks were selected at random, however, and statistically represent the entire population. Using the standard deviation, an upper and lower limit can be placed on the mean value ($\overline{\text{M}}$ct) of the population. For instance, if one is willing to accept a chance of being wrong 15% of the time (85% confidence limit), then

$$\text{upper limit} = \overline{\text{M}}\text{ct} + Z\left(\frac{\sigma}{\sqrt{n}}\right) \tag{13.3}$$

where $\sigma/\sqrt{n}$ represents the standard error factor.

TABLE 13.4 Risk-Upper-Limit Variations

Risk (%)	Confidence (%)	Z	Upper limit (min)
5%	95%	1.65	65.72
10%	90%	1.28	64.89
15%	85%	1.04	64.35
20%	80%	1.84	63.89

The Z value is obtained from Appendix C.3, where 0.1492 is close to 15% and reflects a Z of 1.04. Thus,

$$\text{upper limit} = 62 + (1.04)\left(\frac{16}{\sqrt{50}}\right) = 64.35 \text{ min}$$

This means that the upper limit is 64.4 minutes at a confidence level of 85%, or that there is an 85% chance that $\overline{\text{M}}\text{ct}$ will be less than 64.4. Variations in risk and upper limit values are shown in Table 13.4. If a specified $\overline{\text{M}}\text{ct}$ limit is established for the design of equipment (based on mission and operational requirements), and it is known (or assumed) that maintenance times are normally distributed, then one would have to compare the results of predictions or measurements (e.g., 64.35 minutes) accomplished during the development process with the specified value to determine the degree of compliance.

When considering probability distributions in general, the time dependency between probability of repair and the time allocated for repair can usually be expected to produce a probability density function in one of three common distribution forms (normal, exponential, and log-normal) as illustrated in Figure 13.5.[3]

1. The *normal* distribution applies to the relatively straightforward maintenance tasks and repair actions (simple removal and replacement tasks) that consistently require a fixed amount of time to complete with little variation.
2. The *exponential* distribution applies to maintenance tasks involving part substitution methods of failure isolation in large systems that result in a constant repair rate.
3. The *log-normal* distribution applies to most maintenance tasks and repair actions comprised of several subsidiary tasks of unequal frequency and time duration.

As indicated, the maintenance task times for many systems and equipments do not always fit the normal curve. There may be a few representative maintenance actions where repair times are extensive, causing a skew to the right. This is particularly true

[3] Although past experience has indicated that repair times usually follow the normal and log-normal distributions, sometimes the exponential distribution is assumed for the sake of convenience in modeling, particularly in reliability modeling where the exponential distribution is assumed when dealing with failure rates. Such as assumption is sometimes misleading and presumes that zero repair times are the most frequent. In any event, the analyst must take care to ensure that the correct repair-time distribution is selected in analyzing maintainability data.

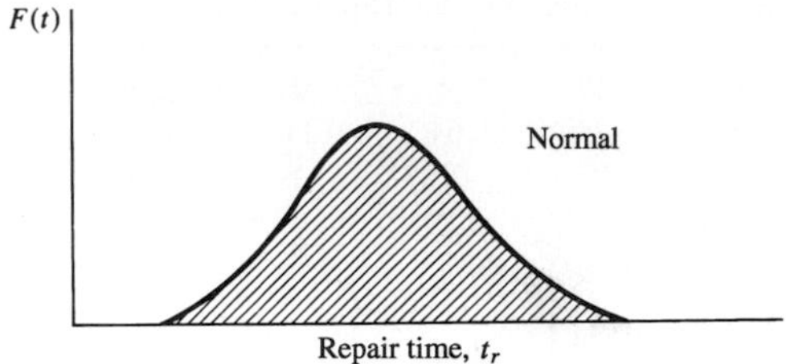

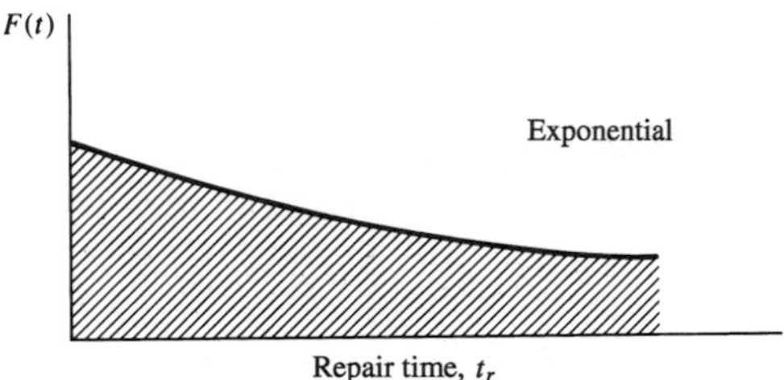

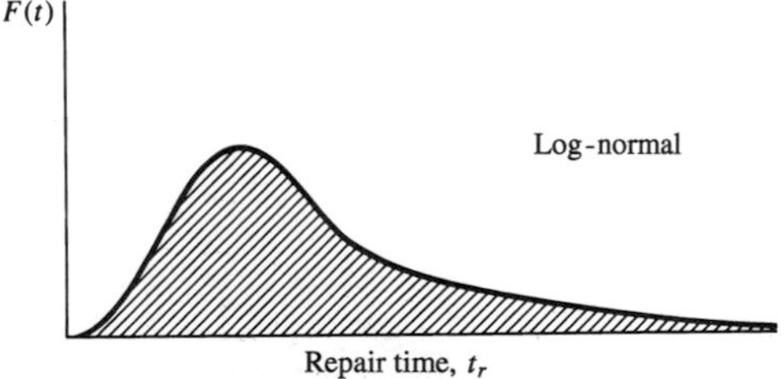

Figure 13.5 Repair time distributions.

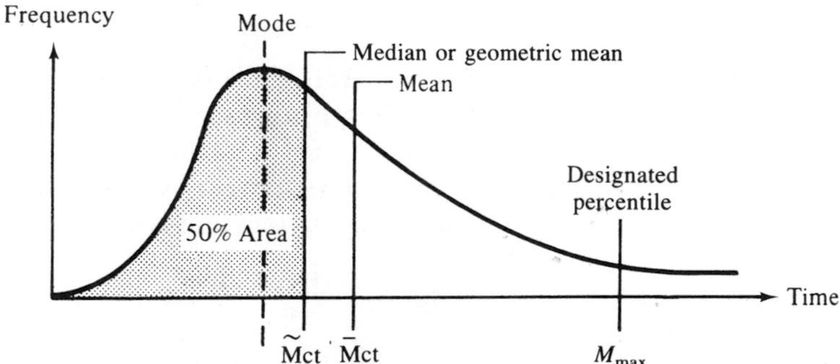

Figure 13.6 Log-normal distribution of repair times.

for electronic equipment, where the distribution of repair times often follows a log-normal curve as shown in Figure 13.6. Derivation of the specific distribution curve for a set of maintenance task times is accomplished using the same procedure as given in the preceding paragraphs. A frequency table is generated, and a histogram is plotted.

TABLE 13.5 Corrective Maintenance Repair Times (min)

55	28	125	47	58	53	36	88
51	110	40	75	64	115	48	52
60	72	87	105	55	82	66	65

A sample of 24 corrective maintenance repair actions for a typical electronic equipment item is presented in Table 13.5. Using the data in the table, the mean is determined as

$$\overline{\text{Mct}} = \frac{\sum_{i=1}^{n} \text{Mct}_i}{n} = \frac{1{,}637}{24} = 68.21 \text{ min}$$

When determining the mean corrective maintenance time ($\overline{\text{Mct}}$) for a specific sample population of maintenance repair actions, the use of Equation 13.4, which has wider application, is appropriate and is

$$\overline{\text{Mct}} = \frac{\Sigma\,(\lambda_i)(\text{Mct}_i)}{\Sigma\,\lambda_i} \tag{13.4}$$

where λ_i is the failure rate of the individual (ith) element of the item being measured, usually expressed in failures per system operating hour. Equation 13.4 calculates $\overline{\text{Mct}}$ as a weighted average using reliability factors.

It should be noted that $\overline{\text{Mct}}$ considers only *active* maintenance time or the time that is spent working directly on the system. Logistics delay time and administrative delay time are not included. Although all elements of time are important, the $\overline{\text{Mct}}$ factor is primarily oriented to a measure of the supportability characteristics in equipment design.

Mean preventive maintenance time ($\overline{\text{Mpt}}$). Preventive maintenance refers to the actions required to retain a system at a specified level of performance. It may include such functions as periodic inspection, servicing, scheduled replacement of critical items, calibration, overhaul, and so on. $\overline{\text{Mpt}}$ is the mean (or average) elapsed time to perform preventive or scheduled maintenance on an item and is expressed as

$$\overline{\text{Mpt}} = \frac{\Sigma\,(\text{fpt}_i)(\text{Mpt}_i)}{\Sigma\,\text{fpt}_i} \tag{13.5}$$

where fpt_i is the frequency of the individual (ith) preventive maintenance action in actions per system operating hour, and Mpt_i is the elapsed time required for the ith preventive maintenance action.

Preventive maintenance may be accomplished while the system is in full operation or could result in downtime. In this instance, the concern is for preventive maintenance actions that result in system downtime. Again, $\overline{\text{Mpt}}$ includes only *active* system maintenance time, not logistic delay and administrative delay times.

TABLE 13.6 Calculation for $\widetilde{\text{Mct}}$

Mct_i	LOG Mct_i	$(\text{LOG Mct}_i)^2$	Mct_i	LOG Mct_i	$(\text{LOG Mct}_i)^2$
55	1.740	3.028	64	1.806	3.262
28	1.447	2.094	1	2.041	4.248
123	2.097	4.397	48	1.681	2.826
47	1.672	2.796	52	1.716	2.945
58	1.763	3.108	60	1.778	3.161
53	1.724	2.972	72	1.857	3.448
36	1.556	2.421	87	1.939	3.760
88	1.945	3.783	105	2.021	4.084
51	1.708	2.917	55	1.740	3.028
110	2.041	4.166	82	1.914	3.663
40	1.602	2.566	66	1.819	3.309
75	1.875	3.516	65	1.813	3.287
Total				43.315	78.785

Median active corrective maintenance time ($\widetilde{\text{Mct}}$). The median maintenance time is that value which divides all of the downtime values so that 50% are equal to or less than the median and 50% are equal to or greater than the median. The median will usually give the best average location of the data sample. The median for a normal distribution is the same as the mean, whereas the median in a log-normal distribution is the same as the geometric mean, illustrated in Figure 13.6. $\widetilde{\text{Mct}}$ (which is also equivalent to MTTR_g) is calculated as

$$\widetilde{\text{Mct}} = \text{antilog}\frac{\sum_{i=1}^{n} \log \text{Mct}_i}{n} = \text{antilog}\frac{\Sigma\,(\lambda_i)(\log \text{Mct}_i)}{\Sigma\,\lambda_i} \tag{13.6}$$

For illustrative purposes, the maintenance time values in Table 13.5 are presented in the format illustrated in Table 13.6. The median is computed as

$$\widetilde{\text{Mct}} = \text{antilog}\frac{\sum_{1}^{24} \log \text{Mct}_i}{24}$$

$$= \text{antilog}\frac{43.315}{24} = \text{antilog}\,1.805 = 63.8 \text{ min}$$

Median active preventive maintenance time ($\widetilde{\text{Mpt}}$). The median active preventive maintenance time is determined using the same approach as for calculating $\widetilde{\text{Mct}}$. $\widetilde{\text{Mpt}}$ is expressed as

$$\widetilde{\text{Mpt}} = \text{antilog}\frac{\Sigma\,(\text{fpt}_i)(\log \text{Mpt}_i)}{\Sigma\,\text{fpt}_i} \tag{13.7}$$

Mean active maintenance time ($\overline{\text{M}}$). $\overline{\text{M}}$ is the mean or average elapsed time required to perform scheduled (preventive) and unscheduled (corrective) maintenance. It excludes logistics delay time and administrative delay time, and is expressed as

$$\overline{\text{M}} = \frac{(\lambda)(\overline{\text{Mct}}) + (\text{fpt})(\overline{\text{Mpt}})}{\lambda + \text{fpt}} \tag{13.8}$$

where λ is the corrective maintenance rate or failure rate, and fpt is the preventive maintenance rate.

Maximum active corrective maintenance time (M_{max}). M_{max} can be defined as that value of maintenance downtime below which a specified percentage of all maintenance actions can be expected to be completed. M_{max} is related primarily to the log-normal distribution, and the 90th or 95th percentile point is generally taken as the specified value as shown in Figure 13.6. It is expressed as

$$\text{M}_{\text{max}} = \text{antilog}\,(\overline{\log \text{Mct}}) + Z\sigma_{\log \text{Mct}_i}) \tag{13.9}$$

where $\overline{\log \text{Mct}}$ is the mean of the logarithms of Mct_i, Z the value corresponding to the specific percentage point at which M_{max} is defined (see Table 13.4, −1.65 for 95%), and

$$\sigma_{\log \text{Mct}_i} = \sqrt{\frac{\sum_{i=1}^{n} (\log \text{Mct}_i)^2 - \left(\sum_{i=1}^{n} \log \text{Mct}_i\right)^2 / n}{n - 1}} \tag{13.10}$$

or the standard deviation of the sample logarithms of average repair times, Mct_i.

For example, M_{max} at the 95th percentile for the data sample in Table 13.5 is determined as

$$\text{M}_{\text{max}} = \text{antilog}\,[\log \overline{\text{Mct}} + (1.65)\,\sigma_{\log \text{Mct}_i}] \tag{13.11}$$

where, referring to Equation (13.10) and Table 13.6,

$$\sigma_{\log \text{Mct}_i} = \sqrt{\frac{78.785 - (43.315)^2/24}{23}} = 0.163$$

Substituting the standard deviation factor and the mean value into Equation 13.11, gives

$$\begin{aligned} \text{M}_{\text{max}} &= \text{antilog}\,[\log \overline{\text{Mct}} + (1.65)(0.163)] \\ &= \text{antilog}\,(1.805 + 0.269) = 119 \text{ min} \end{aligned}$$

If maintenance times are distributed log-normally, M_{max} cannot be derived directly by using the observed maintenance values. However, by taking the logarithm of each repair value, the resulting distribution becomes normal, facilitating usage of the data in the manner identical to the normal case.

Logistics delay time (LDT). Logistics delay time refers to that maintenance downtime, which is expended as a result of waiting for a spare part to become available, waiting for the availability of an item of test equipment in order to perform maintenance, waiting for transportation, waiting to use a facility required for maintenance, and so on. LDT does not include active maintenance time but does constitute a major element of total maintenance downtime (MDT).

Administrative delay time (ADT). Administrative delay time refers to that portion of downtime during which maintenance is delayed for reasons of an administrative nature (e.g., personnel assignment priority, labor strike, organizational constraint, etc.). ADT does not include active maintenance time but often constitutes a significant element of total maintenance downtime (MDT).

Maintenance downtime (MDT). Maintenance downtime constitutes the total elapsed time required (when the system is not operational) to repair and restore a system to full operating status, or to retain a system in that condition. MDT includes mean active maintenance time ($\overline{M}$), logistics delay time (LDT), and administrative delay time (ADT). The mean or average value is calculated from the elapsed times for each function and the associated frequencies (similar to the approach used in determining $\overline{M}$).

In summary, Figure 13.7 illustrates the relationships of the various downtime factors within the context of the overall time domain.

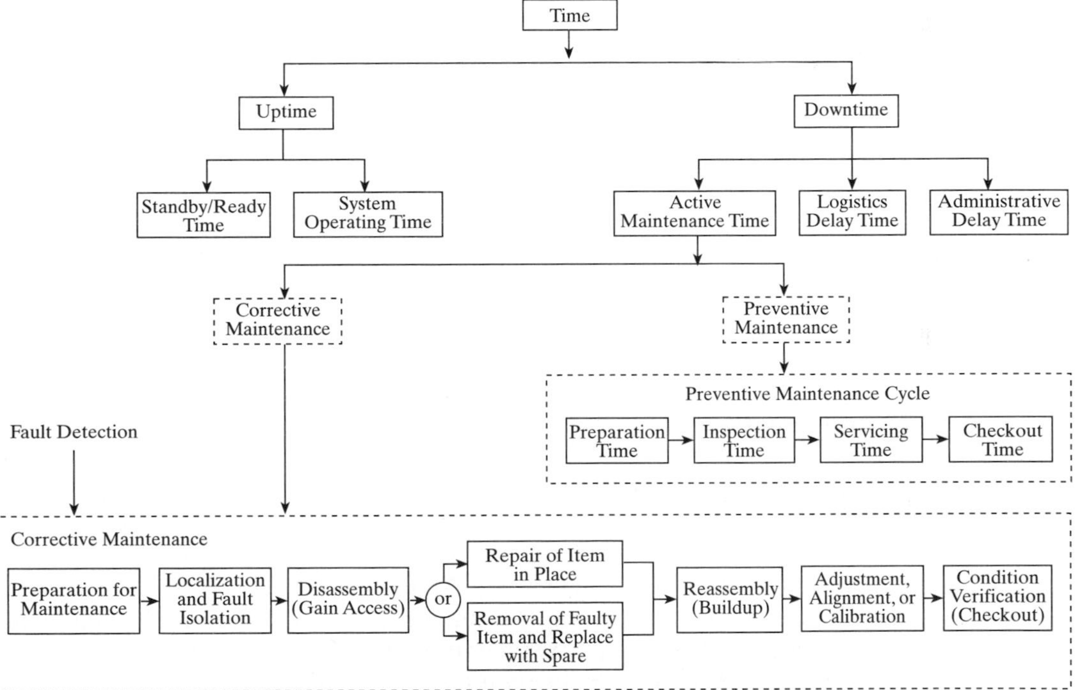

Figure 13.7 Composite view of uptime/downtime factors.

Maintenance Labor-Hour Factors

The maintainability factors covered in the previous paragraphs relate to elapsed times. Although elapsed times are extremely important in the performance of maintenance, one must also consider the maintenance labor-hours expended in the process. Elapsed times can be reduced (in many instances) by applying additional human resources in the accomplishment of specific tasks. However, this may turn out to be an expensive trade-off, particularly when high skill levels are required to perform tasks that result in less overall clock time. In other words, maintainability is concerned with the *ease* and *economy* in the performance of maintenance. As such, an objective is to obtain the proper balance among elapsed time, labor time, and personnel skills at a minimum maintenance cost.

When considering measures of maintainability, it is not only important to address such factors as $\overline{\text{M}}\text{ct}$ and MDT, but it is also necessary to consider the labor-time element. Thus, some additional measures must be employed, such as the following:

1. Maintenance labor-hours per system operating hour (MLH/OH)
2. Maintenance labor-hours per cycle of system operation (MLH/cycle)
3. Maintenance labor-hours per month (MLH/month)
4. Maintenance labor-hours per maintenance action (MLH/MA)

Any of these factors can be specified in terms of mean values. For example, $\overline{\text{MLH}_c}$ is the mean corrective maintenance labor-hours, expressed as

$$\overline{\text{MLH}_c} = \frac{(\Sigma\, \lambda_i)(\text{MLH}_i)}{\Sigma\, \lambda_i} \tag{13.12}$$

where λ_i is the failure rate of the ith item (failures/hour), and MLH_i is the average maintenance labor-hours necessary to complete repair of the ith item.

Additionally, the values for mean preventive maintenance labor-hours and mean total maintenance labor-hours (to include preventive and corrective maintenance) can be calculated on a similar basis. These values can be predicted for each level of maintenance and are employed in determining specific support requirements and associated cost.[4]

Maintenance Frequency Factors

Section 12.2 covers the measures of reliability, with MTBF and λ being key factors. Based on the discussion thus far, it is obvious that reliability and maintainability are very closely related. The reliability factors, MTBF and λ, are the basis for determining the frequency of corrective maintenance. Maintainability deals with the characteristics in system design pertaining to minimizing the corrective maintenance requirements for the system when it assumes operational status. Thus, in this area, reliability and maintainability requirements for a given system must be compatible and mutually supportive.

In addition to the corrective maintenance aspect of system support, maintainability also deals with the characteristics of design which minimize (if not eliminate) preventive maintenance requirements. Sometimes, preventive maintenance require-

[4] Note that labor-hours can be expended in the accomplishment of some preventive maintenance actions without causing any system downtime. Accordingly, it may be possible to accomplish certain scheduled maintenance activities on the system while it is in an operational mode.

ments are added with the objective of improving system reliability (e.g., reducing failures by specifying selected component replacements at designated times). However, the introduction of preventive maintenance can turn out to be costly if not carefully controlled. Further, the accomplishment of too much preventive maintenance (particularly for complex systems or products) often has a degrading effect on system reliability as failures are frequently induced in the process. Hence, an objective of maintainability is to provide the proper balance between corrective maintenance and preventive maintenance at least overall cost.

Mean time between maintenance (MTBM). As described in Section 12.2, MTBM is the mean or average time between all maintenance actions (corrective and preventive) and can be calculated as

$$\text{MTBM} = \frac{1}{1/\text{MTBM}_u + 1/\text{MTBM}_s} \tag{13.13}$$

where MTBM_u is the mean interval of unscheduled (corrective) maintenance and MTBM_s is the mean interval of scheduled (preventive) maintenance. The reciprocals of MTBM_u and MTBM_s constitute the maintenance rates in terms of maintenance actions per hour of system operation. MTBM_u should approximate MTBF, assuming that a combined failure rate is used that includes the consideration of primary inherent failures, dependent failures, manufacturing defects, operator and maintenance induced failures, and so on (refer to Table 12.1). The maintenance frequency factor, MTBM, is a major parameter in determining system achieved availability, Equation 12.17, and operational availability, Equation 12.18, as described in Section 12.2.

Mean time between replacement (MTBR). MTBR, a factor of MTBM, refers to the mean time between item replacement and is a major parameter in determining spare part requirements. On many occasions, corrective and preventive maintenance actions are accomplished without generating the requirement for the replacement off a component part (refer to Figure 12.11). In other instances, item replacements are required, which, in turn, necessitates the availability of a spare part and an inventory requirement. Additionally, higher levels of maintenance support (i.e., intermediate and depot levels) may also be required.

In essence, MTBR is a significant factor, applicable in both corrective and preventive maintenance activities involving item replacement, and is a key parameter in determining logistic support requirements. A maintainability objective in system design is to maximize MTBR (or minimize the number of component replacements where possible).

Maintenance Cost Factors

For many systems/products, maintenance cost constitutes a major segment of total life-cycle cost. Further, experience has indicated that maintenance costs are significantly affected by design decisions made throughout the early stages of system development. Thus, it is essential that total life-cycle cost be considered as a major design parameter beginning with the definition of system requirements (refer to Section 2.6, Figure 2.11).

Of particular interest in this chapter is the aspect of *economy* in the performance of maintenance actions. Maintainability is directly concerned with the characteristics of system design that will ultimately result in the accomplishment of maintenance at minimum overall cost.

When considering maintenance cost, the following cost-related indices may be appropriate as criteria in system design:

1. Cost per maintenance action ($/month)
2. Maintenance cost per system operating hour ($/OH)
3. Maintenance cost per month ($/month)
4. Maintenance cost per mission or mission segment ($/mission)
5. The ratio of maintenance cost to total life-cycle cost

The various aspects of maintenance and related costs are discussed in detail in Chapter 17.

Related Maintenance Factors

It is evident from the system maintenance concept in Section 3.4 that there are several additional factors that are closely related to and highly dependent on the maintainability measures described. These include various logistics factors, such as the following:

1. Supply responsiveness or the probability of having a spare part available when needed, spare part demand rates, supply lead times for given items, levels of inventory, and so on
2. Test and support equipment effectiveness (reliability and availability of test equipment), test equipment use, system test thoroughness, and so on
3. Maintenance facility availability and use
4. Transportation modes, times between maintenance facilities, and frequency
5. Maintenance organizational effectiveness and personnel efficiency
6. Data and information processing capacity, time, and frequency

There are numerous other logistics factors that should also be specified, measured, and controlled if the ultimate mission of the system is to be fulfilled. For instance, specifying a 15-minute $\overline{\text{M}}\text{ct}$ requirement for a prime element of the system is highly questionable if there is a low probability of having a spare part available when required (resulting in a long logistics delay-time possibility); specifying specific maintenance labor-hour requirements may not be appropriate if the maintenance organization is not properly staffed or available to perform the required function(s); specifying system test-time requirements may be inappropriate if the predicted reliability of the test equipment is less than the reliability of the item being tested; and so on.

There are many examples where the interactions between the prime system and its elements of support are critical, and both areas must be considered in the establishment of system requirements during conceptual design. Maintainability, as a characteristic in design, is closely related to the area of system support since the results of maintainability directly affect maintenance requirements. Thus, when specifying maintainability factors, one should also address the qualitative and quantitative requirements for system support to determine the effects of one area on another.

Referring to Figure 13.8, there is an up-down relationship. To comply with the specified maintenance time requirements, the appropriate support infrastructure must be in place. This, in turn, may have a feedback effect on overall system effectiveness and cost. Additionally, if the support infrastructure is not ideal, this may require some additional modifications in the design of the prime elements of the system (e.g., the incorporation of additional redundancy in design) or some added redundancies in the support structure (e.g., additional spares, personnel, test equipment). The results could turn out to be costly. Logistic support factors are discussed further in Chapter 15.

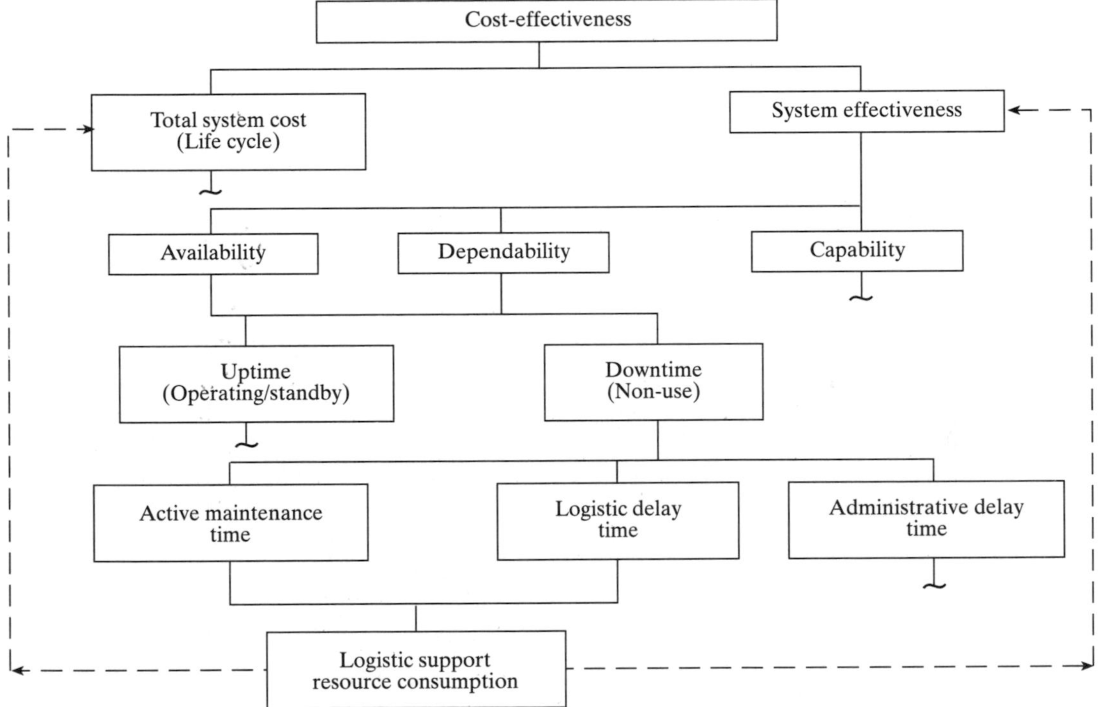

Figure 13.8 The relationship between maintenance downtime and logistics factors (refer to Figure 2.9).

13.3 MAINTAINABILITY IN THE SYSTEM LIFE CYCLE

Maintainability, like reliability, must be an inherent consideration within the overall systems engineering process beginning during the conceptual design phase. Referring to Figure 13.9 (which is an extension of Figure 2.2), qualitative and quantitative maintainability requirements are developed through the accomplishment of feasibility

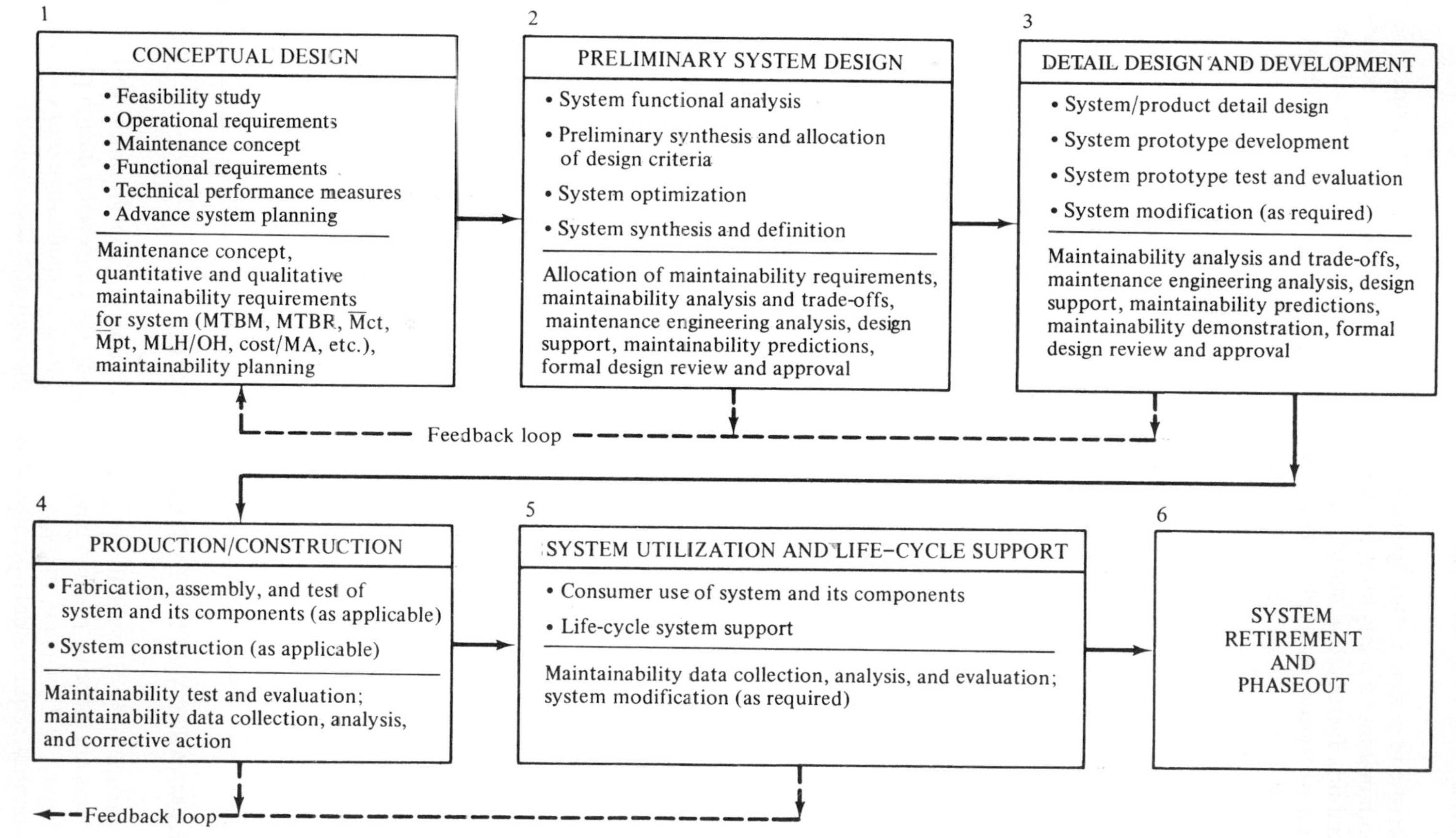

Figure 13.9 Maintainability requirements in the system life cycle (refer to Figure 2.2).

analysis (Section 3.2), the development of operational requirements and the maintenance concept (Section 3.4), and the identification and prioritization of TPMs (Section 3.5). The applicable measures of maintainability must be established, their importance with respect to other system metrics must be delineated, and the requirements for design must be identified through the development of the appropriate DDPs (refer to Figure 2.9).

Given the specification of maintainability requirements in Block 1 of the figure (also included in the type A System Specification—refer to Section 3.8), the next step is to transition from the functional analysis described in Sections 3.6 and 4.1 (i.e., the identification of maintenance functions in particular) and down to the subsystem level and below as part of the system allocation process discussed in Section 4.2. Having defined the basic requirements, one evolves through the iterative process of system synthesis, analysis, and evaluation (refer to Figure 2.10). To facilitate this process, there are several different analysis methods/tools that can be effectively used in preliminary and detail design (i.e., Blocks 2 and 3 of Figure 13.9). As design progresses and physical models are developed, the ongoing evaluation process occurs, and maintainability demonstration testing is included as part of the overall system test and evaluation activity described in Chapter 6.

System Requirements

Every system is developed in response to a need, or to fulfill some anticipated function. The effectiveness with which the system fulfills this function is the ultimate measure of its utility and its value to the customer. This effectiveness is a composite of performance, maintainability, and other factors. In essence, maintainability constitutes a major factor in determining the usefulness of the system.

Maintainability requirements for a system, specified in both quantitative and qualitative terms, are defined as part of the overall system operational requirements and the maintenance concept described in Section 3.4. Of particular interest is the following:

1. Definition of system performance factors, the mission profile, and system use requirements (use conditions, duty cycles, and how the system is to be operated).
2. Definition of the operational life cycle (i.e., the anticipated time that the system will be in the inventory and in operational use).
3. Definition of the basic system maintenance and support concept (i.e., the anticipated levels of maintenance, maintenance responsibilities, major functions at each level, and the prime elements of logistic support at each level—type of spares, test equipment, personnel skills, facilities, etc.).
4. Definition of the environment in which the system is expected to operate and be maintained (e.g., temperature, humidity, vibration, etc). This should include a range of values as applicable and should cover all transportation, handling, and storage modes.

Given the foregoing information, one must determine the extent to which maintainability characteristics should be incorporated in the system design. If the operational requirements specify that the system must function 90% of the time and the estimated reliability is low, then the system maintainability requirements may be rather stringent in order to maintain the overall 90% availability. Conversely, if the estimated reliability is high (resulting in few failures), the specified maintainability requirements may be different. Further, if the maintenance concept dictates that only the organization and depot levels of maintenance are allowed, then the maintainability requirements may be different from what would likely be specified with three levels of maintenance planned (to include intermediate maintenance). In any event, quantitative and qualitative maintainability requirements must be defined for the system based on the foregoing considerations. Quantitative requirements are usually expressed in terms of MTBM, MTBR, $\overline{\mathrm{M}}$, $\overline{\mathrm{M}}$ct, $\overline{\mathrm{M}}$pt, MLH/OH, \$/MA, or a combination thereof.

In specifying maintainability, one must first identify the major system requirements from the top-level functional flow diagram (as illustrated in Figures 3.12 and 4.4). The functions in the top-level diagram are expanded as necessary for the purposes of system definition, and a functional packaging scheme is identified as illustrated in Figure 4.6. Maintenance functional diagrams are developed as shown in Figure 4.4, significant maintenance functions are defined, and maintainability requirements are established for the appropriate maintenance functions and the various functional packages of the system. This is an iterative process of requirements definition.[5]

Maintainability Allocation

After requirements for the system have been established, it is then necessary to translate these requirements into lower-level design criteria through maintainability allocation. This is accomplished as part of the system requirements allocation process described in Section 4.2.

For the purpose of illustration, it is assumed that System *XYZ* in Figure 4.8 must be designed to meet an inherent availability requirement of 0.9989, a MTBF of 450 hours, and a MLH/OH (for corrective maintenance) of 0.2, and a need exists to allocate $\overline{\mathrm{M}}$ct and MLH/OH to the assembly level. From Equation 12.16, $\overline{\mathrm{M}}$ct is

$$\overline{\mathrm{M}}\mathrm{ct} = \frac{\mathrm{MTBF}(1 - A_i)}{A_i} \tag{13.14}$$

or

$$\overline{\mathrm{M}}\mathrm{ct} = \frac{450(1 - 0.9989)}{0.9989} = 0.5 \text{ hr}$$

Thus, the system's $\overline{\mathrm{M}}$ct requirement is 0.5 hour, and this requirement must be allocated to Units A, B, and C, and the assemblies within each unit. The allocation process is facilitated through the use of a format similar to that illustrated in Table 13.7.[6]

Referring to Table 13.7, each item type and the quantity (Q) of items per system are indicated. Allocated reliability factors are specified in column 3 and the degree to which the failure rate of each unit contributes to the overall failure rate (represented

[5] Refer to Appendix A.1 for additional coverage of functional analysis and the development of maintenance functional block diagrams.

[6] MTBM and $\overline{\mathrm{M}}$pt may be allocated on a comparable basis with MTBF and $\overline{\mathrm{M}}$ct, respectively.

TABLE 13.7 Maintainability Allocation for System ZYZ

1	2	3	4	5	6	7
ITEM	QUANTITY OF ITEMS PER SYSTEM (Q)	FAILURE RATE (λ) × 1,000 hr	CONTRIBUTION OF TOTAL FAILURES $C_f = (Q)(\lambda)$	PERCENT CONTRIBUTION $C_p = C_f \Sigma\, C_f \times 100$	AVERAGE CORRECTIVE MAINTENANCE TIME $\overline{\text{M}}\text{ct}$ (hr)	CONTRIBUTION OF TOTAL CORRECTIVE MAINTENANCE TIME $C_f = (C_f)(\overline{\text{M}}\text{ct})$
1. Unit A	1	0.246	0.246	11%	0.9	0.221
2. Unit B	1	1.866	1.866	84%	0.4	0.746
3. Unit C	1	0.110	0.110	5%	1.0	0.110
Total			$\Sigma\, C_f = 2.222$	100%		$\Sigma\, C_f = 1.077$

$$\overline{\text{M}}\text{ct for System } XYZ = \frac{\Sigma\, C_t}{\Sigma\, C_f} = \frac{1.077}{2.222} = 0.485 \text{ hr (requirement: 0.5 hr)}$$

by C_f) is entered in column 4. The average corrective maintenance time for each unit is estimated and entered in column 6. These times are ultimately based on the inherent characteristics of equipment design, which are not known at this point in the system life cycle. Thus, corrective maintenance times are initially derived using a complexity factor, which is indicated by failure rate. As a goal, the item that contributes the highest percentage to the anticipated total failures (Unit B in this instance) should require a low $\overline{\text{Mct}}$, and those with low contributions may require a higher $\overline{\text{Mct}}$. On certain occasions, however, the design costs associated with obtaining a low $\overline{\text{Mct}}$ for a complex item may lead to a modified approach, which is feasible as long as the end result ($\overline{\text{Mct}}$ at the system level) falls within the quantitative requirement.[7]

The estimated value of C_t for each unit is entered in column 7, and the sum of the contributions for all units can be used to determine the overall system's $\overline{\text{Mct}}$ as

$$\overline{\text{Mct}} = \frac{\Sigma\, C_t}{\Sigma\, C_f} = \frac{1.077}{2.222} = 0.485 \text{ hr} \tag{13.15}$$

In Table 13.7, the calculated $\overline{\text{Mct}}$ for the system is within the requirement of 0.5 hour. The $\overline{\text{Mct}}$ values for the units provide corrective maintenance downtime criteria for design (i.e., DDPs), and the values are included in the applicable design specifications.

Once allocation is accomplished at the unit level, the resultant $\overline{\text{Mct}}$ values can be allocated to the next-lower equipment indenture item. For instance, the 0.4 hour $\overline{\text{Mct}}$ value for Unit B can be allocated to assemblies 1, 2, and 3, and the procedure for allocation is the same as employed in Equation 13.15. An example of allocated values for the assemblies of Unit B is included in Table 13.8.

The $\overline{\text{Mct}}$ value covers the aspect of *elapsed* or *clock time* for restoration actions. Sometimes this factor, when combined with a reliability requirement, is sufficient to establish the necessary maintainability characteristics in design. On other occasions, specifying $\overline{\text{Mct}}$ by itself is not adequate because there may be several design approaches that will meet the $\overline{\text{Mct}}$ requirement but not necessarily in a cost-effective manner. Meeting a $\overline{\text{Mct}}$ requirement may result in an increase in the skill levels of personnel accom-

TABLE 13.8 Unit B Allocation

1	2	3	4	5	6	7
Assembly 1	1	0.116	0.116	6%	0.5	0.058
Assembly 2	1	1.550	1.550	83%	0.4	0.620
Assembly 3	1	0.200	0.200	11%	0.3	0.060
Total			1.866	100%		0.738
$\overline{\text{Mct}}$ for Unit B $= \frac{\Sigma\, C_t}{\Sigma\, C_f} = \frac{0.738}{1.866} = 0.395$ hr (requirement: 0.4 hr)						

[7] In any event, the maintainability parameters are dependent on the reliability parameters. Also, it will frequently occur that reliability allocations are incompatible with maintainability allocations (or vice versa). Hence, a close feedback relationship between these activities is mandatory.

plishing maintenance actions, increasing the quantity of personnel for given maintenance functions, or incorporating automation for manual operations. In each instance there are costs involved; thus, one may wish to specify additional constraints, such as the skill level of personnel at each maintenance level and the maintenance labor-hours per operating hour (MLH/OH) for significant items. In other words, a requirement may dictate that an item be designed such that it can be repaired within a specified elapsed time with a given quantity of personnel possessing skills of a certain level. This will influence design in terms of accessibility, packaging schemes, handling requirements, diagnostic provisions, and so on, and is perhaps more meaningful overall.

The factor MLH/OH is a function of task complexity and the frequency of maintenance. The system-level requirement is allocated on the basis of system operating-hours, the anticipated quantity of maintenance actions, and an estimate of the number of labor-hours per maintenance action. Experience data are used where possible.

Following the completion of quantitative allocations for each indenture level of equipment, all values are included in the functional breakdown illustrated in Figure 4.8. The figure provides an overview of major system design requirements.

Component Selection and Application

The extent to which maintainability characteristics are built into the design depends not only on the design of individual components themselves, but on the way in which they are mounted or located within the system structure. Specific objectives may include the following:

1. The selection of *standardized* components and materials. The goal is to minimize the overall number and different types of components, reducing the quantity and variety of spares/repair part requirements (and associated inventories) while ensuring dependable sources of supply throughout the system life cycle.
2. For repairable items, select those that incorporate built-in *self-test features* and the level and depth of *diagnostics* that will facilitate completing the maintenance cycle in minimum time and with a high degree of confidence. Referring to Figure 13.1, the most time-consuming steps are often in the areas of localization and diagnostics, particularly for electronic equipment.
3. Select items that can be repaired using common and standard tools and test equipment, in readily available facilities, and without requiring high-skilled personnel with extensive training.
4. Ensure that the appropriate degree of *accessibility* is provided that will allow one to identify rapidly the item requiring removal (or gain access to the area requiring repair), enable the quick removal and replacement of the faulty item with a spare, and accomplish the follow-on system checkout to ensure a proper operating state with a minimum of alignment and adjustment (without requiring calibration). Generally, the items that are expected to require the most maintenance should be the most accessible (i.e., high failure-rate items or critical components).
5. Incorporate a modularized functional-packaging approach where, in the event of failure, a faulty item can be removed and replaced quickly, without requiring the

removal and replacement of other items in the process. The interdependency of items on each other should be minimize. Additionally, the faulty items should easily be removable, of a plug-in variety, and with quick-release fasteners. Complete *physical* and *functional interchangeability* is desired.

6. Avoid the selection of short-life components (i.e., critical-useful life items discussed in Section 12.4) and the requirement for preventive maintenance. The goal is to eliminate all requirements for scheduled or preventive maintenance unless justified on the basis of reliability information. This can be facilitated through the accomplishment of a reliability-centered maintenance (RCM) analysis that is described in Section 13.4.
7. Incorporate the proper amount of labeling and the identification of components in repairable items to aid the technician in completing his or her tasks in an effective and efficient manner.

The above areas represent just a few of the concerns in designing for maintainability. The overall objective is to minimize maintenance frequencies, times, and the resouces required in the performance of maintenance tasks.

13.4 MAINTAINABILITY ANALYSIS METHODS

Within the context of maintainability analysis (accomplished as part of the iterative process of system synthesis, analysis, and evaluation), there are several tools that can be effectively used in support of the objectives described throughout this text. Of particular interest here are the trade-offs between reliability and maintainability, reliability-centered maintenance (RCM), repair versus discard analysis (or level of repair analysis), maintainability prediction, and the maintenance task analysis. These are discussed subsequently.

Reliability-Maintainability Trade-Off Evaluation

Suppose that there is a requirement to replace an existing equipment item with a new item for the purpose of improving operational effectiveness. The current need specifies that the equipment must operate 8 hours per day 360 days per year, for 10 years. The existing equipment meets an availability of 0.961, a MTBF of 125 hours, and a $\overline{\text{M}}\text{ct}$ of 5 hours. The new system must meet an availability of 0.990, a MTBF greater than 300 hours and a $\overline{\text{M}}\text{ct}$ not to exceed 5.0 hours. An anticipated quantity of 200 items of equipment is to be procured. Three different alternative design configurations are being considered to satisfy the requirement, and each configuration constitutes a modification of the existing equipment.

Figure 13.10 graphically shows the relationships among inherent availability (A_i), MTBF, and $\overline{\text{M}}\text{ct}$ and illustrates the allowable area for trade-off. The selected configuration must reflect the reliability and maintainability characteristics represented by the shaded area. The existing design is not compatible with the new requirement. Three alternative design configurations are being considered. Each configuration meets the

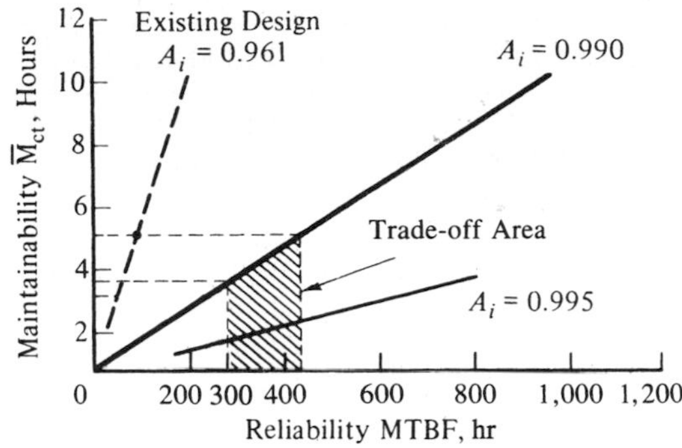

Conf.	A_i	MTBF	$\overline{M}$ct
Existing	0.961	125	5.0
Alt. A	0.991	450	4.0
Alt. B	0.990	375	3.5
Alt. C	0.991	320	2.8

Other configurations are eligible for consideration as long as the effectiveness parameters fall within the trade-off areas

Figure 13.10 Reliability-maintainability trade-off.

availability requirement, with configuration A having the highest estimated reliability MTBF and configuration C reflecting the best maintainability characteristics with the lowest $\overline{M}$ct value. The objective is to select the best of the three configurations on the basis of cost.

When considering cost, there are costs associated with research and development (R&D) activity, investment or manufacturing costs, and operation and maintenance (O&M) costs.[8] For instance, improving reliability or maintainability characteristics in design will result in an increase in R&D and investment (manufacturing) cost. In addition, experience has indicated that such improvements will result in lower O&M cost, particularly in the areas of maintenance personnel and support cost and the cost of spare/repair parts. Thus, initially the analyst may look at only these categories. If the final decision is close, it may be appropriate to investigate other categories. A summary of partial cost data is presented in Table 13.9.

As shown in Table 13.9, the delta costs associated with the three alternative equipment configurations are included for R&D and investment. Maintenance personnel and support costs, included as part of the O&M cost, are based on estimated operating time for the 200 items of equipment throughout the required 10-year period of use (i.e., 200 items of equipment operating 8 hours per day, 360 days per year, for 10 years) and the reliability MTBF factor. Assuming that the average cost per maintenance action is $100, maintenance personnel and support costs are determined by multiplying this factor by the estimated quantity of maintenance actions, which is determined from total operating time divided by the MTBF value. Based on the delta values presented in Table 13.9, configuration A satisfies the system availability, reliability, and maintainability requirements with the least cost.

[8] Only those *delta costs* considered significant for this evaluation are included here.

TABLE 13.9 Cost Summary (Partial Costs)

CATEGORY	CONF. *A*	CONF. *B*	CONF. *C*	REMARKS
R and D cost Reliability design Maintainability design	 $17,120 2,109	 $15,227 4,898	 $12,110 7,115	High reliability parts, packaging, accessibility
Investment cost manufacturing (200 Systems)	$3,422,400	$3,258,400	$3,022,200	$17,112/Equipment A; $16,292/Equipment B; $15,111/Equipment C
O&M cost Maintenance personnel and support Spare and repair parts	 $1,280,000 342,240	 $1,536,000 325,840	 $1,800,000 302,220	12,800 Maint. Action/Equipment A; 15,360 Maint. Action/Equipment B; 18,800 Maint. Action/Equipment C; 10% of manufacturing cost for spares
Total	$5,063,869	$5,140,365	$5,143,645	

Reliability-Centered Maintenance (RCM)

Reliability-centered maintenance (RCM) is a systematic approach to develop a focused, effective, and cost-efficient preventive maintenance program and control plan for a system or product. This technique is best initiated during the early design process and evolves as the system is developed, produced, and deployed. However, this technique can also be used to evaluate preventive maintenance programs for existing systems with the objective of continuous product/process improvement in mind.[9]

The RCM technique was developed in the 1960s, primarily through the efforts of the commercial airline industry.[10] It can be accomplished using a structured decision tree which leads the analyst through a "tailored" logic approach to delineate the most

[9] Quite often during system development, the designer, when selecting components, will not be knowledgeable of the specific reliability physics-of-failure characteristics of an item. So, to ensure the proper level of reliability for the system, the designer will select a component and indicate the need for preventive maintenance (e.g., a periodic item replacement) "just in case!" This has, for many systems in the past, resulted in the over-specification of maintenance requirements (i.e., the specification of too much preventive maintenance, which is unjustified from a reliability perspective and can be costly).

[10] A maintenance steering group (MSG) was formed in the 1960s which undertook the development of this technique. The result was a document called "747 Maintenance Steering Group Handbook: Maintenance Evaluation and Program Development (MSG-1)" published in 1968. This effort, focused toward a particular aircraft, was next generalized and published in 1970 as "Airline/Manufacturer Maintenance Program Planning Document-MSG2." The MSG-2 approach was further developed and published in 1978 as "Reliability Centered Maintenance," Report Number A066-579, prepared by United Airlines, and in 1980 as "Airline/Manufacturer Maintenance Program Planning Document-MSG3." The MSG-3 report has been revised and is currently available as "Airline/Manufacturer Maintenance Program Development Document (MSG-3), 1993." These reports are available from the Air Transport Association.

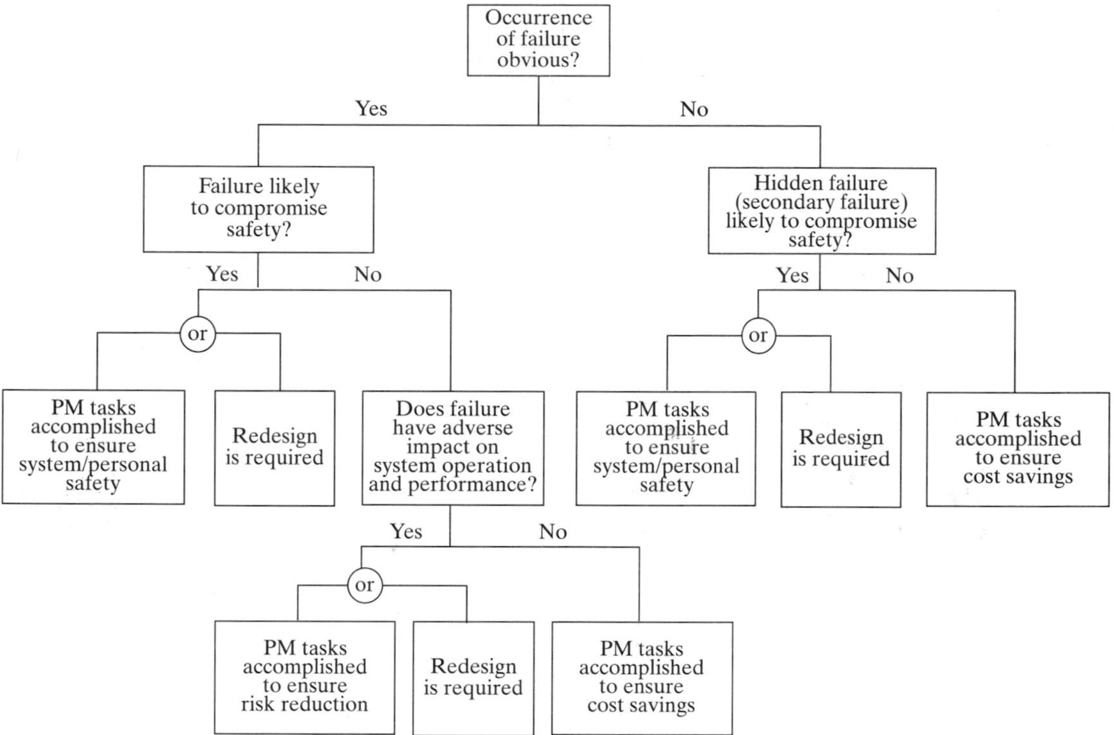

Figure 13.11 Simplified RCM decision logic.

applicable preventive maintenance tasks (their nature and frequency). Referring to Figure 13.11, a simplified top-down RCM decision-logic structure is presented, where system safety is a prime consideration along with performance and cost. By following the steps, one is led to either the specification of preventive maintenance or a recommendation for redesign.

With the RCM analysis directed toward the establishment of a cost-effective preventive maintenance program, a necessary prerequisite is the accomplishment of the FMECA described in Section 12.4. Referring the Figure 12.27, one of the outputs from the FMECA could lead to the identification of a preventive maintenance requirement—one that can be justified on the basis of reliability information. Figure 13.12 shows how the RCM analysis model can be integrated with some of the other tools discussed herein.[11]

[11] RCM decision logics, with some variations, have also been included in: (a) MIL-STD-2173(AS)—*Reliability-Centered Maintenance Requirements for Naval Aircraft, Weapons Systems, and Support Equipment;* (b) AMC-P-750-2—*Guide to Reliability-Centered Maintenance;* and (c) *J. Moubray, Reliability-Centered Maintenance,* (London: Butterworth-Heinemann, 1991).

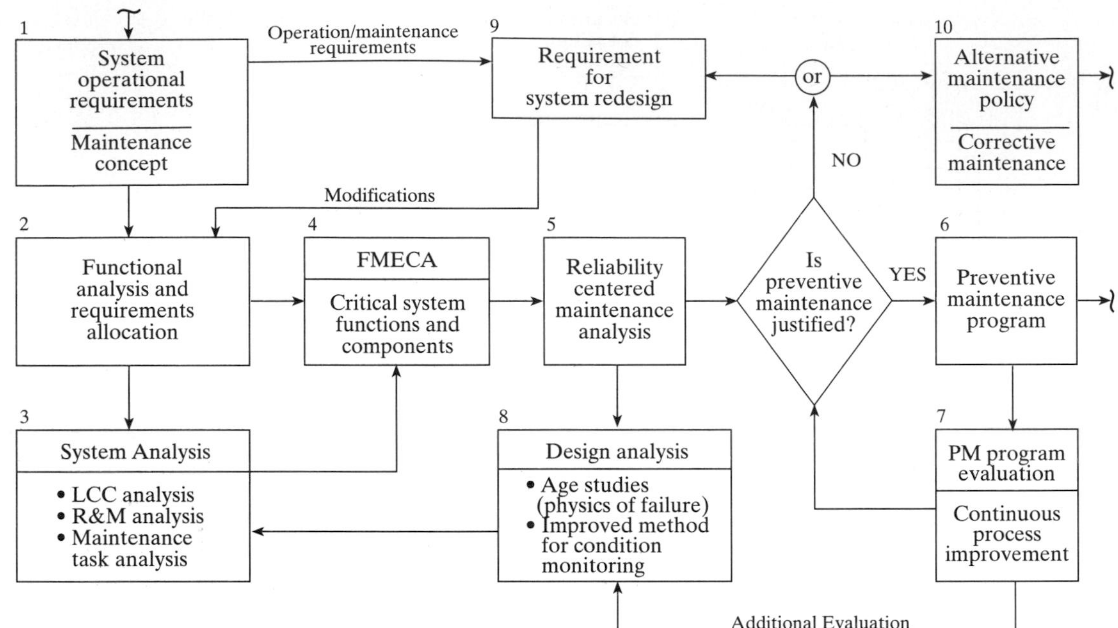

Figure 13.12 The RCM analysis model and its interfaces.

Repair versus Discard Analysis

In expanding the maintenance concept to establish criteria for system design, it is necessary to determine whether it is economically feasible to repair certain assemblies or to discard them when failures occur. If the decision is to accomplish repair, it is appropriate to determine the maintenance level at which the repair should be accomplished (i.e., intermediate maintenance or supplier/depot maintenance). This example illustrates a level-of-repair analysis based on life-cycle-cost criteria, which may be accomplished during conceptual or preliminary system design.

Suppose that a computer system is to be distributed in quantities of 65 throughout three major geographical areas. The system will be used to support both scientific and management functions within various industrial firms and government agencies. Although the actual system use will vary from one consumer organization to the next, an average use of 4 hours per day (for a 360-day year) is assumed.

The computer system is currently in the early development stage, should be in production in 18 months, and will be operational in 2 years. The full complement of 65 computer systems is expected to be in use in 4 years and will be available through the 8 year of the program before system phaseout commences. The system life cycle, for the purposes of the analysis, is 10 years.

Based on early design data, the computer system will be packaged in major units with a built-in test capability that will enable the isolation of faults to the unit level. Faulty units will be removed and replaced at the organizational level (i.e., customer's facility), and sent to the intermediate maintenance shop for repair. Unit repair will be accomplished through assembly replacement, and assemblies will be either repaired or discarded. A total of 15 assemblies are being considered, and the requirement is to justify the assembly repair or discard decision on the basis of life-cycle cost criteria. The operational requirements, maintenance concept, and program plan are illustrated in Figure 13.13.

The stated problem primarily pertains to the analysis of 15 major assemblies of the given computer system configuration to determine whether the assemblies should

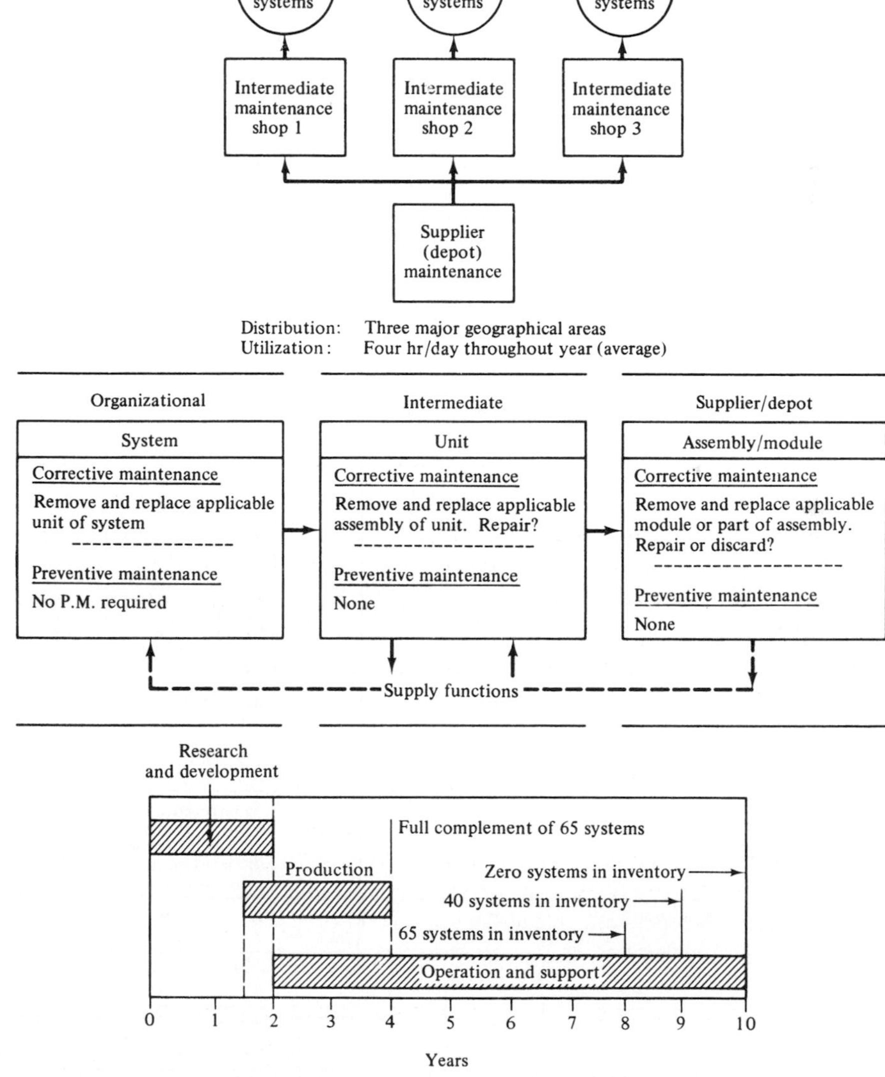

Figure 13.13 Basic system concepts.

be repaired or discarded when failures occur. In other words, the various assemblies will be individually evaluated in terms of (1) assembly repair at the intermediate level of maintenance, (2) assembly repair at the supplier or depot level of maintenance, and (3) disposing of the assembly. Life-cycle costs, as applicable to the assembly level, will be

Evaluation Criteria	*Repair at Intermediate Cost* ($)	*Repair at Supplier Cost* ($)	*Discard at Failure Cost* ($)	*Description and Justification*
1. Estimated acquisition costs for assembly A-1 (to include R and D cost and production cost)	550/Assy. or 35,750	550/Assy. or 35,750	475/Assy. or 30,875	Acquisition cost includes all applicable costs allocated to assembly A-1 based on a requirement of 65 systems. Assembly design and production are simplified in the discard area.
2. Unscheduled maintenance costs	6,480	8,100	Not applicable	Based on the 8-year useful system life, 65 systems, a utilization of 4 hrs/day, a failure rate (λ) of 0.00045 for assembly A-1, and a $\overline{Mct}$ of 2 hr, the expected number of maintenance actions is 270. When repair is accomplished, two technicians are required on a full-time basis. The labor rates are $12/hr for intermediate maintenance and $15/hr for supplier maintenance.
3. Supply support spare assemblies	3,300	4,950	128,250	For intermediate maintenance 6 spare assemblies are required to compensate for transportation time, the maintenance queue, TAT, etc. For supplier/depot maintenance 9 spare assemblies are required. 100% spares are required in the discard case.
4. Supply support spare modules or parts for assembly repair	6,750	6,750	Not applicable	Assume $25 for materials per repair action.
5. Supply support inventory management	2,010	2,340	25,650	Assume 20% of the inventory value (spare assemblies, modules, and parts).
6. Test and support equipment	5,001	1,667	Not applicable	Special test equipment is required in the repair case. The acquisition and support cost is $25,000 per installation. The allocation for assembly A-1 per installation is $1,667. No special test equipment is required in the discard case.
7. Transportation and handling	Not applicable	2,975	Not applicable	Transportation costs at the intermediate level are negligible. For supplier maintenance, assume 340 one-way trips at $175/100 lb One assembly weighs 5 pounds.
8. Maintenance training	260	90	Not applicable	Delta training cost to cover maintenance of the assembly is based on the following: Intermediate – 26 students, 2 hrs each, $200/student week. Supplier – 9 students, 2 hrs each, $200/student week.
9. Maintenance facilities	594	810	Not applicable	From experience, a cost estimating relationship of $0.55 per direct maintenance laborhour is assumed for the intermediate level, and $0.75 is assumed for the supplier level.
10. Technical data	1,250	1,250	Not applicable	Assume 5 pages for diagrams and text covering assembly repair at $250/page.
11. Disposal	270	270	2,700	Assume $10/assembly and $1/module or part as the cost of disposal
Total estimated cost	61,665	64,952	187,475	

Figure 13.14 Repair versus discard evaluation (Assembly A-1).

estimated and employed in the alternative selection process. Total overall computer system costs have been determined at a higher level, and are not included in this example.

Given the information in the problem statement, the next step is to develop a cost breakdown structure (CBS) and to establish evaluation criteria. The evaluation criteria include consideration of all costs in each applicable category of the CBS, but the emphasis is on operation and support (O&S) costs as a function of acquisition cost. Thus, the research and development cost and the production cost are presented as one element, whereas various segments of O&S costs are identified individually. Figure 13.14 presents evaluation criteria, cost data, and a brief description and justification supporting each category. The information shown in the figure covers only one of the 15 assemblies but is typical for each case.

The data presented in Figure 13.14 represent assembly A-1, used in a manner illustrated in Figure 13.13. The next step is to employ the same criteria to determine the recommended repair-level decision for each of the other 14 assemblies (i.e., assemblies 2 to 15). Although acquisition costs, reliability and maintainability factors, and certain logistics requirements are different for each assembly, many of the cost-estimating relationships are the same. The objective is to be *consistent* in analysis approach and in the use of input cost factors to the maximum extent possible and where appropriate. The summary results for all 15 assemblies are presented in Table 13.10.

Referring to Table 13.10, note that the decision for assembly A- 1 favors repair at the intermediate level; the decision for assembly A-2 is repair at the supplier or depot level; the decision for assembly A-3 is not to accomplish repair at all but to discard the assembly when a failure occurs; and so on. The table reflects recommended

TABLE 13.10 Summary of Repair-Level Costs

	Maintenance Status			
Assembly number	Repair at intermediate cost ($)	Repair at supplier cost ($)	Discard at failure cost ($)	Decision
A-1	61,665	66.702	187,475	Repair—intermediate
A-2	58,149	51,341	122,611	Repair—supplier
A-3	85,115	81,544	73,932	Discard
A-4	85,778	78,972	65,071	Discard
A-5	66,679	61,724	95,108	Repair—supplier
A-6	65,101	72,988	89,216	Repair—intermediate
A-7	72.223	75,591	92,114	Repair—intermediate
A-8	89.348	78,204	76,222	Discard
A-9	78,762	71,444	89,875	Repair—supplier
A-10	63,915	67,805	97,212	Repair—intermediate
A-11	67,001	66,158	64,229	Discard
A-12	69,212	71,575	82,109	Repair—intermediate
A-13	77,101	65,555	83,219	Repair—supplier
A-14	59,299	62,515	62,005	Repair—intermediate
A-15	71,919	65,244	63,050	Discard
Policy cost	1,071,267	1,037,362	1,343,449	Repair—supplier

policies for each individual assembly. In addition, the overall policy decision, when addressing all 15 assemblies as an integral package, favors repair at the supplier.

Before arriving at a final conclusion, the analyst should reevaluate each situation where the decision is close. Referring to Figure 13.14, it is clearly uneconomical to accept the discard decision; however, the two repair alternatives are relatively close. Based on the results of the various individual analyses, the analyst knows that repair-level decisions are highly dependent on the unit acquisition cost of each assembly and the total estimated number of replacements over the expected life cycle (i.e., maintenance actions based on assembly reliability). The trends are illustrated in Figure 13.15, where the decision tends to shift from discard to repair at the intermediate level as the unit acquisition cost increases and the number of replacements increases (or the reliability decreases).[12]

In instances where the individual analysis result lies close to the crossover lines in Figure 13.15, the analyst may wish to review the input data, the assumptions, and accomplish a sensitivity analysis involving the high-cost contributors. The purpose is to assess the risk involved and verify the decision. This is the situation for assembly A-l, where the decision is close relative to repair at the intermediate level versus repair at the supplier's facility (refer to Figure 13.14).

After reviewing the individual analyses of the 15 assemblies to ensure that the best possible decision is reached, the results in Table 13.10 are updated as required. Assuming that the decisions remain basically as indicated, the analyst may proceed in either of two ways. First, the decisions in the table may be accepted without change, supporting a *mixed* policy with some assemblies being repaired at each level of maintenance and other assemblies being discarded at failure. With this approach, the ana-

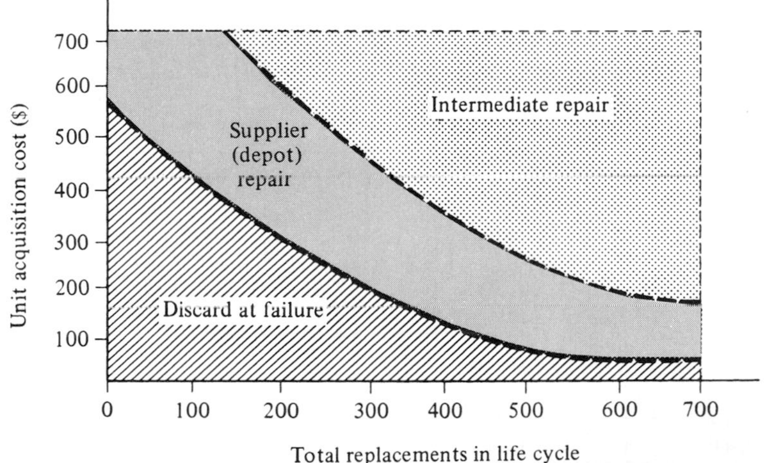

Figure 13.15 Economic screening criteria.

[12] The curves projected in Figure 13.15 are characteristic for this particular analysis and will vary with changes in operational requirements, system use, the maintenance concept, production requirements, and so on.

lyst should review the interaction effects that could occur (i.e., the effects on spares, use of test and support equipment, maintenance personnel use, etc.). In essence, each assembly is evaluated individually based on certain assumptions, the results are reviewed in the context of the whole, and possible feedback effects are assessed to ensure that there is no significant impact on the decision.

A second approach is to select the overall least-cost policy for all 15 assemblies treated as an entity (i.e., assembly repair at the supplier or depot level of maintenance). In this case, all assemblies are designated as being repaired at the supplier's facility, and each individual analysis is reviewed in terms of the criteria in Figure 13.15 to determine the possible interaction effects associated with the single policy. The result may indicate some changes to the values in Table 13.10.

Finally, the output of the repair-level analysis must be reviewed to ensure compatibility with the initially specified system maintenance concept. The analysis data may either directly support and be an expansion of the maintenance concept, or the maintenance concept will require change as a consequence of the analysis. If the latter occurs, other facets of system design may be significantly impacted. The consequences of such maintenance concept changes must be thoroughly evaluated before arriving at a final repair-level decision. The basic level of repair analysis procedure is illustrated in Figure 13.16.

Maintainability Prediction

Maintainability prediction involves an early assessment of the maintainability characteristics in system design and is accomplished periodically at different stages in the design process. Through the review of design data, predictions of the MTBM, $\overline{\text{M}}\text{ct}$, $\overline{\text{M}}\text{pt}$, MLH/OH, and so on, are made and are compared with the initially specified requirements identified in the maintainability allocation process. Areas of noncompliance are evaluated for possible design improvement.

Maintainability prediction, in the broadest sense, includes the early quantitative estimation of maintenance elapsed-time factors, maintenance labor-hour factors, maintenance frequency factors, and maintenance cost factors. The accomplishment of this requires that the engineer or analyst review design data, layouts, component-part lists, reliability factors, and supporting data with the intent of identifying anticipated maintenance tasks and the resources required for task completion. Maintainability prediction not only requires the estimation of task times and frequencies, but requires a qualitative assessment of the design characteristics for supportability.

Prediction of $\overline{\textbf{M}}\textbf{ct}$. Prediction of mean corrective maintenance time ($\overline{\text{M}}\text{ct}$) may be accomplished using a system element breakdown, as illustrated in Figure 4.8, and determining maintenance tasks and the associated elapsed times in progressing from one element to another. The breakdown covers subsystems, units, assemblies, subassemblies, and parts. Maintainability characteristics such as localization, isolation, accessibility, repair, and checkout as incorporated in the design are evaluated and identified with one of the functional levels and down to each component part. Times applicable to each part (assuming that every part will fail at some point) are combined to

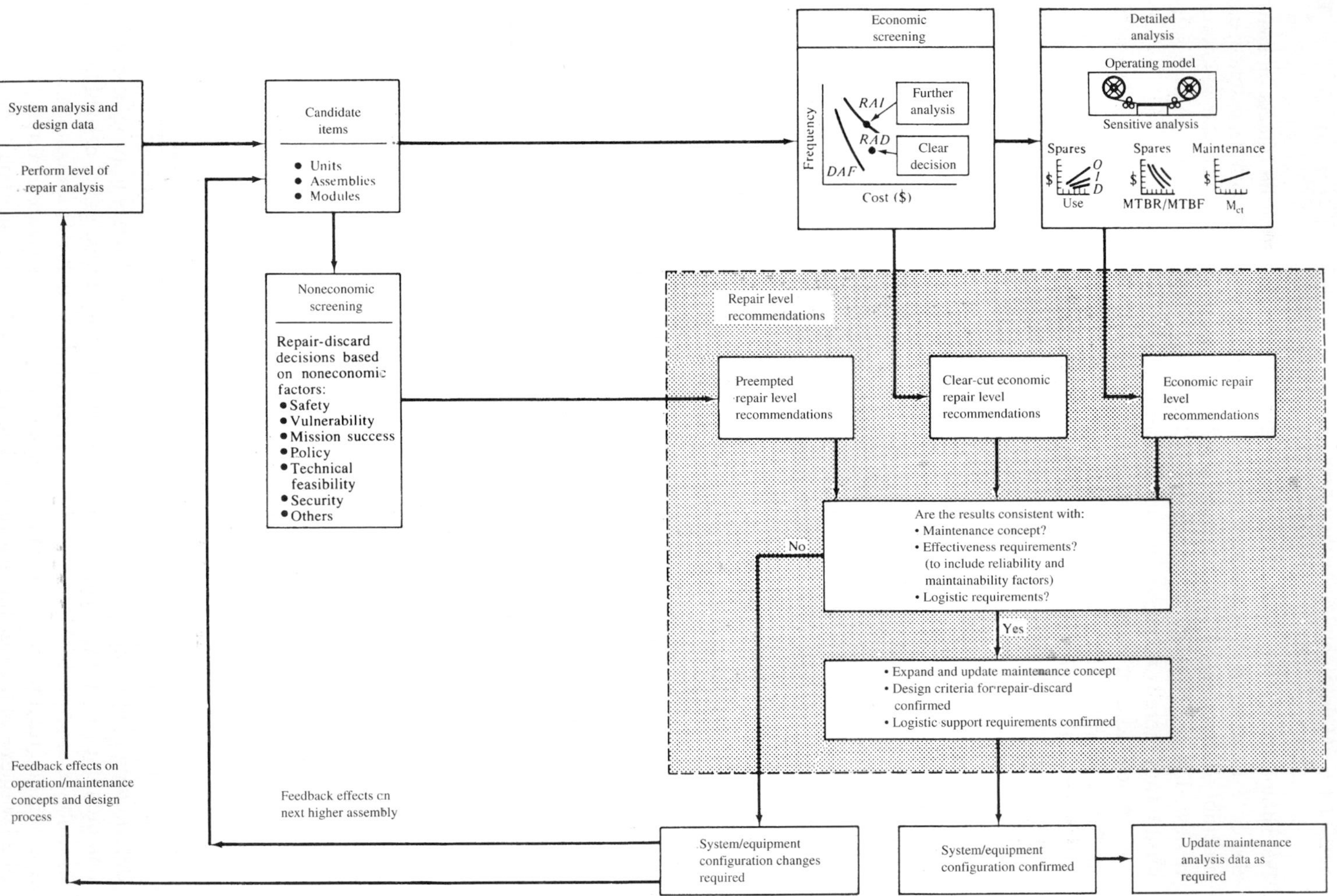

Figure 13.16 Level-of-repair analysis procedure (refer to Figure 2.2).

TABLE 13.11 Maintainability Prediction Worksheet (Assembly 4)

Part category	λ	N	$(N)(\lambda)$	Maintenance Times (Hr)							$(N)(\lambda)(Mct_i)$
				Loc	Iso	Acc	Ali	Che	Int	Mct_i	
Part A	0.161	2	0.322	0.08	0.08	0.14	0.01	0.01	0.11	0.370	0.119
Part B	0.102	4	0.408	0.01	0.05	0.12	0.01	0.02	0.12	0.330	0.134
Part C	0.021	5	0.105	0.03	0.04	0.11	—	0.01	0.14	0.330	0.034
Part D	0.084	1	0.084	0.01	0.03	0.10	0.02	0.03	0.11	0.300	0.025
Part E	0.452	9	1.060	0.02	0.04	0.13	0.02	0.03	0.08	0.390	1.299
Part F	0.191	8	1.520	0.01	0.02	0.11	0.01	0.02	0.07	0.240	0.364
Part G	0.022	7	0.154	0.02	0.05	0.15	—	0.05	0.15	0.420	0.064
	Total		6.653							Total	2.039

N = Quantity of parts
λ = Failure rate
Loc = Localization
Iso = Isolation
Acc = Access
Ali = Alignment
Che = Checkout
Int = Interchange
Mct_i = Maintenance cycle time

For determination of $\overline{MLH}_c$ enter labor hours for maintenance times.

provide factors for the next-higher level. A sample data format for an assembly is presented in Table 13.11.

Referring to Table 13.11, the frequency factors are represented by the failure rate (λ) for each part determined from reliability prediction data and the elapsed times required for fault localization, isolation, disassembly to gain access, and so on, are noted. Maintenance task times are usually estimated from experience and data obtained on similar systems already in use. The summation of the various individual times constitutes a maintenance cycle time (i.e., Mct_i) as illustrated in Figure 13.1.

Similar data prepared on each assembly in the system are combined as illustrated in Table 13.12, and the factors are computed to arrive at the predicted $\overline{\mathrm{Mct}}$.

TABLE 13.12 Maintainability Prediction Data Summary

Work sheet No.	Item designation	WORK SHEET FACTOR	
		$\Sigma\,(N)(\lambda)$	$\Sigma\,(N)(\lambda)(\mathrm{Mct}_i)$
1	Assembly 1	7.776	3.021
2	Assembly 2	5.328	1.928
3	Assembly 3	8.411	2.891
4	Assembly 4	6.653	2.039
5	Assembly 5	5.112	2.576
⋮	⋮	⋮	⋮
13	Assembly 13	4.798	3.112
Grand Total		86.476	33.118

$$\overline{\mathrm{Mct}} = \frac{\Sigma\,(N)(\lambda)(\mathrm{Mct}_i)}{\Sigma\,(N)(\lambda)} = \frac{33.118}{86.486} = 0.382 \text{ hr}$$

Prediction of $\overline{\mathbf{Mpt}}$. Prediction of preventive maintenance time may be accomplished using a method similar to the corrective maintenance approach discussed earlier. Preventive maintenance tasks are estimated along with frequency and task times. An example is presented in Table 13.13.

TABLE 13.13 Preventive Maintenance Data Summary

Description of preventive Maintenance task	Task frequency $(\mathrm{fpt}_i)(N)$	Task time (Mpt_i)	Product $(\mathrm{fpt}_i)(N)(\mathrm{Mpt}_i)$
1. Lubricate	0.115	5.511	0.060
2. Calibrate	0.542	4.234	0.220
⋮	⋮	⋮	⋮
31. Service	0.321	3.315	0.106
Grand total	13.260		31.115

$$\overline{\mathrm{Mpt}} = \frac{\Sigma\,(\mathrm{fpt}_i)(N)(\mathrm{Mpt}_i)}{\Sigma\,(\mathrm{fpt}_i)(N)} = \frac{31.115}{13.260} = 2.346 \text{ hr}$$

Prediction of maintenance resource requirements. Maintenance resources in this instance include the personnel and training requirements, test and support equipment, supply support (e g., spare and repair parts), transportation and handling requirements, facilities, computer software, and data needed in the accomplishment of maintenance actions. For instance, what test equipment is required to accomplish fault isolation? What personnel types and skill level are required to repair the faulty item? What type of handling equipment is necessary to transport an item to the intermediate maintenance shop? In essence, the elapsed-time element alone does not provide an adequate prediction of the maintainability characteristics in design. One needs to predict the time element and the resources required.

Maintenance Task Analysis (MTA)

The maintenance task analysis (MTA) constitutes the process of evaluating a given system configuration to determine the following[13]:

1. Identify the resources required for the sustaining maintenance and support of the system throughout its planned life cycle. Such resource requirements may include the quantity and skill levels of personnel needed for maintenance, spares/repair parts and associated inventories, tools and test equipment, transportation and handling requirements, maintenance facilities and associated capital assets, data, and computer resources.
2. Provide an assessment of the configuration relative to the incorporation of maintainability characteristics in design, both in the design of the prime mission-related elements of the system and in the design of the maintenance and support infrastructure. The objective is to ensure that the system design configuration at the time is compatible with the TPMs and the DDPs defined in Section 3.5, and that the required resource requirements for support are minimized. As discussed in Chapter 2, the system make-up should not only include the prime mission-related elements (i.e., operational hardware, software, and operating personnel) but the elements of the support network as well (refer to Figure 3.5)—there are two sides of the balance as conveyed in Figure 12.12.

The MTA may be accomplished at a gross level during the conceptual design phase when there is enough definition of specific elements of the system (evolving from the maintenance concept—refer to Section 3.4), or when an existing repairable item is being considered for incorporation into the overall design (i.e., a COTS item). As the design progresses during the preliminary and detail design phases and the system configuration becomes better defined, the analyst may accomplish a MTA through a review of available design data, drawings, component part and material lists, technical reports, and so on. The analyst will evaluate what he or she sees from the data, projected in the context of what is anticipated in the system utilization phase.

[13] A complete description of the MTA, as described herein, is included in B. S. Blanchard, *Logistics Engineering and Management*, 5th ed. (Upper Saddle River, N. J. Prentice Hall, Inc., 1998).

In conducting the MTA, the functional analysis described in Sections 3.6 and 4.1 provides the necessary foundation. For the purposes of illustration, a manufacturing plant has been selected and is described through the abbreviated functional block diagram shown in Figure 13.17. The appropriate metrics (i.e., TPMs) should be identified and allocated to each functional block. Each operational function can then be viewed in terms of *go/no-go* criteria, with the development of maintenance functions evolving from the operating functions. This process is illustrated in Figure 13.18.

The maintenance functions can then be broken down into subfunctions, duties, groups of activity, tasks, subtasks, and so on. With the "whats" described in terms of identified tasks, the analyst identifies and evaluates the various alternative approaches for task accomplishment. The results lead to the identification of the "hows" and the specific resources required for maintenance.

Referring to Figure 13.17, it is suspected that the "System Inspection and Test" activity (Function 13) constitutes a potential problem area and that it is desired to accomplish a MTA for the purposes of evaluation. Given that this function fails frequently, the analyst may commence with the preparation of an abbreviated logic flow diagram as shown in Figure 13.19. Starting with a typical symptom of failure (one that occurs most often), a step-by-step *go/no-go* approach can be delineated, with the critical tasks being identified by the numbers just above each block. The selected "path" covers the corrective maintenance cycle to include diagnostics, repair, and condition verification.

The next step is to evaluate the critical tasks identified in Figure 13.19. Referring to Figure 13.20 (sheet 2 is an extension of sheet 1), each of the selected tasks is evaluated and the required maintenance resources are identified (i.e., task times and sequences in Blocks 12 and 13, task frequency in Block 14, personnel quantitities and skill-level requirements in Blocks 15–18, spares and replacement parts in Blocks 8-11 [sheet 2], test and support equipment in Blocks 12–15, and facility requirements in Block 16.[14]

Given the information presented in Figure 13.20 (sheets 1 and 2), the analyst may wish to question a few specific areas of concern. For example:

1. With the extensive resources required for the repair of Assembly A-7 (e.g., the variety of special test and support equipment, the necessity for a "cleanroom" facility for maintenance, the extensive amount of time required for the removal and replacement of CB-1A5, etc.), it may be feasible to identify Assembly A-7 as being nonrepairable. In other words, investigate the feasibility of whether the assemblies of Unit B should be classified as "repairable" or "discard at failure."
2. Referring to Tasks 1 and 2, a "built-in test" capability exists at the organizational level for fault isolation to the subsystem. However, fault isolation to the unit requires a Special System Tester (0-2310B), and it takes 25 minutes of testing plus a highly skilled (supervisory skill) individual to accomplish the function. In essence, one should investigate the feasibility of extending the built-in test down

[14] B. S. Blanchard, D. Verma, and E. L. Peterson, *Maintainability: A Key to Effective Serviceability and Maintenance Management* (New York: John Wiley & Sons, Inc. 1995).

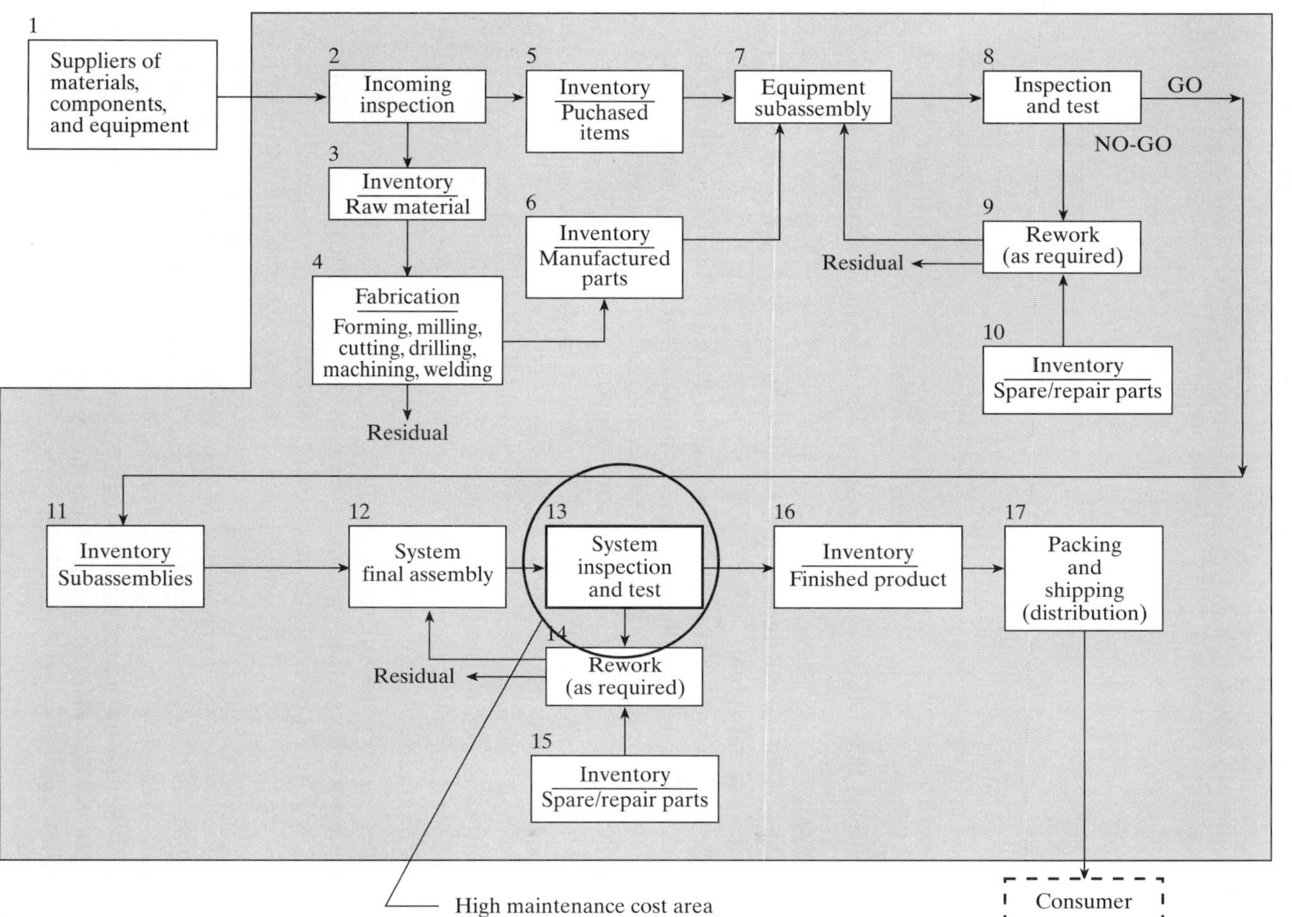

Figure 13.17 Function flow block diagram of a manufacturing operation.

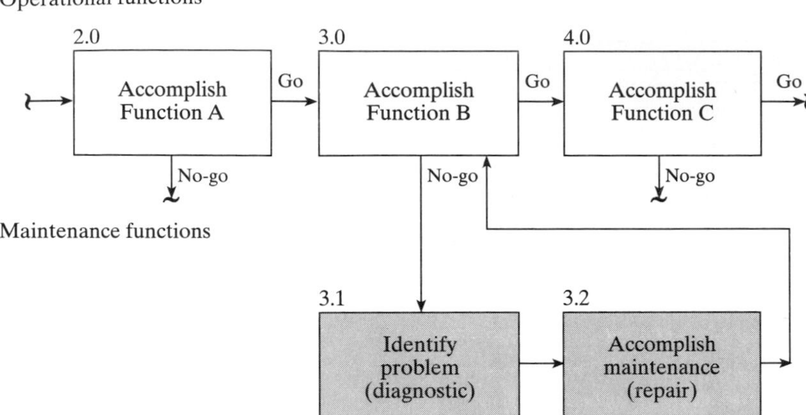

Figure 13.18 The identification of maintenance functions from operational functions.

to the unit level and eliminate the need for the special system tester and the high-skill level individual.

3. The physical removal and replacement of Unit B from the system takes 15 minutes, which seems rather extensive. Although perhaps not a major item, it would be worthwhile investigating whether the removal/replacement time can be reduced (to less than 5 minutes for example).
4. Referring to Tasks 10-15, a special cleanroom facility is required for maintenance. Assuming that the various assemblies of Unit B are repaired (versus being classified as "discard at failure"), then it would be worthwhile to investigate changing the design of these assemblies such that a cleanroom environment is not required for maintenance. In other words, can the expensive maintenance facility requirement be eliminated?
5. There is an apparent requirement for a number of new "special" test equipment/tool items; i.e., special system tester [0-2310B], special system tester [I-8891011-A], special system tester [I-8891011-8], C.B, test set [D-2252-A], special extractor tool [EX20003-4], and special extractor tool [EX45112-6]. Usually, these *special* items are limited as to general application for other systems, and are expensive to acquire and maintain. Initially, one should whether or not these items can be eliminated. If test equipment/tools are required, can *standard* items be used (in lieu of special items)? Also, if the various special testers are required, can they be integrated into a "single" requirement? In other words, can a single item be designed to replace the three special testers and the C.B. test set? Reducing the overall requirements for special test and support equipment is a major objective.
6. Referring to Task 9, there is a special handling container for the transportation of Assembly "A-7". This may impose a problem in terms of the availability of the container at the time and place of need. It would be preferable if normal packaging and handling methods could be used.

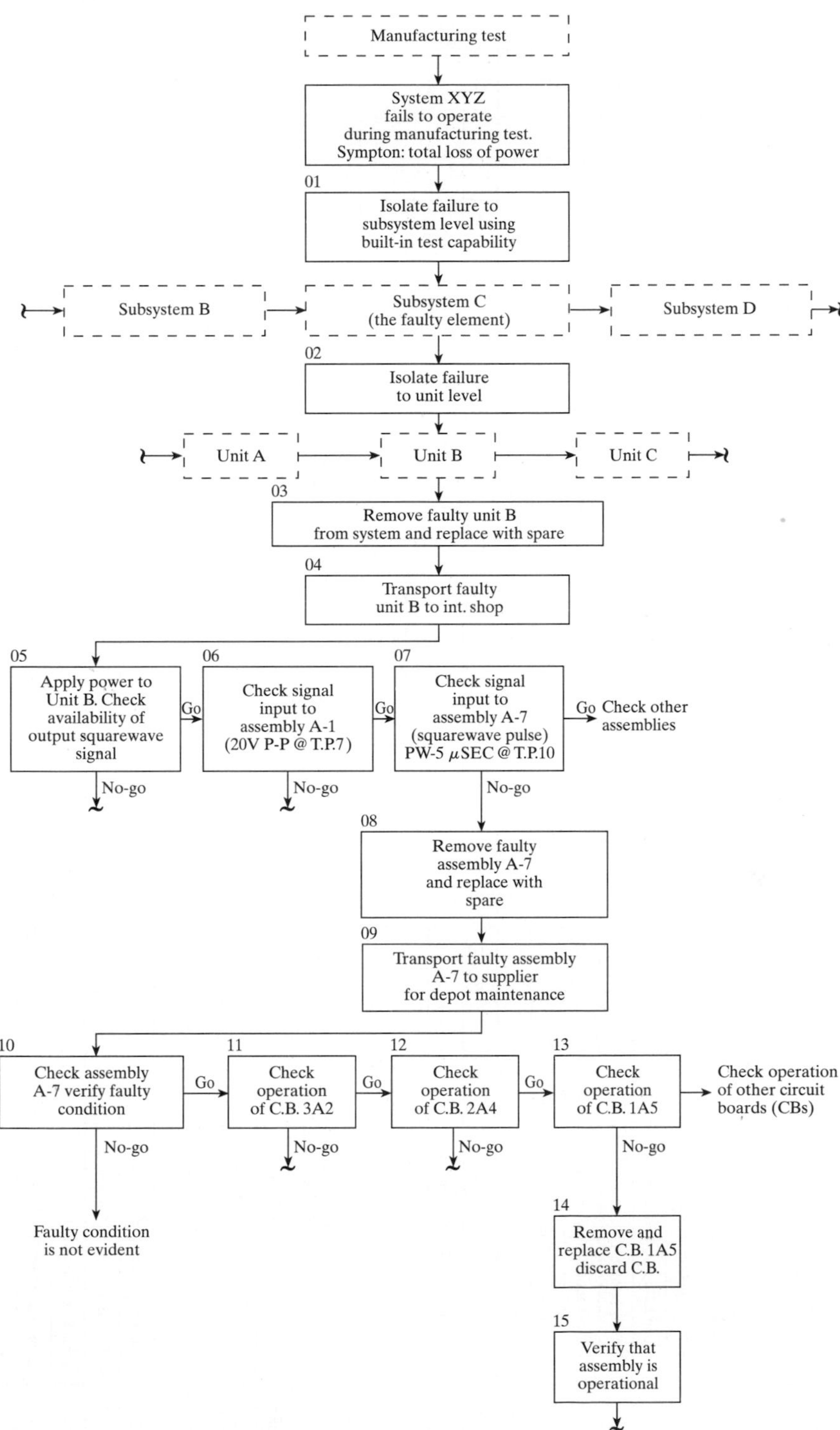

Figure 13.19 Abbreviated logic flow diagram.

1. System XYZ	2. Item Name/Part No. Manufacturing Test/A4321	3. Next Higher Assy. Assembly and Test	4. Description of requirements
5. Req. No. 01	6. Requirement Diag. /Repair	7. Req. Freq. 0.00450	8. Maint. Level Org/Inter.
9. Ma. Cont. No. A12B100			During manufacturing and test of Product 12345 (Serial No. 654), System XYZ failed to operate. The symptom of failure was "loss of total power output." Requirements: troubleshoot and repair system.

10. Task Number	11. Task Description	12. Elapsed Time-Minutes (2 4 6 8 10 12 14 16 18 20 22 24 26 28 30 32 34 36 38)	13. Total elapsed time	14. Task Freq.	Personnel Man-Min 15. B	16. I	17. S	18. Total
01	Isolate failure to subsystem level		5	0.00450	5	–	–	**5**
	(Subsystem C is faulty)							
0.2	Isolate failure to unit level	①	25		–	25	–	**25**
	(Unit B is faulty)	②					25	**25**
0.3	Remove unit B from system and	(2nd cycle)	15		15	–	–	**15**
	replace with a spare unit B							
0.4	Transport faulty unit to int. shop		30		30	–	–	**30**
0.5	Apply power to faulty unit, check	(3nd cycle)	20		–	20	–	**20**
	for output for squarewave signal							
0.6	Check signal input to assembly		15		–	15	–	**15**
	A-1 (20v P-P @ T.P. 7)							
0.7	Check signal input to assembly	(4th cycle)	20		–	20	–	**20**
	A-7 (Squarewave, PW-5 μsec @ T.P.2)							
0.8	Remove faulty A-7 and replace		10		10	–	–	**10**
0.9	Transport faulty assembly A-7 to	14 Calendar days in transit						
	supplier for depot maintenance							
10	Check A-7 and verify faulty cond.	(5th cycle)	25		–	–	25	**25**
11	Check operation of CB-3A2		15		–	–	15	**15**
12	Check operation of CB-2A4		10		–	–	10	**10**
13	Check operation of CB-1A5	(6th cycle)	20		–	–	20	**20**
14	Remove and replace faulty CB-1A5	(7th cycle)	40		–	40	–	**40**
	Discard faulty circuit board							
15	Verify that assembly is operational		15		–	–	15	**15**
	and return to inventory							
		Total	**265**	**0.00450**	**60**	**120**	**110**	**290**

(a)

Figure 13.20 Maintenance task analysis (sheet 1).

1. Item Name/Part No. Manufacturing Test/A4321			1. Req. No. 01	3. Requirement Diagnostic Troubleshooting and Repair		4. Req. Freq. 0.00450	5. Maint. Level.: Organization, Intermediate, Depot	6. Ma. Cont. No. A12B100
7. Task Number	8. Qty. per Assy.	Replacement Parts		Test and Support/Handling Equipment			16. Description of Facility Requirements	17. Special Technical Data Instructions
		9. Parts Nomenclature / 11. Part Number	10. Rep. Freq.	12. Qty.	13. Item Nomenclature / 15. Item Part Number	14. Use Time (min)		
01	–	————	—	1	Built-in test equip. A123456	5	————	Organizational maintenance
02	–	————	—	1	Special system tester O–2310B	2.5	————	↓
03	1	Unit B B180265X	0.01866	1	Standard tool kit STK–100–B	15	————	↓
04	–	————	—	1	Standard cart (M–10)	30	————	Intermediate maintenance
05	–	————	—	1	Special system tester I–8891011–A	20	————	↓
06	–	————	—	1	Special system tester I–8891011–A	15	————	↓
07	–	————	—	1	Special system tester I–8891011–A	20	————	↓
08	1	Assembly A–7 MO–2378A	0.00995	1	Special extractor tool EX20003–4	10	————	Refer to special removal instructions
09	–	————	—	1	Container, special handling T–300A	14 days	————	Normal trans. environment
10	–	————	—	1	Special system tester I–8891011–A	25	Clean room environment	Supplier (depot) maintenance
11	–	————	—	1	C.B. test set D–2252–A	15	↓	↓
12	–	————	—	1	C.B. test set D–2252–A	10	↓	↓
13	–	————	—	1	C.B. test set D–2252–A	20	↓	↓
14	1	CB–1A5 GDA–221056C	0.00450	1	Special extractor tool/EX45112–6; standard tool kit STK–200	40	↓	↓
15	–	————	—	1	Special system tester I–8891011–A	15	↓	Return operating assy. to inventory

(b)

Figure 13.20 *(continued)* Maintenance task analysis (sheet 2).

7. Referring to Task 14, the removal and replacement of C8-1A5 takes 40 minutes and requires a high-skilled individual to accomplish the maintenance task. Assuming that Assembly A-7 is repairable, it would be appropriate to simplify the circuit board removal/replacement procedure by incorporating plug-in components, or at least simplify the task to allow one with a basic skill level to accomplish.

The MTA evolves from the defined maintenance concept (Section 3.4) and the functional analysis (Sections 3.6 and 4.1) and uses, as an input, data from the FMECA, RCM, reliability and maintainability predictions, and the level of repair analysis as necessary. Figure 3.21 includes a sample illustration showing the relationships of these various analyses. With the accomplishment of the MTA covering all of the repairable elements of the system, the total maintenance and support infrastructure can be further defined through the development of a detailed maintenance plan. From this plan, one can proceed with the necessary provisioning actions, procurement, and the subsequent acquisition of spare parts, test equipment, facilities, data, and so on. The MTA constitutes an essential input to the support requirements described in Chapter 15.

Evaluation of Alternative Design Configurations Using Multiple Criteria

Company ABC is developing a large system made up of several subsystems. Subsystem *XYZ* is to be procured from an outside supplier, and there are three different configurations being considered for selection. Each of the three candidate configurations represents existing design (i.e., COTS items), with some new design required to ensure compatibility with the overall system (i.e., design of the interface). The evaluation criteria includes such parameters as performance, operability, effectiveness, design characteristics, support requirements, schedule, and life-cycle cost. Both qualitative and quantitative factors, which were derived from the applicable TPMs (from Section 3.5), are covered in the evaluation process. Figure 13.22 identifies the evaluation parameters, with items 3, 4, 6 to 9, and 11 being of particular interest for maintainability.

In accomplishing the evaluation, the analyst commences with the development of a list of evaluation parameters. The 11 selected parameters in Figure 13.22 have been prioritized in terms of the relative levels of importance, and weighing factors have been assigned accordingly. The QFD process (described in Section 3.5), the Delphi method, or some equivalent evaluation technique, may be used to help establish the weighing factors.

For each of the 11 parameters, the analyst may wish to develop a special checklist, including the criteria for the evaluation of the three proposed configurations. For instance, a sample checklist supporting item 11 ("cost-life cycle") is presented in Figure 13.23. Using this and similar checklists for each factor, the three supplier proposals are evaluated independently. Base rating values from 0 to 10 are applied according to the degree of compatibility with the desired goals. If a "highly desirable" evaluation is realized, a rating of 10 is assigned.

The base rate values are multiplied by the weighing factors to obtain a score. The total score is then determined by adding the individual scores for each configuration.

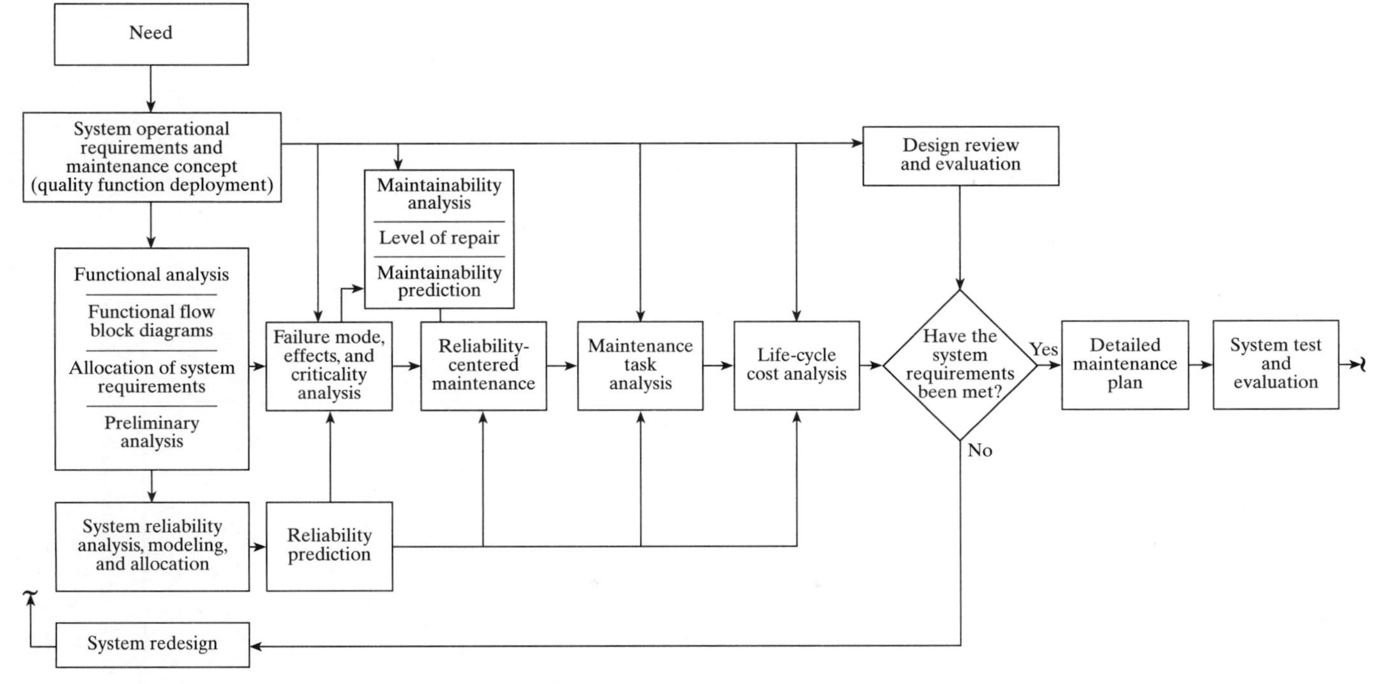
Need
System operational requirements and maintenance concept (quality function deployment)
Functional analysis
Functional flow block diagrams
Allocation of system requirements
Preliminary analysis
System reliability analysis, modeling, and allocation
Reliability prediction
Failure mode, effects, and criticality analysis
Maintainability analysis
Level of repair
Maintainability prediction
Reliability-centered maintenance
Maintenance task analysis
Life-cycle cost analysis
Design review and evaluation
Have the system requirements been met?
Yes
No
Detailed maintenance plan
System test and evaluation
System redesign

Figure 13.21 The MTA and supporting tools/models.

Item	Evaluation parameter	Weighting factor	Configuration A		Configuration B		Configuration C	
			Base rate	Score	Base rate	Score	Base rate	Score
1	Performance—input, output, accuracy, range, compatibility	14	6	84	9	126	3	42
2	Operability—simplicity and ease of operation	4	10	40	7	28	4	16
3	Effectiveness—Ao, MTBM, $\overline{M}ct$, $\overline{M}pt$, MDT, MLH/OH	12	5	60	8	96	7	84
4	Design characteristics—reliability, maintainability, human factors, supportability, producibility, interchangeability	9	8	72	6	54	3	27
5	Design data—design drawings, specifications, logistics data, operating and maintenance procedures	2	6	12	8	16	5	10
6	Test aids—common and standard test equipment, calibration standards, maintenance and diagnostic computer programs	3	5	15	8	24	3	9
7	Facilities and utilities—space, weight, volume, environment, power, heat, water, air conditioning	5	7	35	8	40	4	20
8	Spare/repair parts—part type and quantity, standard parts, procurement time	6	9	54	7	42	5	30
9	Flexibility/growth potential—for reconfiguration, design change acceptability	3	4	12	8	24	6	18
10	Schedule—research and development, production	17	7	119	8	136	9	153
11	Cost—life cycle (R & D. investment, O & M)	25	10	250	9	225	5	125
Subtotal				753		811		534
Derating factor (development risk)				113 15%		81 10%		197 20%
Grand Total		100		640		730		427

Figure 13.22 Evaluation summary (three alternative designs).

Since some new "interface" design is required in each instance, a special derating factor is applied to cover the risk associated with a failure on the part of the supplier to meet the initially specified requirement. In this particular example, configuration B is the preferred choice based on the criteria selected.

Rating (Points)	Evaluation Criteria—Item 11: Cost–Life Cycle
9–10	The supplier has justified his design on the basis of life-cycle cost and has included a complete life-cycle cost analysis in his proposal (i.e., cost breakdown structure, cost profile, etc.).
7–8	The supplier has justified his design on the basis of life-cycle cost but did not include a complete life-cycle cost analysis in his proposal.
5–6	The supplier's design has not been based on life-cycle cost; however, he plans to accomplish a complete life-cycle cost analysis and has described the approach, model, etc., that he proposes to use in the analysis process.
3–4	The supplier's design has not been based on life-cycle cost, but he intends to accomplish a life-cycle cost analysis in the future. No description of approach, model, etc., was included in his proposal.
0–2	The subject of life-cycle cost (and its application) was not addressed at all in the supplier's proposal.

Figure 13.23 Checklist for single criterion evaluation (example) (refer to Figure 13.22 for other evaluation criteria).

13.5 DESIGN REVIEW AND EVALUATION

Review of the extent to which maintainability is incorporated in system design is accomplished as an inherent part of the process described in Sections 3.9, 4.6, and 5.7. The characteristics of the system (and its elements) are evaluated in terms of the initially specified maintainability requirements for the system. If the requirements appear to have been met, the design is approved and the program enters the next phase. If not, the appropriate changes are initiated for corrective action.

In accomplishing a maintainability review, a checklist may be developed to facilitate the review process. An abbreviated representative list, not to be considered as being all inclusive, is presented below (it should be noted that the answer to those questions that are applicable should be *yes*).

1. Have maintainability quantitative and qualitative requirements for the system been adequately defined and specified?
2. Are the maintainability requirements compatible with other system requirements? Are they realistic?
3. Are the maintainability requirements compatible with the system maintenance concept?
4. Has the proper level of accessibility been provided in the design to allow for the easy accomplishment of repair or item replacement? Are access requirements compatible with the frequency of maintenance? Accessibility for items requiring frequent maintenance should be greater than that for items requiring infrequent maintenance.

5. Is standardization incorporated to the maximum extent possible throughout the design? In the interest of developing an efficient supply support capability, the number of different types of spares should be held to a minimum.
6. Is functional packaging incorporated to the maximum extent possible? Interaction affects between modular packages should be minimized. It should be possible to limit maintenance to the removal of one module (the one containing the failed part) when a failure occurs and not require the removal of two, three, or four different modules.
7. Have the proper diagnostic test provisions been incorporated into the design? Is the extent or depth of testing compatible with the level of repair analysis?
8. Are modules and components having similar functions electrically, functionally, and physically interchangeable?
9. Are the handling provisions adequate for heavy items requiring transportation (e.g., hoist lugs, lifting provisions, handles, containers, etc.)?
10. Have quick-release fasteners been used on doors and access panels? Have the total number of fasteners been minimized? Have fasteners been selected based on the requirement for standard tools in lieu of special tools?
11. Have adjustment, alignment, and calibration requirements been minimized (if not eliminated)?
12. Have servicing and lubrication requirements been held to a minimum (if not eliminated)?
13. Are assembly, subassembly, module, and component labeling requirements adequate? Are the labels permanently affixed and unlikely to come off during a maintenance action or as a result of environmental conditions?
14. Have all system maintainability requirements been met?

13.6 MAINTAINABILITY DEMONSTRATION

Maintainability demonstration is conducted as part of the system test and evaluation effort discussed in Chapter 6. Specifically, maintainability demonstration is accomplished as part of Type 2 testing to verify that qualitative and quantitative maintainability requirements have been achieved. It also provides for the assessment of various logistic support factors related to, and having an impact on, maintainability parameters and item downtime (e.g., test and support equipment, spare/repair parts, technical data, personnel, maintenance policies).

Maintainability demonstration is usually accomplished during the latter part of detail design, and should be conducted in an environment that simulates, as closely as practical, the operational and maintenance environment planned for the item. The maintainability demonstration may vary considerably depending on system requirements and the test objectives. Two representative approaches are described in this section to provide an idea of the steps involved.

Demonstration Method 1

Maintainability demonstration method 1 follows a sequential test approach that is similar to the reliability test described in Section 12.6. Two different sequential test plans are employed to demonstrate $\overline{M}ct$ and M_{max} (for corrective maintenance). An accept decision for the equipment under test is reached when that decision can be made for both test plans. The test plans assume that the underlying distribution of corrective maintenance task times is log-normal. The sequential test plan approach allows for a quick decision when the maintainability of the equipment under test is either far above or far below the specified values of $\overline{M}ct$ and M_{max}.

Maintainability testing is accomplished by simulating faults in the system and observing the task times and logistic resources required to correct the situation. It involves the following steps:

1. A failure is induced in the equipment without knowledge of the test team. The induced failure should not be evident in any respect other than that normally resulting from the simulated mode of failure. In other words, the technician(s) scheduled to perform the maintenance demonstration shall not be given any hints (through visual or other evidence) as to where the failure is induced.
2. The maintenance technician will be called on to check out the equipment operationally. At some point in the checkout procedure, a symptom of malfunction is detected.
3. Once a malfunction has been detected, the maintenance technician(s) will proceed to accomplish the necessary corrective maintenance tasks (i.e., fault localization, isolation, disassembly, remove and adjustment and alignment, and system checkout—see Figure 13.1). In the performance of each step, the technician should follow approved maintenance procedures and use the proper test and support equipment. The maintenance tasks performed must be consistent with the maintenance concept and the specified levels of maintenance appropriate to the demonstration. Replacement parts required to perform repair actions being demonstrated shall be compatible with the spare and repair parts recommended for operational support.
4. While the maintenance technician is performing the corrective tasks (commencing with the identification of a malfunction and continuing until the equipment has been returned to full operational status), a test recorder collects data on task sequences, areas of task difficulty, and task times. In addition, the adequacies and inadequacies of logistic support are noted. Was the right type of support provided? Were there test delays because of inadequacies? Was there an overabundance of certain items and a shortage of others? Did each specified element of support do the job in a satisfactory manner? Were the test procedures adequate? These and related questions should be in the mind of the data recorder during the observation of a test.

This maintainability test cycle is accomplished n times, where n is the selected sample size. For the sequential test, the number of demonstrations could possibly

TABLE 13.14 Corrective Maintenance Task Allocation

(1) Item	(2) Quantity (Q)	(3) Failures/item % 1,000 hr (λ)	(4) Total failures $(Q)(\lambda)$	(5) % Contribution	(6) Allocated maint. tasks for demonstration
Unit A	1	0.48	0.48	21	21
Unit B	1	1.71	1.71	76	76
Unit C	1	0.06	0.06	3	3
Total			2.25	100	100

extend to 100, assuming that a continue-to-test decision prevails. Thus, in preparing for the test, a sample size of 100 demonstrations should be planned. The selected tasks should be representative and based on the expected percent contribution toward total maintenance requirements. Those items with high failure rates will fail more often and require more maintenance and logistic resources; hence, they should appear in the demonstration to a greater extent than items requiring less maintenance.

The task selection process is accomplished by proportionately distributing the 100 tasks among the major functional elements of a system. Assuming that three units compose a system, the 100 tasks may be allocated as illustrated in Table 13.14. Referring to the table, the elements of the system and the associated failure rates (from reliability data) are listed. The percentage contribution of each item to the total anticipated corrective maintenance (column 5) is computed as

$$\text{item percent contribution} = \frac{Q\lambda}{\Sigma\, Q\lambda} \times 100 \qquad (13.16)$$

This factor is used to allocate the tasks proportionately to each unit. In a similar manner, the 21 tasks within Unit A can be allocated to assemblies within that unit, and so on. When the allocation is completed, there may be one task assigned to a particular assembly and the assembly may contain several components, the failure of which reflects different failure modes (e.g., no output, erratic output, low output, etc.). Through a random process, one of the components in the assembly will be selected as the item where the failure is to be induced, and the method by which the failure is induced is specified.

With the tasks identified and listed in random order, the demonstration proceeds with the first task, then the second, third, and so on. The criteria for accept-reject decisions are illustrated in Figure 13.24. Task times (Mct_i) are measured and compared with the specified $\overline{\text{Mct}}$ and M_{max} values. When the demonstrated time excceds the specified value, an event is noted along the ordinate of the graph and problem areas are described. Testing then continues until the event line either enters the reject region or the accept region.

An example of the demonstration test score sheet is illustrated in Table 13.15. The accept-reject numbers support the decision lines in Figure 13.24 (refer to the $\overline{\text{Mct}}$ curve). In this instance, 29 tasks were completed before an accept decision was

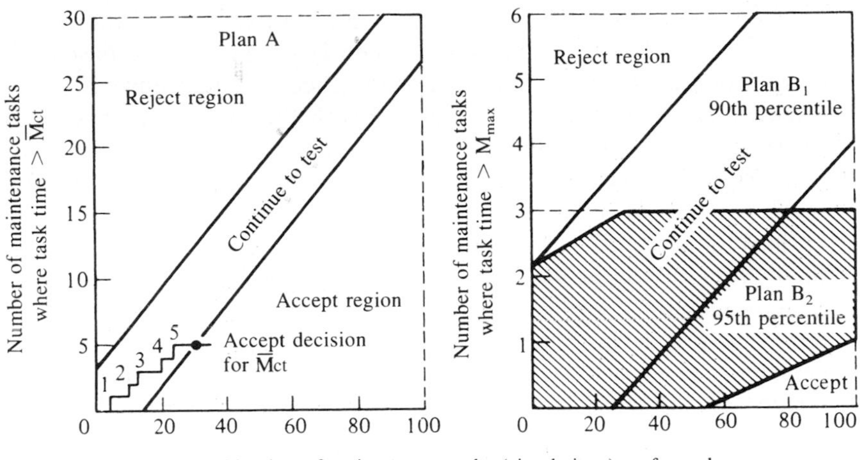

Figure 13.24 Graphical representation of maintainability demonstration plans. *Source:* MIL-STD-471, *Maintainability Verification, Demonstration, Evaluation,* Washington, D.C.: Department of Defense.

TABLE 13.15 Demonstration Source Sheet

REQMT: $\overline{M}ct$ = 0.5 hr = 30 min — Plan A

Maint. task no.	Task time Mct_i	Cum no. $Mct_i > \overline{M}ct$	Accept when cum ≤ then	Reject when cum > than
1	12 Min	0	—	—
2	6	0	—	—
3	18	0	—	—
4	32	1	—	—
5	19	1	—	5
6	27	1	—	6
7	108	2	—	6
8	6	2	—	6
9	14	2	—	7
10	47	3	—	7
11	28	3	—	7
12	19	3	0	7
13	4	3	0	8
14	24	3	0	8
15	78	4	1	8
⋮	⋮	⋮	⋮	⋮
24	20	4	3	11
25	127	5	4	11
26	21	5	4	12
27	13	5	4	12
28	28	5	4	12
29	8	5	5	12
Accept for $\overline{M}ct$				

reached. The sequential test requires that both the $\overline{\text{M}}\text{ct}$ criteria and the M_{max} criteria be met before the equipment is fully acceptable. M_{max} may be based on either the 90th or 95th percentile, depending on the specified system requirement and the test plan selected. If one test plan is completed with the event line crossing into the accept region, testing will continue until a decision is made in the other test plan.

Referring to Figure 13.24, the criteria for sequential testing specifies that the minimum number of tasks possible for a quick decision for $\overline{\text{M}}\text{ct}$ is 12 (test plan A). For M_{max} at the 90th percentile, the least number of tasks possible is 26 (test plan B_1), whereas the figure is 57 for the 95th percentile (test plan B_2). Thus, if the maintainability of an item is exceptionally good, demonstrating the complete sample of 100 tasks may not be necessary, thus saving time and cost. Conversely, if the maintainability of an item is marginal and a continue-to-test decision prevails, the test program may require the demonstration of all 100 tasks. If truncation is reached, the equipment is acceptable for $\overline{\text{M}}\text{ct}$ if 29 or less tasks do not exceed the specified $\overline{\text{M}}\text{ct}$ value. Comparable factors for M_{max} are 5 or less for the 90th percentile and 2 or less for the 95th percentile.

Demonstration Method 2

This method is applicable to the demonstration of $\overline{\text{M}}\text{ct}$, $\overline{\text{M}}\text{pt}$, and $\overline{\text{M}}$. The underlying distribution of maintenance times is not restricted (no prior assumptions), and the sample size constitutes 50 corrective maintenance tasks for $\overline{\text{M}}\text{ct}$ and 50 preventive maintenance tasks for $\overline{\text{M}}\text{pt}$. $\overline{\text{M}}$ is determined analytically from the test results for $\overline{\text{M}}\text{ct}$ and $\overline{\text{M}}\text{pt}$. M_{max} can also be determined if the underlying distribution is assumed to be lognormal. This method offers the advantage of a fixed sample size, which facilitates the estimation of test costs.

The method involves the selection and performance of maintenance tasks in a similar manner as described for demonstration method 1. Tasks are selected based on their anticipated contribution to the total maintenance picture, and each task is performed and evaluated in terms of maintenance times and required logistic resources. Illustration of this method is best accomplished through an example. It is assumed that a system is designed to meet the following requirements and must be demonstrated accordingly:

$$\overline{\text{M}} = 75 \text{ min}$$

$$\overline{\text{M}}\text{ct} = 65 \text{ min}$$

$$\overline{\text{M}}\text{pt} = 110 \text{ min}$$

$$\text{M}_{\text{max}} = 120 \text{ min}$$

$$\text{Producer's risk } (\alpha) = 20\%$$

This test is accomplished and the data collected are presented in Table 13.16. The determination of $\overline{\text{M}}\text{ct}$ (upper confidence limit) is based on the expression

$$\text{Upper limit} = \overline{\text{M}}\text{ct} + Z\left(\frac{\sigma}{\sqrt{n_c}}\right) \tag{13.17}$$

TABLE 13.16 Maintenance Test-Time Data

Demonstration task number	Observed time Mct_i	$Mct_i - \overline{Mct}$ $(Mct_i - 62)$	$(Mct_i - \overline{Mct})^2$
1	58	−4	16
2	72	+10	100
3	32	−30	900
⋮	⋮	⋮	⋮
50	48	−14	196
Total	3105		15,016

where

$$\overline{Mct} = \frac{\sum_{i=1}^{n_c} Mct_i}{n_c} = \frac{3{,}105}{50} = 62.1 \quad \text{(assume 62)}$$

and $Z = 0.84$ from Table 13.4 Then,

$$\sigma = \sqrt{\frac{\sum_{i=1}^{n_c} (Mct_i - \overline{Mct})^2}{n_c - 1}} = \sqrt{\frac{15{,}016}{49}} = 17.5$$

where n_c is the corrective maintenance sample size of 50

$$\text{Upper limit} = 62 + \frac{(0.84)(17.5)}{\sqrt{50}} = 64.07 \text{ min}$$

The completed $\overline{Mct}$ statistic is compared to the corresponding accept-reject criteria, which is to accept if

$$\overline{Mct} + Z\left(\frac{\sigma}{\sqrt{n_c}}\right) \leq \overline{Mct} \text{ (specified)} \qquad (13.18)$$

and reject if

$$\overline{Mct} + Z\left(\frac{\sigma}{\sqrt{n_c}}\right) > \overline{Mct} \text{ (specified)} \qquad (13.19)$$

Applying demonstration test data, it can be seen that 64.07 minutes (the upper value of $\overline{Mct}$ derived by test) is less than the specified value of 65 minutes. Therefore, the system passes the $\overline{Mct}$ test and is accepted.

For preventive maintenance the same approach is used. Fifty (50) preventive maintenance tasks are demonstrated and task times (Mpt_i) are recorded. The sample mean preventive downtime is

$$\overline{Mpt} = \frac{\Sigma\, Mpt_i}{n_p} \qquad (13.20)$$

The accept-reject criteria is the same as stated in Equations 13.18 and 13.19, except that preventive maintenance factors are used. That is, accept if

$$\overline{\mathrm{M}}\mathrm{pt} + Z\left(\frac{\sigma}{\sqrt{n_p}}\right) \leq 110 \text{ min}$$

and reject if

$$\overline{\mathrm{M}}\mathrm{pt} + Z\left(\frac{\sigma}{\sqrt{n_p}}\right) > 110 \text{ min}$$

Given the test values for $\overline{\mathrm{M}}$ct and $\overline{\mathrm{M}}$pt, the calculated mean maintenance time is

$$\overline{\mathrm{M}} = \frac{(\lambda)(\overline{\mathrm{M}}\mathrm{ct}) + (\mathrm{fpt})(\overline{\mathrm{M}}\mathrm{pt})}{\lambda + \mathrm{fpt}} \tag{13.21}$$

where

λ = corrective maintenance rate or the expected number of corrective maintenance tasks occurring in a designated period

fpt = preventive maintenance rate or the expected number of preventive maintenance tasks occurring in the same period

Using test data, the resultant value of $\overline{\mathrm{M}}$ should be equal to or less than 75 minutes. Finally, $\mathrm{M}_{\max}$ is determined from

$$\mathrm{M}_{\max} = \text{antilog}\,(\overline{\log \mathrm{Mct}} + Z\sigma_{\log \mathrm{Mct}_i}) \tag{13.22}$$

The calculated value should be equal to or less than 120 for acceptance. An example of calculation for $\mathrm{M}_{\max}$ is presented in Section 13.2. If all the demonstrated values are better than the specified values, following the criteria defined previously, the system is accepted. If not, some retest or redesign may be required, depending on the seriousness of the problem.

Maintainability Assessment

The true assessment of system maintainability in an operational environment is accomplished through an effective data collection, analysis, and evaluation capability, as described in Section 6.5. The maintenance event report illustrated in Figure 6.6 is designed for the collection of maintenance data throughout the system operational life cycle when failures occur. Maintenance tasks, task sequences, task times, and logistics factors are recorded. Major areas of deficiency are noted and should be corrected through the initiation of changes and the appropriate system modification(s).

QUESTIONS AND PROBLEMS

1. Define *maintainability*. How does it differ from *maintenance*? Provide some examples.
2. Why is maintainability important in system design? When in the life-cycle process should it be considered? Why?

3. What are the quantitative measures of maintainability (discuss measures for hardware, software, personnel, and facilities)?
4. What is the significant difference between MTBF and MTBM? Between MTBF and MTBR? Between MTBR and MTBM?
5. Corrective-maintenance task times were observed as given in the table.

Task time (min)	Frequency	Task time (min)	Frequency
41	2	37	4
39	3	25	10
47	2	36	5
35	5	31	7
23	13	13	3
27	10	11	2
33	6	15	8
17	12	29	8
19	12	21	14

(a) What is the range of observations?
(b) Using a class interval width of 4, determine the number of class intervals. Plot the data and construct a curve. What type of distribution is indicated by the curve?
(c) What is the geometric mean of the repair times?
(d) What is the standard deviation of the sample data?
(e) What is the M_{max} value? Assume 90% confidence level.
(f) What is the $\overline{\text{Mct}}$?

6. Corrective-maintenance task times were observed as given in the table.

Task time (min)	Frequency	Task time (min)	Frequency
35	2	25	12
17	6	19	10
12	2	21	12
15	4	23	13
37	1	29	8
27	10	13	3
33	3	9	1
31	6	—	—

(a) What is the range of observations?
(b) Assuming 7 classes with a class interval width of 4, plot the data and construct a curve. What type of distribution is indicated by the curve?
(c) What is the mean repair time?
(d) What is the standard deviation of the sample data?
(e) The system is required to meet a mean repair time of 25 minutes at a stated confidence level of 95%. Do the data reveal that the specification requirements will be met? Why?

7. With a specified inherent availability of 0.990 and a calculated MTBF of 400 hr, what is the $\overline{\text{Mct}}$?

8. Calculate as many of the following parameters as you can with the information given.

DETERMINE		GIVEN
A_i	MTBM	$\lambda = 0.004$
A_a	MTBF	Total operation time = 10,000 hr
A_o	$\overline{\text{M}}$	Mean downtime = 50 hr
$\overline{\text{M}}\text{ct}$	MTTR_g	Total number of maintenance actions = 50
M_{max}		Mean preventive maintenance time = 6 hr
		Mean logistics plus administrative time = 30 hr

9. Find the upper 90% confidence limit for the population mean repair time based on a normally distributed sample of 70 repair actions, given $\bar{x}$ = 80 minutes and

$$\sum_{1}^{N} (x_1 - \bar{x})^2 = 18{,}000$$

10. What is maintainability allocation? How is it accomplished?

11. Recommend either of the following maintenance concepts based on the data provided.

—Concept 1. Discard all component-board assemblies on failure.
—Concept 2. Repair all component-board assemblies on failure.
—Concept 3. Discard component-board assemblies on failure when the replacement cost is less than $100 each, and repair all other component boards.

(a) Quantity data
Total number of assemblies: 10,000
Number of items under $100 unit cost: 5,000
Number of items over $100 unit cost: 5,000 (average cost per unit is $300)

(b) Failure-rate data
The items under $100 unit cost will fail on the average of every 36 months. The failure rates, respectively, are 0.00048 and 0.00016 per item; or 2.40 and 0.80 per class.

(c) The service life of the system is 10 years. The system is operated 2,080 hours per year.

(d) An investment of $1,000,000 is necessary for equipment to repair any or all of the subassemblies. The equipment life is 10 years, and it will have no scrap value at the end of that time.

(e) Labor costs are $3.00 for each repair. Component spares cost $2.00 per repair.

12. What is the purpose of maintainability prediction? When are maintainability predictions accomplished in the life cycle?

13. Given the following data, calculate the achieved availability.
(a) $\overline{\text{M}}\text{ct}$ = 0.5 hour.
(b) MTBM_u = 2.0 hours.
(c) $\overline{\text{M}}\text{pt}$ = 2.0 hours.
(d) MTBM_s = 1,000 hours.

14. A $\overline{\text{M}}\text{ct}$ of 1 hour has been assigned to system ABC. Allocate this time to the assembly level of the system using the data given in the table.

Assembly	Quantity	λ	$\overline{\text{M}}\text{ct}$
A	1	0.05	_____ ?
B	2	0.16	_____ ?
C	1	0.27	_____ ?
D	1	0.12	_____ ?

15. Is MDT in itself a true measure of equipment design for maintainability? If so, explain. If not, why not?

16. How are preventive maintenance requirements determined? How are they justified? What is likely to occur if they are not properly justified?

17. How are the FMECA and the RCM analysis related (if at all)?

18. What is the purpose of the MTA? What information is provided through the MTA? How should it be applied?

19. What is the purpose of the repair-level analysis? When should it be accomplished? How does it relate to the MTA (if at all)?

20. Refer to Figure 13.22. How are the criteria for evaluation selected and how are the weighting factors determined?

21. What is the purpose of maintainability demonstration? What elements are demonstrated?

22. When selecting tasks for maintainability demonstration, how does the selection process result in a representative sample of what is to be expected in an operational situation?

23. What criteria are used in the selection of personnel for the demonstration of maintenance tasks? Define the criteria for the selection of test and support equipment.

24. Refer to maintainability demonstration method 2, and determine whether the system will meet the $\overline{\text{M}}\text{ct}$ requirement if the assumed producer's risk is 5%.

25. The $\overline{\text{M}}\text{ct}$ requirement for an equipment item is 65 minutes and the established risk factor is 10%. A maintainability demonstration is accomplished and yields the results given in the table for the 50 tasks demonstrated. (Task times are in minutes.)

39	57	70	51	74	63	66	42	85	75
42	43	54	65	47	40	53	32	50	73
64	82	36	63	68	70	52	48	86	36
74	67	71	96	45	58	82	32	56	58
92	91	75	74	67	73	49	62	64	62

Did the equipment item pass the maintainability demonstration?

26. The $\tilde{M}pt$ requirement for an equipment is 120 minutes. A maintainability demonstration is accomplished and yields the results given in the table for the 50 tasks demonstrated. (Task times are in minutes.) Did the equipment item pass the maintainability demonstration? The risk factor is 10%.

150	120	133	92	89	115	122	69	172	161
144	133	121	101	114	112	181	78	112	91
82	131	122	159	135	108	95	67	118	103
78	93	144	152	136	86	113	102	65	115
113	101	94	129	148	118	102	106	117	115

14 DESIGN FOR USABILITY (HUMAN FACTORS)

Most of the material presented in the earlier chapters was concerned with the hardware and software elements of a system. For the system to be complete, one also needs to address the human being and the interfaces between the human and other system elements. Good hardware and software design alone will not guarantee good usability. Consideration must be given to the user's anthropometric characteristics (i.e., human physical dimensions), sensory factors (i.e., sight, hearing, etc.), physiological factors (i.e., impact from environmental forces), psychological factors (i.e., needs, expectations, attitude, motivation, etc.), and their interrelationships. System design requires the proper integration of human factors (along with hardware, software, facilities, data, and other elements) throughout the system acquisition process as illustrated in Figure 2.4.

The purpose of this chapter is to deal briefly with the human as an integral part of the system. In this context, the activities discussed within this broad subject area include the identification of functions and tasks that are to be performed by the human, the establishment and allocation of human factors design criteria, the accomplishment of human factors analysis, the identification of personnel and training requirements, and the conduct of personnel test and evaluation activities. A survey of the literature indicates that such activities may also be classified and included under such general categories as *ergonomics, engineering psychology, systems psychology*, or *human engineering*.

14.1 DEFINITION AND EXPLANATION OF HUMAN FACTORS

Requirements for the *human*, as part of a system design, are initially derived through the definition of system operational requirements, the maintenance concept, and the accomplishment of a top-level functional analysis (refer to Figure 2.4, Blocks 0.1 and 0.2). A description of the mission(s) to be performed is an essential first step in the requirements definition process. Based on this, a functional analysis is completed where operational and maintenance functions are identified indicating the "whats;" i.e., what must the system do (Sections 3.6 and 4.1). Through the subsequent process of synthesis, analysis, and evaluation, a specific design approach may be selected for determining the "hows" (i.e., the manner by which the functions will be accomplished). This leads to the identification of *human* requirements for the system, as shown in Figure 3.14.[1]

Given the functions that have been identified and allocated to the human, the next step is to break these down into *job operations, duties, tasks, subtasks*, and so on, as illustrated in Figure 14.1. A brief description of each is presented subsequently.[2]

1. *Job operation.* Completion of a function normally includes a combination of duties and tasks. A job operation may involve one or more related groups of duties, and may require one or more individuals in its accomplishment (i.e. positions in a given specialty field). An example of a job operation is operate a motor vehicle or accomplish scheduled maintenance.
2. *Duty.* Defined as a set of related tasks within a given job operation. For instance, when considering the operating of a motor vehicle, a set of related tasks may include (a) driving the motor vehicle in traffic on a daily basis, (b) registering the motor vehicle yearly, (c) servicing the motor vehicle as required, and (d) accomplishing vehicle preventive maintenance on a periodic basis.
3. *Task.* Constitutes a composite of related activities (informational, decision, and control activities) performed by an individual in accomplishing a prescribed amount of work in a specified environment. A task may include a series of closely associated operations, maintenance inspections, and so on. Relative to driving a motor vehicle, examples of tasks are (a) applying the appropriate pressure on the accelerator in order to maintain the desired vehicle speed, (b) shifting gears as necessary in order to maintain engine rpm, or (c) turning the steering wheel as required to enable the motor vehicle to move in the desired direction.
4. *Subtask.* Depending on the complexity of the situation, a task may be broken down into subtasks to cover discrete actions of a limited nature. A subtask may constitute the shifting of gears from first to second (in the motor vehicle example), a machine adjustment, or a similar act.
5. *Task element.* Task elements may be categorized as to the smallest logically definable facet of activity (based on perceptions, decisions, and control actions) that requires individual behavioral responses in completing a task or a subtask. For example, the identification of a specific signal level on a display, a decision per-

[1] Referring to Figure 3.14, it should be noted that human activities and tasks may also evolve from the hardware life cycle as lower elements of the system are defined. In other words, through the analysis of a unit of hardware a human requirement may be identified at a lower level.

[2] A hierarchy of behavioral and related terms is discussed further in several references in Appendix F.

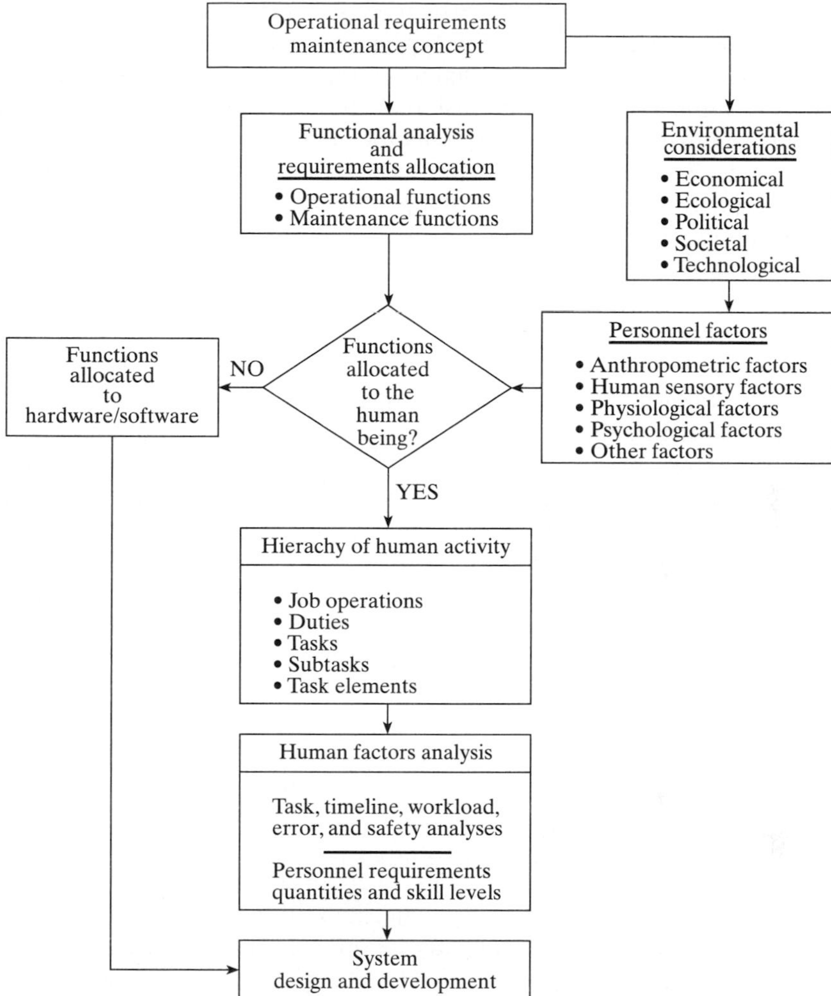

Figure 14.1 Human factors requirements.

taining to a single physical action, the actuation of a switch on a control panel, and the interpretation of a go/no-go signal might be classified as a task element. This is the lowest category of activity where job behavioral characteristics are identified and evaluated.

The hierarchy outlined earlier represents a breakdown from the top system-level functions described in Chapter 3 to the smallest element of activity involving human performance. In developing this breakdown, it is not always easy to separate functions from tasks, tasks from subtasks, and so on. However, one must proceed from the top-level function(s) down to the level necessary to establish the proper human-machine interface. As with an equipment hierarchy, this breakdown constitutes a logical division of human activity which serves as a basis in establishing human factors requirements in design.

With the functional analysis results and the basic hierarchy of human activity identified (at least, on a preliminary basis), one needs to address some of the environmental and personnel factors that will ultimately influence work accomplishment. Figure 14.1 illustrates the relationship of these elements in the process of defining the overall human factors requirements for the system. The environmetal considerations noted in the figure constitute those factors that impose a constraint on the system development activity as an entity. Personnel factors may be categorized in terms of anthropometric factors, human sensory factors, physiological factors, and psychological factors. These categories are briefly described subsequently.

Anthropometric Factors

When establishing basic design requirements for the system, particularly regarding human activities, one obviously must consider the physical dimensions of the human body. The weight, height, arm reach, hand size, and so on, are critical when designing operator stations, consoles, control panels, accesses for maintenance purposes, and the like. Further, body dimensions will vary somewhat from a static position to a dynamic condition. *Static* measurements pertain to the human subject in a rigid standardized position, whereas *dynamic* measurements are made with the human in various working positions and undergoing continuous movement. As movement occurs, body measurements will change. Thus, when considering various system design alternatives, the designer must be aware of the different human profiles that are likely to occur in the performance of operator and maintenance activities associated with the system.

To aid the designer in obtaining information on body measurements, there are two basic sources of data to consider: (1) anthropometric surveys, in which measurements of a sample of the population have been made; and (2) experimental data derived from simulating the operating conditions peculiar to the system being developed. The designer may have access to either source of data, or a combination of both. In many instances, however, the acquisition of meaningful experimental data is quite costly. Hence, static measurements are often used for the purposes of design guidance.

Figures 14.2 and 14.3 illustrate examples of static body measurements in the standing and sitting positions, respectively.[3] Anthropometric measurements are usually provided in percentiles, ranges, and means (or medians). The data presented represent the 5th and 95th percentiles for men and women for a given sample population. Although the designer will not be able to cover all possible sizes or profiles, he or she can cover most situations as well as decide on where to cut off. The ultimate design configuration must accommodate most of the population if operator and maintenance activities are to be accomplished efficiently. Those individuals having extreme measurements (e.g., the small percentage of the population below the 5th percentile and above the 95th percentile) may be able to perform the necessary operator or mainte-

[3] The factors presented in Figures 14.2 and 14.3 were taken from MIL-STD-1472, Military Standard, *Human Engineering Design Criteria for Military Systems, Equipment and Facilities* (Washington, D.C.: Department of Defense). Another good source for anthropometric data is Sanders, M. S. and E. J. McCormick, *Human Factors in Engineering Design*, 7th ed. (New York: McGraw-Hill, 1992).

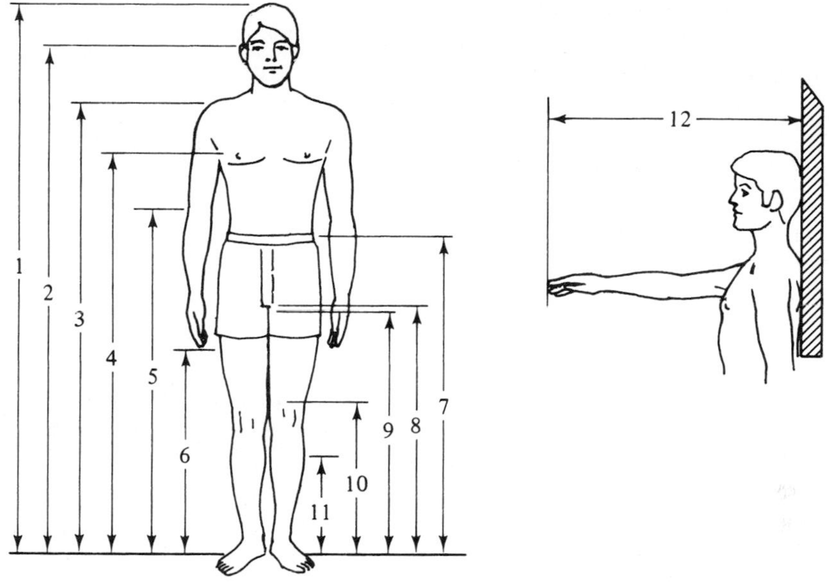

Factors	Percentile values (cm)			
	5th Percentile		95th Percentile	
	Men	Women	Men	Women
Weight (kg)	57.4	46.4	91.6	76.5
Standing body dimensions				
1. Stature	163.8	152.4	185.6	172.2
2. Eye height	152.1	142.7	173.3	160.1
3. Shoulder height	133.6	123.0	154.2	143.4
4. Chest height		110.0		127.3
5. Elbow height	104.8		120.0	
6. Fingertip height	61.5		73.2	
7. Waist height	97.5	93.1	115.2	110.1
8. Crotch height	76.3	66.4	91.8	81.4
9. Gluteal furrow height		66.2		79.4
10. Kneecap height	47.5		58.6	
11. Calf height	31.1		40.6	
12. Functional reach	72.7	67.7	90.9	80.4

Figure 14.2 Anthropometric data—standing body dimensions.

nance functions; however, the results are likely to be inefficient due to the introduction of various human stresses causing early personal fatigue and possible system failure.

In designing a system, work-space requirements must be established for both operator personnel in the performance of operating functions and for maintenance personnel in the accomplishment of maintenance tasks. Figures 14.4 and 14.5 provide examples of the design with anthropometric factors in mind.

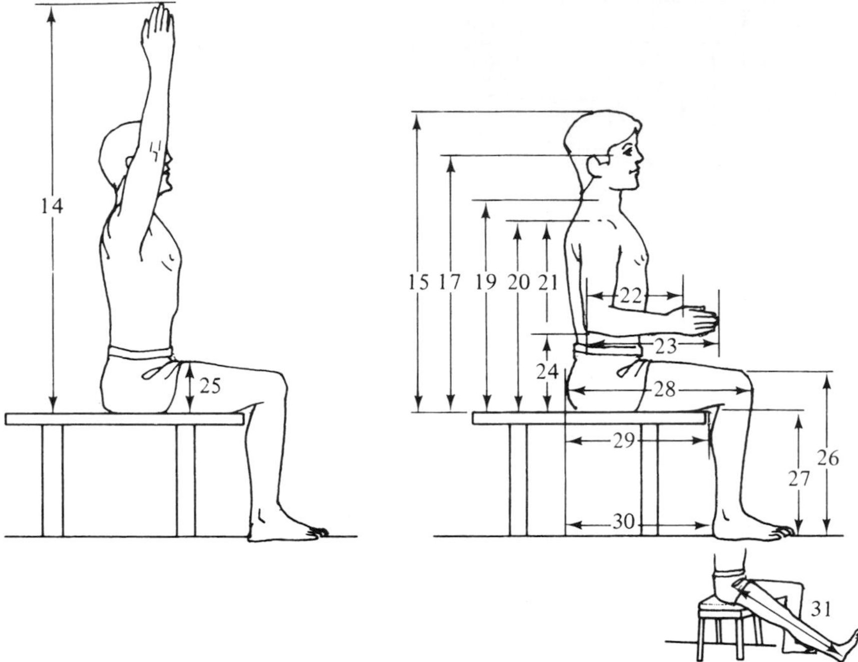

Factors	Percentile values (cm)			
	5th Percentile		95th Percentile	
	Men	Women	Men	Women
Seated body dimensions				
14. Vertical arm reach, sitting	128.6		147.8	
15. Sitting height, erect	84.5	78.4	96.9	90.9
16. Sitting height, relaxed	82.5	76.9	94.8	89.7
17. Eye height, sitting erect	72.8	68.7	84.6	78.8
18. Eye height, sitting relaxed	70.8	67.2	82.5	77.6
19. Mid-shoulder height	57.1	53.7	67.7	62.5
20. Shoulder height, sitting	54.2		65.4	
21. Shoulder-elbow length	33.8	30.2	40.2	36.2
22. Elbow-grip length	32.6		37.9	
23. Elbow-fingertip length	44.3	38.9	51.9	45.7
24. Elbow rest height	17.5	18.7	28.0	26.9
25. Thigh clearance height		10.4		14.6
26. Knee height, sitting	49.7	43.7	58.7	51.6
27. Popliteal height	40.6	38.0	50.0	44.1
28. Buttock-knee length	54.9	52.0	64.3	61.9
29. Buttock-popliteal length	45.8	43.4	54.5	52.6
30. Buttock-heel length	46.7		56.4	
31. Buttock-heel length (diagonal)	103.9		120.4	

Figure 14.3 Anthropometric data—seated body dimensions.

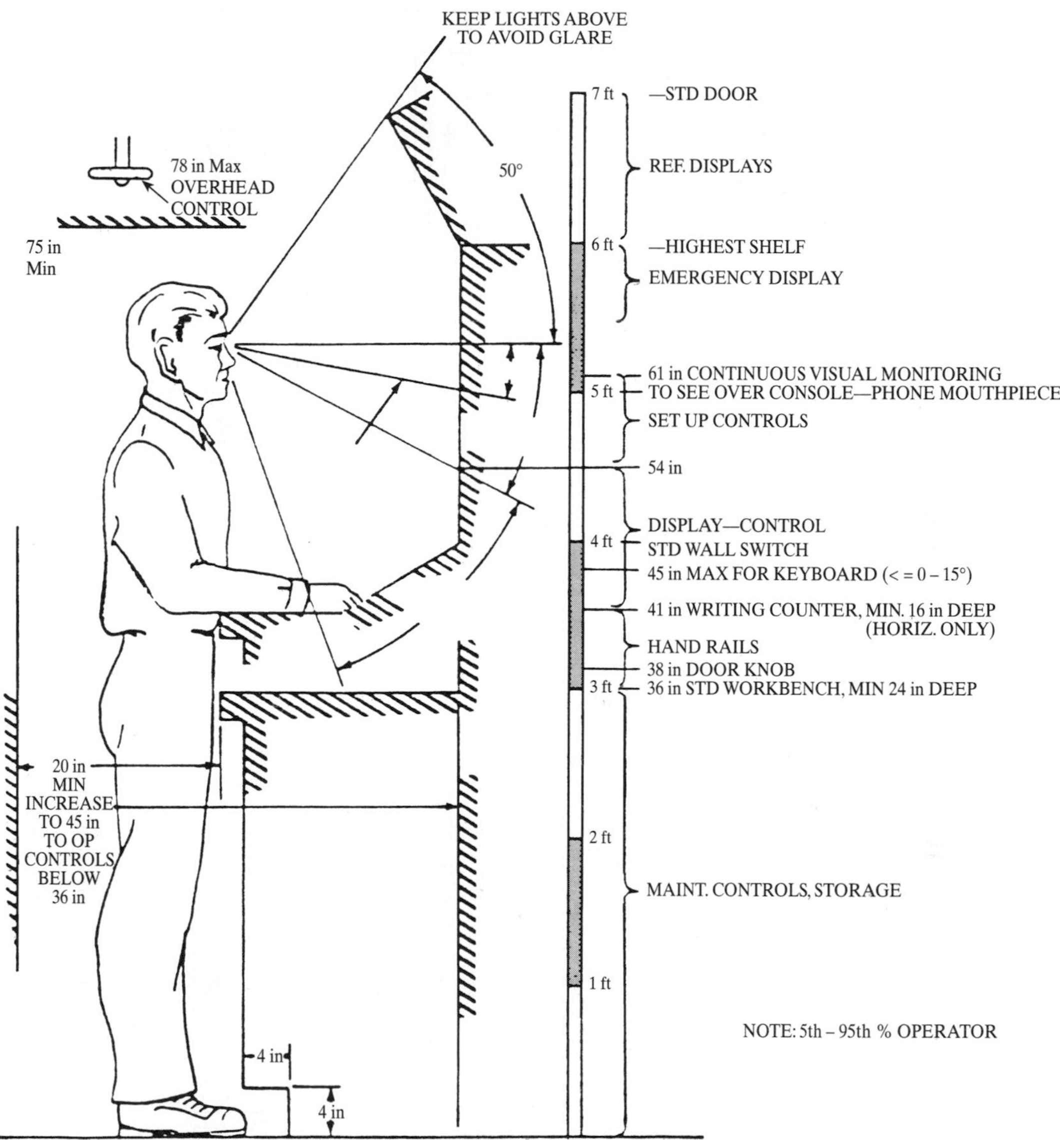

Figure 14.4 Suggested parameters for standard operator workplaces. *Source:* H. P. Van Cott, and R. G. Kincade, *Human Engineering Guide to Equipment Design*, rev. ed. Washington, D.C.: U.S. Government Printing Office, 1972.

The application of anthropometric data in design involves many considerations. The human body and work-space dimensions are significant. However, one should consult the literature before proceeding further because the limited material in this section is provided for illustrative purposes only. There exist criteria covering many facets of console and control panel design, work-space design, seat design, work surface design, and so on. Also, associated with body dimensions are the aspects of force and

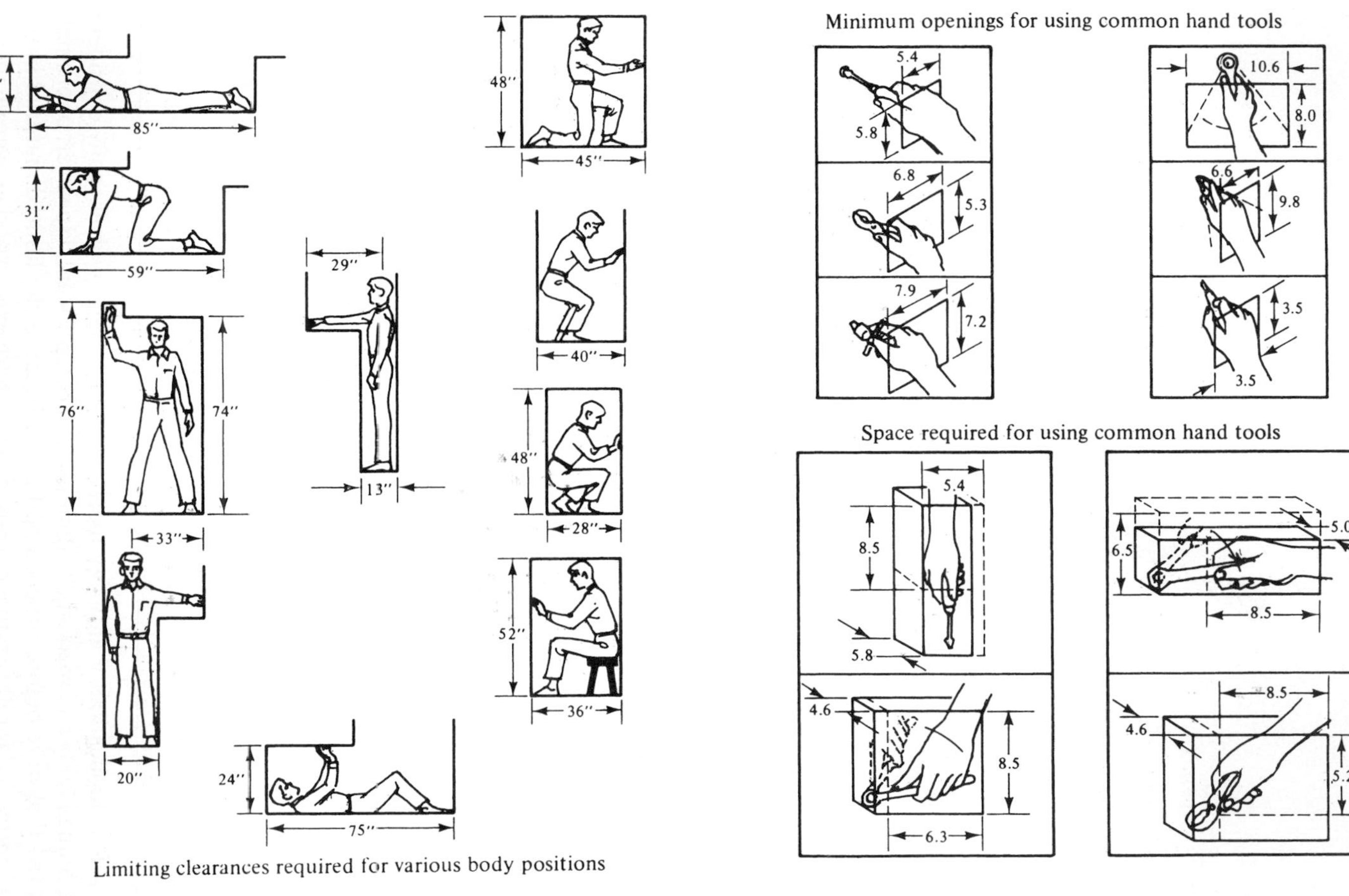

Figure 14.5 Anthropometric data pertaining to work space requirements. *Source:* NAV-SHIPS 94324, *Maintainability Design Criteria Handbook for Designers of Shipboard Electronics Equipment.* Naval Ship Systems Command, U.S. Navy, Washington, D.C.: 1964.

weight-lifting capacity. A human being can exert more force, with less fatigue, if the system (i.e., the segment of the system where the interface exists) is designed properly. The amount of force that can be exerted is determined by the position of the body and the members applying the force, the direction of application, and the object to which the force is applied. Weight-lifting capacity, which is closely related to force, is highly dependent on the individual's weight, size, and position. The field of anthropometry deals primarily with the science and technique of human measurements; however, there are numerous other factors that are directly affected by design decisions.[4]

Human Sensory Factors[5]

In dealing with the human-machine interface in system design, one must be cognizant of certain human sensory capacities. Although most of the human senses may be affected through system design, factors pertaining to vision (sight) and hearing (noise) are of particular significance.

Vision. Vision or sight, as it pertains to system design, is usually limited to the desired vertical and horizontal fields, as illustrated in Figure 14.6. The designer should consider the specified degrees of eye and head rotation as the maximum allowable values in the design of operator consoles and control panels. Requirements outside of these recommended limits will result in operator inefficiency and system failures.

Within the broad field of view illustrated in the figure, the human eye may see different objects from different angles. Sight is stimulated by the electromagnetic radiation of certain wavelengths, or the visible portion of the electromagnetic spectrum. The eye sees different lines (i.e., parts of the spectrum) with varying degrees of brightness. Relative to color, one can usually perceive all colors while looking straight ahead. However, color perception begins to decrease as the viewing angle decreases. Therefore, when designing consoles (or panels) with color-coded meters or color warning light displays, one must consider the placement of such relative to the operator's field of view. Figure 14.7 illustrates the limits of color vision.

Finally, the satisfactory performance of tasks is highly dependent on the level of illumination. Not only must the designer be concerned with field of view and color, but the proper level of illumination is an obvious necessity. Illumination levels will vary somewhat depending on the task to be performed. Table 14.1 identifies suggested levels of illumination for designated items.

Hearing. In human activities, the designer needs to address both the requirements for oral communication and the aspects of noise. Of particular concern is the effect of noise on the performance of work. Noise is generally regarded as a distractor and a deterrent when considering efficiency of work accomplishment. As the noise level

[4] K. Kroemer, H. Kroemer, and K. Kroemer-Elbert, *Ergonomics: How to Design for Ease and Efficiency* (Upper Saddle River, N. J.: Prentice Hall, Inc., 1994). Also refer to Appendix F for additional human factors references.

[5] NAVSHIPS 94324. *Maintainability Design Criteria Handbook for Designers of Shipboard Electronic Equipment* (Washington, D.C.: Naval Ship Systems Command, Department of the Navy, 1964).

TABLE 14.1 Specific Task Illumination Requirements

Illumination levels		
	Foot-candles*	
Type of task	Recommended	Minimum
1. Business machine operation	100	50
2. Console surface	50	30
3. Dials, gages, and meters	50	30
4. Factory assembly—general	50	30
5. Factory assembly—precise	300	200
6. Inspection tasks—extra fine	300	200
7. Inspection tasks—rough	50	30
8. Office work, general	70	50
9. Ordinary seeing tasks	50	30
10. Panels	50	30
11. Passageways and stairways	20	10
12. Reading, large print	30	10
13. Reading, small type	70	50
14. Repair work—general	50	30
15. Repair work—instrument	200	100
16. Testing—extra fine	200	100
17. Testing—rough	50	30
18. Transcribing and tabulation	100	50

*As measured at the task object or 30 in, above the floor.
This information (in part) was extracted from MIL-STD-1472, Military Standard, *Human Engineering Design Criteria for Military Systems, Equipment, and Facilities* (Washington, D.C.: Department of Defense).

increases, a human being begins to experience discomfort, and both productivity and efficiency decrease. Further, oral communication becomes ineffectual or impossible. When the noise level approaches 120 dB, a human being will usually experience a physical sensation in some form; and at levels above 130 dB, pain can occur.

Another major factor in determining the extent to which noise is a distracter is its character—whether steady or intermittent. In situations where the noise is steady, an individual can adapt to it, and work efficiency may not be significantly compromised. If the noise level is too high (even though steady), the individual will probably experience permanent injury through loss of hearing. Conversely, when the noise is intermittent, the individual is usually distracted regardless of the intensity. In this instance, a greater effort is required to maintain job efficiency and the possibility of early fatigue occurs.

The designer, when dealing with the human being as a component of the system, must identify the operator and maintenance tasks that are to be accomplished by people. These tasks must be evaluated in terms of the environment in which the tasks are to be performed, and the noise generated by the system (or generated externally) must

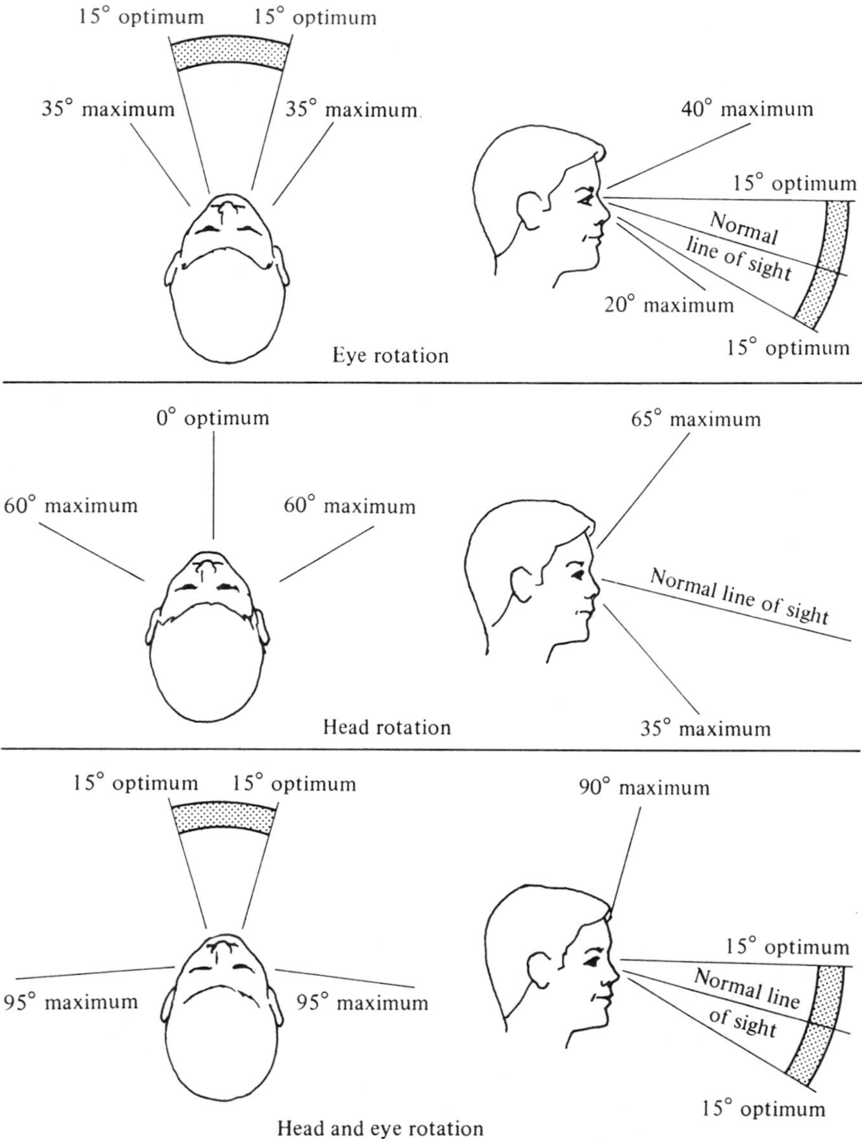

Figure 14.6 Vertical and horizontal visual field.

be maintained at a level where human efficiency is maximized. The desired intensity level will likely fall within a range 50 to 80 dB; however, the designer should consult the available literature to establish the proper design criteria for the variety of applications that are likely to occur.

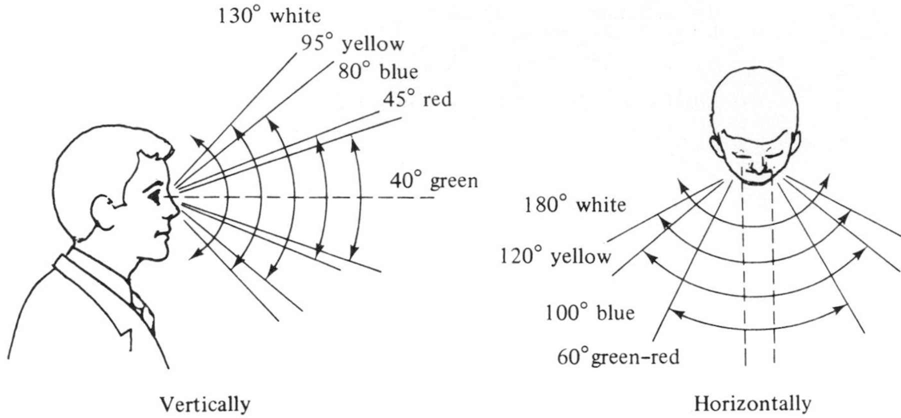

Figure 14.7 Approximate limits of color differentiation.

Physiological Factors[6]

The study of physiology is obviously well beyond the scope of this text; however, some recognition must be given to the effects of environmental stresses on the human body while performing system tasks. *Stress* refers to any aspect of external activity or the environment acting on the individual (who is performing a system tasks) in such a manner as to cause a degrading effect. Stress may result in physiological effects and psychological effects, and *strain* is often the consequence of stress measured in terms of one or more physical characteristics of the human body. Some of the causes of stress are the following:

1. *Temperature extremes.* Experience has indicated that certain temperature extremes are detrimental to work efficiency. As the temperature increases above the comfort zone (e.g., 55°F to 75°F), mental processes slow down, motor response is slower, and the likelihood of error increases. Conversely, as the temperature is lowered (e.g., 50°F and lower), physical fatigue and stiffening of the extremities begins. For this reason, the designer needs to be aware of the environmental profiles for the system, particularly with regard to anticipated system use in the arctic or in the tropics.
2. *Humidity.* Heat and humidity are usually significant factors in causing a reduction in the operational efficiency of personnel. A human being can tolerate much higher temperatures with dry air than if the air were humid.
3. *Vibration.* High levels of vibration will often affect human proficiency. Figure 14.8 illustrates desired criteria for various values of frequency and acceleration. The designer must consider these factors to permit safe conditions for the accomplishment of system operational and maintenance activities.

[6] Additional coverage of physiological factors is presented in M. S. Sanders, and E. J. Mc-Cormick, *Human Factors in Engineering and Design*, 7th ed. (New York: McGraw-Hill Book, Co. 1992).

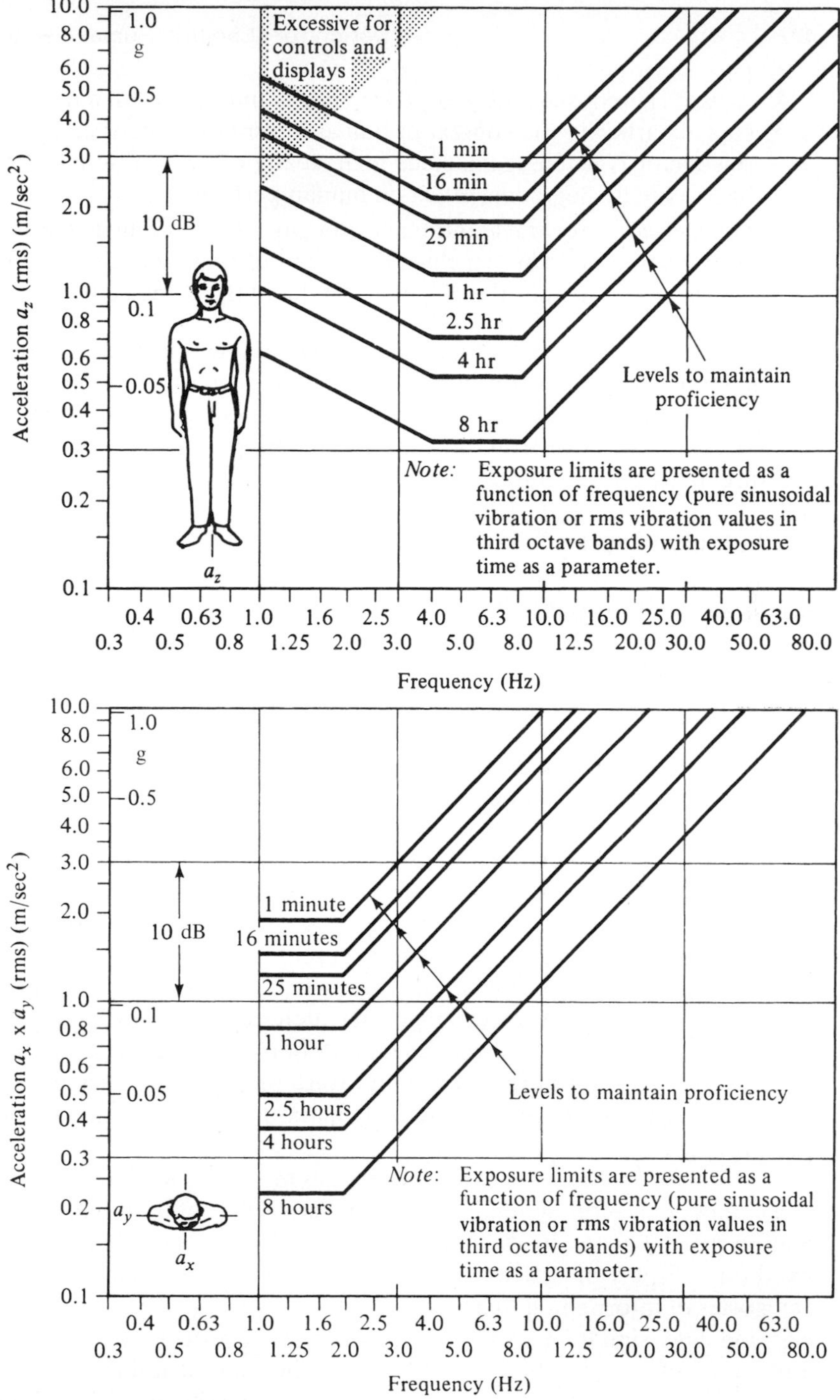

Figure 14.8 Vibration exposure criteria for longitudinal (upper curve) and transverse (lower curve) directions with respect to body axis. *Source:* MIL-STD-1472, Military Standard, *Human Engineering Design Criteria for Military Systems, Equipment, and Facilities.* Washington, D.C.: Department of Defense.

4. *Noise.* The consequences of noise and its impact on human efficiency was discussed earlier in the context of human sensory factors (i.e., hearing). However, this factor is listed again, since high steady or intermittent levels of noise may have a highly degrading effect on human performance.
5. *Other factors.* Several additional factors may (to varying degrees) cause stress on the human body. These include the effects of radiation, the effects of gas or toxic substances in the air, the effects of sand and dust, and so forth.

The external stress factors listed previously will normally result in individual human strain. Strain may, in turn, have an impact on any one or more of the human body systems (i.e., circulatory system, digestive system, nervous system, respiratory system, etc.). Measures of strain include such common parameters as pulse rate, blood pressure, body temperature, oxygen consumption, and the like.

Psychological Factors

Psychological factors pertain to the human mind and the aggregate of emotions, traits, and behavior patterns as they relate to job performance. All other conditions may be optimum from the standpoint of performing a task in an efficient manner; however, if the individual operator (or maintenance technician) lacks initiative, motivation, dependability, self-confidence, communication skills, and so on, the probability of performing in an efficient manner is low.

Psychological factors may be highly influenced by physiological factors. In general, one's attitude, initiative, motivation, and so on, is dependent on the needs and expectations of the individual. Fulfillment of these needs and expectations is a function of the organizational environment within which the individual performs and the leadership characteristics of supervisory personnel in that organization. If the individual operator does not perceive the opportunity for on-the-job growth, or if the managerial style of his or her immediate supervisor is not directly supportive of personal goals and objectives, job performance may deteriorate accordingly.

The psychological aspects of human performance are many and varied, and the reader is advised to review the literature on behavioral science in order to gain some insight as to the cause-and-effect relationships pertaining to good versus poor performance.

With the preceding basic characteristics in mind, the human must now be viewed as a component of the system. A key issue here is that of *information processing*. Information may be generally classified as *static* or *dynamic* (i.e. that which is constant and fixed on a display or that which is continually changing). Further, there are different categories of information to include quantitative information, qualitative information, warning and signal information, identification information, graphic information, alphanumeric and symbolic information, time-phased information, and so on. The designer must be cognizant of the requirements for the system, and must understand the human's capacities and abilities in this area.[7]

[7] M. S. Sanders, and E. J. McCormick, *Human Factors in Engineering and Design*, 7th ed. (New York: McGraw-Hill Book Co., 1992).

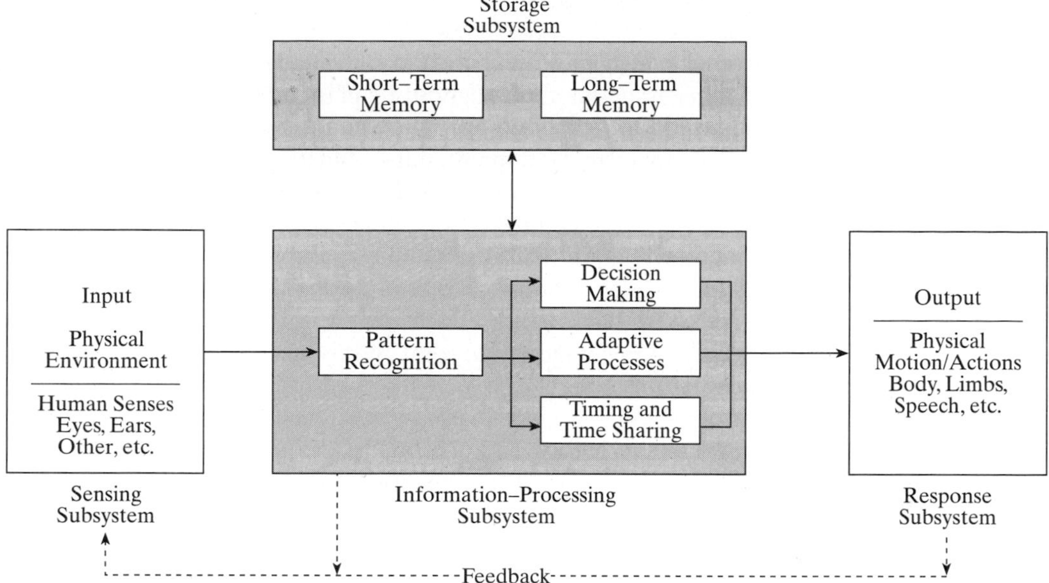

Figure 14.9 The human information processing subsystem (simplified).

Figure 14.9 portrays a simple information processing model where four basic human subsystems are identified. The *sensing* subsystem responds to specific types of energy identified through the human senses (i.e., vision, hearing, feeling, etc.). This provides the *stimulus* to initiate some form of action. The *information-processing* subsystem addresses the human's capacity to receive and process information. Of particular interest is the amount of information that the human can transmit (often expressed in terms of "bits"), and the rate with which he or she can transmit it. The *storage* subsystem refers to the human memory and its capacity, or the ability to retrieve data and facilitate the information processing. Finally, there is the *response* subsystem, which leads to the accomplishment of some function/task through a combination of physical motions (i.e., the output from the model). Inherent within this model is the *feedback*, which permits responses to be accurate in terms of the original input.[8]

In summary, the designer needs to understand not only the characteristics dealing with anthropometry, sensory factors, and the impact of external factors on the human, but the capacity and abilities of the human relative to the information processing requirement illustrated in Figure 14.9.

[8] Figure 14.9 constitutes a modified version of Figure 2.1 in H. P. Van Cott, and R. G. Kinkade, *Human Engineering Guide to Equipment Design*, Rev. ed. (Washington, D.C.: U.S. Government Printing Office, 1972).

14.2 THE MEASURES IN HUMAN FACTORS

An overall objective is to design a system such that it can be operated and maintained in an effective and efficient manner, throughout its planned life cycle, and in response to the needs of the customer. Referring to Figure 12.12, the *effectiveness* side of the spectrum infers that all operating and maintenance functions should be accomplished in a specified manner, within the desired period, and without inducing human errors in the process. The goal is to maximize performance, availability, dependability, and whatever other system-level goals that have been established. At the same time, to be *efficient* infers that such functions must be accomplished at minimum cost. Thus, the life-cycle cost side of the balance is important, particularly because personnel costs often account for a large percentage of the total. With this in mind, some measures of concern may include the following:

1. The number of personnel required in the *operation* of the system in a designated period, throughout its planned life cycle, or in terms of cycles of operation. A metric could be stated as the total number of labor hours per hour of system operation or per mission.
2. The number of personnel required in the *maintenance* of the system in a designated period, in terms of mission, or in terms of hours of system operation. A common measure is Maintenance Labor Hours Per Operating Hour (refer to maintenance labor hour factors in Section 13.2).
3. The operator personnel cost per hour of system operation ($/hour).
4. The maintenance personnel cost per hour of system operation, per mission segment, or per month (refer to maintenance cost factors in Section 13.2).
5. The number of operator or maintainer errors per period or mission segment. This may include both errors in procedure that result in having no detrimental effect on the other elements of the system, and the errors that cause the induction of problems in the other elements. In any event, a "system" failure occurs.

In addressing personnel requirements in general, an objective is to design the system such that it can be operated and maintained, with an existing group of personnel available to perform the functions desired, and where personnel entry skill-level requirements are held to a minimum. For instance, an objective may be to design the system such that it can be successfully operated by an individual with a *basic* skill level, where this level constitutes a person with the following characteristics:

1. Age: 18 to 21 years.
2. Education level: high school graduate.
3. General reading/writing level: ninth grade.
4. Experience: no regular work experience prior to training.
5. General qualifications: after a limited amount of familiarization training on the system, this individual can accomplish normal operator functions, involving system actuation, manipulation of controls, communication, and the recording of

data. The individual is able to follow clearly presented instructions where interpretation and decision making are not necessary. This individual will normally require close supervision.

The objective here is to establish system design criteria that will promote simplicity in operation (and simplicity in the performance of maintenance). The accomplishment of such will tend to minimize personnel training costs, reduce the probability of personnel-induced system failures, and minimize overall life-cycle cost.

In addition to measures at the system-level, there are numerous metrics associated with the human as an entity. For example, the anthropometric factors (i.e., static and dynamic measures of the human body) provide design-to guidance where the human is involved in performing system functions, particularly in workstation design. The volume and rate of information processing in *bits*, or *bits per unit of time*, serve as measures in the design of communication subsystems. The allocated time to accomplish a given function could be extremely important in the design of a process. When measuring the impact of external environmental conditions on the human, the degree of *strain* can be measured by observing the human's pulse rate, blood pressure, body temperature, oxygen consumption, or a combination of these. In essence, the measures can be numerous and varied, and all that are applicable for a given system need to be traceable from the requirements at the system level described in Section 3.4.

14.3 HUMAN FACTORS IN THE SYSTEM LIFE CYCLE

Human factors, like reliability and maintainability, must be an inherent consideration within the overall systems engineering process, beginning with the conceptual design phase. Referring to Figure 14.10 (which is an extension of Figure 2.2), qualitative and quantitative requirements pertaining to the human must be developed through the accomplishment of feasibility analysis (Section 3.2, the development of operational requirements and the maintenance concept (Section 3.4), and the identification and prioritization of TPMs (Section 3.5). As the top-level functions are initially identified for the system, trade-off studies are conducted with the intent of identifying functions/tasks that will be (1) accomplished solely by the human; (2) accomplished automatically through the use of equipment or software only, and where there is no anticipated human involvement; or (3) accomplished through a combination of people, equipment, software, facilities, and the like. The requirements for people, as elements of the system, are identified, leading to the development of the appropriate DDPs (refer to Figure 2.9).

Given the specification of the requirements for human factors in Block 1 of the figure (also included in the type A system specification—refer to Section 3.8), the next step is to transition from the functional analysis described in Sections 3.6 and 4.1 and down to the subsystem level and below as part of the system allocation process discussed in Section 4.2. The iterative process of synthesis, analysis, and evaluation is accomplished, addressing the human interfaces with other elements of the system on a continuing basis. This process is facilitated through the effective utilization of selected

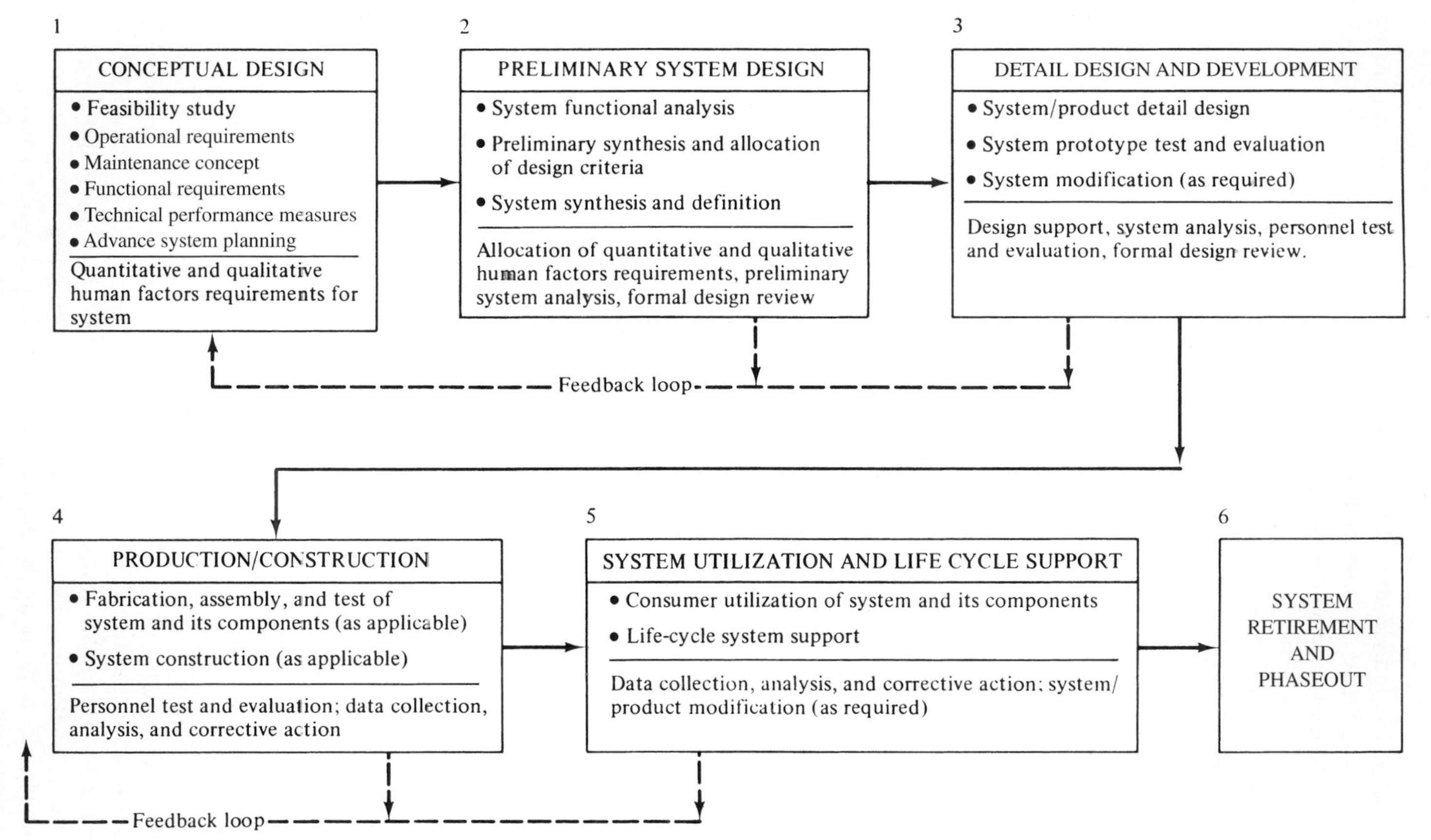
1
CONCEPTUAL DESIGN
• Feasibility study
• Operational requirements
• Maintenance concept
• Functional requirements
• Technical performance measures
• Advance system planning
Quantitative and qualitative human factors requirements for system
2
PRELIMINARY SYSTEM DESIGN
• System functional analysis
• Preliminary synthesis and allocation of design criteria
• System synthesis and definition
Allocation of quantitative and qualitative human factors requirements, preliminary system analysis, formal design review
3
DETAIL DESIGN AND DEVELOPMENT
• System/product detail design
• System prototype test and evaluation
• System modification (as required)
Design support, system analysis, personnel test and evaluation, formal design review.
Feedback loop
4
PRODUCTION/CONSTRUCTION
• Fabrication, assembly, and test of system and its components (as applicable)
• System construction (as applicable)
Personnel test and evaluation; data collection, analysis, and corrective action
5
SYSTEM UTILIZATION AND LIFE CYCLE SUPPORT
• Consumer utilization of system and its components
• Life-cycle system support
Data collection, analysis, and corrective action; system/product modification (as required)
6
SYSTEM RETIREMENT AND PHASEOUT
Feedback loop

Figure 14.10 Human factors in the system life cycle (refer to Figure 2.2).

methods/tools throughout the preliminary and detail design phases (i.e., Blocks 2 and 3 of Figure 14.10). As design progresses and physical models are developed, the ongoing evaluation process occurs and personnel test and evaluation is accomplished as part of the overall system test and evaluation activity described in Chapter 6.

System Requirements

The human, as an element of the system, is first visualized during the initial description of the need and the accomplishment of the feasibility analysis (refer to Sections 3.1 and 3.2). The customer (or user) identifies his or her role as an inherent part of the system from the beginning, whether one is dealing with a transportation vehicle, a computer workstation, a communications capability, or a health-care facility. Through the development of system operational requirements (Section 3.4), operating functions are further delineated, leading to the functional analysis and the identification of all of the major functions in which the human is involved (refer to Section 3.6 and Figure 3.14). Operating functions will subsequently lead to the development of maintenance functions.

At this early stage in conceptual design, the designer must conduct some initial trade-off studies to determine the extent to which the human will be involved in the operation of the system, considering the criteria discussed in Section 14.1 (i.e., the anthropometric factors, human sensory factors, information processing capabilities, and so on). Depending on the specific functions to be accomplished, the human may be totally involved 100% of the time, or partially involved with some degree of automation incorporated to facilitate accomplishment of the mission. In any event, human factors must be addressed initially at the system level (through early system analysis activities), and later at the subsystem level and below with the design of the various system elements and their interfaces.

Requirements Allocation

Referring to Figure 3.13, specific design-to criteria should be established for the "system operator." For example, it may be stated that—*the system shall be designed such that it can be successfully operated by no more than* x *number of personnel (between the 95th percentile man and the 5th percentile woman), with basic skills not exceed* y *level, throughout the mission, in time* z, *and without introducing any errors in the process.* Additional measures may include any one or more of the factors introduced in Section 14.2. As is the case for reliability and maintainability, specific requirements covering personnel must be tailored to the mission scenario, and are derived through the development of TPMs described in Section 3.5. The results could be included in a breakdown of requirements for the human such as that presented for the equipment in Figure 4.8.

Given the requirements for the system, the appropriate design criteria and DDPs can be established for its various elements (i.e., equipment, software, facilities, etc). The guidelines presented in Figures 14.2 through 14.8 provide an example for a start. Referring to Figure 14.4, in the design of rack-type consoles specific criteria can be established covering the placement of components based on anthropometric and sensory factors. In the design of control panels, there are specific guidelines pertaining to the type, sequencing, and location of switches, knobs, or other operator controls; there are guidelines for the type of guages or readout devices depending on the nature of the information desired; there are guidelines pertaining to labeling and color coding; there are guidelines pertaining to the transfer of information through audio and/or visual means; and so on. For assistance in this area, the reader is advised to review some of the literature in Appendix F.

14.4 HUMAN FACTORS ANALYSIS METHODS

Within the context of human factors analysis (accomplished as an iterative part of system synthesis, analysis, and evaluation), there are several tools that can be effectively used in support of the objectives described throughout this text. Of particular interest herein are the operator task analysis (OTA), development of operational sequence diagrams (OSDs), error analysis, system safety/hazard analysis, and the use of mock-ups for the purpose of evaluation. The relationships between some of these activities are illustrated in Figure 14.11.

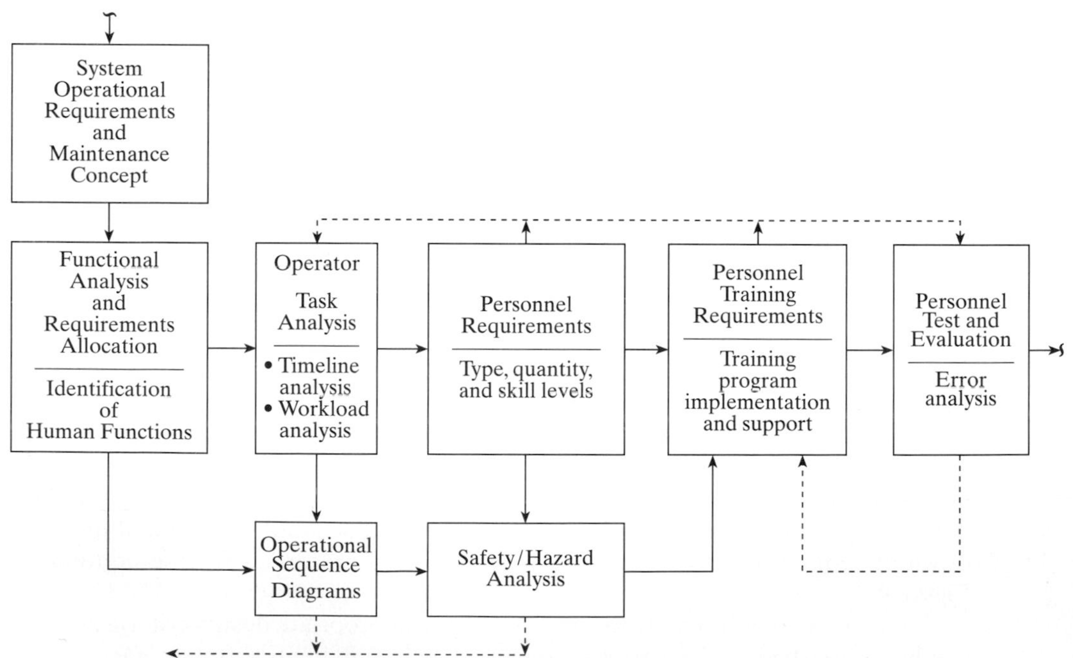

Figure 14.11 The application and relationships of selected tools/methods used for human factors in design.

Operator Task Analysis (OTA)

This facet of analysis involves a systematic study of the human behavior characteristics associated with the completion of system tasks. In accomplishing a task analysis, the following steps are applied in most instances and Figure 14.12 presents a format that is often used:

1. Identify system operator functions and establish a hierarchy of these functions in terms of job operations, duties, tasks, subtasks, and task elements as described in Section 14.1.
2. For each function involving the human element, determine the specific information necessary for operator personnel decisions. Such decisions may lead to the actuation of a control, the monitoring of a system condition, or equivalent. Information required for decision making may be presented in the form of a visual display or an audio signal of some type.
3. For each action, determine the adequacy of the information fed back to the human as a result of control activations, operational sequences, and so on.
4. Determine the time requirements, frequency of occurrence, accuracy requirements, and the criticality of each action (or series of actions) accomplished by the human.[9]
5. Determine the impact of the environmental and personnel factors and constraints on the human activities identified in item 1.
6. Determine the human skill-level requirements for all operator personnel actions, group and assign tasks to a designated workstation, and determine the total quantity and skill levels of personnel required for the system.[10]

Figure 14.12 can be used as an aid in accomplishing the OTA. Referring to the figure, each subtask (or equivalent) is identified in column 3. This may involve operating, checking, adjusting, troubleshooting, or conducting some similar activity. The action stimulus, or event, that instigates performance of the subtask is noted in column 4. This may include initiating a command, an out-of-tolerance display indication, a symptom of failure, or equivalent. The required action, or response, is noted in column 5. This may include a control movement, a voice communication command, or some other human response. The feedback or indication of the adequacy of the response is included in column 6. Column 7 indicates the initial classification of the subtask along with other similar subtasks, and column 8 is used to enter potential sources of human error. Column 9 is divided into two sections, with *allowable* time being the period within which the subtask must be completed and *necessary* time being the actual time

[9] Depending on the results of this part of the analysis, it may be appropriate to revisit the FMECA to determine possible impacts on the system overall (refer to Section 12.4).

[10] The approach here is similar to what is accomplished for the assignment and grouping of maintenance tasks analyzed through the maintenance task analysis (MTA) described in Section 13.4. In certain cases, it may be appropriate to accomplish the OTA and MTA jointly on an integrated basis, particularly when dealing with the determination of personnel quantities and skill levels.

Function (1)	Operate aircraft power plant and system controls
Task (2)	Control jet engine operation

Subtask (3)	Action Stimulus (4)	Required Action (5)	Feedback (6)	Task Classification (7)	Potential Errors (8)	Time (9)		Work station (10)	Skill Level (11)
						Allowable (9a)	Necessary (9b)		
3.1 Adjust engine rpm	4.1 Engine rpm on tachometer	5.1 Depress throttle control downward	6.1 Increase in indicated tachometer rpm	7.1 Operator task, aircraft commander	8.1 a. Misread tachometer b. Fail to adjust throttle to proper rpm	9a.1 10 sec	9b.1 7 sec	10.1 aircraft commander seat	11.1 Low

Figure 14.12 Sample format for task allocation and analysis. *Source:* H. P. Van Cott, and R. G. Kincade, *Human Engineering Guide to Equipment Design,* rev. ed. Washington, D.C.: U.S. Government Printing Office, 1972.

required. A time constraint may be initially specified through the allocation process discussed in Section 14.3. The geographical location or the prospective workstation where the subtask is to be accomplished is noted in column 10. Finally, the anticipated skill level required for subtask completion is included in column 11.

The purpose of the task analysis is to ensure that each *stimulus* is tied to a *response* (and that each response is directly related to a stimulus). Further, the individual human motions are analyzed on the basis of dexterity, mental and motor skill requirements, stress and strain characteristics of the human performing the task/subtask, and so no The objective is to (1) identify those areas of system design where potential human-machine problems exist, (2) identify the necessary personnel quantities and skill-level requirements for operating the system, and (3) identify requirements for personnel training. The OTA format, as shown in Figure 14.12, can be extended to include the performance of a *timeline* analysis and a *workload* analysis. The timeline analysis addresses those tasks that can be accomplished sequentially and those in parallel, and the workload analysis leads to the identification of the quantity of personnel required.

Operational Sequence Diagram (OSD)

The OSD can be used to aid in evaluating the *flow of information* from the point when the operator first becomes involved with the system to the completion of the mission. Information flow in this instance pertains to operator decisions, operator control activities, and the transmission of data. Figure 14.13 presents an example of an OSD format.

Stemming from the functional analysis and the OTA, *operator* actions can be identified in a top-down sequential format as shown in the figure. In this instance, there are two system operators required, each assigned to a workstation. The objective is to show how the information flows between the two operators and their respective workstations, presented in time as one progresses from the top down. The symbology included in the figure reflects the "actions" that occur. This, in turn, leads to the requirements in design for specific types of displays, the incorporation of electrical or mechanical means for information processing, the incorporation of automation, and so on. Through an evaluation of these sequences of operation, and considering the human characteristics described in Section 14.1, one can assess the adequacy of design regarding the interfaces between the human and other elements of the system (i.e., control panel design in this instance). Results from the generation of OSDs can aid in the development of personnel training requirements.

Error Analysis

An error occurs when a human action exceeds some limit of acceptability, where the limits of acceptable performance have been defined. Errors may be broken down into errors of *omission* when a human fails to perform a necessary task and errors of *commission* when a task is performed incorrectly (i.e., selection, sequence, or time errors). The possible causes of error are due to the following:

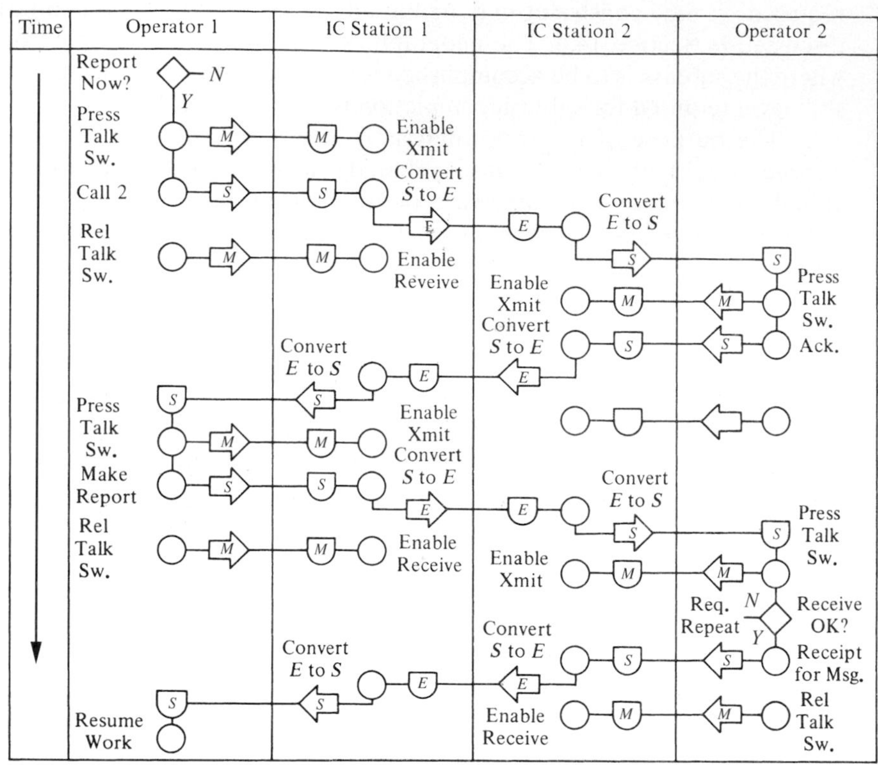

Notes on Operational Sequence Diagram

Symbols

◇ = Decision

○ = Operation

⇨ = Transmission

⊔ = Receipt

D = Delay

□ = Inspect, Monitor

▽ = Store

Links

M = Mechanical or Manual

E = Electrical

V = Visual

S = Sound etc.

Stations or subsystems are shown by columns; sequential time progresses down the page.

Figure 14.13 Operational sequence diagram (example). *Source:* MIL-H-46855, Military Specification, *Human Engineering Requirements for Military Systems, Equipment and Facilities.* Washington, D.C.: Department of Defense.

1. Inadequate work space and work layout—poor workstation design relative to seating, available space, activity sequences, and accessibility to system elements
2. Inadequate design of facilities, equipment, and contol panels for human factors—inadequate displays and readout devices, poor layout of controls, and lack of proper labeling

3. Poor environmental conditions—inadequate lighting, high or low temperatures, and high noise level
4. Inadequate training, job aids, and procedures—lack of proper training, and poorly written operating and maintenance procedures
5. Poor supervision—lack of communications, no feedback, and lack of good planning resulting in overtime

The error analysis can be accomplished analytically in conjunction with the OTA, MTA, during the development of OSDs, and physically as part of the system test and evaluation effort discussed in Chapter 6. The objective is to select a human with the appropriate skills and training, simulate the operation of the system by undertaking a series of functions/tasks, and measure the number of errors that occur in the process. Using an Ishikawa "cause-and-effect" diagram approach, as conveyed in Figure 12.25, can aid in determining the causes of the errors (i.e., how was the error introduced in the system)? Additionally, it is important to note the effects of the error on other elements of the system and on the system as an entity. Thus, it is important that the error analysis be closely integrated with both the FMECA (Section 12.4) and the System Safety/Hazard Analysis discussed subsequently.

Safety/Hazard Analysis

The safety/hazard analysis is closely aligned with the FMECA described in Section 12.4. Safety pertains to both personnel and the other elements of the system with personnel being emphasized herein. The safety/hazard analysis generally includes the following basic information[11]:

1. *Description of hazard.* System operational requirements and the system maintenance concept are reviewed to identify possible hazardous conditions. Past experience on similar systems or products utilized in comparable environments serves as a good starting point. Hazardous conditions may include acceleration and motion, electrical shock, chemical reactions, explosion and fire, heat and temperature, radiation, pressure, moisture, vibration and noise, and toxicity.
2. *Cause of hazard.* Possible causes should be described for each identified hazard. In other words, what events are likely to occur in creating the hazard?
3. *Identification of hazard effects.* Describe the effects of each identified hazard on both personnel and equipment. Personnel effects may include injuries, such as cuts, bruises, broken bones, punctures, heat exhaustion, asphyxiation, trauma, and respiratory or circulatory damage.
4. *Hazard classification.* Hazards may be categorized according to their impact on personnel and equipment as follows:

[11] Two references in this area are (1) W. Hammer, *Occupational Safety Management and Engineering*, 4th ed. (Upper Saddle River, N. J.: Prentice Hall, Inc., 1988); and (2) H. E. Roland, and B. Moriarty, *System Safety Engineering and Management*, 2nd ed. (New York: John Wiley & Sons, Inc., 1990).

   a. *Negligible hazard (category I).* Such conditions as environment, personnel error, characteristics in design, errors in procedures, or equipment failures that will not result in significant personnel injury or equipment damage.
   b. *Marginal hazard (category II).* Such conditions as environment, personnel error, characteristics in design, errors in procedures, or equipment failures that can be controlled without personnel injury or major system damage.
   c. *Critical hazard (category III).* Such conditions as environment, personnel error, characteristics in desiring errors in procedures, or equipment failures that will cause personnel injury or major system damage, or that will require immediate corrective action for personnel or system survival.
   d. *Catastrophic hazard (category IV).* Such conditions as environment, personnel error, characteristics in design, errors in procedures, or equipment failures that will cause death or severe injury, or complete system loss.

5. *Anticipated probability of hazard occurrence.* Through statistical means, estimate the probability of occurrence of the anticipated hazard frequency in terms of calendar time, system operation cycles, equipment operating hours, or equivalent.
6. *Corrective action or preventive measures.* Describe the action(s) that can be taken to eliminate or minimize (through control) the hazard. It is hoped that all hazardous conditions will be eliminated; however, in some instances it may only be possible to reduce the hazard level from category IV to one of the lesser critical categories.

The safety/hazard analysis serves as an aid in initially establishing design criteria and as an evaluation tool for the subsequent assessment of design for safety. Although the format may vary somewhat, a safety analysis may be applied in support of requirements for both industrial safety and system/product safety.

Mockups

Quite often, and before the development of the engineering and prototype models described in Section 5.6, there is a need to visualize an equipment layout, a panel design, and other elements of the system in a three-dimensional context. This is particularly true when designing workstations and associated space needs, and in the identification and location of switches, knobs, displays, and related controls in the design of a console control panel. To facilitate the visualization objective, the designer may wish to develop one or a series of *mockups* to demonstrate a concept. Mockups can be easily constructed with cardboard, wood, sheet metal, actual system components, or various combinations of such. The idea is to replicate the final configuration as economically as possible, to the desired degree of accuracy, and as early in the system life cycle as practicable. The use of mockups can either facilitate the modeling of a system configuration using CAD methods, or take the place of a CAD presentation when the design is simple and the use of CAD is too expensive.

14.5 PERSONNEL AND TRAINING REQUIREMENTS

Personnel requirements for the system may be categorized in terms of *operator* personnel and *maintenance* personnel. Operator personnel requirements are derived through the human factors analysis, particularly through the generation of the detailed operator task analysis and operational sequence diagrams. Maintenance personnel requirements evolve from the maintainability analysis, the supportability analysis, and the detail maintenance task analysis.

In defining these requirements for the system, personnel numbers and skill levels are determined for all human activities by function, job operation, duty, task, and so on. These requirements, initially identified in small increments, are combined on the basis of similarities and complexity level. Individual position requirements are identified, and worksheets are prepared specifying the duties and tasks to be performed by each individual. From this information, personnel quantities and skill level requirements are identified for the system as an entity.

Given the personnel requirements (in terms of positions) for the system, the next step is to identify available resources and those individuals who can be employed to operate and maintain the system. These selected individuals are evaluated relative to their current skills as compared to the skill levels necessary for the system. The differences dictate the needs for training.

For the sake of illustration, let us assume that as a result of the human factors analysis the system will require x individuals with a *basic* skill, y individuals with an *intermediate* skill, and z individuals with a *high* skill, and that these skill level classifications are broadly defined as follows[12]:

Basic skill level. A basic skill level is assumed to require an individual between 18 and 21 years of age, a high school graduate with a ninth-grade general reading and writing level and having no regular work experience. After a limited amount of familiarization training on the system, this individual can accomplish simple operator functions involving system actuation, manipulation of controls, communications, and the recording of data. The individual is able to follow clearly presented instructions where interpretation and decision making are not required. Close supervision of this individual is normally required.

Intermediate skill level. An intermediate skill level normally requires an individual over 21 years of age, with approximately 2 years of college or equivalent course work in a technical institute, has had some specialized training on similar systems in the field, and has had from 2 to 5 years of work experience. Personnel in this classification can perform relatively complex tasks, where the interpretation of data and some decision making may be required, and can accomplish simple on-site preventive maintenance tasks. This individual requires little supervision.

[12] As indicated, these classifications are defined in rather broad terms to illustrate a concept. In certain instances it may be necessary to elaborate further, particularly regarding complex functions, to provide a good basis for determining specific training needs.

High skill level. A high skill level normally requires an individual with 2 to 4 years of formal college or equivalent course work in a technical institute, who has taken several specialized training courses in various related fields and possesses 10 years or more of related on-the-job experience. An individual in this classification may be assigned to train and supervise basic and intermediate skill-level personnel, and is in the position to interpret procedures, accomplish complex tasks, and make major decisions affecting system operating policies. Further, this individual is qualified to accomplish (or supervise) all on-site preventive maintenance requirements for the system.

Given requirements for the system, it is assumed that all available resource personnel currently possess the necessary qualifications for entry into positions requiring a basic skill level; that is, with some general familiarization training on the system, they can usually perform the level of activity noted for an individual with a basic skill.

On the basis of these assumptions, it is necessary to develop a training program that will

1. Train entry-level personnel in the fundamentals of system operation to fill the positions where a *basic skill* level is required.
2. Train entry-level personnel in the performance of operator and maintenance functions to the extent necessary to satisfy the *intermediate skill-level* requirements for the system.
3. Train entry-level personnel in the performance of operator and maintenance functions to the extent necessary to satisfy the *high skill-level* requirements for the system.

Training may be accomplished through a combination of formal structured programs and on-the-job (OJT) training. Training requirements include both the training of personnel initially assigned to the system and the training of replacement personnel throughout the life cycle as required due to attrition. In addition, training must cover both operator activities and maintenance activities.

In designing training programs, the specific requirements may vary considerably depending on the complexity of the human functions to be performed in the operation and maintenance of the system. The level of complexity is a function of system design as discussed in the earlier sections of this chapter. If there are many human functions of a highly complex nature, then the training requirements will likely be extensive and the associated cost will be high. Conversely, if the functions to be performed by the human being are relatively simple, then training requirements (and associated cost) will be minimal. In any event, a training program should be designed and tailored to meet the personnel needs for the system. The program content must be at the appropriate level to provide future system operators and maintenance personnel with the tools necessary to enable the performance of their respective functions in an effective and efficient manner.

With this objective in mind, a formal training plan is often developed for the system during the detail system/product design phase. This plan should include a statement of training objectives, a description of the training program in terms of modules and content, a proposed training schedule, a description of required training aids and data, and an estimate of anticipated training cost(s). The training plan must be implemented in time to support system operation and maintenance activities as the system is distributed for use.

14.6 DESIGN REVIEW AND EVALUATION

Review of the design for the incorporation of human engineering characteristics is accomplished as an inherent part of the process described in Sections 3.9, 4.6, and 5.7. The characteristics of the system (and its elements) are evaluated in terms of the initially specified human factors requirements for the system. If the requirements appear to be fulfilled, the design is approved as is. If not, the appropriate changes are initiated for corrective action.

In support of conducting a human factors review, one may wish to develop a design review checklist, including questions such as those listed subsequently.

1. Have the qualitative and quantitative requirements for human factors in design been adequately defined and specified? Have they been allocated from the top down?
2. Has a top-level system analysis been accomplished to identify those functions that are to be completed by the human? Have these functions been broken down into job operations, duties, tasks, subtasks, and task elements to the extent practicable?
3. Have the appropriate interfaces been defined between the human and the other elements of the system (i.e., equipment, software, and facilities)?
4. Does the system design adequately reflect the proper consideration of anthropometric, human sensory, physiological, and psychological factors?
5. Does the design reflect consideration of the abilities and capacities of the human in dealing with information processing requirements (see Figure 14.9)?
6. Have those tasks to be accomplished by the human been justified through the OTA? Can they be traced from the functional analysis (refer to Sections 3.6 and 4.1)?
7. Have the personnel quantities and skill levels been defined for the system? Has the system been designed such that it can be successfully operated by an individual with *basic* skills? Have the number of operating personnel been optimized to the extent practicable?
8. Have the interfaces between the human and other elements of the system been properly defined and justified through the development of OSDs? Are the OTA and OSDs compatible? Do they justify and lead into the development of training requirements?
9. Has a system safety/hazard analysis been completed? Is it compatible with the FMECA (refer to Section 12.4)?
10. Has an error analysis been conducted? Do the results feed back to the FMECA and the safety/hazard analysis? Have the results been provided as an input in the development of personnel training requirements?

11. Have the environments been adequately defined for each area where system functions/tasks are to be accomplished by the human? Are they optimal?
12. In the design of control panels and displays, are the controls standardized? Are the controls sequentially positioned? Are they placed according to frequency or criticality of use? Is control spacing adequate? Is control labeling adequate? Have the proper control and display relationships been incorporated? Are the proper type of panel switches used? Is control panel lighting adequate? Has the overall design been justified through the OSDs?
13. In the area of safety, have system/product hazards from heat, cold, thermal change, barometric change, humidity change, shock vibration, light, mold, bacteria, corrosion, rodents, fungi, odors, chemicals, oils, greases, handling and transportation, and so on been eliminated?
14. Have fail-safe provisions been incorporated in the design?
15. Have protruding devices been eliminated or are they suitably protected?

The items covered certainly are not intended as being all-inclusive but merely represent a sample of possible interest areas (it should be noted that the answer to those questions that are applicable should be *yes*). A good design review should address these and related issues.

14.7 PERSONNEL TEST AND EVALUATION

Chapter 6 addresses the overall requirements for system test and evaluation in a fairly comprehensive manner. Included in the system test and evaluation effort is the human element of the system. As equipment is being evaluated for performance, reliability, maintainability, and so on, the operator and the maintenance technician should be evaluated in terms of human performance efficiency.

During the latter part of the detail design phase and throughout the subsequent phases of the life cycle (refer to Figure 14.10), there is a personnel test and evaluation activity. The basic objective of this activity is to (1) monitor human performance in fulfilling system operator and maintenance requirements; (2) assess operator and maintenance tasks in terms of elapsed time, step sequences, dexterity requirements, areas of difficulty, and human error; (3) identify problem-areas and initiate recommendations for corrective action; and (4) modify personnel selection and training requirements as necessary for compatibility with any system changes that may occur. Thus, the process described in Chapter 6 is followed, but the emphasis is on the human being.

QUESTIONS AND PROBLEMS

1. Define what is meant by *human factors*. When in the system life cycle are human factors considered? Why is it important to consider human factors in design?
2. What is meant by *anthropometry*, and how does it relate to system design? How do human *physiology* and *psychology* relate to system design? Are they important? Why?

3. Describe the *human sensory factors*. How do they relate to system design?
4. Describe the possible impact(s) of physiological and psychological factors on system operation and the successful completion of a defined mission. Provide some examples.
5. What effects on the human are likely to occur if the operator in performing his or her function is undertrained for the job? Overtrained? How would these effects likely influence system operation?
6. Refer to Figure 14.9. Describe some of the considerations pertaining to the human that need to be addressed in the design of an information processing capability.
7. Identify some of the measures that you might apply as "design-to" requirements for the human elements of the system. Allocate these factors to the subsystem level.
8. What is the purpose of the OTA? From what source does it evolve? What information is included? How might the results be used in system design? Provide an example (or two). How does the OTA relate to the MTA (if at all)?
9. From what source can a *timeline* and *workload* analysis be developed? What might be the objectives in accomplishing these analyses?
10. Select a system of your choice and accomplish an OTA on a major element of that system.
11. What is the purpose of the OSD? How does it relate to the OTA? How might the results be used in system design? Provide an example (or two).
12. In the system selected for Question 10, identify a major human-equipment interface area and develop an OSD. Show how the OSD might evolve from the OTA.
13. What is the purpose of the system *safety/hazard analysis*? What information is included? How does this analysis relate to the FMECA (if at all)?
14. What is the purpose of an *error analysis*? Briefly describe the steps involved? When can an error analysis be accomplished? How does the error analysis relate to the OTA? OSD? FMECA? Safety/hazard analysis?
15. What are the possible uses and benefits of a *mockup*?
16. How are personnel quantities and skill-level requirements defined for the system? Identify the steps involved. In analyzing the results, what would you look for in identifying potential problem areas?
17. How are personnel training requirements determined? What input data are required?
18. What steps would you take in the design of a training program?
19. How would you measure the effectiveness of a training program?
20. What is the purpose of personnel test and evaluation? When are the requirements initially defined? When is test and evaluation accomplished? What would you expect from the results? What steps would you take if you should identify a problem as a result of the test and evaluation effort.
21. How does the test and evaluation effort described in Section 14.7 relate to the maintainability demonstration described in Section 13.6 (if at all)?

15 Design for Supportability (Serviceability)

Major systems and products have been planned, designed and developed, produced, and delivered to the customer with little consideration given to the aspects of maintenance and supportability. Supportability pertains to the infrastructure that includes all activities such as (1) the procurement and acquisition of people, materials, equipment, software, facilities, and data required for the initial distribution and sustaining support of the system throughout its planned life cycle; (2) the services and training that are required to assist the customer or user in the operation of the system; (3) providing the necessary day-to-day maintenance and support to ensure that the system is "operational" on a continuing basis; and (4) providing the mechanism for the retirement, disposal, or recycling of materials that are no longer useful. It is this infrastructure that is needed to ensure that the system will successfully accomplish its mission effectively, efficiently, and with complete consumer satisfaction.

Although the issues of maintenance and logistic support have been discussed in the earlier chapters, the emphasis has primarily been directed toward the consideration of reliability, maintainability, and usability characteristics in the design of the prime mission-related elements of the system. The objective has been to ensure that the major elements of the system can be effectively and efficiently supported throughout their planned life cycle. However, in accomplishing this objective, one needs to design the maintenance and support infrastructure as well to ensure that it can be "responsive" in an effective and efficient manner. From a systems engineering perspective, this area of activity is critical and must be in place if the overall objectives discussed in this text are to be accomplished.

15.1 DEFINITION AND EXPLANATION OF SUPPORTABILITY

Supportability, as defined herein, refers to the inherent characteristics of design and installation that enable the effective and efficient maintenance and support of the system throughout its planned life cycle. It is closely related to reliability and maintainabilty in that to acquire good supportability requires the incorporation of reliability and maintainability characteristics in the design (i.e., accessibility, good diagnostics, interchangeability, modularization, labeling, etc.).[1]

To realize the objective of designing a system for supportability, one needs to address not only the incorporation of reliability and maintainability characteristics in the design of the prime elements of the system, but the design of the support infrastructure is critical in that it must be responsive to the demands exhibited by these prime elements. It is this support infrastructure that is primarily being addressed in this chapter. Thus, it is appropriate to commence with some discussion of this infrastructure and the elements within.

Figure 15.1 illustrates the basic activities of concern and the relative flow from one area to another. Planning is initiated, design and development is accomplished,

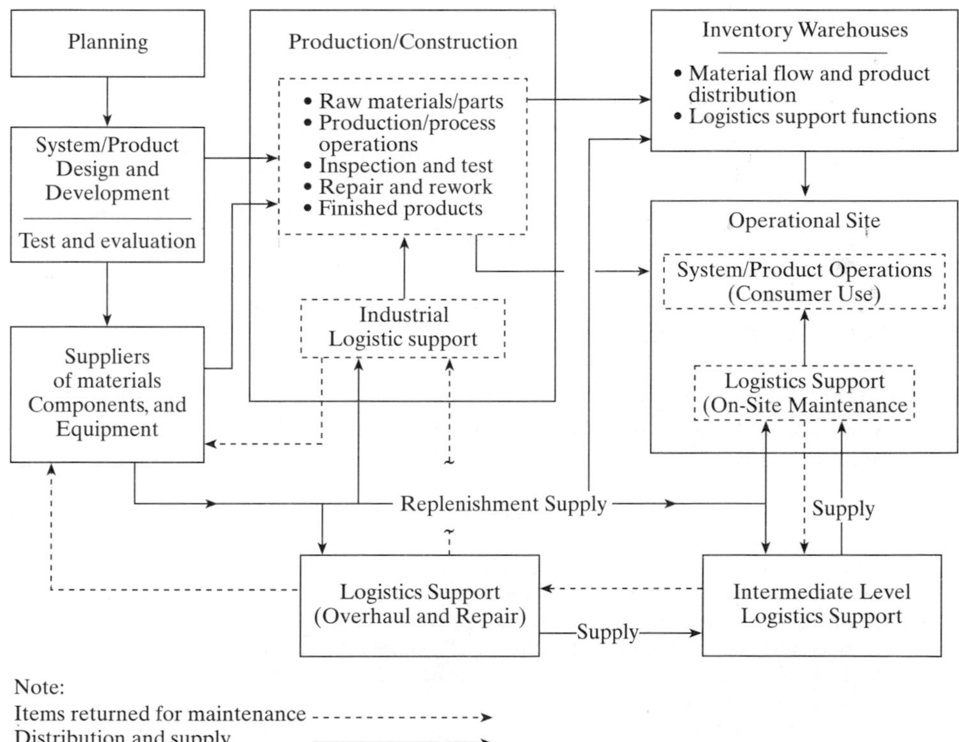

Figure 15.1 The flow of activities and materials for system operation and maintenance.

[1] In the commercial (nondefense) sector, the term *serviceability* is often used in a similar context.

items are produced with suppliers providing the needed materials, system elements are delivered and installed at the customer's operational site, and system use occurs. This *outward* flow of activities and materials constitutes one segment of the overall logistics cycle (i.e., the procurement, distribution, transportation, warehousing, and ultimate delivery of the system to the customer [user].[2]

Given that the system is "operational," then there is a *reverse* flow when failures occur and items need to be returned for intermediate-level and depot/producer-level maintenance. The maintenance cycle is amplified further in Figure 3.5. In support of maintenance activities, there is a need for properly trained personnel, spares/repair parts and associated inventories, test and support equipment, facilities, a transportation capability, computer resources, and data. The appropriate resources are assigned at each level with the requirements at the organizational level being supported by the capability at the intermediate level, activities at the intermediate level being supported by the depot/producer capability, and so on. This maintenance network constitutes another segment of the logistics cycle (i.e., an element of *integrated logistic support* [ILS].[3]

Because *systems* are addressed throughout this text, the concepts and principles associated with ILS have been adopted herein and are considered applicable to any type of a system, and the structure in Figure 3.5 will serve as a reference for discussion purposes in subsequent sections of this chapter. With this in mind, logistics may be defined as "a disciplined, unified, and iterative approach to the management and technical activities necessary to (1) integrate support considerations into system and equipment design; (2) develop support requirements that are related consistently to readiness objectives, to design, and to each other; (3) acquire the required support; and (4) provide the required support during the operational phase at minimum cost."[4] The first two objectives in this definition, being design related, pertain to *supportability*.

15.2 THE ELEMENTS OF SUPPORT

Figure 15.2 identifies eight basic elements of maintenance and support, which, in turn, can be related directly to the activities associated within the support infrastructure presented in Figure 3.5. Although these elements are shown as independent entities, the

[2] In the commercial sector, *logistics* can be defined as the "process of planning, implementing, and controlling the efficient, cost-effective flow and storage of raw materials, in-process inventory, finished goods, and related information from the point of origin to the point of consumption for the purpose of conforming to customer requirements." This definition, from the Council of Logistics Management, is basically oriented to the "business" aspects pertaining to the physical distribution of *consumable* items and not systems. Nevertheless, many of the principles and concepts applied in the commercial sector are also applicable in the acquisition of systems and their components. Refer to the literature on business logistics or logistics management in Appendix F.

[3] In the defense sector, ILS can be defined as a "management function that provides the initial planning, funding, and controls which help to assure that the ultimate consumer (or user) will receive a system that will not only meet performance requirements, but one that can be expeditiously and economically supported throughout its programmed life cycle." This definition, which addresses the maintenance and support of *systems*, can be applied in both the defense and commercial sectors.

[4] DSMC, *Integrated Logistic Support Guide* (Fort Belvoir, Va.: 1986).

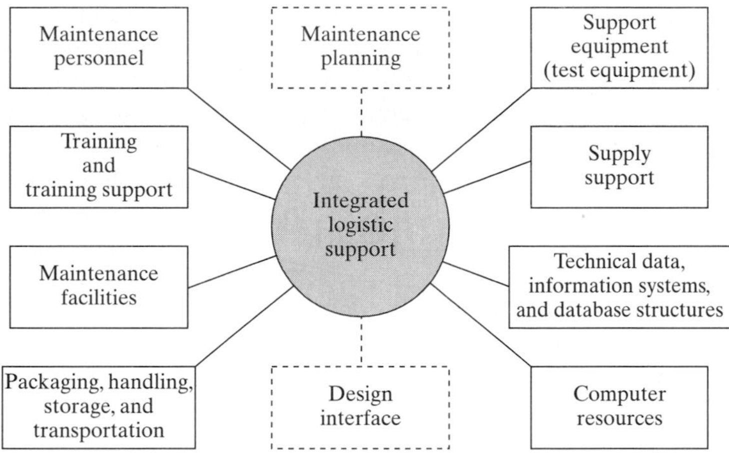

Figure 15.2 The elements of support.

key issue is that of *integration*. As these elements are all interrelated, it is essential that they be addressed as a major subsystem of the end product. It may be necessary (as part of the system synthesis, analysis, and evaluation process) to trade off certain decisions in one area in favor of another. In any event, the end result should convey a total integrated approach. The elements are noted as follows:

1. *Maintenance personnel* required for the installation, checkout, and sustaining maintenance and support of the system throughout its planned life cycle. This includes personnel at all levels (refer to Figure 3.6), mobile teams, operators at test facilities and calibration laboratories, and so on. Personnel requirements are based on the results of the maintenance task analysis discussed in Section 13.4.
2. *Training and training support* required for all system operator and maintenance personnel throughout the system life cycle (refer to Section 14.5). Personnel requirements are identified through the operator task analysis (Section 14.4) and the maintenance task analysis (Section 13.4) respectively. The objective is to assess the personnel quantities and skills levels required for the new system and to develop the capabilities of existing personnel to meet these needs. Formal training may be completed through a combination of regular classroom instruction, periodic seminars and workshops, computer-based education, instruction through the use of video tapes, correspondence courses, or on-the-job training (OJT). Training support pertains to the required training facilities and associated resources, training aids (mockups, simulators, special devices, and software), and trailing procedures and manuals.
3. *Supply support*, which includes all spares (repairable units, assemblies, modules), repair parts (nonrepairable components), consumables (liquids, lubricants, disposable items), special supplies, and related inventories necessary to support the prime mission-oriented elements of the system, maintenance computers and software, test and support equipment, transportation and handling equipment, training equipment, and facilities. The requirements for spares/repair-parts/

consumables are initially identified through the maintenance task analysis (Section 13.4) and later through the supportability analysis (Section 15.5). This category includes the initial provisioning (and subsequent reprovisioning) documentation, procurement activities, warehousing, and the distribution of materials to the proper locations (refer to Figures 3.5 and 15.1).

4. *Support equipment*, which includes all tools, condition monitoring equipment, diagnostic and checkout equipment, special test equipment, metrology and calibration equipment, maintenance fixtures and stands, and special handling equipment required to support all scheduled and unscheduled maintenance actions accomplished at each level of maintenance (refer to Figures 3.5 and 3.6). These requirements are initially derived from the maintenance task analysis (Section 13.4) and later through the supportability analysis (Section 15.5).
5. *Computer resources*, which cover all computers and associated software necessary to support scheduled and unscheduled activites at each level of maintenance. This may include any condition monitoring programs, diagnostic tapes, and so on. Any requirements for the processing of maintenance data (refer to Section 6.5), the implementation of a computer-integrated-maintenance-management (CIMM) capability, and so on, could fall within this category.
6. *Packaging, handling, storage, and transportation*, which includes all equipment, special provisions, containers (reusable and disposable), and supplies necessary to support the packaging, preservation, storage, handling, and/or transportation of the prime mission-related elements of the system, personnel, spares/repair parts, support equipment, technical data, and mobile facilities. This category basically covers the initial and sustaining transportation requirements in support of the distribution of materials and maintenance cycle illustrated in Figure 15.1 (i.e., the *outward* and *reverse* flows discussed in Section 15.1). The primary modes of transportation include *air, highway, railway, waterway*, and *pipeline*.[5]
7. *Maintenance facilities* includes all facilities required to support scheduled and unscheduled maintenance actions at all levels of maintenance (refer to Figure 3.5). Physical plant, portable buildings, mobile vans, housing, intermediate-level maintenance shops, calibration laboratories, and special repair shops (depot, overall, suppliers) must be considered. Capital equipment and utilities (heat, power, energy requirements, environmental controls, communications, etc.) are generally included as part of facilities.
8. *Technical data, information systems, and database structures* includes system installation and checkout procedures, operating and maintenance instructions, inspection and calibration procedures, overhaul instructions, facilities data, modification instructions, engineering design data (specifications, drawings, materials and parts lists, digital data), supplier data, and logistics provisioning and pro-

[5] The area of *transportation*, a major element of logistics in the commercial sector (refer to footnote 2), is covered extensively in a number of the "business" logistics references in Appendix F. Two good sources are J. J. Coyle, E. J. Bardi, and J. L. Cavinato, *Transportation*, 4th ed. West Publishing Co., (St. Paul, Minn. 1992); and N. A. Glaskowsky, D. R. Hudson, and R. M. Ivie, *Business Logistics*, 3rd ed. (Orlando: The Dryden Press, Harcourt Brace Jovanovich Co., 1992).

curement data that are necessary in the performance of system development, production, operation, maintenance, and retirement functions. Such data should not only cover the prime mission-oriented elements of the system but the other elements of the support infrastructure as well (i.e., test and support equipment, transportation and handling equipment, training equipment, and facilities). Included within this category are the information system capabilities, and associated databases, that allow for the implementation of effective *electronic data interchange* (EDI) processes and the requirements associated with *Continuous Acquisition and Life-Cycle Support* (CALS).

For large-scale systems, the logistic support requirements throughout the life cycle may be significant. The prime mission-oriented segment of the system must be designed with support in mind, and the various elements of logistic support must be designed to be compatible with the prime mission equipment. Further, these different elements of logistic support interact with each other and the effects of these interactions must be reviewed and evaluated continually. A major decision or a change involving any one of these elements could significantly impact other elements and the system as a whole.

Conversely, the logistics requirements for relatively small systems (or products) may entail only the functions of product distribution to the customer and initial system installation and checkout, whereas the sustaining life-cycle maintenance support will be minimal. In this instance, the emphasis on logistic support (and the design for supportability) will not be as great as for large complex systems. The specific requirements must be tailored accordingly.

The objective is to develop a system reflecting the proper balance between the prime mission elements of that system and its related support. This balance deals with establishing the proper relationships between system performance characteristics, reliability and maintainability characteristics, supportability characteristics, economic factors, and so on (refer to Figure 12.12). To attain such a balance requires the consideration of maintenance and logistic support throughout the system life cycle, particularly in the early phases of planning and conceptual design when major decisions affecting the overall system configuration are made. This emphasis in the early phases of the life cycle is provided through design for supportability, an inherent part of the systems engineering process.

15.3 THE MEASURES OF SUPPORTABILITY

The quantitative factors of maintenance and support can vary significantly depending on the type of system, its mission, and the criteria employed in system evaluation. These factors pertain to supply (spares), transportation, test equipment, maintenance organizations, facilities, and the like. Thus, one should deal with the measures associated with the elements defined in Section 15.2. A few of these measures are described subsequently.

Supply Support Measures

Supply support includes the spare parts and the associated inventories necessary for the accomplishment of unscheduled and scheduled maintenance actions. At each maintenance level, one must determine the type of spare part (by manufacturing part number) and the quantity of items to be purchased and stocked. Also, it is necessary to know how often various items should be ordered and the number of items that should be procured in a given purchasing transaction.

Spare part requirements are initially based on the system maintenance concept (Section 3.4) and are subsequently defined and justified through the maintenance task analysis (Section 13.4) and the supportability analysis (Section 15.5). Essentially, spare-part quantities are a function of demand rates and include the consideration of the following:

1. Spares and repair parts covering actual item replacements occurring as a result of corrective and preventive maintenance actions. Spares are considered as being major replacement items which are repairable, whereas repair parts are nonrepairable smaller components.
2. An additional stock level of spares to compensate for repairable items in the process of undergoing maintenance. If there is a backup (lengthy queue) of items in the intermediate shop or at the depot awaiting repair, these items will not be available as recycled spares for subsequent maintenance actions; thus, the inventory is further depleted (beyond expectations) and a stock-out condition may result. In addressing this problem, it becomes readily apparent that the test equipment capability, personnel, and facilities directly impact the maintenance turnaround times and the quantity of additional spare items needed.
3. An additional stock level of spares and repair parts to compensate for the procurement lead times required for item acquisition. For instance, prediction data may indicate that 10 maintenance actions requiring the replacement of a certain item will occur within a 6-month period, and it takes 9 months to acquire replacements from the supplier. What additional repair parts will be necessary to cover the operational needs and yet compensate for the long supplier lead time? The added quantities will vary depending on whether the item is designated as repairable or will be discarded at failure.
4. An additional stock level of spares to compensate for the condemnation or scrapage of repairable items. Repairable items returned to the intermediate shop or depot are sometimes condemned (i.e., not repaired) because, through inspection, it is decided that it is not economically feasible to repair the item. Condemnation will vary, depending on equipment use, handling, environment, and organizational capability. An increase in the condemnation rate will generally result in an increase in spare part requirements.

In reviewing the foregoing considerations, of particular significance is the determination of spares requirements covering item replacements in the performance of corrective maintenance resulting from system failure. Major factors involved in this process are (1) the reliability of the item to be spared, (2) the quantity of items used, (3) the required probability that a spare will be available when needed, (4) the criticality of item application regarding mission success, and (5) cost. Use of the reliability and probability factors are illustrated in the examples presented.

Probability of success with spares availability considerations. Assume that a single component with a reliability of 0.8 is used in a unique system application and that one backup spare component is purchased. Determine the probability of system success having a spare available in time, t, (given that failures occur randomly and are exponentially distributed).

This situation is analogous to the case of an operating component and a parallel component in standby (i.e., standby redundancy) discussed in Section 12.2. The applicable expression is Equation 12.27 stated as

$$P = e^{-\lambda t} + (\lambda t)e^{-\lambda t} \tag{15.1}$$

With a component reliability of 0.8, the value of λt is 0.223. Substituting this value into Equation 15.1 gives a probability of success of

$$\begin{aligned} P &= e^{-0.223} + (0.223)e^{-0.223} \\ &= 0.8 + (0.223)(0.8) = 0.9784 \end{aligned}$$

Assume next that the component is supported with two backup spares (where all three components are interchangeable). The probability of success during time, t, is determined from

$$P = e^{-\lambda t} + (\lambda t)e^{-\lambda t} + \frac{(\lambda t)^2 e^{-\lambda t}}{2!}$$

or

$$P = e^{-\lambda t}\left[1 + \lambda t + \frac{(\lambda t)^2}{2!}\right] \tag{15.2}$$

With a component reliability of 0.8, and a value of λt of 0.223, the probability of success is

$$\begin{aligned} P &= 0.8\left[1 + 0.223 + \frac{(0.223)^2}{(2)(1)}\right] \\ &= 0.8(1.2479) = 0.9983 \end{aligned}$$

Thus, adding another spare component results in one additional term in the Poisson expression. If two spare components are added, two additional terms are added, and so forth.

The probability of success for a configuration consisting of two operating components, backed by two spares, with all components being interchangeable can be found from

$$P = e^{-2\lambda t}\left[1 + 2\lambda t + \frac{(2\lambda t)^2}{2!}\right] \tag{15.3}$$

With a component reliability of 0.8 and $\lambda t = 0.223$

$$P = e^{-0.446}\left[1 + 0.446 + \frac{(0.446)^2}{(2)(1)}\right]$$

$$= 0.6402(1 + 0.446 + 0.0995) = 0.9894$$

These examples illustrate the computations used in determining system success with spare parts for three simple component configuration relationships. Various combinations of operating components and spares can be assumed, and the system success factors can be determined by using

$$1 = e^{-\lambda t} + (\lambda t)e^{-\lambda t} + \frac{(\lambda t)^2 e^{-\lambda t}}{2!} + \frac{(\lambda t)^3 e^{-\lambda t}}{3!} + \cdots + \frac{(\lambda t)^n e^{-\lambda t}}{n!} \tag{15.4}$$

Equation 15.4 can be simplified into the general Poisson expression (refer to Appendix B.2) as

$$f(x) = \frac{(\lambda t)^x e^{-\lambda t}}{x!} \tag{15.5}$$

The objective is to determine the probability of x failures occurring if an item is placed in operation for t hours, and each failure is corrected (through item replacement) as it occurs. With n items in the system, the number of failures in t hours will be $n\lambda t$, and the general Poisson expression becomes

$$f(x) = \frac{(n\lambda t)^x e^{-n\lambda t}}{x!} \tag{15.6}$$

To facilitate calculations, a cumulative Poisson probability graph is derived from Equation 15.6 and presented in Figure 15.3. The ordinate value can be viewed as a confidence factor. Several simple examples will be presented to illustrate the application of Figure 15.3.

Probability of mission completion. Suppose that one must determine the probability that a system will complete a 30-hour mission without a failure when the system has a known mean life of 100 hours. Let

$$\lambda = 1 \text{ failure}/100 \text{ hours or } 0.01 \text{ failure/hour}$$

$$t = 30 \text{ hours}$$

$$n = 1 \text{ system}$$

$$n\lambda t = (1)(0.01)(30) = 0.3$$

Enter Figure 15.3 where $n\lambda t$ is 0.3. Procced to the intersection where r equals zero and read the ordinate scale, indicating a value of approximately 0.73. Thus, the probability of the system completing a 30-hour mission is 0.73.

Assume that the system identified above is installed in an aircraft and there are 10 aircraft scheduled for a 15-hour mission. Determine the probability that at least seven systems will operate for the duration of the mission without failure as

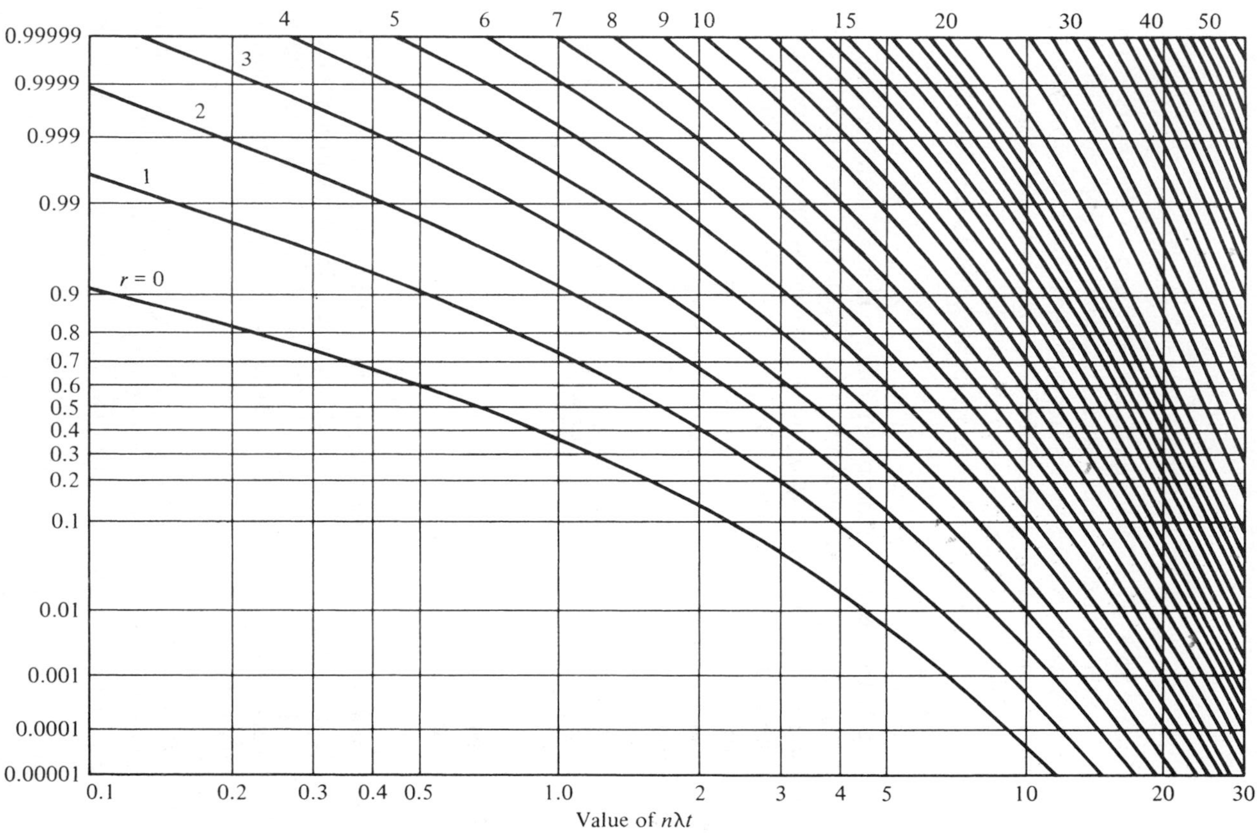

Figure 15.3 Spares determination using Poisson cumulative probabilities. *Source:* NAVAIR 00-65-502/NAVORD OD 41146, *Reliability Engineering Handbook,* Naval Air Systems Command and Naval Ordnance Systems Command. Washington, DC.

$$n\lambda t = (10)(0.01)(15) = 1.5$$

and

$$r = 3 \text{ failures or less (allowed)}$$

Enter Figure 15.3, where $n\lambda t$ equals 1.5. Proceed to the intersection where r equals 3 and read the ordinate scale, indicating a value of approximately 0.92. Thus, there is a 92% confidence that at least seven systems will operate successfully out of 10. If an 80% operational reliability is specified (i.e., eight systems must operate without failure), the confidence factor decreases to about 82%.

Although the graph in Figure 15.3 provides a simplified solution, the use of Equation 15.4 is preferable for more accurate results.

Spares quantity determination. Spare part quantity determination is a function of the probability of having a spare part available when required, the reliability of the item in question, the quantity of items used in the system, and so on. An expression, derived from the Poisson distribution, useful for spare-part-quantity determination is

$$P = \sum_{n=0}^{n=s} \left[\frac{(R)[-\ln R]^n}{n!} \right] \tag{15.7}$$

where

P = probability of having a spare of a particular item available when required
S = number of spare parts carried in stock
R = composite reliability (probability of survival); $R = e^{-K\lambda t}$
K = quantity of parts used of a particular type
$\ln R$ = natural logarithm of R

In determining spare part quantities, one should consider the level of protection desired (safety factor). The protection level is the P value in Equation 15.7. This is the probability of having a spare available when required. The higher the protection level, the greater the quantity of spares required. This results in a higher cost for item procurement and inventory maintenance. The protection level, or safety factor, is a hedge against the risk of a stock depletion.

When determining spare part quantities, one should consider system operational requirements (e.g., system effectiveness, availability) and establish the appropriate level at each location where corrective maintenance is accomplished. Different levels of corrective maintenance may be appropriate for different items. For instance, spares required to support prime equipment components that are critical to the success of a mission may be based on one factor; high-value or high-cost items may be handled differently than low-cost items; and so on. In any event, an optimum balance between stock level and cost is required.

Figure 15.4 (sheets 1 and 2) presents a nomograph that simplifies the determination of spare part quantities using Equation 15.7. The nomograph not only simplifies solutions for basic spare part availability questions, but provides information that can aid in the evaluation of alternative design approaches in terms of spares and in

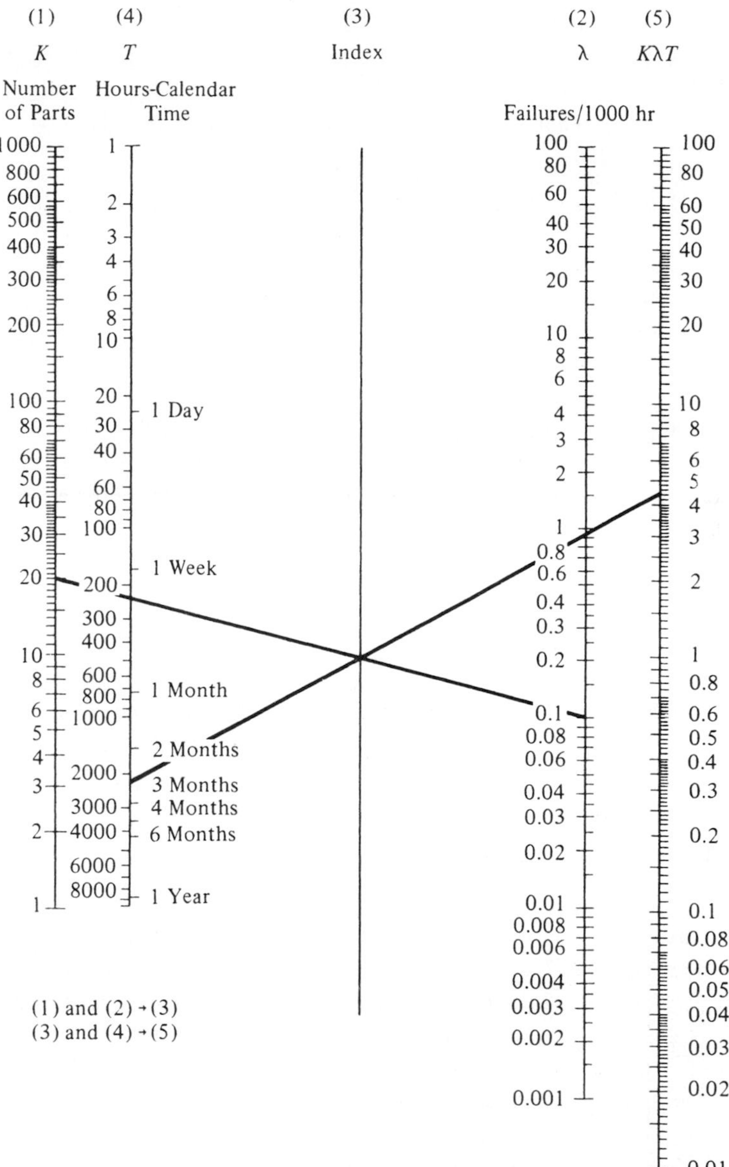

Figure 15.4A (Sheet 1) Spare part requirement nomograph. *Source:* NAVAIR 00-65-502/NAVORD OD 41146, *Reliability Engineering Handbook,* Naval Air Systems Command and Naval Ordnance Systems Command. Washington, DC.

the determination of provisioning cycles. The following examples illustrate the use of this nomograph.

Suppose that a piece of equipment contains 20 parts of a specific type with a failure rate (λ) of 0.1 failure per 1,000 hours of operation. The equipment operates 24

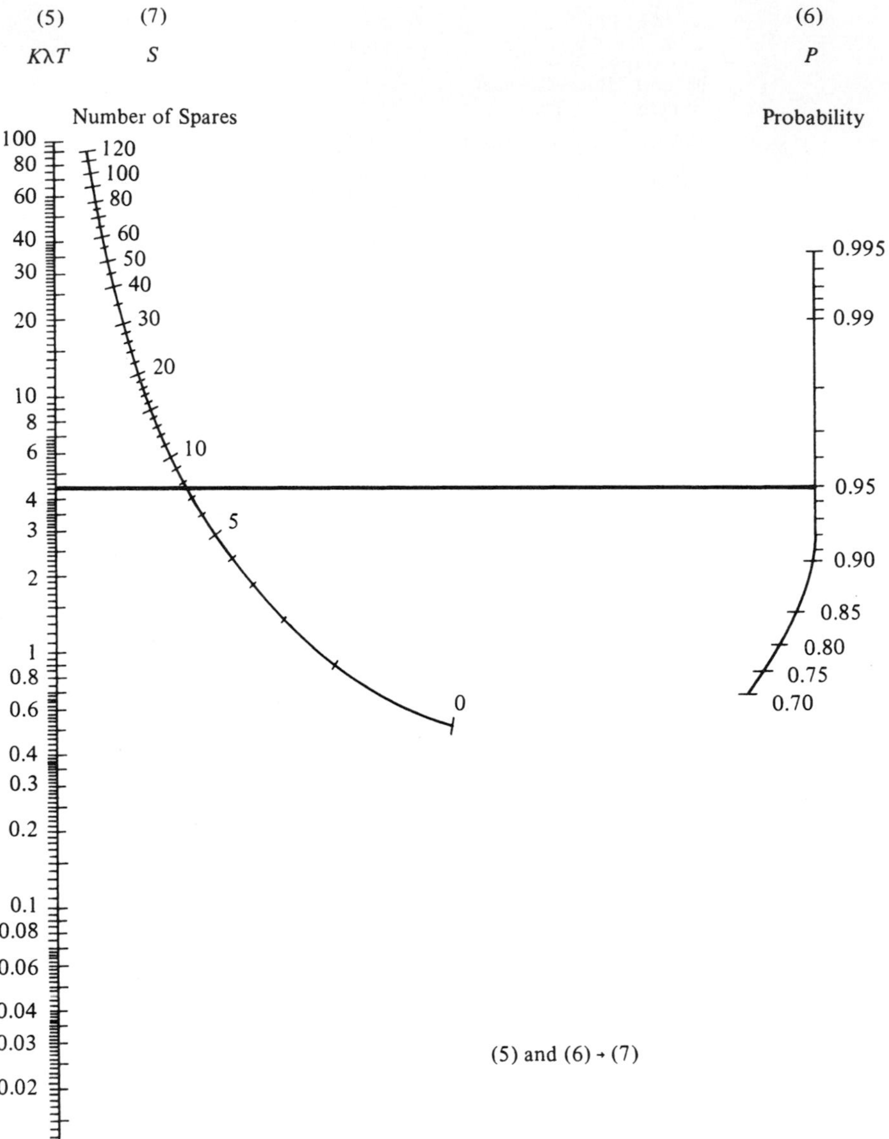

Figure 15.4B (Sheet 2) *(continued).*

hours a day and spares are procured and stocked at 3-month intervals. How many spares should be carried in inventory to ensure a 95% chance of having a spare part when required? Let

$$K = 20 \text{ parts}$$

$$\lambda = 0.1 \text{ failure/1,000 hours}$$

$$T = 3 \text{ months}$$
$$K\lambda T = (20)(0.0001)(24)(30)(3) = 4.32$$
$$P = 95\%$$

Using the nomograph in Figure 15.4 as illustrated, approximately eight spares are required.

As a second example, suppose that a particular part is used in three different items of equipment (A, B, and C). Spares are procured every 180 days. The number of parts used, failure rate, and the equipment operating hours per day are given in Table 15.1.

TABLE 15.1 Data for Spares Inventory

Item	K	Failures per 1000 hours	Operating hours per day
A	25	0.10	12
B	28	0.07	15
C	35	0.15	20

The number of spares that should be carried in inventory to ensure a 90% chance of having a spare available when required is calculated as follows:

1. Determine the procuct of K, λ, and T as

$$A = (25)(0.0001)(180)(12) = 5.40$$
$$B = (28)(0.00007)(180)(15) = 5.29$$
$$C = (35)(0.00015)(180)(20) = 18.90$$

2. Determine the sum of the $K\lambda T$ values as

$$\Sigma\, K\lambda T = 5.40 + 5.29 + 18.90 = 29.59$$

3. Using sheet 2 of the nomograph (Figure 15.4), construct a line from $K\lambda T$ value of 29.59 to the point where P is 0.90. The approximate number of spares required is 36.

Inventory system considerations. The overall inventory requirements for spares and repair parts must be addressed in addition to evaluating specific demand situations. Although too much inventory on hand may be responsive in meeting the demand requirements and all related contingencies, the cost of maintaining the inventory may be high. There is a great deal invested (i.e., resources tied up) and the possibility of obsolescence may be great, particularly if system design changes are being introduced. Conversely, too little inventory increases the risk of stock depletion, the possibility of the system not being in an operational state, and high costs as a result. An optimum balance (not too much or too little) must be sought between the inventory on hand, the procurement frequency, and the procurement quantity, as is modeled in Section 9.2.[6]

[6] Although the value and objectives of the *just-in-time* approach to inventory are recognized, the practical applications of such are not always possible when considering the distribution of system elements, criticality of need, available transportation, and so on.

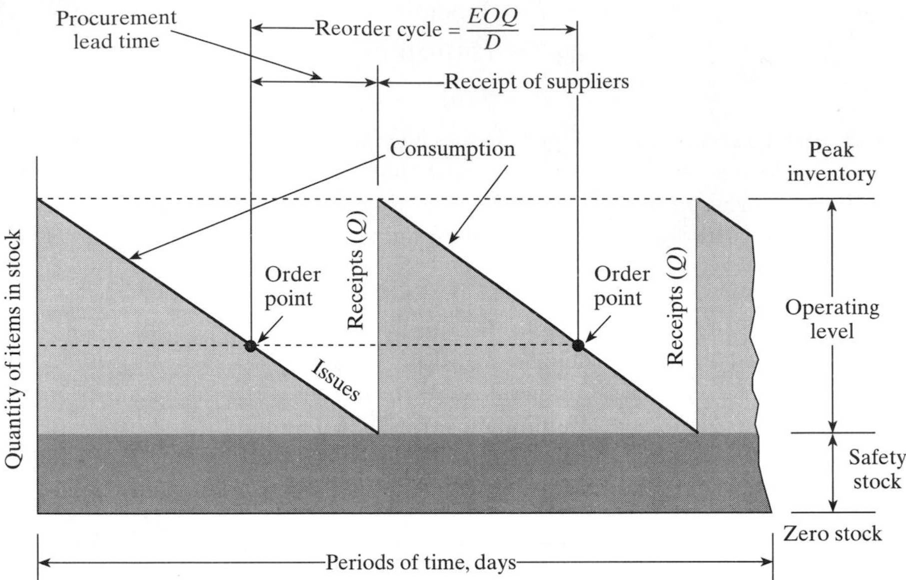

Figure 15.5 Extension of the theoretical inventory cycle (refer to Figure 9.11).

Figure 9.11 shows the general deterministic inventory process (constant demand and constant procurement lead time). This illustration is a theoretical representation of an inventory cycle for a given item. Figure 15.5 is an extension of this figure, identifying some of the more traditional factors in inventory control, such as the following:

1. *Operating level* is the quantity of material items required to support normal system operations in the interval between orders and the arrival of successive shipments.
2. *Safety level* is the additional stock required to compensate for unexpected demands, repair and recycle times, procurement lead time, and unforeseen delays.
3. *Procurement cycle* (or reorder cycle) is interval of time between successive procurement orders.
4. *Procurement lead time* is the span of time from the date of procurement to receipt of the order in the inventory. This includes (a) administrative lead time from the date that a decision is made to initiate an order to the receipt of the order at the supplier, (b) production lead time or the time from receipt of the order by the supplier to completion of the manufacture of the item ordered, and (c) delivery lead time from completion of manufacture to receipt of the item in the inventory.
5. *Procurement level* is the stock level when a replenishment order is initiated for additional stock of the spare/repair parts.

Figure 15.6 presents a situation that is more realistic, where the demand or the procurement lead time are random variables. The necessity for safety stock is evident from this illustration, as a "stock-out" condition may cause the system to be nonoper-

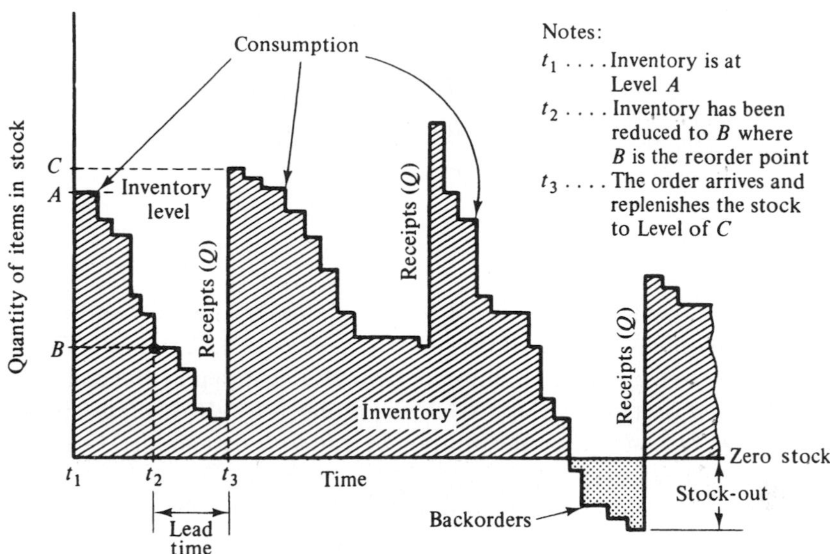

Figure 15.6 Representation of an actual inventory cycle.

ational which may be very costly. In any event, the overall objective is to have the needed amount and type of spares available for the lowest total cost.

Procurement costs vary with the number of orders placed. The economic inventory principle involves a trade-off between the placing of many orders, resulting in high material acquisition costs and the placing of orders less frequently while maintaining a higher level of inventory, causing high inventory maintenance and carrying costs. The economic order principle equates the "cost to order" to the "cost to hold", and the point at which the combined costs are at a minimum indicates the desired size of order.

The optimization model in Section 9.2 can be adapted to this simple inventory situation. To do so means that the derivation based on Figure 9.11 must be modified. This is accomplished by setting $R = \infty$ (instantaneous replenishment), $C_s = \infty$ (no shortages permitted), and adding safety stock. With these modifications, Figure 9.11 takes on the form of Figure 15.5.

Neglecting the cost of holding safety stock, the total cost per period (day) can be found from Equation 9.32, which is stated as

$$TC = C_iD + \frac{C_pD}{Q} + \frac{C_h}{2Q(1 - D/R)}[Q(1 - D/R) + L - DT]^2 + \frac{C_s(DT - L)^2}{2Q(1 - D/R)}$$

With $R = \infty$ and $C_s = \infty$

$$TC = C_iD + \frac{C_pD}{Q} + \frac{C_h}{2Q}(Q + L - DT)^2 \tag{15.8}$$

Because $L = DT$, plus safety stock that is neglected, Equation 15.8 reduces further to

$$TC = C_iD + \frac{C_pD}{Q} + \frac{C_hQ}{2} \tag{15.9}$$

Optimization is accomplished by taking the derivative of TC with respect to Q and setting the result equal to zero as

$$\frac{dTC}{dQ} = -\frac{C_p D}{Q^2} + \frac{C_h}{2} = 0$$

Solving for the optimal value of Q gives

$$Q^* = \sqrt{\frac{2C_p D}{C_h}} \tag{15.10}$$

This also could have been found from Equation 9.43 with R and C_s set equal to ∞.

The optimum value of the procurement level, L^*, is simply D times T plus safety stock, where the cost of holding safety stock is justified by the contribution it makes to reducing shortages. An approach to setting a level of safety stock is presented in Appendix B.3, where the distribution of lead time demand, LTD, is developed through Monte Carlo analysis.

As an example of the application of Equation 15.10, assume that the annual demand for a spare part is 1,000 units. The cost per unit is \$6.00 delivered in the field. Procurement cost per order placed is \$10.00 and the cost of holding one unit in inventory for one year is estimated to be \$1.32.

The economic procurement quantity may be found by substituting the appropriate values into Equation 15.10 as

$$Q^* = \sqrt{\frac{2(\$10)(1{,}000)}{\$1.32}} = 123 \text{ units}$$

Total cost may be expressed as a function of Q by substituting the costs and various values of Q into Equation 15.9. The result is shown in Table 15.2. The tabulated total cost value for $Q = 123$ is the minimum-cost quantity for the condition specified. Total cost as a function of Q is illustrated in Figure 15.7.[7]

TABLE 15.2 Inventory Cost as a Function of Q

Q	$C_i D$	$\frac{C_p D}{Q}$	$\frac{C_h Q}{2}$	TC
0	\$6,000	\$ ∞	\$ 0	\$ ∞
50	\$6,000	200	33	6,233
100	\$6,000	100	66	6,166
123	\$6,000	81	81	6,162
150	\$6,000	67	99	6,165
200	\$6,000	50	132	6,182
400	\$6,000	25	264	6,289
600	\$6,000	17	396	6,413

[7] The total cost found does not include the cost of safety stock, which must be separately justified, as indicated.

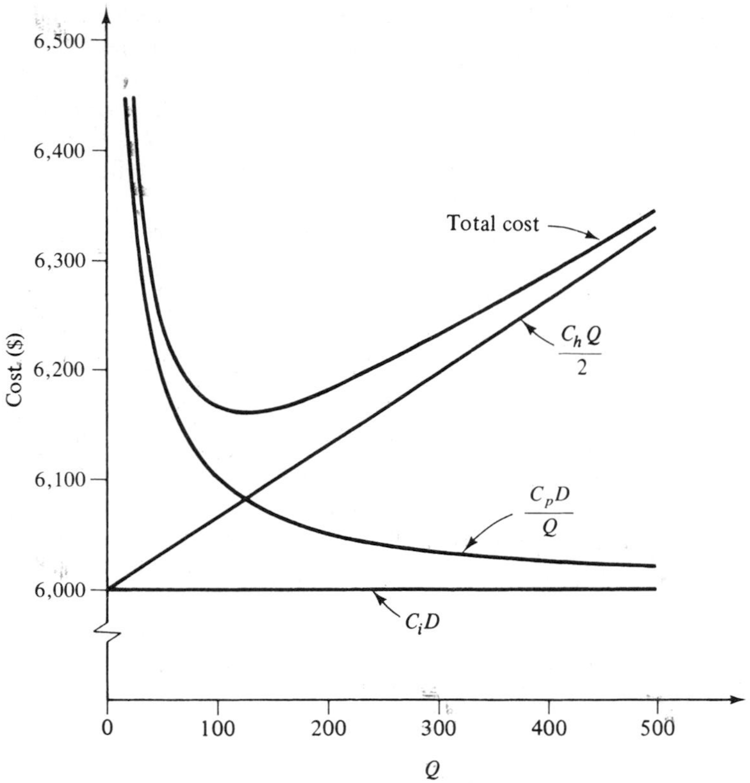

Figure 15.7 Inventory cost as a function of Q.

The economic procurement model of Equation 15.10 is generally applicable in instances where there are relatively large quantities of common spares and repair parts. However, it may be feasible to employ other models of procurement for major high-value items and for those items considered to be particularly critical to mission success.

High-value items are those components with a high unit procurement cost that should be purchased on an individual basis. The dollar value of these components is usually significant and may even exceed the total value of the hundreds of other spares and repair parts in the inventory.

Often a relatively small number of items may represent a large percentage of the total inventory value. Because of this, it may be preferable to maintain a given quantity of these items in the inventory to compensate for repair and recycle times, pipeline and procurement lead times, and so on, and order new items on a one-for-one basis as failures occur and as a spare is withdrawn from the inventory. Thus, where it may appear to be economically feasible to purchase many of these items in a given purchase transaction, only a small quantity of items may be actually procured, owing to the cost of tying up too much capital and the resultant high inventory maintenance cost.[8]

[8] The classification of high-value items will vary with the program and may be established at a certain dollar value (i.e., all components whose initial unit cost exceeds x dollars are considered as high-value items).

Another consideration in the spares acquisition process is that of *criticality*. Some items are considered more critical than others in terms of impact on mission success. For instance, the lack of a $10 item may cause a system to be inoperative, while the lack of a $1,000 item might not cause a major problem. The criticality of an item is generally based on its function in the system and not necessarily on its acquisition cost. The degree of criticality may be determined through failure mode, effects, and criticality analysis (FMECA), described in Section 12.4.

The spares acquisition process may also vary somewhat between items of a comparable nature if the usage rates (i.e., duty cycles) are significantly different. Fast-moving items may be procured locally near the point of usage (such as an intermediate maintenance shop), whereas slower-moving items stocked at the depot may be acquired from a remotely-located supplier if the pipeline and procurement lead times are not as critical.

Test and Support Equipment Measures

The general category of test and support equipment may include a wide spectrum of items, such as precision electronic test equipment, mechanical test equipment, ground handling equipment, special jigs and fixtures, maintenance stands, and the like. These items, in varying configurations and mixes, may be assigned to different levels of maintenance and geographically dispersed throughout the country (or world). However, regardless of the nature and application, the objective is to provide the right item for the job intended, at the proper location, and in the quantity required.

Because of the likely diversification of the test and support equipment for any given system, it is difficult to specify quantitative measures that can be universally applied. Certain measures are appropriate for electronic test equipment, other measures are applicable to ground-handling equipment, and so on. Further, the specific location and application of a given item of test equipment may also result in different measures. For instance, an item of electronic test equipment used in support of onsite organizational maintenance may have different requirements than a similar item of test equipment used for intermediate maintenance accomplished in a remote shop.

Although all the test and support equipment requirements at each level of maintenance are considered to be important relative to successful system operation, the testers or test stations in the intermediate and depot maintenance facilities are of particular concern, since these items are likely to support several system elements at different customer locations. That is, an intermediate maintenance facility may be assigned to provide the necessary corrective maintenance support for a large number of system elements dispersed throughout a wide geographical area. This means that a variety of items (all designated for intermediate-level maintenance) will arrive from different operational sites at differents times.

When determining the specific test equipment requirements for a shop, one must define (1) the type of item that will be returned to the shop for maintenance; (2) the test functions to be accomplished, including the performance parameters to be measured as well as the accuracies and tolerances required for each item; and (3) the anticipated frequency of test functions per unit of time. The type and frequency of item returns (i.e., shop arrivals) is based on the maintenance concept and system reliability data. The distribution of arrival times for a given item is often a negative exponential with the number of items arriving within a given period following a Poisson distribution, as presented in Appendix B.2. As items arrive in the shop, they may be processed immediately or there may be a waiting line, or queue, depending on the availability of the test equipment and the personnel to perform the required maintenance functions.

When evaluating the test process itself, one should calculate the anticipated test equipment use requirements (i.e., the total amount of on-station time required per day, month, or year). This can be estimated by considering the repair-time distributions for the various items arriving in the shop. However, the ultimate elapsed times may be influenced significantly, depending on whether manual, semiautomatic, or automatic test methods are employed.

Given the test equipment use needs (from the standpoint of total test station time required for processing shop arrivals), it is necessary to determine the anticipated availability of the test equipment configuration being considered for the application. As defined in Equation 12.18, availability is a function of reliability and maintainability. Thus, one must consider the MTBM and MDT values for the test equipment itself. Obviously, the test equipment configuration should be more reliable than the system component being tested. Also, in instances where the complexity of the test equipment is high, the logistic resources required for the support of the test equipment may be extensive (e.g., the frequent requirement to calibrate an item of test equipment against a secondary or primary standard in a *clean-room* environment). There may be a requirement to determine the time that the test equipment will be available to perform its intended function.

The final determination of the requirements for test equipment in a maintenance facility is accomplished through an analysis of various alternative combinations of arrival rates, queue length, test station process times, or quantity of test stations. In essence, one is dealing with a single-channel or multichannel queuing situation using the queuing models discussed in Chapter 10. As the maintenance configuration becomes more complex, involving many variables (some of which are probabilistic in nature), Monte Carlo analysis may be appropriate (refer to Appendix B.3). In any event, there may be a number of feasible servicing alternatives, and a preferred approach is sought.

Organizational Measures

The measures associated with a maintenance organization are basically the same as those factors which are typical for any organization. Of particular interest relative to logistic support are the following:

1. The direct maintenance labor time for each personnel category, or skill level, expended in the performance of system maintenance activities. Labor time may be broken down to cover both unscheduled and scheduled maintenance individually. Labor time may be expressed in:
   a. Maintenance labor-hours per system operating hour (MLH/OH).

   b. Maintenance labor-hours per mission cycle (or segment of a mission).
   c. Maintenance labor-hours per month (MLH/Mo).
   d. Maintenance labor-hours per maintenance action (MLH/MA).
2. The indirect labor time required to support system maintenance activities (i.e., overhead factor).
3. The personnel attrition rate or turnover rate (in percent).
4. The personnel training rate or the worker-days of formal training per year of system operation and support.
5. The number of maintenance work orders processed per unit of time (e.g., week, month, or year), and the average time required for work-order processing.
6. The average administrative delay time, or the average time from when an item is initially received for maintenance to the point when active maintenance on that item actually begins.

When addressing the total spectrum of logistics (and the design for supportability), the organizational element is critical to the effective and successful life-cycle support of a system. The right personnel quantities and skills must be available when required, and the individuals assigned to the job must be properly trained and motivated. As in any organization, it is important to establish measures dealing with organizational effectiveness and productivity (refer to Section 14.5).

Transportation and Packaging Measures

Referring to Figures 3.5 and 15.1, it is evident that the area of *transportation* constitutes a major ingredient of the maintenance and logistic support infrastructure. Transportation requirements include the movement of human and material resources, in support of both operational and maintenance activities, from one location to another. When evaluating the effectiveness of transportation, one must deal with such factors as the following:

1. Transportation route, both national and international (distances, number of nationalities, customs requirements, political and social factors, and so on).
2. Transportation capability or capacity (volume of goods transported, number of loads, ton-miles per year, frequency of transportation, modes of transportation, legal forms, and so on).
3. Transportation time (short-haul versus long-haul time, mean delivery time, time per transportation leg, and so on).
4. Transportation cost (cost per shipment, cost of transportation per mile/kilometer, cost of packaging and handling, and so on).

The basic modes of transportation include *railroad, highway, waterway, air*, and *pipeline*. Additionally, there are various combinations of these modes that can be categorized as *intermodal*, as shown in Figure 15.8. Depending on the geographical location of system elements and the applicable suppliers (i.e., the need for transportation),

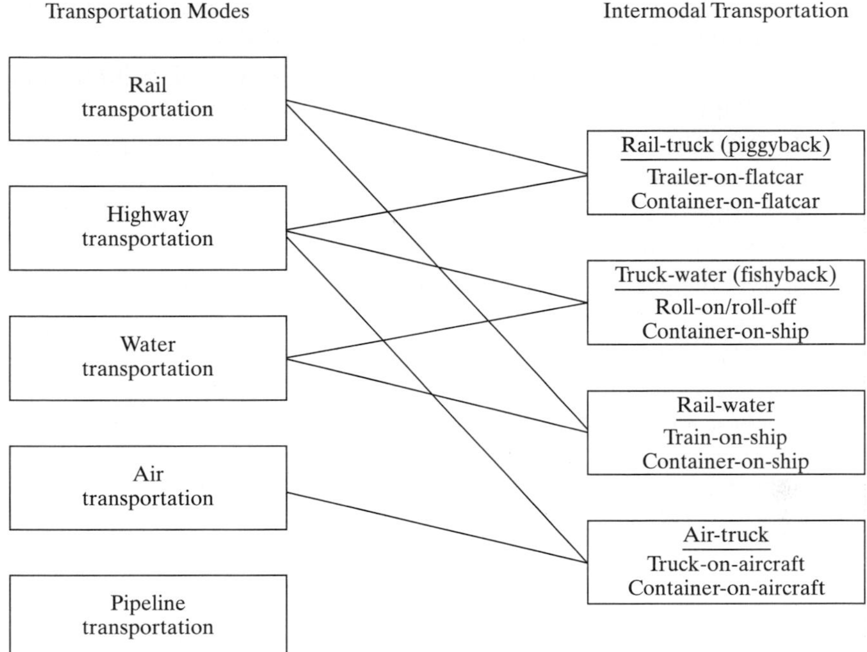

Figure 15.8 The various forms of transportation.

the degree of urgency in terms of delivery-time requirements, etc., alternative routing requirements can be identified. Through the process of conducting trade-offs considering time factors, available modes, and cost, a recommended transportation route is established. This may include transportation by a specific mode or through the use of an intermodal approach.[9]

Given the possible modes of transportation and the proposed routes that have been identified to support the *outward* and *reverse* flows pertaining to operational and maintenance activities (refer to Figure 3.5), along with the various environments in which an item may be subjected when being transported from one location to another, the key issue becomes that of *packaging* design or the design of an item for *transportability* or *mobility*. Products that are to be transported must be designed in such a way to eliminate damage, possible degradation, and so on. The following questions should be addressed[10]:

[9] This is an oversimplification of the transportation issue. One should consider many different factors in determining a specific transportation route to include the mode and intermodal possibilities, the legal forms of transportation (common, contract, and private), the various rate structures, the many governing legislative acts and regulations, and the overall structure of the transportation industry in general. A good reference is N. A. Glaskowsky, D. R. Hudson, and R. M. Ivie, *Business Logistics,* 3rd ed. (Orlando: The Dryden Press, Harcourt Brace Jonanovich Publishing Co., 1992). Also, refer to the literature on business logistics listed in Appendix F.

[10] It is not uncommon to find that a product/item in transit will be subjected to a harsh or rigorous environment—one that is more rigorous than will be experienced throughout the accomplishment of its operational mission(s).

1. Does the package (in which the item being transported is contained) incorporate the desired strength and material characteristics?
2. Can it stand rough handling or long-term storage without degradation (refer to Chapter 12)?
3. Does the package provide adequate protection against various environmental conditions such as rain, vibration and shock, temperature, sand/dust, salt spray)?
4. Is the package compatible with existing transportation and handling methods? Does it possess good stacking qualities?
5. Has the package been designed for safety (i.e., to discourage pilferage or theft)?
6. Can the packaging materials be recyled for additional use. If not, will the materials meet the requirements for disposability (refer to Chapter 16)?

The design for *transportability* must be considered along with the requirements pertaining to performance, reliability, maintainability, usability, and the many other factors identified in Figures 2.9 and 2.14.

Facility Measures

Facilities are required to support activities pertaining to the accomplishment of active maintenance tasks, providing warehousing functions for spares and repair parts, and providing housing for related administrative functions. Although the specific quantitative measures associated with facilities may vary significantly from one system to the next, the following factors are considered to be relevant in most instances:

1. Item process time or turnaround time (TAT) (i.e., the elapsed time necessary to process an item for maintenance, returning it to full operational status).
2. Facility use (i.e., the ratio of the time used to the time available for use, percentage use in terns of space occupancy, and so on).
3. Energy use in the performance of maintenance (i.e., unit consumption of energy per maintenance action, cost of energy consumption per increment of time or per maintenance action, and so on).
4. Total facility cost for system operation and support (i.e., total cost per month, cost per maintenance action, and so on).

Summary

In summary, the measures of logistic support are numerous and will vary depending on the system and the nature of its mission. For some systems, supply support may represent a major portion of the total logistics capability, and the measures associated with supply support are critical in the system design and development process. In other cases, facility support may turn out to be predominant, particularly if the system requires the design and construction of a unique maintenance building (or equivalent).

15.4 LOGISTIC SUPPORT IN THE SYSTEM LIFE CYCLE

Logistics, and the design for supportability, must be addressed from the beginning. Although many of the activities within the overall logistics domain are accomplished later on in the life cycle (e.g., distribution and system support), the ultimate success in their accomplishment downstream is highly dependent on early design and management decisions made during the conceptual and preliminary design phases. In other words, decisions pertaining to the initial selection of technologies, the choice of materials, equipment packaging schemes, levels of repair, and so on, can have a great impact on the maintenance and support infrastructure. Thus, supportability (along with reliability, maintainability, and human factors) must be an inherent consideration within the overall systems engineering process.[11]

Referring to Figure 15.9 (which is an extension of Figure 2.2), qualitative and quantitative supportability requirements are developed through the accomplishment of the feasibility analysis (Section 3.2), the development of the operational requirements and the maintenance concept (Section 3.4), and the identification and prioritization of TPMs (Section 3.5). The applicable measures of supportability must be established, their importance with respect to other system metrics must be delineated, and the requirements for design must be identified through the development of the appropriate DDPs (refer to Figure 2.9). Such requirements may be specified not only for the prime mission-oriented elements of the system but for the maintenance and support infrastructure as well (refer to Figure 3.5).

As discussed in Section 15.1, supportability is being addressed within the context of *integrated logistic support* (ILS). ILS is basically a management function implemented with the objective of developing and producing a system that can be effectively and efficiently supported throughout its planned life cycle (see footnote 3). Thus, one of the initial activities is that of planning and the identification of logistics activities throughout the life cycle. A brief description of these activities is provided subsequently.

Logistic Support Planning

Logistic support planning begins during the early conceptual phase of a program through the development of a preliminary integrated logistic support plan (Figure 15.9, Block 1) and extends through the preparation of a formal integrated logistic support plan during preliminary design (Block 2). The preliminary plan leads directly into the final ILS plan. Although the level and depth of planning may vary significantly depending on the type and size of the system, the planning approach proposed herein is still considered to be appropriate provided that the result is properly tailored to the need.

[11] For additional discussion of supportability and logistics from a *systems engineering* approach, refer to B. S. Blanchard, *Logistics Engineering and Management*, 5th ed. (Upper Saddle River, N.J. Prentice Hall, Inc., 1998).

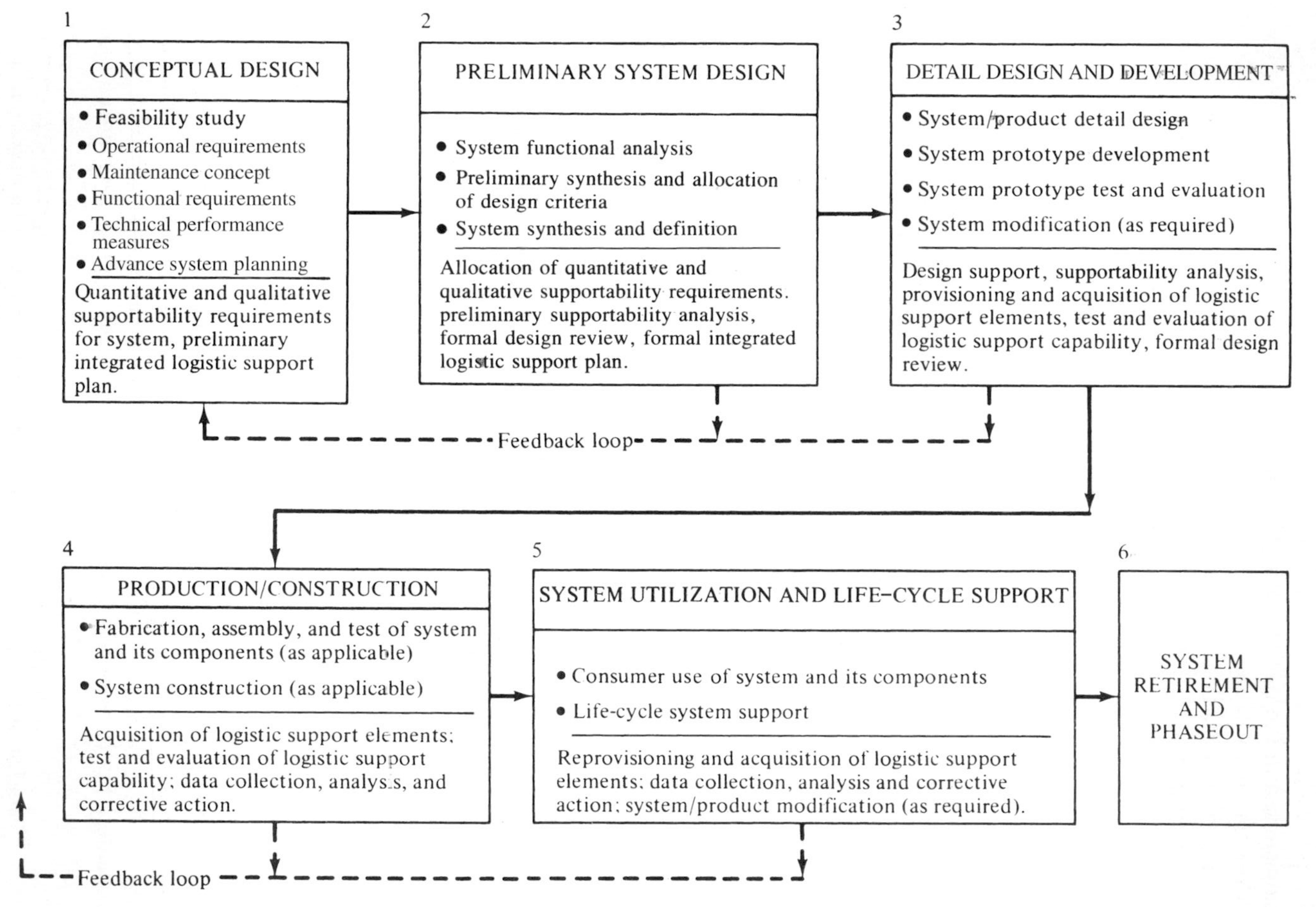

Figure 15.9 Logistic support in the system life cycle (refer to Figure 2.2).

Preliminary integrated logistic support plan. The preliminary ILS plan should address the overall approach to system support as well as the functions to be performed throughout system design as related to supportability. Specifically, this plan should cover the following:

1. Definition of system operational requirements and the system maintenance concept (logistic support requirements must be based on the mission profile, the anticipated geographical area where the system is to be located, the environment in which it is intended to operate, and the maintenance concept as defined in Section 3.4)
2. Quantitative and qualitative supportability design requirements for the system (TPMs and DDPs)
3. Functional analysis and the allocation of supportability requirements to the subsystem level and beyond as appropriate (Sections 3.6 and 4.1)
4. Supportability analysis (i.e., maintenance task analysis, level-of-repair analysis, life-cycle cost analysis, etc.)
5. Design of new logistic support elements as required (e.g., new test and support equipment, new facilities, training equipment, etc.)
6. Formal design reviews (Sections 3.9, 4.6, and 5.7)
7. Early provisioning and the acquisition of selected logistic support elements (i.e., the plan for the procurement of long-lead-time items to include spares and repair parts, test and support equipment, and so on, as necessary to meet program schedules)
8. Test and evaluation of logistic support elements (Chapter 6)

The preliminary ILS plan covers those functions in the early phases of the life cycle that directly interface with detail day-to-day design activities, reliability activities, maintainability activities, and so on. This plan may be included as part of the system engineering management plan (SEMP) discussed in Chapter 18 and must lead directly into the formal ILS plan.

Formal integrated logistic support plan (ILSP). The formal ILS plan is initially prepared during preliminary design. It evolves directly from the preliminary ILS plan and covers logistic support activities that deal with the detailed requirements for the system and the provisioning, procurement, and acquisition, and the maintaining of a support capability for the system throughout its planned life cycle. This plan directly ties into the systems engineering process because knowledge of the support requirements is acquired directly from the design process. The formal ILS plan usually includes the following individual sections[12]:

[12] Although the formal ILS plan is initially prepared during the latter stages of preliminary design, it needs to be updated periodically if it is to reflect specific support requirements being defined on a continuing basis through detail design. For the purposes of this text, the ILS plan is to be considered as a "living" document. An ILSP may include a supportability analysis plan, and/or a reliability and maintainability plan depending on the program.

1. *Maintenance plan.* A detailed plan for maintenance is developed from the maintenance concept, the ongoing maintenance planning effort accomplished throughout system design, and the results of the maintainability and supportability analyses. This plan includes a description of the levels of maintenance required for sustaining system life-cycle support, the major functions to be performed at each level, the organizations responsible for performing the required maintenance, and a summary of the logistic support resources required. The maintenance plan is based on a given system design configuration, while the maintenance concept reflects a before-the-fact design approach to system support. The information contained in this plan serves as a *technical baseline* for all other sections of the formal ILS plan.
2. *Supply support plan.* A supply support plan is developed to include the following:
   - **a.** A summary listing of spares/repair parts, and consumables for each level of maintenance. This list is based on the results of the supportability analysis.
   - **b.** Procedures covering the procurement and handling (packaging and transportation) of spares, repair parts, and consumables.
   - **c.** Definition of warehousing and the accountability functions associated with maintenance support. This includes initial cataloging, stocking, inventory maintenance and control, establishing procurement cycles, and the disposition of residual assets.
   - **d.** Procedures for data collection, analysis, and the updating of spares and repair-part demand factors necessary to improve procurement cycles and reduce waste.
3. *Test and support equipment plan.* A detailed plan is developed to cover the acquisition of test and support equipment, and should include the following:
   - **a.** A summary listing of test and support equipment requirements for each level of maintenance. This listing is based on the results of the supportability analysis.
   - **b.** Procedures covering the acquisition of newly designed support equipment and the procurement of off-the-shelf items from supplier inventories (i.e., COTS items).
   - **c.** Procedures covering support equipment/prime equipment compatibility testing and the subsequent distribution of test and support equipment to the geographical point of need.
   - **d.** Procedures for data collection, analysis, and the evaluation of test and support equipment to assess testing accuracy and thoroughness, test equipment utilization, reliability of the test equipment, and test equipment maintenance.
4. *Personnel and training plan.* A plan should be developed to cover the following:
   - **a.** Operator training (training of system operators) to include the type and level of training, the proposed schedule for training, basic entry-skill requirements, and a brief course outline.
   - **b.** Maintenance training for each level of maintenance to include the type and depth of training, the proposed schedule for training, basic entry requirements, and a brief course outline.

 c. Training equipment, devices, aids, and data required to support operator and maintenance training activity.
 d. Proposed approach for the accomplishment of periodic training for new personnel or for the recycling of existing personnel for upgrading purposes.

 Personnel quantities and skill-level requirements, particularly regarding the accomplishment of maintenance functions, are identified through maintenance task analysis (MTA) and the supportability analysis. The differences between the required skills and the existing skills of currently available personnel will lead to the definition of training requirements (i.e., the process of upgrading individuals to higher-level skills to successfully operate and maintain the system [refer to Sections 14.5 and 15.2]).

5. *Facilities plan.* A facilities plan is developed to identify all real property and equipment to support system testing (Types 3 and 4), training, operation, and maintenance functions. The plan must contain sufficient qualitative and quantitative information to allow facility planners to the following:
 a. Initially assess and allocate requirements. The supportability analysis identifies maintenance activities that will lead to the establishment of facility requirements for maintenance support. This includes the requirements associated with capital equipment, utilities, and so on.
 b. Analyze existing facilities to determine adequacies or deficiencies.
 c. Determine requirements for new facilities or modifications to existing facilities.
 d. Estimate the cost of construction (and capital equipment) or modification projects required to meet the needs of the system, and establish the appropriate facility development schedule.
6. *Data plan.* A data plan should be prepared to include[13]:
 a. A summary of technical data requirements (installation, operating, maintenance, servicing, inspection, calibration, overhaul instructions) for each level of maintenance for the system and its components.
 b. Technical data development schedule (major milestones).
 c. A procedure for data verification and validation.
 d. Database for data verification and validation.
7. *Computer resources plan.* A detailed listing of computer hardware and maintenance software programs, and the procedures for modifying such programs.
8. *Retirement plan.* Throughout the system life cycle, faulty nonrecoverable items will be phased out of the inventory as maintenance actions occur. Further, the entire system and its components will ultimately be phased out upon retirement after the system has fulfilled its intended purposes. In the past, the disposal of residual items has not been very well planned, system components have been inappropriately discarded and left to the whims of nature, and the results have had a costly impact on the environment. Thus, it is essential that one address the areas of disposability, possible candidates and procedures for item recycling, and methods for material decomposition. The accomplishment of material recycling

[13] Data may constitute a database in digital format, a CAD/CAM/CALS data package, computer disks, drawings, parts lists, procedures, technical manuals, supplier manuals, or various combinations of these.

and/or disposal often requires a relatively high level of logistic support. The intent of this plan is to identify the magnitude of effort required in this final phase of the life cycle, and to define the anticipated logistic support resources needed (refer to Chapter 16).

The formal ILS plan provides the necessary guidance for life-cycle system support. This guidance is developed through system design activities and the decisions pertaining to supportability. If the system is to be effectively and economically supported, it is essential that the logistic support elements covered in the plan be addressed at the early stages of the life cycle through development of the maintenance concept, and expanded subsequently through the iterative process involving reliability, maintainability, and supportability analyses. Thus, logistic support planning is a significant ongoing activity, and the early phases of this planning effort must be closely integrated with the system engineering management plan (SEMP).

Design for Supportability

There are many contributing factors that must be considered to obtain effective results. The system must be able to perform certain designated functions, it must be reliable, it must be maintainable, it must be operable, it must be economically feasible, and so on. Relative to the aspect of logistic support, the major level of activity occurs during the system use phase, when the customer (user) is operating the system on a day-to-day basis. However, the greatest impact on system support is realized during the early stages of system design, when decisions are made that significantly affect the type and extent of support to be provided. Thus, it is essential that logistic support be a major consideration during the early system design and development process (i.e., design for supportability must take place). When design for supportability occurs, one must consider the following:

1. The design of the prime-mission–oriented elements of the system to ensure that the inherent characteristics in the design reflect a configuration that can be easily and economically supported.
2. The design of the various elements of logistic support to ensure that these elements are compatible with the prime mission elements of the system, compatible with each other, and reflect an overall economical approach to system support.

The design of the prime-mission–oriented elements of the system for supportability is accomplished largely through the reliability and maintainability efforts described in Chapters 12 and 13, whereas the design of the logistic support elements of the system is emphasized in this chapter. The interaction effects between these two major segments of the system can be significant. Dealing with one of these areas without considering the other on an integrated basis can prove to be costly. For instance, the following problems often arise when only prime mission equipment design is accomplished without the proper integration of elements of support (this list is not all-inclusive but represents some of the areas of concern):

1. Individual items of test equipment are not compatible with the tasks scheduled to be accomplished at the maintenance level assigned. Items incorporating insufficient test capability will not allow for a positive verification of the system condition. This will often lead to the accomplishment of maintenance based on a "best-guess" approach which, in turn, results in the consumption of valuable resources. Conversely, items incorporating more test capability than what is necessary are costly, both from the standpoint of acquisition cost and from the standpoint of requiring more maintenance (in themselves) than what otherwise would be required.
2. Individual items of test equipment are not as reliable as the elements of the system requiring the test support. This point tends to destroy the validity of the test being conducted, particularly in situations when a failure occurs and it is difficult (if not impossible) to determine whether the failure is in the system itself or in the test equipment. Such situations usually result in more total system downtime than desired.
3. Spares and repair parts stocked at each level of maintenance are not compatible with the removal and replacement requirements or the repair-discard decisions anticipated. This usually results in excessive system downtime, the possibility of costly inventory requirements, or the added high costs associated with the increase in transportation and handling requirements necessary to obtain the needed item(s).
4. The personnel quantities and skill levels are not adequate to perform the maintenance functions that are assigned at a given level. Personnel performing maintenance tasks without the necessary skills will likely induce problems into the system and cause an increase in the maintenance requirements overall. Personnel possessing skills greater than what are actually required will tend to induce problems because of carelessness or seek employment elsewhere (resulting in cost to the organization).
5. The facilities are not adequate for the operating and maintenance tasks scheduled to be performed. This often leads to poor maintenance practices, the possible introduction of hazards because of unsafe conditions, excessive system downtime, or a combination thereof.
6. Maintenance procedures and supporting technical data do not track the tasks that are to be performed at a given maintenance level. Further, such procedures are not prepared to the correct skill levels of the personnel performing the tasks. Inadequate data coverage tends to lead to maintenance-induced faults in the system because the assigned personnel are performing tasks without the necessary guidance. Conversely, procedures that provide more information than what is actually required will encourage the accomplishment of maintenance beyond that which is considered to be economically feasible.

These six points are symptomatic of the problems that can occur, particularly for large systems, when inadequate consideration is given to maintenance and logistic support. The decisions associated with system support often turn out to be significant in terms of life-cycle cost, and it may be necessary to make a change in one element of support in order to reach a cost-effective solution for the system as a whole. This, in turn, may necessitate a change in the prime mission equipment design, as well as requiring changes in the other elements of support for the purposes of compatibility. In essence, a change in any given area may require changes in many other areas, and the interactions that occur between the prime equipment and the elements of logistic support are numerous.

When designing for system supportability, it is essential that logistic support be considered at program inception. This is accomplished through the development of the maintenance concept as described in Section 3.4. The design must be compatible with the levels of maintenance identified, the basic functions to be performed at each level, the anticipated personnel skills at each level, the test and support equipment approach conveyed, and the quantitative requirements specified. These requirements, where appropriate, are allocated to various subsystems, units, or other parts of the system. As design progresses, alternative support approaches are evaluated and a preferred configuration is selected. This iterative process involving the evaluation of alternatives is accomplished through the supportability analysis (SA), which is directly tied into the overall systems analysis activity. The SA constitutes the integration and application of various analytical techniques to define and optimize (to the extent possible) the logistic support requirements for a given system. The output of the SA provides (1) an analytical assessment of the system in terms of its supportability, and (2) a database for the identification, provisioning, and acquisition of those logistic support elements required for system support during the utilization phase. The results of the SA are used in the development of the formal ILS plan.

The logistic support activities in the system design process are highlighted in Figure 15.9, Blocks 1, 2, and 3. These activities are an integral part of the systems engineering process described in Part II and are paramount in meeting the objective of design for supportability.

Logistics in the Production/Construction Phase

Logistics in the production phase includes two basic areas of activity. The first deals with the production process itself and the flow of materials from suppliers to the producer's manufacturing facility, and from the manufacturing facility to the customer. This aspect of logistics, often classified under the term of *business logistics* or *industrial logistics,* covers the overall physical distribution of the prime mission equipment and consumable products. Specific functions include packaging, transportation and handling, warehousing and inventory control, and related activities involved in the transition of items from the producer to the consumer (refer to footnote 2).

The second area relates to the production and distribution of the elements of logistics as described in Section 15.2 required for the sustaining support of the system throughout its useful life. This includes test and support equipment, spares and repair parts, training equipment, data, and the like. The specific requirements in this area are based on the supportability analysis (SA), specifications are prepared, and production begins.

As the production process evolves, it is essential that the item being produced is of high quality and performs satisfactorily. For instance, if the proper level of quality control is not maintained in the production of test equipment, the test equipment will, in all probability, not provide the desired accuracy in testing or be reliable. If the spares are not manufactured to the same specifications and quality standards as those comparable items in the prime equipment with the same part numbers, the system may not function properly when replacements occur as a result of maintenance actions. In essence, the elements of logistics must be treated in the production process using the same specifications and controls as those imposed for the prime equipment.

Logistics for Operating Systems

The major elements of logistics, applicable to the sustaining support of a system or product being used by the customer in the field, are described in Section 15.2. These elements are required to support the ongoing scheduled and unscheduled maintenance activities that are necessary to keep the system or product in proper operating condition. The prime objective in this phase is to effectively and efficiently manage all logistic support resources such that the right type of support is available at the right location and in a timely manner. A major requirement toward meeting this objective is to be able to evaluate and assess the effectiveness of the overall logistic support capability, and to initiate corrective action in areas where problems exist. The assessment process is accomplished through the development of a data collection and analysis capability. Data may be collected using the forms illustrated in Figures 6.5 and 6.6.

15.5 SUPPORTABILITY ANALYSIS (SA)[14]

The supportability analysis (SA) constitutes the integration and application of different analytical techniques/methods to solve a wide variety of problems. The SA, in its application, is the process employed on an iterative basis throughout system development that addresses the aspect of supportability in design. As such, it is an inherent part of the systems engineering process and uses the results of reliability analysis (models, FMECA/FTA, predictions), maintainability analysis (RCM, level of repair analysis, MTA, predictions), and human factors analysis (OTA, OSD, error analysis, safety/hazard analysis, training requirements) described in Chapters 12, 13, and 14 respectively.[15]

[14] The scope of activity defined in this section is sometimes identified by the term *logistic support analysis* (LSA), *maintenance analysis* (MA), *maintenance engineering analysis* (MEA), *maintenance level analysis* (MLA), or something equivalent. However, the objectives are the same in each instance.

[15] For defense systems, the SA is covered in MIL-HDBK-502, "DOD Handbook-Acquisition Logistics," Department of Defense, Washington, D.C., May 1997. The data requirements, evolving from the SA (or "Logistics Management Information"), are specified in MIL-PRF-49506, "Performance Specification for Logistics Management Information," Department of Defense, Washington, D.C., Novemeber 1996. The emphasis herein is on the SA process.

The SA process, which follows the approach for systems analysis illustrated in Figure 3.15, leads to the identification of specific maintenance and support resources as shown in Figure 15.10. More specifically:

1. The SA aids in the evaluation of prime equipment design characteristics in terms of logistic support requirements. This includes consideration of alternative equipment packaging schemes, test approaches, accessibility features, transportation and handling provisions, and so on. Logistic support resource requirements are estimated for each alternative being considered and a preferred approach is selected.
2. The SA aids in the evaluation of alternative repair policies allowable within the constraints dictated by the maintenance concept. For example, the SA supports repair-discard decisions, and decisions pertaining to where repair is to be accomplished. In this context, the SA is closely interrelated with the maintainability analysis described in Chapter 13.
3. The SA aids in the evaluation of two or more off-the-shelf equipment items being considered for a single system application (i.e., COTS items). Assuming that new design is not appropriate and outside supplier sources are being considered, which item (when installed as part of the system) will reflect the least overall logistics burden? The SA facilitates the evaluation effort which, in turn, affects procurement decisions in system acquisition.
4. Given an assumed or fixed design configuration for the prime mission segment of the system, the SA aids (through reliability and maintainability analyses and predictions) in determining the specific logistic support resource requirements for that configuration. Once design data are available, it is possible to determine the type and quantity of test and support equipment, the type of quantity of spares and repair parts, personnel quantities and skill levels, transportation and handling requirements, facility requirements, and technical data needs.
5. During system test and evaluation and when the system is in operational use by the customer, the SA aids (through application of operations analysis and logistics modeling techniques) in the measurement of the overall effectiveness of the system and its associated support. Problem areas readily become apparent and the SA can assist in the evaluation of alternatives for corrective action, including improvements in the use and applications of support resources.

The SA is developed on an iterative basis throughout system definition and design. Basically, the analysis effort stems from the maintenance concept and is ultimately dependent on engineering data, reliability and maintainability analyses and predictions, cost data, and so on. The analyst, in developing SA data, reviews engineering documentation with the objective of estimating future support requirements. Thus, the SA includes coverage of the following:

1. All significant repairable items (i.e., system, subsystem, assembly, and subassembly). Items defined through the level-of-repair analysis as being "repairable" are evaluated to determine the type and extent of support necessary.

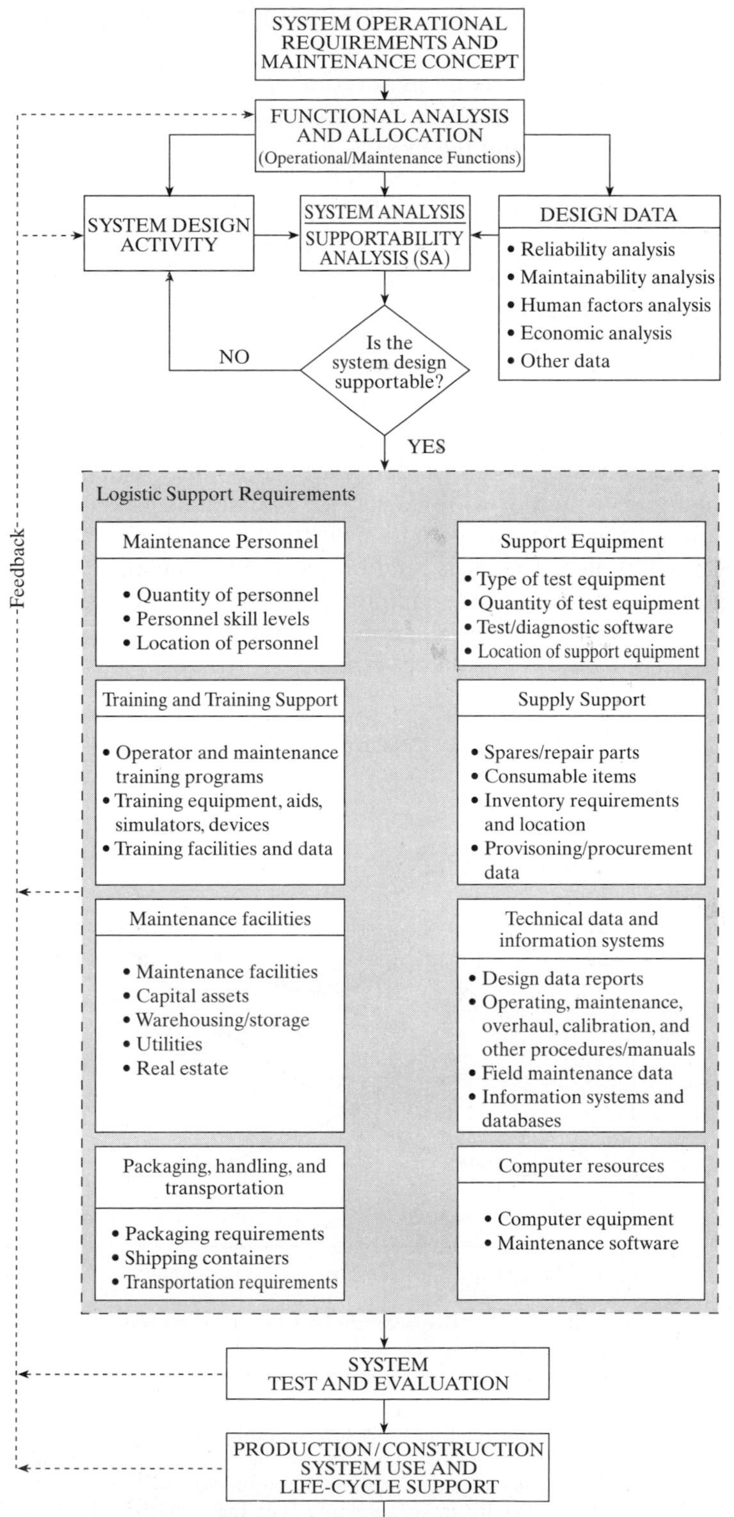

Figure 15.10 The identification of specific resource requirements from the LSA.

2. All maintenance requirements (i.e., troubleshooting, remove and replace, servicing, alignment and adjustment, functional test and checkout, inspection, calibration, overhaul, and the like).

SA data are used to support design decisions and to serve as the basis for the subsequent provisioning and acquisition of specific logistic support elements needed for the operation and maintenance of the system. The accomplishment of these objectives requires the generation of certain qualitative and quantitative factors dealing with the various elements of logistics discussed earlier. Typical SA data (or "Logistics Management Information") cover the factors identified in Table 15.3.[16]

The SA data format (i.e., output data package) is by no means fixed. It varies from program to program and is tailored to the specific information desired and the time in the life cycle when needed. Early in the life cycle, major trade-offs are accomplished and include a wide variety of individual evaluations at a gross level. For instance, the analyst may wish to evaluate the effects of system fault isolation capability on spares cost. The basic relationship is illustrated in Figure 15.11 and the evaluation process requires prime equipment design data and supply support data.

A second example is illustrated in Figure 15.12. This figure indicates the overall relationships between corrective maintenance and preventive maintenance. Quite

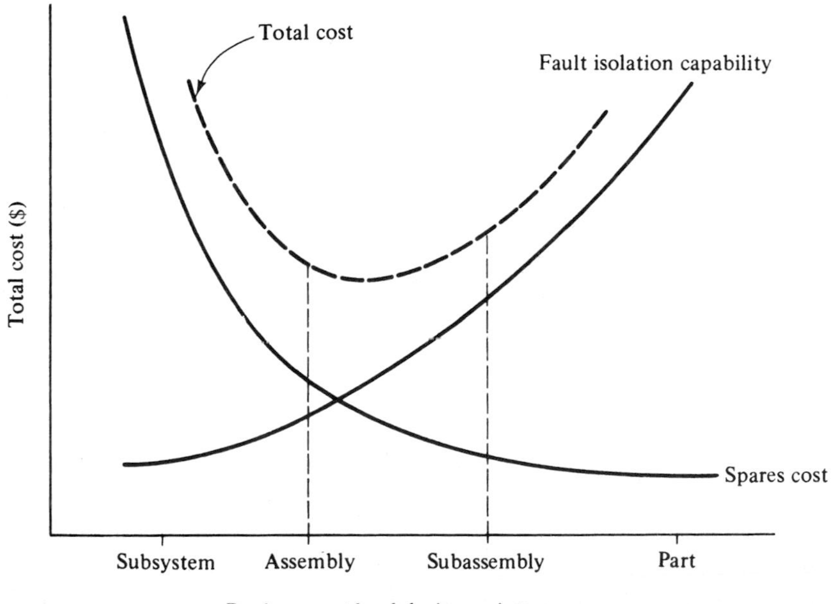

Figure 15.11 Cost-effectiveness of system diagnostic capability.

[16] A maintenance task analysis is often accomplished within the scope of the SA in addition to what might be required in a maintainability program (see Section 13.3). For a detailed description of the procedures for accomplishing maintenance analyses, refer to B. S. Blanchard, *Logistics Engineering and Management*, 5th ed. (Upper Saddle River, N.J.: Prentice Hall, Inc., 1998).

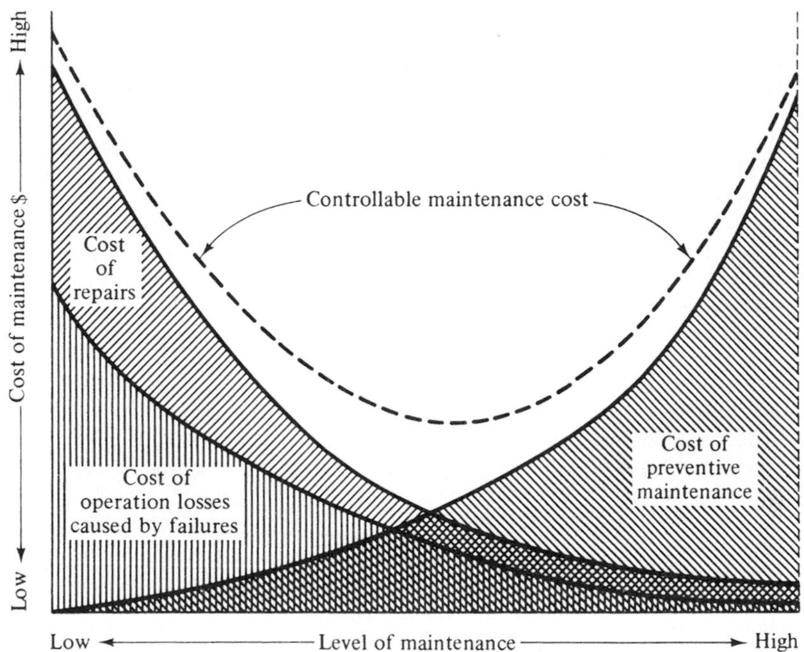

Figure 15.12 Cost-effectiveness evaluation: corrective maintenance versus preventive maintenance.

often, the question arises as to how much maintenance should be planned and the nature of such maintenance. To evaluate this requires knowledge of many of the factors listed in Table 15.3, although the level of detail required may not be extensive. In any event, the evaluation is accomplished, a preferred approach is selected, and the system design reflects the results.

In essence, the early analysis efforts must provide the right information expeditiously as the designer is required to make many decisions in a relatively short period of time. If this information is not available in a timely manner, the decisions will be made anyway and the results may not prove to be beneficial from the standpoint of system supportability.

As system development progresses, the SA effort assumes a more comprehensive approach. Additional design data are available and the analyst is able to define support needs in a relatively precise manner. SA data are used to evaluate the design configuration (as proposed at the time) in terms of the measures described in Sections 15.3. Further, SA data are used to identify specific logistic support resource requirements.

Figure 15.10 (and Table 15.3) shows the SA output leading to the definition of support requirements. Actually, the process is somewhat more complex. For instance, one might refer to the measures of supply support discussed in Section 15.3. To determine spares requirements, the analyst must review design data and identify operating and maintenance functional requirements, equipment packaging schemes, repair levels, fault isolation capability, and so on. After identifying the specific items that are to be replaced during maintenance actions, it is necessary to estimate demand rates and

TABLE 15.3 Logistics Management Information (LMI)

1. Maintenance levels
2. Maintenance tasks and levels
   —Tasks sequence(s)
   —Task time
   —Task frequency
3. Supply support
   —Repair levels and location
   —Quantity and type of spares, repair parts, and consumables
   —Critical items
   —Replacement frequency
   —Condemnation rate
   —Wear-out rate
   —Inventory level and location
   —Safety stock level
   —High-value items
   —Shelf life
   —Spares and repair part availability
   —Procurement lead time
   —Inventory cost (order cost, material cost, holding cost)
   —Provisioning cycle
4. Test and support equipment
   —Quantity, type, and location
   —Utilization rate
   —Availability and reliability
   —Maintenance requirements
   —Cost (R&D, production, operation, maintenance)
5. Transportation and handling equipment
   —Quantity, type, and location
   —Containers (disposable and reusable)
   —Packing and shipping
   —Transportation cost
6. Personnel and training
   —Personnel quantity and skill level requirements and maintenance location
   —Attrition rate
   —Learning curve(s)
   —Indirect labor (overhead)
   —Initial training requirements
   —Replacement training requirements
   —Training aids (equipment and data)
   —Personnel costs (direct and indirect)
   —Training costs (direct and indirect)
7. Facilities
   —Maintenance and training facility requirements (space and layout)
   —Storage requirements
   —Capital equipment
   —Tooling and special handling equipment
   —Environmental requirements
   —Utility requirements (power, light, heat, water, telephone, etc.)
   —Facility utilization
   —Facilities cost (R&D, construction, operation, maintenance, taxes, energy utilization, etc.)
8. Technical data
   —Technical manual requirements (operating and maintenance instructions, overhaul procedures, etc.)
   —Logistics provisioning data
   —Maintenance reporting
9. Additional factors
   —Availability (A_o, A_a, A_i)
   —MTBM, MTBR, MTBF, R, λ, fpt
   —$\overline{\text{M}}\text{ct}$, $\overline{\text{M}}\text{pt}$, $\overline{\text{M}}$, MDT, M_{max}
   —MLH/OH, MLH/MA, MLH/year
   —Turnaround time (TAT)
   —Maintenance actions (MA/year)
   —Logistic support cost factors

the various quantitative parameters associated with the anticipated inventories. These factors are evaluated, an optimum approach is selected, and a formal spares and repair parts list is developed for the system. This, in turn, leads to the provisioning and acquisition of the required spares. The process is illustrated in Figure 15.13. Note that the SA includes the analytical steps involved in determining supply support requirements, not the aspects of provisioning and acquisition.

Figure 15.14 covers the development process for test and support equipment in a manner similar to the illustration for supply support (see Figure 15.13). Referring to the logistics measures in Section 15.3, one needs to identify the items to be tested, the parameters to be measured for each item tested, and the frequency of test. Loading studies are accomplished and alternative configurations are evaluated. The results are

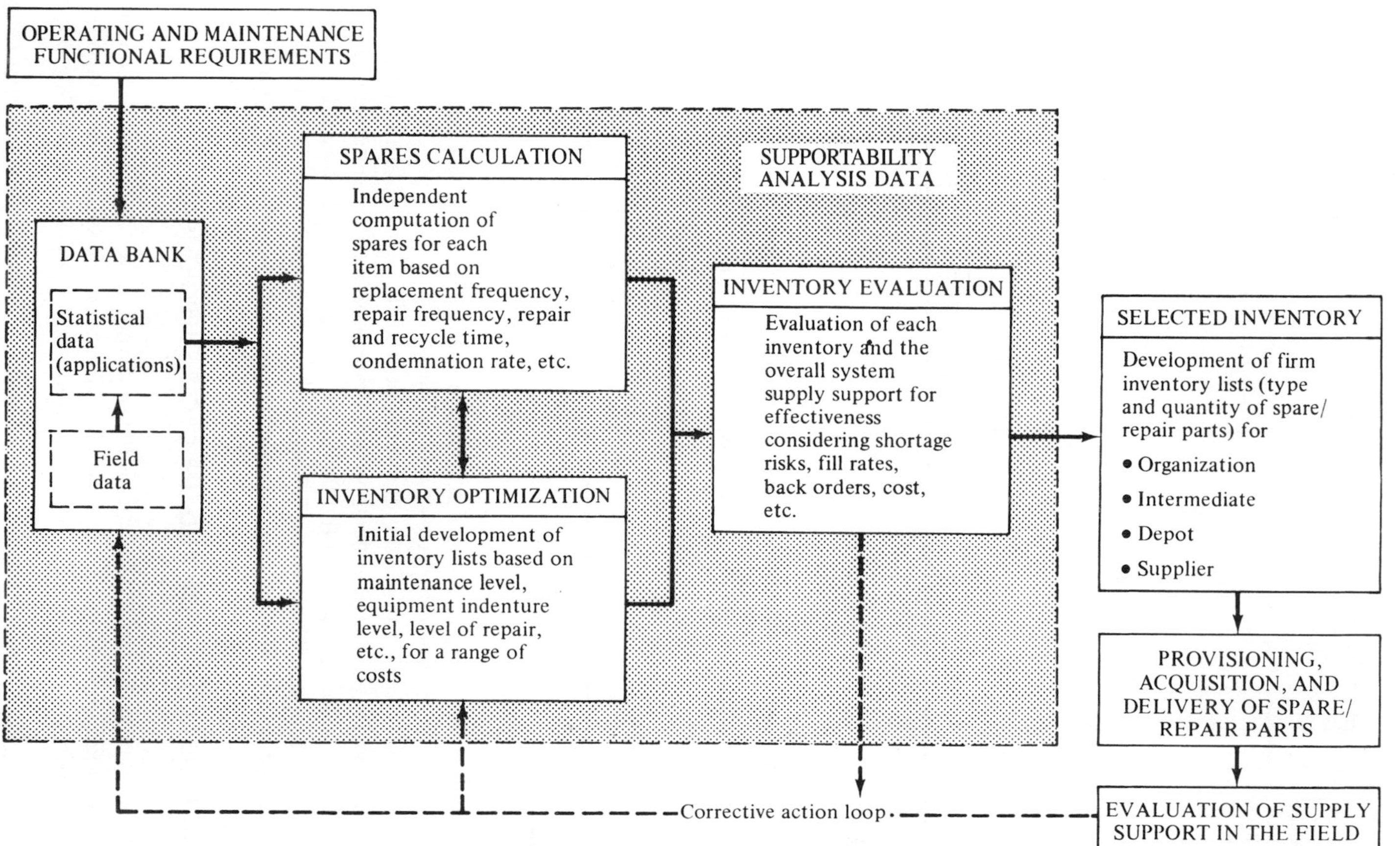

Figure 15.13 Spare and repair parts development process.

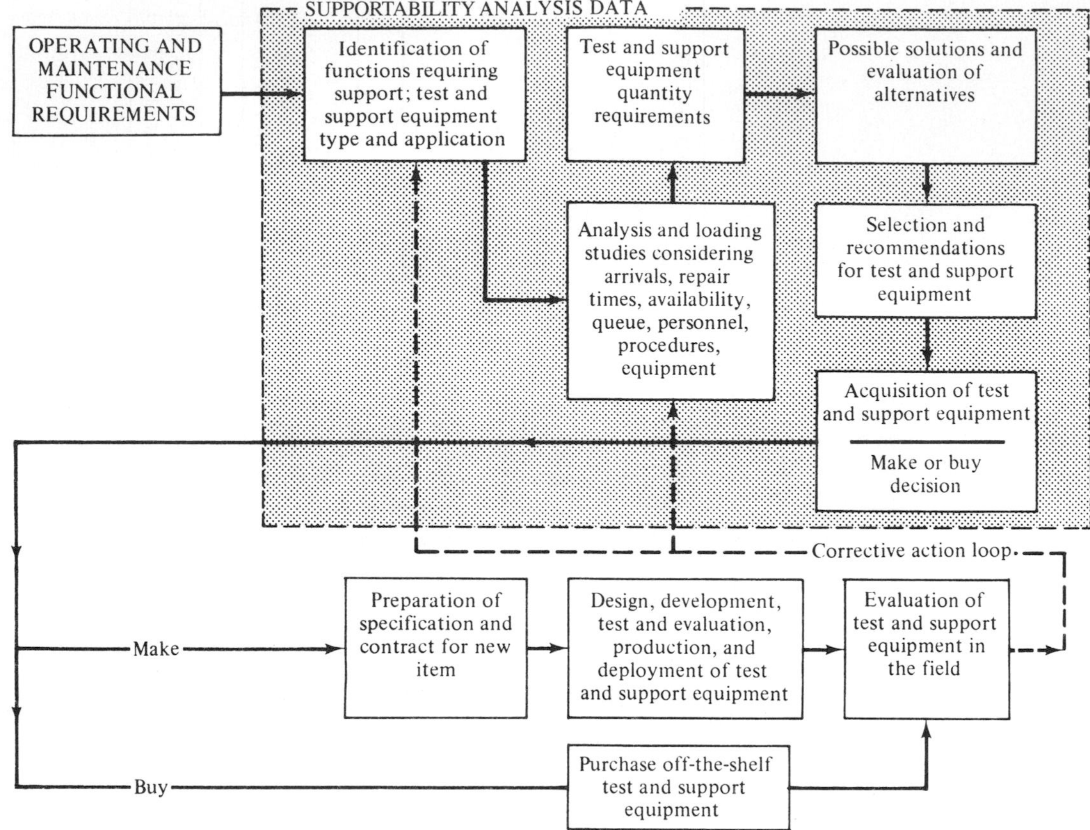

Figure 15.14 Test and support equipment development process.

presented in the form of a list of recommended test and support equipment. The SA deals with the analytical aspects of this process.

Although only two of the elements of logistic support are covered in Figures 15.13 and 15.14, the same basic approach is applicable to the other elements. The SA provides a database, covering the factors in Table 15.3, that is used for requirements definition purposes. The analysis is iterative in nature, and is employed throughout the early phases of the life cycle to promote supportability in system design.

15.6 CONTINUOUS ACQUISITION AND LIFE-CYCLE SUPPORT (CALS)[17]

Inherent within the field of logistics is the requirement for an extensive amount of data (as described in Section 15.5), particularly for complex systems and large-scale programs. With the ever increasing need to address supportability in design, it becomes

[17] The terms *computer-aided logistic support* and *computer-aided acquisition and logistic support* have also been applied to this activity. A good reference is MIL-HDBK-59A, Military Handbook, *Computer-Aided Acquisition and Logistic Support* (CALS) (Washington, D.C.: Department of Defense).

essential that such data (which provides an assessment of system design for supportability) be processed and made available to the design community in a timely manner. In other words, as the SA activity evolves commencing during conceptual design, a design evaluation is accomplished, the results are recorded, and recommendations for possible corrective action need to be fed back to designers on a realtime basis.

In the past, design evaluation and feedback associated with logistic support have been almost nonexistent, primarily because of inabilities to collect, process, and evaluate appropriate data in a timely manner. In recognition of this deficiency, combined with the advent of computers and their use in the design process through computer-aided design (see Section 4.4), the Department of Defense is addressing this problem through the advent of CALS. Basically, the objective is to develop and convert existing databases to a digital format and to expedite the processing of data within government organizations, within the supplier organizations, and between the government and its various contractors. Of further significance is the processing of information to enhance the design-for-supportability objectives discussed throughout this chapter.

15.7 TOTAL PRODUCTIVE MAINTENANCE (TPM)

Total productive maintenance, a concept introduced by the Japanese in the early 1970s, represents an "integrated life-cycle approach to the maintenance and support of a manufacturing plant."[18] Experience indicated that many of the factories in being had been operating at less than full capacity, productivity was low in general, and the costs of factory operations were high. Further, a large portion of the total cost of doing business was due to "production losses." These costs, in turn, were transferred to the cost of the product being manufactured.[19] In other words, the cost of many of the products being delivered for consumer use was high because of the cost of manufacturing, which, in turn, had a negative impact in the commercial marketplace. Thus, an intensive effort was initiated in Japan to increase productivity and reduce the costs of manufacturing, with the issue of *maintenance* being a "target" of opportunity.

Although the issue of maintenance was the initial "motivator," the implications relative to implementing the concepts and principles of TPM are far more reaching. In essence, TPM is an approach for improving the overall effectiveness and efficiency of a manufacturing plant. More specifically, the objectives are to

1. Maximize the overall effectiveness of manufacturing equipment and processes. This pertains to maximizing the availability of the production process through

[18] S. Nakajima, (ed.), *TPM Development Program: Implementing Total Productive Maintenance*, (Portland: English trans. Productivity Press, 1989). Subsequently, there have been numerous additional references on TPM published by Productivity Press. Also, refer to T. Wireman, *Total Productive Maintenance: An American Approach*, (New York: Industrial Press, Inc., 1991).

[19] According to a survey conducted by R. K. Mobley in the 1980s, between 15% and 40% of the total cost of the finished goods produced could be attributed to the maintenance costs in the factory. Where in the light process industries (e.g., food-related products) this percentage was relatively low, the percentage was high in heavy industries (e.g., iron and steel, pulp and paper). Refer to R. K. Mobley, *An Introduction to Predictive Maintenance* (New York: Van Nostrand Reinhold, 1990).

improvement of equipment reliability and maintainability, with the goal of minimizing downtime (refer to Chapters 12 and 13).

2. Establish a life-cycle approach in the accomplishment of preventive maintenance. This pertains to the application of a reliability-centered maintenance (RCM) approach to justify the requirements for preventive maintenance (refer to Section 13.4).
3. Involve all operating departments/groups within a manufacturing plant organization in the planning for and subsequent implementation of a maintenance program (i.e., representation from engineering, operations, testing, marketing, and maintenance). The objective is to gain full *commitment* throughout the manufacturing organization.
4. Involve employees from the plant manager to the workers on the floor. There must be commitment from the top down in the organizational hierarchy.
5. Initiate a program based on the promotion of maintenance through "motivation management" and the development of autonomous small-group activities (i.e., the accomplishment of maintenance through a "team" approach). The operator must assume greater responsibility for the operation of his or her own equipment, and must be trained to detect equipment abnormalities, understand the cause-and-effect relationships, and accomplish minor repair. The operator is then supported by other members of the team for the accomplishment of higher-level maintenance activities. The application of the FMECA and the Ishikawa diagram can help in identifying cause-and-effect relationships (refer to Figures 12.24 and 12.25).

The measure of total productive maintenance (TPM) can be expressed in terms of *overall equipment effectiveness* (OEE), which is a function of *availability, performance rate*, and *quality rate* as

$$\text{OEE} = (\text{availability})(\text{performance rate})(\text{quality rate}) \tag{15.11}$$

where

$$\text{Availability } (A) = \frac{\text{loading time} - \text{downtime}}{\text{loading time}} \tag{15.12}$$

with *loading time* referring to the total time available for the manufacture of products (expressed in days, months, or minutes) and *downtime* referring to stoppages resulting from breakdown losses (failures), losses because of setups/adjustments, and so forth

and

$$\text{Performance rate } (P) = \frac{(\text{output})(\text{actual cycle time})}{(\text{loading time} - \text{downtime})} \times \frac{(\text{ideal cycle time})}{(\text{actual cycle time})} \tag{15.13}$$

with *output* referring to the number of products produced; *ideal cycle time* referring to the ideal or designed cycle time per product (e.g., minutes per product); and *actual cycle time* referring to the actual time experienced (minutes per product). This metric addresses the process and rate of performance.

and

$$\text{Quality rate } (Q) = \frac{\text{input} - (\text{quality defects} + \text{startup defects} + \text{rework})}{\text{input}} \tag{15.14}$$

with *input* referring to the total number of products processed and *defects* referring to the total number of failures for one reason or another.

In essence, one is dealing with the manufacturing plant as a *system*, with the OEE factor being a measure of system effectiveness, as discussed in Section 12.2. As such, the manufacturing plant can be defined in functional terms, metrics can be allocated for each functional element, high-cost contributors can be identified, cause-and-effect relationships can be established, and recommendations for design improvement can be initiated as appropriate. Thus, although this subject is being addressed in this chapter (because of the fact that *maintenance* was the initial "driver"), many of the concepts and principles discussed throughout Part II of this text are applicable to the design, development, construction, operation, and support of a manufacturing plant.

The TPM approach first became popular in the United States in the late 1980s, particularly with those companies that are producing products in a highly competitive international environment and desire to reduce product costs. In the early 1990s, the American Institute of Total Productive Maintenance (AITPM) was established, annual conferences have been scheduled, and monthly newsletters have been published. Although the terminology is different, the goals and objectives of TPM are consistent with the concepts and principles of ILS and the design for system supportability.[20]

15.8 SUPPORTABILITY TEST AND EVALUATION

The ultimate objective in designing for supportability is to ensure that the proper level of logistic support is provided for the system in a timely and economical manner throughout its life cycle. Specific quantitative and qualitative goals are established during conceptual design, analyses are completed throughout detail design and development, a fixed design configuration is assumed based on these analyses, a prototype system is produced, and test and evaluation is accomplished to compare the results with the initially specified requirements.

System test and evaluation and its general requirements are discussed in Chapter 6. The test and evaluation effort associated with logistic support is accomplished through a combination of individual demonstrations (Type 2 testing), an integrated systems test at a designated test site (Type 3 testing), and an operational test in a customer environment (Type 4 testing). The most information can be derived through the latter, whereas individual demonstrations provide limited results. However, if problems do occur, it is far preferable to make changes as early in the life cycle as possible. Thus, it is important to begin with the assessment of the logistic support capability through Type 2 testing when prototype equipment first becomes available.

[20] The Japanese established an initial OEE benchmark of 85%. As this text went to press, a few companies have attained this goal; however, the measured OEE for most of those companies who have applied TPM principles (to varying degrees) has been between 55% and 65%.

Much of the early evaluation of the logistic support elements for the system can be accomplished in conjunction with other testing activities. For instance the results of reliability testing as is described in Chapter 12 will provide some indication of the failures that are likely to occur, causing maintenance actions and the demand for logistics resources. To an even greater extent, the results of maintainability demonstration, as discussed in Chapter 13, will provide an initial assessment of the following:

1. The most likely maintenance tasks that will be accomplished throughout the life cycle, task sequences, and task times.
2. The personnel skill levels that will be required for maintenance task accomplishment, the labor-hours expended in task completion, complexity, dexterity factors, and error rates.
3. The removals and replacements required in accomplishing corrective and preventive maintenance (spare and repair part requirements).
4. The compatibility between the test and support equipment, the transportation and handling equipment, and the prime-mission–oriented equipment. Will the item perform the intended function effectively and efficiently?
5. The procedures that will be used in the accomplishment of corrective and preventive maintenance.

Maintainability demonstration is primarily intended to evaluate the supportability characteristics inherent in the prime mission equipment. As such, this requires the availability of the right elements of the logistic support capability for the system, and much valuable information can be acquired.

System test and evaluation, accomplished as Type 3 and 4 testing, provides the first real opportunity to look at the system as a whole. Prime equipment and all the required elements of logistic support are delivered to an operational test site, integrated, and scheduled through a series of mission profiles. The logistic support provided is in accordance with the requirements of the formal ILS plan. Throughout the accomplishment of the scheduled mission exercises, system operational, performance, and supportability characteristics are measured and evaluated. Areas of deficiency are noted and corrective action is initiated as appropriate. In essence, the same procedure as that discussed in Chapter 6 is employed, except that the major emphasis here is on the elements and measures of logistic support treated in Sections 15.2 and 15.3.

QUESTIONS AND PROBLEMS

1. Define *supportability*. How does it relate to *reliability, maintainability*, and *usability*?
2. Define *logistics*. What are the elements of logistic support? How would you compare the *business logistics* approach with *integrated logistic support* (ILS)? Identifiy some commonalities/differences.
3. How does logistics relate to systems engineering?
4. Briefly describe logistics activities in each phase of the life cycle.

5. Why is it important to consider logistics during the early planning and conceptual design phase of the life cycle?
6. Refer to Figure 15.1. Identify and describe some of the *metrics* associated with the different activities conveyed through the figure.
7. What factors should be considered in determining supply support requirements? Test and support equipment requirements? Personnel requirements? Training requirements? Transportation and handling requirements? Facility requirements? Technical data requirements?
8. Assuming that a single component with a reliability of 0.85 is used in a unique application in the system and that there is one backup spare component, determine the probability of system success having a spare available it time, *t*, when required.
9. Assuming that the component in Problem 8 is supported with two backup spares, determine the probability of system success.
10. Determine the probability of system success (having a spare available) for a configuration consisting of two operating components backed by two spares (assume that the components are interchangeable with a reliability of 0.875).
11. There are 10 systems located at a site scheduled to perform a 20-hour mission. The system has a predicted MTBF of 100 hours. What is the probability that at least eight of these systems will operate for the duration of the mission without failure?
12. An equipment contains 30 parts of the same type. The part has a predicted failure frequency of 10,000 hours. The equipment operates 24 hours a day and spares are procured at 90-day intervals. How many spares should be carried in the inventory to ensure a 95% probability of success?
13. Determine the economic order quantity (EOQ) for spares inventory replenishment when the cost per unit is $100; the cost of preparing for a shipment and sending a truck to the warehouse is $25; the estimated cost of managing an item in inventory is 25% of the inventory value; and the annual demand is 200 units.
14. What is the total cost per period given that the demand is 4 units; the cost per unit is $12.00; the procurement cost is $16.00; and the inventory management cost is $2.00?
15. Suppose that an equipment contains 40 parts of a specific type with a failure rate of 0.3 failures per 1,000 hours of operation. The equipment operates 24 hours a day and spares are procured and stocked at 1-week intervals. How many spares should be carried in stock to ensure a 90% probability of success?
16. What are the major considerations in determining an economic order quantity (EOQ)? Should the EOQ principle be applied in all instances when spares/repair parts are required (explain)?
17. Some spare parts have a higher priority (in importance) than others. What factor(s) determines this priority? What considerations should be addressed in the procurement of such items?
18. Assume that you have an intermediate-level shop full of test equipment. What steps would you take to ensure the adequacy of this equipment over time?
19. How could the maintenance requirements for test and support equipment affect the availability of the prime elements of the system?
20. Assume that you have been requested to set up a maintenance organization for system support. What factors would you consider? Describe the steps involved. How would you later evaluate this organization in terms of effectiveness (identify the metrics that you would use)?
21. How would you determine the transportation requirements for the maintenance shop that you have organized in Problem 20?

22. What considerations would you address in selecting a specific mode of transportation, (or combination of modes)?
23. How can *packaging* design impact the reliability of a system (if at all)?
24. Describe what is meant by the *supportability analysis* (SA). What is its purpose? What type of information is included? When in the life cycle is it accomplished? What is the purpose of the *logistics management information* (LMI)? How does it relate to the SA?
25. In accomplishing the SA, there are several design tools/techniques that can be effectively applied in fulfilling its objective(s). Identify a few of these and describe their application.
26. Briefly describe CALS, its objective, and its application.
27. Describe *total productive maintenance* (TPM). What are the objectives of TPM? Metrics? How does it relate to ILS (if at all)?
28. Calculate the OEE given that: loading time is 460 minutes/day; downtime is 100 minutes; ideal cycle time is 10 minutes/product; actual cycle time is 15 minutes/product; the number of products scheduled for production is 22; and the number of products actually produced is 20. If this value turned out to be significantly less than your "benchmark" value, what steps would you take to correct the situation?
29. How are the elements of logistics evaluated through testing? Can any information of value be gained from other testing? If so, please identify.
30. What is the purpose of the ILSP and what is included? How does it relate to the SEMP? Reliability and maintainability plans? Test and evaluation master plan (TEMP)? How does it apply to the (a) production/construction phase (if at all); (b) utilization and support phase; and (c) the retirement and material disposal phase?

16 Design for Producibility and Disposability

Producibility and disposability are design-dependent parameters related to each other in subtle ways. This relationship is traceable to "bringing into being" and "ceasing to be."[1] The degree of order exhibited by a system in being is a remarkable manifestation of system design, an overview of which is given in Chapter 2. Producibility is usually given priority over disposability, because it is internal to the firm and its profitability. But disposability (an externality) is becoming a product characteristic of increasing interest to producers and consumers alike. System and product disposal (along with waste from the production or construction process) requires attention during both product and process design.

In this chapter, both producibility and disposability are discussed as important design concerns. They are linked insofar as possible, realizing that measures and metrics for these parameters are different. Likewise, models for the evaluation of the system-level impacts of these design concerns are not yet fully developed. Accordingly, producibility and disposability are presented in this chapter in a descriptive rather than a normative manner, with an emphasis on their place in and relationship to the systems engineering process.

16.1 INTRODUCTION TO PRODUCIBILITY AND DISPOSABILITY

Market opportunities are expanding for the production of "environmentally friendly" or "green" products, systems, and structures. Although an externality to the producer, impacts on the environment are of concern to people and societies worldwide. Pro-

[1] To bring an entity into being is to achieve a high degree of organization and order; to cease being is to return to a state of disorganization and disorder, as in the concept of entropy.

ducibility, conversely, is internal to the manufacturer. It is not directly visible to the customer. Nevertheless, both producibility and disposability are factors on which the ultimate net value of a product or operational system depends.

Technological and Ecological Services

It is through production or construction that technology is put to use by humans. "Green" manufacturing is not a goal that can be achieved easily, however. It should be an objective adopted by producers to reduce the environmental impact of their products and production operations continuously. Real environmental improvement requires a system life-cycle approach to guide design decisions and operational policies.

Technical systems are the source for technological services.[2] These services generally involve the substitution of energy for human effort. They include delivering diverse foods independent of season or local climate, delivering potable water, modifying the climate in buildings, delivering communications capability, removing wastes, and so on. Civilization, especially in advanced countries, has developed an extensive infrastructure for the delivery of technological services, including electrical transmission lines, communication links, highways and airports, homes and public buildings, sewer systems, and so forth.

Humans also depend on ecosystem services. Ecosystem services are those functions of ecosystems that people desire, including the maintenance of atmospheric balance, flood control, carbon storage, production of food and fiber, and maintenance of water quality. Although these services are often taken for granted, people are beginning to realize that many functions of natural systems are not immutable. They can be affected by human actions and effected through potent technologies.

The definition of an ecosystem service is a matter of individual and societal perception; it depends on subjective evaluation. However, people most familiar with natural systems are not usually in positions of power and are not usually consulted by those who are. Balancing technological and natural systems will be exceedingly difficult if the operational characteristics of the natural systems are not as well understood as the physical aspects of the technological system.

Responding to societal needs, regulatory agencies such as the U.S. Enviromental Protection Agency, the Food and Drug Administration, and the Occupational Safety and Health Agency were formed by the government to assess real or anticipated environmental quality losses. Because regulatory decisions represent a means for incorporating externalities into designs for products, systems, and services, they have far reaching impacts and must be made carefully and wisely.

Producing for Environmental Quality

With the rapid growth in population and widespread manufacturing on a finite planet, the generation of wastes associated with the production, use, and disposal of goods and services is of great concern. These byproducts of market transactions impact environ-

[2] The notion of technical systems is introduced in Section 1.5.

mental quality. The losses to environmental quality resulting from pollution of the air, water, and soil by effluents from manufacturing are one example of externalities. Noise pollution from factories, aircraft, trucks, and cars is another. Externalities have no direct markets; clean air, rivers, or low decibel noise cannot be bought or sold in a market place.

The loosely regulated market has poorly internalized environmental quality concerns. There are two reasons for this problem. First, environmental quality loss is often not seen or understood at the time the transaction occurs. Second, losses are usually insignificant for a single market transaction. They arise from the aggregate, cumulative effects of the manufacture, use, and disposal of products. Degradation over the long term of common property resources, such as air and water, is difficult to assess now.

Designers have always had the opportunity to consider environmental concerns. Such concerns are reflected in the market, where many products are made to be recycled, are biodegradable, and are free from toxic substances. In the past, the existence of environmental effects or constraints imposed on the manufacturing process has not been a major issue for designers. Perhaps this is because the waste and pollution generated from manufacturing processes are considered to be byproducts of manufacturing and, hence, are assumed not to affect the product's producibility and quality. Consequently, the costs incurred become part of the fixed cost of the producer and are not linked to the product realization process. As a result, little effort has been made to design "green" products and associated "environmentally-friendy" manufacturing processes.

Factors Promoting Green Engineering

The reuse value of recoverable materials and components from most products at the time of disposal is usually small in relation to initial production cost. Accordingly, it is difficult to persuade manufacturers and the first buyer to make changes to the product that would improve the eventual ease of recovery of these materials. This is especially true if those changes incur cost or have even a small actual or imagined effect on the initial performance, functionality, or affordability of the product.

However, this situation is changing. Most leading companies have increasingly taken initiatives to improve the environmental performance of their products and processes to reduce undesirable impact. What are the changes in environmental consciousness that favor movement away from "end-of-pipe" solutions toward environmentally-oriented product design and production? The following major factors are now recognized as primary drivers toward this newer thinking:

1. *Competitive differentiation.* Traditionally, marketplace decisions have been based on a products performance, quality, and price. In the world-class competition of today, some nonprice factors are being included in the company's competitive strategy. All signs indicate that environmental impacts, one of the nonprice factors, became the most important issue in the commercial world during the 1990s.
2. *Customer consciousness.* Most people believe that a healthy natural environment not only enhances their quality of life, but assures them that the quality of life

will be sustained. Increasingly, customers are becoming concerned about the environmental quality of the products they use and the quality of the environment in which they live. Consumer awareness of potential damage inflicted on the environment by abuses of technology is beginning to create pressure to develop green products and clean manufacturing operations. Regulatory bodies tend to respond to this change in attitudes by developing legislation for use against those that neglect environmental issues.

3. *Governmental regulatory pressures.* Pending environmental legislation is expected to impose environmentally-related restrictions on industrial and consumer product manufacturers. To address environmental problems, regulations inspired by the "polluter pays principle" are emerging. Motivated by environmental laws and customers' environmental consciousness, manufacturers are forced to consider seriously how to produce greener products and to maintain competition at the same time.
4. *Profitability improvement.* An eco-design approach to systems can have a significant impact on profitability through savings in manufacturing and other operating costs, by waste elimination-related strategies, and through increased market share. A growing trend toward safer product design and production satisfies a company's responsibility to provide a safer, healthy working environment for its employees. Incorporation of such values into business activities contributes to a positive feeling about their workplace.
5. *International standards.* Coordinated by the International Standards Organization (ISO), many manufacturers are participating in a worldwide effort to establish standards for environmental stewardship over the full product life cycle.[3]

Environmentally Conscious Design and Manufacturing (ECDM)

An evolutionary design paradigm, which starts with the consideration of environmental impacts caused by products and product-related processes during the product and process design stage, is known as *environmentally conscious design and manufacturing* (ECDM). ECDM has been increasingly characterized in the last decade within the domain of industrial ecology.[4]

The ultimate aim of ECDM is to reduce the environmental impact or to improve the environmental friendliness of products, processes, and related activities. A significant difference between ECDM and waste management or pollution prevention is that the ECDM approach is proactive. Its objective is to reduce environmental impact in the early stages of product design. Production designers and manufacturing engineers (as well as managers) have a shared responsibility for planning and implementing environmentally safe and environmentally-friendly manufacturing operations. Much can

[3] Refer to ISO 14001, "Environmental Management System" (EMS). This standard was developed recently and is being introduced in many companies and implemented in conjunction with ISO 9000-9004 *Quality Standards* (ANSI/ASQC Q90-Q94).

[4] T. E. Graedel, and B. R. Allenby, *Industrial Ecology* (Upper Saddle River, N.J.: Prentice Hall, Inc., 1995).

be gained by careful analysis of products, their design, the materials used, and the manufacturing processes and practices employed.

ECDM can be generally divided into two categories: design for environment (DFE) and environmental management (EM). The DFE approach is the proactive activity that aims at prevention of environmental impacts, whereas EM is remedial in nature. Accordingly, it is ECDM that is most effective as a system design strategy when implemented over the life cycle.

16.2 PRODUCIBILITY AND DISPOSABILITY IN THE SYSTEM LIFE CYCLE[5]

As conveyed in Figures 2.9 and 2.14, *producibility* and *disposability* constitute two major areas of consideration in the design of a system/product and in the implementation of the systems engineering process. The principles discussed in this chapter can be applied in a manner similar to those for reliability, maintainability, human factors, and supportability (Chapters 12 to 15, respectively). Referring to Figure 2.2, specific design-to requirements need to be established and specified early during the conceptual design phase, functional analysis and requirements allocation are accomplished in preliminary design, trade-offs and design optimization occur, and so on. The objective is to influence the design such that system elements and components can be produced effectively and efficiently and that, when retired, can be disposed of efficiently and without causing detrimental impact(s) on the environment.

Concurrent Life-Cycle Interfaces

Referring to Figure 16.1, which is an extension of Figure 5.1, there are four basic life cycles of which to be concerned. The initial emphasis in designing the *product* for producibility and for disposability is applied in the top series of activities (i.e., the first life cycle in this instance). However, there are major interfaces involving the design of the production process (the second life cycle) and in the design of the disposal/recycling capability (the fourth life cycle), and these must be addressed on a concurrent basis. The design of the system support capability (the third life cycle) must also be addressed, as discussed in Chapter 15.[6]

When designing the product with producibility/disposability criteria in mind, the production process must also be designed to support the objectives imposed on the product. It is not uncommon for a system/product to be designed with the desired characteristics incorporated and then to be subjected to an existing production process, with product modifications being required for the purposes of compatibility. In other words, a manufacturer, with good intent relative to product design, may not wish to

[5] Some of the material and figures in this section and in Section 16.5 were adapted from research presented in the unpublished dissertation by M. J. Goan, "An Integrated Approach to Environmentally Conscious Design and Manufacturing," Virginia Tech, Blacksburg, Va. June 1996.

[6] The principles of *concurrent engineering*, discussed in Chapter 5, must be implemented here if the overall objectives of system engineering are to be realized.

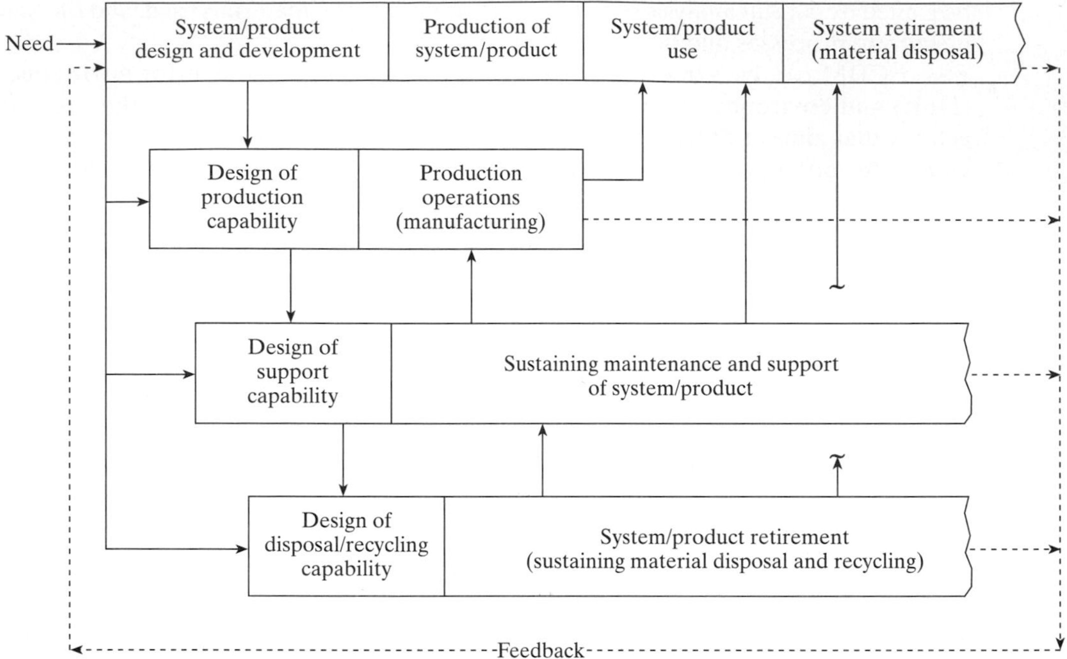

Figure 16.1 The life cycles and their interrelationships (refer to Figure 5.1).

modify the existing factory for economic reasons; thus, the initial product design may be impacted in a negative manner. The same concern exists when considering the product design for disposability (the first life cycle) and the design of the disposal/recycling capability (the fourth life cycle). Further, the design of the disposal/recycling capability must consider the requirements for material disposal and/or recycling resulting from the maintenance and support activities illustrated in Figure 16.1 (i.e., the third life cycle). Thus, the four life cycles must be addressed from the top down, in an integrated manner, and with the appropriate feedback provisions.

Within the context of the activities in Figure 16.1, design synthesis, analysis, and evaluation are required throughout the product design and development cycle, as illustrated in Figure 2.7. During the ECDM process, producers must be able to predict and compare the environmental impacts of various product and process design alternatives. If environmental requirements for the product/system are specified well, design alternatives can be evaluated against them along with other requirements. In addition to design for performance and functionality, designers are constantly faced with choosing among materials and processes, as well as among end-use characteristics, such as the degree of energy efficiency and recyclability.

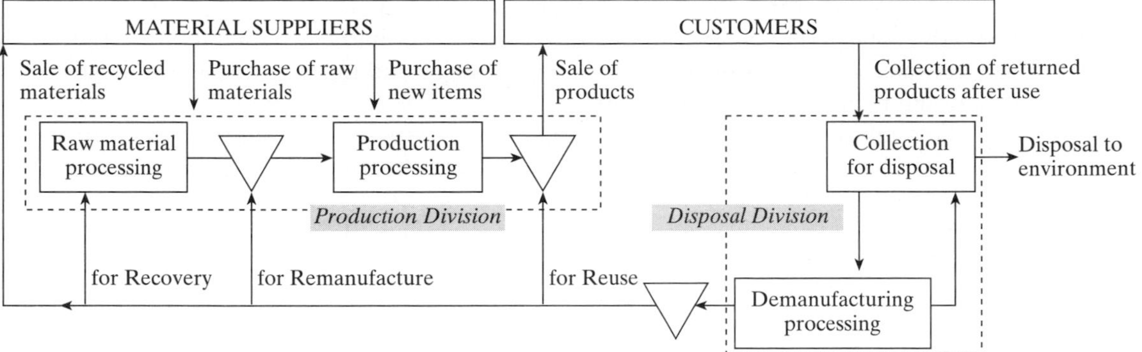

Figure 16.2 A production system with demanufacturing and disposal.

Manufacturing and Demanufacturing Issues[7]

In a production system with demanufacturing capability, the basic structure of the demanufacturing component consists of two major divisions as shown in Figure 16.2: a production division and a disposal division. In the production division, manufacturing purchases raw materials or components from outside suppliers, and then manufactures and distributes specified types of products for sale to consumers. The production division is classified into three sectors: (1) recyclable products for producing the same type of products inside the plant, (2) recyclable products for producing different kinds of products inside the plant, and (3) wastes to be discarded. Accordingly, a production system with demanufacturing capacity actually contains the closed-looped flow (for division 1) or the open-loop flow (for division 2), or both, inside the plant.

In a typical product development process, the product configuration established and product attributes are processed through several manufacturing operations. Returned (or recycled) products are processed by demanufacturing operations. Recycling and demanufacturing, which are the product's major or end-of-life activities, should be integrated into the design and life-cycle processes. ECDM results will affect downstream process planning for product end-of-life and, in turn, processes planning has an effect on green product design.

It is now recognized that demanufacturing plays an important role in the green product concept. Manufacturing companies respond to social and regulatory pressures for environmental issues in several different ways. These responses are directed at decreasing disposal costs by demanufacturing the recycled products received from consumers for reuse, remanufacture, and recycling. Also, these company responses have

[7] *Demanufacturing* is used in this chapter instead of the term *disassembly*. Demanufacturing includes consideration of planning and processing related to recycling as a whole. Disassembly, viewed as a recycling activity as part of demanufacturing, is employed to reuse, refurbish, remanufacture, and recycle. Further, it should be noted that many of the principles related to *design for disassembly* are similar to those for maintainability (refer to Chapter 13).

an impact on the firm's strategy for green product design and development. Accordingly, the collection, demanufacture, processing, recycling, and disposal of used products recycled from consumers present major design and manufacturing challenges to traditional manufacturing operations.

Demanufacturing operations are essential processes for the purposes of reuse, demanufacturing, recovery, and so forth. These are described subsequently:

1. *Reuse* is the highest form of waste reduction and has the potential to increase the products end-of-life value. Reuse of a product may or may not require total disassembly. Reuse may be most easily justified in the case of components with high manufacturing costs, long innovation cycles, and long lifetimes.
2. *Remanufacturing* is the refurbishing or partially rebuilding of a product returned from the market (by collection) at the end of its life to give it the functionality of a new product. Remanufacturing requires disassembly efforts that contribute to the demanufacturing cost. The recycled components and parts may be used for reproducing either the same or different products, according to closed-loop or the open-loop recycling concept.
3. *Recovery* from products to obtain raw materials or reuseable components is an important means of reducing disposal volume and cost.

The efforts listed previously are important ways to reduce product life-cycle cost. However, these activities contribute to demanufacturing costs. The significant demanufacturing costs for recycling and reuse include costs for retrieval, separation, disassembly, sorting, storage, transportation, identification, testing, reprocessing, and remarking. The value of the recycled or remanufactured product is a function of the intrinsic value of its components/parts and materials, as opposed to the cost of recycling and disposal.

Costs of recycling and disposal can be significantly affected by product design decisions. Design for demanufacturing, design for disassembly, design for serviceability, or supportability and design for recycling are fundamental to effective demanufacturing and recycling strategies and programs. Both manufacturing and demanufacturing processes are affected by design decisions, especially when the environmental impacts caused by the processes are considered. The product realization process is complicated by the nature of economic, logistic, and environmental considerations. To explain the relationships between value and cost throughout a products life-cycle, a cost/value flow diagram is presented next.

A Life-Cycle Cost/Value Flow Diagram

Integration approaches are required in constructing a cost-benefit analysis model, as shown in Figure 16.3. A cost/value flow diagram provides basic concepts of costs and values that vary throughout a products life cycle. It also depicts a life-cycle cost/value, economic flow component that explains the complexity of economic, logistic, and environmental considerations for manufacturing systems with demanufacturing/recycling capacity.

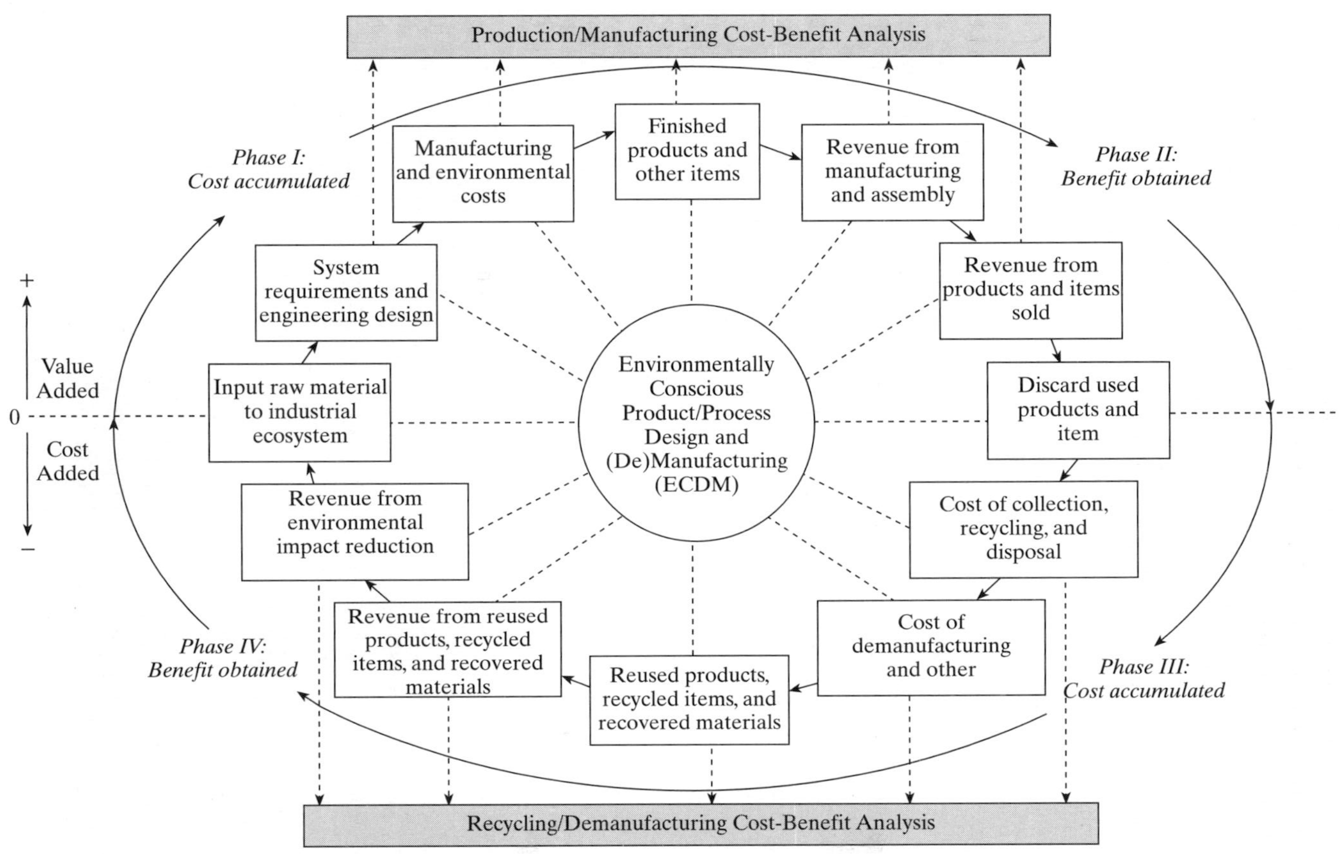

Figure 16.3 Life-cycle cost and benefit flow for ECDM.

Under traditional manufacturing engineering and life-cycle costing, production/manufacturing cost-benefit analysis is employed in Phases I and II of Figure 16.3. Cost is committed during manufacturing of the product and related items, including environmental cost (Phase I). Value is added to the products and items as they are fashioned from raw materials. The extent of the added value is determined by the production/manufacturing costs arising from capital equipment, design engineering, purchases, labor, processing operations, and so on. During Phase II, a benefit is obtained only when the finished products or items are sold. The value decreases because the products and items depreciate physically or functionally during their life as a result of use, environmental stress, and by competition from newer products. For capital products, such as vehicles, a net salvage value may be realized. But, for most small items, the net value is zero or negative reflecting removal or disposal costs. Thus, traditional product life cycle and manufacturing cost-benefit analysis does not consider the cost because of environmental effects and the potential revenue from recycling.

A more complete economic cycle includes the lower half of the flow diagram in Figure 16.3. Products and items can be recycled in the salvaged condition. Collection, sorting, disassembly, contaminant removal, and material recovery necessary during the recycling and demanufacturing process. These activities add to costs, while extending the actual net value of the product and the items recycled from consumers and industry (Phase III). In Phase IV, recycled items and recovered materials are sold to the original or other product manufacturers. Other savings in the overall four-phase life-cycle result from the reuse of recycled items and recovered materials, as well as from the reduction of environmental impacts, including waste reduction energy saving.

Environmental costs are difficult to specify accurately, but show up as environmental loading costs, landfill fees, disposal costs, waste handling charges, and general environmental liability. Nevertheless, recycling makes good economic sense in many instances in addition to its environmental benefit. In such a complete economic cycle, the opportunities and feasibility of item recycling and materials recovery should be considered throughout product and process design. These considerations should be subjected to a cost-benefit analysis model employed to select design alternatives for the product and process.

16.3 DESIGN FOR PRODUCIBILITY

Producibility is not synonymous with manufacturability. Producibility is a more encompassing term, which includes not only the goal of ease of manufacture, but also the ability of the designed entity to be manufactured, packaged, and shipped (the physical distribution dimension). Producibility, like manufacturability, is a characteristic of design. It can be improved by proper design decisions incorporating life-cycle considerations.[8]

[8] The term *producibility*, as defined in this context, is closely related to the *design for transportability and packaging* described in Section 15.3.

Measures of Producibility

It is difficult to express producibility in a generic way or in quantitative terms by mathematical measures. However, it is possible to partition the concept into the measurable areas of manufacturing on the one hand and marketing on the other.

Manufacturability measures. The more manufacturable a product is, the more quickly (and inexpensively) it may be produced. Therefore, manufacturability has been measured in terms of traditional industrial engineering methods. An example of manufacturability may be *manufacturing lead time* (MLT) or the time needed for a product to be in the manufacturing process. For example, a product must be processed on n_m machines and the average operational time, T_O, is required for each machine. Setup requirements for the these machines are T_{SU} and non-operational time for the product is T_{NO}. If the number of units produced per batch is Q, then the MLT can be expressed as

$$\text{MLT} = \Sigma\,(T_{SUi}/Q_{TOi} + T_{Oi} + T_{NOi}), \qquad i = 1 \text{ to } n_m \tag{16.1}$$

For a product-oriented measurement, a common metric is the average production time per unit TP, which is simply the batch time per machine divided by the number of units per batch. With the preceding measurements already defined and a zero defect rate, T_p can be expressed as

$$T_P = [\text{batchtime/machine}]/Q = T_{SU}/Q + T_O \tag{16.2}$$

The processes and methods for manufacturing will determine how quickly the product is manufactured. Note, however, that the cost of the processes are not a factor in these measures; these costs will affect the overhead rates and must be included in the life-cycle analysis for process design trade-off studies. A process may be twice as fast as another, but cost three times as much per product unit or per hour. Accordingly, more than just the process time must be considered.

These considerations embody a systems approach to design because the product and its manufacturing processes in the producer's facilities are considered when calculating are done. In the development stage of the system life cycle, these values are purely theoretical. Once production begins, these values become significant in terms of cost and profit. Continual improvement in the manufacturing process must be considered to improve the manufactureability of the product. This type of program could be planned for, similar to the manner that "Reliability Growth Plans" account for such improvements in the life cycle (refer to Section 12.4). Thus, "design for manufacturability" (DFM) is highly desirable before major process design is completed.

Subjective measures can be used to determine manufacturability. Industrial science methods use surveys and direct observation for determining the ease of manufacture of a process. Sometimes, a lack of manufacturability can result in operator injuries over time. A common example of this occurrence is the appearance of "carpal tunnel syndrome" in the poultry industry and in data entry operations.

Market measures. Marketing measures, as defined herein, refers primarily to the flow of finished goods (i.e., the product) from the manufacturing plant to the ultimate consumer (refer to the outward flow in Figure 15.1). This aspect of product distribution, often considered and tied closely with design for producibility, is also included as a major element of logistic support (i.e., the *packaging, handling, storage*, and *transportation* elements, identified in Figure 15.2.)

Referring to Section 15.3, the primary measures of a "market" nature include the following:

1. The time that it takes to move a product from the source of manufacture to the ultimate customer (i.e., user). This includes the elements of time associated with packaging and handling, inprocess warehousing and storage, and transportation. Given a demand for a product, how long does it take for delivery and installation?
2. The cost of processing an item from the source of manufacture to the customer. This includes the cost of packaging, handling, storage, and transportation. Given a demand for a product, how much does it cost for delivery and installation?

In addition to the design of a product for producibility, the product must be designed so that it can be economically packaged (using standard and conventional materials), easily handled (using standard handling methods), and transportable (using commercial transportation provisions) without causing degrading effects on either the product itself or on the environment. Accordingly, the requirements for producibility must be closely coordinated with the requirements for supportability discussed in Chapter 15.

Manufacturability in Product Design

Manufacturing processes can generally be classified into five categories.

1. *Forming processes.* Processes in which an original shape is created from a molten or gaseous state, or from solid particles of an undefined shape.
2. *Deforming processes.* Processes that convert the original shape of solid to another shape without changing its mass or material composition.
3. *Removing processes.* Processes in which material removal occurs during the process itself.
4. *Joining processes.* Processes that unite individual work pieces to make subassemblies or final products.
5. *Material properties modification processes.* Processes that purposely change the material properties of a work piece to achieve characteristics without changing its shape.

These categories are applied to a variety of engineering materials, which can be generally classified into metals, ceramics, plastics (polymers), and composites. The selection of materials depends on the product design and should involve environmental considerations. Sustainable and recyclable materials are always considered first in

TABLE 16.1 Matching Materials and Processes (Overview)

Materials	Processes				
	Forming	Deforming	Removing	Joining	Modifying
Metals	Widely used	Widely used	Widely used	Widely used	Widely used
Ceramics	Widely used	Not used	Seldom used	Not used	Not used
Polymers	Widely used	Seldom used	Seldom used	Seldom used	Not used
Composites	Widely used	Not used	Seldom used	Seldom used	Not used

green production design. Consequently, the different processes are applied to different engineering material as summarized in Table 16.1.

The selection of which process to apply to a particular material is influenced by several factors (e.g., the lot size of the parts to be made and the physical properties of the materials) that affect process attributes (e.g., cost, production rate, flexibility, and part quality), and is especially influenced by the environmental loads caused by manufacturing processes. Selection of an appropriate process often depends on a trade-off (balance) of all process attributes (or criteria) including the potential environmental impacts. Table 16.2 gives an example based on engineering requirements for making a desirable product in an efficient, ecological, and profitable manner.

TABLE 16.2 Some Characteristics of Manufacturing Processes

Attribute	Processes				
	Forming	Deforming	Removing	Joining	Modifying
Costs	High tooling/ low labor cost	High tooling/low labor cost	Medium tooling/ high labor cost	Low capacity/ high labor cost	Medium to high capital/ low labor cost
Environmental costs	Expert judgment	Expert judgment	Expert judgment	Expert judgment	Expert judgment
Production rate	High	High	Medium (milling) to low (grinding)	Medium (welding) to low (adhesive)	Low
Part quality	Medium to low	Medium to low	Medium to high	Medium to low	High
Flexibility	Low	Low	High	High	Medium to high
Environmental factors	Inputs/outputs	Inputs/outputs	Inputs/outputs	Inputs/outputs	Inputs/outputs

There are a few high level design for manufacturability principles of general importance. These are the following:

1. *Use gravity.* It is easier to work with lighter components, with up and down movements, with snap fit in vertical slot openings, and so on.

2. *Use fewer parts.* An increase in the number of parts means an increase in design and manufacturing cost. Non existent components cost nothing to purchase, assemble, and test.
3. *Design for ease of fabrication.* Design parts so that (a) tolerances are compatible with the assembly method employed and (b) fabrication costs are compatible with targeted product costs. This eliminates part rejections or tolerance failures during assembly.
4. *Reduce nonstandard parts.* The use of common and standard parts eliminates the development costs associated with designing and manufacturing.
5. *Add more functionally per part.* The goal is to accomplish the functions required with fewer parts, or to allocate more functions per part. This is an objective of the functional analysis and packaging requirement described in Section 4.1.

In the assembly area, a few high level guidelines are of general importance. These are the following:

1. *Employ automatic inserters.* Specify parts that can be automatically sequenced and inserted using DIP inserters, a variable center device (VCD), axial part inserters, or selective compliance assembly robot arm (SCARA) insertion robots. If not, minimize setups and reorientation.
2. *Employ "preoriented" parts.* Parts that cannot be supplied "preoriented" in reels, tubes, or matrix arrays for easy insertion should be avoided. The key to successful automatic insertion is "preoriented" parts.
3. *Minimize sudden and frequent changes in assembly direction.* Following a unidirectional assembly sequence, such as design for top-down assembly, is usually the most desirable.
4. *Maximize process compliance.* Process compliance consists of designing with standard parts, standard processes, ease of assembly, and so on. Use processes that are easy to install and maintain; for example, use snap fits instead of fasteners.
5. *Maximize accessibility.* Designers should provide adequate clearances for disassembly or for accessing the part for future repair replacement.
6. *Minimize handling.* There are two aspects to minimize handling: design of parts for ease of feeding (insertion) and design of parts so that they are easy to grasp, manipulate, or orient.
7. *Avoid flexible components.* Flexible components are generally difficult to handle and assemble.

Informal guidelines have been developed for design with consideration for manufacturability. Mechanical design, predating the industrial revolution, probably has the most universally understood guidelines. Some of these are the following:

1. *Assemble to a foundation.* This method allows for automated assembly by griping to a foundation. The foundation must be designed for accurate machine posi-

tioning, since a large tolerance on the foundation location will be added to the assembled components.

2. *Assemble from as few positions as possible.* Repetitive machinery is more reliable with fewer components. Reliability of the production equipment will be reduced with an increase in components. This practice (monodirectional access) also improves maintainability.
3. *Make parts independently replaceable.* Subcomponents of an assembly should not require removal of other components to get at the faulty ones. This assembly practice also improves maintainability.
4. *Order assembly so that most reliable goes in first, with the least reliable last.* This guideline concerns the testing of a product before shipping. If a particular component or subassembly requires a significant portion of the final test, production time devoted to troubleshooting is minimized.
5. *Assure commonality in design.* Commonalty in design attempts to reduce the types of subcomponents in a system. The more standardized a product is, the less overhead is associated with supporting the variety of the parts before assembly. Also, the variety of tools used to assemble these types of parts can be reduced.

16.4 MODELING MANUFACTURING PROGRESS

Production and related operations require a coordinated and integrated set of activities that are often repeated over time. This repetition makes possible improvements in the production process such as a reduction in the time to produce a unit, an increase in the rate at which selected activities are performed and a corresponding increase in the number of units produced, and a reduction in the cost per unit of output.

Because an objective of systems engineering is to implement a procedure that will lead to the product/process improvement of systems on a continuing basis, it is necessary that one develop the appropriate analytical techniques/tools for purposes of *evaluation*. Referring to Figure 2.7, a set of requirements for the system are established, a design configuration is defined for the purpose of synthesis, an analysis is accomplished, and the system is evaluated in terms of the initially specified requirements. This process of synthesis, analysis, and evaluation is accomplished on an iterative basis, with the objective of realizing some improvement on each iteration. One of the greatest areas of opportunity for improvement is in product manufacturing where there are often many activities that are repetitive in nature.

It is appropriate to develop a *production progress model* that can be used not only to evaluate the current status of a production capability but to enable the prediction of future events. An objective is to be able to predict (in graphical and mathematical terms) possible future reductions in production times and costs resulting from the learning that occurs through repetitive activities. The production progress model is sometimes referred to as a *learning curve* model.[9]

[9] Refer to W. J. Fabrycky, P. M. Ghare, and P. E. Torgersen, *Applied Operations Research and Management Science* (Upper Saddle River, N.J.: Prentice Hall, Inc., 1984), Chap. 11.

The development of learning curves is of great interest, particularly when considering the potential costs associated with the production of units downstream. Learning generally occurs within an individual or within an organization as a function of the number of units produced. It is commonly accepted that the amount of time required to complete a given task or produce a unit of output will be less each time that the task is undertaken. The unit time will decrease at a decreasing rate, and this time reduction will follow a predictable pattern.

Empirical evidence supporting the concept of a *learning curve* was first noted in the aircraft industry years ago. The reduction in direct labor hours required to build an aircraft was observed and found to be predictable. Since then, the learning curve has found applications in other industries as a means for adjusting costs for items produced beyond the first one.

Most learning curves are based on the assumption that the direct labor hours needed to complete a unit of product will decrease by a constant percentage each time the production quantity is doubled. A typical rate of improvement in the aircraft industry is estimated to be 20% between doubled quantities. This establishes, for example, an 80% learning function and means that the direct labor-hours needed to build the second aircraft will be 80% of the hours required to build the first aircraft. The fourth aircraft will require 80% of the hours that the second required, the eight aircraft will require 80% of the fourth, and so on.[10]

Given the preceding relationships, the application of learning curves has found wide usage in recent years in a variety of industries where multiple quantities of "like" items are being produced. Although the 80% factor in the aircraft industry may be assumed as being characteristic for this particular industry, other percentage factors may be more realistic for different industries depending on system complexities, the manufacturing processes used, and so on. For instance, a 90% learning curve may be more realistic in the production of sophisticated electronic equipment, whereas a 70% learning curve may be appropriate in the production of mechanical items. In any event, learning curves have been developed and adapted to fit the situation at hand, and these curves have subsequently been used in the prediction of production costs. An example of the application of learning curves used in the estimation of manufacturing costs is discussed further in Chapter 17.

16.5 DESIGN FOR DISPOSABILITY

Examples of "green" products, "clean" processes, and the "eco-factory" are beginning to appear across a wide spectrum of industries, driven by the so-called green revolution. The green revolution derives from Industrial Ecology, where environmentally conscious design and manufacturing (ECDM) plays a central role in green product realization. Because the nature of product realization is cross-functional, interdisciplinary teams are needed to pursue the activities of design, manufacturing, utilization, support, and disposal/recycling.

[10] Learning curve derivation and application is found in W. J. Fabrycky, and B. S. Blanchard, *Life-Cycle Cost and Economic Analysis* (Upper Saddle River, N.J.: Prentice Hall, Inc., 1991).

Disposability Issues and Industrial Ecology

Waste and emissions caused by the supply chain result in serious impacts to the local and global environment, such as global warming and acid rain. Accordingly, disposability must be recognized as an important design-dependent parameter in product design and development under the ECDM paradigm. Decisions about product design, production planning, process selection, logistics, inventory disposition, and recycling will increase in importance because of customer preferences and governmental regulations.

The relationship between the supply and environmental chains is the domain of industrial ecology in which manufacturing enterprises adopt environmental strategies to discharge their environmental responsibility. Consequently, there is a need to implement new design and manufacturing ideas to deal quickly with opportunities for reducing environmental impacts caused by the supply chain. The evolutionary ECDM paradigm is such an ecologically conscious strategy.

A green product may be defined as one that, at the ultimate end of its useful life, passes through disassembly and other reclamation processes to reuse nonhazardous and renewable materials. ECDM strategies deliberately attempt to reduce the ecological impacts of industrial activity without sacrificing quality, cost, reliability, performance, or energy-use efficiency. However, efforts leading to environmental friendliness result in complicated problems. The general ECDM paradigm requires new design and specified analysis method and methodology to shift from reactive end-of-pipe treatment to integrated, multidisciplinary, proactive design activities. Although the explicit consideration of environmental objectives and constraints in the product development cycle can lead to a green technology, the question as to what is green depends on how environmental problems and system boundaries are defined.

The concepts of design for environment (DFE) are also employed in design of the manufacturing processes in the factory, called *eco-factory*. An eco-factory's initiatives focus primarily on various manufacturing strategies and technologies. The major emphasis for manufacturing technology is on production systems, restoration systems, and control and assessment. Eco-factory technology points toward the protection of the environment and the effective use of the world's resources by industry. In a sense, the eco-factory is viewed as physical system derived on the basis of ECDM approach to implement the concept of industrial ecology. Various technologies within each of these groupings are shown in Figure 16.4

Manufacturing Systems with Recycling Application

Recycling has been recently recognized as one of the effective means for solving environmental problems. Therefore, in the design of future manufacturing systems, the recycling of materials used should be considered.

Recycling of products to obtain raw materials or reuseable components is an important means of reducing disposal costs and increasing total product value. In demanufacturing processes, demanufacturing involves refurbishing or partially rebuilding a product returned at the end of its life. The objective is to give it the func-

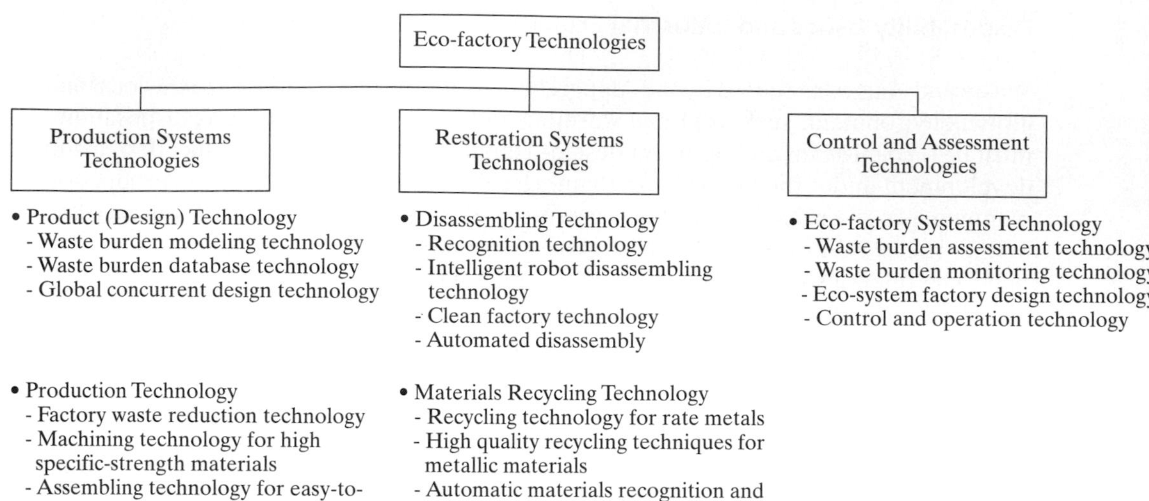

Figure 16.4 Some technology categories for the eco-factory.

tionality of a new product. This strategy has been identified by several companies as a major source of cost saving and competitive advantage. Effective recycling and demanufacturing require ease of product recovery through disassembly or separation, performance and market acceptance of the recycled materials and components, and the environmentally responsible disposal of nonrecoverables remaining after recycling.

Recycling activities could occur during the production stage. One of the features in a plastic extrusion production process is that waste materials are created, reground, and mixed with virgin plastic raw material for use in subsequent (extrusion) production cycles. The recycling of these waste materials helps to reduce the need to use the more expensive virgin raw materials. However, the yield from the overall extrusion process decreases somewhat.

Reuse in demanufacturing are the highest form of waste reduction, and it has the potential to reduce the end-of-life cost of products. Reuse may be easily justified in the case of components with high manufacturing costs, long innovation cycles, and long lifetimes.

Ecology-Based Manufacturing

In the domain of life-cycle design, one of the important issues in manufacturing is the harmonization of manufacturing activities with the global ecology. This is a new concept of an ecologically conscious manufacturing system, namely, an ecology-based production system (or eco-factory).

The basic requirements for an eco-factory are low-energy consumption, limited use of rare natural resources, and recycling. An eco-factory can be evaluated by the reduction in its propensity to damaging the environment, by no decline in productivity, economy of the manufacturing process, and by high value-added products. In the eco-factory, the manufacturing system consists of a product design technology that aims to develop and design products with a low ecological load, and a production technology that aims for low ecological load. Assuming an "eco-factory" approach could lead to complete implementation of the concept of integrated design (i.e., the design of products that are conceived not only to be assembled and used, but also to be totally and efficiently disassembled and recycled). This implies that the a closed-loop factory should be evaluated both by environmental (ecological) criteria and by conventional criteria, such as economy, productivity, performance, and marketability.

Environmental loads have seldom been considered in conventional manufacturing activities. A closed eco-manufacturing system that aims for reduced consumption of energy and resources, and incorporates recycling processes, will be more desirable and necessary in the future.

16.6 INTEGRATION OF PRODUCIBILITY AND DISPOSABILITY

The system-oriented approach for achieving a balanced industrial eco-system requires that designers consider eco-product design and eco-manufacturing processes simultaneously in the early design stage. Life-cycle analysis (LCA) is a guide to evaluate the environmental consequences of a product or process over its entire life, from raw material acquisition to final disposal. Life-cycle assessment has been classified as both a conceptual and a technical framework that may be used to evaluate the environmental performance of a product, process, or activity over the entire product life cycle; from the first stage of procuring raw materials to the final stage of waste management. Recycling analysis within a closed-loop system requires information from other activities, such as material selection analysis, manufacturing processes analysis, and product utilization analysis.

Product and Process Eco-Design

Much attention is now being paid to the integration of design and manufacturing, not to mention integration with demanufacturing and recycling. The impact of environmental issues on product development ranges from production planning and product production to product disassembly and recycling, both economically and ecologically. However, none of the preceding consider the life-cycle design problem as a whole throughout the product/material life cycle. In many ways, the previous types of problems and solutions are simply parts (or subproblems) of the whole DFE problem. Systematic integration of environmental considerations into life-cycle engineering and design is needed.

Ecology-oriented product development is not limited to production and marketing. The entire product life cycle must be considered. Beyond green product design, companies have to consider the environmental impact of the product-related processes and logistic support. Therefore, an integrative approach is desired that considers three elements (i.e., products, processes, and logistic support) during the whole life cycle, including the recycling phase (see Figure 16.1).

Within the context of eco-design of products and processes, the ECDM approach seeks to discover product innovations that will result in reducing harmful environmental impacts at any or all stages of the life cycle, while satisfying cost and performance as well as quality objectives. ECDM should be practiced in balance against other design considerations to get the "best" design. For ECDM to be implemented and integrated effectively into the eco-product development process, the following key elements are required throughout the life-cycle stages: life-cycle synthesis, life-cycle analysis, and life-cycle evaluation, to determine the best alternative with balance among the various design considerations.

The ECDM System Problem

Figure 16.5 summarizes the ECDM-related problems that were dealt with in this chapter. Total product value is generally the goal. System design evaluation or design optimization are essential procedures in eco-product development to maximize the products total value. In the context of ECDM, the products total life-cycle value is created from production activities and demanufacturing efforts. Both production and demanufacturing are significantly affected by product design and the production system by which they are realized. Either may contribute to an increase or decrease in the value of a product.

From a product design perspective, total product value is determined by performance functionality, quality, environmental compatibility, and economic viability. From a manufacturing process perspective, the coordination of disassembly, demanufacturing, and recycling operations, with each other and with other manufacturing facilities, are complex tasks. There appears to be a need for analytical models that will allow designers and engineers to gain insight into product-design and manufacturing process issues that related to the environment. In addition, life-cycle assessment is developed and employed as a tool to evaluate the environmental burden of manufacturing processes and of products. The results of LCA are an important factor in the evaluation of the environmental impacts of a product and its manufacturing process.

Life-cycle cost analysis (LCCA) is used for an economic assessment of the environmental issues. The life-cycle approach is one of the significant features of the ECDM framework. Another feature involves product/materials-oriented and system-oriented approaches. These approaches led to the integration of DEE, LCA, and LCCA into the ECDM concept. Life-cycle costing is discussed further in Chapter 17.

QUESTIONS AND PROBLEMS

1. Define *producibility*. Why is it important? When in the system life cycle should it be addressed?
2. Identify and describe some of the measures of producibility.
3. How does producibility relate to *reliability, maintainability, human factors*, and *supportability*?
4. Define *disposability*. Why is it important? When in the system life cycle should it be addressed? How does it relate to *producibility*?

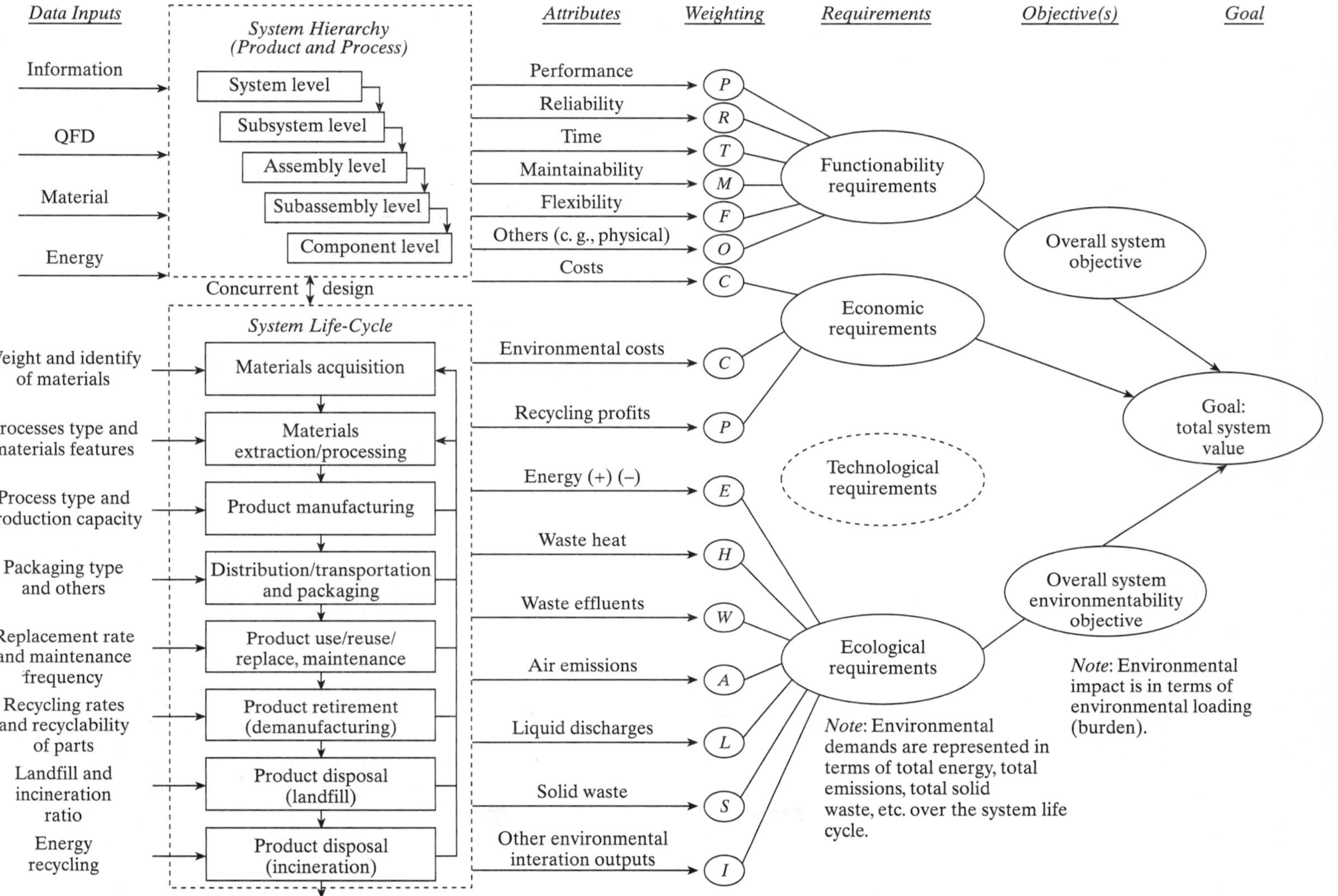

Figure 16.5 A goal-oriented flow model for producibility and disposability.

5. Identify and describe some of the measures of disposability.
6. Describe what is meant by *environmental quality*.
7. Describe in your own words what is meant by *environmentally conscious design and manufacturing* (ECDM).
8. Refer to Figure 16.1. Describe each of the life cycles shown, the activities within, and their interrelationships.
9. Refer to Figure 2.10. Describe how these activities might apply (if at all) to the life cycles shown in Figure 16.1.
10. What impact might the results of the functional analysis (described in Section 4.1) have on producibility?
11. What is meant by *demanufacture*? Provide some examples.
12. If you were tasked with the objective of reducing *waste* in your manufacturing process, what goals would you establish for the design of the product to be produced?
13. How does producibility relate to *mobility* and *transportability*?
14. Identify and describe some of the manufacturing processes that need to be addressed to enhance producibility (pick at least five).
15. Describe what is meant by a *learning curve*. Provide a simple illustration of a 70% and a 80% learning curve. Under what conditions can learning curves be applied?
16. Describe what is meant by the *eco-factory*. What elements should be included?
17. Define what is meant by *green engineering*.
18. Assume that you have been assigned by your organization to develop a retirement and material disposal plan. Prepare a topical outline of the plan (i.e., what is to be included). Develop a flow-process identifying the activities (and their interfaces) that you would specify in the implementation of such a plan.

17 DESIGN FOR AFFORDABILITY (LIFE-CYCLE COST)

Many systems have been planned, designed, produced, and operated with little concern for economic issues and the total cost of the system over its intended life cycle; i.e., *life-cycle cost* (LCC). Referring to Figure 2.9, although the "technical factors" of the balance are usually addressed, the "economic factors" are often neglected. In instances where costs have been introduced, they have been viewed in a fragmented manner. The costs associated with activities such as research, design, testing, production or construction, system use and support, and material disposal have been isolated and addressed independently at various stages in the life cycle. They have not been viewed on an *integrated* basis.

This chapter addresses cost and economic factors under the general theme of *design for affordability*. An introduction to life-cycle costing is followed by a focus on costs as they occur throughout the system life cycle. Then, in the main body of this chapter, the life-cycle cost analysis process, incorporating 12 steps and a companion example, describes the preferred methodology. Finally, the applications and benefits of life-cycle costing are summarized.

17.1 INTRODUCTION TO LIFE-CYCLE COSTING

Current trends indicate that, in general, the complexity of systems is increasing, and many of those systems in use today are not meeting the needs of the customer (i.e., user) in terms of performance, effectiveness, and overall cost. New technologies are being introduced on a continuing basis, whereas the duty cycles for many systems in use are being extended. The length of time that it takes to develop and acquire a new system needs to be reduced, the industrial base is rapidly changing, and available resources

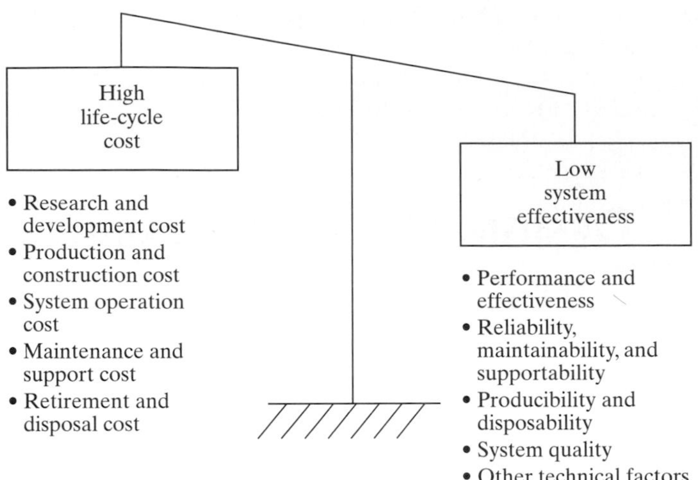

Figure 17.1 The imbalance between "cost" and "system effectiveness."

are dwindling. These trends, combined with past practices in system design and development, have tended to create an *imbalance* between the "cost" side of the spectrum and the "effectiveness" side, as illustrated in Figure 17.1 (an extension of Figure 2.9).

Referring to the economic side of the balance in Figure 17.1, experience has indicated that not only have the acquisition costs associated with a new system been rising, but the costs of operating and maintaining systems already in use have been increasing alarmingly. This is due primarily to a combination inflation and cost growth from causes such as the following:[1]

1. Cost growth resulting from engineering changes occurring throughout the design and development of a system or product (for the purposes of improving performance, adding capability, etc.)
2. Cost growth resulting from changing suppliers in the procurement of system components
3. Cost growth resulting from system production or construction changes
4. Cost growth resulting from changes in the logistic support capability
5. Cost growth resulting from initial estimating inaccuracies and changes in estimating procedures
6. Cost growth resulting from unforeseen problems.

At a time when considerable system cost growth is being experienced, budget allocations for many categories of systems are decreasing from year to year. The net result is that less money is available for acquiring and operating new systems and in maintaining and supporting systems that are already in being. The available funds for

[1] It has been noted that cost growth resulting from these various causes has ranged from 5 to 10 times the rate of inflation over the past several decades.

projects (i.e.. buying power), when inflation and cost growth are considered, are decreasing rapidly.

The current economic situation is further complicated by some additional problems related to the actual determination of system or product cost:

1. Total system cost is often not visible, particularly those costs associated with system operation and support. The cost visibility problem can be related to the "iceberg effect" illustrated in Figure 17.2. One must not only address system acquisition cost, but other costs as well.
2. In estimating cost, individual factors are often improperly applied. For instance, costs are identified and often included in the wrong category; variable costs are treated as fixed costs (and vice versa); indirect costs are treated as direct costs; and so on.
3. Existing accounting procedures do not always permit a realistic and timely assessment of total cost. In addition, it is often difficult (if not impossible) to determine costs on a functional basis.
4. Budgeting practices are often inflexible regarding the shift in funds from one category to another or from year to year, to facilitate improvements in system acquisition and use.

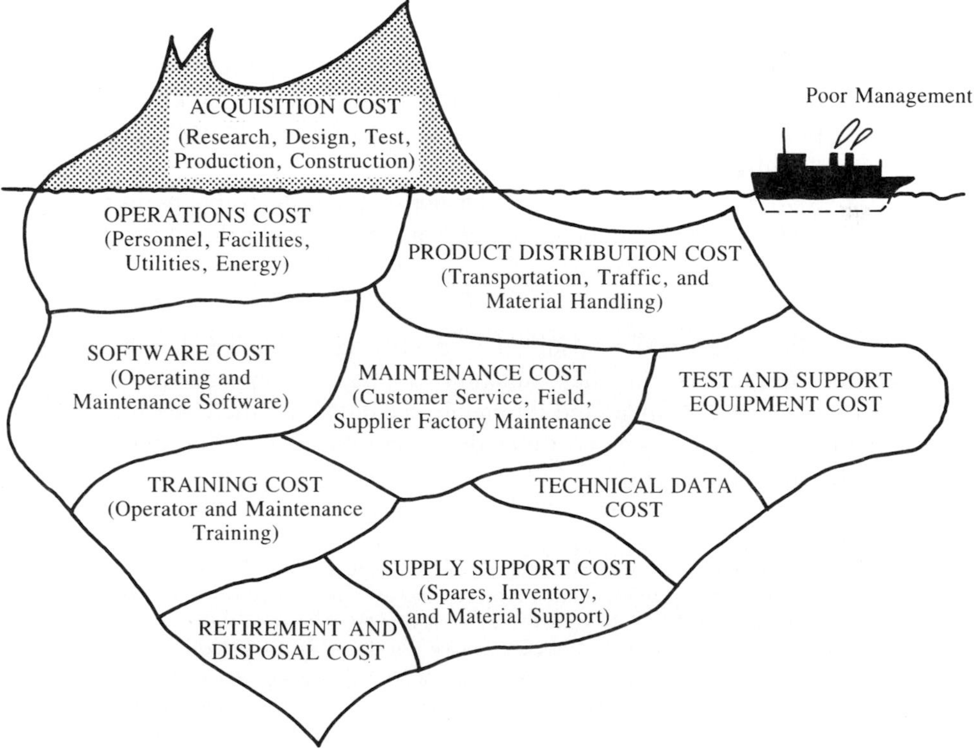

Figure 17.2 Total cost visibility.

The current trends of inflation and cost growth, combined with these additional problems, have caused inefficiencies in the utilization of valuable resources. Systems and products have been developed that are not cost-effective. Further, it is anticipated that conditions will become worse unless an increased degree of cost consciousness is assumed in day-to-day activities. Design for economic feasibility must address all aspects of *life-cycle cost*, not just segments thereof.

Life-cycle cost refers to all costs associated with the system as applied to the defined life cycle. The life cycle and the major functions associated with each phase are illustrated in Figure 2.2. The life cycle, tailored to the specific system being addressed, forms the basis for life-cycle costing. In general, life-cycle cost includes the following:[2]

1. *Research and development cost.* Initial planning; market analysis; feasibility studies; product research; engineering design; design documentation; software; test and evaluation of engineering models; and associated management functions
2. *Production and construction cost.* Industrial engineering and operations analysis; manufacturing (fabrication, assembly, and test); facility construction; process development; production operations; quality control; and *initial* logistic support requirements (e.g., initial consumer support, the manufacture of spare parts, the production of test and support equipment, etc.)
3. *Operation and support cost.* Customer or user operations of the system in the field; product distribution (marketing and sales, transportation, and traffic management); and sustaining logistic support throughout the system or product life cycle (e.g., customer service, maintenance activities, supply support, test and support equipment, transportation and handling, technical data, facilities, system modifications, etc.)
4. *Retirement and disposal cost.* Disposal of nonrepairable items throughout the life cycle; system/product retirement; material recycling; and applicable logistic support requirements

Life-cycle cost is determined by identifying the applicable functions in each phase of the life-cycle, costing these functions, applying the appropriate costs by function on a year-to-year schedule, and ultimately accumulating the costs for the entire span of the life cycle. Life-cycle cost includes all producer, supplier, customer (user), maintainer, and related costs.[3]

[2] Describing the system/product life cycle may appear to be rather elementary; however, experience has indicated that many different interpretations may exist about what constitutes the life cycle. Because this establishes the major reference point for life-cycle costing, it is essential that a *common understanding* prevail as to what is meant by the life cycle and what is included (or excluded).

[3] It should be noted that *all* life-cycle costs may be difficult (if not impossible) to predict and measure. For instance, some indirect costs caused by the interaction effects of one system on another, social costs, and so on may be impossible to quantify. Thus, the emphasis should relate primarily to those costs that can be *directly* attributed to a given system or product. Further, there needs to be an improvement in the traceability of costs back to the actual *causes* for such.

17.2 COST EMPHASIS IN THE SYSTEM LIFE CYCLE

Experience indicates that a large percentage of the total cost for many systems is the direct result of the downstream activities associated with the operation and support of these systems, whereas the commitment of these costs is based on engineering and management decisions made in the early conceptual and preliminary design stages of the life cycle (refer to Figure 2.11). The costs associated with the different phases of the life cycle are interrelated, and a decision made in any one phase can have an impact on the other phases. Further, the costs associated with customer, producer, and supplier activites are all interrelated. Thus, in addressing the economic issues, one must look at total cost in the context of the overall life cycle (refer to Figure 2.2), and particularly during the early stages of advance planning and conceptual design. As is the case pertaining to the other characteristics of design discussed in the earlier chapters, a cost target can be (1) initially specified for the system (as a "design-to-cost" TPM); (2) allocated to a lower-level function or element of the system; and (3) measured for the purposes of system evaluation as one progresses through the life cycle. The emphasis here is on *total life cycle cost* and not just individual cost elements. This is essential if one is to properly assess the risks associated with the day-to day decision-making process.[4]

Figure 17.3 reflects a characteristic life-cycle cost commitment curve as related to actions occurring during the various phases of the life cycle (refer to Figure 2.11).

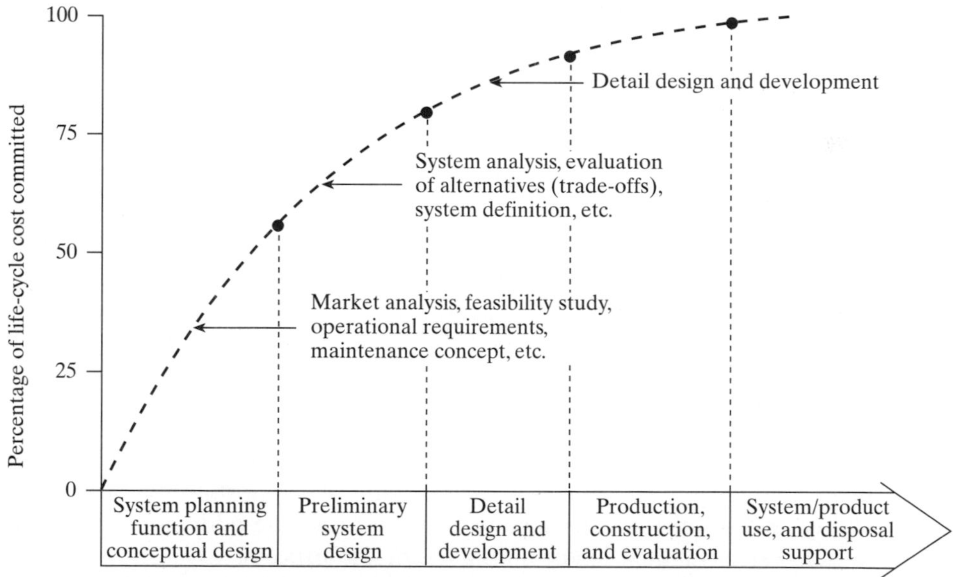

Figure 17.3 Actions affecting life-cycle cost (example).

[4] Inherent within the conduct of a systems engineering program is the preparation and implementation of a *risk management plan*. The initial identification and assessment of risks, along wiht the subsequent actions required for risk abatement, are key factors that must be addressed.

The system/product life-cycle phases presented in Figure 2.2 are translated to reflect emphasis on the early planning and design stages of a program. As illustrated, approximately 60% of the projected life-cycle cost is *committed* by the end of the system planning and conceptual design stage, even though actual project expenditures are relatively minimal at this point in time. This curve varies with the individual system; however, it does convey a trend relative to the affects of decisions on ultimate life-cycle cost.

The initial step in a life-cycle cost analysis is to establish cost targets or firm goals; i.e., one or more quantitative TPM figures-of-merit to which the system or product should be designed, produced (or constructed), and supported for a designated period of time. Second, these cost targets are then allocated to specific subsystems or elements as design constraints or criteria (refer to Figure 4.8). With the progression of design, various alternative configurations are evaluated in terms of compliance with the allocated targets, and a preferred approach is selected. As the system or product continues to evolve through various stages of development, life-cycle cost estimates are made and the results are compared against the initially specified targets (refer to Figure 5.12). Areas of noncomplicance are noted and corrective action is initiated where appropriate. Cost emphasis throughout the system/product life cycle is illustrated in Figure 17.4 and discussed subsequently.

Conceptual Design (Figure 17.4, Block 1)

Experience has shown that a major portion of the projected life-cycle cost for a given system stems from the consequences of decisions made during early planning and as part of system conceptual design. These decisions deal with system operational requirements, performance and effectiveness factors, the maintenance concept, the system configuration (system packaging schemes and levels of diagnostics), quantity or items to be produced, customer utilization factors, logistic support policies, and so on. Such decisions, made as a result of a market analysis or a design feasibility study, actually guide subsequent design and production activities, product distribution functions, and the various aspects of sustaining system support. Thus, if ultimate life-cycle costs are to be minimized in designing for economic feasibility, it is essential that a high degree of cost emphasis be applied in the early stages of system/product development.

In the early stages of planning and conceptual design, quantitative cost figures-of-merit should be established as requirements to which the system or product is to be designed, tested, produced (or constructed), operated, and supported. A *design-to-cost* (DTC) TPM may be established as a system design requirement along with performance, effectiveness, reliability, maintainability, supportability, producibility, disposability, and so on (refer to Section 3.5). Cost must be addressed on a *proactive* basis, rather than a *resultant* factor during the design process.

Design-to-cost figures-of-merit can be specified in terms of life-cycle cost at the system level. However, DTC parameters sometimes are established at a lower level to facilitate improved cost visibility and closer control throughout the life cycle. For example,

1. *Design to unit acquisition cost.* A factor that includes only research and development cost and production or construction cost

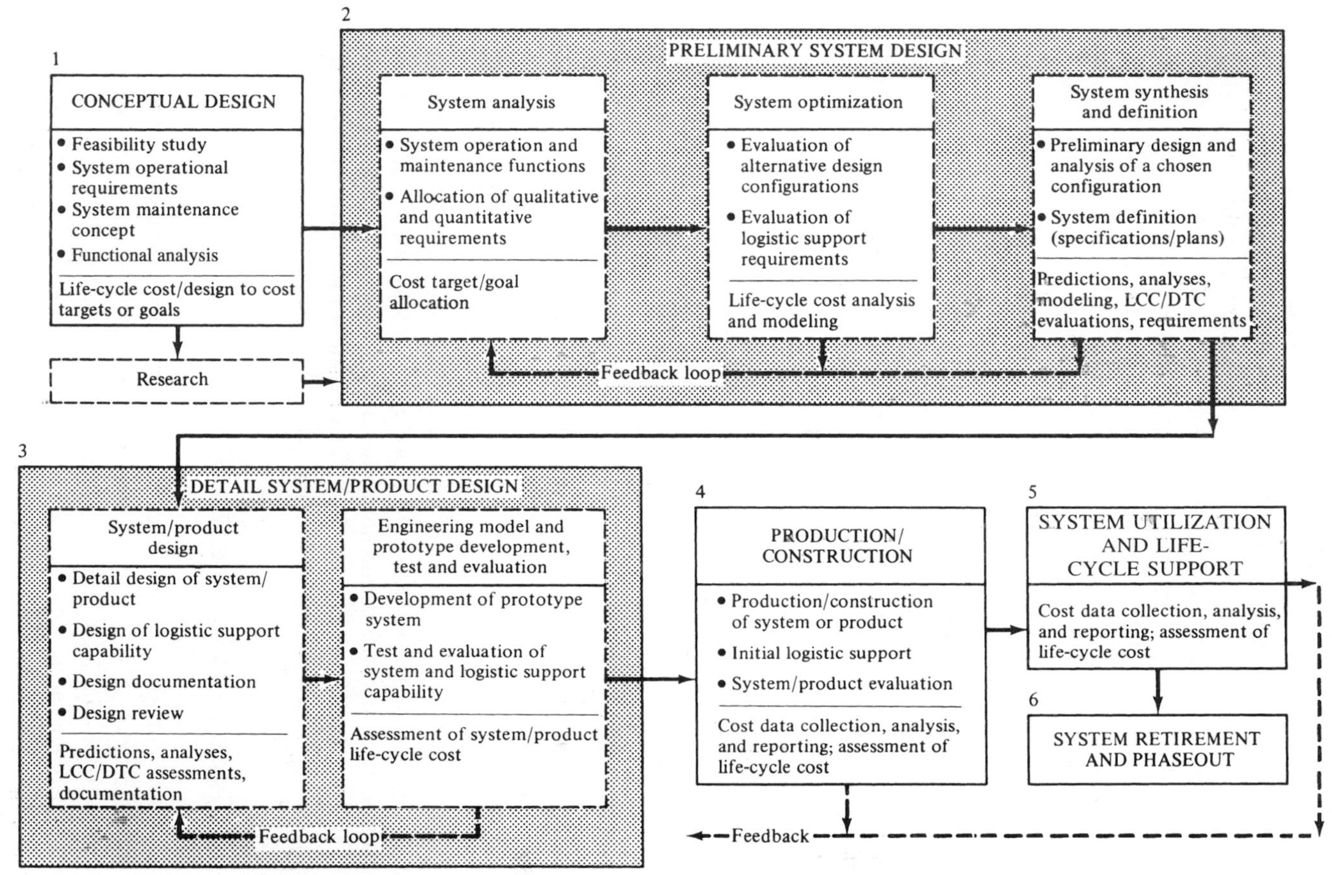
1
CONCEPTUAL DESIGN
• Feasibility study
• System operational requirements
• System maintenance concept
• Functional analysis
Life-cycle cost/design to cost targets or goals
Research
2
PRELIMINARY SYSTEM DESIGN
System analysis
• System operation and maintenance functions
• Allocation of qualitative and quantitative requirements
Cost target/goal allocation
System optimization
• Evaluation of alternative design configurations
• Evaluation of logistic support requirements
Life-cycle cost analysis and modeling
System synthesis and definition
• Preliminary design and analysis of a chosen configuration
• System definition (specifications/plans)
Predictions, analyses, modeling, LCC/DTC evaluations, requirements
Feedback loop
3
DETAIL SYSTEM/PRODUCT DESIGN
System/product design
• Detail design of system/ product
• Design of logistic support capability
• Design documentation
• Design review
Predictions, analyses, LCC/DTC assessments, documentation
Engineering model and prototype development, test and evaluation
• Development of prototype system
• Test and evaluation of system and logistic support capability
Assessment of system/product life-cycle cost
Feedback loop
4
PRODUCTION/ CONSTRUCTION
• Production/construction of system or product
• Initial logistic support
• System/product evaluation
Cost data collection, analysis, and reporting; assessment of life-cycle cost
5
SYSTEM UTILIZATION AND LIFE-CYCLE SUPPORT
Cost data collection, analysis, and reporting; assessment of life-cycle cost
6
SYSTEM RETIREMENT AND PHASEOUT
Feedback

Figure 17.4 System or product life-cycle process (refer to Figure 2.2).

2. *Design to unit operation and support cost.* A factor that includes only operation and maintenance support cost

When considering only a single segment of life-cycle cost, one must be sure that decisions are not based on that one segment alone, without considering the overall effects on total life-cycle cost. For example, one can propose a given design configuration on the basis of a low unit acquisition cost, but the projected operation and support cost, and life-cycle cost, for that configuration may be considerably higher than necessary. Ideally, acquisition cost should not be addressed without consideration for operation and support cost (and vice versa), and both segments of cost must be viewed in terms of life-cycle cost.

Preliminary System Design (Figure 17.4, Block 2)

With the quantitative cost requirements established, the next step involves an iterative process (i.e.. optimization, synthesis, and system/product definition). The criteria defined in Block 1 are initially allocated, or apportioned, to various segments of the system to establish guidelines for the design or the procurement of the applicable element(s).

Referring to Figure 4.8, allocation is accomplished from the system level down to the level necessary to provide adequate control. The factors projected reflect the target cost per individual unit (i.e., a single system or product in a total population), and are based on system operational requirements, the system maintenance concept, and so on.

As the design process evolves, various alternative approaches are considered in arriving at a preferred system configuration. Life-cycle cost analyses are accomplished in evaluating each possible candidate with the objective of (1) ensuring that the candidate selected is compatible with the established cost targets, and (2) determining which of the various candidates being considered is preferred from an overall cost-effectiveness standpoint. Numerous trade-off studies are accomplished, using life-cycle cost analysis as an evaluation tool, until a preferred design configuration is chosen. Areas of compliance are justified, and noncompliant approaches are discarded. This is an iterative process, which is inherent within the overall model illustrated in Figure 2.10, with the necessary feedback and corrective action illustrated by Block 2 of Figure 17.4.

Detail System Design (Figure 17.4, Block 3)

As system or product design is further refined and design data become available, the life-cycle cost analysis involves the evaluation of specific design characteristics (as reflected by design documentation and engineering or prototype models), the prediction of cost-generating variables, the estimation of costs, and the projection of life-cycle cost as a cost profile. The results are compared with the initial requirement and corrective action is taken as necessary. Again, this is an iterative process but at a lower level than what is accomplished during preliminary system design.

Production/Construction, Utilization, Support, Retirement, and Material Disposal (Figure 17.4, Blocks 4–6)

Cost concerns in these latter stages of the system product life cycle involve a data collection, analysis, and assessment function. Through proper assessment, the high-cost contributors can be easily identified. It is hoped that valuable information is gained and used for the purposes of product improvement and for the development of good historical data for future applications.

In summary, life-cycle costing is applicable in all phases of system design and development, production, construction, operational use, logistic support, retirement and material disposal. Cost emphasis is created early in the life cycle by establishing quantitative cost factors as *requirements*. As the life cycle progresses, cost is employed as a major parameter in the evaluation of alternative design configurations and in the selection of a preferred approach. Subsequently, cost data are generated based on established design and production characteristics and used in the development of life-cycle cost projections. These projections are then compared with the initial requirements to determine degree of compliance and the ultimate necessity for corrective action. In essence, life-cycle costing evolves from a series of rough estimates to a relatively refined methodology, and is employed as a management tool for decision-making purposes.

17.3 THE LIFE-CYCLE COST ANALYSIS PROCESS

Throughout the system life cycle, there are many decisions required of both a technical and nontechnical nature. Most of these decisions, particularly those at the earlier stages, have life-cycle implications and directly impact life-cycle cost. It may initially be perceived that a specific decision will not directly affect life-cycle cost; however, the indirect effects on cost may turn out to be very significant.

For each specific problem where there are possible alternative solutions and a decision is required to select a preferred approach, there is an overall analysis process that one usually follows, either intuitively or on a formal basis. This process is comparable with the generic engineering analysis approach illustrated in Figure 3.15. Specifically, one should (1) define the need for analysis, (2) identify the ground rules and constraints, (3) establish the analysis approach, (4) select a model to facilitate the analysis and evaluation process, (5) acquire the appropriate data for each alternative being considered, (6) evaluate the alternatives, and (7) recommend a proposed solution in response to the original problem. Inherent within this process is the identification of the appropriate analytical techniques/tools and the selection of a model that is responsive to the need.[5]

[5] Such techniques may include the use of simulation, linear and dynamic programming, queuing analysis, accounting methods, networking, or various combinations of these and other tools as described in Part III (Chapters 7 to 11).

Referring to Figure 17.4 (which is an extension of Figure 2.2, there are numerous possible applications where life-cycle cost analysis methods can be applied. More specifically, LCC analysis may be used in the evaluation of the following:

1. Alternative technology approaches in accomplishing a feasibility analysis (refer to Section 3.3)
2. Alternative system operational profiles or mission scenarios (refer to Section 3.4)
3. Alternative maintenance and logistic support concepts (refer to Section 3.4)
4. Alternative approaches in considering various design methods in response to the functional analysis (i.e., trade-offs pertaining to the accomplishment of functions by the human or through automation, the selection of COTS items versus the design of new items, the appropriate mix of hardware versus software, and so on) (refer to Sections 4.1–4.3)
5. Alternative system design configurations relative to packaging schemes, levels of repair, diagnostic routines, built-in versus external test, component selection and standardization, reliability versus maintainability, and so on
6. Alternative sources of supply and procurement schemes
7. Alternative production approaches (i.e., continuous versus discontinuous production, internal production versus outsourcing, quantity of production lines, number of inventory points and levels of inventory, inspection and test alternatives, etc.)
8. Alternative product distribution channels, packaging methods, transportation and handling approaches, warehouse and storage locations, and so on (refer to Chapter 15)
9. Alternative maintenance and logistic support policies (i.e., levels of maintenance performed, inhouse versus contracted or third-party maintenance, alternative warranty and guarantee policies, etc.) (refer to Chapter 15)
10. Alternative product disposal or recycling policies (i.e., various degrees of disposability/recyclability, alternative logistic support configurations for disposal or recycling functions, etc.) (refer to Chapter 16)

Although there is nothing novel relative to addressing such problems in the development of new systems (the re-engineering of existing systems), the emphasis here is to accomplish the necessary analysis and evaluation functions on the basis of *life-cycle cost*, and not just the short-term costs associated with design and procurement activites.

In conducting a life-cycle cost analysis, there are certain steps that the analyst should perform to acquire the desired results. For the purposes of illustration, the steps outlined in Figure 17.5 will serve as a basis for further discussion.[6]

[6] The specific steps in accomplishing a life-cycle cost analysis (and the depth of analysis) may vary somewhat from one application to the next and must be tailored to the problem at hand. The approach presented in Figure 17.5 is included to provide a basic understanding of the overall process and items that need to be considered in performing a LCC analysis.

1. *Define system requirements.* Define system operational requirements and the maintenance concept, identify the applicable technical performance measures (TPMs), and describe the system in functional terms (functional analysis at the system level).
2. *Describe the system life cycle and identify the activities in each phase.* Establish a "baseline" for the development of a cost breakdown structure and the estimation of cost for each year of the projected life cycle.
3. *Develop a cost breakdown structure* (CBS). Provide a top-down/bottom-up structure, to include all categories of cost, for the purposes of the initial allocation of costs (top-down) and the subsequent collection and summarization of costs (bottom-up). All life-cycle activities must be covered within.
4. *Identify data input requirements.* Identify the input data requirements and the possible sources from where such can be acquired. The type and amount of data will depend on the nature of problem being addressed, the phase of the system life cycle, and on the depth of analysis.
5. *Establish the costs for each category in the CBS.* Develop the appropriate cost-estimating relationships and estimate the costs for each category in the CBS on a year-by-year basis.
6. *Select a cost model for the purposes of analysis and evaluation.* Select (or develop) a computer-based model to facilitate the life-cycle cost analysis process. The model must be sensitive to the specific system being evaluated.
7. *Develop a cost profile and summary.* Construct a cost profile (i.e., cost stream) showing the "flow" of costs over the entire life cycle, and provide a summary identifying the cost for each category in the CBS and the percentage contribution in terms of the total.
8. *Identify the high-cost contributors and establish cause-and-effect relationships.* Highlight those functions, elements of the system, or segments of a process that should be investigated for possible areas of design improvement.
9. *Conduct a sensitivity analysis.* Evaluate the model, input-output data relationships, and the results of the "baseline" analysis to ensure that (1) the overall LCC analysis approach is valid; and (2) the model itself is well constructed and sensitive to the problem at hand. Given this, the sensitivity analysis can aid in identifying major areas of risk (as part of a risk analysis).
10. *Construct a Pareto diagram and identify priorities for problem resolution.* Conduct a Pareto analysis, construct a Pareto diagram, and identify priorities for problem resolution (i.e., those problems that require the most management attention).
11. *Identify feasible alternatives for design evaluation.* Having developed an approach for the LCC evaluation of a given single design configuration, it is now appropriate to extend the LCC analysis through the evaluation of multiple design alternatives.
12. *Evaluate feasible alternatives and select a preferred approach.* Develop a cost profile for each of the alternatives being evaluated, compare the alternatives considering the time value of money, construct a break-even analysis, and select a preferred design approach.

Figure 17.5 The basic steps in a life-cycle cost analysis.

17.4 A LIFE-CYCLE COST ANALYSIS EXAMPLE

The life-cycle cost analysis process outlined in the previous section is implemented in accordance with the 12 steps in Figure 17.5. In this section, each step will be addressed in greater depth with the aid of an example application involving the evaluation of two candidate communication system design alternatives being considered in response to a specific customer need.

Define System Requirements (Figure 17.5, Step 1)

Independent of the nature of the problem, an initial or first step is to define a "baseline" terms of system expectations, operational requirements, the maintenance concept, the basic functions that the system must perform and the appropriate TPMs. This provides the overall framework within which the problem can be defined. The need for defining operational requirements and the maintenance concept is particularly important and a necessary input in determining the downstream costs associated with activities in the system utilization and sustaining support phase of the life cycle.

For the purposes of illustration, it is assumed that a metropolitan area plans to upgrade its overall communications capability by procuring new communication equipment for installation in patrol aircraft, helicopters, ground vehicles, certain designated ground facilities, and in a central communications control facility. Two alternative system configurations are being considered and evaluated. The objective is to define system operational requirements, the maintenance concept, and required program planning information so that the system configuration with the lowest life-cycle cost can be acquired.[7]

The communication system network is shown in Figure 17.6. There is an overall requirement for 80 equipment installations. Proposed maintenance and support facilities are also shown. The new communication equipment should be adaptable for use in all designated installations (i.e., interchangeable). The only configuration differences should be in the interface connections, mounting fixtures, and so on.

System planning activity has identified six operational objectives to be met by the new equipment. Although these objectives pertain largely to the equipment, the communication system is composed of the equipment packages collectively, together with the proposed maintenance support capability. The six objectives are the following:

Objective 1. The equipment is to be installed in three low-flying light aircraft (10,000 ft. or less) in quantities of one per aircraft. This will enable communication with loitering helicopters dispersed within a 200-mile radius and with the central communication control facility. It is anticipated that each aircraft will fly 15 times per month with an average flight duration of 3 hours. The equipment use requirement is 1.1 hours of operation for every hour of aircraft operation, which includes air time plus some ground time. It is assumed that all functions are fully operational throughout this time period. The equipment must exhibit a MTBM of at least 500 hours and a $\overline{\mathrm{M}}\mathrm{ct}$ not to exceed 15 minutes.

Objective 2. The equipment is to be installed in each of five helicopters and will enable communication with patrol aircraft within a 200-mile range, other helicopters within a 50-mile radius and with the central communications control facility. It is anticipated that each helicopter will fly 25 times per month, with an average flight duration of 2 hours. The use requirement is 0.9 hour of equipment operation for every flight hour of helicopter operation. The equipment must meet a 500-hour MTBM requirement and a $\overline{\mathrm{M}}\mathrm{ct}$ of 15 minutes or less.

Objective 3. The equipment is to be installed in each of 50 police patrol vehicles (one per vehicle) and will enable communication with other vehicles within a range of 25 miles, and with the central communications control facility from a range of 50

[7]This illustration is based on a realistic case study involving a life-cycle cost analysis conducted in evaluating a given communication system design. This case study was described in detail in W. J. Fabrycky, and B. S. Blanchard, *Life-Cycle Cost and Economic Analysis* (Upper Saddle River, N.J.: Prentice-Hall, Inc., 1991), chap. 12, and selected segments have been extracted for illustration herein.

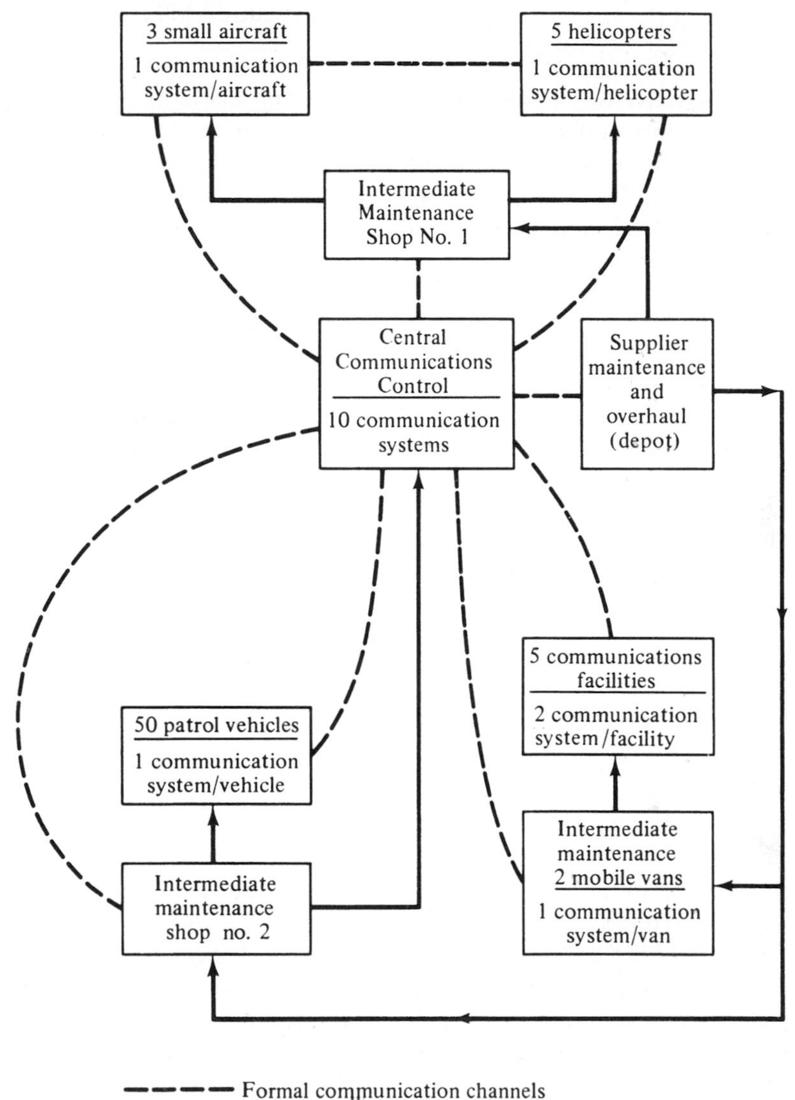

Figure 17.6 Communication system network.

miles or less. Each vehicle will be in operation on the average of 5 hours per day, 5 days per week, and will be used 100% during that time. The required MTBM is 400 hours, and the $\overline{\text{M}}\text{ct}$ should be less than 30 minutes.

Objective 4. Two installations are to be made in each of five fixed communications facilities optimally located throughout the metropolitan geographical area This will enable communication with patrol vehicles within a range of 25 miles and the central communications control facility from a range of up to 50 miles. Equipment use requirements are 120 hours per month, the required MTBM is 200 hours, and the $\overline{M}ct$ should not exceed 60 minutes.

Objective 5. Ten installations are to be made in the central communications control facility and will enable communication with patrol aircraft and loitering helicopters within a range of 400 miles, the five fixed communication facilities within a range of 50 miles, and with patrol vehicles within a 50-mile radius. In addition, each will be able to communicate with the intermediate maintenance facilities. The average equipment use requirement is 3 hours per day for 360 days per year. The MTBM requirement is 200 hours, and the $\overline{M}ct$ should not exceed 45 minutes.

Objective 6. Two mobile vans will be used to support intermediate-level maintenance at the five fixed communications facilities. Each van will incorporate one communication package, which will be used an average of 2 hours per day for a 360-day year. The required MTBM is 400 hours, and the $\overline{M}ct$ is 15 minutes.

To achieve the stated operational objectives, there is a need to acquire equipment that will meet certain performance and effectiveness requirements (e.g., voice transmission range, clarity of message, MTBM, $\overline{M}ct$, etc.). Advanced program planning indicates that a full complement of units must be in operation 5 years after the start of the program, and that this capability must be maintained through the 11th year.

Based on a review of the available sources of supply, there is no known existing system that will completely fulfill the need; but there are two new candidate design configurations that should suffice, assuming that all design goals are met. The objective is to evaluate each configuration in terms of its life-cycle cost and to recommend the preferred configuration.

Before the identification and development of cost elements, the baseline configuration for the communication equipment and its maintenance concept should be described in detail. This configuration (including units, assemblies, and modules) is developed from conceptual design considerations to be fairly representative of each candidate being considered. The maintenance concept can be defined as a series of statements and/or illustrations that include criteria covering maintenance levels, support policies, effectiveness factors (e.g., maintenance time constraints, turnaround times, transportation times), and basic logistic support requirements. The maintenance concept is a prerequisite to the system or product design, whereas a detailed maintenance plan reflects the results of design and is used for the acquisition of the logistics elements required for the sustaining life-cycle support of the system in the field. The maintenance concept is illustrated by Figure 17.7.

The specific functions scheduled to be accomplished on the equipment at each level of maintenance are noted. In the event of a malfunction, fault isolation is performed to the applicable unit by using the built-in test capability (i.e., Unit A, B, or C). Units are removed and replaced at the organizational level and sent to the intermediate maintenance shop for corrective maintenance. At the intermediate shop, units are repaired through assembly or part replacement, and assemblies are repaired through module replacement. In case of fault isolation to assembly 5 in Unit C, the faulty assembly is sent to the depot, where repair is accomplished through module replacement.

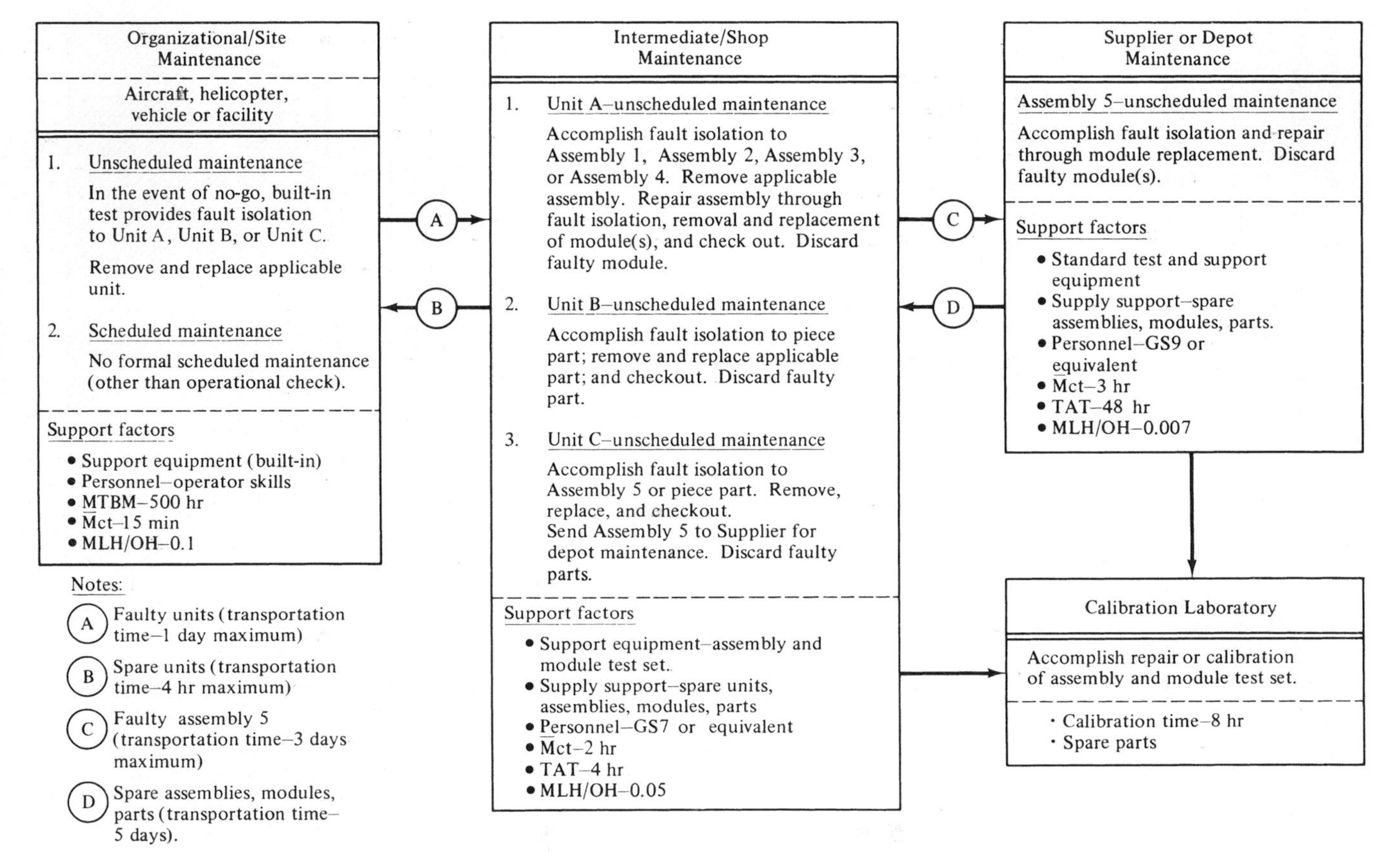

Figure 17.7 System maintenance concept (repair policy).

The maintenance concept illustrated in Figure 17.7 identifies the functions that are anticipated for each level of maintenance, the effectiveness requirements in terms of maintenance frequency and times, and the major elements of logistic support to include personnel skill levels, test and support equipment, supply support requirements, and facilities. This information is not only required as an input to the system design process, but serves as the basis for determining operation and support costs.

Describe the System Life Cycle and Identify Activities in Each Phase (Figure 17.5, Step 2)

Given the operational requirements as a baseline, the next step is to describe the system life cycle and to identify the major activities in each phase. For the communication system requirements, the major activities and milestones include *research and development, production*, and *system operation and support*, as shown in Figure 17.8. Regarding the area of operation and support, the assumed buildup, distribution, and

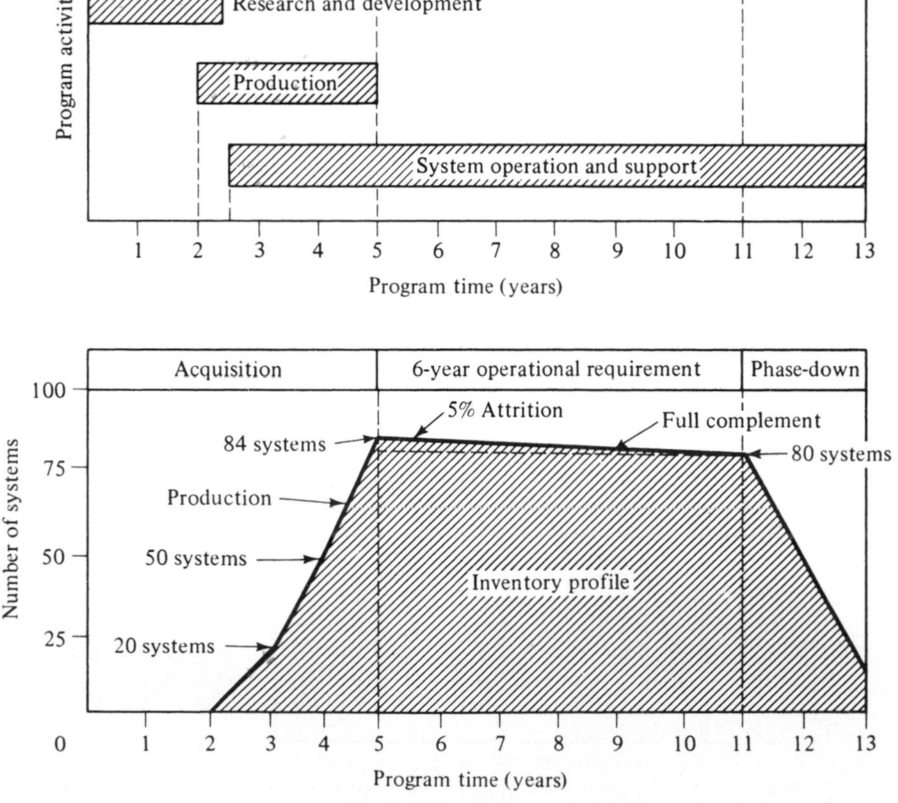

Figure 17.8 Program plan and profile.

the projected inventory profile are noted. This information serves as a basis for the development of a cost breakdown structure (CBS).[8]

Develop a Cost Breakdown Structure (CBS) (Figure 17.5, Step 3)[9]

Referring to Figure 2.9, the cost side of the spectrum needs to be broken down into specific cost categories. Figure 17.9 shows a "generic" CBS, which may be used as a frame of reference. This, in turn, should be expanded and tailored to the system configuration being evaluated, as shown in Figure 17.10.

The cost breakdown structure links objectives and activities with resources, and constitutes a logical subdivision of cost by functional activity area, major element of a system, and one or more discrete classes of common or like items. The cost breakdown structure, which is usually adapted or tailored to meet the needs of each individual program, should exhibit the following characteristics:

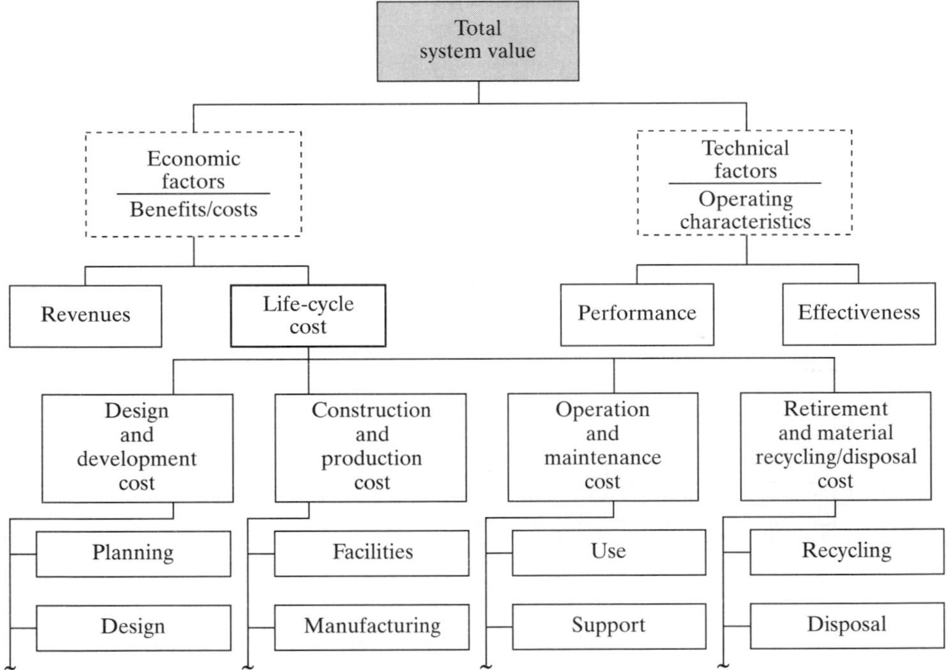

Figure 17.9 A generic cost breakdown structure (refer to Figure 2.9).

[8] It is not uncommon for changes to occur in the projected life cycle, distribution rates, inventory profiles, and so on. However, the analyst needs to make some initial assumptions at this point and then account for possible variations through a sensitivity analysis conducted later.

[9] In some instances, a *summary* work breakdown structure (WBS) may be equated to a cost breakdown structure (CBS) as long as ALL life-cycle activities, material requirements, and so on, are included. On programs where a *contract* WBS (CWBS) is negotiated, only those work units that apply to the contract are included, and the other areas of activity are not addressed. In these instances, although there may be some similarities between the CWBS and the CBS, the CBS prevails because it addresses the total spectrum of cost.

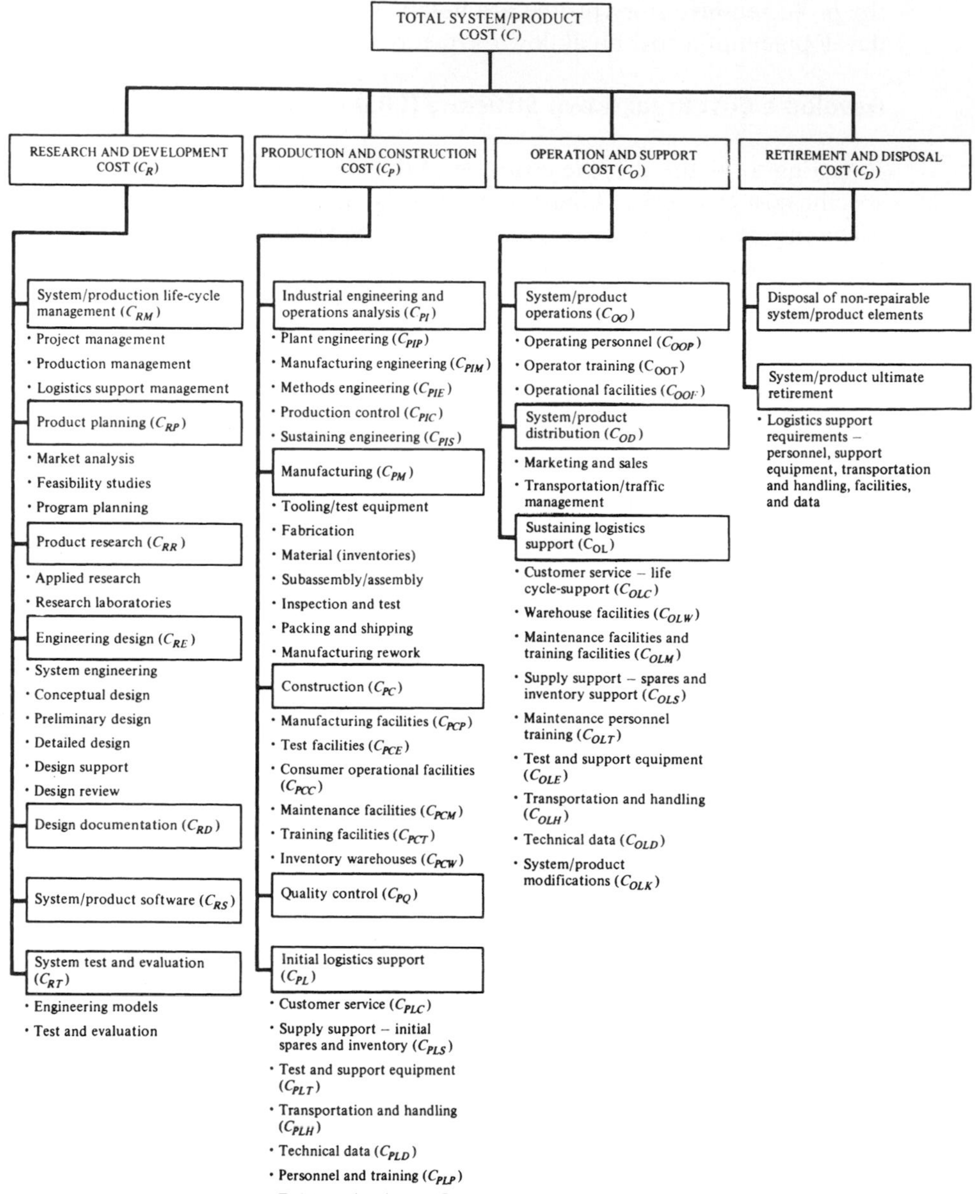

Figure 17.10 Cost breakdown structure for the communication system.

1. All life-cycle costs should be considered and identified in the cost breakdown structure. This includes research and development cost, production and construction cost, operation and system support cost, and retirement and disposal cost.[10]

[10] This does not mean to imply that all cost categories are relevant to all analyses. The objective is to include all life-cycle costs and then identify those categories that are considered significant relative to the problem at hand.

2. Cost categories are generally identified with a significant function, level of activity, or with a major item of material. Cost categories in the CBS must be well defined, and managers, engineers, accountants, and others must have the same understanding of what is included in a given cost category and what is not included. Cost omissions and doubling (i.e., counting the same cost in two or more categories) must be avoided.[11]
3. Costs must be broken down to the level necessary to provide management with the visibility required in evaluating various facets of system design and development, production, operational use, and support. Management must be able to identify high-cost areas and cause-and-effect relationships.
4. The CBS and the categories defined should be coded in a manner to facilitate the analysis of specific areas of interest while virtually ignoring other areas. For example, the analyst may wish to investigate supply support costs as a function of engineering design or distribution costs as a function of manufacturing, independent of other aspects of the system. The CBS should be designed with this objective in mind.
5. The CBS and the categories defined should be coded in such a manner as to enable the separation of producer costs, supplier costs, and consumer costs in an expeditious manner.
6. When related to a particular program, the cost structure should be directly compatible (through cross-indexing, coding, etc.) with planning documentation, the work breakdown structure, work packages, the organization structure, PERT-CPM and PERT-COST scheduling networks, Gantt charts, and so on. Costs that are reported through various management information systems must be compatible and consistent with those comparable cost factors in the CBS.

Referring to Figure 17.10, the cost categories identified are too broad to ensure any degree of traceability, accountability, or control. The analyst can not readily determine what is and what is not included, nor can he or she validate that the proper parameter relationships have been used in determining the specific cost factors that are inputted into the cost breakdown structure. Thus, the analyst requires a much more indepth description of each cost category, the methods for calculating the costs in each category, and the assumptions upon which the costs are based. Figure 17.11 presents an example of a description of three of the categories identified in the CBS.[12]

In summary, the CBS constitutes a *functional* breakdown of costs. It involves *all* costs related to customer, contractor, supplier, and consumer (user) activities over the

[11] On completion of a LCC analysis, all costs within the CBS should be *traceable* back to the applicable functional blocks identified through the functional analysis described in Sections 3.6 and 4.1. The objective is to determine the costs associated with the accomplishment of a given *function*.

[12] An example of a CBS, with the cost categories described, along with the methods for calculating the costs in each category, is presented in W. J. Fabrycky, and B. S. Blanchard, *Life-Cycle Cost and Economic Analysis* (Upper Saddle River, N.J.: Prentice Hall, Inc., 1991), appendix B; and B. S. Blanchard, *Logistics Engineering and Management* 5th ed. (Upper Saddle River, N.J.: Prentice Hall, Inc., 1998), appendix E.

Cost Category (Reference Figure 17.10)	Method of Determination (Quantitative Expression)	Cost Category Description and Justification
Spare/repair parts cost (C_{OLS})	$C_{OLS} = [C_{SO} + C_{SI} + C_{SD} + C_{SS} + C_{SC}]$ C_{SO} = Cost of organizational spare/repair parts C_{SI} = Cost of intermediate spare/repair parts C_{SD} = Cost of depot spare/repair parts C_{SS} = Cost of supplier spare/repair parts C_{SC} = Cost of consumables $C_{SO} = \sum_{N_{MS}} \Big[(C_A)(Q_A) + \sum_{i=1} (C_{Mi})(Q_{Mi}) + \sum_{i=1} (C_{Hi})(Q_{Hi})\Big]$ C_A = Average cost of material purchase order ($/order) Q_A = Quantity of purchase orders C_M = Cost of spare item i Q_M = Quantity of i items required or demand C_H = Cost of maintaining spare item i in the inventory ($/$ value of the inventory) Q_H = Quantity of i items in the inventory N_{MS} = Number of maintenance sites C_{SI}, C_{SD}, and C_{SS} are determined in a similar manner.	Initial spare/repair part costs are covered in C_{PL}. This category includes all replenishment spare/repair parts and consumable materials (e.g., oil, lubricants, fuel, etc.) that are required to support maintenance activities associated with prime equipment, operational support and handling equipment, test and support equipment, and training equipment at each level (organizational, intermediate, depot, supplier). This category covers the cost of purchasing; the actual cost of the material itself; and the cost of holding or maintaining items in the inventory. Costs are assigned to the applicable level of maintenance. Specific quantitative requirements for spares (Q_M) are derived from the Supportability Analysis (SP) discussed in Chapter 15. These requirements are based on the criteria described in Chapter 3. The optimum quantity of purchase orders (Q_A) is based on the EOQ criteria described in Section 15.3. Support equipment spares are based on the same criteria used in determining spare part requirements for prime equipment.
Maintenance facilities cost C_{OLM}	$C_{OLM} = [(C_{PPM} + C_U) \times (\%\,\text{allocation})(N_{MS})]$ C_{PPM} = Cost of maintenance facility support ($/site) C_U = Cost of utilities ($/site) N_{MS} = Number of maintenance sites *Alternate approach* $C_{OMF} = [(C_{PPO})(N_{MS})(S_O)]$ C_{PPO} = Cost of maintenance facility space ($/square foot/site). Utility cost allocation is included. S_O = Facility space requirements (square feet) Determine C_{OMF} for each appropriate echelon of maintenance.	Initial acquisition (construction) cost for maintenance facilities is included in C_{PCM}. This category covers the annual recurring costs associated with the occupancy and support (repair, modification, paint, etc.) of maintenance shops at all level throughout the system life cycle. On some occasions, a given maintenance shop will support more than one (1) system, and in such cases, associated costs are allocated proportionately to each system concerned.
Engineering design (C_{RE})	$C_{RE} = \sum_{i=1}^{N} C_{RE_i}$ C_{RE_i} = Cost of specific design activity i N = Number of design activities	Includes all initial design effort associated with system/equipment definition and development. Specific areas include system engineering; design engineering (electrical, mechanical, drafting); reliability, maintainability, and human factors (Chapters 12, 13, and 14); functional analysis and allocation (Chapter 4); supportability analysis (Chapter 15); components; producibility; standardization; safety; etc. Design modifications are covered in C_{OLK}.

Figure 17.11 Sample breakout of cost categories and estimating relationships.

entire life cycle. Variable and fixed costs, direct and indirect costs, recurring and non-recurring costs, inflationary and other cost-growth factors, etc., must be included. It provides the necessary *visibility* to the depth required for engineering and management decisions. If such visibility is not apparent, the applicable categories in the CBS may be extended downward as required (or summarized upward if such visibility in not required). Finally, the CBS may be utilized initially to facilitate a top-down allocation of costs stemming from a given design-to-cost (DTC) requirement and later as a mechanism for a bottom-up summarization of cost for the purposes of evaluation (refer to Section 17.2).

Identify Data Input Requirements (Figure 17.5, Step 4)

While the initial presumption is that the completion of a life-cycle cost analysis requires a lot of input data, the actual requirements in this area depend on the phase in which the analysis is accomplished and the depth of the analysis performed. A LCC analysis can be accomplished at the system (or subsystem) level in conceptual design with very little actual input data. The analyst, using past experience and intuition, can accomplish his or her objective by making rough estimates, with enough accuracy to adequately support the top-level design decisions made at this point in the life cycle. What is required is a thorough understanding of the LCC analysis process (i.e., the steps involved and what to look for), some knowledge pertaining to how the system will be operated and maintained by the customer in the field, a feeling for some of the major interrelationships between activities and costs, and a knowledge of applicable cost-estimating relationships (CERs). Accordingly, an analyst who has had some past experience in completing LCC analyses and who has a good understanding of system requirements should be able to complete successfully the necessary task in a timely manner.[13]

As one progresses through the life cycle and the system configuration becomes better defined, then a LCC analysis may be accomplished to a greater depth (Figure 17.4, Blocks 2 and 3). Referring to the CBS in Figure 17.10, the analyst may proceed with an estimation of costs for the various categories shown, delving down to the depth required to provide the desired visibility. In areas where the costs appear to be high, a greater depth may be required to identify the high-cost "drivers." In the event that the current CBS is not adequate in terms of cost-traceability, then the appropriate categories need to be extended to a greater depth. For example, if the system in question is *transportation intensive*, then it may be necessary to expand Category C_{OD} to a lower level. This often involves an iterative process of analysis, feedback, a more indepth analysis, and so on, which will ultimately dictate the amount of input data required.

[13] In conducting a LCC analysis, there is a tendency to begin by requesting a lot of data from various parts of the organization (e.g., design data, reliability and maintainability data, production data, logistics data, etc.). This can result in the development of too much data, which may take a long time to acquire. The results can be costly. Futher, the responsible design engineer/manager who requested that the LCC analysis be accomplished in the first place needs feedback as soon as possible and cannot afford to wait too long. Thus, it is important to first understand the problem, know what input data are required, and complete the analysis in a timely manner.

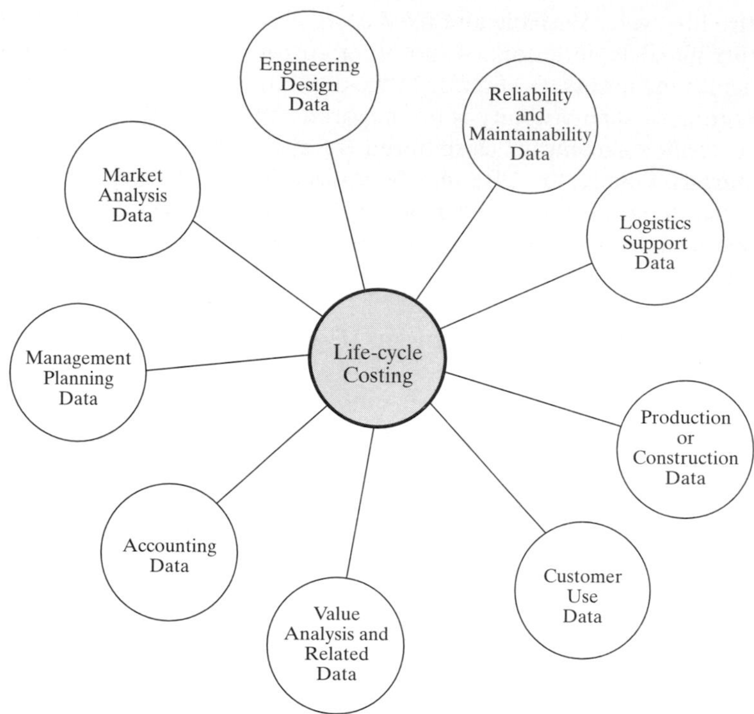

Figure 17.12 Possible sources of data for a life-cycle cost analysis.

As the life-cycle cost analysis effort becomes more complex the data-input requirements will expand. For example, in conducting an evaluation of an existing system configuration currently in use (and where the objective is to identify highest contributors, determine cause-and-effect-relationships, and provide specific recommendations for design improvement) it may be appropriate to solicit data from several different sources, accomplish an indepth analysis, study the interrelationships between system elements/activities and costs, and develop some recommendations through a *continuous product/process improvement effort*. In such instances, the analyst may solicit data from a number of different sources such as those illustrated in Figure 17.12.[14]

Establish the Costs for Each Category in the CBS (Figure 17.5, Step 5)

In conducting a LCC analysis in the early phases of a program (where the greatest impact on life-cycle cost can be realized), available input data may be limited because of the lack of design definition at that time. Thus, the cost analyst must rely primarily on the use of various cost-estimating techniques in the development of cost data, as indicated us Figure 17.13. As the system design progresses, more complete design infor-

[14] It is assumed that with the system well defined and operating, the required data are readily available, and that the costs of acquiring such are minimal.

mation becomes available, and the analyst is able to develop cost estimates by comparing the characteristics of the new system with similar systems where historical data are recorded. The generation of cost data is based on *analogous* estimating methods. Finally, as the system design configuration becomes firm, design data (to include drawings and layouts, parts and material lists, specifications, predictions, etc., as conveyed in Figure 17.12) are produced that will enable the development of good engineering and manufacturing cost estimates. Further, the results from available reliability, maintainability, supportability, and disposability analyses can be used to aid in the prediction of operation, support, retirement, and material disposal/recycling costs.

Referring to Figure 17.13, the most general approach for estimating costs during the early phases of a program (in the absence of good input data) is through the use of *parametric* methods. Parametric cost-estimating relationships (CERs) are basically "rules of thumb," developed from past experience, which relate various cost factors to explanatory variables of one form or another. These explanatory variables usually reflect the characteristics of the system such as performance, effectiveness factors, physical features, or even other cost elements. For example, from past experience a cost factor may be specified in terms of a unit of speed for a vehicle, a mile of range for a radar system, a unit of weight per mile for a transportation system, a unit of space for a facility, a unit of volume for materials or liquids, or in terms of some cost-to-cost factor. Cost-estimating relationships may take different forms (i.e., continuous or discontinuous, mathematical or nonmathematical, linear or nonlinear, etc.) and should relate to the specified TPMs for a given system.

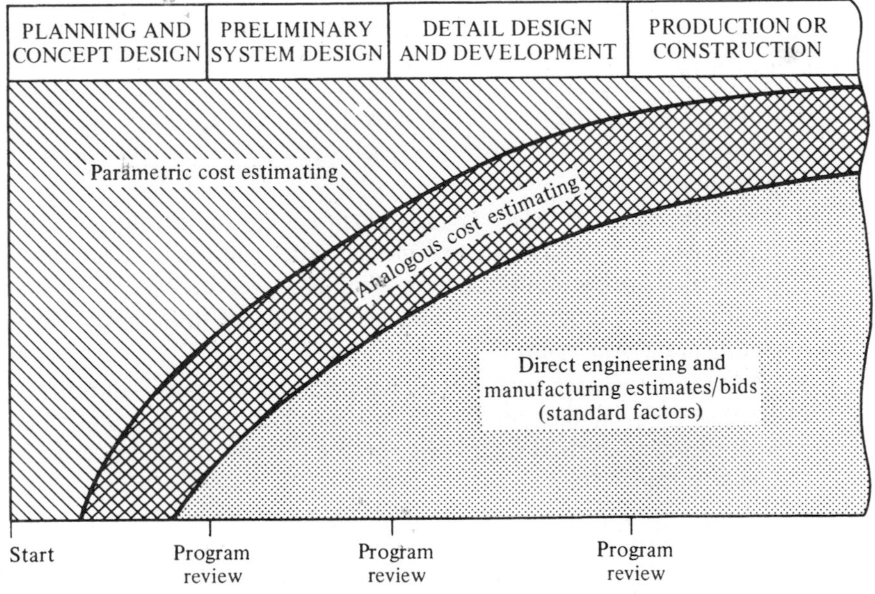

Figure 17.13 Cost estimation by program phase.

Activity-based costing (ABC). To be effective in total cost management (and in the accomplishment of cost-effectiveness analyses) requires full-cost visibility allowing for the traceability of all costs back to the activities, processes, or products that generate these costs. In traditional accounting structures employed in most organizations, a large percentage of the total cost can not be traced back to the "causes." For example, "overhead" or "indirect" costs that often constitute greater than 50% of the total include management costs, supporting organization costs, and other costs that are difficult to trace and assign to specific objects. With these costs being allocated across the board, it is impossible to identify the actual "causes" and to pinpoint the *true* high-cost contributors. As a result, the concept of *activity-based costing* has been introduced in recent years.[15]

Activity-based costing (ABC) is a methodology directed toward the detailing and assignment of costs to the items that cause them to occur. The objective is to enable the "traceability" of *all* applicable costs to the process or product that generates these costs. The ABC approach allows for the initial allocation and later assessment of costs by function, and was developed to deal with the shortcomings of the traditional management accounting structure where large overhead factors are assigned to all elements of the enterprise across the board without concern for whether they directly apply or not. More specifically, the principles of ABC are noted.

1. Cost are directly traceable to the applicable cost-generating process, product, or a related object. Cause-and-effect relationships are established between a cost factor and a specific process or activity.
2. There is no distinction between direct and indirect (or overhead) costs. Although 80% to 90% of all costs are traceable, those nontraceable costs are not allocated across the board, but are allocated directly to the organizational unit(s) involved in the project.
3. Costs can be easy allocated on a *functional* basis (i.e., the functions identified in Sections 3.6 and 4.1). It is relatively easy to develop cost-estimating relationships in terms of the cost of activities per some activity measure (i.e., the cost per unit output).
4. The emphasis in ABC is on "resource consumption" (versus "spending"). Processes and products consume activities, and activities consume resources. With resource consumption being the objective, the ABC approach facilitates the evaluation of day-to-day decisions in terms of their impact on resource consumption downstream.
5. The ABC approach fosters the establishment of "cause-and-effect" relationships and, as such, enables the identification of the "high-cost contributors." Areas of risk can be identified with some specific activity and the decisions that are being made within.

[15] Activity-based costing is covered further in J. R. Canada, W. G. Sullivan, and J. A. White, *Capital Investment Analysis for Engineering and Management*, 2nd ed. (Upper Saddle River, N.J.: Prentice Hall, Inc., 1996); and P. T. Kidd, *Agile Manufacturing: Forging New Frontiers* (Reading, Mass.: Addison-Wesley, 1994).

6. The ABC approach tends to eliminate some of the cost-doubling (or double-counting) that occurs when attempting to differentiate on what should be included as a "direct" cost or as an "indirect" cost. By not having the necessary visibility, there is the potential of including the same costs in both categories.

Implementation of the ABC approach, or something of an equivalent nature, is essential if one is to do a good job of total cost management. Costs are tied to objects and viewed over the long term, and the use of such facilitates the life-cycle cost analysis process.

Developing cost data. In developing cost data for a life-cycle cost analysis, the cost analyst should initially investigate all possible data sources to determine what is available for direct application in support of analysis objectives. If the required data are not available, the use of parametric cost estimating techniques may be appropriate. However, one should first determine what can be derived from existing data banks, initial system planning data, supplier documentation, reliability and maintainability predictions, supportability analyses, test data, field data, and so on. Some of these data sources are discussed subsequently.

1. *Existing data banks.* Actual historical information on existing systems, similar in configuration and function to the item(s) being developed, may be used when applicable. Often it is feasible to employ such data and apply adjustment factors as necessary to compensate for any differences in technology, configuration, projected operational environment, and time frame. Included in this category of existing data are standard cost factors which have been derived from historical experience that can be applied to specific functions or activities. Standard cost factors may cover such areas as the following:
   - **a.** The cost of engineering labor—dollars per labor hour for principal engineer, senior engineer, technician, and so on
   - **b.** The cost of manufacturing labor by classification—dollars per labor hour per classification
   - **c.** Overhead rate—dollars per direct labor cost (or percentage)
   - **d.** Training cost—dollars per student week
   - **e.** Shipping cost—dollars per pound per mile
   - **f.** The cost of fuel—dollars per gallon
   - **g.** The cost of maintaining inventory—percent of the inventory value per year
   - **h.** The cost of facilities—dollars per cubic foot of occupancy
   - **i.** The cost of material *x*—dollars per pound or per foot

   These and comparable factors, where actual quantitative values can be directly applied, are usually established from known rates and costs in the marketplace, and are a direct input to the analysis. However, care must be exercised to ensure that the necessary inflationary and deflationary adjustments are incorporated on a year-to-year basis.

2. *Advance system/product planing data.* Advanced planning data for the system or product being evaluated usually includes market analysis data, definition of system operational requirements and the maintenance concept, the results of technical feasibility studies, and program management data. The cost analyst needs information pertaining to the proposed physical configuration and major performance features of the system, the anticipated mission to be performed and associated utilization factors, system effectiveness parameters, the geographical location and environmental aspects of the system, the maintenance concept and logistic support philosophy, and so on. This information serves as the baseline from which all subsequent program activities evolve. If the basic information is not available, the analyst must make some assumptions and proceed accordingly. These assumptions must then be thoroughly documented.
3. *Individual cost estimates, predictions, and analyses.* Throughout the early phases of a program, cost estimates are usually generated on a somewhat continuing basis. These estimates may cover research and development activities, production or construction activities, or system operating and support activities. Research and development activities, which are basically nonrecurring in nature, are usually covered by initial engineering cost estimates or by cost-to-complete projections. Such projections primarily reflect labor costs and include inflationary factors, cost growth resulting from design changes, and so on.

   Production cost estimates are often presented in terms of both nonrecurring costs and recurring costs. Nonrecurring costs are handled in a manner similar to research and development costs. Conversely, recurring costs are frequently based on individual manufacturing cost standards, value engineering data, industrial engineering standards, and so on. Quite often, the individual standard cost factors that are used in estimating recurring manufacturing costs are documented separately, and are revised periodically to reflect labor and material inflationary effects, supplier price changes, effects of learning curves, and so on.

   System operating and support costs are based on the projected activities throughout the operational use and support phase of the life cycle and are undoubtedly the most difficult to estimate. Operating costs are a function of system or product mission requirements and use factors. Support costs are basically a function of the inherent reliability and maintainability characteristics in the system design and the logistics requirements necessary to support all scheduled and unscheduled maintenance actions throughout the programmed life cycle. Logistic support requirements include maintenance personnel and training, supply support (spares, repair parts, and inventories), test and support equipment, transportation and handling, facilities, and certain facets of technical data. Thus, individual operation and support cost estimates are based on the predicted frequency of maintenance or the mean of the maintenance (MTBM) factor, and on the logistic support resources required when maintenance actions occur. These costs are derived from reliability and maintainability prediction data (refer to Chapters 12 and 13), supportability analysis (SA) data (refer to Chapters 15), and other supporting information, all of which is based on system/product engineering design data.

4. *Supplier documentation.* Proposals, catalogs, design data, and reports covering special studies conducted by suppliers (or potential suppliers) may be used as a data source when appropriate. Quite often, major elements of a system are either procured off the shelf or developed through a subcontracting arrangement of some type. Various potential suppliers will submit proposals for consideration, and these proposals may include not only acquisition cost factors but (in some instances) life-cycle cost projections. If supplier cost data are used, the cost analyst must become completely knowledgeable as to what is and is not included. Omissions or the double counting of costs must not occur.
5. *Engineering test and field data.* During the latter phases of system development and production and when the system or product is being tested or is in operational use, the experience gained represents the best source of data for actual analysis and assessment purposes (refer to Section 6.5). Such data are collected and used as an input to the life-cycle cost analysis. Also, field data are used to the extent possible in assessing the life-cycle cost impact that may result from any proposed modifications on prime equipment, software, or the elements of logistic support.

These five main sources of data identified above for life-cycle costing purposes are presented in a summary manner to provide an overview as to what the cost analyst should look for. In pursuing the data requirements further, the analyst will find that a great deal of experience has been gained in determining research and development and production/construction costs. However, few historical cost data are currently available in the operations and support area. Accounting for operation and support costs has been lacking in the past, but this situation should ultimately rectify itself as the emphasis on life-cycle costing continues to increase.[16]

Select a Cost Model for the Purposes of Analysis and Evaluation (Figure 17.5, Step 6)

After the establishment of the cost breakdown structure, it is necessary to develop a model of some type to facilitate the life-cycle cost evaluation process. The model may be a simple series of relationships or a complex set of computer subroutines, depending on the phase of the system life cycle and the nature of the problem at hand. In any case, the criteria for model selection and application discussed in Section 4.4 and in Chapter 7 are applicable.

Life-cycle costing itself includes a compilation of a variety of cost factors, reflecting the many different types of activities indicated by the CBS. The objective in using a model is to evaluate a system in terms of total life-cycle cost, as well as the various individual segments of cost. Figure 17.14 illustrates this point. Total system life-cycle

[16] The subject of cost estimation is a significant area and should be investigated further. For additional material, refer to P. F. Ostwald, *Engineering Cost Estimating*, 3rd ed. (Upper Saddle River, N.J.: Prentice Hall, Inc., 1992); R. D. Stewart, and R. M. Wyskida, *Cost Estimator's Reference Manual*, 2nd ed. (New York: John Wiley & Sons, Inc., 1995); and G. J. Thuesen, and W. J. Fabrycky, *Engineering Economy*, 8th ed. (Upper Saddle River, N.J.: Prentice Hall, Inc., 1993).

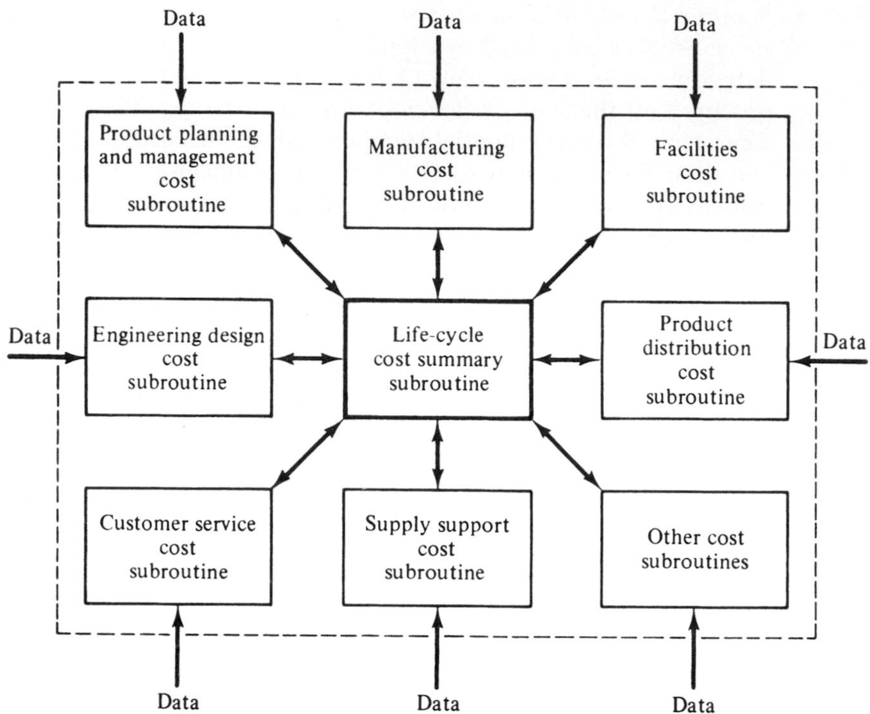

Figure 17.14 Life-cycle model configuration. *Note:* Although not shown here, many of the subroutines actually interact with each other in terms of data input and output.

cost is compiled through the use of accounting techniques in the life-cycle cost summary subroutine, whereas major areas of cost (which constitute input data to the LCC summary subroutine) can be viewed on an individual basis.

The subroutines that reflect segments of cost may be structured differently depending on the system element covered. For instance, supply support costs may be compiled through the application of spare and repair part demand factors and inventory techniques discussed in Chapter 15. This, in turn, may require factors derived through the use of a reliability model. On the other hand, engineering design costs may be extracted directly from a proposal or a set of engineering cost projections. The relationship here is analogous to the relationship of the models illustrated in Figure 4.12.

Develop a Cost Profile and Summary (Figure 17.5, Step 7)

In developing a cost profile, there are different approaches that one may follow. However, the following steps are suggested as a start:

1. Identify all activities throughout the life cycle that will generate costs in one form or another. This includes functions associated with planning, research and development, test and evaluation, production or construction, distribution, system use and support, and retirement and material disposal (refer to Figures 2.2 and 17.8).

2. Relate each activity identified in item 1 to a specific function (refer to the functional analysis in Sections 3.6 and 4.1), and to a specific cost category in the cost breakdown structure (refer to Figures 17.9 and 17.10). All functions and program activities should fall in one or more categories in the CBS, and there should be a "traceability" of requirements from the functional analysis to the CBS.
3. Develop a format for the recording of costs for each activity in the CBS and for each year in the life cycle. Figure 17.15 conveys a basic format that is often used in the presentation of costs at the top level, and Figure 17.16 shows an example of a worksheet that can serve to help identify costs by functional element.
4. Through application of the cost estimating methods discussed earlier, compute the costs for each CBS category and for each year in the life cycle. Considerations pertaining to the effects of learning curves, cost growth because of changes in suppliers and the development of new contactural agreements, changes in procurement price levels, changes in reliability and maintainability factors that influence system operating and maintenance support costs, and so on, should be addressed as appropriate. The analyst may commence with the prediction of costs for each year in terms of *constant* dollars (i.e., today's dollars) because it will allow for the direct comparison of various activity levels from year to year. Also, using constant dollars tends to assure consistency in accomplishing comparative studies.
5. Within each category in the CBS, the individual cost elements are next projected to include the appropriate inflationary factors. The modified values constitute a new cost stream and reflect realistic costs as they are anticipated for each year in the future (i.e., expected 1999 costs in 1999, 2000 costs in 2000, etc.). This cost stream represents a *budgetary* profile, which is used for management decision making.
6. Construct a third profile considering the "time value of money" or the presentation of costs in terms of the *present equivalent* (PE). When evaluating two or more alternative system configurations, each will include different levels of activity conducted in different years, different design approaches, and different maintenance and support requirements. Although all of the options being considered may be in full compliance with the specified system requirements, no two alternatives will be identitical in terms of component makeup, use of materials, and so on. Thus, when individual profiles are developed for each alternative and compared, must be related on an "equivalent" basis, considering the time value of money.[17]

Determining production costs. Production costs are heavily dependent on system operational requirements and the desired buildup of components in the inventory; i.e., the demand for the system and its elements by the customer. Referring to inventory profile in Figure 17.8, the assumptions made that pertain to the "front-end" of the profile are critical in determining the costs later. Further, this phase of the profile stems from

[17] The principles and concepts of economic analysis are included in Chapter 8. Additionally, an example of the comparison of two alternative design configurations is presented as part of item 12, Figure 17.5.

PROGRAM ACTIVITY	COST CATEGORY DESIGNATION	COST BY PROGRAM YEAR ($)													TOTAL COST (CONSTANT $)	TOTAL COST (ACTUAL $)	TOTAL COST (PE $)	% CONTRIBUTION
		1	2	3	4	5	6	7	8	9	10	11	12	13				
Alternative A 1. Research and development a. Life-cycle management b. Product planning (1) Feasibility studies (2) Program planning 2. 3. Others	 C_R C_{RM} C_{RP} —																	
Total cost (constant)																		
Total cost (actual)																		
Total cost (PE)																		
Alternative B 1. Research and development 2.	 C_R																	

Figure 17.15 Cost collection worksheet.

Functions / Parts	A	B	C					X	Y	Z	Total cost ($)	%
1	2	4	6					27	50	4	93	
2	5	7	19					2	6	23	62	
3												
Total												
% Total												
High or low												

Figure 17.16 Functional-cost analysis matrix.

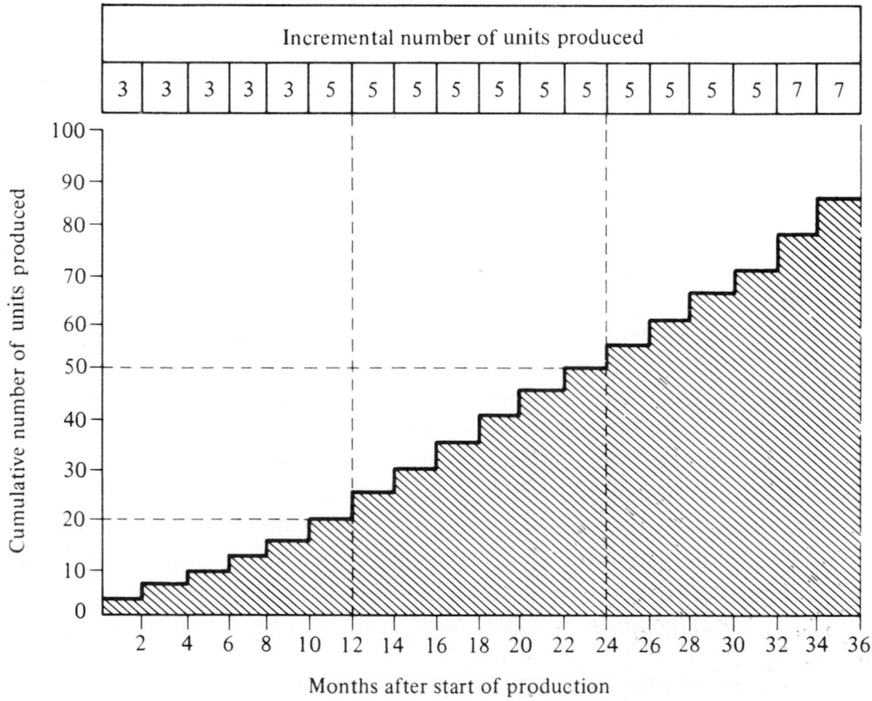

Figure 17.17 Communication system production requirements.

the activities represented by the first life cycle shown in Figure 16.1. Thus, in completing a LCC analysis, a more indepth plan is considered as being essential, even though the information provided may be preliminary in nature. Figure 17.17 shows the planned production buildup for the communication system elements shown in Figure 17.6.

Application of learning curves. When performing a repetitive activity or process, learning occurs, and the experience gained often results in reduced cost. Although learning and the associated cost variations occur at different activity points throughout the life cycle, the greatest impact of learning on cost is realized in the production of large quantities of a given item. In such instances, the cost of the first unit produced is generally higher than the cost of the 25th unit, which may be higher than the cost of the 50th unit, and so on. This is primarily because of job familiarization by the workers in the production facility, development of more efficient methods for item fabrication and assembly, the use of more efficient tools, and improvement in overall management. The effects of learning generally result in the largest portion of any cost savings taking place relatively early in a uniform production run, with a leveling off occurring later.

Learning curves are commonly derived on the basis of assuming a constant percentage cost savings for each doubling of the quantity of production units. For example, 80% *unit* learning curve implies that the second unit costs 80% of the first, the fourth unit costs 80% of the second, the eighth units costs 80% of the fourth, and so on. This learning curve is then applied to find the production cost profile.

Unit learning curves may vary considerably, depending on the expected magnitude of the cost savings estimated for the 5th unit, the 10th unit, the 20th unit, and so on. Because of product complexity, an 80% unit cost reduction may not be realized until the production of the 10th unit. In this case, an 80% learning curve will be based on the 10th, 20th, 40th, and 80th units as being the major milestones for cost measurement. Thus, we still may use an 80% learning curve, but the cost factors will be different. The analyst should evaluate the complexity of tasks in the production process and attempt to determine the type of unit learning curve that is most appropriate for the situation at hand. A variety of unit learning curves, which are idealistic, is shown in Figure 17.18a.

Sometimes in the application of learning curves it may be more appropriate to use a cumulative average learning curve. If it turns out that the projected average cost of producing the first 20 units is 80% of the average cost of producing the first 10 units, the process follows an 80% *cumulative average* learning curve.

Learning curves can also cover material cost, although the percentage(s) may be different. Although labor costs relate to both the required personnel skill levels and labor-hours to accomplish a given function, material costs may vary as a function of factory tooling, material scrapage rates (or percent of raw material used after the fabrication process), procurement methods, and inventory policies. Again, there are many factors involved, and different learning curves may be applied depending on the specific situation.

In the application of learning curves the cost analyst must ensure that the production process is indeed continuous and relatively devoid of design changes, manufacturing changes in producing the product, or organization changes that will ultimately cancel out the effects of a learning curve altogether, if not create a negative learning curve where producing the 10th unit is more costly than the first unit. Figure 17.18b shows a set of learning curves that have been adjusted to reflect actual experience. Idealistic learning curves of 80% and 90% were developed initially and then adjusted to reflect the expected "unstable" conditions that are likely to occur in the producer's factory.

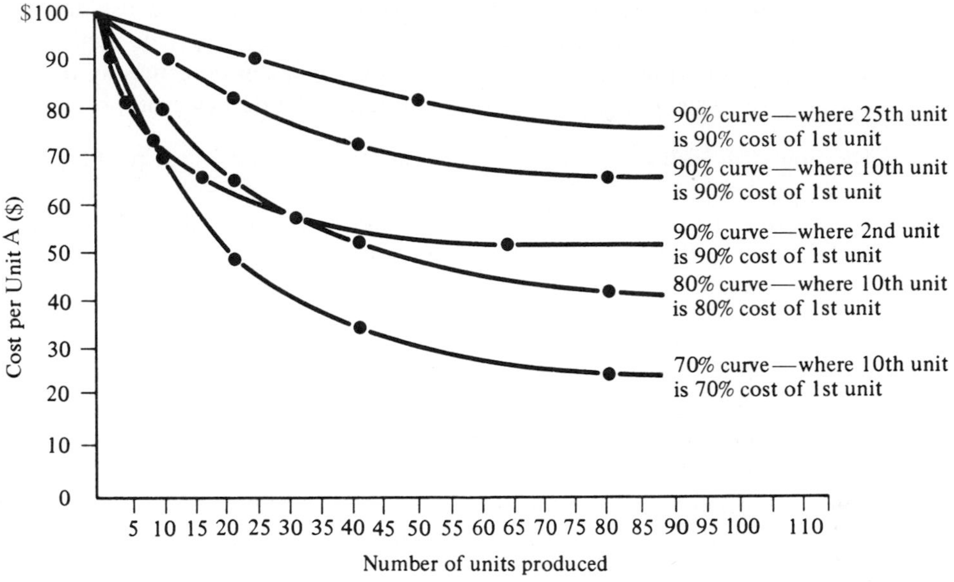

$100
90
80
70
60
50
40
30
20
10
0
Cost per Unit A ($)
5 10 15 20 25 30 35 40 45 50 55 60 65 70 75 80 85 90 95 100 110
Number of units produced
90% curve—where 25th unit is 90% cost of 1st unit
90% curve—where 10th unit is 90% cost of 1st unit
90% curve—where 2nd unit is 90% cost of 1st unit
80% curve—where 10th unit is 80% cost of 1st unit
70% curve—where 10th unit is 70% cost of 1st unit

(a)

Learning curve for Alternative B adjusted for inflation, modifications, management policy changes, etc.
Learning curve for Alternative A adjusted for inflation, modification, management policy changes, etc.
$10,000
9,000
8,000
7,000
6,000
5,000
4,000
3,000
2,000
1,000
Cost per unit ($)
90% Learning curve for Alternative B
80% Learning curve for Alternative A—where 20th system is 80% cost of the 1st system
0 5 10 15 20 25 30 35 40 45 50 55 60 65 70 75 80 85
Number of units produced

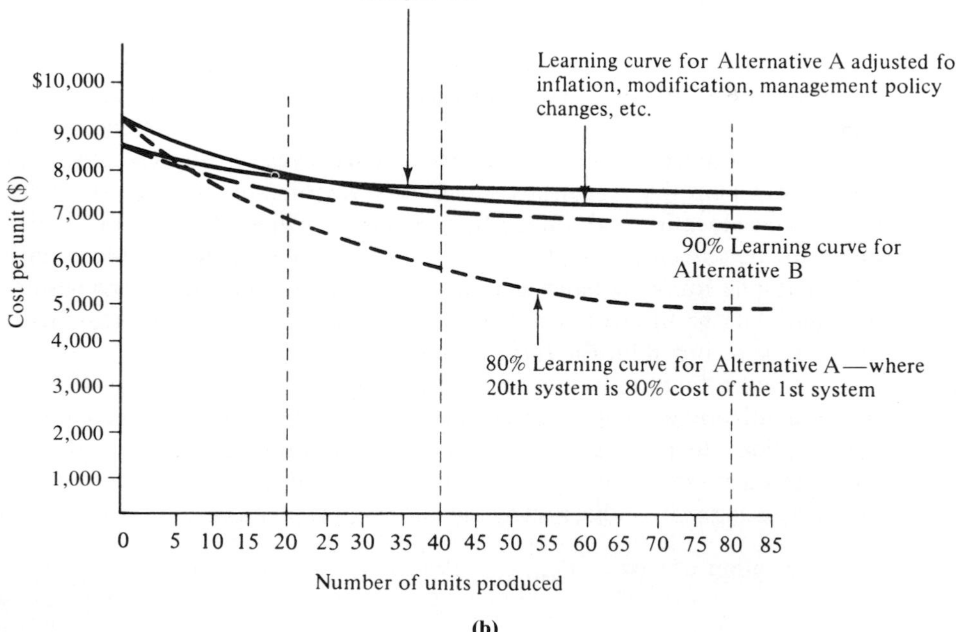

(b)

Figure 17.18 **(a)** Idealistic learning curves (70%, 80%, and 90%). **(b)** Learning curves adjusted based on past experience of the producer.

In summary, the proper application of learning curves is considered significant in life-cycle costing. Although the effects of learning curves may be minimal for research and development activities, they may be significant in determining production costs and follow-on system or product operation and maintenance support costs.[18]

Dealing with inflation. When developing time-phased cost profiles, the reality of inflation should be considered for each future year in the life cycle. During the past several decades, inflation has been a significant factor in the rising costs of products and services and in the reduction of the purchasing power of the dollar. Inflation is a broad term covering the general increase(s) in the unit cost of an item or activity and is related primarily to labor and material costs, as follows:

1. Inflation factors applied to labor costs are due to salary and wage increases, cost of living increases, and increases in overhead rates due to the rising costs of personnel fringe benefits, retirement benefits, insurance, and so on. Inflation factors should be determined for different categories of labor (i.e., engineering labor, technician labor, manufacturing labor, construction labor, customer service personnel labor, management labor, etc.) and should be estimated for each year in the life cycle.
2. Inflation factors applied to material costs are due to material availability (or unavailability), supply and demand characteristics, the increased costs of material processing, and increases in material handling and transportation costs. Inflation factors will often vary with each type of material and should be estimated for each year in the life cycle.

Increasing costs of an inflationary nature often occur as a result of new contract provisions with suppliers, new labor agreements and union contracts, revisions in procurement policies, shifts in sources of supply, the introduction of engineering changes, program schedule shifts, changes in productivity levels, changes in item quantities, and for other comparable reasons. Also, inflation factors are influenced to some extent by geographical location and competition. When reviewing the various causes of inflation, one must be extremely careful to avoid overestimating and double counting for the effects of inflation. For instance, a supplier's proposal may include provisions for inflation, and unless this fact is noted, there is a chance that an additional factor for inflation will be included for the same reason.

Inflation factors should be estimated on a year-to-year basis if at all possible. Because inflation estimates may change considerably with general economic conditions at the national level, cost estimates far out in the future (i.e., 5 years and more) should be reviewed at least annually and adjusted as required. Inflation factors may be established by using price indices or by the application of a uniform escalation rate.

Summarization of costs. Figure 17.19 refects the results of collecting and summarizing the costs recorded in Figure 17.15. Cost streams may be plotted first, based on the actual inflated (or "budgetary") costs projected through the life cycle. The important issue here is to identify any areas (for any given category of the CBS) where there may

[18] Learning curves are discussed in detail in W. J. Fabrycky, P. M. Ghare, and P. E. Torgersen, *Applied Operations Research*, (Upper Saddle River, N.J.: Prentice Hall Inc., 1984), chap. 11.

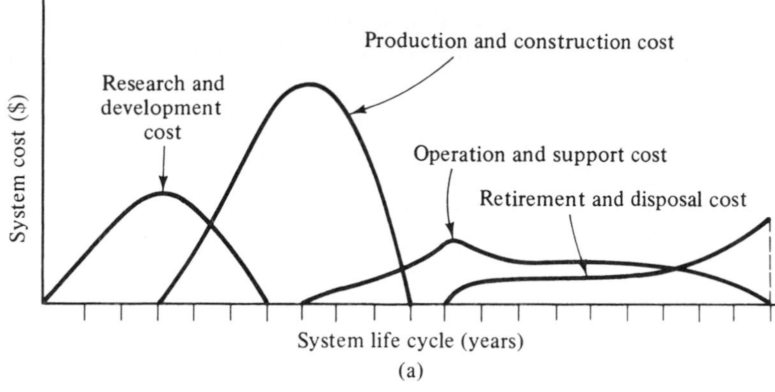

(a)

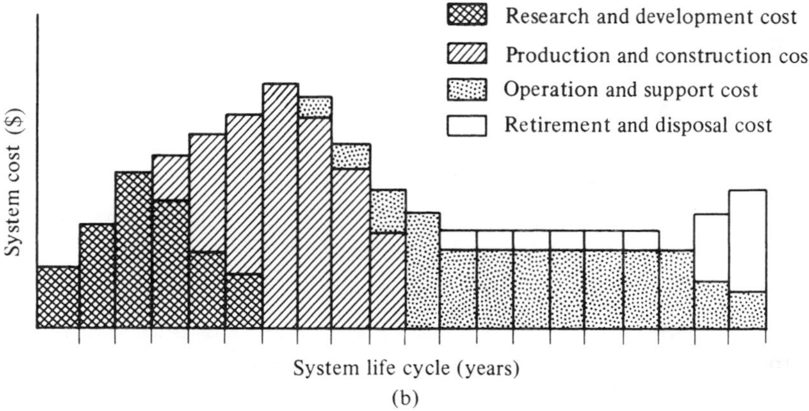

(b)

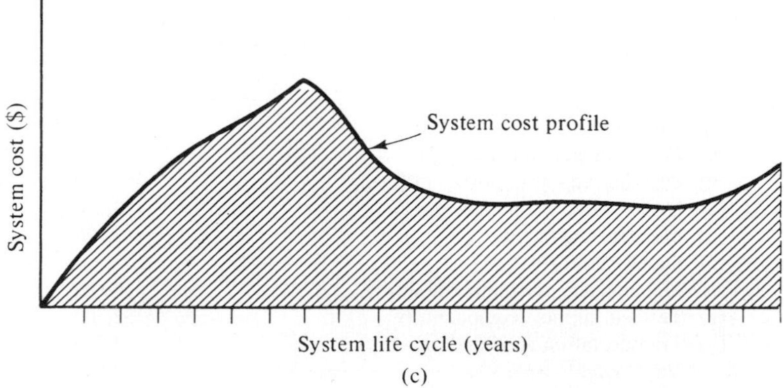

(c)

Figure 17.19 Development of cost profile.

be significant changes from one year to the next and to investigate the possible "causes" accordingly. Conversely, if the purpose is to compare alternative profiles, then a second cost stream should be plotted to reflect present-equivalent costs.

Table 17.1 reflects a summary listing of costs and a breakdown showing the percent contribution of each major category in the CBS to the total. For a single design configuration, the objective is to identify the high-cost contributors (from a budgetary perspective) and to compare these on a relative basis with other categories of cost for this configuration and with similar cost areas for other systems where the costs are presented in a like manner. When comparing alternative configurations as part of a LCC

TABLE 17.1 Life-Cycle Cost Summary

Cost category	Cost ($)	% of Total
1. Research and development cost (C_R)		
(a) System/product management (C_{RM})	247,071	9.8
(b) Product planning (C_{RP})	37,090	1.5
(c) Engineering design (C_{RE})	237,016	9.4
(d) Design data (C_{RD})	48,425	1.9
(e) System test and evaluation (C_{RT})	67,648	2.7
Subtotal	637,250	25.3
2. Production and construction cost (C_P)		
(a) System/product management (C_{RM})	112,497	4.5
(b) Industrial engineering and operations analysis (C_{PI})	68,200	2.8
(c) Manufacturing—recurring (C_{PMR})	635,100	25.3
(d) Manufacturing—nonrecurring (C_{PMN})	48,900	1.9
(e) Quality control (C_{PQ})	78,502	3.1
(f) Initial logistics support (C_{PL})		
(1) Supply support—initial (C_{PLS})	48,000	1.9
(2) Test and support equipment (C_{PLT})	70,000	2.8
(3) Technical data (C_{PLD})	5,100	0.2
(4) Personnel training (C_{PLP})	46,400	1.8
Subtotal	1,112,699	44.3
3. Operation and support cost (C_0)		
(a) Operating personnel (C_{OOP})	24,978	0.9
(b) Distribution—transportation (C_{ODT})	268,000	10.7
(c) Unscheduled maintenance (C_{OLA})	123,130	4.9
(d) Maintenance facilities (C_{OLM})	4,626	0.2
(e) Supply support (C_{OLS})	295,920	11.8
(f) Maintenance personnel training (C_{OLT})	9,100	0.4
(g) Test and support equipment (C_{OLE})	32,500	1.3
(h) Transportation and handling (C_{OLH})	5,250	0.2
Subtotal	763,504	30.4
Grand total	2,513,453	100%

analysis, costs should be broken down both in terms of inflated budgetary values and present value. This will allow for a comparison of alternatives on an equivalent basis and in terms of the anticipated resource requirements for the applicable categories.

Identify the High-Cost Contributors and Establish Cause-and-Effect Relationships (Figure 17.5, Step 8)

Referring to Table 17.1, there are certain categories of cost that stand out as being *high-cost contributors*. An objective at this point is to be able to determine some of the major "causes" for these high costs. For example, 25.3% of the total projected life-cycle cost is in the area of "manufacturing-recurring (C_{PMR})"; 11.8% of the cost is in the area of "supply support (C_{OLS})"; and 10.7% is in the area of "distribution-transportation (C_{ODT})." As an initial step, the analyst needs to refer back to the CBS and review the underlining input assumptions and the parameter-cost relationships that were established when estimating these costs initially. The question is: Are there any underlying assumptions that are questionable or relationships that may be considered as being invalid and that may have a major impact on the projected output results? If it turns out that the "traceability" from the output (e.g., the $295,920 or 11.8%) back to specific input factors is not apparent, then the analyst may wish to break the cost category (C_{OLS}) down into subcategories to gain the additional visibility needed.

Assuming that the proper level of traceability is apparent, the analyst should then relate the applicable high-cost factors in the CBS back to the specific "function(s)" that is being performed, which "causes" this high cost. This may, in turn, lead to a particular element of the system that fails frequently and consumes a lot of resources to maintain, a process that requires people with high skills to complete, or the identification of a method of transportation that is costly to operate. The objective is to identify specific design-related characteristics that tend to "drive" the costs of the system from a life-cycle perspective, whether it be the prime mission-related elements of the system or an element of the support infrastructure.

Having identified the causes, the question is: Are there any alternative system design approaches that can be implemented that will allow for at least the same level of system electiveness but at less overall cost? Feasible candidate solutions should be evaluated in terms of life-cycle cost, and major areas of improvement may be realized through the modification process described in Section 6.6. This process of evaluation, identifying the high-cost contributors, determining the cause-and-effect relationships, and modifying the system for improvement can be accomplished on a continuing basis.

Conduct a Sensitivity Analysis (Figure 17.5, Step 9)

In the performance of a life-cycle cost analysis, there may be a few key areas where the results might be "suspect" because of inadequate data input, poor prediction data, invalid assumptions in the beginning, and so on. Identifying the high-cost areas from Table 17.1 provides a good place to start. There are a comple of questions that need to be addressed: How sensitive are the results of the analysis to possible variations of these uncertain input factors? To what extent can certain input parameters be varied without causing the configuration being evaluated to be discarded for no longer being feasible?

Referring to Table 17.1, the analyst should select the high contributors (those that contribute more than 10% of the total cost), identify the critical input factors that directly impact cost (from Step 8), change some of these factors at the input stage, and determine the changes in cost at the output. Such input variations can be accomplished over a designated range, considering appropriate distributions, and so on. For example, key input parameters in the analysis of the communication system include the system *operating time* and the *MTBM*. Using the life-cycle cost model (from Step 6), the analyst may apply a multiple factor to the operating time and MTBM values and determine the delta cost associated with each variation. From this, trend curves are projected, as illustrated in Figure 17.20.

Through evaluation of the information presented, it is evident that a *small* variation in operating time and MTBM may cause a significant delta increase in cost at the output. This provides an indication that there is a high degree of *risk* associated with making decisions based on the results. Although the results in Figure 17.20 are in terms of the total life-cycle cost figure-of-merit (i.e., \$2,513,453), it may be appropriate to assess the results of each variation at the second or third indenture of the CBS

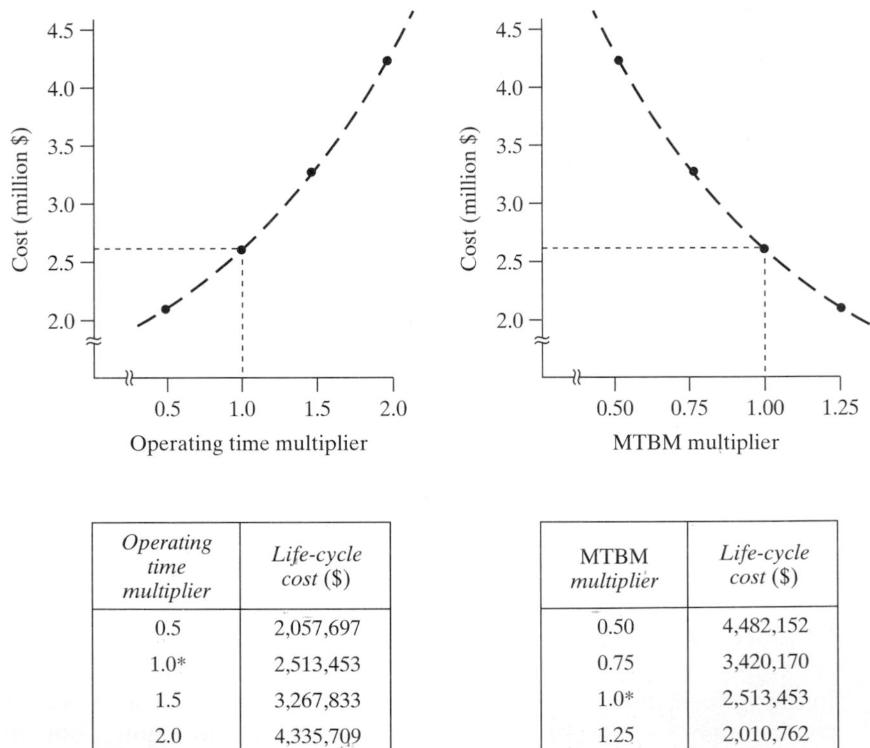

Operating time multiplier	*Life-cycle cost* (\$)
0.5	2,057,697
1.0*	2,513,453
1.5	3,267,833
2.0	4,335,709

MTBM *multiplier*	*Life-cycle cost* (\$)
0.50	4,482,152
0.75	3,420,170
1.0*	2,513,453
1.25	2,010,762

*Baseline configuration.

Figure 17.20 Results of a sensitivity analysis (partial).

in Figure 17.10. Such input variations may have a greater impact on some categories than on others.

On identifying the high-risk areas (as measured by the magnitude of the delta cost at the output stage), the analyst should make every effort to reduce this risk by improving the input data to the greatest extent possible. This may, in turn, lead to a revised MTBM prediction, a more indepth description of system operational requirements, a more indepth maintenance task analysis (MTA), and so on. The most significant challenge here is that the analyst needs to be familiar with the various parameter interrelationships, their impact on cost, the sensitivities of the model in the identification of cause-and-effect relationships, and the potential areas of risk associated with the LCC analysis results.

Construct a Pareto Diagram and Identify Priorities for Problem Resolution (Figure 17.5, Step 10)

The problem areas identified in Steps 8 and 9 should be evaluated and prioritized in terms of degree of importance. High-cost/high-risk areas should receive the most management attention and, as these are addressed, others may rise to the top. A Pareto diagram, illustrated in Figure 17.21, may be constructed to show the relative importance of different problem areas that (either directly or indirectly) cause high costs. "Importance" factors are based not only on high-cost areas, but on criticality as it pertains to the system and its ability to accomplish its designated mission.[19]

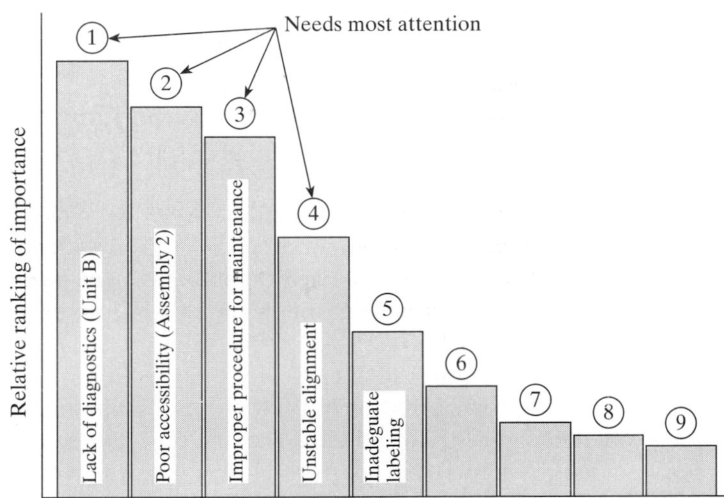

Figure 17.21 A Pareto ranking of major problem areas.

[19] The FMECA (Section 12.4) and the Ishikawa diagram (Figure 12.25) can be used to help identify "cause-and-effect" relationships, and the results conveyed in Figure 17.21 should consider the results from the Pareto analysis shown in Figure 12.26. Although the examples in Figure 17.21 are primarily maintenance related, the results from such an analysis should identify areas of concern across the board (i.e., a process-related problem, a system operational problem, an organizational problem, and so on).

Identify Feasible Alternatives for Design Evaluation (Figure 17.5, Step 11)

Having acquired the necessary visibility relative to the most critical problems and their causes, the next step is to investigate possible alternative ways by which the applicable functions can be accomplished. Design trade-offs are conducted with the objective of selecting a few feasible candidates for further evaluation. The goal is to improve the balance illustrated in Figure 17.1 by continuing to meet the effectiveness requirements while, at the same time, reducing the overall life-cycle cost. Figure 17.22 shows a projected life-cycle cost profile for each of three candidate solutions being considered.

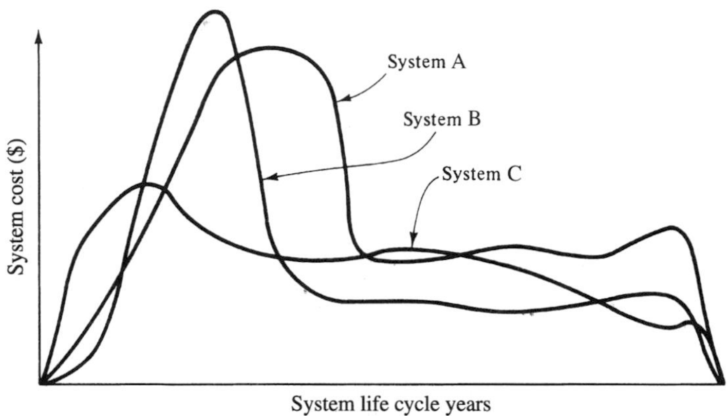

Figure 17.22 Alternative life-cycle cost profiles being considered for evaluation.

Evaluate Feasible Alternatives and Select Preferred Approach (Figure 17.5, Step 12)

The first 10 steps (Figure 17.5, Steps 1–10) address the life-cycle cost analysis approach as it applies to a single system configuration. Step 11 introduces the requirement to address alternatives and to evaluate each on an equivalent basis. It is now appropriate to extend the LCC analysis to cover this area where multiple options are being considered.[20]

For the communication system example (refer to Step 1 in Figure 17.5), there are two candidates being considered to fulfill the operational and maintenance requirements illustrated in Figures 17.6 and 17.7. The first objective is to ensure that both meet the specified requirements (i.e., performance, unit design-to-cost goal of $35,000, MTBM of 500 hours, etc.). Given this, the question is: Which candidate is preferred assuming that the cost of capital is 10%?

Following the steps identified in Figure 17.5, the cost profile for each alternative was developed in terms of both inflated (budgetary) cost and present-value cost for the

[20] Several methods for evaluating alternatives on an equivalent basis using money-flow modeling are presented in Sections 8.3 and 8.4. Methods based on economic optimization are presented in Sections 9.2 and 9.3.

purposes of comparison. The projected costs, on a year-to-year basis, are shown in Table 17.2.[21]

For the communication requirement, Table 17.2 summarizes the costs for each of the two alternatives, along with the percentage contribution of the various cost categories in terms of the total. The cost categories reflected in the table are extracted from the CBS in Figure 17.10 and represent those where sigificant values have been recorded through the analysis effort. The two alternatives, presented in terms of the initially specified requirements, are identified in Figure 17.23.

By inspection, it appears that Alternative A is preferred over Alternative B. However, prior to making a final decision on which alternative to select, the analyst should accomplish a break-even analysis to determine the point in time when Alternative A assumes a point of preference (i.e., becomes less costly). Referring to Figure 17.24 it appears that this point is approximately 4 years and 10 months after program start.

Because the resultant values in Table 17.3 are relatively close, the analyst may wish to accomplish a sensitivity analysis with the objective of identifying possible areas of risk (refer to Figure 17.5, Step 9). With the variation of factors, as an input to the LCC analysis, the break-even point in Figure 17.24 may shift either inward or outward, which may have an impact on the risk(s) associated with the decision-making process.

In this instance, it is assumed that Alternative A is the preferred approach, particularly since the break-even point is early enough in the planned life cycle of 13 years (refer to Figure 17.8). Given the selection of A, the analyst may then convert the present-value cost figures back into inflated budgetary values in order to gain visibility as to the potential high-cost contributors in terms of expected resource consumption in the future. Referring to Table 17.3, it appears that the categories of "manufacturing-recurring (C_{PMR})" at 25.3%, "supply support (C_{OLS})" at 11.8%, and "distribution-transportation (C_{ODT})" at 10.7% constitute the high-cost contributors and are worthy of some additional investigation (refer to Figure 17.5, Step 8). The question is: Can the design of the configuration selected (i.e., Alternative A) be improved with the goal of further reducing the projected life cycle cost?

17.5 LIFE-CYCLE COST APPLICATIONS AND BENEFITS

Inherent within the activities identified in Figure 17.4, there are many different and varied design and management decisions that are made and can have a significant impact on life-cycle cost. In particular, those early system-level decisions made during the conceptual and preliminary design phases will have a significant influence on the activities and associated costs of system operations, maintenance and support, and retirement and material disposal (refer to Figures 17.2 and 17.3). Thus, it is essential that designers, support personnel, managers, and others in the decision-making loop consider the impact of their day-to-day actions on total life-cycle cost. In other words, the greatest benefits can be gained if the issues of economics and cost were addressed from the beginning in the design and development of new systems.

[21] The conversion of the inflated (budgetary) cost stream to a present-equivalent cost stream is accomplished by using Equation 8.2 in Chapter 8, supplemented by use of the 10% interest table in appendix D.

TABLE 17.2 Cost Allocation by Program Year

Program activity	Cost category designation	Cost by program year													Total actual cost ($)
		1	2	3	4	5	6	7	8	9	10	11	12	13	
A. Alternative A															
1. Research and development	C_R	162,712	194,357	280,181	—	—	—	—	—	—	—	—	—	—	637,250
2. Production/construction	C_P	—	85,400	307,000	360,270	360,270									1,112,699
3. Operation and support	C_O	—	—	23,388	56,792	90,954	90,954	90,954	90,954	90,954	90,954	90,954	34,884	11,762	763,504
Total actual cost (budgetary)	C	162,712	279,757	610,569	417,062	450,983	90,954	90,954	90,954	90,954	90,954	90,954	34,884	11,762	$2,513,453
Total present-equivalent cost	$C_{10\%}$	147,921	231,191	458,720	284,853	280,015	51,344	46,659	42,430	38,574	35,063	31,879	11,114	3,407	$1,663,200
B. Alternative B															
1. Research and development	C_R	166,124	173,317	255,616	—	—	—	—	—	—	—	—	—	—	595,057
2. Production/construction	C_P	—	96,850	305,140	361,510	379,357									1,142,857
3. Operation and support	C_O	—	—	26,682	64,721	104,037	104,037	104,037	104,037	104,037	104,037	104,037	40,007	13,258	872,927
Total actual cost (budgetary)	C	166,124	270,167	587,438	426,231	483,394	104,037	104,037	104,037	104,037	104,037	104,037	40,007	13,258	$2,610,841
Total present-equivalent cost	$C_{10\%}$	151,023	223,266	441,342	291,116	300,139	58,729	53,371	48,533	44,122	40,106	36,465	12,746	3,841	$1,704,799

TABLE 17.3 Life-Cycle Cost Breakdown

	Alternative A		Alternative B	
Cost Category	Cost ($)	% of Total	Cost ($)	% of Total
1. Research and development cost (C_R)				
(a) System/product management (C_{RM})	247,071	9.8	240,823	9.2
(b) Product planning (C_{RP})	37,090	1.5	15,960	0.7
(c) Engineering design (C_{RE})	237,016	9.4	238,119	9.1
(d) Design data (C_{RD})	48,425	1.9	39,624	2.3
(e) System test and evaluation (C_{RT})	67,648	2.7	60,531	2.3
Subtotal	637,250	25.3	595,057	22.8
2. Production and construction cost (C_P)				
(a) System/product management (C_{RM})	112,497	4.5	112,497	4.3
(b) Industrial engineering and operations analysis (C_{PI})	68,200	2.8	66,000	2.5
(c) Manufacturing—recurring (C_{PMR})	635,100	25.3	657,500	25.2
(d) Manufacturing—nonrecurring (C_{PMN})	48,900	1.9	55,100	2.1
(e) Quality control (C_{PQ})	78,502	3.1	72,990	2.8
(f) Initial logistics support (C_{PL})				
(1) Supply support—initial (C_{PLS})	48,000	1.9	57,020	2.2
(2) Test and support equipment (C_{PLT})	70,000	2.8	70,000	2.7
(3) Technical data (C_{PLD})	5,100	0.2	5,350	0.2
(4) Personnel training (C_{PLP})	46,400	1.8	46,400	1.8
Subtotal	1,112,699	44.3	1,142,857	43.8
3. Operation and support cost (C_0)				
(a) Operating personnel (C_{OOP})	24,978	0.9	24,978	0.9
(b) Distribution—transportation (C_{ODT})	268,000	10.7	268,000	10.7
(c) Unscheduled maintenance (C_{OLA})	123,130	4.9	151,553	0.2
(d) Maintenance facilities (C_{OLM})	4,626	0.2	5,705	0.2
(e) Supply support (C_{OLS})	295,920	11.8	365,040	14.0
(f) Maintenance personnel training (C_{OLT})	9,100	0.4	9,100	0.4
(g) Test and support equipment (C_{OLE})	32,500	1.3	42,251	1.6
(h) Transportation and handling (C_{OLH})	5,250	0.2	6,300	0.2
Subtotal	763,504	30.4	872,927	33.4
Grand total	2,513,453	100%	2,610,841	100%

At the same time, many benefits can be gained through the application of life-cycle costing methods in the evaluation and subsequent improvement of systems that are already in the inventory and in operational use. The identification of high-cost contributors and the follow-on system modification for improvement (realizing a reduction in projected life-cycle cost), implemented on an iterative and continuing basis, can result in many benefits in this day and age where resources are limited and there is a great deal of international competition. Success in this area is heavily dependent on the

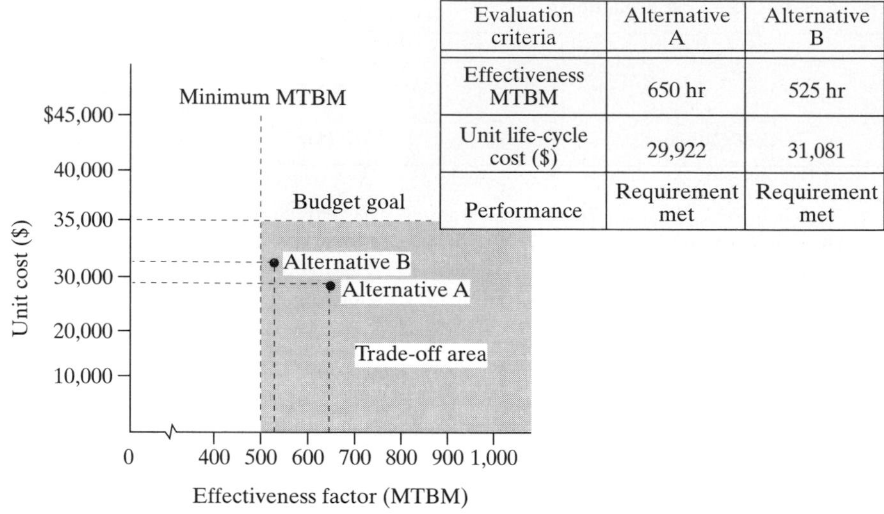

Evaluation criteria	Alternative A	Alternative B
Effectiveness MTBM	650 hr	525 hr
Unit life-cycle cost ($)	29,922	31,081
Performance	Requirement met	Requirement met

Figure 17.23 Effectiveness versus unit cost.

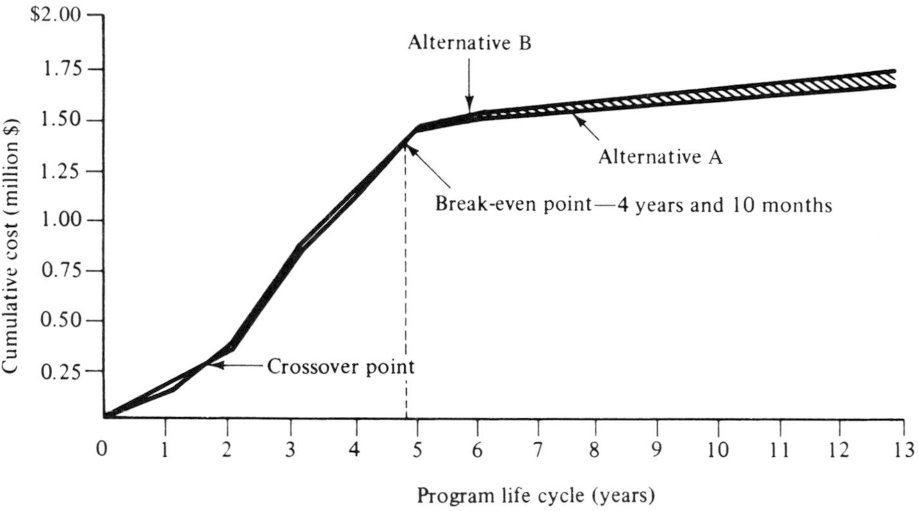

Figure 17.24 Break-even analysis.

availability of good historical data and having the visibility needed for the implementation of a *continuous product/process improvement* capability. Regarding *visibility*, can you answer the following questions:

1. For your system, can you identify the high-cost contributors from a *functional* perspective? What is it costing to perform certain critical functions over the life cycle?
2. Can you identify the *causes* for these high-cost areas? What elements of the system, or segments of a process, are "driving" these costs?

3. What are the relationships between the highest areas and functions that are *critical* regarding accomplishment of the system mission?
4. What are the *high-risk* areas?

The life-cycle cost analysis process can be used as an aid in solving a wide variety of problems. Figure 17.25 highlights a few areas where one can apply the steps identified in Figure 17.5; however, the process reflected by the steps identified in the figure must be properly "tailored" to the problem at hand. In essence, one can accomplish a LCC analysis on a single sheet of paper in a few minutes, or can go to great depths in terms of data manipulations. The results of such an analysis must be to the proper depth, must be timely, and must be responsive to designer/manager involved in the decision-making process. The important issue here is to *think economics* and to *think life cycle*, the former being within the context of the latter.

In summary, the LCC analysis approach:

1. Forces long-range planning versus the more traditional short-term thinking and decision-making processes. As a result, decisions can be based on more complete information with less risk involved.
2. Forces total cost visibility and the identification of the high-cost system elements, equipment, processes, and so on. This aids in pinpointing the specific functional areas where resource consumption is high and modifications for improvement are desired.
3. Enables a better understanding of the interrelationships between different system elements and categories of cost. The interaction effects become more visible through a life-cycle cost sensitivity analysis.

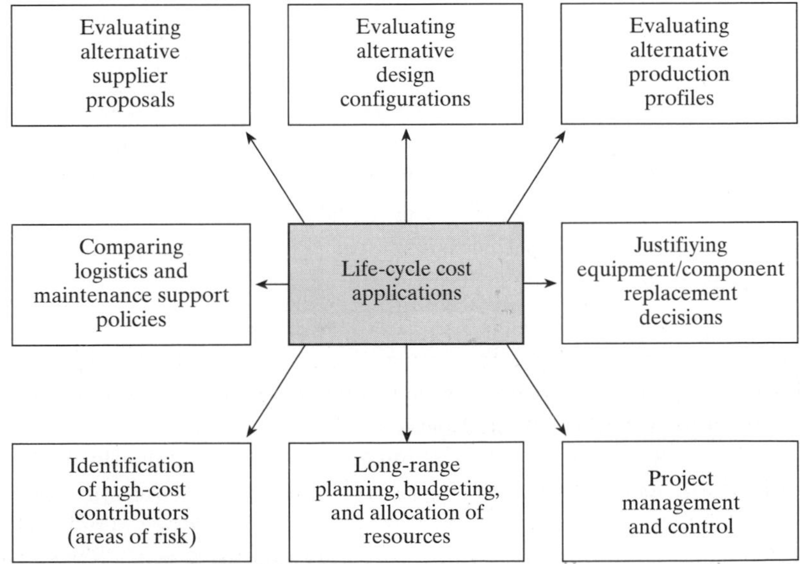

Figure 17.25 Life-cycle cost applications.

4. Aids in the early identification of potential high-risk areas, the quantification of such, and in the subsequent elimination of the possible causes of risk.
5. Allows for better overall resource management because of the long-term visibility that is provided.

Although the benefits are numerous, there are also some major impediments. Our current "thought processes," accounting practices, budgetary cycles, organizational objectives, and politically driven activities are more oriented to the "short term!" Further, on many occasions the visibility from a LCC analysis is not desired due to the fear of exposure for one reason of another.

To be successful in this area requires that the proper *organizational environment* be established that will allow it to happen. There must be a commitment to "life-cycle thinking" from the top down; the right type of data must be collected and available; and the analyst must have direct access to all applicable areas of activity. Given this, it is important to get "involved" by understanding the process, applying it to a known entity and evaluating the results, and establishing some cost-estimating relationships that can be applied to future LCC analyses.

QUESTIONS AND PROBLEMS

1. Define *life-cycle cost* (LCC) in your own words. What is included (excluded)?
2. What is meant by DTC? When in the system life cycle should it be applied? How does it relate to the TPMs identified in Section 3.5 (if at all)?
3. Why is life-cycle costing important?
4. What is meant by *functional costing?* How does it relate to the functional analysis described in Sections 3.6 and 4.1?
5. Describe the steps involved in accomplishing a life-cycle cost analysis.
6. What is the cost *breakdown structure* (CBS)? What are its purposes? What is included (or excluded)? How does it relate to a *work breakdown structure* (WBS)?
7. How does the CBS relate to the functional analysis (if at all)?
8. Describe some of the more commonly used cost estimating methods. Under what conditions should they be applied? Provide some examples.
9. What is *activity-based costing* (ABC)? Why is it important (if at all)? What are some of the differences between this method of costing and some of the more conventional approaches?
10. Refer to Figure 17.9. Under what conditions should *revenues* be treated as an element of cost?
11. Refer to Figure 17.10. Would you consider the CBS in the figure to be adequate for all applications where a LCC analysis is required? If so, why? If not, why not?
12. Refer to Figure 17.13. What is meant by *parametric cost estimating?* How would you develop cost-estimating relationships?
13. Describe what is meant by a *learning curve*? How are learning curves developed? How can they be applied? What are some of the cautions that need to be addresses in applying a learning curve?

14. In this chapter, three different cost profiles are identified. What are they? Under what conditions can each be applied?

15. Refer to Table 17.1. Why is it sometimes beneficial to present the results of a LCC analysis in this type of a format?

16. What is meant be a *sensitivity* analysis? What type of information may be derived from the accomplishment of such? Why is the accomplishment of a sensitivity analyst important?

17. What is a *Pareto analysis?* What benefits can it provide?

18. Calculate the anticipated life-cycle cost for your personal automobile.

19. Referring to the following table, which alternative would you select? Why?

	Configuration A		Configuration B	
Cost category	Present cost	% of total	Present cost	% of total
Research and development	$70,219	7.8	$53,246	4.2
Management	9,374	1.1	9,252	0.8
Engineering design	45,552	5.0	28,731	2.3
Test and evaluation	12,176	1.4	12,153	0.9
Technical data	3,117	0.3	3,110	0.2
Production	407,814	45.3	330,885	26.1
Construction	45,553	5.1	43,227	3.4
Manufacturing	362,261	40.2	287,658	22.7
Operations and maintenance	422,217	46.7	883,629	69.4
Operations	37,811	4.2	39,301	3 1
Maintenance	382,106	42.5	841,108	66.3
Maintenance personnel	210,659	23.4	407,219	32.2
Spares/repair parts	103,520	11.5	228,926	18.1
Test equipment	47,713	5.3	131,747	10.4
Transportation	14,404	1.6	51,838	4.1
Maintenance training	1,808	0.2	2,125	0.1
Facilities	900	0.1	1,021	Neg
Field data	3,102	0.4	18,232	1.4
Phaseout and disposal	2,300	0.2	3,220	0.3
Grand total	$900,250	100%	$1,267,760	100%

20. Refer to the table in Problem 19. What would be the likely impact on LCC if

(a) The system MTBF is decreased?

(b) The $\overline{\text{M}}$ct is increased?

(c) The MLH/OH is increased?

(d) System use is increased?

(e) The fault-isolation capability in the system were inadequate?

What steps would you take to reduce these costs?

21. The BAF Corporation is considering the possibility of introducing Product Y into the market. A market analysis indicates that the corporation could sell all of the products that it can produce at a price of $200 each for at least 10 years into the future. The product is not repairable and is discarded at failure.

To manufacture the product, the corporation needs to invest in some capital equipment. Based on a survey of potential sources, there are three alternatives considered feasible to meet the need.

(a) *Configuration A* includes a machine that is semiautomatic (requiring a part-time machine operator), will produce 395 products per year, and can be purchased at a price of $15,000. The expected reliability (MTBF) of this machine is 165, and the anticipated cost per corrective maintenance action is $200. The average machine operating cost is $3.00 per hour, and the estimated salvage value after 10 years of operations is $1,000.

(b) *Configuration B* includes a machine that is automatic, will produce 525 products per year, and can be purchased at a price of $28,000. The expected reliability (MTBF) of this machine is 270, and the anticipated cost per corrective maintenance action is $300. Preventive maintenance is required every six (a) months, and the cost per maintenance action is $250. The average machine operating cost is $0.60 per hour, and the estimated salvage value after 10 years of operations is $2,500.

(c) *Configuration C* includes a machine that is automatic, will produce 500 products per year, and can be purchased at a price of $23,000. The expected reliability (MTBF) of this machine is 195, and the anticipated cost per corrective maintenance action is $225. Preventive maintenance is required every ninety (90) days, and the cost per maintenance action is $200. The average machine operating cost is $0.50 per hour, and the estimated salvage value after 10 years of operation is $2,200.

The anticipated machine use will be 8 hours per day and 270 days per year. Machine output (when considering a normal "buildup" rate) is expected to be 0.5 during the first year of operation and at full capacity during year 2 and on. The manufacturing cost (materials and labor associated with material procurement, material handling, quality control, inspection and test, and packaging) for Product Y is $50 using Machine A, $30 using Machine B, and $40 using Machine C. The expected allocated distribution cost (transportation, warehousing, etc.) is $10 per item. Assume that the interest rate is 10%.

Which configuration would you select? Why?

21. A need has been identified which will require the addition of a new communications capability to a ground transportation system. This new communications capability, referred to as System XYZ, must be developed, and there are two different supplier configurations being considered for procurement. Based on the information provided: (a) compute the life-cycle cost for each of the two configurations; (b) plot the applicable cost streams (discounted and undiscounted); (c) accomplish a break-even analysis; and (d) select a preferred approach. In computing present-equivalent costs, assume a 15% discount factor.

System XYZ is to be installed in a transportation vehicle, and the total number of systems in operational use for each year of the projected life cycle is noted. It is assumed that the vehicles will be distributed, with the first 12 systems in Area A, the next 12 systems in Area B, and so on.

Year number									
1	2	3	4	5	6	7	8	9	10
0	0	10	20	40	60	60	60	35	25

It is assumed that each System XYZ will be used on the average of 4 hours per day, 365 days per year. The vehicle operator is assigned to operate several different systems throughout the accomplishment of a mission, and it is assumed that 1% of his time is allocated to System XYZ.

System XYZ is a newly designed entity, and each of the two candidate configurations is packaged in Units A to C, as illustrated.

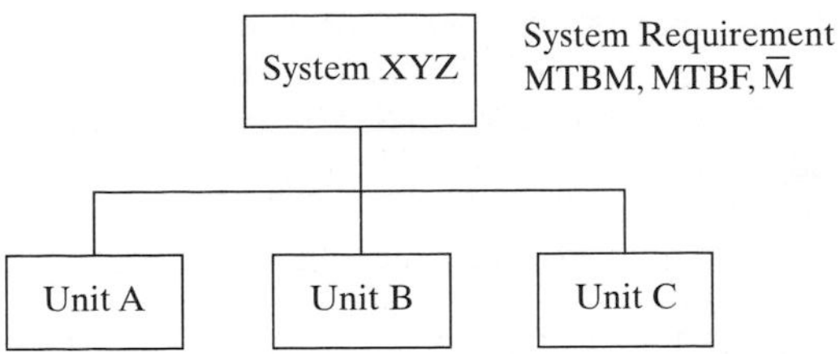

The predicted reliability and maintainability factors associated with each of the two candidate configurations are noted in the following table. Relative to the maintenance concept, two levels of maintenance are assumed (with an intermediate-level shop in each of the five geographical areas). Each of the two configurations incorporates a built-in self-test capability that enables rapid system checkout and fault isolation to the unit level. No external support equipment is required for organizational maintenance on the vehicle. In the event of a "no-go" condition, fault isolation is accomplished to the unit and the applicable unit is removed, replaced with a spare, and the faulty unit is sent to the intermediate-level maintenance shop for corrective maintenance. Unit repair is accomplished through module replacement, with the modules being discarded at failure (i.e., the modules are assumed to be "non-repairable"). Scheduled (preventive) maintenance is accomplished for configuration A (Unit A) and configuration B (Unit B) , as noted in the table, in the intermediate-level shop every 6 months. No supplier-level (or depot) maintenance is required; however, the supplier does provide backup supply and support functions as required.

Parameter*	Configuration A	Configuration B
System level (organizational maintenance)		
MTBM	195 hr	249 hr
$MTBM_u$ (or MTBF)	267 hr	377 hr
$\bar{M}$	30 min	30 min
Unit level (intermediate maintenance)		
Unit A		
MTBM	382 hr	800 hr
$MTBM_u$	800 hr	800 hr
$MTBM_s$	730 hr	—
$\overline{Mct}$	5 hr	5 hr
$\overline{Mpt}$	16 hr	—
Unit B		
MTBM	500 hr	422 hr
$MTBM_u$	500 hr	1,000 hr
$MTBM_s$		730 hr
$\overline{Mct}$	4 hr	5 hr
$\overline{Mpt}$	—	12 hr
Unit C		
MTBM	2,000 hr	2,500 hr
$\overline{Mct}$	2 hr	3 hr

*Assume that $MTBM_u$ = MTBF. When there is no scheduled maintenance, $MTBM_u$ = MTBM.

The requirements for System XYZ dictate the program profile in the following figure. Assume that life-cycle costs are broken down into the three categories represented by the blocks in the program profile (i.e., design and development, production, and operations and maintenance).

Year number

1 2 3 4 5 6 7 8 9 10

Design and development

Production

Operations and maintenance

In an attempt to simplify the problem, the following additional factors are assumed:

1. Design and development costs for System XYZ (to include labor and material).
   Configuration A: $80,000 ($50,000/year 1 and $30,000/year 2)
   Configuration B: $100,000 ($70,000/year 1 and $30,000/year 2)
2. Design and development costs for special support equipment at the intermediate level of maintenance.
   Configuration A: $30,000 ($20,000/year 1 and $10,000/year 2)
   Configuration B: $23,000 ($17,000/year 1 and $6,000/year 2)
3. System XYZ models for operational use are produced and delivered in the year before the identified need (i.e., 10 models are produced and delivered in year 2, etc.). The production costs for each System XYZ (Units A–C) are
   Configuration A: $21,000
   Configuration B: $23,000
4. Special support equipment is required at each intermediate maintenance shop (for the corrective maintenance of units) at the start of the year when System XYZ operational models are distributed (i.e., Area "A" at the beginning of year 3). In addition, a backup set of special support equipment is required at the supplier location when the first intermediate shop becomes operational. Special support equipment is produced and delivered at a cost of
   Configuration A support equipment: $13,000
   Configuration B support equipment: $12,000
5. Spare units are required at each intermediate-level maintenance shop at the time of activation. Assume that one (1) Unit A, one (1) Unit B, and one (1) Unit C constitute a set of spares, and that the cost of a set is equivalent to the cost of a production system (i.e., $21,000 for Configuration A and $23,000 for Configuration B). Also, assume that a set

of spares is stocked at the supplier's facility at the time when the first intermediate shop is activated.

Additional spares constitute components (i.e., assemblies, modules, parts, etc.). Assume that the material costs are $250 per corrective maintenance actions, and $100 per preventive maintenance action. The cost factors include amortized inventory maintenance costs.

6. Maintenance facilities, as defined here, include the supporting resources required for System XYZ (i.e., at the intermediate shop), above and beyond spares/inventories, personnel, and data. A burden rate of $1 per direct maintenance manhour associated with the prime equipment is assumed.
7. Maintenance data include the preparation and distribution of maintenance reports, failure reports, and related data associated with each maintenance action. Maintenance data costs are assumed to be $25 per maintenance action.
8. For each maintenance action at the system level, one low-skilled technician at $20 per direct maintenance manhour is required on a full-time basis. It is assumed that this is an average value, applied throughout the life cycle, and it includes direct, indirect, and inflationary factors. The $\overline{M}$ is 30 minutes for each of the two configurations.
9. For each corrective maintenance action involving Unit A, B, or C, two technicians are required on a full-time basis (i.e., the duration of the $\overline{M}ct$ value). One low-skilled technician at $20 per hour and one (1) high-skilled technician at $30 per hour are required. Direct, indirect, and inflationary factors are considered in these average values.
10. For each preventive maintenance action involving units (Configurations A and B), one (1) high-skilled technician at $30 per hour is required on a full-time basis (i.e., the duration of the $\overline{M}pt$ value).
11. For the operation of System *XYZ*, the allocated cost for the operator is $40 per hour.

In solving this problem, be sure to state all assumptions in a clear and concise manner.

23. Referring to the figure, there are four feasible configurations identified in the "trade-off" area. Which one would you select? Why?

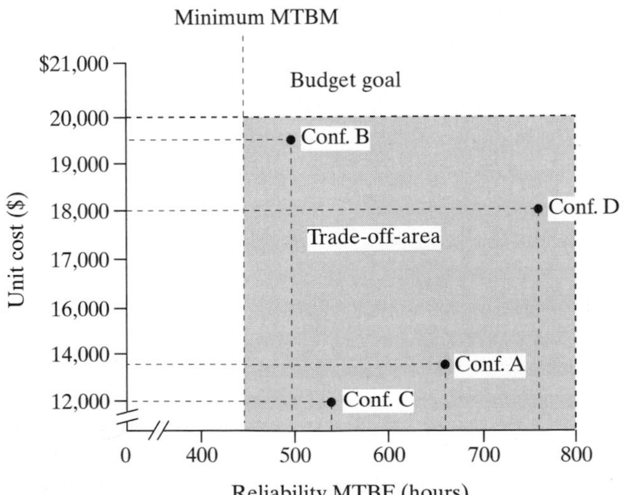

24. The figure results from a breakeven analysis, influenced by a sensitivity analysis. What steps would you take to reduce the identified "risk?"

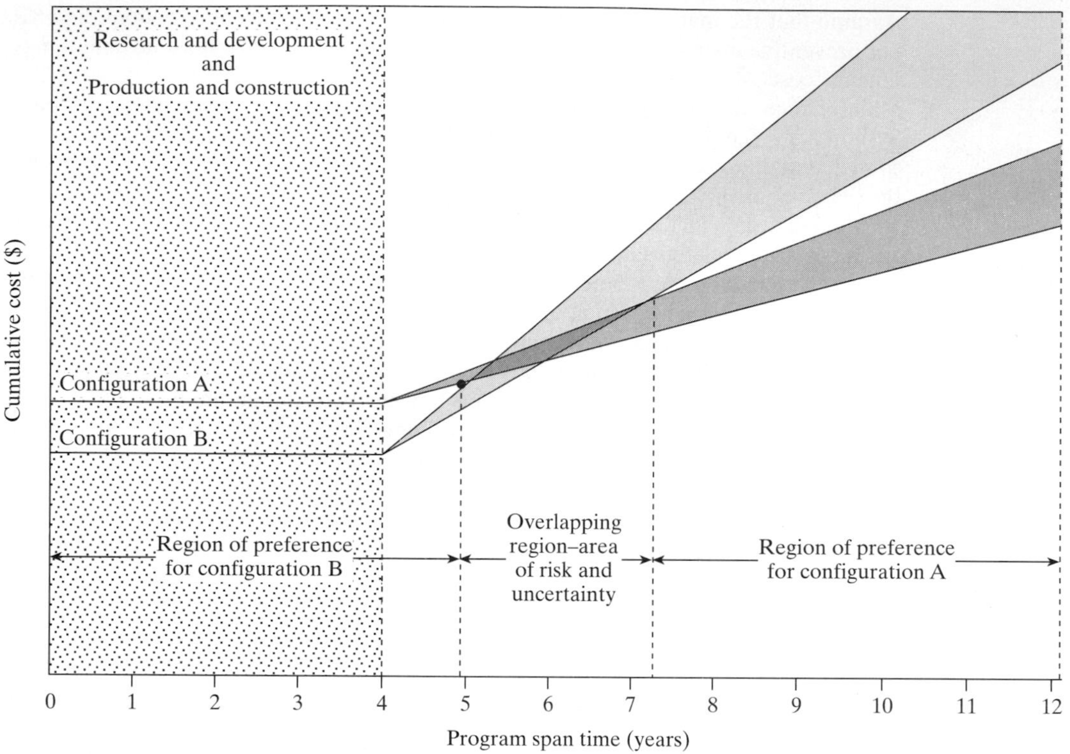

25. Select a system of your choice and accomplish a life-cycle cost analysis.

26. Describe some of the benefits derived from accomplishing a life-cycle cost analysis. What are some of the impediments? How would you go about removing some of these impediments?

PART V
SYSTEMS ENGINEERING MANAGEMENT

18 SYSTEMS ENGINEERING PLANNING AND ORGANIZATION

Successful application of the principles and concepts of systems engineering involves both the *technological* and the *management* sides of the spectrum, as illustrated in Figure 2.16. The earlier parts of the text concentrate on the technology issues, while this part addresses the management issues. One may think that a given system technical approach is the best. However, the objectives of such may never be realized without good initial planning, the subsequent establishment of the proper organizational environment that will allow it to happen, and the follow-on management and controls that are necessary to ensure effective and efficient program implementation.

Systems engineering management involves planning, organizing, staffing, monitoring, and controlling the process of designing, developing, testing, and producing a system that will fully meet the needs of the customer (i.e., the process described in Part II). It provides the necessary overview function to ensure that all needed engineering disciplines and related specialties are properly integrated (i.e., the traditional engineering design disciplines, those described in Part IV, and all supporting disciplines). It ensures that the system being developed contains the proper mix of hardware, software, facilities, personnel, data, and so on. It causes the selection and proper integration of the appropriate tools to facilitate completion of the day-to-day design process (i.e., the analytical tools/models described in Parts II and III). The objective of systems engineering management is to provide the right item, at the right location, at the right time, and with a minimum expenditure of human and physical resources.[1]

[1] Systems engineering management is covered in two chapters. This chapter addresses the issues of early planning and organization for systems engineering. Program functions and tasks, stemming from Figures 2.2 and 2.4, are described; scheduling and cost projections are developed; and personnel requirements are identified. Chapter 19 presents the subsequent day-to-day program management and control activities needed to ensure that the initial planning is properly implemented.

18.1 SYSTEMS ENGINEERING PLANNING

Referring to Figure 2.2, early systems engineering planning starts with the identification of a need and the definition of program requirements. Although this may appear to be rather basic, every program is different, and it is essential that systems engineering requirements be tailored accordingly. The principles and concepts described throughout this text are applicable to all programs. Only the nature and depth of application may vary from one program to the next.

The successful implementation of systems engineering program requirements is dependent on (1) employing a top-down approach in the development of the system in question, (2) the early integration of design and related supporting activities, (3) viewing the requirements in terms of the entire system life cycle, and (4) the preparation of the necessary requirements and planning documentation from the beginning. As early system concepts are generated and feasibility studies are conducted to determine possible alternative technical solutions in response to a given design problem, the appropriate level of planning must be initiated to ensure that the ideas generated through analysis are properly consumated and integrated into a final product configuration in a cost-effective manner.

Referring to Figure 3.2, the critical program documentation includes a combination of specifications and plans, with the system specification (type A) covering the overall *technical* requirements for the system as an entity and the system engineering management plan addressing the *management* requirements. The "specification" side of the spectrum is discussed in Section 3.8, and the "planning" requirements are addressed in this chapter.

18.2 SYSTEM ENGINEERING MANAGEMENT PLAN (SEMP)

The SEMP is the key management document covering those activities and milestones necessary to accomplish the objectives discussed throughout this text. The SEMP directly supports the program management plan (PMP), which, for discussion purposes, is considered to be the top management document for a given program or project. The objectives of the SEMP are to provide the structure, policies, and procedures to foster the integration of the various engineering-related activities needed for system design and development. Referring to Figure 18.1, the SEMP is developed during the conceptual design phase to provide the integration of a number of individual design-related program plans (e.g., reliability program plan, maintainability program plan, etc.) and to promote the necessary communication links with other top-level planning documentation (e.g., configuration management plan, test and evaluation master plan, total quality management plan, etc.).[2]

[2] It should be emphasized that planning requirements, the number and type of plans, the preparation and availability of such, and so on, will vary from one program to the next. Figure 18.1 conveys what might be considered to be the requirements for a large system project. Conversely, all of the planning areas noted in the figure may be integrated into a single, relatively small document for a small system project, assuming that they are applicable initially.

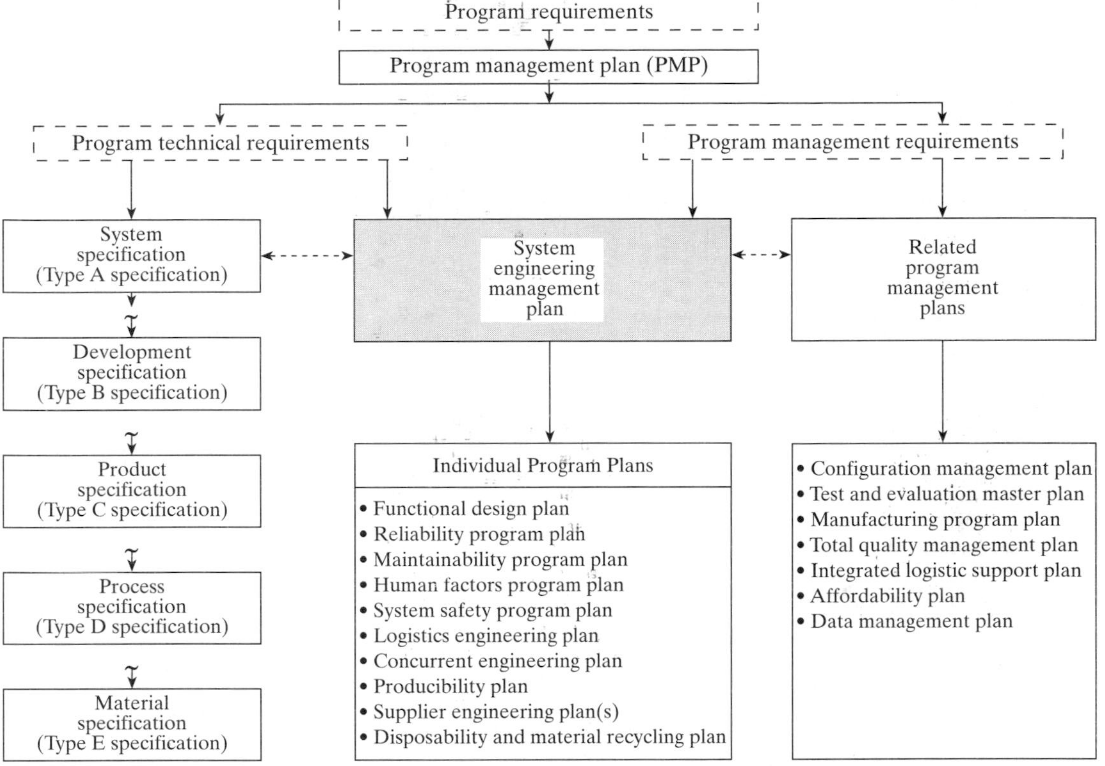

Figure 18.1 Program plans and specifications (refer to Figure 3.2).

Figure 18.2 presents a "generic" approach as to what information might be included in the SEMP. Part I includes a description of program requirements, a statement of work (SOW), a work breakdown structure (WBS), a description of systems engineering tasks, the necessary schedules and cost projections, an organizational structure, and all other management functions necessary in carrying out a successful program in response to systems engineering objectives. Part II describes the systems engineering process, which is covered in Chapters 2 to 6 of this text, or the steps that need to be effectively planned and managed through the activities identified in Part I of the SEMP. Part III covers the integration of the appropriate key engineering disciplines that are required for implementation of the steps described in Part II. Thus, the proposed plan includes a description of the management requirements for a program, the activities to be managed, and the resources required for program implementation.[3]

Figure 18.3 conveys a different format for the material that might be included in a SEMP. This outline is basically organized around the systems engineering process and promotes the combining of many of the activities covered in the three parts in Figure 18.2. One must ensure, however, that all systems engineering tasks are properly

[3] The proposed SEMP outline in Figure 18.2 represents a modification of an earlier outline that was included in the *Systems Engineering Management Guide,* Defense Systems Management College, (Fort Belvoir, Va. 1990) and earlier editions.

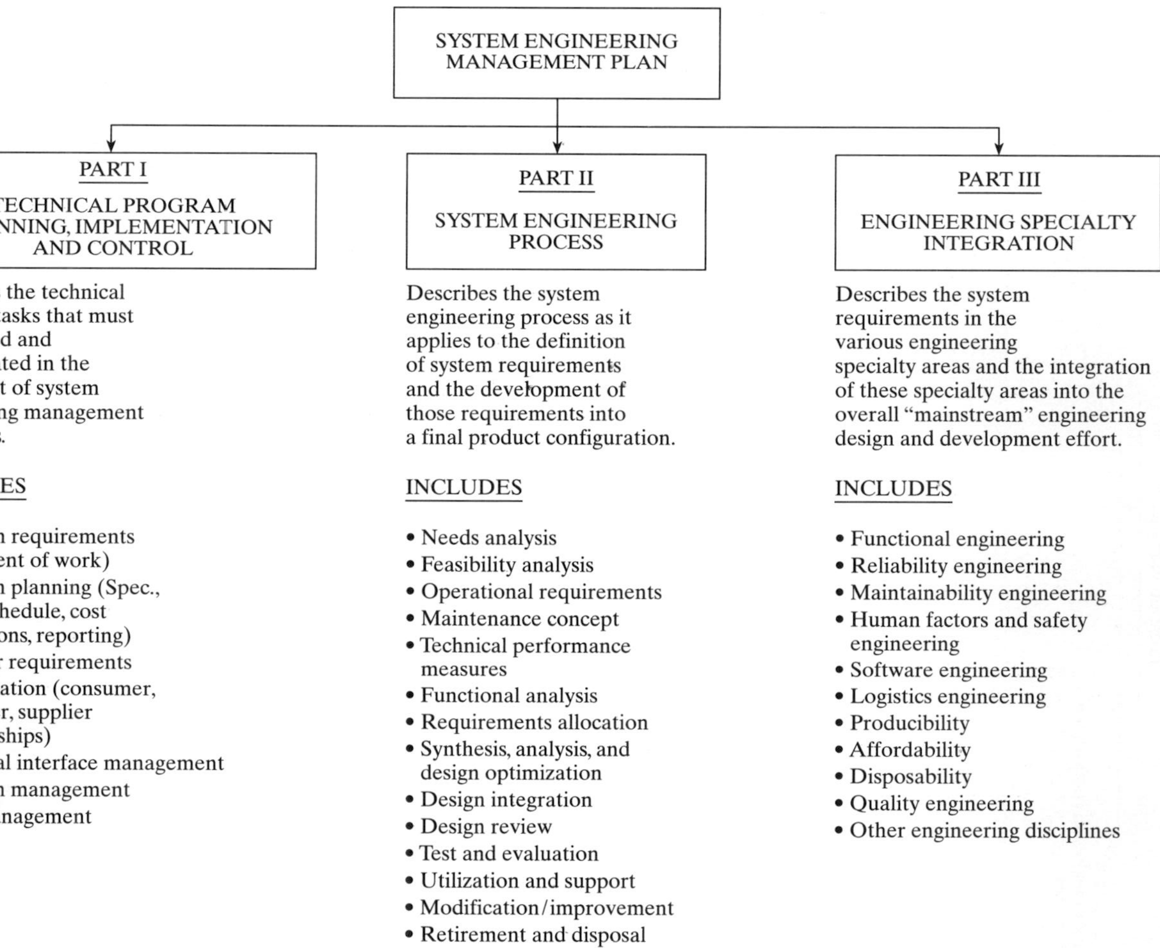

SYSTEM ENGINEERING MANAGEMENT PLAN
PART I
TECHNICAL PROGRAM PLANNING, IMPLEMENTATION AND CONTROL
Describes the technical program tasks that must be planned and implemented in the fulfillment of system engineering management objectives.
INCLUDES
• Program requirements (statement of work)
• Program planning (Spec., WBS, schedule, cost projections, reporting)
• Supplier requirements
• Organization (consumer, producer, supplier relationships)
• Technical interface management
• Program management
• Risk management
PART II
SYSTEM ENGINEERING PROCESS
Describes the system engineering process as it applies to the definition of system requirements and the development of those requirements into a final product configuration.
INCLUDES
• Needs analysis
• Feasibility analysis
• Operational requirements
• Maintenance concept
• Technical performance measures
• Functional analysis
• Requirements allocation
• Synthesis, analysis, and design optimization
• Design integration
• Design review
• Test and evaluation
• Utilization and support
• Modification/improvement
• Retirement and disposal
PART III
ENGINEERING SPECIALTY INTEGRATION
Describes the system requirements in the various engineering specialty areas and the integration of these specialty areas into the overall "mainstream" engineering design and development effort.
INCLUDES
• Functional engineering
• Reliability engineering
• Maintainability engineering
• Human factors and safety engineering
• Software engineering
• Logistics engineering
• Producibility
• Affordability
• Disposability
• Quality engineering
• Other engineering disciplines

Figure 18.2 System engineering management plan—a generic approach.

SYSTEMS ENGINEERING MANAGEMENT PLAN
(Outline, EIA Engineering Standard 632, "Systems Engineering")

1. Title Page, Table of Contents, Applicable Documents
2. Systems Engineering Process
   2.1 Systems Engineering Process Planning—major products and results from process, process inputs, technical objectives, work breakdown structure (WBS), training, standards and procedures, resource allocation, constraints, work authorization, verification planning, subcontractor/supplier technical effort.
   2.2 Requirements Analysis—reliability and maintainability, survivability, electromagnetic compatibility, human engineering, safety, security, producibility, product support (supportability), test and evaluation, integrated diagnostics, transportability, infrastructure support, other areas of functionality.
   2.3 Functional Analysis and Allocation—approach, methods, procedures tools.
   2.4 Synthesis — factor-dependent approaches and methods, use of leverage options (COTS/NDI, open systems architecture, reuse).
   2.5 Systems Analysis And Control—approach, methods, procedures, and tools (trade studies, system cost-effectiveness analysis, risk management, configuration management, interface management, data management, SE master schedule, technical performance measurement, subcontractor/supplier control, requirements traceability).
3. Transitioning Critical Technologies—activities, risks, and criteria for selecting technologies and for the transitioning of these technologies.
4. Integration of Systems Engineering Effort—organization and integration of design disciplines and related activities (concurrent engineering).
5. Implementation Tasks—technology verifications, process proofing, manufacturing engineering test articles, development test and evaluation, generation and reuse of software, sustaining engineering and problem solution support, other systems engineering implementation tasks.
6. Additional Systems Engineering Activities—long-lead items, engineering tools, design-to-cost, value engineering, system integration, other methods and controls.
7. Notes and Appendicies

Figure 18.3 System engineering management plan—a tailored approach.

related to each other, to the applicable elements of the work breakdown structure, to the appropriate schedules and cost projections, and so on.[4]

Statement of Work (SOW)

The SOW is a narrative description of the work required for a given project. Regarding the SEMP, it must be developed from the overall project SOW described in the PMP, and it should include the following:

1. A summary statement of the tasks to be accomplished.
2. An identification of the input requirements from other tasks. These may include the results from other tasks accomplished within the project, tasks completed by the customer, or tasks accomplished by a supplier.

[4] The source of this outline is the interim standard, IS-632, "Systems Engineering," Electronic Industries Association, 2001 Pennsylvania Avenue N.W., Washington, D.C., 1994.

3. References to applicable specifications (to include the System A Specification), standards, procedures, and related documentation as necessary for the completion of the defined scope of work. These references should be identified as key requirements in the documentation tree described in Figure 3.2.
4. A description of the specific results to be achieved. This may include deliverable equipment, software, design data, reports, or related documentation, along with the proposed schedule of delivery.

In preparing a SOW, the following general guidelines are considered to be appropriate.

1. The SOW should be relatively short and to the point (not to exceed two pages), and must be written in a clear and precise manner.
2. Every effort must be made to avoid ambiguity, and the possibility of misinterpretation by the reader.
3. Describe the requirements in sufficient detail to ensure clarity, considering both practical applications and possible legal interpretations. Do not *underspecify* or *overspecify*.
4. Avoid unnecessary repetition and the incorporation of extraneous material and requirements. This could result in unnecessary cost.
5. Do not repeat detailed specifications and requirements that are already covered in the applicable referenced documentation.

The SOW will be read by many different individuals with a variety of backgrounds (e.g., engineers, accountants, contract managers, schedulers, lawyers), and there must be no unanswered questions as to the scope of work desired. It forms a basis for the definition and costing of detailed tasks, for the establishment of subcontractor and supplier requirements, and so on.

Systems Engineering Program Tasks[5]

Systems engineering, as defined throughout this text, covers a broad spectrum of activity. It may even appear that the "systems engineer," or the "systems engineering organization," does everything. Although this is not practical, the fulfillment of systems engineering objectives does require some involvement, either directly or indirectly, in almost every facet of program activity. The challenge is to identify those functions (or tasks) that deal with the overall *system* as an entity and, when successfully completed, will have a positive impact on the many related and subordinate tasks that must be accomplished. Although there are variations from one program to the next, the fol-

[5] Refer to B. S. Blanchard, *System Engineering Management*, 2nd ed. John Wiley & Sons, Inc., (New York: 1998), for a more indepth coverage of program tasks, input-output requirements, and related material.

lowing tasks have been identified where strong *leadership* from the systems engineering organization (or a designated alternate activity) is required:

1. *Perform a needs analysis and conduct feasibility studies* (refer to Sections 3.1 and 3.2). These activities should be the responsibility of the systems engineering organization because they deal with the system as an entity and are fundamental in the initial interpretation and subsequent definition of system requirements. A strong interface and good communications with the customer/user are required.
2. *Define system operational requirements, the maintenance concept, and identify and prioritize the technical performance measures* (refer to Sections 3.4 and 3.5). The results of these activities are included in the overall defininition of system-level requirements and are the basis for top-down system design. Trade-off studies are accomplished, requiring an excellent understanding of system interfaces and the objectives of the customer/user.
3. *Accomplish a functional analysis at the system level and allocate requirements to the next lower level* (refer to Section 3.6). It is essential that a good *functional* baseline be defined, from which specific resource requirements will be identified. This "baseline" serves as a common frame of reference used as an input source for many of the engineering and support activities accomplished later on. Although the systems engineering organization may not accomplish the functional analysis in total, a strong leadership role is essential in providing a necessary foundation at the system level.
4. *Prepare system specification, Type A* (refer to Section 3.8). This represents the top technical document for design, and serves as the basis for the development of all subordinate specifications. The systems engineering organization should be responsible for the preparation of this top-level document, and should ensure that all subordinate specifications "track" and are mutually supportive.
5. *Prepare the test and evaluation master plan* (refer to Section 6.3). Because this document reflects the approach for system *validation*, the systems engineering organization should assume a leadership role to ensure that there is connectivity between the prioritized TPMs, the levels of design complexity, the risks associated with pursuing given design approaches, and so on. As specific quantitative requirements are initially defined, a viable plan must be established for the validation of these requirements. The systems engineering organization needs to ensure an integrated test and evaluation effort and that the system specification, the SEMP, and other design plans "talk" to each other.
6. *Prepare the systems engineering management plan* (SEMP). Because this plan constitutes the top-level "integrating" engineering document that ensures systems engineering objectives are fulfilled, it must be prepared, revised as necessary, and implemented under the leadership of the systems engineering organization (refer to Figures 18.1 and 18.2).
7. *Accomplish synthesis, analysis, and evaluation* (refer to Sections 2.5 and 3.7). Although this area of activity is continuous and widespread across the design

community, the systems engineering organization must provide an "oversight" function to ensure that the major day-to-day design decisions are in compliance with the system specification, that the results of trade-off studies are properly documented, and that the design risks have been properly identified and addressed.

8. *Plan, coordinate, and conduct formal design review meetings* (refer to Sections 3.9, 4.6, and 5.7). The conductance of formal design reviews is necessary to ensure that the systems engineering process is being properly implemented; the appropriate *functional* baselines are being well defined (refer to Figure 2.4); good configuration management is being implemented; and that all members of the design team understand the basis for past decisions and are "tracking" the same design database. The systems engineering organization, or designated representative, should "chair" the formal design review meetings.
9. *Monitor and review system test and evaluation activities* (refer to Sections 6.5 and 6.6). This constitutes a systems engineering organization "oversight" function to review the results of test and evaluation (and thus "validation") to ensure that systems requirements are being met. If not, the problems need to be identified rapidly, followed by implementation of the appropriate corrective action.
10. *Coordinate and review all formal design changes and modifications for improvement* (refer to Section 6.6 and Figure 6.9). The systems engineering organization, or designated representative, should "chair" the Change Control Board (CCB) identified in Figure 6.9 to ensure that (1) system requirements (i.e., all TPMs) will still be met should a change be approved; (2) all proposed changes have been evaluated and selected considering impacts on mission criticality, system effectiveness, life-cycle cost, and environmental factors; (3) a comprehensive modification and rework plan has been developed; and (4) the approved changes are incorporated efficiently and in a timely manner. The application of good configuration management is required here.
11. *Initiate and establish the necessary ongoing liaison activities throughout the production/construction, utilization and support, and retirement and material disposal phases* (refer to Chapters 15 and 16). The systems engineering organization must maintain surveillance of production/construction activities to ensure that the system is being produced/constructed as designed. Further, an ongoing level of surveillance of consumer/user operations in the field is essential in order to provide the feedback necessary for the purposes of system "validation."

These 11 basic program tasks constitute an example of what might be appropriate for a typical large-scale program, although the specific requirements may vary from one instance to the next. The goal is to identify tasks that are oriented to the *system* and are *critical* relative to meeting the requirements described in the earlier chapters. The overall objectives are to ensure that (1) the requirements for the system are initially well defined from the beginning; (2) the appropriate characteristics are "designed-in"; and (3) the system has been validated in terms of the initially specified requirements.

Work Breakdown Structure (WBS)[6]

One of the first steps in the program planning process after the generation of the statement of work (SOW) is the development of a WBS. The WBS is a product-oriented family tree that leads to the identification of the functions, activities, tasks, subtasks, work packages, and so on, that must be performed for the completion of a given program. It displays and defines the system (or product) to be developed, produced, operated and supported, and portrays all of the elements of work to be accomplished. The WBS is *not* an organizational chart in terms of project personnel assignments and responsibilities, but does represent an organization of work packages prepared for the purposes of program planning, budgeting, contracting, and reporting.

Figure 18.4 illustrates an approach to the development of the WBS. During the early stages of system planning, a *Summary Work Breakdown Structure* (SWBS) should be prepared to include all elements of activity through the projected system life cycle, working from the top down. Generally, the SWBS includes three levels of activity[7]:

1. *Level 1.* Identifies the total anticipated scope of work related to the design and development, production, distribution, operation, support, and retirement of a system.
2. *Level 2.* Identifies the various projects, or categories of activity, that must be completed in response to program requirements. It also may include major elements of a system or significant project activities (e.g., subsystems, equipment, software, facilities, data, elements of support, program management, etc.). Program budgets are usually prepared at this level.
3. *Level 3.* Identifies the functions, activities, major tasks, or components of the system that are directly subordinate to the Level 2 items. Program schedules are prepared at this level.

As program planning progresses and individual contract negotiations are consummated, the SWBS may be developed further and adapted to a particular contract or procurement action, resulting in a CWBS. Referring to Figure 18.4, the SWBS can be broken down as shown (i.e., a CWBS for the elements of work in the preliminary system design phase, another CWBS to reflect the elements of work in the detail design and development phase, etc.).

Figure 18.5 presents a sample SWBS reflecting all anticipated levels of activity throughout the life cycle of System XYZ. In many instances, when developing a SWBS

[6] For a more indepth discussion of the development of a WBS, one can refer to almost any text on project management. A good reference is H. Kerzner, *Project Management: A Systems Approach to Planning, Scheduling and Controlling* 5th ed. Van Nostrand Reinhold, (New York: 1995). The purpose here is to illustrate how systems engineering activities may be included.

[7] Although it is more conventional to cover only system *acquisition* activites in the SWBS, the approached presented herein initially addresses all activities because the implementation of systems engineering principles and concepts requires that one consider the entire system life cycle and all of the functions/activities that need to be performed. Thus, the development of a SWBS assumes a broad approach, leading to one or more detailed work breakdown structures for the purposes of contracting (i.e., a *contract work breakdown structure* [CWBS].

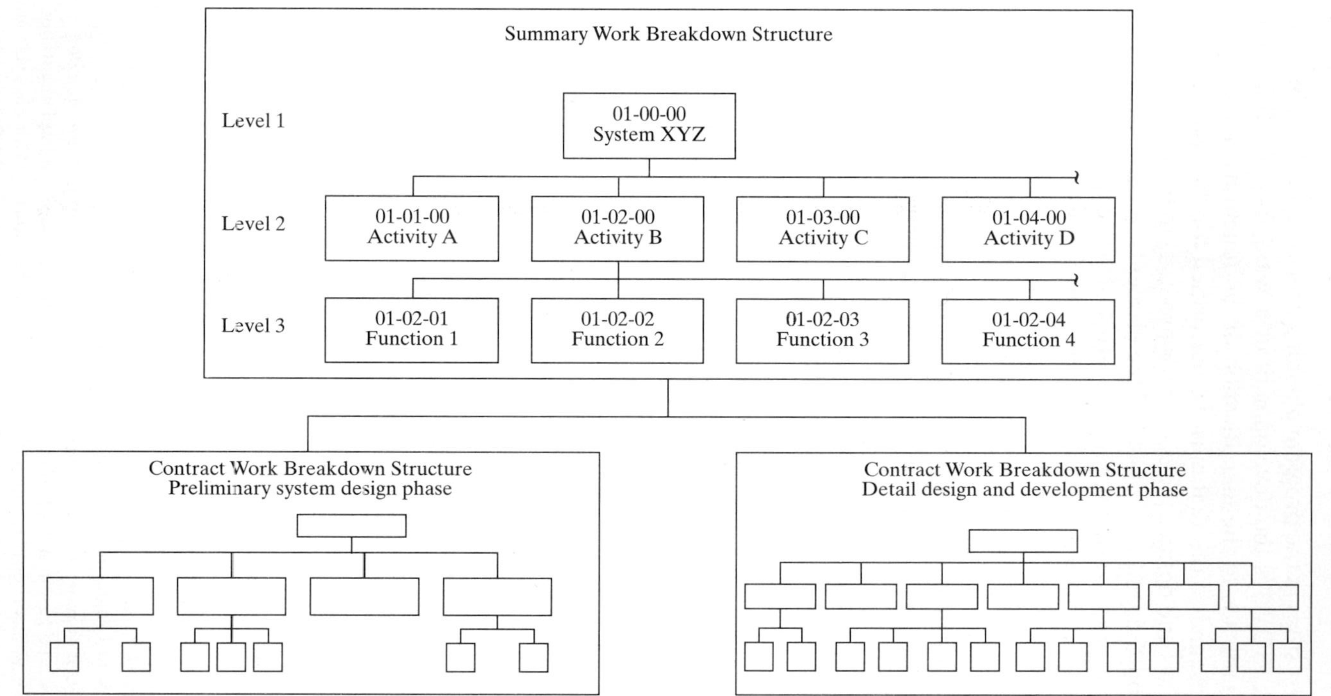

Figure 18.4 Work breakdown structure development (partial).

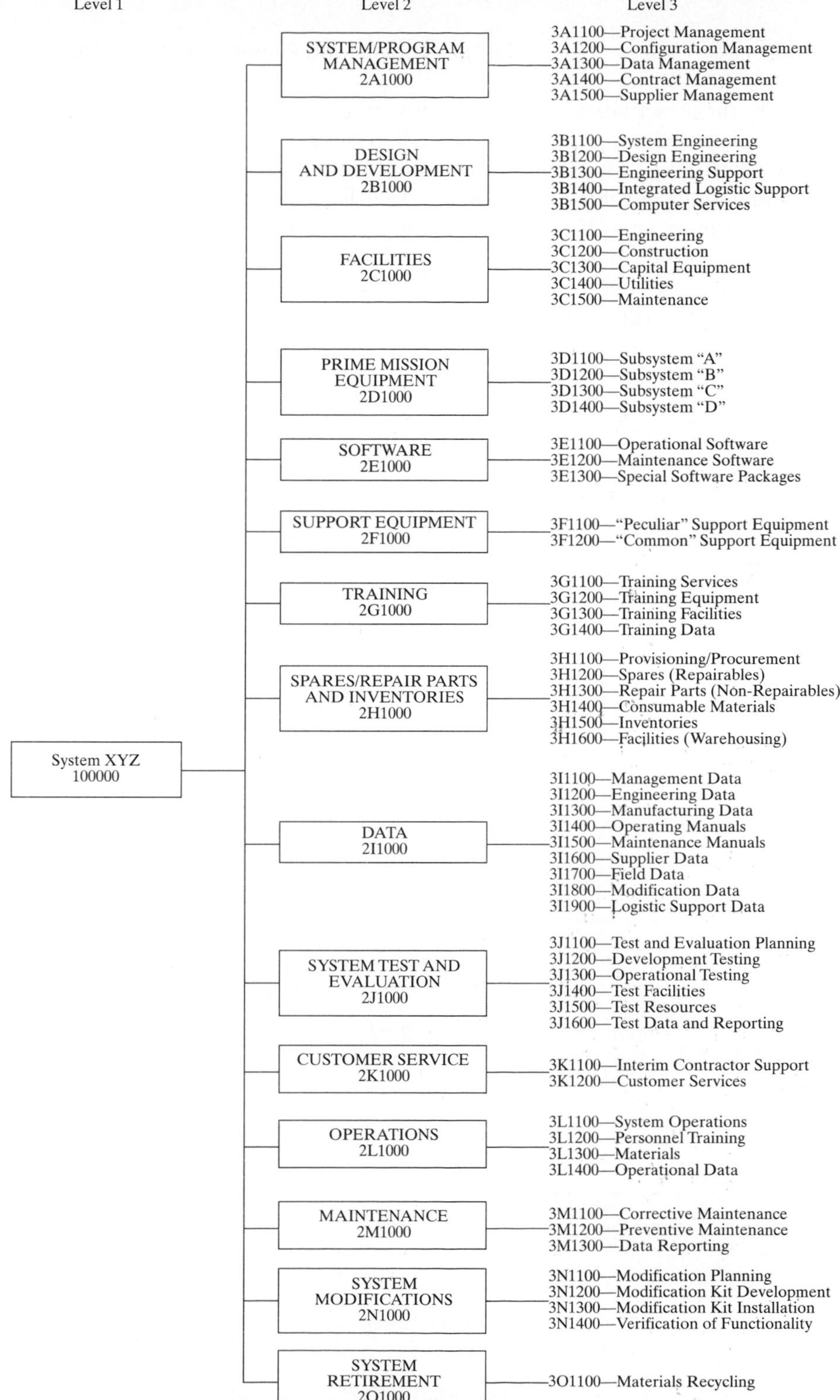

Level 1
Level 2
Level 3
System XYZ
100000
SYSTEM/PROGRAM MANAGEMENT
2A1000
3A1100—Project Management
3A1200—Configuration Management
3A1300—Data Management
3A1400—Contract Management
3A1500—Supplier Management
DESIGN AND DEVELOPMENT
2B1000
3B1100—System Engineering
3B1200—Design Engineering
3B1300—Engineering Support
3B1400—Integrated Logistic Support
3B1500—Computer Services
FACILITIES
2C1000
3C1100—Engineering
3C1200—Construction
3C1300—Capital Equipment
3C1400—Utilities
3C1500—Maintenance
PRIME MISSION EQUIPMENT
2D1000
3D1100—Subsystem "A"
3D1200—Subsystem "B"
3D1300—Subsystem "C"
3D1400—Subsystem "D"
SOFTWARE
2E1000
3E1100—Operational Software
3E1200—Maintenance Software
3E1300—Special Software Packages
SUPPORT EQUIPMENT
2F1000
3F1100—"Peculiar" Support Equipment
3F1200—"Common" Support Equipment
TRAINING
2G1000
3G1100—Training Services
3G1200—Training Equipment
3G1300—Training Facilities
3G1400—Training Data
SPARES/REPAIR PARTS AND INVENTORIES
2H1000
3H1100—Provisioning/Procurement
3H1200—Spares (Repairables)
3H1300—Repair Parts (Non-Repairables)
3H1400—Consumable Materials
3H1500—Inventories
3H1600—Facilities (Warehousing)
DATA
2I1000
3I1100—Management Data
3I1200—Engineering Data
3I1300—Manufacturing Data
3I1400—Operating Manuals
3I1500—Maintenance Manuals
3I1600—Supplier Data
3I1700—Field Data
3I1800—Modification Data
3I1900—Logistic Support Data
SYSTEM TEST AND EVALUATION
2J1000
3J1100—Test and Evaluation Planning
3J1200—Development Testing
3J1300—Operational Testing
3J1400—Test Facilities
3J1500—Test Resources
3J1600—Test Data and Reporting
CUSTOMER SERVICE
2K1000
3K1100—Interim Contractor Support
3K1200—Customer Services
OPERATIONS
2L1000
3L1100—System Operations
3L1200—Personnel Training
3L1300—Materials
3L1400—Operational Data
MAINTENANCE
2M1000
3M1100—Corrective Maintenance
3M1200—Preventive Maintenance
3M1300—Data Reporting
SYSTEM MODIFICATIONS
2N1000
3N1100—Modification Planning
3N1200—Modification Kit Development
3N1300—Modification Kit Installation
3N1400—Verification of Functionality
SYSTEM RETIREMENT
2O1000
3O1100—Materials Recycling

Figure 18.5 Summary work breakdown structure for System XYZ. *Source:* B. S. Blanchard, *System Engineering Management*, 2nd ed. N.Y.: John Wiley & Sons, 1998.

only those activities associated with system design and development are included, which sometimes results in the omission of some critical activities (or elements of the system). Thus, it is preferred that one identify ALL activities first, and then (from this foundation) develop the necessary lower-level structures as necessary.[8]

Referring to Figure 18.5, systems engineering activities are reflected under Category 3B1100, an element of "Design & Development (Category 2B1000)." Although such activities are accomplished throughout all phases of the life cycle, this category serves as a focal point for the establishment of initial budgeting requirements for systems engineering tasks and later for collecting the costs resulting from task completion. As an initial step, the 11 systems engineering program tasks discussed in the previous section can be included within Category 3B1100 of the SWBS, and then can be amplified or adjusted to meet specific program requirements through the development of a CWBS, such as the example presented in Figure 18.6.

In summary, the WBS provides a mechanism for the purposes of initial program bugeting and subsequently for cost collection and reporting progress against individ-

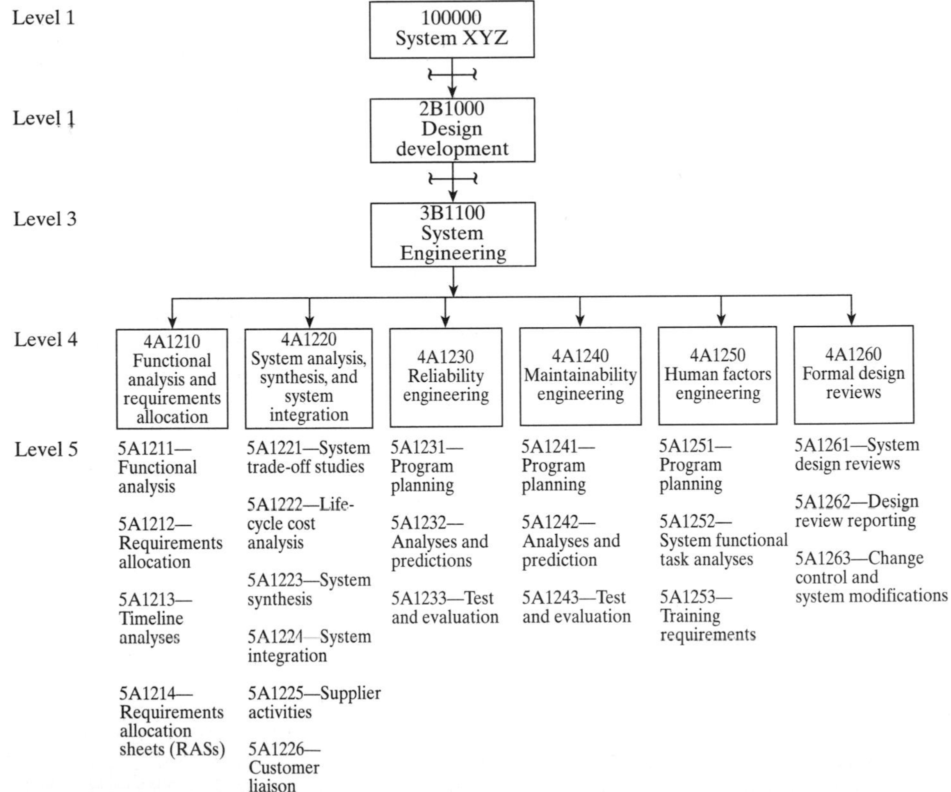

Figure 18.6 CWBS expansion showing systems engineering activities in the preliminary system design phase (partial).

[8] The SWBS, presented in this context, can be related directly to the cost breakdown structure (CBS) discussed in Section 17.3 (refer to Figure 17.10).

ual work packages. It constitutes an excellent project management tool and is used in many programs.

Scheduling of Tasks[9]

Available scheduling methods include the use of *bar charts, milestone charts, program networks, Gantt charts*, or combinations thereof. Figure 18.7 identifies some of the major systems engineering activities, presented in the form of a combined bar/milestone chart.

Although application of the traditional bar or milestone chart is popular, it is often difficult to ensure that the proper interfaces are being addressed as the program manager "tracks" progress against each of the 15 tasks shown. In other words, one may attempt to assess progress for Task 10, "Design Integration," in terms of the scheduled timeline. However, in accomplishing systems engineering activities, there may be numerous inputs from various sources that are required to complete this task in a satisfactory manner. Thus, if this scheduling approach is used, care must be taken to ensure that all of the inputs and outputs are properly defined and can be properly assessed.

A preferred method of scheduling is through the use of program networks such as the program evaluation and review technique (PERT), the critical path method (CPM), or various combinations of these. PERT and CPM are ideally suited for early planning where there are many interfaces (i.e., suppliers providing input from various parts of the world), precise task time data are not readily available, and where the aspects of probability are introduced to aid in defining the risks associated with the many day-to-day design decisions.

In applying the PERT/CPM scheduling approach to a project, one must initially identify all interdependent *activities* and *events* for each applicable phase of the project. Activities refer to continuous levels of effort and events are related to program milestone dates based on management objectives. Managers and programmers work with engineering organizations to define these objectives and identify specific tasks and subtasks. When this is accomplished to the necessary level of detail, networks are developed which start with a summary network covering the key system engineering activities identified in the WBS. An example is illustrated in Figure 18.8. Figure 18.9 lists the activities that are reflected by the lines in the network and Figure 18.10 provides an example of program network calculations.

When constructing networks, one starts with the end objective (e.g., the system is delivered to the customer) and works backward until the beginning or starting event is identified. Each event is labeled, coded, and checked in terms of program timeframe. Activities are then identified and checked to ensure that they are properly sequenced. Activity times are estimated and these times are stated in terms of their probability of occurrence. Some activities can be performed on a concurrent basis, but others must be accomplished in series. For each completed network, there is one beginning event

[9] This chapter presents only a *survey* of key techniques used in system engineering management. Thus, the details associated with different scheduling methods, the development of program/project networks, and so on, are not included herein. PERT/CPM methods are introduced in Section 11.4. Refer to Appendix F for references addressing project management.

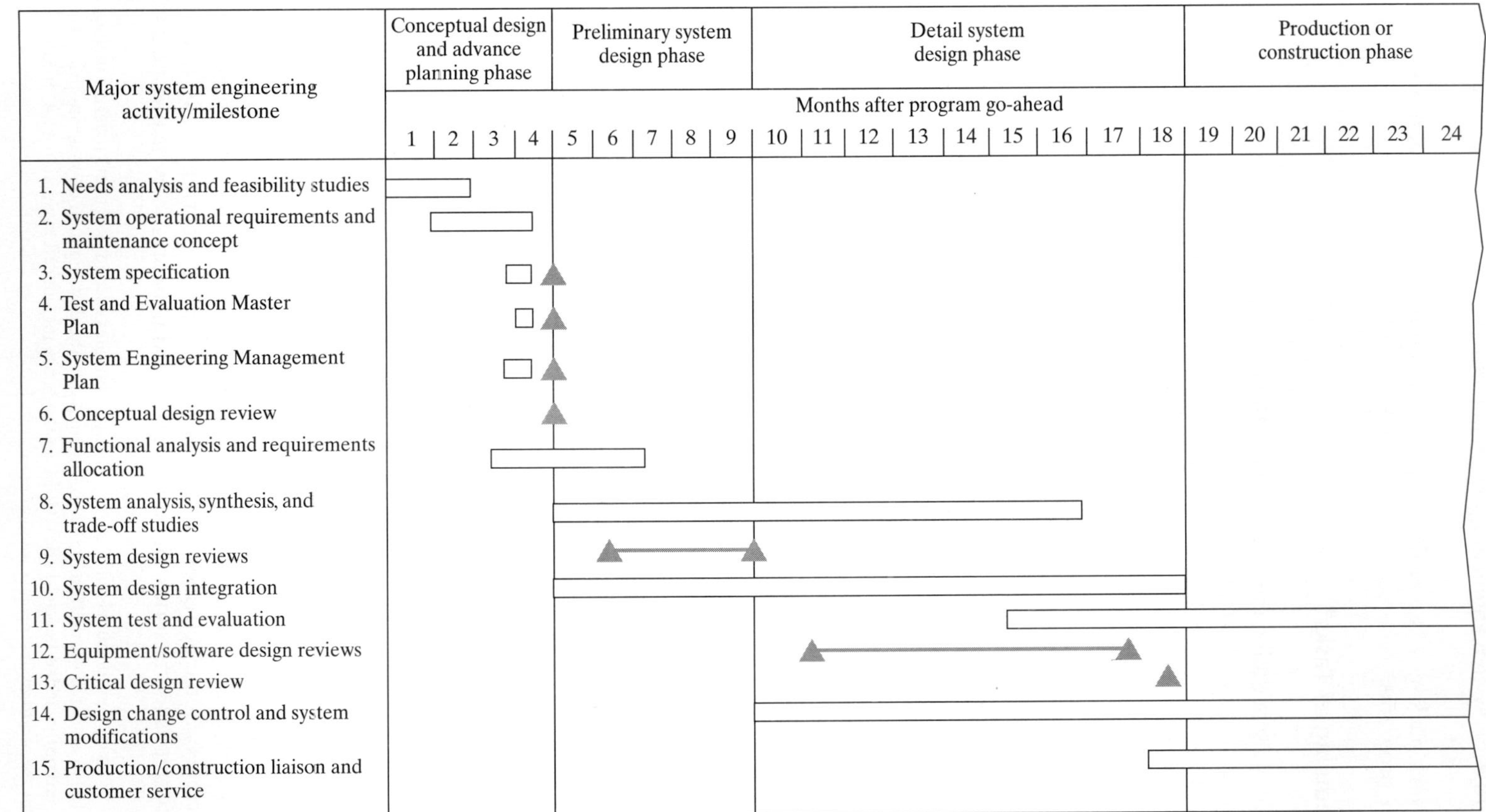

Figure 18.7 Major systems engineering activities and milestones (example).

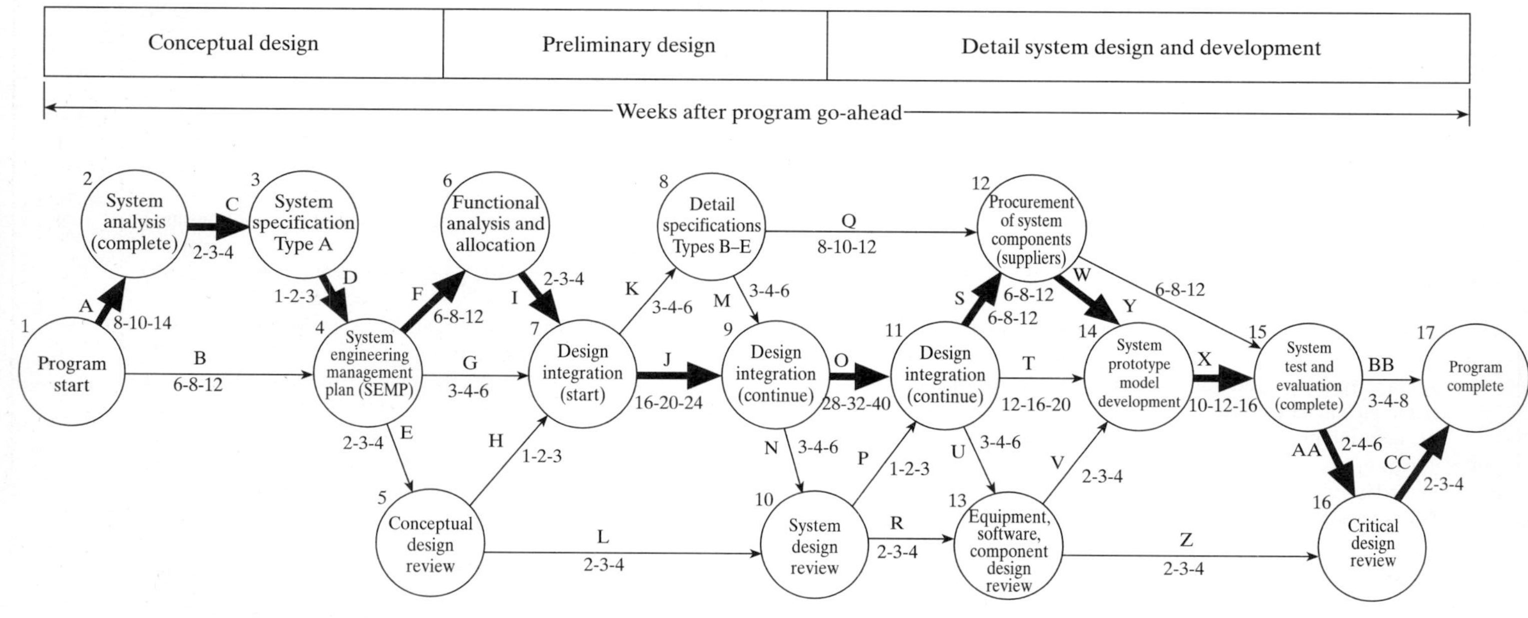

Critical Path: 1-2-3-4-6-7-9-11-12-14-15-16-17

Figure 18.8 Program summary network, *Source:* B. S. Blanchard, *System Engineering Management*, 2nd ed. N.Y.: John Wiley & Sons, 1998.

Activity	Description of Program Activity	Activity	Description of Program Activity
A	Perform needs analysis, conduct feasibility studies, and accomplish systems analysis (operational requirements, maintenance concept, and functional definition of the system).	Q	Identify the appropriate system component suppliers, impose the necessary specification requirements through contracts, and monitor supplier activities.
B	Conduct advance planning, perform initial management functions, and complete the System Engineering Management Plan (SEMP).	R	Conduct the necessary planning and prepare for the Equipment/Software/Component Design Reviews (there may be a series of individual design reviews covering different system components).
C	Prepare the System Specification (Type A).	S	Provide detail design data (as necessary) to support supplier operations.
D	Develop system-level technical requirements for inclusion in the System Engineering Management Plan (SEMP).	T	Develop a prototype model, with associated support, in preparation for system test and evaluation.
E	Prepare system-level design data and supporting materials for the Conceptual Design Review.	U	Prepare design data and supporting materials (as a result of detail design) for the Equipment/Software/Component Design Reviews.
F	Accomplish functional analysis and the allocation of overall system requirements to the subsystem level and below (as required).	V	Translate the results from the Equipment/Software/Component Design Reviews for Incorporation into the prototype model(s) as applicable. The prototype model that is to be utilized in test and evaluation must reflect the latest design configuration.
G	Develop the necessary organizational and related infrastructure in preparation for the accomplishment of the required program design integration tasks.	W	Provide supplier components, with supporting data, for the development of the system prototype to be utilized in test and evaluation activities
H	Translate the results from the Conceptual Design Review to the appropriate design activities (i.e., approved design data, recommendations for improvement/corrective action).	X	Prepare for and conduct System Test and Evaluation (implement the requirements of the Test and Evaluation Master Plan).
I	Translate the results from the functional analysis and allocation activity into specific design criteria required as an input for the design integration process.	Y	Provide test data and logistic support, from the various suppliers, throughout the system test and evaluation phase. Test data are required to cover individual tests conducted at supplier facilities, and logistic support (i.e., spare/repair parts, test equipment, etc.) is necessary to support system testing activities.
J	Accomplish preliminary design and related design integration activities.	Z	Conduct the necessary planning and prepare for the Critical Design Review.
K	Translate the results from system level design into specific requirements at the subsystem level and below. Prepare Development, Process, Product, or Material Specifications as required.	AA	Test results, in the form of either design verification or recommendations for improvement/corrective action, are provided as an input the Critical Design Review.
L	Conduct the necessary planning and prepare for the System Design Review.	BB	Prepare system test and evaluation report.
M	Translate the requirements contained within the various applicable specifications into specific design criteria required as an input for the design integration process.	CC	Translate the results from the Critical Design Review for incorporation into the final system configuration prior to entering the Production or Construction Phase of the Program.
N	Prepare design data and supporting materials (as a result of preliminary design) for the System Design Review.		
O	Accomplish detail design and related design integration activities.		
P	Translate the results from the System Design Review to the appropriate design activities (i.e. approved design data, recommendations for improvement/corrective action).		

Figure 18.9 List of activities in program network.

1	2	3	4	5	6	7	8	9	10	11	12
Event number	Previous number	t_a	t_b	t_c	t_e	s^2	TE	TL	TS	TC	Probability (%)
17	16	2	3	4	3.0	0.111	115.2	115.2	0	110	6.4
	15	3	4	8	4.5	0.694	112.1	115.2	3.1	115	47.9
16	15	2	4	6	4.0	0.444	112.1	112.2	0	120	91.9
	13	2	3	4	3.0	0.111	86.5	112.2	25.7		
15	14	10	12	16	12.3	1.000	108.2	108.2	0		
	12	6	8	12	8.3	1.000	95.9	108.2	12.3		
14	13	2	3	4	3.0	0.111	86.5	95.9	9.4		
	12	6	8	12	8.3	1.000	95.9	95.9	0		
	11	12	16	20	16.0	1.778	95.3	95.9	0.6		
13	11	3	4	6	4.2	0.250	83.5		13.6		
	10	2	3	4	3.0	0.111	53.8		42.1		
12	11	6	8	12	8.3	1.000	87.6	87.6	0		
	8	8	10	12	10.0	0.444	60.8	87.6	26.8		
11	10	1	2	3	2.0	0.111	52.8	79.3	26.5		
	9	28	32	40	32.7	4.000	79.3	79.3	0		
10	9	3	4	6	4.2	0.250	50.8		30.7		
	5	2	3	4	3.0	0.111	21.3		59.0		
9	8	3	4	6	4.2	0.250	35.0	46.6	11.6		
	7	16	20	24	20.0	1.778	46.6	46.6	0		
8	7	3	4	6	4.2	0.250	30.8		15.8		
7	6	2	3	4	3.0	0.111	26.6	26.6	0		
	5	1	2	3	2.0	0.111	20.3	26.6	6.3		
	4	3	4	6	4.2	0.250	19.5	26.6	7.1		
6	4	6	8	12	8.3	1.000	23.6	23.6	0		
5	4	2	3	4	3.0	0.111	18.3		9.3		
4	3	1	2	3	2.0	0.111	15.3	15.3	0		
	1	6	8	12	8.3	1.000	8.3	15.3	7.0		
3	2	2	3	4	3.0	0.111	13.3	13.0	0		
2	1	8	10	14	10.3	1.000	10.3	10.3	0		

Figure 18.10 Example of program network calculations.

and one ending event, with all activities leading to the ending event. Finally, a summary network can be expanded and broken down into lower-level networks (i.e., a network for a reliability engineering program, another network for a maintainability program, etc.).

The application of network scheduling is appropriate for both small- and large-scale projects and is of particular value for a one-of-a-kind system development effort where there are numerous interdependencies, or for those programs where repetitive tasks are not predominant. Networking is readily adaptable to advance planning and forces the precise definition of tasks, task sequences, and task interrelationships. The technique enables management and engineering to predict with some degree of certainty the probable time that it will take to achieve an objective. It also enables the rapid assessment of progress and the detection of problems and delays, and is particularly adaptable to computer methods. The application of PERT/CPM scheduling is particularly appropriate for systems engineering where there are many different and varied activities that must be integrated in a timely manner, and where the early identification of potential areas of risk is critical.

Projecting Costs for Program Tasks

The WBS serves as the basis for the initial identification of functions, activities, tasks, and the grouping of tasks into work packages. Further, the WBS provides the framework against which cost projections are made and budgets are developed for projects. The subsequent identification of the required human and material resources that are needed for task accomplishment is then accomplished and is based on project scheduling requirements.

Referring to Figure 18.8, while the emphasis indicated is on the element of *time*, a cost network can be superimposed upon the PERT/CPM network by estimating the total cost and cost function for each activity line. Figure 18.11 shows a sample activity cost function. Such functions can be generated for the entire network.

When using this method, there is a *time-cost* option that enables management to evaluate alternatives relative to the allocation of resources for activity accomplishment. In many instances, time can be saved by applying more resources, or, conversely, costs may be reduced by extending the time to complete an activity. Time and cost alternatives are evaluated with the objective of selecting the lowest-cost approach within the maximum allowable time limit.

18.3 ORGANIZATION FOR SYSTEMS ENGINEERING

The initial planning for systems engineering begins during the early phases of conceptual design, and evolves through the development of the SEMP described in Section 18.2. To implement this plan successfully requires an organizational structure that will promote, support, and generally enhance the application of systems engineering principles and concepts, on programs. The proper organizational *environment* must be cre-

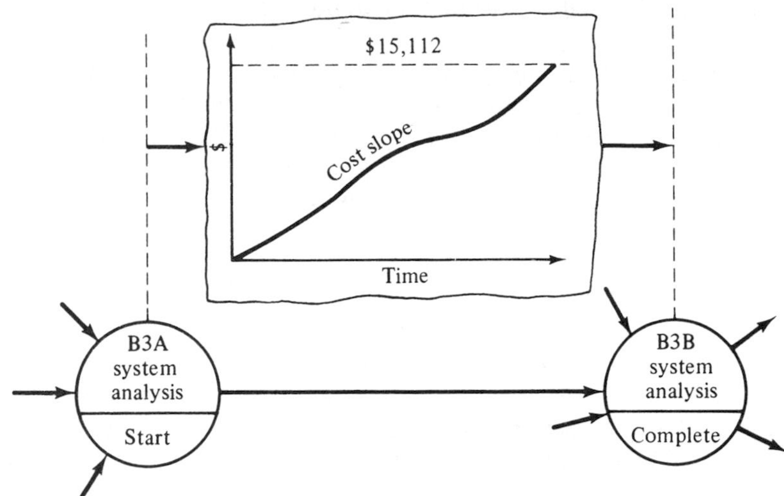

Figure 18.11 Activity-cost function.

ated that will (1) allow for the accomplishment of systems engineering requirements, and (2) will facilitate the implementation of these requirements in an effective and efficient manner.

Organization is the combining of human resources in such a manner as to fulfill a need. Organizations constitute groups of individuals of varying levels of expertise combined into a social structure of some form to accomplish one or more functions. Organizational structures will vary with the functions to be performed, and the results will depend on established goals and objectives, the resources available, the communications and working relationships among the individual participants, the motivation of the personnel, and many other factors. The ultimate objective is to achieve the most effective and efficient use of human, material, and monetary resources through the establishment of decision-making and communications processes designed to accomplish specific objectives.

The fulfillment of systems engineering objectives is highly dependent on the proper mix of resources, the establishment of good communications, and on the development of good interpersonal skills by the participants. The uniqueness of tasks and the many different interfaces that exist requires not only good communication skills, but an understanding of the system as an entity and the many design disciplines that contribute to its development.[10]

[10] The level of discussion of organizational concepts herein is cursory in nature and intended to provide the reader with an overview of some of the key points with respect to systems engineering. A more indepth coverage of organizations, organizational dynamics, and management theory is provided in several of the references in appendix F.

Developing the Organizational Structure

An initial step in the development of any type of an organizational structure is to determine the goals and objectives for the overall company/agency/institution involved, along with the functions and tasks that must be accomplished. Depending on the complexity and size of programs, the structure may assume a pure *functional* model, a *project* orientation, a *matrix* approach, or a combination thereof. Further, the structure may change in context as the system development effort evolves from the conceptual design phase through preliminary system design, detail design and development, production, and so on.

Regarding systems engineering, a prime objective during the early stages of conceptual design is to ensure the proper development of system-level requirements (i.e., the first six tasks described in Section 18.2). These activities are highly *customer/consumer* focused, are directed toward the *system* as an entity, and the accomplishment of such does not require a large organization per se. However, the selection of a few *key* personnel with the appropriate skills, background, and experience levels is essential.

As the program evolves into the preliminary and detail design and development phases, the number of personnel assigned may increase as the design requirements at the subsystem level (and below) may dictate the necessity for including expertise from many different design disciplines (e.g., reliability, maintainability, human factors, logistics, etc.). In this context, the organizational structure may transition from a pure *project* configuration to a mixed *functional-project* or *matrix* approach. As the system and its elements enter the production/construction phase, the organizational structure may shift once again.

In addressing the overall organizational issue, the emphasis herein is intended to stress the accomplishment of the many and varied tasks described in Section 18.2, independent of which organizational element (department, section, or group of personnel) accomplishes the work. Experience has indicated that there are organizational departments or groups located within industrial firms or government agencies that have been designated as "systems engineering" and given the appropriate responsibilities, but who are not actually performing the tasks required. Conversely, there are organizational elements with different identities who are, in actuality, performing the desired functions effectively and efficiently. Further, for small projects, where a single individual must assume different roles, the systems engineering responsibilities may be accomplished by the chief engineer, an electrical engineer, a mechanical engineer, or someone equivalent. On the one hand, the project manager may serve as the "systems engineer," or there may be a designated group of people performing the required tasks.

Consumer, Producer, and Supplier Relationships

To address the subject of "organization for systems engineering" properly, one needs to understand the environment in which systems engineering functions are performed. Although this may vary somewhat depending on the size of the project and the stage of design and development, this discussion is primarily directed to a large project operation, characteristic in the acquisition of many large-scale systems. By addressing large projects, it is anticipated that a better understanding of the role of systems engineering in a somewhat complex envrionment is provided. The reader must, of course, adapt and structure an approach for his/her program.

For a relatively large project, the systems engineering function may appear at several levels as shown in Figure 18.12. The requirements for systems engineering and the responsibility for implementing the 11 tasks described in Section 18.2 lie with the customer (or user) since it is at this level where the *system* is addressed as an entity. The customer may establish a systems engineering organization to accomplish the required tasks, or these tasks may be relegated (in part or in total) to the producer through some form of contractual arrangement. In any event, the responsibility, along with the authority, for accomplishing systems engineering functions must be clearly defined.

In some instances, the customer may assume full responsibility for the overall design and development, production, integration, and installation of the system for operational use. The needs analysis, accomplishment of feasibility studies, definition of operational requirements and the maintenance concept, identification and prioritization of TPMs, preparation of the system specification (type A), preparation of the SEMP, and so on are accomplished within the customer's organization. Top-level functions are defined and specific program requirements (i.e., tasks) are allocated to individual producers, subcontractors, and component suppliers.

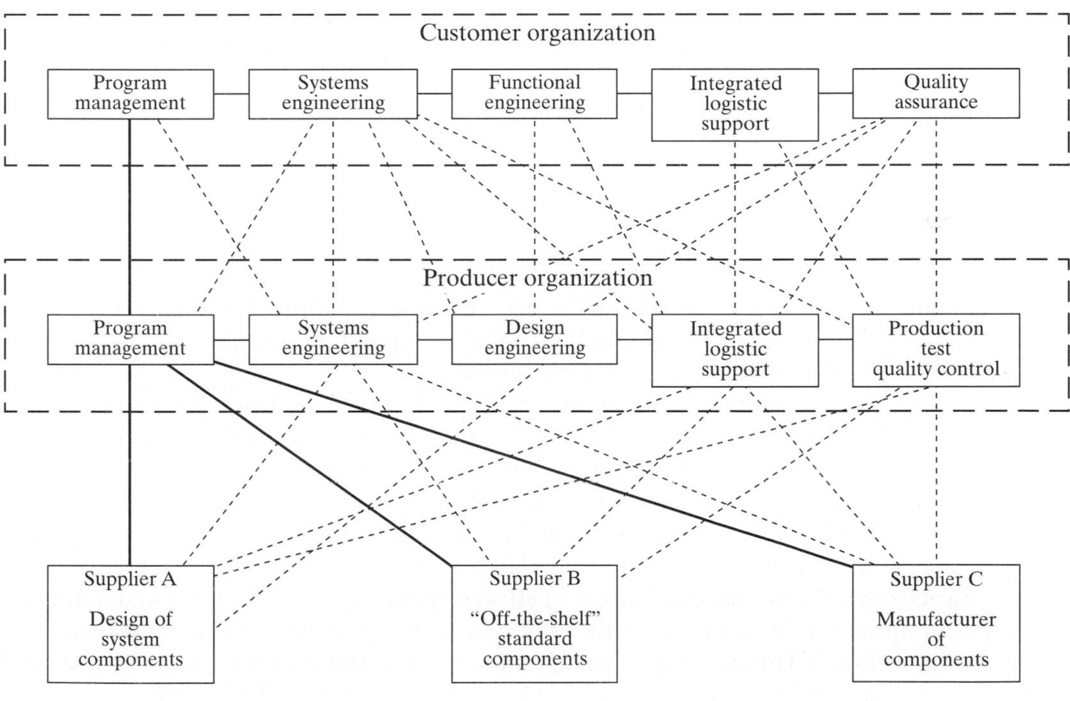

Figure 18.12 Customer, producer, and supplier organizational relationships.

In other cases, although the customer provides the overall guidance in terms of issuing a general statement of work (SOW), or a contractual document of an equivalent nature, the producer (or *prime contractor*) is held responsible for the entire system design and development effort and for completing the 11 tasks described in Section 18.2. Although both the customer and producer have established systems engineering organizations (as shown in Figure 18.12), the basic responsibility for fulfilling the objectives described throughout this text lies with the producer's organization, with supporting tasks being accomplished by individual suppliers as required. To accomplish this, the customer must not only delegate the appropriate level of *responsibility* for completing the functions specified but the necessary *authority* as well. Further, the customer must make accessible all of the necessary input data for the producer to complete successfully the conceptual design tasks noted earlier.[11]

Referring to Figure 18.12, note that there is not only an extensive amount of communications required within each of the customer and producer organizations, but between the the various customer, producer, and supplier organizations as well. Although the solid lines pertain primarily to the more formal program management direction and that of a contractual nature, there are many informal channels of communication that must exist to ensure that the proper dialogue is established between the numerous and varied entities involved in the system development effort. The succesful implementation of a *teaming* or *partnership* approach, along with the fostering of the concurrent engineering principles discussed in Part II, is heavily dependent on good communications (both downward and upward) from the beginning.

Producer Organization and Functions

A primary building block for most organizational patterns is the functional approach, which involves the grouping of functional specialities or disciplines into separately identifiable entities. The intent is to perform similar work within one organizational component. In the pure functional structure, all engineering work is the responsibility of one executive, all manufacturing effort is the responsibility of another executive, and so on. In this case, the same organizational group will accomplish the same type of work for *all* ongoing projects on a concurrent basis.

A partial functional organization, including the identification of major work packages, is illustrated in Figure 18.13. The systems engineering function comes within the overall engineering organization and includes the tasks described in Section 18.2. The organization is responsible for accomplishing these tasks as they apply to all projects. This approach is often desirable for small firms or agencies, since it is easier to manage a homogenous group of similar functions and personnel with comparable backgrounds. Also, the duplication of effort is minimized. On the other hand, for large operations this is not necessarily preferable, owing to impractical centralization of responsibility. Thus, for large multiproduct firms, the pure functional approach is

[11] It is not uncommon for the customer to perform a requirements analysis, prepare a report describing the requirements for a new system, file the report somewhere, and fail to pass on the necessary information later to the responsible producer. Thus, the producer has to generate a new set of requirements, which may, or may not, be consistent with those initially developed by the customer.

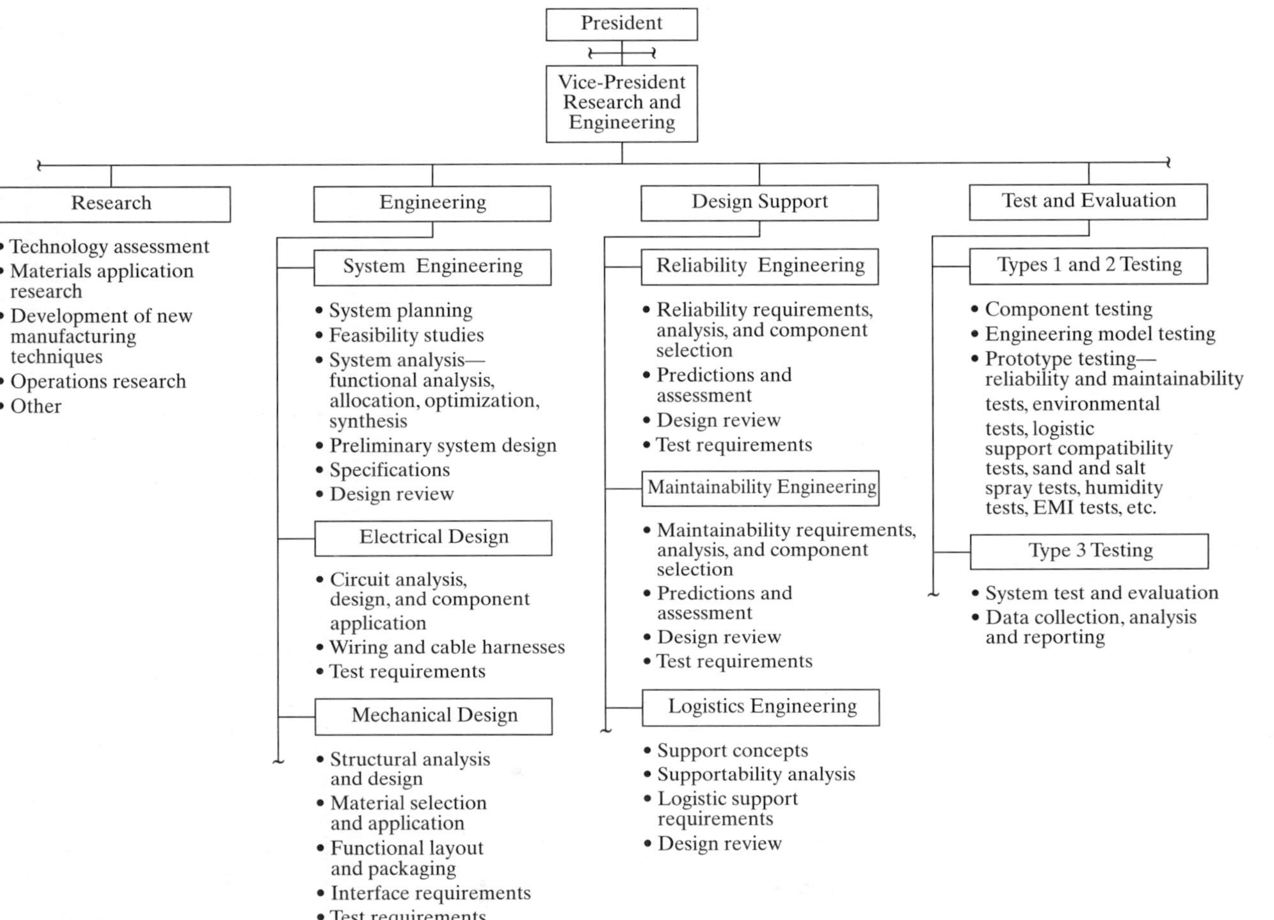

Figure 18.13 Functional organization (partial).

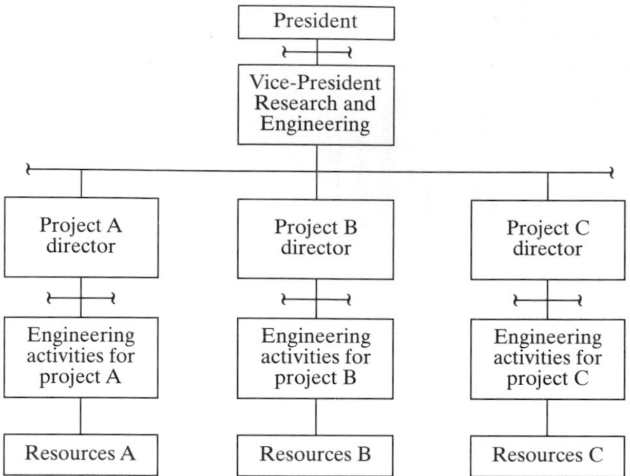

Figure 18.14 Simplified pure project organization.

modified somewhat and organized in terms of individual projects, with some cross-disciplinary activity introduced as appropriate.

Figure 18.14 illustrates a pure project organization that is solely-responsive for the planning, design and development, production, operational use, and support of a unique or single system. It is time limited, and the commitment of the varied skills and resources required is purely for the purpose of accomplishing tasks associated with this particular system. Each project organization will include its own engineering functions, its own testing functions, and its own support functions. Thus, there is a systems engineering group for Project A, another for Project B, and so on. Each group performs the same basic tasks described in Section 18.2. While the pure project organization does provide some benefits in terms of responsiveness to a particular customer, there may be some redundancies within a given company/agency due to the duplication of resources across multiple project-lines.

Figure 18.15 illustrates a matrix configuration, with a mix of the pure functional organization in Figure 18.13 and the project organization in Figure 18.14. For large single-system acquisitions, where the scope of work and funding are adequate, the project configuration may be the preferred approach. However, if there are many inhouse projects where each individually can not justify a separate project organization, the matrix configuration may be preferable. In this instance, it may be appropriate to accomplish some functions through a project structure, whereas others are accomplished by specialized staff groups. In this example, systems engineering tasks are accomplished through a staff function, supporting different projects on an as-required basis. Although the resources are combined within the functional structure in this case, the direct involvement in the day-to-day project activities may be inhibited since the systems engineering activity is not actually a part of the project organization.

Figure 18.16 illustrates a slightly different version of a typical project-staff organizational configuration. In this example, each project includes a systems engineering group while many of the engineering support activities are provided on a task-by-task basis through various staff functions. Because most tasks described in Section 18.2 must be accomplished as an integral part of the system design and development process,

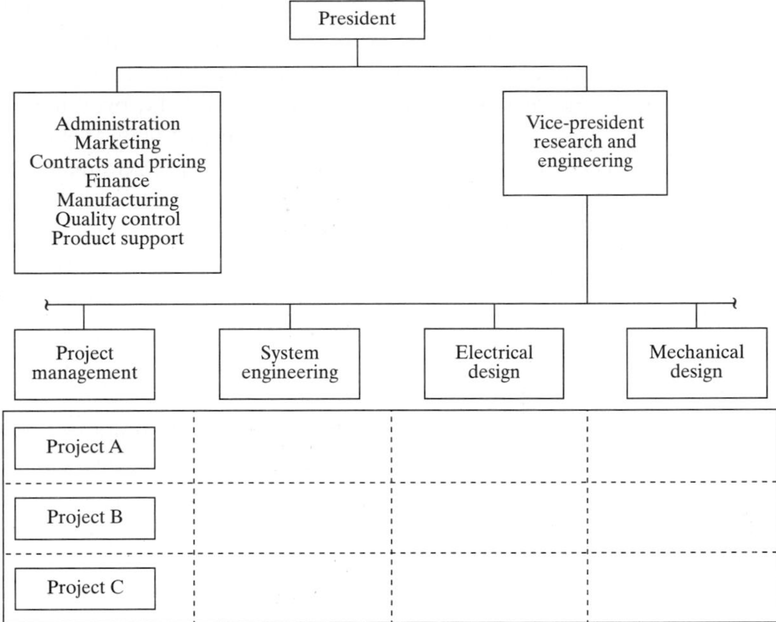

Figure 18.15 Matrix organization.

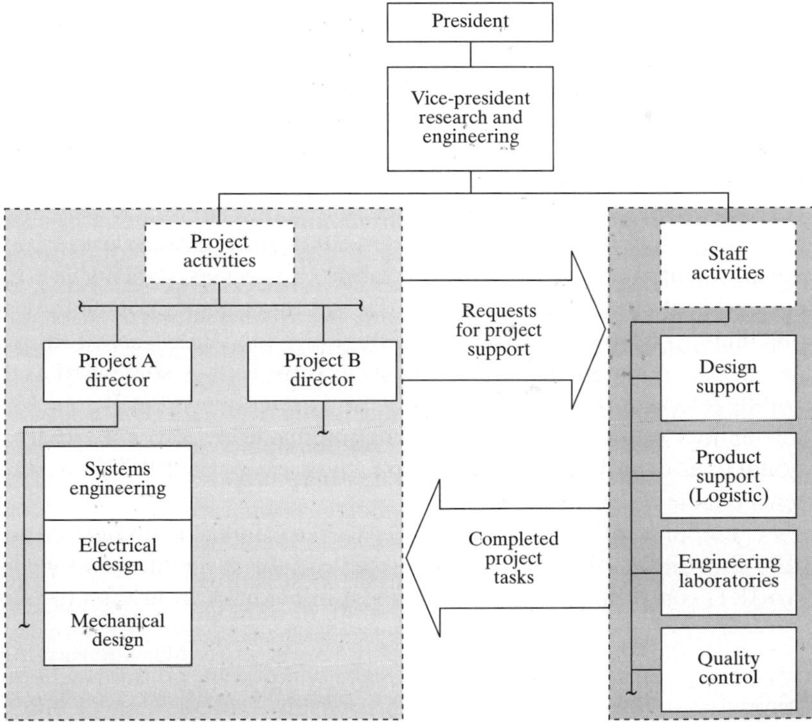

Figure 18.16 Typical project-staff organization.

combined with the fact that the systems engineering activity should assume a leadership role in a "technical" sense, it is preferable that such a function be an inherent part of the project organization. Further, the project type of organization facilitates the communications process between the customer and the producer. The systems engineering activity must provide a leadership role in fostering this communication.

As conveyed earlier, the interfaces between the systems engineering group and other segments of a producer's organization are numerous, particularly when involved in a large system acquisition. Referring to Figure 18.17, the systems engineering group for Project Y is challenged by the fact that there are many other activities with which to deal on an almost day-to-day basis. There is a need to establish a good communications link with the many internal project activities noted; i.e., design engineering, software engineering, the business unit, other project organizations, the logistic support function, production operations, and so on. An example of the required communication network is illustrated in Figure 18.18 (which explains the dotted lines in Figure 18.17).

The ultimate objective in the design and development of any system is to establish a *team* approach, with the appropriate communications, enabling the application of concurrent engineering methods throughout. However, there are often problems when dealing with large programs involving many different functional organizational units. Barriers are developed which tend to inhibit the necessary day-to-day close working relationships, the timely transfer of essential information, and the communications discussed earlier. As there are many companies organized strictly along functional lines (versus the project-staff configuration shown in Figure 18.17), there is a need to ensure that the proper level of communications is maintained between all applicable organizational units, regardless of where they are permanently located within the overall company structure.

With this objective in mind, the Department of Defense (DOD) initiated the concept of *integrated product and process development* (IPPD) in the early 1990s. IPPD can be defined as a "management technique that simultaneously integrates all essential acquisition activities through the use of multidisciplinary teams to optimize the design, manufacturing, and supportability processes."[12] This concept promotes the communications and integration of the key functional areas, as they apply to the various phases of program activity. Although the specific nature of the activities involved and the degree of emphasis exerted will change change somewhat as the system design and development effort evolves, the structure conveyed in Figure 18.19 is maintained throughout to foster the necessary communications across the more traditional functional lines of authority. In this regard, the concept of IPPD is directly in line with systems engineering objectives.

Inherent within the IPPD concept is the establishment of *integrated product teams* (IPTs), with the objective of addressing certain designated and well-defined issues.[13] An IPT, constituting a selected team of individuals from the appropriate disciplines,

[12] DOD Regulation 5000.2, "Mandatory Procedures for Major Defense Acquisition Programs (MDAPs) and Major Automated Information System (MAIS) Acquisition Programs" (Washington, D.C.: Department of Defense, 1996).

[13] The term IPT is also used as a designator for "integrated process team."

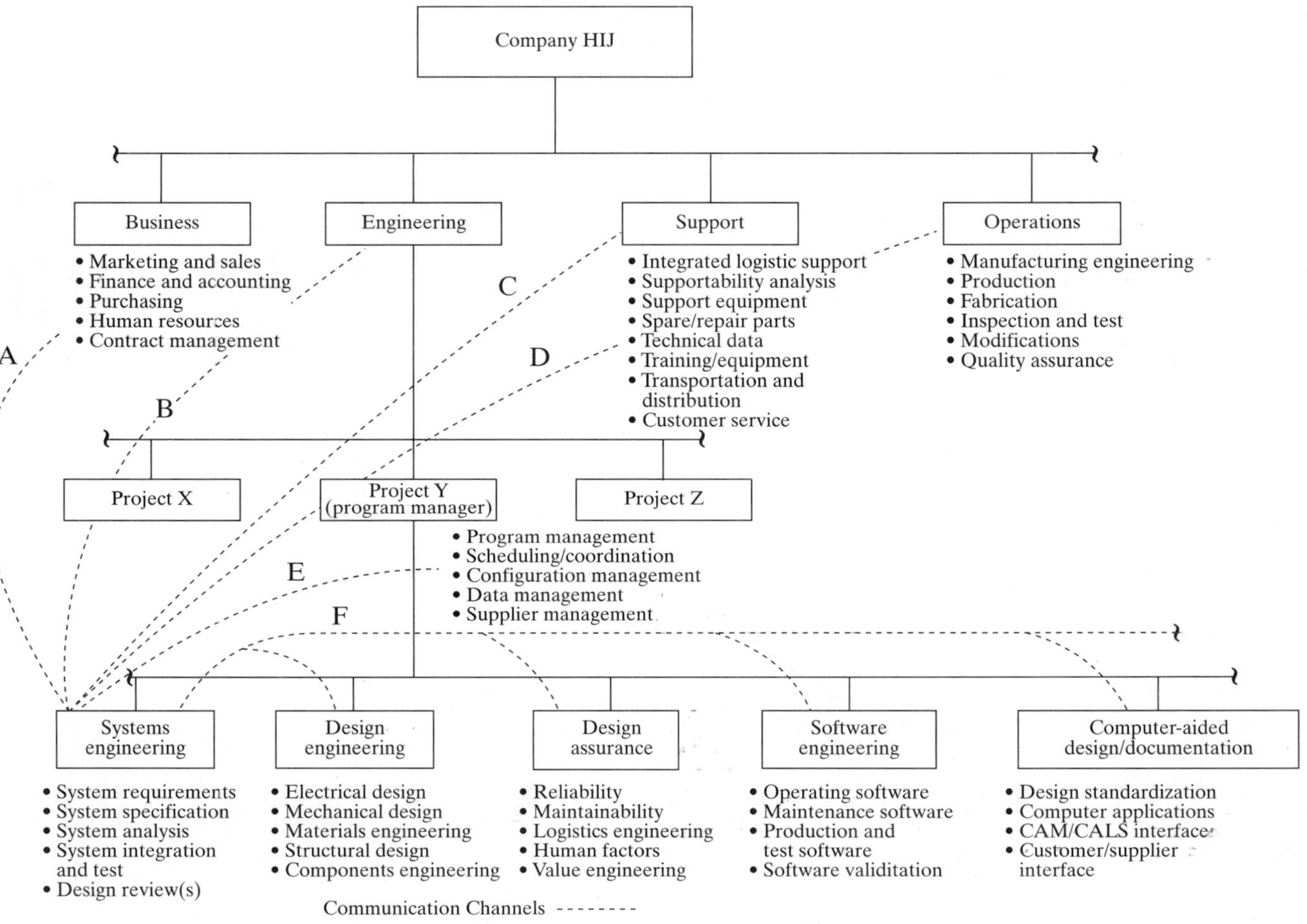

Figure 18.17 Major systems engineering communication links (producer organization).

Communication Channel (Figure 18.17)	Supporting Organization (Interface Requirements)
A	1. *Marketing and sales*—to acquire and sustain the necessary communications with the customer. Supplemental information pertaining to customer requirements, system operational and maintenance support requirements, changes in requirements, outside competition, etc., is needed. This is above and beyond the formal "contractual" channel of communications. 2. *Accounting*—to acquire both budgetary and cost data in support of economic analysis efforts (e.g., life-cycle cost analysis). 3. *Purchasing*—to assist in the identification, evaluation, and selection of component suppliers with regard to technical, quality, and life-cycle cost implications. 4. *Human resources*—to solicit assistance in the initial recruiting and hiring of qualified project personnel for system engineering, and in the subsequent training and maintenance of personnel skills. To conduct training programs for all project personnel across the board relative to system engineering concepts, objectives, and the implementation of program requirements. 5. *Contract management*—to keep abreast of contract requirements (of a technical nature) between the customer and the contractor. To ensure that the appropriate relationship are established and maintained with suppliers as they pertain to meeting the *technical* needs for system design and development.
B	To establish and maintain on-going liaison and close communications with other projects with the objective of transferring knowledge that can be applied for the benefit of Project Y. To solicit assistance from other company-wide functionally-oriented engineering laboratories and departments relative to the application of new technologies in support of system design and development.
C	To provide an input relative to project requirements for system support, and to solicit assistance in terms of the functional aspects associated with the design, development, test and evaluation, production, and sustaining maintenance of a support capability through the planned system lifecycle.
D	To provide an input relative to project requirements for production (i.e., manufacturing, fabrication, assembly, inspection and test, and quality assurance), and to solicit assistance relative to the design for producibility and the implementation of quality engineering requirements in support of system design and development.
E	To establish and maintain close relationships and the necessary on-going communications with such project activities as scheduling (the monitoring of critical program activities through a network scheduling approach); configuration management (the definition of various configuration baselines and the monitoring and control of changes/modifications); data management (the monitoring, review, and evaluation of various data packages to ensure compatibility and the elimination of unnecessary redundancies); and supplier management (to monitor progress and ensure the appropriate integration of supplier activities).
F	To provide an input relative to *system-level* design requirements, and to monitor, review, evaluate, and ensure the appropriate integration of system design activities. This includes providing a *technical* lead in the definition of system requirements, the accomplishment of functional analysis, the conductance of system-level trade-off studies, and the other project tasks.

Figure 18.18 Description of major project interface requirements.

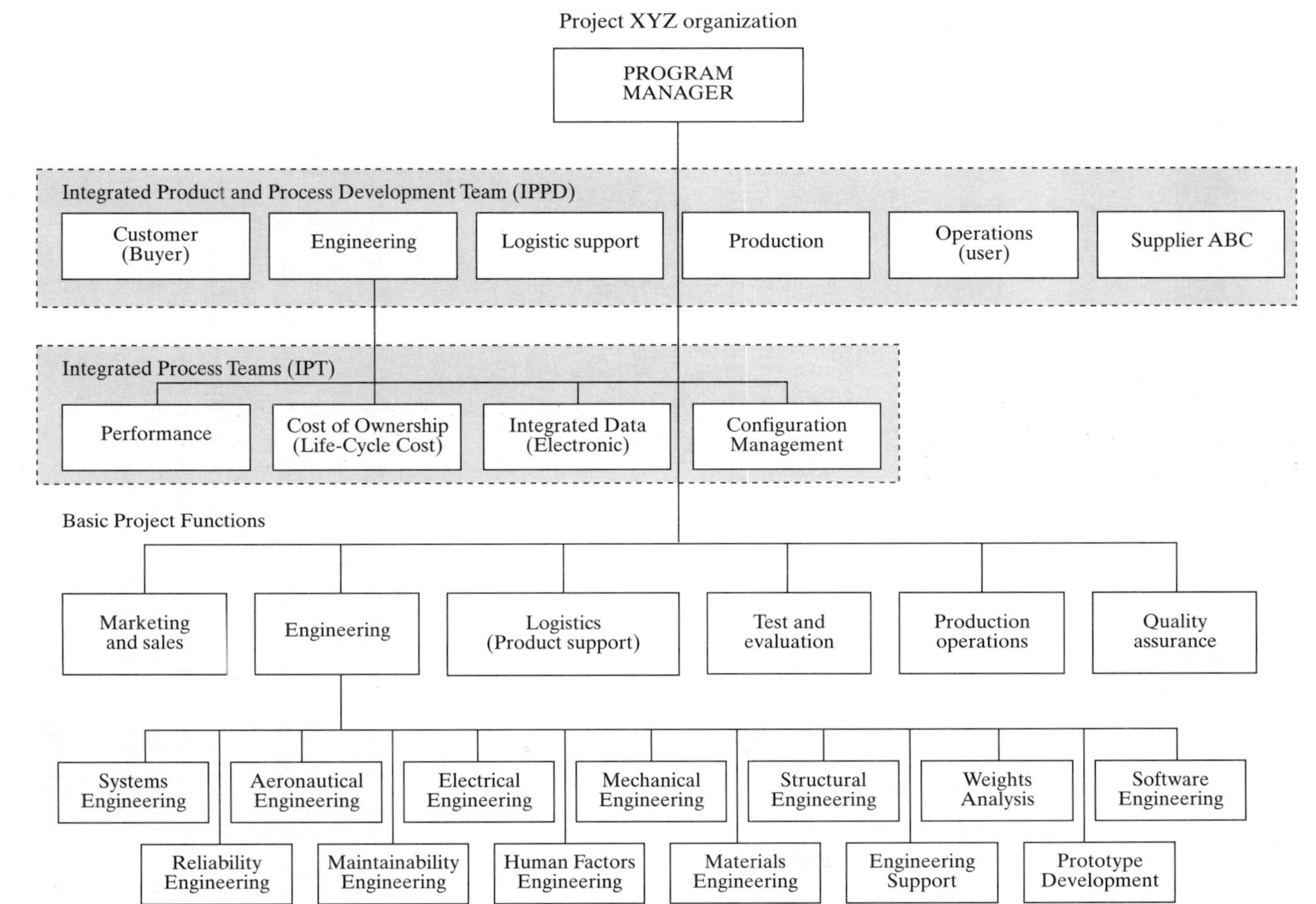

Figure 18.19 Functional organization with IPPD/IPT activities.

may be established to investigate a specific segment of design, a solution for some outstanding problem, design activities that have a large impact on a high-priority TPM, and so on. The objective is to create a *team* of qualified individuals that can effectively work together to solve some problem in response to a given requirement. Further, there may be several different teams established to address issues at different levels in the overall system hierarchical structure (i.e., issues at the system, subsystem, or component level). Referring to Figure 18.19, an IPT may be established to concentrate on those activities that significantly impact selected *performance* factors, *cost-of-ownership, integrated data*, and *configuration management*. There may be another IPT assigned to "track" the *integrated data* environment issue. The objective is to provide the necessary emphasis in critical areas and to reap the benefits of a *team* approach in arriving the best solution possible.

IPTs are often established by the program manager, or by some designated high-level authority in the organization. The representative team members must be well qualified in their respective areas of expertise, must be empowered to make on-the-spot decisions when necessary, must be proactive relative to team participation, success oriented, and be resolved to addressing the problem assigned. The program manager must clearly define the objectives for the team, expectations in terms of results, and the team members must maintain a continuous "up-the-line" communications channel. The longevity of the IPT will depend on the nature of the problem and the effectiveness of the team in progressing toward meeting its objective. Care must be taken to avoid the establishment of too many teams, as the communication processes and interfaces become too complex when there are many teams in place. Additionally, there often are conflicts when it comes to issues of importance and which issue is "traded-off' as a result. Further, as the team ceases to be effective in accomplishing its objectives, it should be disbanded. An established team that has "outlived its usefulness" can be counterproductive.

18.4 SUMMARY

This chapter presented an overview of some of the considerations that need to be addressed early in the life cycle when the initial planning for system acquisition commences. Systems engineering planning is accomplished throughout the conceptual design phase and evolves through the development of SEMP. Given the basic plan, it is then necessary to establish an organizational approach that will enable the proper implementation of this plan. These first two steps of planning and organization are highlighted in this chapter, and lead directly into Chapter 19, which covers the sustaining management and control functions associated with the program as the system evolves through the design and development, production/construction, and subsequent phases.

QUESTIONS AND PROBLEMS

1. When in the life cycle should systems engineering planning be initiated? Why is it important?
2. What is the purpose of SEMP? When in the life cycle should it be developed? How does it relate to each of the following: (a) PMP; (b) reliability program plan; (c) TEMP; (d) ILSP; (e) TQM plan; and (f) configuration management plan?

3. How does the SEMP relate to the system specification (type A)?
4. Select a system of your choice and develop a detailed outline of a SEMP for your system.
5. Identify and describe the tasks that should be included in the implementation of a systems engineering program.
6. What is the purpose of a WBS? What is the difference between a SWBS and a CWBS? What is the difference between a CWBS and a CBS? How does the relate to the functional block diagrams in Chapters 3 and 4?
7. What should be included in a SOW? What cautions should be exercised in preparing a SOW?
8. If you are assigned to manage a systems engineering program, what scheduling method would you select to gain as much visibility as possible in determine your status? Why did you select this method?
9. Assume that you have been assigned to develop a SEMP for a system of your choice. Identify and describe the functions/tasks that need to be accomplished. Then construct a network (PERT/CPM) for the program. This can be a real or hypothetical system.
10. For the network developed in Problem 9, develop a cost structure, relating the costs to each scheduled activity.
11. Identify some of the key objectives in organizing for systems engineering.
12. With systems engineering in mind, describe some of the advantages and disadvantages of a pure *functional* organization structure; a pure *project* organization structure; and a *matrix* organization structure.
13. Assume that you have just been assigned to develop a new systems engineering capability in your company. Construct a hypothetical organizational chart for the company, identify the organizational elements responsible for the accomplishment of systems engineering tasks, and describe the interface relationships with other critical organizational groups.
14. If you were charged with the staffing of a systems engineering organization, what type of an individual would you hire? Describe background and experience expectations, personal characteristics, specific desired skills.
15. As a systems engineering manager, what type of an organizational *environment* would you try to create to accomplish your objectives effectively?
16. What is the objective of an IPPD concept? What is an IPT, and what are its objectives?
17. Refer to Figure 18.17. Assume that you have been appointed as the manager of the systems engineering organization and that the communication processes represented by the dotted lines were not functioning properly. What steps would you take to strengthen these communication links?

19 PROGRAM MANAGEMENT AND CONTROL

Given a comprehensive and in-place organizational structure, the challenge is that of *implementation.* Experience has indicated that the past is replete with examples where excellent planning was accomplished in the beginning, only to find that the subsequent implementation process did not follow the plan and was deficient as a result. Thus, for the successful accomplishment of systems engineering objectives, a two-step management process is necessary: (1) the planning and organization for systems engineering described in Chapter 18; and (2) the subsequent program implementation and control activities described in this chapter.

An initial step in the implementation of any program is to define the specific goals and objectives of the responsible organization(s), and to describe the proposed methods by which these goals and objectives will be met. This information, which evolves from the early planning activity covered in Chapter 18, is then used as a basis for establishing the necessary day-to-day program direction and control tasks. As the program progresses through time, there must be a built-in mechanism by which *evaluation* is accomplished. The results of such aid in identifying the *risks* of possibly not meeting certain requirements. Finally, there must be a provision for feedback and corrective action as required; i.e., a management information system (MIS), which provides the visibility essential for good program management.

19.1 ORGANIZATIONAL GOALS AND OBJECTIVES

The purpose of organization, presented in this context, is to fulfill the requirements pertaining to the design and development, production, and delivery of a system, which is completely responsive to the needs of the customer, in an effective, efficient, and timely manner; i.e., the activities described in the earlier parts of this text. The goals of the organization are directed toward this end.

In addressing the subject of *goals*, one needs to deal with two levels of activity: (1) the goals specified by the TPMs, discussed in Section 3.5, which relate to the characteristics that must be incorporated into the design; and (2) the goals of the organization relative to accomplishing the necessary activities to ensure that the first objective is attained. A basic question is: How effective and efficient is the organization functioning in the accomplishment of this first goal? In this instance, the question can be addressed to the systems engineering organization relative to (1) its impact on the ultimate system/product design configuration itself, and (2) its effectiveness and efficiency in performing the 11 tasks described in Section 18.2.[1]

Regarding the organizational issue, which is the point of emphasis in this chapter, the objective is to develop an overall systems engineering *capability* that will enable the accomplishment of the 11 program tasks (or functions of an equivalent nature) in an effective manner. This, in turn, requires (1) the initial establishment of a desired set of "metrics" against which each of the applicable tasks can be assessed; (2) the development of the necessary processes for the performance of the tasks; and (3) the implementation of a data collection and information capability that will enable management to "track" performance and determine just how well the organization is functioning overall. The overall organizational objective is to establish a disciplined, well-defined approach in the performance of its tasks, establish measurable goals that can be quantitatively expressed and controlled, and initiate a provision for continuous process improvement.

Having identified the systems engineering tasks to be accomplished, along with the metrics associated with each, the manager needs to assess current status of the organization. At the same time, it may be appropriate to compare the results with other comparable entities. In other words, a *benchmarking* approach can be applied to aid in the development of future goals. Benchmarking can be defined as "the ongoing activity of comparing one's own process, product, or service against the best known similar activity, so that challenging by attainable goals can be set and a realistic course of action implemented to efficiently become and remain the best of the best in a reasonable time."[2] The ultimate questions are: Where are we today? How do we compare with others relative to both product and organization (i.e., the competition)? Where would we like to be in the future?

In summary, the organizational goals and objectives must first be responsive to the established TPMs for the system being developed (refer to Section 3.5). Given this, additional goals can be established for the systems engineering organization itself. Through the use of benchmarking, future goals can be identified and an organizational

[1] It is not uncommon to address only the second goal, believing that a given systems engineering organization is doing a good job in accomplishing the 11 identified tasks when, at the same time, the organization (and its personnel) has not impacted the design at all. Thus, when assessing the organization, the second goal must be evaluated in terms of the first.

[2] G. J. Balm, *Benchmarking: A Practitioner's Guide for Becoming and Staying the Best of the Best* (Schaumburg, Il. QPMA Press, 1992). Another good reference is R. C. Camp, *Benchmarking: The Search for Industry Best Practices That Lead to Superior Performance*: (Wis: Milwaukee, ASQC Quality Press, 1989).

"growth" plan can be developed, which, when implemented, can aid in leading to the desired objectives.

19.2 DIRECTION AND CONTROL OF PROGRAM ACTIVITIES

Successful program implementation includes the day-to-day managerial functions associated with the influencing, guiding, and supervising of subordinates in the accomplishment of the organizational activities identified in the SEMP. This requires a continuing awareness, on the part of the manager, of the "specifics" pertaining to the tasks being performed and whether the results are fulfilling the program objectives in an effective manner.

The "controlling" aspect pertains to (1) the ongoing monitoring of program activities to assess the status of the tasks being performed at designated times in the program, and (2) the initiation of corrective action as required to overcome any noted deficiencies in the event that program objectives are not being met. The key here is to develop an appropriate management information system (MIS) that will provide the right information at the right time, and to initiate a feedback and corrective-action capability that is responsive in terms of elapsed time.

Program Review and Reporting Requirements

Periodic program management reviews are usually conducted throughout the system design and development process. The nature and frequency of these will vary with each program and will be a function of design complexity, the number and location of organizations participating in the program, the number of suppliers, the number of problems being encountered, and so on. Systems being developed, incorporating new technologies and where the risks are greater, will require more frequent reviews than for other programs.

The objective of these management-oriented reviews is to determine the status of selected program tasks, assess the degree of progress with respect to fulfilling contractual requirements, review program schedule and cost data, review supplier activities, and coordinate the results of the technical design reviews (refer to Sections 3.9, 4.6, and 5.7). These reviews are directed toward dealing with issues of a "program" nature, and are not intended to duplicate the formal design review process which deals with design-related problems of a "technical" nature. However, of mutual interest is the "tracking" of the critical TPMs, as illustrated in Figure 5.12.

The data system (i.e., the MIS) that includes the information needed for these program reviews must be designed in such a way as to provide the right type of information, to selected locations and in the proper format, in a timely manner and at the desired frequency, with the right degree of reliability, and at the right cost. The nature and the content of the data will be based on a number of factors, both of a technical nature (e.g., TPMs) and of a program nature (e.g., high-volume suppliers). It is not uncommon for one to go ahead and design and develop, or procure, a large computer-based MIS capability with good intent, but that turns out to be too complex, nonresponsive by providing the wrong type of information, and very costly to operate. Such

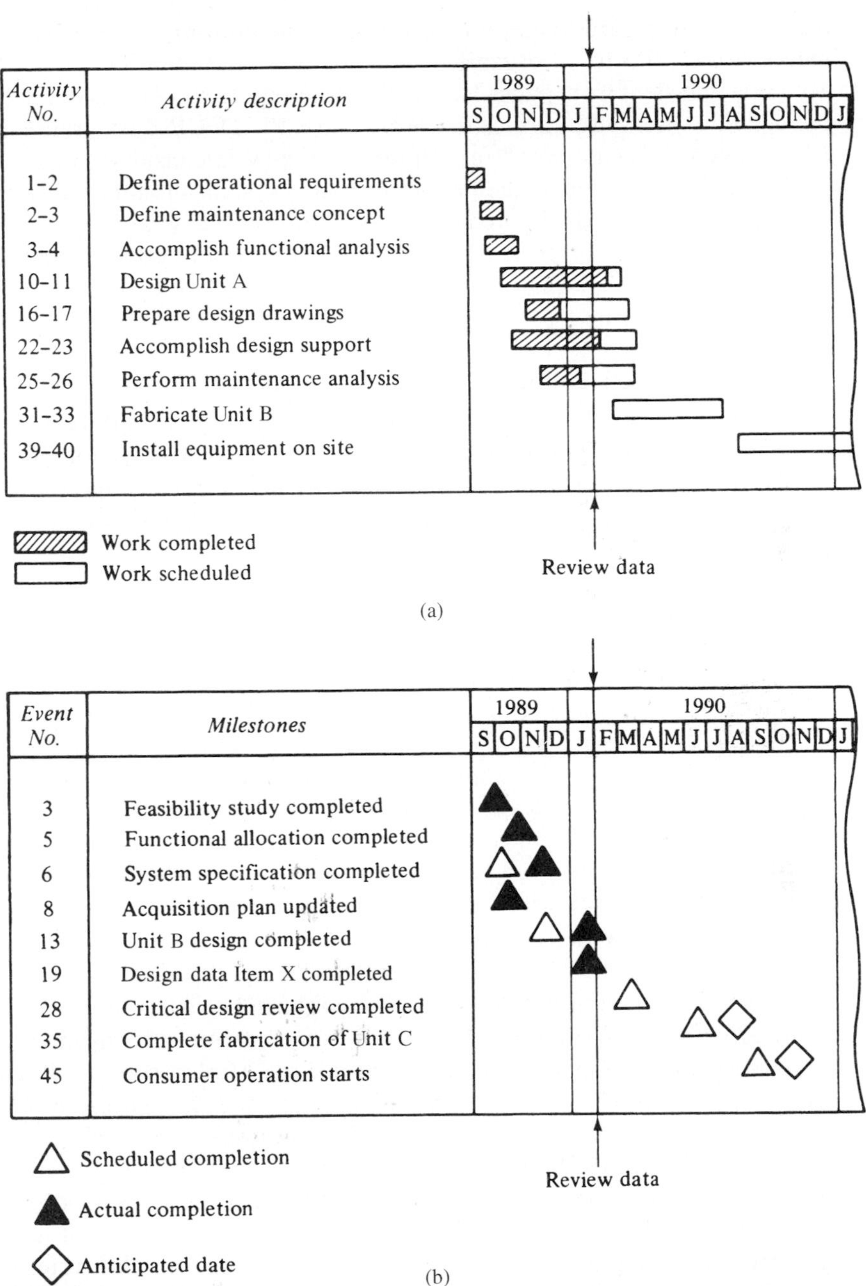

Figure 19.1 Sample bar chart (top) and milestone chart (bottom).

a capability must be tailored to the situation and sensitive to the needs of the particular project.

Regarding the more conventional type of reporting requirements, Figure 19.1 shows the current status of program activities (the sample bar graph at the top) and the status with regard to major milestones (at the bottom). Figure 19.2 relates the elements of a CWBS to an organizational structure, and identifies the appropriate cost accounts that stem from each. This leads to the reporting of costs in Figure 19.3, where the activities are tied to both the CWBS and a PERT/CPM program network. Figure 19.4 shows a projected cost stream based on life-cycle considerations.

The data from a management information system should readily point out existing problem areas. Also, potential areas where problems are likely to occur, if program operations continue as originally planned, should become visible. To deal with these situations, planning should establish a corrective-action procedure that will include the following:

1. Problems are identified and ranked in order of importance.
2. Each problem is evaluated on the basis of the ranking, addressing the most critical ones first. Alternative possibilities for corrective action are considered in terms of (a) effects on program schedule and cost, (b) effects on system performance and effectiveness, and (c) the risks associated with the decision as to whether to take corrective action.
3. When the decision to take corrective action is reached, planning is initiated to take steps to resolve the problem. This may be in the form of a change in management policy, a contractual change, an organizational change, or a system/equipment configuration change.
4. After corrective action has been implemented, some follow-up activity is required to (a) ensure that the incorporated change(s) actually has resolved the problem, and (b) assess other aspects of the program to ensure that additional problems have not been created as a result of the change.

Figure 19.5 identifies three areas that need attention, as detected at program review M1.

Review of Supplier Activities

Much of the discussion in Section 18.3 on organization structure pertains to systems engineering responsibilities and activities at the customer (consumer) and producer levels only, while supplier functions were not really addressed at all. However, realization of the objectives described herein requires not only a top-down but a bottoms-up approach as well (refer to Section 2.4). Decomposition must initially occur from the top down, and then the various components of the system provided by suppliers must be integrated from the bottom up. This *integration* task accomplished as part of the detail design and development phase, constitutes a major "test" of just how well the systems engineering process has been implemented to this point (refer to Chapter 6).

From an organizational perspective, systems engineering requirements must be initially defined by the customer and then allocated to the producer, as conveyed in Figure 18.12. At the same time, the requirements at the producer level must be allocated to the various suppliers of system components (as applicable).

The term *supplier* refers to a broad class of external organizations who provide products, components, materials, or services to the producer. This may range from the

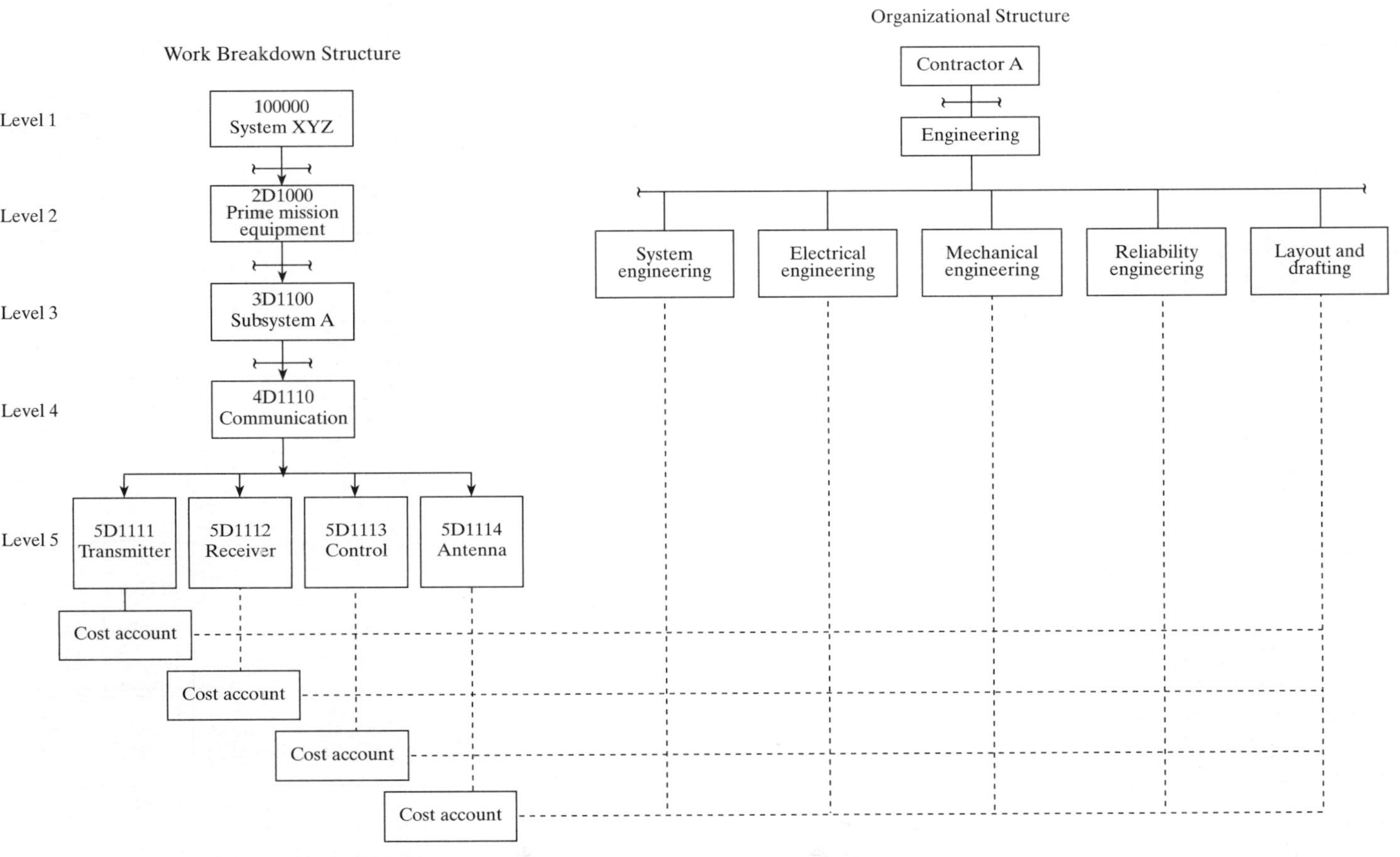

Figure 19.2 The identification of cost accounts through the integration of CWBS and the organizational structure.

PERT TIME-COST STATUS REPORT

PROJECT: System XYZ				CONTRACT NO.: 4B-3015					REPORT DATE: 4/10/90			
Item/identification				*Time status*					*Cost status*			
WBS no.	Project no.	Beginning event no.	Ending event no.	Exp. elap. time (t_e) (weeks)	Earliest completion date (D_E)	Latest completion date (D_L)	Slack $D_L - D_E$ (weeks)	Actual date completed	Cost. est. ($)	Actual cost to date ($)	Latest revised est. ($)	Overrun (underrun) ($)
3A1100	814763	36	37	5.2	3/4/86	4/11/90	7.8	4/5/90	2,000	1,950	1,950	(50)
2A3100	824823	239	240	3.0	5/15/86	4/30/90	–3.2		3,500	3,650	4,000	500
3A5400	833110	312	313	4.4	6/15/86	8/2/90	9.6		5,660	3,550	5,660	0

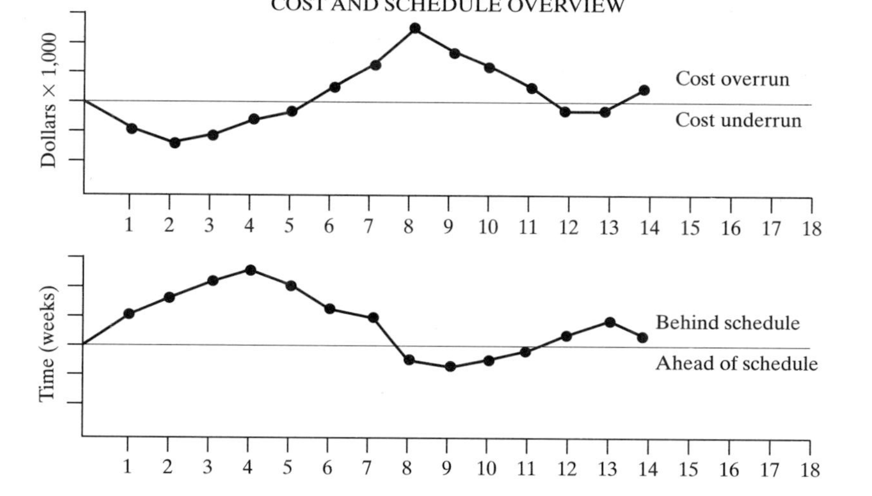

Figure 19.3 A status report relating cost to the PERT/CPM network and the CWBS.

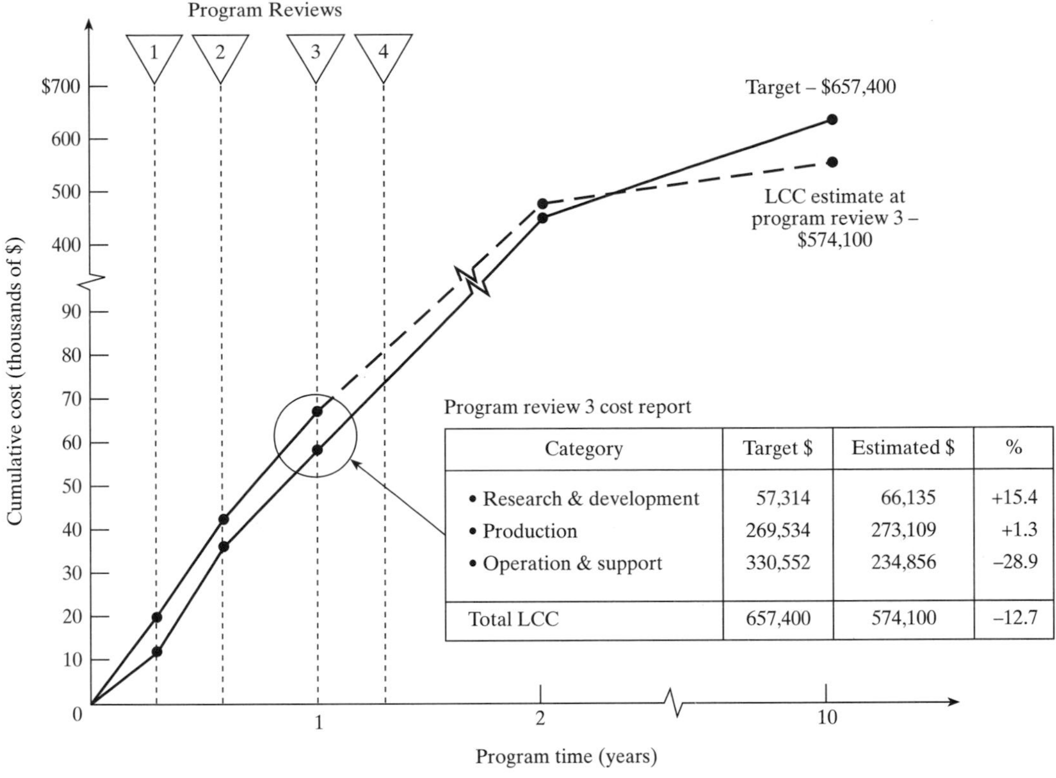

Category	Target $	Estimated $	%
• Research & development	57,314	66,135	+15.4
• Production	269,534	273,109	+1.3
• Operation & support	330,552	234,856	–28.9
Total LCC	657,400	574,100	–12.7

Figure 19.4 Project life-cycle cost projection at a major program review point.

delivery of a major subsystem, configuration item, unit, or down to a small component part. Referring to Figure 18.12, suppliers may provide services to include (1) the design, development, and manufacture of a major element of the system; (2) the production and distribution of items already designed (providing a manufacturing source); (3) the distribution of commercial and standard components from an established inventory (i.e., COTS items); and/or (4) the implementation of a process in response to some functional requirement. For many large programs, suppliers provide a large number of the elements that make up the system (e.g., more than 50% of the components in some instances), as well as the spares and repair parts that are required to support maintenance activities.

Referring to Figure 19.6, for large programs there may be a *layering* of suppliers as shown, with one of more component suppliers providing services to the supplier of a major subsystem or a configuration item. Further, recent trends in the economy have favored the practices of increased "outsourcing," or the process of seeking external sources of supply as compared to the completion of work internally within a producer's organization. This, combined with the current emphasis in *globalization*, has resulted in suppliers being selected from many different geographical centers located throughout the world.

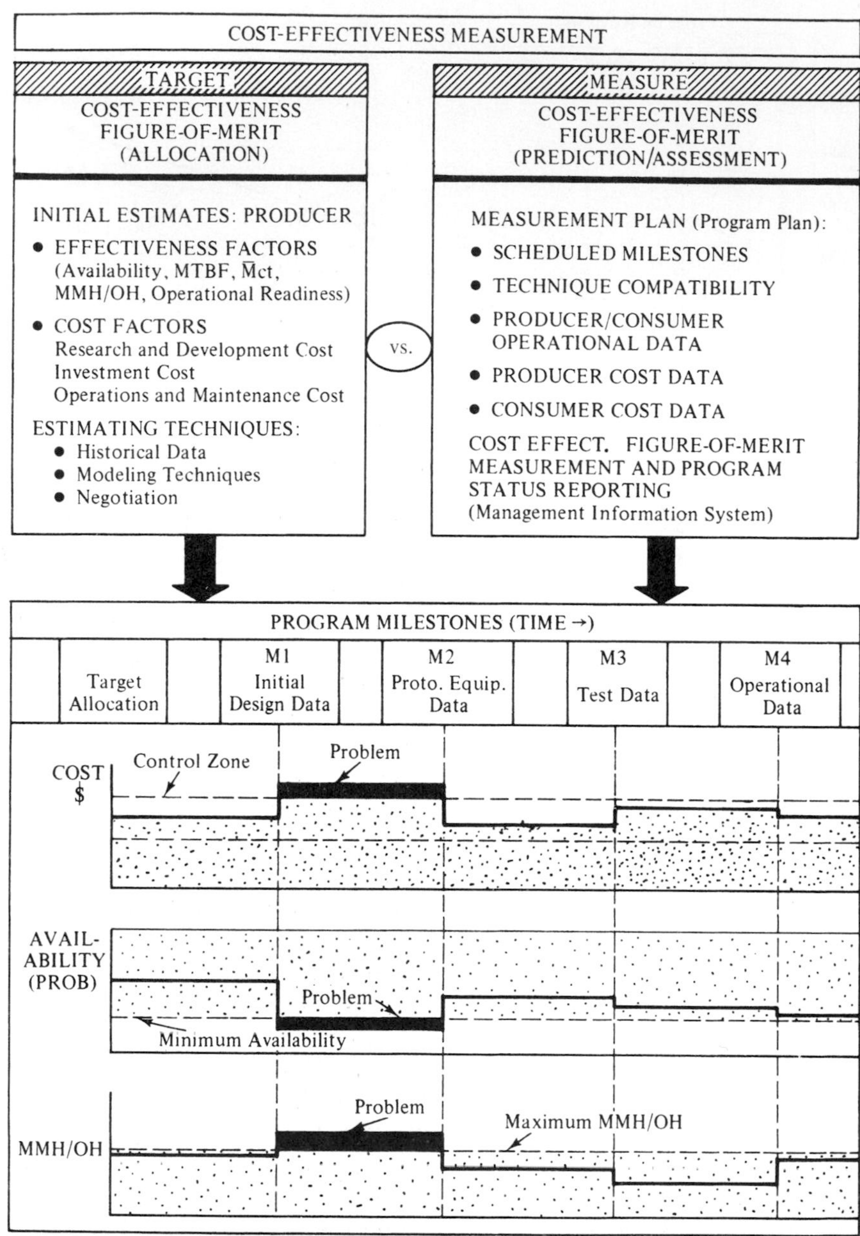

Figure 19.5 Parameter measurement and evaluation.

With the involvement of many different suppliers in the design, development, manufacture, and support of systems, there is an ever-increasing need for the implementation of systems engineering practices and techniques, more so now than ever before. Major suppliers, as key participants in the design and development process,

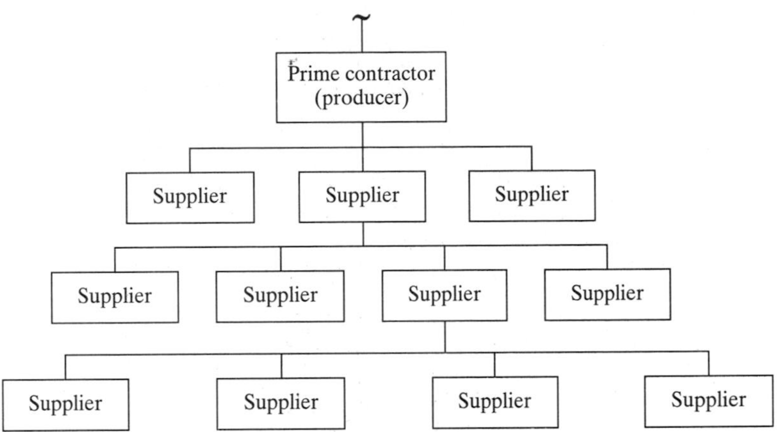

Figure 19.6 A typical structure involving the "layering" of suppliers.

must be involved from the beginning. The SEMP must include coverage of supplier functions and activities (refer to Figure 18.1). The system specification (type A), described in Section 3.8, must provide a good functional baseline, from which the various lower-level specifications can be developed. In essence, major suppliers must be brought into the early design process, participate as members of the design team, and must be committed to implementation of the systems engineering process.

Referring to Figure 3.2, there is a need to develop a "specification tree" showing a hierarchical relationship of specifications as they apply to various elements of the system. The purpose is to provide a structure for the traceability of requirements (from the top down) and to indicate *precedence* (i.e., which document takes precedence in the event of a possible conflict). Figure 19.7, which is an extension of Figure 3.2, identifies supplier requirements for a typical large program. As shown, there are suppliers involved in the design and development of a subsystem (type B specification), the delivery of commercial off-the-shelf products (type C specification), the development of a process (type D specification), and in the delivery of materials (type E specification).[3]

Qualitative and quantitative "design-to" criteria must be determined and included in each of the specifications identified in Figure 19.7. These requirements, which evolve from the TPMs, are determined primarily through the allocation process described in Section 4.2 (refer to Figure 4.8). Trade-offs are conducted within the producer's organization to ensure a proper "balance" across the board, and that the applicable functional interfaces between the numerous elements that are being provided by different suppliers are well defined. From a systems engineering perspective, it is necessary to ensure that the many supplier-delivered components actually fit together when combined during system integration.

With the specifications well defined, the producer will generally prepare a *request for proposal* (RFP), solicit a response from all qualified suppliers in terms of a formal proposal, and proceed through the process of supplier evaluation and ultimate source selection. Qualified proposals will be evaluated, graded, and the winner will be

[3] The reader may wish to review the various categories of specifications described in Section 3.8.

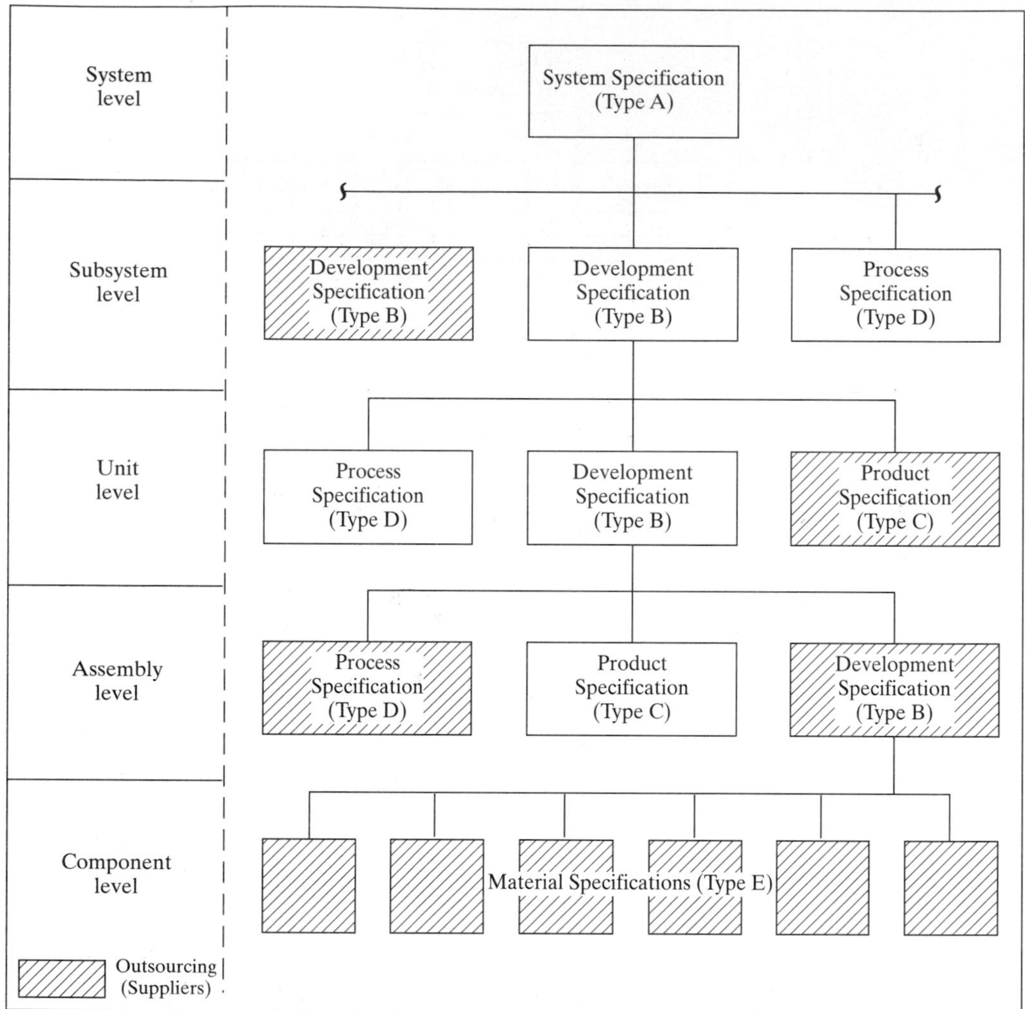

Figure 19.7 A "specification tree" showing supplier requirements (example) (refer to Figure 3.2).

awarded a contract. The systems engineering organization must be an integral part of this process to ensure that the requirements included in both the SEMP and the system specification (type A) are properly extended downward to the degree necessary. Figure 19.8 provides a sample checklist that can be utilized to facilitate the supplier evaluation process.[4]

Subsequent to supplier selection and the awarding of a contract, the day-to-day communication processes described in Chapter 18 must be implemented as illustrated

[4] The supplier identification, solicitation, review, evaluation, and selection process (along with a detailed checklist) is included in B. S. Blanchard, *System Engineering Management*, 2nd ed. (New York: John Wiley & Sons, Inc., 1998), chapter 8 and appendix D.

SUPPLIER EVALUATION CHECKLIST

D.1 GENERAL CRITERIA

D.2 PRODUCT DESIGN CHARACTERISTICS

D.2.1 TECHNICAL PEFORMANCE MEASURES
D.2.2 TECHNOLOGY APPLICATIONS
D.2.3 PHYSICAL CHARACTERISTICS
D.2.4 EFFECTIVENESS FACTORS
1. RELIABILITY
2. MAINTAINABILITY
3. HUMAN FACTORS
4. SAFETY FACTORS
5. SUPPORTABILITY/SERVICEABILITY
6. QUALITY FACTORS
D.2.5 PRODUCIBILITY FACTORS
D.2.6 DISPOSABILITY FACTORS
D.2.7 ENVIRONMENTAL FACTORS
D.2.8 ECONOMIC FACTORS

D.3 PRODUCT MAINTENANCE AND SUPPORT INFRASTRUCTURE

D.3.1 MAINTENANCE AND SUPPORT REQUIREMENTS
D.3.2 DATA/DOCUMENTATION
D.3.3 WARRANTY/GUARANTEE PROVISIONS
D.3.4 CUSTOMER SERVICE
D.3.5 ECONOMIC FACTORS

D.4 SUPPLIER QUALIFICATIONS

D.4.1 PLANNING/PROCEDURES
D.4.2 ORGANIZATIONAL FACTORS
D.4.3 AVAILABLE PERSONNEL AND RESOURCES
D.4.4 DESIGN APPROACH
D.4.5 MANUFACTURING CAPABILITY
D.4.6 TEST AND EVALUATION APPROACH
D.4.7 MANAGEMENT CONTROLS
D.4.8 EXPERIENCE FACTORS
D.4.9 PAST PERFORMANCE
D.4.10 MATURITY
D.4.11 ECONOMIC FACTORS

Figure 19.8 Supplier evaluation checklist.

in Figure 19.9 (which is an extension of Figure 18.12). The producer's systems engineering organization must provide a leadership role from a *technical* perspective, particularly because the various supplier organizations selected may not necessarily include a similar activity, or any one individual who is knowledgeable of the principles and concepts described herein.

In summary, the role of the *supplier* is critical in meeting the goals and objectives of systems engineering. While the supplier's contribution may seem trivial if all is going well, the lack of a needed supplier response can be damaging if not addressed.

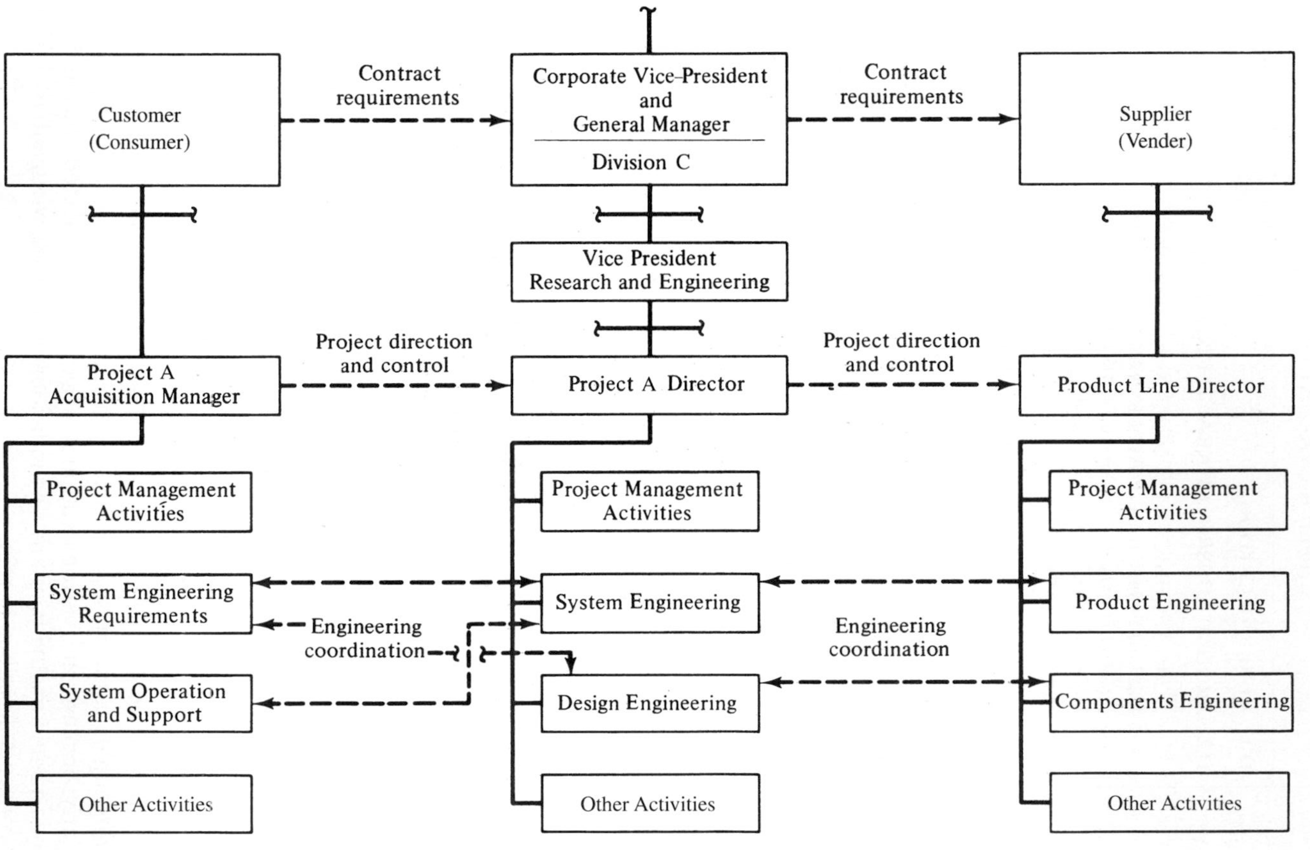

Figure 19.9 Consumer-producer-supplier organizational relationships (refer to Figure 18.12).

19.3 PROGRAM EVALUATION FACTORS

Having established certain systems engineering objectives and the recommended tasks that need to be performed, the questions are: To what extent is your organization completing these tasks electively and efficiently? Does your management understand the principles and concepts of systems engineering? Is there a commitment from the top down toward the implementation of the systems engineering process? Have the appropriate internal standards, measurable goals, and processes been established for the successful accomplishment of the stated objectives?

Although many questions of this nature can be asked, the objective is to assess the systems engineering organization's *level of maturity*, where it may "fit" in the hierarchical structure when comparing it with other organizations in a similar area, and where there are weaknesses that need to be addressed. In other words, although the benchmarking process discussed earlier aids in establishing certain goals, there is a need to develop a model to assist in the evaluation of the organization's current capability. This, in turn, leads to the development of a plan for improvement.

System Engineering Capability Maturity Model (SE-CMM)

In response to the preceding, there was a concerted effort initiated in the mid 1990s to develop a model that will address the organizational assessment issue. Although there are numerous models being used to varying degrees, two specific projects stand out. Under the auspices of Carnegie Melon University, and through the efforts of many individuals from industry, government, and academia, the "Systems Engineering Capability Maturity Model" (SE-CMM) was developed and first made available for use in 1994.[5] A second effort involved the development of the "Systems Engineering Capability Assessment Model" (SECAM), through the efforts of the Capability Assessment Working Group of the International Council On System Engineering (INCOSE), with its first version being released in 1994.[6] Although both models are considered effective in their application, a brief description of the SE-CMM has been selected to provide an overview of the objectives of organizational assessment.

The SE-CMM describes the essential *elements* of an organization's systems engineering process that must exist to ensure that the principles and practices of *good systems engineering* are being implemented. The model does not specify a particular process or sequence of activities, but provides a reference for comparing the actual systems engineering practices being implemented by an organization against these essential elements. These essential elements have been broken down into 17 *process areas* (PAs), against which organizations may be evaluated. More specifically, is the organization:

[5] SECMM-95-01, "A Systems Engineering Capability Maturity Model (SE-CMM)," Version 1.1 (Pittsburgh: 1995).

[6] A good reference that provides a historical basis for the SECAM and its applications is B. A. Andrews, and E. R. Widmann, "A Synopsis of Metrics and Observations from Systems Engineering Process Assessments Conducted Using the INCOSE SECAM," Proceedings, *Sixth Annual International Symposium Of The INCOSE*, vol. 1, International Council On Systems Engineering, (Seattle: 1996).

1. Analyzing candidate solutions in response to a need?
2. Deriving and allocating requirements?
3. Evolving system architecture?
4. Integrating design disciplines?
5. Integrating system elements?
6. Understanding customer/consumer needs and expectations?
7. Verifying and validating the system?
8. Ensuring good quality?
9. Managing configurations?
10. Managing risk?
11. Monitoring and controlling technical effort?
12. Defining its system engineering process and activities?
13. Improving its system engineering processes?
14. Managing product line evolution?
15. Managing system engineering support environment?
16. Providing ongoing skills and knowledge?
17. Providing coordination with suppliers?

In evaluating a given organization, an assessment can be made to determine the extent to which it is addressing each of the above 17 process areas. A questionnaire has been developed, with specific questions supporting each area. A numerical scale is used to rate the degree of compliance, numerical values are combined and compiled, and an overall rating is calculated.[7]

From the results of the combined assessment pertaining to how well the organization is complying with the objectives in the 17 process areas, a *capability index* is determined and the organization is placed in any one of six specified *levels of maturity*, as illustrated in Figure 19.10. As shown, the step-function illustrated in the figure evolves from "Level 0" where there is no evidence of systems engineering activity being performed, to "Level 5" where all of the appropriate systems engineering activities are being effectively implemented and where there is a *continuous process improvement* mechanism in place.

Given the overall results from the organization's evaluation using the SE-CMM, the manager may wish to evaluate the data further to determine which of the 17 processes are the weakest. Figure 19.11 shows a comparitive presentation which, in turn, can lead to areas for improvement.

Organizational Development and Improvement

Although the SE-CMM method may not be the only tool used for organizational assessment, the conceptual approach used should be applicable for any systems engineering organization. Review of Figure 19.11, and an analysis of the results from the

[7] A good reference covering the questionnaire development and application is SPC-95075-CMC, "Systems Engineering Maturity And Benchmarking," Version 01.00.06 (Herndon, Va. 1996).

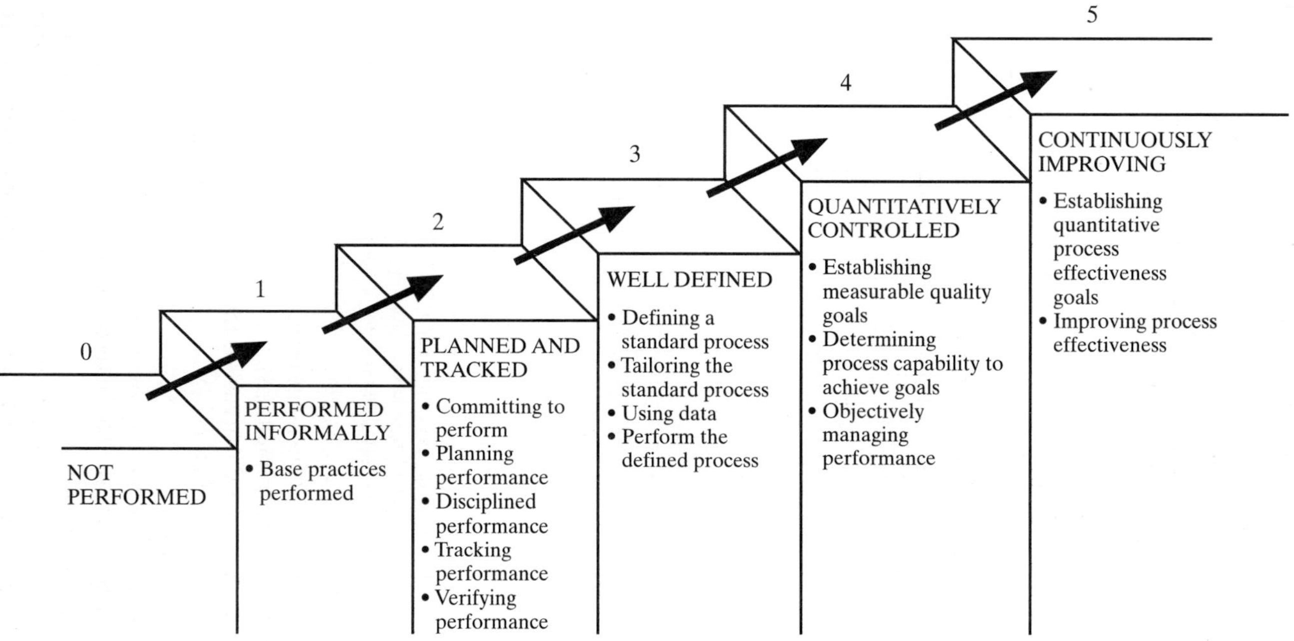

Figure 19.10 Improvement path for systems engineering process capability. *Source:* SECMM-95-01, "A Systems Engineering Capability Maturity Model (SE-CMM)," Version 1.1. Pittsburgh: Software Engineering Institute, Carnegie Melon University, November 1995.

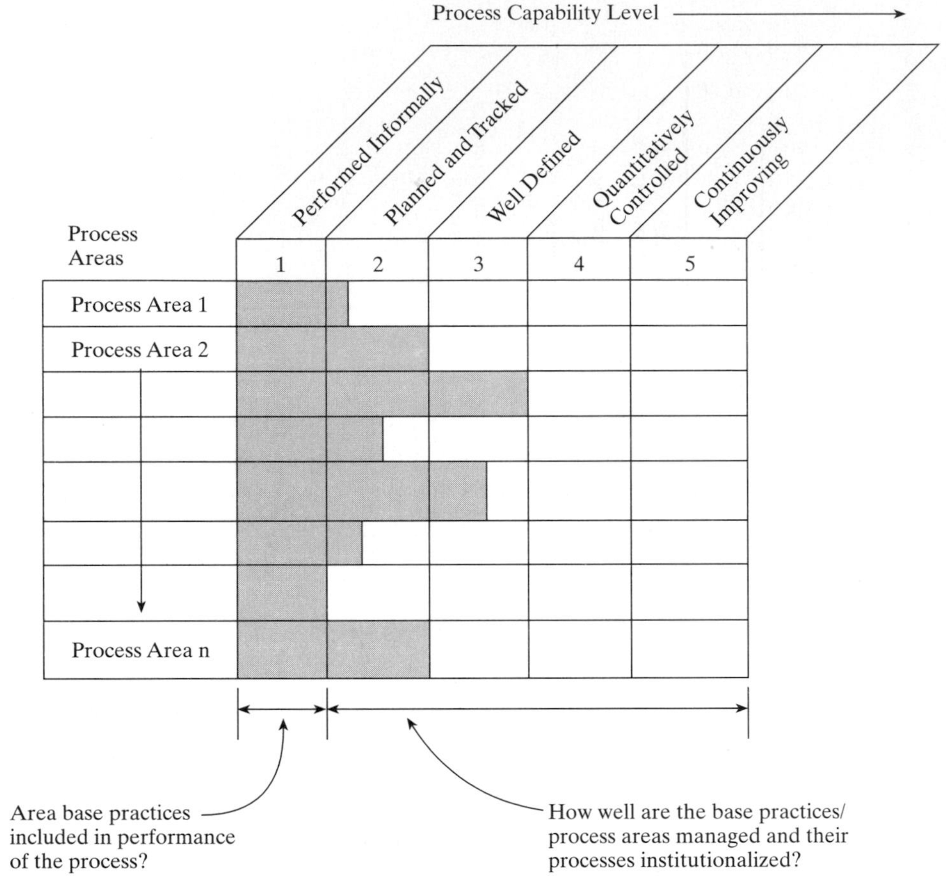

Figure 19.11 Determining process capabilities and areas for improvement. *Source:* SECMM-95-01, "A Systems Engineering Capability Maturity Model (SE-CMM)," Version 1.1. Pittsburgh: Software Engineering Institute, Carnegie Melon University, November 1995.

questionnaire, should lead to the identification of areas of weakness (i.e., areas where process improvement is recommended).

Having identified potential areas for improvement, there are two steps that need to be addressed:

1. Determining ways for the improvement of internal processes *within* the systems engineering organization. This encompasses evaluating alternative methods of doing business, determining the requirements for changing the existing procedures and processes, and assessing the impact of such changes on the other processes. A change in any one process should not have a negative impact on any other process.
2. Determining the possible impact(s) of changes in the processes followed by the systems engineering organization on the higher-level organizational structure or

on any other related function (i.e., the customer, other organizational groups, suppliers, etc). The proper environment must be established within the overall organizational structure for the proposed changes from item 1 to result in an improvement.

Proposed changes within the systems engineering organization cannot be initiated in a *vacuum*. There must be a mutual commitment throughout the organization and, particularly, regarding the program manager and his or her staff. Nevertheless, there must be a vehicle through which organizational improvement can be initiated on a continuous basis.

19.4 PROGRAM RISK MANAGEMENT

Referring to Figure 18.2, an essential part of a systems engineering effort is the implementation of a risk management program and the development of a plan. The element of *risk* can be defined as the potential that something will go wrong as a result of one or a series of events. It can be measured as the combined effect of the probability of occurrence and the assessed consequences given that occurrence. The potential for risk becomes increasingly higher as complexities and new technologies are introduced into the design of new systems. Risk, as used herein, refers to the potential of not meeting a specified technical and/or program requirement (i.e., not meeting a specified TPM, a schedule, or program cost requirement).

Risk management is an organized method for identifying and measuring risk and for selecting and developing options for handling risk. Risk management should not be considered as a separate program thrust by itself, but should be an inherent part of any sound management activity. The systems engineering organization is particularly interested in those facets that deal with *technical* risk. Risk management includes the following basic activities:

1. *Risk assessment.* This involves the ongoing review of technical design and/or program management decisions, and the identification of potential areas of risk.
2. *Risk analysis.* This includes conducting an analysis to determine the probability of events and the consequences associated with their occurrence. The purpose of risk analysis is to identify the cause(s), the effects, and the magnitude of the risk perceived, and to identify alternative approaches tor risk avoidance. There are many tools that are available and can be used as an aid in conducting risk analyses; for example, scheduling network analysis, life-cycle cost analysis, FMECA, the Ishikawa cause-and effect or "fishbone" diagram, hazard analysis, and trade-off studies in varying forms.
3. *Risk abatement.* This involves the techniques and methods developed to reduce (if not eliminate) or control the risk. A plan must be implemented for the handling of risk.

One of the first steps in risk management is the identification of the potential areas of risk. Although there is some degree of risk associated with any program area of activity where decisions are being made, one needs to identity those in which the potential consequences of failure could be significant. Program areas of risk may include funding, schedule, contract relationships, political, and technical. Technical risks relate primarily to the potential of not meeting a design requirement, not being able to produce an item in multiple quantities, or not being able to support a product in the field. Design engineering risks can be tied directly to the technical performance measures (TPMs) identified in Section 3.5 and Figure 3.10. These TPMs, which reflect critical factors in design, can be prioritized to reflect relative degrees of importance.

Given the identification of "performance" characteristics to which the system is to be designed (i.e.. those parameters which require monitoring on a regular basis), the next step is to evaluate these by indicating possible causes for failure. In the event of a failure to meet a specific design requirement, what are the possible causes and what are the probabilities of occurrence? Although the output measure being monitored may be a high-priority TPM, the cause for a possible failure may be the result of a misapplication of a new technology in design, a schedule delay on the part of a major supplier, a cost overrun, or a combination of these.

The causes are evaluated independently to determine the degree to which they can impact the TPM(s) being monitored. Sensitivity analyses are conducted, using various analytical models as appropriate, to determine the magnitude of the potential risk. This, in turn, will lead to the classification of factors in terms of "high," "medium," or "low" risk. These classifications of risk are then addressed within the program management review and reporting structure. High-risk items are monitored to a greater extent, with a higher priority relative to initiating a risk abatement plan, than low-risk items.

To facilitate the risk management implementation process, it is often feasible to develop a model of some type. One approach is to address risk in terms of two major variables: the probability of failure (P_f), and the effect or consequence of that failure (C_f). Consequences may be measured on the basis of technical performance, cost, or schedule. Mathematically, this model can be expressed as[8]

$$\text{Risk factor } (RF) = P_f + C_f - (P_f)(C_f) \tag{19.1}$$

where P_f is the probability of failure and C_f is the consequence of failure. The quantitative relationships of the parameters are shown in Figure 19.12.

To illustrate the model application, with Figure 19.12 as the prime source of information, consider the following system design characteristics:[9]

1. System design uses off-the-shelf hardware with minor modifications to the software.
2. The design is relatively simple involving the use of standard hardware.

[8] This model was adapted from the procedure in the document, *Systems Engineering Management Guide*, Defense Systems Management College (Fort Belvoir, Va. 1986). Although the later issues of this document do cover the various aspects of risk management, this particular model has been deleted for reason unknown to the authors.

[9] In applying this model to a specific application, it is recommended that Figure 19.12 be replaced with some form of a checklist tailored, with the appropriate weighting factors, to the specific system and in support of the TPMs specified for that system.

Magnitude	Maturity Factor (P_M)		Complexity Factor (P_C)		Dependency Factor (P_D)
	Hardware *PMhw*	Software *PMsw*	Hardware *PChw*	Software *PCsw*	
0.1	Existing	Existing	Simple design	Simple design	Independent of existing system, facility, or associate contractor
0.3	Minor redesign	Minor redesign	Minor increases in complexity	Minor increases in complexity	Schedule dependent on existing system, facility, or associate contractor
0.5	Major change feasible	Major change feasible	Moderate increase	Moderate increase	Performance dependent on existing system performance, facility, or associate contractor
0.7	Technology available, complex design	New software similar to existing	Significant increase	Significant increase/major increase in no. of modules	Schedule dependent on new system schedule, facility, or associate contractor
0.9	State of art some research complete	State of art never done before	Extremely complex	Extremely complex	Performance dependent on new system schedule, facility, or associate contractor

Magnitude	Technical Factor $(C_f) - (C_t)$	Cost Factor (C_C)	Schedule Factor (C_S)
0.1 (low)	Minimal or no consequences, unimportant	Budget estimates not exceeded, some transfer of money	Negligible impact on program, slight development schedule change compensated by available schedule slack
0.3 (minor)	Small reduction in technical performance	Cost estimates exceed budget by 1 to 5	Minor slip in schedule (less than 1 month), some adjustment in milestone required
0.5 (moderate)	Some reduction in technical performance	Cost estimates increased by 5% to 20%	Small slip in schedule
0.7 (significant)	Significant degradation in technical performance	Cost estimates increased by 20% to 50%	Development schedule slip in excess of 3 months
0.9 (high)	Technical goals cannot be achieved	Cost estimates increased in excess of 50 percent	Large schedule slip that affects segment milestones or has possible effect on system milestones

(1) Risk factor $= P_f + C_f - (P_f)(C_f)$

(2) $P_f = (a)(P_{Mhw}) + (b)(P_{Msw}) + (c)(P_{Chw}) + (d)(P_{Csw}) + (e)(P_D)$

where

P_{Mhw} = Probability of failure due to degree of hardware maturity

P_{Msw} = Probability of failure due to degree of software maturity

P_{Chw} = Probability of failure due to degree of hardware complexity

P_{Csw} = Probability of failure due to degree of software complexity

P_D = Probability of failure due to dependency on other items

and where: a, b, c, d, and e are weighing factors whose sum equals one.

(3) $C_f = (f)(C_t) + (g)(C_C) + (h)(C_S)$

where:

C_t = Consequence of failure because of technical factors

C_C = Consequence of failure because of changes in cost

C_S = Consequence of failure changes in schedule

and where: f to h are weighing factors whose sum equals one.

Figure 19.12 A mathematical model for risk assessment. *Source: Systems Engineering Management Guide*, Fort Belvoir, VA: Defense Systems Management College, 1986.

3. The design requires software of somewhat greater complexity.
4. The design requires a new data base to be developed by a supplier (subcontractor).

The characteristics of the system suggest that there is potential risk associated with the software development task. Using the criteria in Figure 19.12 (and applying the weighing factors as indicated), the probability of failure (P_f) is calculated as follows:

$$
\begin{aligned}
P_{M_{hw}} &= 0.1, \text{ or } (a)(P_{M_{hw}}) = (0.2)(0.1) = 0.02 \\
P_{M_{sw}} &= 0.3, \text{ or } (b)(P_{M_{sw}}) = (0.1)(0.3) = 0.03 \\
P_{C_{hw}} &= 0.1, \text{ or } (c)(P_{C_{hw}}) = (0.4)(0.1) = 0.04 \\
P_{C_{sw}} &= 0.3, \text{ or } (d)(P_{C_{sw}}) = (0.1)(0.3) = 0.03 \\
P_D &= 0.9, \text{ or } (e)(P_D) = (0.2)(0.9) = \underline{0.18} \\
& \qquad\qquad\qquad\qquad\qquad\qquad\qquad\quad 0.30
\end{aligned}
$$

Given the preceding criteria, P_f of this item is 0.30.

If the consequence of the item's failure due to technical factors causes problems of a corrective nature, but the correction results in an 8% cost increase and a 2-month schedule slippage, the C_f is calculated as follows:

$$
\begin{aligned}
C_t &= 0.3, \text{ or } (f)(C_t) = (0.4)(0.3) = 0.12 \\
C_c &= 0.5, \text{ or } (g)(C_c) = (0.5)(0.5) = 0.25 \\
C_s &= 0.3, \text{ or } (h)(C_s) = (0.1)(0.3) = \underline{0.03} \\
& \qquad\qquad\qquad\qquad\qquad\qquad\qquad\quad 0.40
\end{aligned}
$$

Based on the preceding (using the weighing factors indicated), the C_f factor is 0.40 and, from Equation (19.1), the calculated risk factor (RF) is 0.580. This can be classified within the category of medium risk, as noted in Figure 19.13. In this instance, the risk is primarily associated with the system software and the reliance on a supplier.

A similar approach can be applied in performing a risk analysis on all other applicable parameters. The net result is the development of a list of critical items, presented in order of priority, that require special management attention. Risk reports are prepared at different times (i.e., frequency of distribution) depending on the nature of the risk. High-risk items require frequent reporting and special management attention. while low-risk items can be handled through the normal program review, evaluation, and reporting process.

For items classified under "high" and "medium" risk, a risk abatement plan should be implemented. This is a formal approach for eliminating (if possible), reducing, or controlling risk. The accomplishment of this may involve one or a combination of the following:

1. Provide increased management review of the problem area(s), and initiate the necessary corrective action through an internal allocation or shift in resources.
2. Hire outside consultants or specialists to help resolve existing design problems.
3. Implement an extensive testing program with the objective of better isolating the problem and eliminating possible causes.

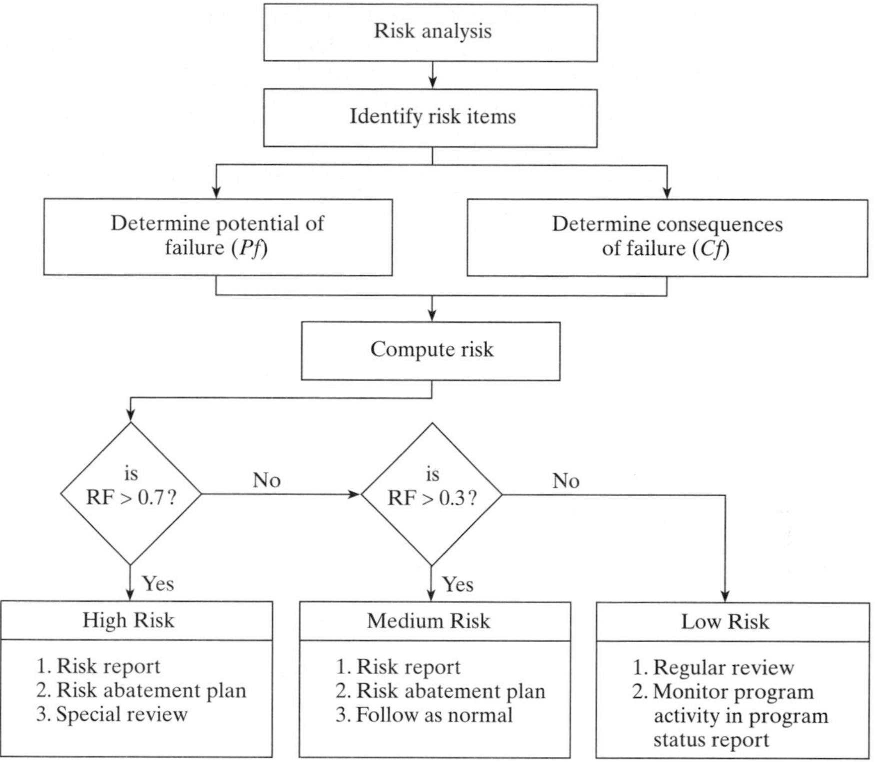

Figure 19.13 Risk analysis and reporting procedure.

4. Initiate special research and development activities, conducted in parallel to provide a "fall-back" position.

The purpose of the risk abatement plan is to *highlight* those areas where special management attention is required. The identification of technical risks is of particular interest regarding system engineering, since the fulfillment of design objectives is very dependent on the proper and expeditious handling of these risks. In this respect, risk management should be an inherent facet of system engineering management.

QUESTIONS AND PROBLEMS

1. The text refers to a two-step process in the implementation of systems engineering requirements. Identify these steps and describe how they relate to each.
2. Assume that you have just been assigned to develop a systems engineering organization. What steps would you follow in developing the goals and objectives for this organization?
3. Describe what is meant by *benchmarking*. For your organization in Problem 2, describe the steps that you would follow in establishing the appropriate "benchmarks?"
4. As a manager of a newly established systems engineering organization, you need to develop a "management information system "(MIS) to provide you with the visibility needed to

properly manage the organization on a day-by-day basis. Describe the steps that you would follow in accomplishing this objective? What information would you include (type, format, frequency, etc.)? What reports should be provided?

5. Refer to the bar chart in Figure 19.1. What areas should be investigated based on the results of the program review?
6. Refer to the cost projection in Figure 19.4. Although the overall estimate looks favorable in terms of the projected life-cycle cost, the results from Program Review 3 are not favorable. What would you do (if anything) to ensure that the ultimate projection is a realistic one and is maintained?
7. Referring to Figure 19.5, there are three problem areas noted. What steps would you take in correcting (i.e., eliminating) each of the problems?
8. In the design and development of a system, decisions have to be made as to whether an item should be developed and produced "inhouse" or considered for "outsourcing." What factors should be addressed in the source-selection process?
9. Assume that a decision has been made in favor of "outsourcing," and that you need to develop a "request for proposal." What information/data would you include in the RFP?
10. Why is the development of a "specification tree" important?
11. Refer to Figure 19.7. How would you determine the specific TPMs that should be included in each of the specifications noted?
12. Explain why it is critical that the functional interfaces (reflected by the supplier activities noted in Figure 19.7) be well defined from the beginning.
13. Refer to Figure 19.8. Develop a supplier evaluation checklist "tailored" to your own specific needs.
14. Assume that you are responsible for the activities of many different suppliers located throughout the world. What steps would you take to ensure that the proper level of integration is being accomplish throughout and on a continuous basis?
15. Assume that you are a manager of a systems engineering organization and that you need to establish an evaluation procedure for assessing just how well your organization is carrying out its functions. What steps would you take in accomplishing this objective?
16. Develop an "assessment model" tailored to your particular organization.
17. Describe what is meant by *risk*. How can it be measured? What aspects of risk would be of particular interest to a systems engineering manager?
18. Develop a detailed outline for a risk management plan. How would this plan relate to the SEMP?
19. Refer to Figure 19.21, assume that as a result of a risk analysis, your calculated RF factor is 0.8. What would you do (if anything)?

PART VI
APPENDICES

CONTENTS

Analysis and Checklist Methods

A critical early step in implementing the systems engineering process is *functional analysis* and the definition of the system in "functional" terms. Functions are intitially identified as part of defining the need and requirements for the system. System operational requirements and the maintenance concept are then defined, and the functional analysis is accomplished to establish a baseline, from which the resource requirements for the system are identified; i.e., equipment, software, people, facilities, data, the various elements of support, etc. The functional analysis is initiated during the conceptual design phase and is described in Chapter 3, Section 3.6. As the design and development process continues, the functional analysis is accomplished to a greater depth during the preliminary system design phase, as described in Chapter 4, Section 4.1. Appendix A.1 is included herein to provide guidance as to the detailed steps involved in accomplishing a functional analysis and, more specifically, the development of functional flow block diagrams.

Evolving from the definition of system requirements and accomplishment of the functional analysis is the establishment of *design criteria*, or the identification of specific characteristics of design and the design-dependent parameters that must be incorporated into the design to meet the overall requirements for the system. These criteria constitute an *input* to the design process. As the design and development process continues, there is a need for *evaluation* on a periodic basis; i.e., the conduct of both informal and formal design reviews at discrete times as the system configuration becomes better defined. The requirements for "design review" are described in Sections 3.9, 4.6, and 5.7, and the scheduling of such is illustrated in Figure 5.10. In support of the evaluation process, design review *checklists* are often utilized as an aid in highlighting the areas that should be addressed. Figure 5.9 shows an abbreviated checklist. The objective in Appendix A.2 is to include a procedure for the development of detailed checklists and supporting questions that can be used to supplement the top-level checklist, such as the one presented in Figure 5.9.

A.1 FUNCTIONAL ANALYSIS

Functional analysis includes the process of translating top-level system requirements into specific qualitative and quantitative design requirements. Given the identified need for a system, supported by the definition of operational requirements and the maintenance concept for that system (described in Chapter 3), it is necessary to translate this information into meaningful design criteria. This translation task constitutes an iterative process of breaking down system-level requirements into successive levels of detail, and a convenient mechanism for communicating this information is the *functional flow diagrams.*

Functional flow diagrams (or functional block diagrams) are developed to describe the system in functional terms. These are developed to reflect both *operational* and *support* activities as they occur throughout the system life cycle, and they are structured in a manner that illustrates the hierarchal aspects of a system (see Figure 3.11, Chapter 3). Some key features of the overall functional flow diagram process are noted as follows:

1. The functional block diagram approach should include coverage of *all* activities throughout the system life cycle, and the method of presentation should reflect proper activity sequences and interface interrelationships.
2. The information included within the functional blocks should be concerned with *what* is required before looking at *how* it should be accomplished.
3. The process should be flexible to allow for expansion if additional definition is required or reduction if too much detail is presented. The objective is to progressively and systematically work down to the level where resources can be identified with *how* a task should be accomplished (refer to Figure 4.1).

In the development of functional flow diagrams, some degree of standardization is necessary (for communications) in defining the system. Thus, certain basic practices and symbols should be used, whenever possible, in the physical layout of functional diagrams. The paragraphs below provide some guidance in this direction.

1. *Function block.* Each separate function in a functional diagram should be presented in a single box enclosed by a solid line. Blocks used for reference to other flows should be indicated as partially enclosed boxes labeled "Ref." Each function may be as gross or detailed as required by the level of functional diagram on which it appears, but it should stand for a definite, finite, discrete action to be accomplished by equipment, personnel, facilities, software, or any combination thereof. Questionable or tentative functions should be enclosed in dotted blocks.
2. *Function numbering.* Functions identified in the functional flow diagrams at each level should be numbered in a manner which preserves the continuity of functions and provides information with respect to function origin throughout the system. Functions on the top-level functional diagram should be numbered 1.0, 2.0, 3.0, and so on. Functions which further indenture these top functions should contain the same parent identifier and should be coded at the next decimal level for

each indenture. For example, the first indenture of function 3.0 would be 3.1, the second 3.1.1, the third 3.1.1.1, and so on. For expansion of a higher-level function within a particular level of indenture, a numerical sequence should be used to preserve the continuity of function. For example, if more than one function is required to amplify function 3.0 at the first level of indenture, the sequence should be 3.1, 3.2, 3.3, ..., 3.*n*. For expansion of function 3.3 at the second level, the numbering shall be 3.3.1, 3.3.2, ..., 3.3.*n*. Where several levels of indentures appear in a single functional diagram, the same pattern should be maintained. While the basic ground rule should be to maintain a minimum level of indentures in any one particular flow, it may become necessary to include several levels to preserve the continuity of functions and to minimize the number of flows required to functionally depict the system.

3. *Functional reference.* Each functional diagram should contain a reference to its next higher functional diagram through the use of a reference block. For example, function 4.3 should be shown as a reference block in the case where the functions 4.3.1, 4.3.2, ..., 4.3.*n*, and so on, are being used to expand function 4.3. Reference blocks shall also be used to indicate interfacing functions as appropriate.
4. *Flow connection.* Lines connecting functions should indicate only the functional flow and should not represent either a lapse in time or any intermediate activity. Vertical and horizontal lines between blocks should indicate that all functions so interrelated must be performed in either a parallel or series sequence. Diagonal lines may be used to indicate alternative sequences (cases where alternative paths lead to the next function in the sequence).
5. *Flow directions.* Functional diagrams should be laid out so that the functional flow is generally from left to right and the reverse flow, in the case of a feedback functional loop, from right to left. Primary input lines should enter the function block from the left side; the primary output, or *go* line, should exit from the right, and the *no-go* line should exit from the bottom of the box.
6. *Summing gates.* A circle should be used to depict a summing gate. As in the case of functional blocks, lines should enter or exit the summing gate as appropriate. The summing gate is used to indicate the convergence, or divergence, or parallel or alternative functional paths and is annotated with the term AND or OR. The term AND is used to indicate that parallel functions leading into the gate must be accomplished before proceeding to the next function, or that paths emerging from the AND gate must be accomplished after the preceding functions. The term OR is used to indicate that any of several alternative paths (alternative functions) converge to, or diverge from, the OR gate. The OR gate thus indicates that alternative paths may lead or follow a particular function.
7. *Go and no-go paths.* The symbols G and $\overline{G}$ are used to indicate go and no-go paths, respectively. The symbols are entered adjacent to the lines leaving a particular function to indicate alternative functional paths.
8. *Numbering procedure for changes.* Additions of functions to existing data should be accomplished by locating the new function in its correct position without regard to sequence of numbering. The new function should be numbered using the first unused number at the level of indenture appropriate for the new function.

With the objective of illustrating how some of these general guidelines are employed, Figures A.1 through A.7 are included to present a few simple applications.

1. Figure A.1 provides an example of the basic format used in the development of functional block diagrams in general.
2. Figure A.2 shows a manufacturing capability (top level or Blocks 1.0–7.0), an expansion of the design function (Block 2.0), and an expansion of the operating functions of the manufacturing plant (Blocks 5.1.1–5.1.16).
3. Figure A.3 shows two levels of operational flow diagrams for a space system.
4. Figure A.4 shows a maintenance functional flow diagram for the space system that evolves from the operation flow in Figure A.3.
5. Figure A.5 shows two levels of operational flow diagrams and two levels of maintenance flow diagrams for a radar system.
6. Figure A.6 shows two levels of operational flow diagrams and a maintenance functional flow diagram for an automotive system.
7. Figure A.7 shows two levels of operational flow diagrams and two levels of maintenance flow diagrams for a lawn mowing system.

Although these sample block diagrams do not cover the selected system entirely, it is hoped that the material is presented in enough detail to provide an appropriate level of guidance for the development of functional block diagrams.

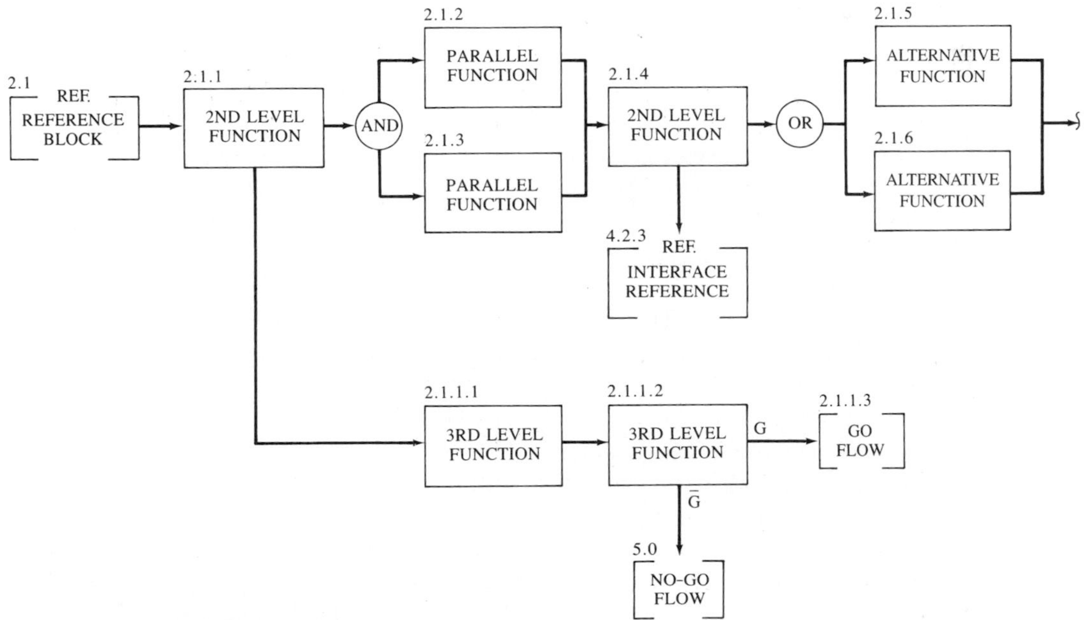

Figure A.1 General format for the development of functional block diagrams.

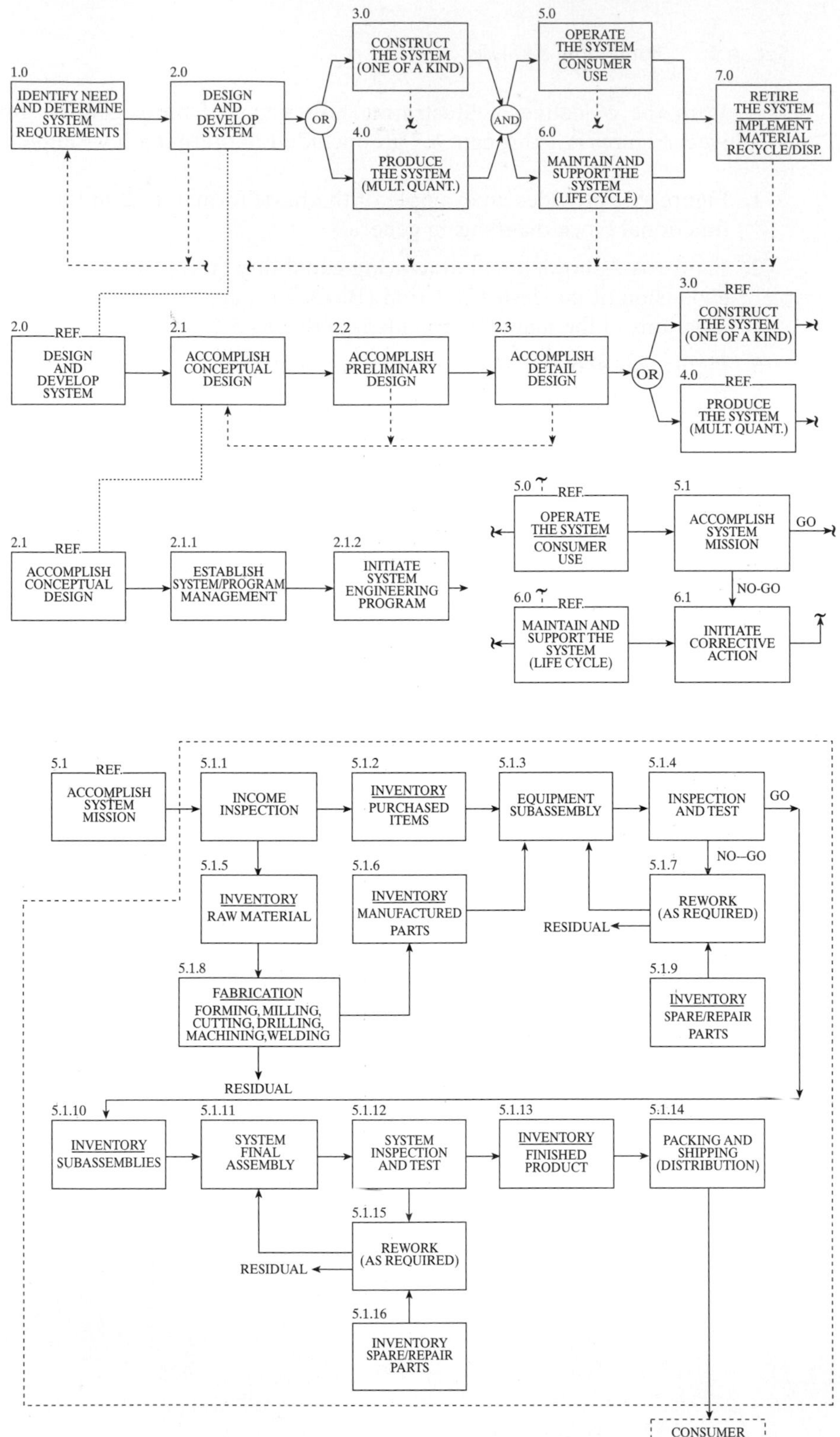

Figure A.2 A manufacturing capability.

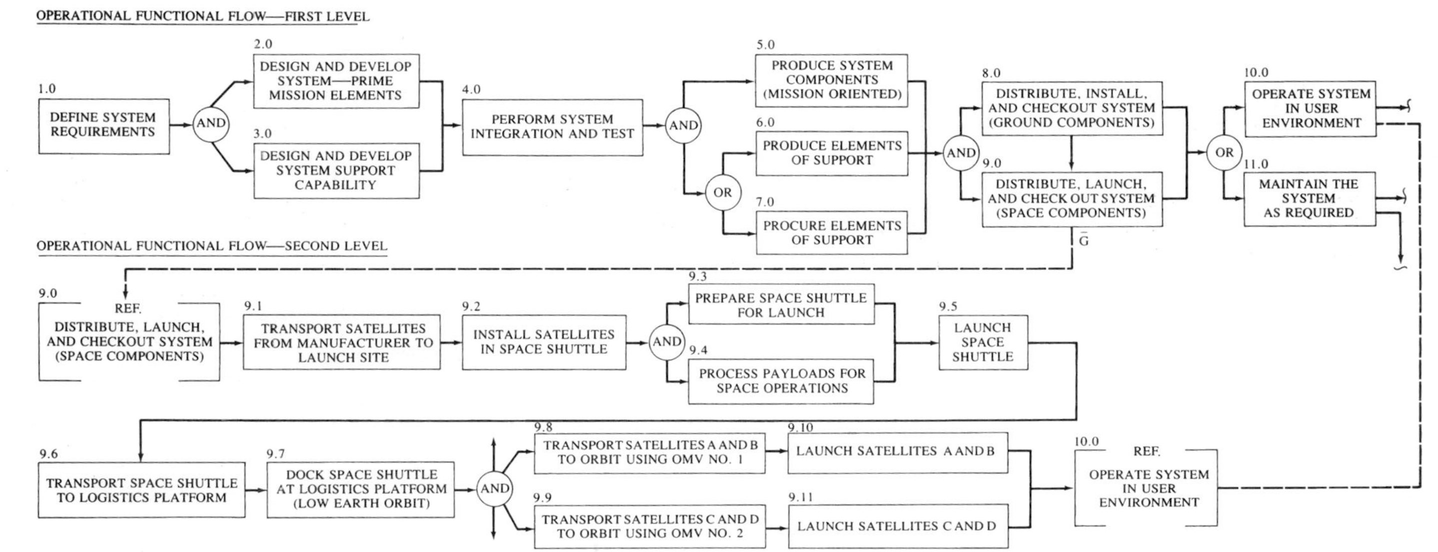
OPERATIONAL FUNCTIONAL FLOW—FIRST LEVEL
1.0
DEFINE SYSTEM REQUIREMENTS
AND
2.0
DESIGN AND DEVELOP SYSTEM—PRIME MISSION ELEMENTS
3.0
DESIGN AND DEVELOP SYSTEM SUPPORT CAPABILITY
4.0
PERFORM SYSTEM INTEGRATION AND TEST
AND
OR
5.0
PRODUCE SYSTEM COMPONENTS (MISSION ORIENTED)
6.0
PRODUCE ELEMENTS OF SUPPORT
7.0
PROCURE ELEMENTS OF SUPPORT
AND
8.0
DISTRIBUTE, INSTALL, AND CHECKOUT SYSTEM (GROUND COMPONENTS)
9.0
DISTRIBUTE, LAUNCH, AND CHECK OUT SYSTEM (SPACE COMPONENTS)
G
OR
10.0
OPERATE SYSTEM IN USER ENVIRONMENT
11.0
MAINTAIN THE SYSTEM AS REQUIRED
OPERATIONAL FUNCTIONAL FLOW—SECOND LEVEL
9.0
REF.
DISTRIBUTE, LAUNCH, AND CHECKOUT SYSTEM (SPACE COMPONENTS)
9.1
TRANSPORT SATELLITES FROM MANUFACTURER TO LAUNCH SITE
9.2
INSTALL SATELLITES IN SPACE SHUTTLE
AND
9.3
PREPARE SPACE SHUTTLE FOR LAUNCH
9.4
PROCESS PAYLOADS FOR SPACE OPERATIONS
9.5
LAUNCH SPACE SHUTTLE
9.6
TRANSPORT SPACE SHUTTLE TO LOGISTICS PLATFORM
9.7
DOCK SPACE SHUTTLE AT LOGISTICS PLATFORM (LOW EARTH ORBIT)
AND
9.8
TRANSPORT SATELLITES A AND B TO ORBIT USING OMV NO. 1
9.9
TRANSPORT SATELLITES C AND D TO ORBIT USING OMV NO. 2
9.10
LAUNCH SATELLITES A AND B
9.11
LAUNCH SATELLITES C AND D
10.0
REF.
OPERATE SYSTEM IN USER ENVIRONMENT

Figure A.3 Operational flow diagram for a space system.

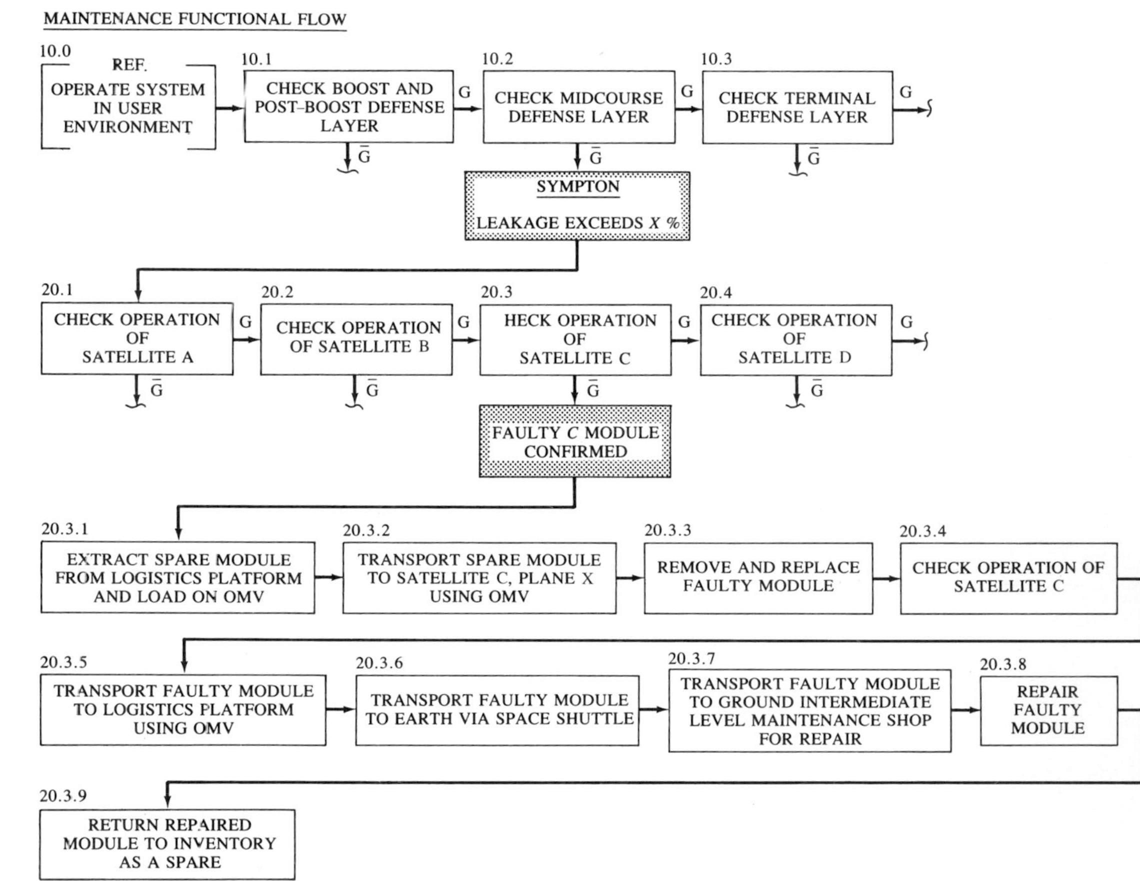

Figure A.4 Space system maintenance flow (example).

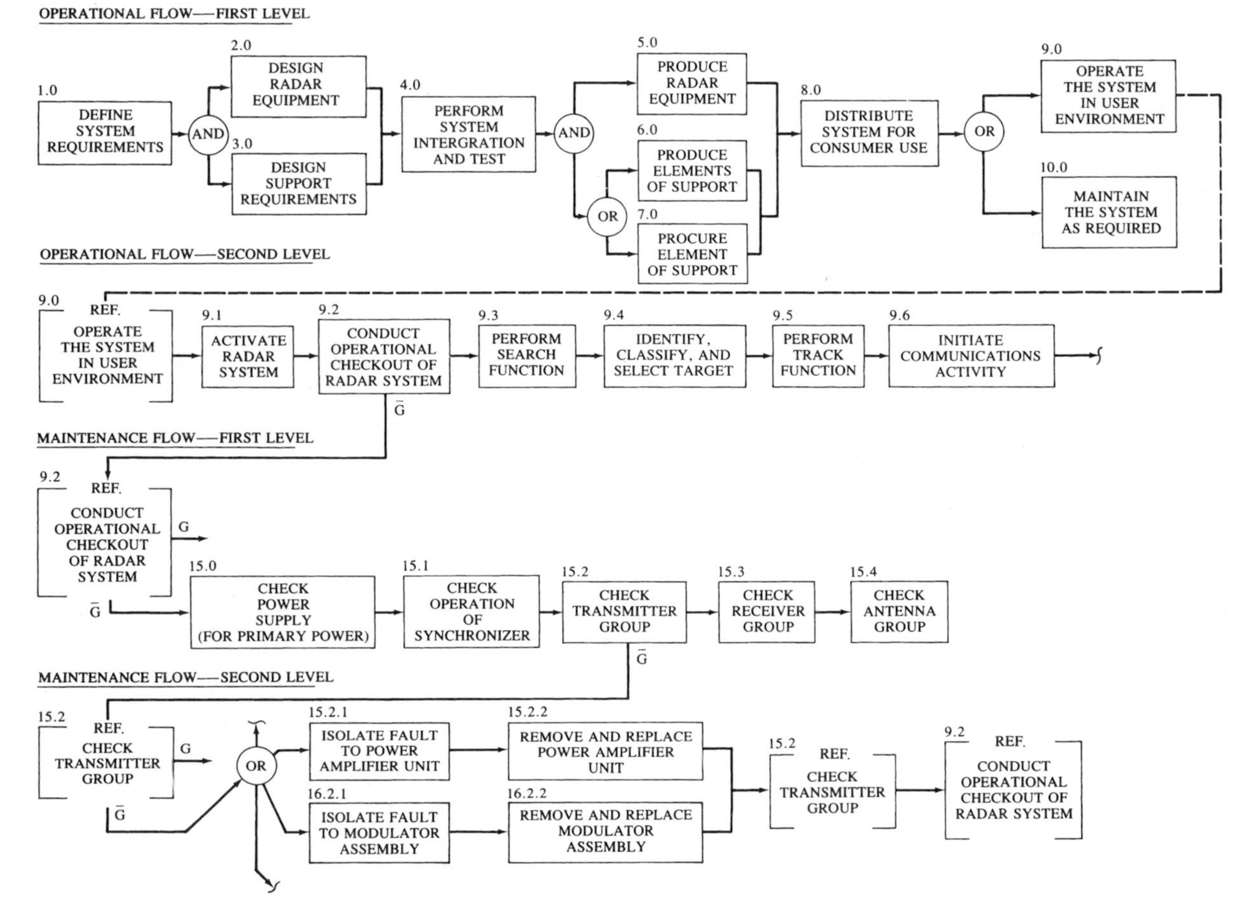

Figure A.5 Radar system flow (example).

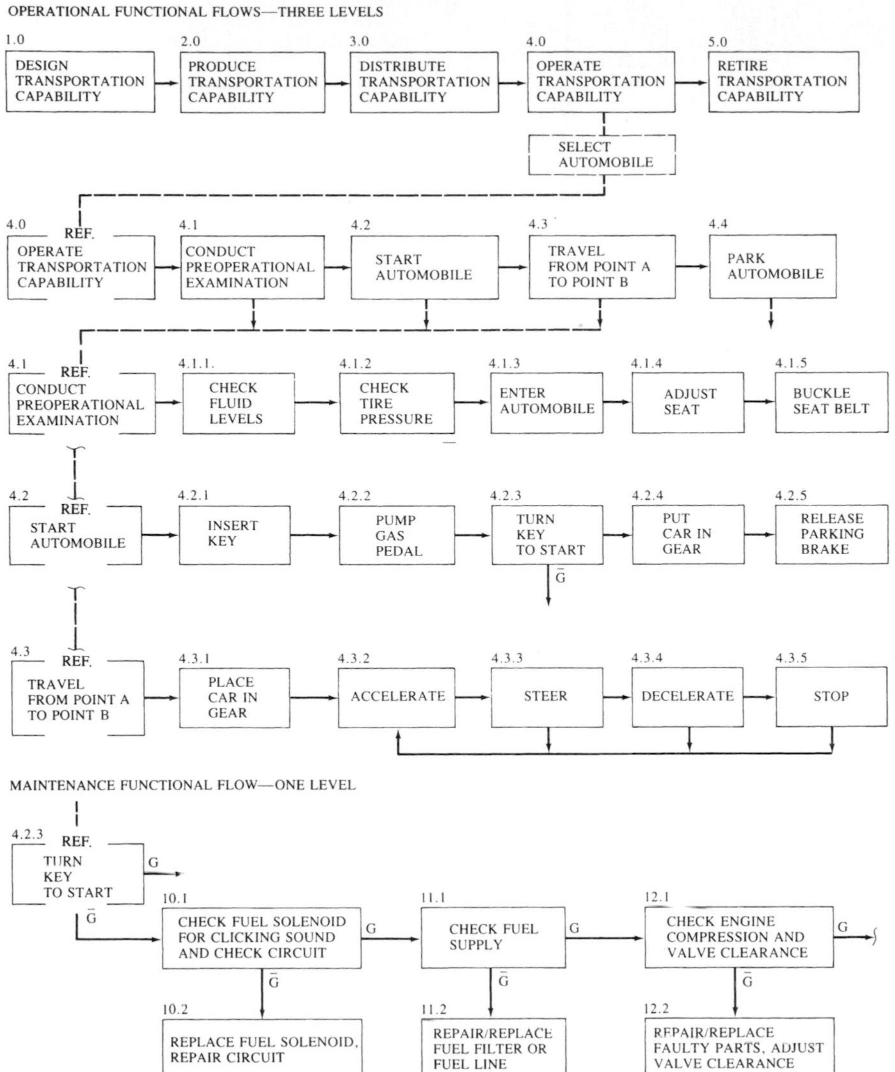

Figure A.6 Operational and maintenance flow diagram for an automobile.

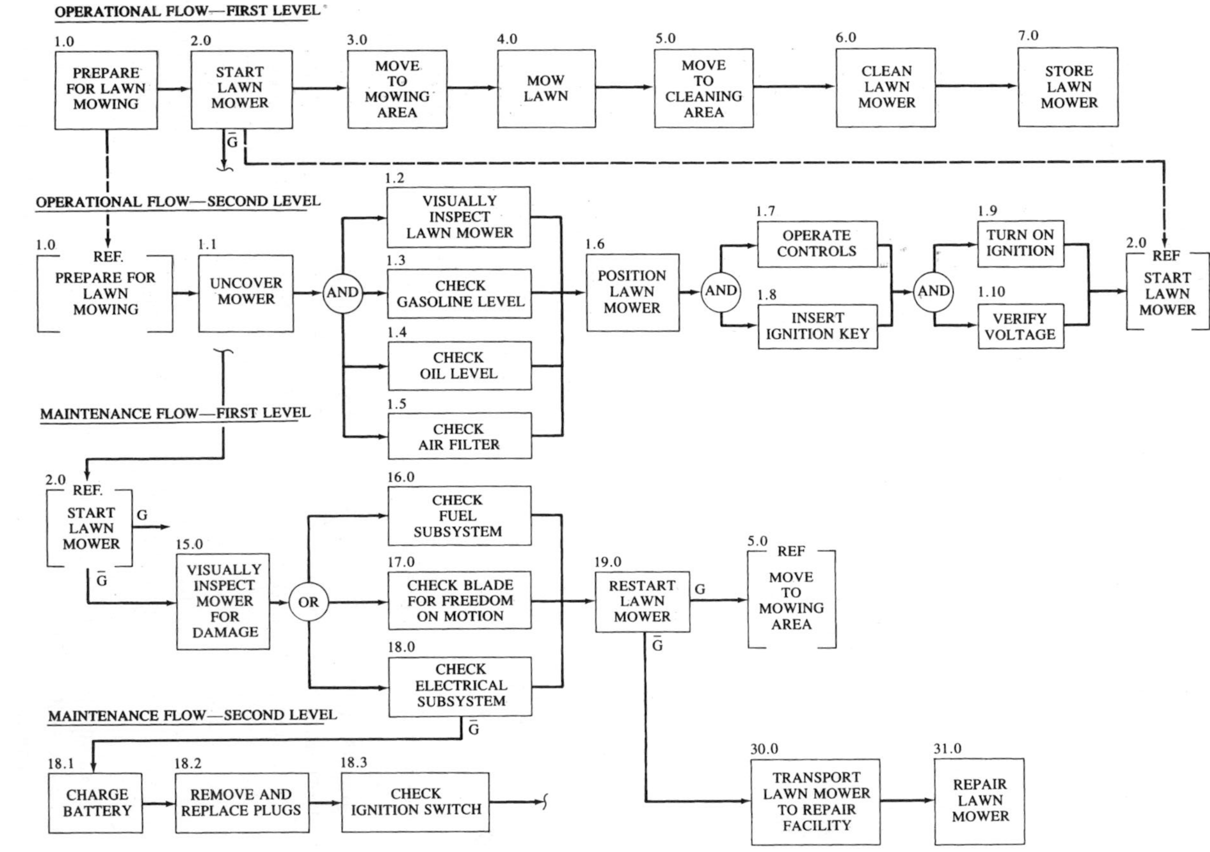

Figure A.7 Operational and maintenance flow diagram for a lawn mowing system.

A.2 DESIGN REVIEW CHECKLISTS

Throughout the system design and development process, there are many occasions when a design review and evaluation (in some form) is appropriate. Referring to Sections 3.9, 4.6, and 5.7, formal design reviews are conducted at discrete times in the life cycle, evolving from a conceptual design review at the system level to a critical design review at the detailed level before going into production (refer to Figure 5.10). During the day-to-day design participation and integration process, there may be a series of informal mini-reviews directed to specific critical areas in the design. In the evaluation of supplier proposals and/or alternative COTS items, each candidate may be evaluated in terms of both the internal characteristics in design and the proposed support infrastructure.

To facilitate the evaluation process, design review *checklists* are often developed to aid in the selection of criteria upon which a review may be based. Referring to Figure 5.9, an abbreviated design review checklist is presented as an example. Topical areas of importance are selected at a broad level (e.g., system operational requirements defined?), and additional areas of concern are identified at the subsystem level and below (e.g., accessibility adequate?). Each of the topic areas listed should be supported with more detailed criteria, presented in the form of one or more questions. For example, the following questions under "accessibility" might be applicable: Are access requirements compatible with the frequency of need? are access openings required for maintenance at least 7.9 inches by 7.2 inches? can access doors be opened without requiring the use of tools? Similar questions can be developed for each topic in Figure 5.9.

The development of checklists can be accomplished following the general steps illustrated in Figure A.8. The first step is to identify those characteristics in design that are considered to be important and their relative level(s) of importance. This, of course, stems back to the definition of system requirements and the mission(s) that the system must accomplished; i.e., the definition of operational requirements, maintenance concept, the identification and prioritization of TPMs, and so on, as shown in Figure 3.1.

Based on the prioritized TPMs, and the subsequent allocation of such, specific design criteria are developed. Referring to Figure 2.9, design characteristics and applicable design-dependent parameters (DDPs) are noted and aligned to specific levels in the overall system hierarchical structure, with the top-level parameters leading to the second level, the second-level parameters leading down to the third level, and so on. Given the hierarchy of design considerations, the next step is to develop specific questions that will lead the "evaluator" to addressing the most important characteristics of design.

In developing questions, the objective is to be as specific as possible and yet tailored to the system, subsystem, equipment, facility, or item of software being reviewed. Further, the design, or structure, of such questions should be presented in a similar manner. For example, in addressing each topic identified in Figure 5.9, the preferred response should be "YES." To go one step further, items 5 (packaging and mounting), 19 (software), and 22 (testability) have been developed through the more detailed

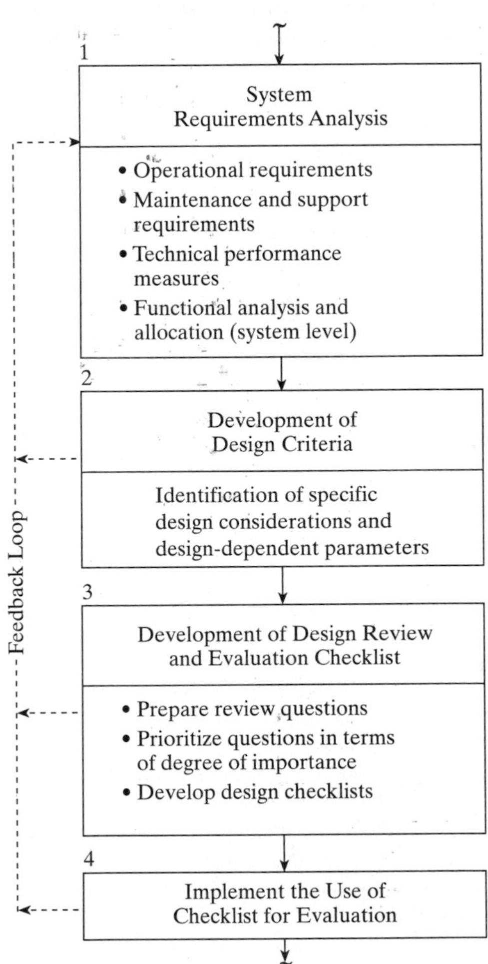

Figure A.8 Development of design review checklists (refer to Figure 3.1).

questions presented in Figure A.9. These, in turn, can be broken down into even more detailed questions as necessary.

To provide the right amount of design emphasis in the proper areas, the applicable questions should be weighted in terms of the level of importance. If *accessibility* is more important than *handling*, for example, then the questions pertaining to accessibility should be weighted heavier than the others. This can be accomplished by adding more questions to the most important areas, or by handling such differences through the establishment of weighting factors such as shown in the evaluation summary in Figure 13.22.

5.0 Packaging and Mounting

5.1 Is the packaging design attractive from the standpoint of consumer appeal (color, shape, size)?

5.2 Is functional packaging incorporated to the maximum extent possible? Interaction effects between packages should be minimized. It should be possible to limit maintenance to the removal of one module (the one containing the fail part) when a failure occurs and not require the removal of two, three, or four modules in order to resolve the problem.

5.3 Is the packaging design compatible with level of repair analysis decisions? Repairable items are designed to include maintenance provisions such as test points, accessibility, and plug-in components. Items to be discarded upon failure should be encapsulated and relatively low in cost. Maintenance provisions within the disposable module are not required.

5.4 Are disposable modules incorporated to the maximum extent practical? It is highly desirable to reduce overall support through a no-maintenance-design concept as long as the items being discarded are relatively high in reliability and low in cost.

5.5 Are plug-in modules and components used to the maximum extent possible (unless the use of plug-in components significantly degrades the equipment reliability)?

5.6 Are accesses between modules adequate to allow for hand grasping?

5.7 Are modules and components mounted such that the removal of any single item for maintenance will not require the removal of other items? Component stacking should be avoided where possible.

5.8 In areas where module stacking is necessary because of limited space, are the modules mounted in such a way that access priority has been assigned in accordance with the predicted removal and replacement frequency? Items that require frequent maintenance should be more accessible.

5.9 Are modules and components, not of a plug-in variety, mounted with four fasteners or less? Modules should be securely mounted, but the number of fasteners should be held to a minimum.

5.10 Are shock-mounting provisions incorporated where shock and vibration requirements are excessive?

5.11 Arc provisions incorporated to preclude installation of the wrong module?

5.12 Are plug-in modules and components removable without the use of tools? If tools are required, they should be of the standard variety.

5.13 Are guides (slides or pins) provided to facilitate module installation?

5.14 Are modules and components labeled?

5.15 Are module and component labels located on top or immediately adjacent to the item and in plain sight?

5.16 Are the labels permanently affixed so that they will not come off during a maintenance action or as a result of environment? Is the information on the label adequate? Disposable modules should be so labeled. In equipment racks, are the heavier items mounted at the bottom of the rack? Unit weight should decrease with the increase installation height.

5.17 Are operator panels optimally positioned? For personnel in the standing position, panels should be located between 40 and 70 inches above the floor. Critical or precise controls should be between 48 and 64 inches above the floor.

Figure A.9 Typical design review questions.

19.0 Software

19.1 Have all system software requirements for operating and maintenance functions been identified? Have these requirements been developed through the system-level functional analysis to provide traceablity?

19.2 Is the software complete in terms of scope and depth of coverage?

19.3 Is the software compatible relative to the equipment with which it interfaces? Is operating software compatible with maintenance software? With other elements of the system?

19.4 Are the language requirements for operating software and maintenance software compatible?

19.5 Is all software adequately covered through good documentation (logic functional flows, coded programs, and so on)?

19.6 Has the software been adequately tested and verified for accuracy (performance), reliablity, and maintainability?

22.0 Testability

22.1 Have self-test provisions been incorporated where appropriate?

22.2 Is reliability degradation due to the incorporation of built-in test minimized? The BIT capability should not significantly impact the reliability of the overall system.

22.3 Is the extent or depth of self-testing compatible with the level of repair analysis?

22.4 Are self-test provisions automatic?

22.5 Have direct fault indicators been provided (a fault light, an audio signal, or a means of determining that a malfunction positively exists)? Are continuous condition monitoring provisions incorporated where appropriate?

22.6 Are test points provided to enable checkout and fault isolation beyond the level of self-test? Test point for fault isolation within an assembly should not be incorporated if the assembly is to be discarded at failure. Test point provisions must be compatible with the level of repair analysis.

22.7 Are test points accessible? Accessibility should be compatible with the extent of maintenance performed. Test points on the operator's front panel are not required for a depot maintenance action.

22.8 Are test points functionally and conveniently grouped to allow for sequential testing (following a signal flow), testing of similar functions. or frequency of use when access is limited?

22.9 Are test points provided for a direct test of all replaceable items?

22.10 Are test points adequately labeled? Each test point should be identified with a unique number, and the proper signal or expected measured output should be specified on a label located adjacent to the test point.

22.11 Are test points adequately illuminated to allow the technician to see the test point number and labeled signal value?

22.12 Can the component malfunctions that could possibly occur be detected through a no-go indication at the system level? Are false alarm rates minimized? This is a measure of test thoroughness.

22.13 Will the prescribed maintenance software provide adequate diagnostic information?

Figure A.9 *(continued)* Typical design review questions.

As a final step, the questions are then combined and included in the development of the applicable checklist. The checklist is then applied to assist the evaluator in addressing the right issues to the extent required based on degrees of importance. The use of checklists provides many benefits, to include serving as a "reminder" for the evaluator to look for certain desired features in reviewing a given design configuration.

B PROBABILITY THEORY AND ANALYSIS

Some models will give satisfactory results if variation is not incorporated. But, these models usually apply to physical phenomena where certainty is generally observed. Models formulated to analize and evaluate human-made systems must incorporate probabilistic elements to be useful in the system engineering process. Acordingly, this appendix presents Probability Concepts and Theory, Probability Distribution Models, and an introduction to Monte Carlo Analysis.

B.1 PROBABILITY CONCEPTS AND THEORY

If one tosses a coin. the outcome will not be known with certainty until either a head or a tail is observed. Prior to the toss, one can assign a probability to the outcome from knowledge of the physical characteristics of the coin. One may know that the diameter of an acorn ranges between 0.80 and 3.20 cm, but the diameter of a specific acorn to be selected from an oak tree will not be known until the acorn is measured. Experiments such as tossing a coin and selecting an acorn provide outcomes called *random events*. Most events in the decision environment are random and probability theory provides a means for quantifying these events.

The Universe and the Sample

The terms *universe* and *population* are used interchangeably. A *universe* consists of all possible objects, stages, and events within an arbitrarily defined boundary. A universe may be finite or it may be infinite. If it is finite, the universe may be very large or very small. If the universe is large, it may sometimes be assumed to be infinite for computational purposes. A universe need not always be large; it may be defined as a dozen events or as only one object. The relative usefulness of the universe as an entity will be paramount in its definition.

A *sample* is a part or portion of a universe. It may range in size from 1 to one less than the size of the universe. A sample is drawn from the population, and observations are made. This is done either because the universe is infinite in size or scope or because the population is large and/or inaccessible as a whole. The sample is used because it is smaller, more accessible, and more economical, and because it suggests certain characteristics of the population.

It is usually assumed that the sample is typical of the population in regard to the characteristics under consideration. The sample is then assessed, and inferences are made in regard to the population as a whole. To the extent that the sample is representative of the population, these inferences may be correct. The problem of selecting a representative sample from a population is an area in statistics to which an entire chapter might be devoted.

Subsequent discussion assumes that the sample is a *random sample*; that is, one in which each object or state or event that consititutes the population has an equally likely chance or probability of being selected and represented in the sample. It is rather simple to state this definition; it may be much more difficult to implement it in practice.

The Probability of an Event

A measure of the relative certainty of an event, before the occurrence of the event, is its probability. The usual representation of a probability is a number $P(A)$ assigned to the outcome A. This number has the following property: $0 \leq P(A) \leq 1$, with $P(A) = 0$ if the event is certain not to occur and $P(A) = 1$ if the event is certain to occur.

Because probability is only a measure of the certainty (or uncertainty) associated with an event, its definition is rather tenuous. The concept of relative frequency is sometimes employed to establish the number $P(A)$. Sometimes probabilities are established a priori. Other times they are simply a subjective estimate. Consider the example of tossing a fair coin. In a lengthy series of tosses, the coin may have come up heads as often as tails. Then the limiting value of the relative frequency of a head will be 0.5 and will be stated as $P(H) = 0.5$.

Two definitions pertaining to events are needed in the development of probability theorems.

1. Events A and B are said to be *mutually exclusive* if both cannot occur at the same time.
2. Event A is said to be *independent* of event B if the probability of the occurrence of A is the same regardless of whether or not B has occurred.

The Addition Theorem

The probability of the occurrence of either one or another of a series of mutually exclusive events is the sum of probabilities of their separate occurrences. If a fair coin is tossed and success is defined as the occurrence of either a head or a tail, then the probability of a head or a tail is

$$\begin{aligned} P(H + T) &= P(H) + P(T) \\ &= 0.5 + 0.5 = 1.0 \end{aligned} \tag{B.1}$$

The key to use of the addition theorem is the proper definition of mutually exclusive events. Such events must be distinct one from another. If one event occurs, it must be impossible for the second to occur at the same time. For example, assume that the probability of having a flat tire during a given period on each of four tires on an automobile is 0.3. Then the probability of having a flat tire on any of the four tires during this time period is not given by the addition of these four probabilities. If $P(T_1) = P(T_2) = P(T_3) = P(T_4) = 0.3$ are the respective probabilities of failure for each of the four tires, then

$$\begin{aligned} P(T_1 + T_2 + T_3 + T_4) &\neq P(T_1) + P(T_2) + P(T_3) + P(T_4) \\ &\neq 0.3 + 0.3 + 0.3 + 0.3 = 1.2 \end{aligned}$$

This cannot be true because the failure of tires is not mutually exclusive. During the time period established, two or more tires may fail, whereas in the example of coin tossing, it is not possible to obtain a head and a tail on the same toss.

The Multiplication Theorem

The probability of occurrence of independent events is the product of the probabilities of their separate events. Implicit in this theorem is the successful occurrence of two events simultaneously or in succession. Thus, the probability of the occurrence of two heads in two tosses of a coin is

$$\begin{aligned} P(H \cdot H) &= P(H)P(H) \\ &= (0.5)(0.5) = 0.25 \end{aligned} \tag{B.2}$$

The tire-failure problem can now be resolved by considering the probabilities of each tire not failing. The probability of each tire not failing is given by $P(\overline{T}_i) = 0.7$. The probability of no tire failing is then given by

$$\begin{aligned} P[(\overline{T}_1)(\overline{T}_2)(\overline{T}_3)(\overline{T}_4)] &= P(\overline{T}_1)P(\overline{T}_2)P(\overline{T}_3)P(\overline{T}_4) \\ &= (0.7)(0.7)(0.7)(0.7) = 0.2401 \end{aligned}$$

And the probability of a tire failing, or of one or more tires failing, is

$$P(T_1 + T_2 + T_3 + T_4) = 1 - 0.2401 = 0.7599$$

This approach is valid, since the probability of one tire not failing is independent of the success or failure of the other three tires.

The Conditional Theorem

The probability of the occurrence of two dependent events is the probability of the first event times the probability of the second event, given that the first has occurred. This may be expressed as

$$P(W_1 \cdot W_2) = P(W_1)P(W_2 | W_1) \tag{B.3}$$

This theorem is similar to the multiplication theorem, except that consideration is given to the lack of independence between events.

As an example, consider the probability of selecting two successive white balls from an urn containing three white and two black balls. This problem reduces to a calculation of the product of the probability of selecting a white ball times the probability of selecting a second white ball, given that the first attempt has been successful, or

$$P(W_1 \cdot W_2) = \left(\frac{3}{5}\right)\left(\frac{2}{4}\right) = \frac{3}{10}$$

The conditional theorem makes allowances for a change in probabilities between two successive events. This theorem will be helpful in constructing finite discrete probability distributions.

The Central Limit Theorem

Although many real-world variables are normally distributed, this assumption cannot be universally applied. However, the distribution of the means of samples or the sums of random variables approximates the normal distribution provided certain assumptions hold. The central limit theorem states: If x has a distribution for which the moment-generating function exists, then the variable $\bar{x}$ has a distribution that approaches normality as the size of the sample tends toward infinity. The sample size required for any desired degree of covergence is a function of the shape of the parent distribution. Fairly good results have been demonstrated with a sample of $n = 4$ for both the rectangular and triangular distributions.

B.2 PROBABILITY DISTRIBUTION MODELS

The pattern of the distribution of probabilities over all possible outcomes is called a probability distribution. *Probability distribution models* provide a means for assigning the likelihood of occurrences of all possible values. Variables described in terms of a probability distribution are conveniently called *random variables*. The specific value of a random variable is determined by the distribution.

A probability distribution is completely defined when the probability associated with every possible outcome is defined. In most instances, the outcomes themselves are represented by numbers or different values of a variable, such as the diameter of an acorn. When the pattern of the probability distribution is expressed as a function of this variable, the resulting function is called a *probability distribution function*.

An example empirical probability distribution function may be developed as follows. A maintenance mechanic attends four machines and his services are needed only when a machine fails. He would like to estimate how many machines will fail each shift. From previous experience, and using the relative frequency concept of probability, the mechanic knows that 40% of the time only one machine will fail at least once during the shift. Further, 30% of the time two machines will fail, three machines will fail 20% of the time, and all four will fail 10% of the time.

The probability distribution of the number of failed machines may be expressed as $P(1) = 0.4$, $P(2) = 0.3$, $P(3) = 0.2$, and $P(4) = 0.1$. This probability distribution is exhibited in Figure B.1.

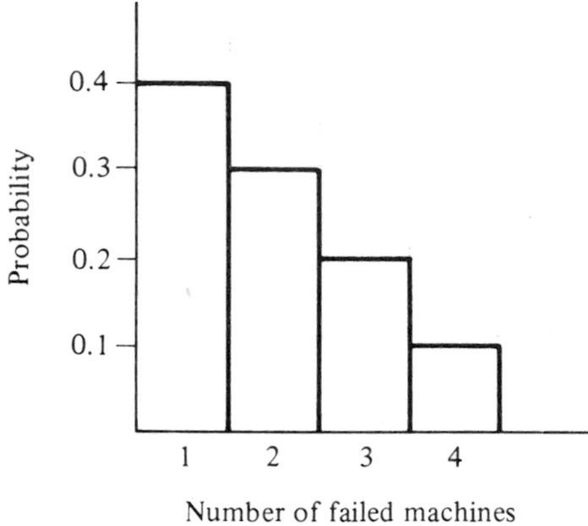

Figure B.1 A probability distribution of the number of failed machines.

The probability distribution function for this case may be defined as

$$p(x) = \frac{5 - x}{10} \qquad \text{if } x = 1, 2, 3, 4$$

$$P(x) = 0 \qquad \text{otherwise}$$

Although the function $P(x) = (5 - x)/10$ uniquely represents the probability distribution pattern for the number of failed machines, the function itself belongs to a wider class of functions of the type $P(x) = (a - x)/b$. All functions of this type indicate similar patterns: yet each pair of numbers (a, b) uniquely defines a specific probability distribution. These numbers (a, b) are called *parameters*.

In the sense that they serve to define the probability distribution function, it is possible to look upon parameters as properties of the distribution function. The choice of representation of parameters is not unique, and the most desirable representation would reflect a measure of the properties of the universe under study. Two most commonly sought measures are the *mean*, an indication of central tendency, and the *variance*, a measure of dispersion.

The probability distribution just presented is discrete in that it assigns probabilities to an event that can only take on integer values. Continuous probability distributions are used to define the probability of the occurrence of an event that may take on values over a continuum. Under certain conditions, it may be desirable to use a continuous probability distribution to approximate a discrete probability distribution. By so doing, tedious summations may be replaced by integrals. In other instances, it may be desirable to make a continuous distribution discrete as when calculations are to be performed on a digital computer. Several discrete and continuous probability distribution models are presented subsequently.

The Binomial Distribution

The binomial distribution is a basic discrete sampling distribution. It is applicable where the probability is sought of exactly x occurrences in n trials of an event that has a constant probability of occurrence p. The requirement of a constant probability of occurrence is satisfied when the population being sampled is infinite in size, or where replacement of the sampled unit takes place.

The probability of exactly x occurrences in n trials of an event that has a constant probability of occurrence p is given as

$$P(x) = \frac{n!}{x!(n-x)!} p^x q^{n-x} \qquad 0 \le x \le n \tag{B.4}$$

where $q = 1 - p$. The mean and variance of this distribution are given by np and npq, respectively.

As an example of the application of the binomial distribution, assume that a fair coin is to be tossed five times. The probability of obtaining exactly two heads is

$$\begin{aligned} P(2) &= \frac{5!}{2!(5-2)!}(0.5)^2(1-0.5)^3 \\ &= 10(0.03125) = 0.3125 \end{aligned}$$

A probability distribution may be constructed by solving for the probability of exactly zero, one, two, three, four, and five heads in five tosses. If $p = 0.5$, as in this example, the resulting distribution is symmetrical. If $p < 0.5$, the distribution is skewed to the right; if $p > 0.5$, the distribution is skewed to the left.

The Uniform Distribution

The uniform or rectangular probablity distribution may be either discrete or continuous. The continuous form of this simple distribution is

$$f(x) = \frac{1}{a} \qquad 0 \le x \le a \tag{B.5}$$

The discrete form divides the interval 0 to a into $n + 1$ cells over the range 0 to n, with $1/(n + 1)$ as the unit probabilities. The mean and variance of the rectangular probability distribution are given as $a/2$ and $a^2/12$ for the continuous case, and as $n/2$ and $n^2/12 + n/6$ for the discrete case.

The general form of the rectangular probability distribution is shown in Figure B.2. The probability that a value of x will fall between the limits 0 and a is equal to

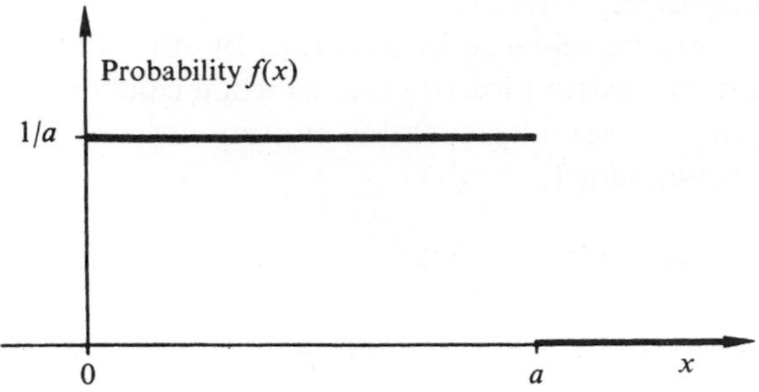

Figure B.2 The general form of the rectangular distribution.

unity. One may determine the probability associated with a specific value of x, or a range of x, by integration for the continuous case. The probability associated with a specific value of x for the discrete distributions of the previous section was found from the functions given. Determination of the probability associated with a range of x required a summation of individual probabilities. This is a fundamental difference in dealing with discrete and continuous probability distributions.

Values drawn at random from the rectangular distribution with x allowed to take on values ranging from 0 through 9 are given in Appendix C, Table C.1. These random rectangular variates may be used to randomize a sample or to develop values drawn at random from other probability distributions as will be illustrated in the last section of this Appendix.

The Poisson Distribution

The Poisson is a discrete distribution useful in its own right and as an approximation to the binomial. It is applicable when the opportunity for the occurrence of an event is large, but when the actual occurrence is unlikely. The probability of exactly x occurrences of an event of probability p in a sample n is

$$P(x) = \frac{(\mu)^x e^{-\mu}}{x!} \qquad 0 \leq x \leq \infty \tag{B.6}$$

The mean and variance of this distribution are equal and given by μ, where $\mu = np$. Cumulative Poisson probabilities are tabulated for values of μ ranging up to 24 in Appendix C, Table C.2.

As an example of the application of the Poisson distribution, assume that a sample of 100 items is selected from a population of items which are 1% defective. The probability of obtaining exactly three defectives in the sample is found from Equation B.6 or Table C.2 as

$$P(3) = \frac{(1)^3(2.72)^{-1}}{3!} = 0.061$$

The Poisson distribution may be used as an approximation to the binomial distribution. Such an approximation is good when n is relatively large, p is relatively small, and in general, $pn < 5$. These conditions were satisfied in the previous example.

The Exponential Distribution

The exponential probability distribution is given by

$$f(x) = \frac{1}{a} e^{-x/a} \qquad 0 \le x \le \infty. \tag{B.7}$$

The mean and variance of this distribution are given by a and a^2, respectively. Its form is illustrated in Figure B.3.

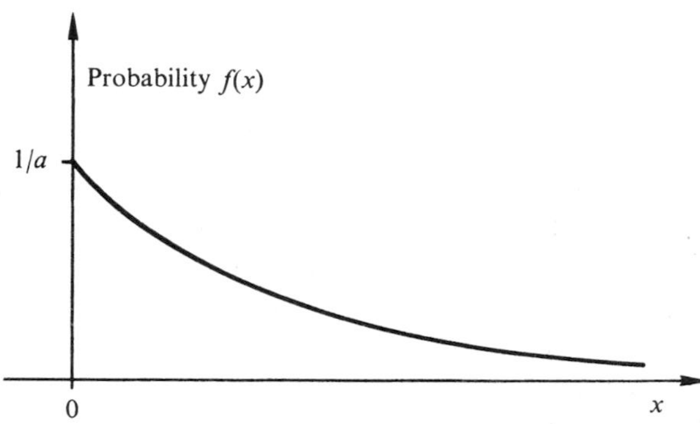

Figure B.3 The general form of the exponential distribution.

As an example of the application of the exponential probability distribution, consider the selection of a light bulb from a population of light bulbs whose life is known to be exponentially distributed with a mean $\mu = 1{,}000$ hour. The probability of the life of this sample bulb not exceeding 1,000 hr would be expressed as $P(x \le 1{,}000)$. This would be the proportional area under the exponential function over the range $x = 0$ to $x = 1{,}000$, or

$$\begin{aligned} P(x \le 1{,}000) &= \int_0^{1{,}000} f(x)\,dx \\ &= \int_0^{1{,}000} \frac{1}{1{,}000} e^{-x/1{,}000} dx \\ &= -e^{-x/1{,}000} \Big|_0^{1{,}000} \\ &= 1 - e^{-1} = 0.632 \end{aligned}$$

Note that 0.632 is that proportion of the area of an exponential distribution to the left of the mean. This illustrates that the probability of the occurrence of an event exceeding the mean value is only $1 - 0.632 = 0.368$.

The Normal Distribution

The normal or Gaussian probability distribution is one of the most important of all distributions. It is defined by

$$f(x) = \frac{1}{\sigma\sqrt{2\pi}} e^{[-(x-\mu)^2/2\sigma^2]} \qquad -\infty \le x \le +\infty \tag{B.8}$$

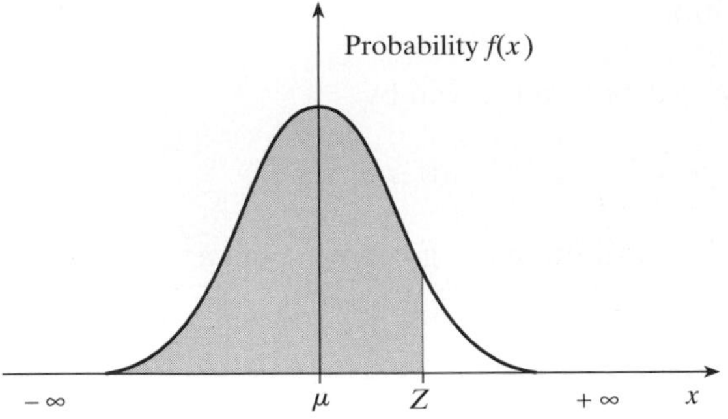

Figure B.4 The normal probability distribution.

The mean and variance is μ and σ^2, respectively. Variation is inherent in nature, and much of this variation appears to follow the normal distribution, the form of which is given in Figure B.4.

The normal distribution is symmetrical about the mean and possesses some interesting and useful properties regarding its shape. Where distances from the mean are expressed in terms of standard deviations, σ, the relative areas defined between two such distances will be constant from one distribution to another. In effect, all normal distributions, when defined in terms of a common value of μ and σ, will be identical in form, and corresponding probabilities may be tabulated. Normally, cumulative probabilities are given from $-\infty$ to any value expressed as standard deviation units as in Appendix C, Table C.3. This table gives the probability from $-\infty$ to Z, where Z is a standard normal variate defined as

$$Z = \frac{x - \mu}{\sigma} \tag{B.9}$$

This is shown as the shaded area in Figure B.4.

The area from $-\infty$ to -1σ is indicated as the shaded area in Figure B.5. From Table C.3, the probability of x falling in this range in 0.1587. Likewise, the area from $-\infty$ to $+2\sigma$ is 0.9773. If the probability of a value falling in the interval -1σ to $+2\alpha$ is required, the following computations are made.

$$\begin{aligned} P(\text{area} - \infty \text{ to } + 2\sigma) &= 0.9773 \\ -P(\text{area} - \infty \text{ to } - 1\sigma) &= 0.1587 \\ P(\text{area} - 1\sigma \text{ to } + 2\sigma) &= 0.8186 \end{aligned}$$

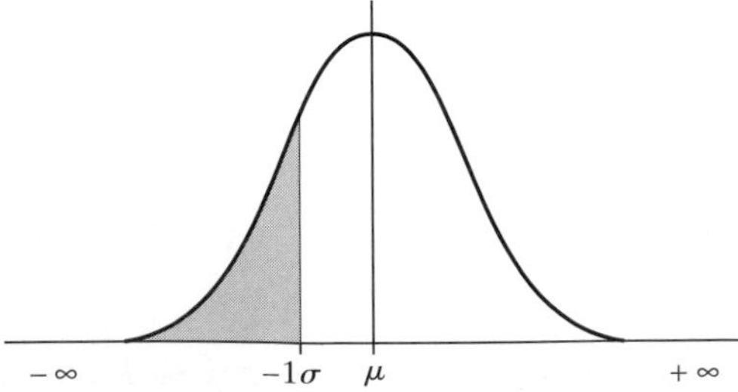

Figure B.5 The area from $-\infty$ to -1σ under the normal distribution.

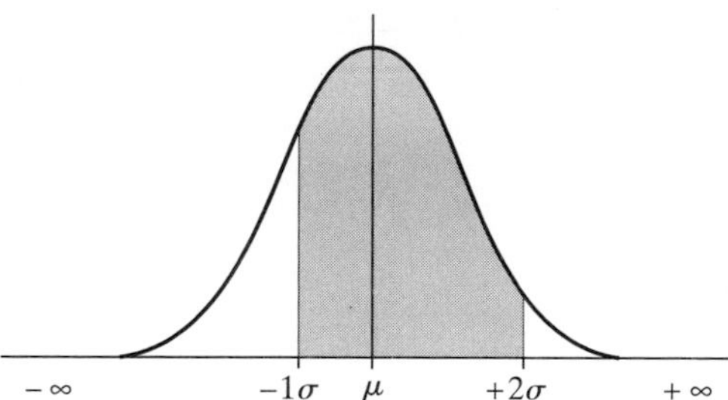

Figure B.6 The area from -1σ to $+2\sigma$ under the normal distribution.

This situation is shown in Figure B.6. The probabilities associated with any normal probability distribution can be calculated by the use of Table C.3.

The Lognormal Distribution

The lognormal probability distribution is related to the normal distribution. If a random variable $Y = \ln X$ is normally distributed with mean μ and variance σ^2, then the random variable X follows the lognormal distribution. Accordingly, the probability distribution function of the random variable X is defined as

$$f(x) = \frac{1}{x\sigma\sqrt{2\pi}} e^{[-(\ln x - \mu)^2/2\sigma^2]}, \qquad x > 0 \tag{B.10}$$

where $\mu \in (-\infty, \infty)$ is called the scale parameter and $\sigma > 0$ is called the shape parameter. The mean and variance of the lognormal distribution are $e^{\mu + \sigma^2/2}$ and $e^{2\mu + \sigma^2}(e^{\sigma^2} - 1)$ respectively.

Figure B.7 shows a lognormal distribution with a constant shape parameter and various scale parameters. Figure B.8 shows a lognormal distribution with a constant scale parameter and various shape parameters.

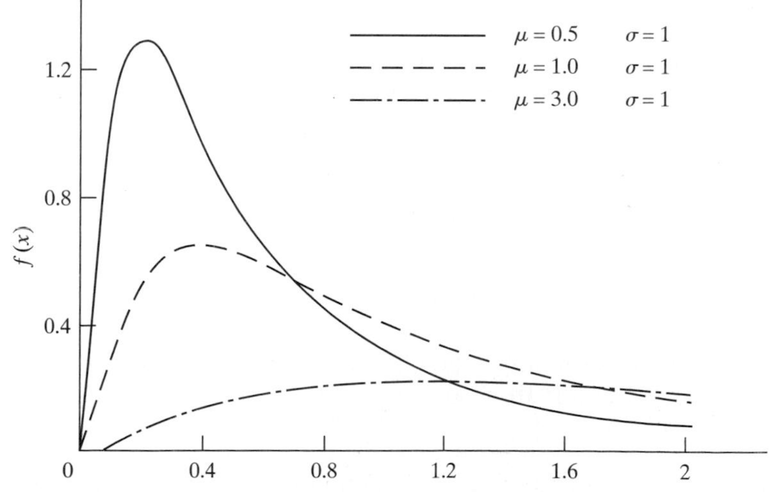

Figure B.7 Longnormal distribution with constant shape parameter.

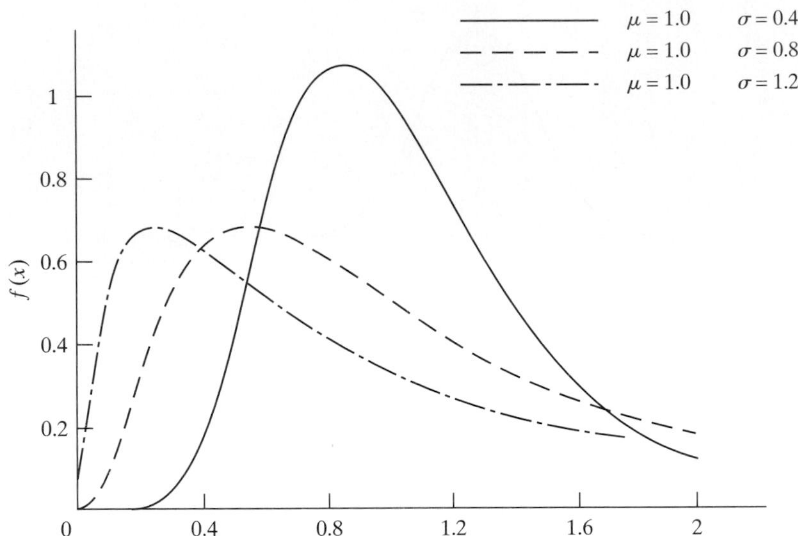

Figure B.8 Longnormal distribution with constant scale parameter.

If $X \sim \ln(\mu, \sigma^2)$ then $\ln X \sim N(\mu, \sigma^2)$. This implies that if n data points $x_1, x_2, \ldots, x_n$ are lognormal, then the logarithms of these data points, $\ln x_1$, $\ln x_2$, ..., $\ln x_n$ will be normally distributed and may be used for parameter estimation, goodness of fit testing, and hypothesis testing.

The Weibull Distribution

The probability distribution function of a random variable X, which follows the Weibull distribution is given by

$$f(x) = \frac{\alpha}{\beta^\alpha} x^{\alpha-1} e^{-(x/\beta)^\alpha}, \qquad x \geq 0 \tag{B.11}$$

where $\alpha > 0$, $\beta > 0$ are the shape and scale parameters respectively, and defined on $[0, \infty)$. A Weibull variate X has mean $\frac{\beta}{\alpha}\Gamma\left(\frac{1}{\alpha}\right)$ and variance $\frac{\beta^2}{\alpha}\left\{2\Gamma\left(\frac{2}{\alpha}\right) - \frac{1}{\alpha}\left[\Gamma\left(\frac{1}{\alpha}\right)\right]^2\right\}$, where $\Gamma(\ )$ is the gamma function

$$\Gamma(z) = \int_0^\infty t^{z-1} e^{-t} dt$$

Weibull distributions with different shape and scale parameters are illustrated in Figures B.9 and B.10, respectively. The Weibull distribution has some interesting characteristics.

1. For $\alpha = 1$, the Weibull distribution is the same as the exponential distribution with parameter β.

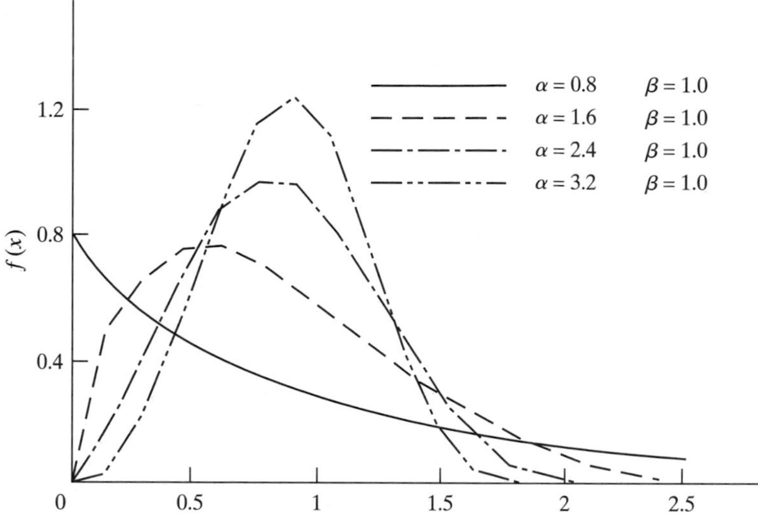

Figure B.9 Weibull distribution with various shape parameters.

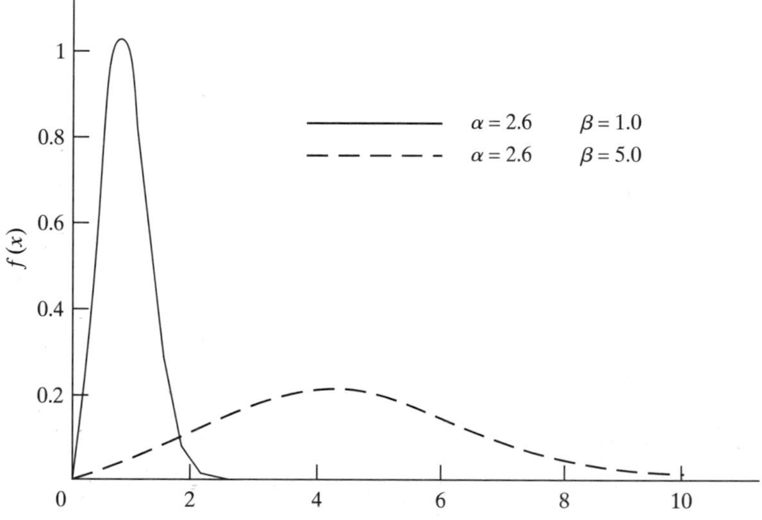

Figure B.10 Weibull distribution with various scale parameters.

2. For $\alpha = 3.4$, the Weibull distribution approximates the normal distribution.
3. If $X \sim$ Weibull (α, β) then $X^a \sim$ exponential$(\beta\alpha)$.

The weibull distribution is frequently used as a time-to-failure model and in reliability analysis, especially in instances where failure data cannot be fitted by the exponential distribution.

B.3 MONTE CARLO ANALYSIS

The decision environment is made up of many random variables. Thus, models used to explain operational systems must often incorporate probabilistic elements. In some cases, formal mathematical solutions are difficult or impossible to obtain from these models. Under such conditions it may be necessary to use a method known as *Monte Carlo analysis*. When applied to an operational system, Monte Carlo analysis provides a powerful means of simulation.

As an introduction to the idea of Monte Carlo analysis, consider its application to the determination of the area of a circle with a diameter of 1 in. Proceed as follows:

1. Enclose the circle of a 1-in. square as shown in Figure B.11.
2. Divide two adjoining sides of the square into tenths, or hundredths, or thousandths, and so on, depending on the accuracy desired.
3. Secure a sequence of pairs of random rectangular variates from Table C.1 in Appendix C.
4. Use each pair of rectangular variates to determine a point within the square and possibly within the circle. This process is illustrated in Table B.1 for 100 trials.
5. Compute a ratio of the number of times a point falls within the circle to the total number of trials. The value of this ratio is an approximate area for the circle expressed as a fraction of the 1 in.^{2} represented by the square. It is 79/100, or 0.79, in this example.

The example just presented has a well-known mathematical solution as follows:

$$A = \pi r^2 = 3.1416(0.50)^2 = 0.7854$$

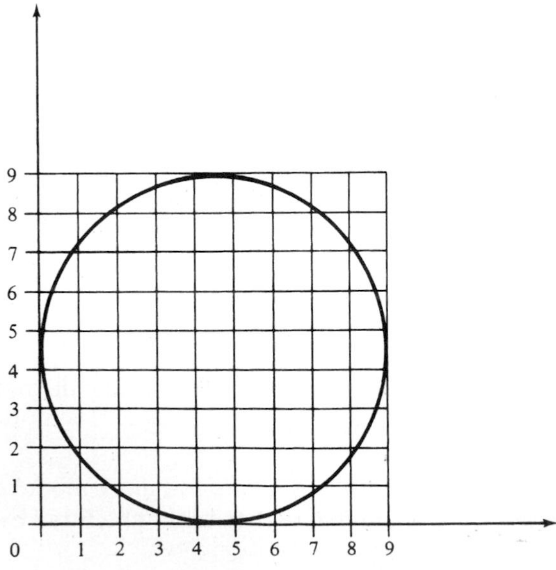

Figure B.11 Area of circle by Monte Carlo analysis.

TABLE B.1 Determining the area of a Circle

Trial	Random number	In	Out
1	73	x	
2	26	x	
3	19		x
4	84	x	
5	81	x	
6	47	x	
7	18		x
8	44	x	
⋮	⋮	⋮	⋮
100	35	x	
Total		79	21

Most models used to explain operational systems incorporate probabilistic elements, and for this reason Monte Carlo analysis is often utilized. A step-by-step procedure used in Monte Carlo analysis is presented next.

Consider the determination of the distribution of the random variable LTD (lead time demand) which is expressed as

$$LTD = \sum^{L} D$$

where D is a random variable representing demand in units per day and L is a random variable expressing lead time in days. The mathematical expression shown above establishes the computational procedure required.

Determine the Distributions

Each random variable in the model refers to an event in the system being studied. Therefore, an important step in Monte Carlo analysis is determining the behavior of these random variables. This involves the development of empirical frequency distributions to describe the relevant variables by the collection of historical data. Once this is done, the frequency distribution for each variable may be studied statistically to ascertain whether it conforms to a known theoretical distribution.

For the example under consideration, assume that data for random variable L have been collected and studied. Suppose that it is concluded that this random variable conforms to the exponential distribution with a mean of 4. This is a distribution model from Equation B.7 expressed as

$$f(L) = \frac{1}{4} e^{-L/4}$$

Likewise, assume that data for random variable D have been collected. Suppose that the resulting frequency distribution is as shown in Figure B.12.

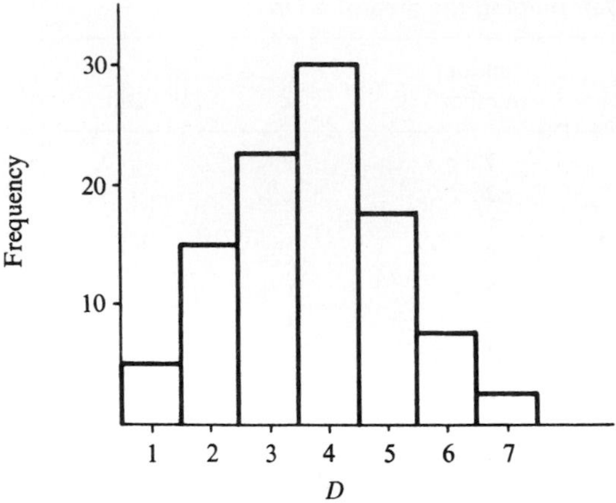

Figure B.12 Distribution of the random variable D.

Develop the Cumulative Distributions

The probability distributions describing the random variables will be theoretical or empirical or both. Those expressed theoretically may be manipulated mathematically in order to obtain the required cumulative probability distributions. The cumulative distribution for random variable L may be expressed as follows:

$$\begin{aligned} F(L) &= \int_0^x f(L)\,dx \\ &= \int_0^x \frac{1}{4} e^{-L/4}\,dx \\ &= -e^{-L/4}\Big|_0^L \\ &= 1 - e^{-L/4} \end{aligned} \tag{B.12}$$

This cumulative exponential distribution is shown in Figure B.13.

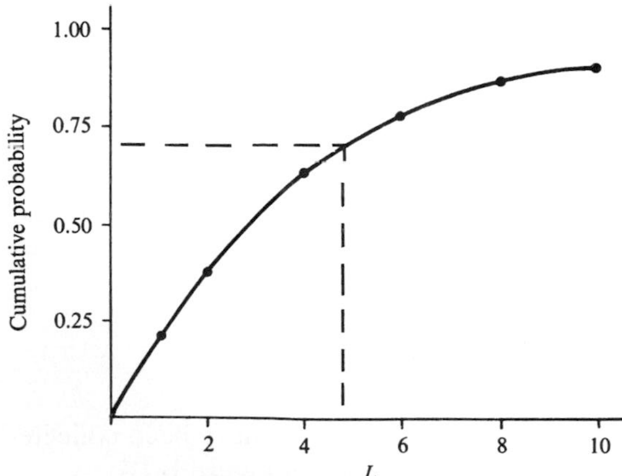

Figure B.13 Cumulative exponential distribution of mean $L = 4$.

Formalize the System Logic

The system under study is usually assumed to operate in accordance with a certain logical pattern. Before beginning the actual Monte Carlo process, it is necessary to formalize the operational procedure by the construction of a model. This may require the development of a step-by-step flow diagram outlining the logic-If the actual simulation process is to be performed on a digital computer, it is mandatory to prepare an accurate logic diagram. From this, the computer can be programmed to pattern the process under study.

Those distributions expressed empirically cannot be converted to cumulative distributions by mathematical means. This is the case for random variable D shown in Figure B.12. The Monte Carlo analysis, however, requires a cumulative distribution for each random variable. Therefore, graphical means must be used. Figure B.14 exhibits the cumulative distribution corresponding to the distribution of Figure B.12. It is developed by summing the probabilities from left to right and plotting the results. This is the same as the mathematical process used for the exponential distribution.

The cumulative distributions of Figure B.13 and Figure B.14 may be used to transform random rectangular variates to values drawn at random from the basic distributions. By this means, specific values are determined for random variable L and random variable D. Thus, a random rectangular variate with a value of 68 gives a value for the random variable L of 4.57, as shown in Figure B.13. Similarly, a random rectangular variate with a value of 654 gives a value for the random variable D of 4, as

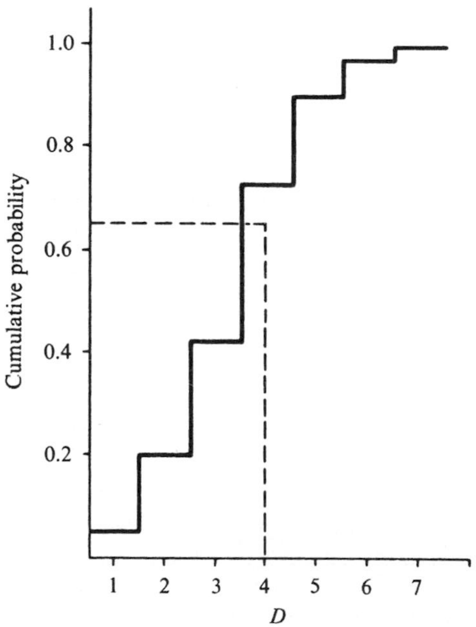

Figure B.14 Cumulative distribution of the random variable D.

shown in Figure B.14. Use of Figure B.13 may be bypassed, since its mathematical equivalent is available. Equation B.12 may be used to transform the random rectangular variate 681 to a random exponential variable directly. The result is

$$0.681 = 1 - e^{-L/4}$$

$$L = 4.572$$

Perform the Monte Carlo Analysis

The example under consideration requires that L random demand variates be added together to form one random LTD value. This is recognized to be the lead-time demand resulting from demand distributed as illustrated in Figure B.12 and lead time distributed exponentially as illustrated in Figure B.13.

To perform the Monte Carlo analysis, L could be arbitrarily rounded upward to the nearest integer. Thus, for the value of L generated in Figure B.13, set $L = 5$ and draw five values at random from the demand distribution, D. Suppose that these are 3, 1, 5, 2, and 4. The first trial then yields a value of LTD of 15. This process is continued numerous times to develop the distribution of LTD as shown in Table B.12. The distribution of LTD would be useful in setting the order level in an inventory system.

TABLE B.2 Determining the Lead-time Demand

Trial	Random number	L rounded	Random number	demand	LTD
1	68	5	28	3	
			03	1	
			81	5	
			18	2	
			54	4	15
2	31	2	92	6	
			63	4	10
3	61	4	05	1	
			90	5	
			33	3	
			82	5	14
4	78	7	62	4	
			14	1	
			89	5	
			37	3	
			17	2	
			19	2	
			94	6	23
⋮	⋮	⋮	⋮	⋮	⋮

C PROBABILITY AND STATISTICAL TABLES

TABLE C.1 RANDOM RECTANGULAR VARIATES

Random variates from the rectangular distribution, $f(x) = \frac{1}{10}$, are presented. (These tables are reproduced with permission from the RAND Corporation, *A Million Random Digits with 100,000 Normal Deviates* [New York: The Free Press, 1955], pp. 130–31.)

TABLE C.2 CUMULATIVE POISSON PROBABILITIES

Cumulative probabilities $\times$ 1,000 for the Poisson distribution are given for μ up to 24. The tabular values were computed from $\Sigma(\mu^x e^{-\mu}/x!)$. (These tables are reproduced from W. J. Fabrycky, P. M. Ghare, and P. E. Torgersen, *Applied Operations Research and Management Science* [Upper Saddle River, N.J.: Prentice Hall, Inc., 1984].)

TABLE C.3 CUMULATIVE NORMAL PROBABILITIES

Cumulative probabilities are given from $-\infty$ to $Z = (x - \mu)/\sigma$ for the standard normal distribution. (Tabular values are adapted with permission from E. L. Grant, and R. S. Leavenworth, *Statistical Quality Control*, 4th ed. [New York: McGraw-Hill Book Co., 1972].)

TABLE C.1 Random Rectangular Variates

14541	36678	54343	94932	25238	84928	30668	34992	69955	06633
88626	98899	01337	48085	83315	33563	78656	99440	55584	54178
31466	87268	62975	19310	28192	06654	06720	64938	67111	55091
52738	52893	51373	43430	95885	93795	20129	54847	68674	21040
17444	35560	35348	75467	26026	89118	51810	06389	02391	96061
62596	56854	76099	38469	26285	86175	65468	32354	02675	24070
38338	83917	50232	29164	07461	25385	84838	07405	38303	55635
29163	61006	98106	47538	99122	36242	90365	15581	89597	03327
59049	95306	31227	75288	10122	92687	99971	97105	37597	91673
67447	52922	58657	67601	96148	97263	39110	95111	04682	64873
57082	55108	26992	19196	08044	57300	75095	84330	92314	11370
00179	04358	95645	91751	56618	73782	38575	17401	38686	98435
65420	87257	44374	54312	94692	81776	24422	99198	51432	63943
52450	75445	40002	69727	29775	32572	79980	67902	97260	21050
82767	26273	02192	88536	08191	91750	46993	02245	38659	28026
17066	64286	35972	32550	82167	53177	32396	34014	20993	03031
86168	32643	23668	92038	03096	51029	09693	45454	89854	70103
33632	69631	70537	06464	83543	48297	67693	63137	62675	56572
77915	56481	43065	24231	43011	40505	90386	13870	84603	73101
90000	92887	92668	93521	44072	01785	27003	01851	40232	25842
55809	70237	10368	58664	39521	11137	20461	53081	07150	11832
50948	64026	03350	03153	75913	72651	28651	94299	67706	92507
27138	59012	27872	90522	69791	85482	80337	12252	83388	48909
03534	58643	75913	63557	25527	47131	72295	55801	44847	48019
48895	34733	58057	00195	79496	93453	07813	66038	55245	43168
57585	23710	77321	70662	82884	80132	42281	17032	96737	93284
95913	24669	42050	92757	68677	75567	99777	49246	93049	79863
12981	37145	95773	92475	43700	85253	33214	87656	13295	09721
62349	64163	57369	65773	86217	00135	33762	72398	16343	02263
68193	37564	56257	50030	53951	84887	34590	22038	40629	29562
56203	82226	83294	60361	29924	09353	87021	08149	11167	81744
31945	23224	08211	02562	20299	85836	94714	50278	99818	62489
68726	52274	59535	80873	35423	05166	06911	25916	90728	20431
79557	25747	55585	93461	44360	18359	20493	54287	43693	88568
05764	29803	01819	51972	91641	03524	18381	65427	11394	37447
30187	66931	01972	48438	90716	21847	35114	91839	26913	68893
30858	43646	96984	80412	91973	81339	05548	49812	40775	14263
85117	38268	18921	29519	33359	80642	95362	22133	40322	37826
59422	12752	56798	31954	19859	32451	04433	62116	14899	38825
73479	91833	91122	45524	73871	77931	67822	95602	23325	37718
83648	66882	15327	89748	76685	76282	98624	71547	49089	33105
19454	91265	09051	94410	06418	34484	37929	61070	62346	79970
49327	97807	61390	08005	71795	49290	52285	82119	59348	55986
54482	51025	12382	35719	66721	84890	38106	44136	95164	92935
30487	19459	25693	09427	10967	36164	33893	07087	16141	12734
42998	68627	66295	59360	44041	76909	56321	12978	31304	97444
03668	61096	26292	79688	05625	52198	74844	69815	76591	35398
45074	91457	28311	56499	60403	13658	81838	54729	12365	24082
58444	99255	14960	02275	37925	03852	81235	91628	72136	53070
82912	91185	89612	02362	93360	20158	24796	38284	55328	96041

TABLE C.1 *(continued)* Random Rectangular Variates

44553	29642	20317	69470	57789	27631	68040	73201	51302	66497
01914	36106	71351	69176	53353	57353	42430	68050	47862	61922
00768	37958	69915	17709	31629	49587	07136	42959	56207	03625
29742	67676	62608	54215	97167	07008	77130	15806	53081	14297
07721	20143	56131	56112	23451	48773	38121	74419	11696	42614
99158	07133	04325	43936	83619	77182	55459	28808	38034	01054
97168	13859	78155	55361	04871	78433	58538	78437	14058	79510
07508	63835	83056	74942	70117	91928	10383	93793	31015	60839
68400	66460	67212	28690	66913	90798	71714	07698	31581	31086
88512	62908	65455	64015	00821	23970	58118	93174	02201	16771
94549	31145	62897	91582	94064	14687	47570	83714	45928	32685
02307	86181	44897	60884	68072	77693	83413	61680	55872	12111
28922	89390	66771	39185	04266	55216	91537	36500	48154	04517
73898	85742	97914	74170	10383	16366	37404	73282	20524	85004
66220	81596	18533	84825	43509	16009	00830	13177	54961	31140
64452	91627	21897	31830	62051	00760	43702	22305	79009	15065
26748	19441	87908	06086	62879	99865	50739	98540	54002	98337
61328	52330	17850	53204	29955	48425	84694	11280	70661	27303
89134	85791	73207	93578	62563	37205	97667	61453	01067	31982
91365	23327	81658	56441	01480	09677	86053	11505	30898	82143
54576	02572	60501	98257	40475	81401	31624	27951	60172	21382
39870	60476	02934	39857	06430	59325	84345	62302	98616	13452
82288	29758	35692	21268	35101	77554	35201	22795	84532	29927
57404	93848	87288	30246	34990	50575	49485	60474	17377	46550
22043	17104	49653	79082	45099	24889	04829	49097	58065	23492
61981	00340	43594	22386	41782	94104	08867	68590	61716	36120
96056	16227	74598	28155	23304	66923	07918	15303	44988	79076
64013	74715	31525	62676	75435	93055	37086	52737	89455	83016
59515	37354	55422	79471	23150	79170	74043	49340	61320	50390
38534	33169	40448	21683	82153	23411	53057	26069	86906	49708
41422	50502	40570	59748	59499	70322	62416	71408	06429	70123
38633	80107	10241	30880	13914	09228	68929	06438	17749	81149
48214	75994	31689	25257	28641	14854	72571	78189	35508	26381
54799	37862	06714	55885	07481	16966	04797	57846	69080	49631
25848	27142	63477	33416	60961	19781	65457	23981	90348	24499
27576	47298	47163	69614	29372	24859	62090	81667	50635	08295
52970	93916	81350	81057	16962	56039	27739	59574	79617	45698
69516	87573	13313	69388	32020	66294	99126	50474	04258	03084
94504	41733	55936	77595	55959	90727	61367	83645	80997	62103
67935	14568	27992	09784	81917	79303	08616	83509	64932	34764
63345	09500	40232	51061	09455	36491	04810	06040	78959	41435
87119	21605	86917	97715	91250	79587	80967	39872	52512	78444
02612	97319	10487	68923	58607	38261	67119	36351	48521	69965
69860	16526	41420	01514	46902	03399	12286	52467	80387	10561
27669	67730	53932	38578	25746	00025	98917	18790	51091	24920
59705	91472	01302	33123	35274	88433	55491	27609	02824	05245
36508	74042	44014	36243	12724	06092	23742	90436	33419	12301
13612	24554	73326	61445	77198	43360	62006	31038	54756	88137
82893	11961	19656	71181	63201	44946	14169	72755	47883	24119
97914	61228	42903	71187	54964	14945	20809	33937	13257	66387

TABLE C.2 Cumulative Poisson Probabilities × 1,000

μ \ x	0	1	2	3	4	5	6	7	8	9	10	11	12	13	14
0.1	905	995	1,000												
0.2	819	982	999	1,000											
0.3	741	963	996	1,000											
0.4	670	938	992	999	1,000										
0.5	607	910	986	998	1,000										
0.6	549	878	977	997	1,000										
0.7	497	844	966	994	999	1,000									
0.8	449	809	953	991	999	1,000									
0.9	407	772	937	987	998	1,000									
1.0	368	736	920	981	996	999	1,000								
1.1	333	699	900	974	995	999	1,000								
1.2	301	663	879	966	992	998	1,000								
1.3	273	627	857	957	989	998	1,000								
1.4	247	592	833	946	986	997	999	1,000							
1.5	223	558	809	934	981	996	999	1,000							
1.6	202	525	783	921	976	994	999	1,000							
1.7	183	493	757	907	970	992	998	1,000							
1.8	165	463	731	891	964	990	997	999	1,000						
1.9	150	434	704	875	956	987	997	999	1,000						
2.0	135	406	677	857	947	983	995	999	1,000						
2.2	111	355	623	819	928	975	993	998	1,000						
2.4	091	308	570	779	904	964	988	997	999	1,000					
2.6	074	267	518	736	877	951	983	995	999	1,000					

TABLE C.2 *(continued)* Cumulative Poisson Probabilities × 1,000

μ \ x	0	1	2	3	4	5	6	7	8	9	10	11	12	13	14
2.8	061	231	469	692	848	935	976	992	998	999	1,000				
3.0	050	199	423	647	815	916	966	988	996	999	1,000				
3.2	041	171	380	603	781	895	955	983	994	998	1,000				
3.4	033	147	340	558	744	871	942	977	992	997	999	1,000			
3.6	027	126	303	515	706	844	927	969	988	996	999	1,000			
3.8	022	107	269	473	668	816	909	960	984	994	998	999	1,000		
4.0	018	092	238	433	629	785	889	949	979	992	997	999	1,000		
4.2	015	078	210	395	590	753	867	936	972	989	996	999	1,000		
4.4	012	066	185	359	551	720	844	921	964	985	994	998	999	1,000	
4.6	010	056	163	326	513	686	818	905	955	980	992	997	999	1,000	
4.8	008	048	143	294	476	651	791	887	944	975	990	996	999	1,000	
5	007	040	125	265	440	616	762	867	932	968	986	995	998	999	1,000
6	002	017	062	151	285	446	606	744	847	916	957	980	991	996	1,000
7	001	007	030	082	173	301	450	599	729	830	901	947	973	987	994
8	000	003	014	042	100	191	313	453	593	717	816	888	936	966	983
9	000	001	006	021	055	116	207	324	456	587	706	803	876	926	959
10		000	003	010	029	067	130	220	333	458	583	697	792	864	917
11		000	001	005	015	038	079	143	232	341	460	579	689	781	854
12		000	001	002	008	020	046	090	155	242	347	462	576	682	772
13			000	001	004	011	026	054	100	166	252	353	463	573	675
14				000	002	006	014	032	062	109	176	260	358	464	570
15				000	001	003	008	018	037	070	118	185	268	363	466

TABLE C.2 *(continued)* Cumulative Poisson Probabilities × 1,000

μ \ x	15	16	17	18	19	20	21	22	23	24	25	26	27	28	29
7	998	999	1,000												
8	992	996	998	999	1,000										
9	978	989	995	998	999	1,000									
10	951	973	986	993	997	998	999	1,000							
11	907	944	968	982	991	995	998	999	1,000						
12	844	899	937	963	979	988	994	997	999	999	1,000				
13	764	835	890	930	957	975	986	992	996	998	999	1,000			
14	669	756	827	883	923	952	971	983	991	995	997	999	999	1,000	
15	568	664	749	819	875	917	947	967	981	989	994	997	998	999	1,000

μ \ x	0	1	2	3	4	5	6	7	8	9	10	11	12	13	14
16					000	001	004	010	022	043	077	127	193	275	368
17					000	001	002	005	013	026	049	085	135	201	281
18						000	001	003	015	015	030	055	092	143	208
19						000	001	002	004	009	018	035	061	098	150
20							000	001	002	005	011	021	039	066	105
21								000	001	003	006	013	025	043	072
22								000	001	002	004	008	015	028	048
23									000	001	002	004	009	017	031
24										000	001	003	005	011	020

TABLE C.2 *(continued)* Cumulative Poisson Probabilities × 1,000

μ \ x	15	16	17	18	19	20	21	22	23	24	25	26	27	28	29
16	467	566	659	742	812	868	911	942	963	978	987	993	996	998	999
17	371	468	564	655	736	805	861	905	937	959	975	985	991	995	997
18	287	375	469	562	651	731	799	855	899	932	955	972	983	990	994
19	215	292	378	469	561	647	725	793	849	893	927	951	969	980	988
20	157	221	297	381	470	559	644	721	787	843	888	922	948	966	978
21	111	163	227	302	384	471	558	640	716	782	838	883	917	944	963
22	077	117	169	232	306	387	472	556	637	712	777	832	877	913	940
23	052	082	123	175	238	310	389	472	555	635	708	772	827	873	908
24	034	056	087	128	180	243	314	392	473	554	632	704	768	823	868

μ \ x	30	31	32	33	34	35	36	37	´38	39	40	41	42	43	44
16	999	1,000													
17	999	999	1,000												
18	997	998	999	1,000											
19	993	996	998	999	999	1,000									
20	987	992	995	997	999	999	1,000								
21	976	985	991	994	997	998	999	999	1,000						
22	959	973	983	989	994	996	998	999	999	1,000					
23	936	956	971	981	988	993	996	997	999	999	1,000				
24	904	932	953	969	979	987	992	995	997	998	999	999	1,000		

TABLE C.3 Cumulative Normal Probabilities

Z	0.09	0.08	0.07	0.06	0.05	0.04	0.03	0.02	0.01	0.00
−3.5	0.00017	0.00017	0.00018	0.00019	0.00019	0.00020	0.00021	0.00022	0.00022	0.00023
−3.4	0.00024	0.00025	0.00026	0.00027	0.00028	0.00029	0.00030	0.00031	0.00033	0.00034
−3.3	0.00035	0.00036	0.00038	0.00039	0.00040	0.00042	0.00043	0.00045	0.00047	0.00048
−3.2	0.00050	0.00052	0.00054	0.00056	0.00058	0.00060	0.00062	0.00064	0.00066	0.00069
−3.1	0.00071	0.00074	0.00076	0.00079	0.00082	0.00085	0.00087	0.00090	0.00094	0.00097
−3.0	0.00100	0.00104	0.00107	0.00111	0.00114	0.00118	0.00122	0.00126	0.00131	0.00135
−2.9	0.0014	0.0014	0.0015	0.0015	0.0016	0.0016	0.0017	0.0017	0.0018	0.0019
−2.8	0.0019	0.0020	0.0021	0.0021	0.0022	0.0023	0.0023	0.0024	0.0025	0.0026
−2.7	0.0026	0.0027	0.0028	0.0029	0.0030	0.0031	0.0032	0.0033	0.0034	0.0035
−2.6	0.0036	0.0037	0.0038	0.0039	0.0040	0.0041	0.0043	0.0044	0.0045	0.0047
−2.5	0.0048	0.0049	0.0051	0.0052	0.0054	0.0055	0.0057	0.0059	0.0060	0.0062
−2.4	0.0064	0.0066	0.0068	0.0069	0.0071	0.0073	0.0075	0.0078	0.0080	0.0082
−2.3	0.0084	0.0087	0.0089	0.0091	0.0094	0.0096	0.0099	0.0102	0.0104	0.0107
−2.2	0.0110	0.0113	0.0116	0.0119	0.0122	0.0125	0.0129	0.0132	0.0136	0.0139
−2.1	0.0143	0.0146	0.0150	0.0154	0.0158	0.0162	0.0166	0.0170	0.0174	0.0179
−2.0	0.0183	0.0188	0.0192	0.0197	0.0202	0.0207	0.0212	0.0217	0.0222	0.0228
−1.9	0.0233	0.0239	0.0244	0.0250	0.0256	0.0262	0.0268	0.0274	0.0281	0.0287
−1.8	0.0294	0.0301	0.0307	0.0314	0.0322	0.0329	0.0336	0.0344	0.0351	0.0359
−1.7	0.0367	0.0375	0.0384	0.0392	0.0401	0.0409	0.0418	0.0427	0.0436	0.0446
−1.6	0.0455	0.0465	0.0475	0.0485	0.0495	0.0505	0.0516	0.0526	0.0537	0.0548
−1.5	0.0559	0.0571	0.0582	0.0594	0.0606	0.0618	0.0630	0.0643	0.0655	0.0668
−1.4	0.0681	0.0694	0.0708	0.0721	0.0735	0.0749	0.0764	0.0778	0.0793	0.0808
−1.3	0.0823	0.0838	0.0853	0.0869	0.0885	0.0901	0.0918	0.0934	0.0951	0.0968
−1.2	0.0985	0.1003	0.1020	0.1038	0.1057	0.1075	0.1093	0.1112	0.1131	0.1151
−1.1	0.1170	0.1190	0.1210	0.1230	0.1251	0.1271	0.1292	0.1314	0.1335	0.1357
−1.0	0.1379	0.1401	0.1423	0.1446	0.1469	0.1492	0.1515	0.1539	0.1562	0.1587
−0.9	0.1611	0.1635	0.1660	0.1685	0.1711	0.1736	0.1762	0.1788	0.1814	0.1841
−0.8	0.1867	0.1894	0.1922	0.1949	0.1977	0.2005	0.2033	0.2061	0.2090	0.2119
−0.7	0.2148	0.2177	0.2207	0.2236	0.2266	0.2297	0.2327	0.2358	0.2389	0.2420
−0.6	0.2451	0.2483	0.2514	0.2546	0.2578	0.2611	0.2643	0.2676	0.2709	0.2743
−0.5	0.2776	0.2810	0.2843	0.2877	0.2912	0.2946	0.2981	0.3015	0.3050	0.3085
−0.4	0.3121	0.3156	0.3192	0.3228	0.3264	0.3300	0.3336	0.3372	0.3409	0.3446
−0.3	0.3483	0.3520	0.3557	0.3594	0.3632	0.3669	0.3707	0.3745	0.3783	0.3821
−0.2	0.3859	0.3897	0.3936	0.3974	0.4013	0.4052	0.4090	0.4129	0.4168	0.4207
−0.1	0.4247	0.4286	0.4325	0.4364	0.4404	0.4443	0.4483	0.4522	0.4562	0.4602
−0.0	0.4641	0.4681	0.4721	0.4761	0.4801	0.4840	0.4880	0.4920	0.4960	0.5000

TABLE C.3 *(continued)* Cumulative Normal Probabilities

Z	0.00	0.01	0.02	0.03	0.04	0.05	0.06	0.07	0.08	0.09
+0.0	0.5000	0.5040	0.5080	0.5120	0.5160	0.5199	0.5239	0.5279	0.5319	0.5359
+0.1	0.5398	0.5438	0.5478	0.5517	0.5557	0.5596	0.5636	0.5675	0.5714	0.5753
+0.2	0.5793	0.5832	0.5871	0.5910	0.5948	0.5987	0.6026	0.6064	0.6103	0.6141
+0.3	0.6179	0.6217	0.6255	0.6293	0.6331	0.6368	0.6406	0.6443	0.6480	0.6517
+0.4	0.6554	0.6591	0.6628	0.6664	0.6700	0.6736	0.6772	0.6808	0.6844	0.6879
+0.5	0.6915	0.6950	0.6985	0.7019	0.7054	0.7088	0.7123	0.7157	0.7190	0.7224
+0.6	0.7257	0.7291	0.7324	0.7357	0.7389	0.7422	0.7454	0.7486	0.7517	0.7549
+0.7	0.7580	0.7611	0.7642	0.7673	0.7704	0.7734	0.7764	0.7794	0.7823	0.7852
+0.8	0.7881	0.7910	0.7939	0.7967	0.7995	0.8023	0.8051	0.8079	0.8106	0.8133
+0.9	0.8159	0.8186	0.8212	0.8238	0.8264	0.8289	0.8315	0.8340	0.8365	0.8389
+1.0	0.8413	0.8438	0.8461	0.8485	0.8508	0.8531	0.8554	0.8577	0.8599	0.8621
+1.1	0.8643	0.8665	0.8686	0.8708	0.8729	0.8749	0.8770	0.8790	0.8810	0.8830
+1.2	0.8849	0.8869	0.8888	0.8907	0.8925	0.8944	0.8962	0.8980	0.8997	0.9015
+1.3	0.9032	0.9049	0.9066	0.9082	0.9099	0.9115	0.9131	0.9147	0.9162	0.9177
+1.4	0.9192	0.9207	0.9222	0.9236	0.9251	0.9265	0.9279	0.9292	0.9306	0.9319
+1.5	0.9332	0.9345	0.9357	0.9370	0.9382	0.9394	0.9406	0.9418	0.9429	0.9441
+1.6	0.9452	0.9463	0.9474	0.9484	0.9495	0.9505	0.9515	0.9525	0.9535	0.9545
+1.7	0.9554	0.9564	0.9573	0.9582	0.9591	0.9599	0.9608	0.9616	0.9625	0.9633
+1.8	0.9641	0.9649	0.9656	0.9664	0.9671	0.9678	0.9686	0.9693	0.9699	0.9706
+1.9	0.9713	0.9719	0.9726	0.9732	0.9738	0.9744	0.9750	0.9756	0.9761	0.9767
+2.0	0.9773	0.9778	0.9783	0.9788	0.9793	0.9798	0.9803	0.9808	0.9812	0.9817
+2.1	0.9821	0.9826	0.9830	0.9834	0.9838	0.9842	0.9846	0.9850	0.9854	0.9857
+2.2	0.9861	0.9864	0.9868	0.9871	0.9875	0.9878	0.9881	0.9884	0.9887	0.9890
+2.3	0.9893	0.9896	0.9898	0.9901	0.9904	0.9906	0.9909	0.9911	0.9913	0.9916
+2.4	0.9918	0.9920	0.9922	0.9925	0.9927	0.9929	0.9931	0.9932	0.9934	0.9936
+2.5	0.9938	0.9940	0.9941	0.9943	0.9945	0.9946	0.9948	0.9949	0.9951	0.9952
+2.6	0.9953	0.9955	0.9956	0.9957	0.9959	0.9960	0.9961	0.9962	0.9963	0.9964
+2.7	0.9965	0.9966	0.9967	0.9968	0.9969	0.9970	0.9971	0.9972	0.9973	0.9974
+2.8	0.9974	0.9975	0.9976	0.9977	0.9977	0.9978	0.9979	0.9979	0.9980	0.9981
+2.9	0.9981	0.9982	0.9983	0.9983	0.9984	0.9984	0.9985	0.9985	0.9986	0.9986
+3.0	0.99865	0.99869	0.99874	0.99878	0.99882	0.99886	0.99889	0.99893	0.99896	0.99900
+3.1	0.99903	0.99906	0.99910	0.99913	0.99915	0.99918	0.99921	0.99924	0.99926	0.99929
+3.2	0.99931	0.99934	0.99936	0.99938	0.99940	0.99942	0.99944	0.99946	0.99948	0.99950
+3.3	0.99952	0.99953	0.99955	0.99957	0.99958	0.99960	0.99961	0.99962	0.99964	0.99965
+3.4	0.99966	0.99967	0.99969	0.99970	0.99971	0.99972	0.99973	0.99974	0.99975	0.99976
+3.5	0.99977	0.99978	0.99978	0.99979	0.99980	0.99981	0.99981	0.99982	0.99983	0.99983

D INTEREST FACTOR TABLES

TABLES D.1 TO D.8 INTEREST FACTORS FOR ANNUAL COMPOUNDING

Values for each of the seven interest formulas derived in Chapter 8 are given for interest rates from 8 to 30%. (These tables are reproduced from W. J. Fabrycky, G. J. Thuesen, and D. Verma, *Economic Decision Analysis* [Upper Saddle River, N.J.: Prentice Hall, Inc., 1998].)

TABLE D.1 8% Interest Factors

n	Single payment		Equal-payment series				Uniform gradient-series factor
	Compound-amount factor	Present-worth factor	Compound-amount factor	Sinking-fund factor	Present-worth factor	Capital-recovery factor	
	To find F given P $F/P, i, n$	To find P given F $P/F, i, n$	To find F given A $F/A, i, n$	To find A given F $A/F, i, n$	To find P given A $P/A, i, n$	To find A given P $A/P, i, n$	To find A given G $A/G, i, n$
1	1.080	0.9259	1.000	1.0000	0.9259	1.0800	0.0000
2	1.166	0.8573	2.080	0.4808	1.7833	0.5608	0.4808
3	1.260	0.7938	3.246	0.3080	2.5771	0.3880	0.9488
4	1.360	0.7350	4.506	0.2219	3.3121	0.3019	1.4040
5	1.469	0.6806	5.867	0.1705	3.9927	0.2505	1.8465
6	1.587	0.6302	7.336	0.1363	4.6229	0.2163	2.2764
7	1.714	0.5835	8.923	0.1121	5.2064	0.1921	2.6937
8	1.851	0.5403	10.637	0.0940	5.7466	0.1740	3.0985
9	1.999	0.5003	12.488	0.0801	6.2469	0.1601	3.4910
10	2.159	0.4632	14.487	0.0690	6.7101	0.1490	3.8713
11	2.332	0.4289	16.645	0.0601	7.1390	0.1401	4.2395
12	2.518	0.3971	18.977	0.0527	7.5361	0.1327	4.5958
13	2.720	0.3677	21.495	0.0465	7.9038	0.1265	4.9402
14	2.937	0.3405	24.215	0.0413	8.2442	0.1213	5.2731
15	3.172	0.3153	27.152	0.0368	8.5595	0.1168	5.5945
16	3.426	0.2919	30.324	0.0330	8.8514	0.1130	5.9046
17	3.700	0.2703	33.750	0.0296	9.1216	0.1096	6.2038
18	3.996	0.2503	37.450	0.0267	9.3719	0.1067	6.4920
19	4.316	0.2317	41.446	0.0241	9.6036	0.1041	6.7697
20	4.661	0.2146	45.762	0.0219	9.8182	0.1019	7.0370
21	5.034	0.1987	50.423	0.0198	10.0168	0.0998	7.2940
22	5.437	0.1840	55.457	0.0180	10.2008	0.0980	7.5412
23	5.871	0.1703	60.893	0.0164	10.3711	0.0964	7.7786
24	6.341	0.1577	66.765	0.0150	10.5288	0.0950	8.0066
25	6.848	0.1460	73.106	0.0137	10.6748	0.0937	8.2254
26	7.396	0.1352	79.954	0.0125	10.8100	0.0925	8.4352
27	7.988	0.1252	87.351	0.0115	10.9352	0.0915	8.6363
28	8.627	0.1159	95.339	0.0105	11.0511	0.0905	8.8289
29	9.317	0.1073	103.966	0.0096	11.1584	0.0896	9.0133
30	10.063	0.0994	113.283	0.0088	11.2578	0.0888	9.1897
31	10.868	0.0920	123.346	0.0081	11.3498	0.0881	9.3584
32	11.737	0.0852	134.214	0.0075	11.4350	0.0875	9.5197
33	12.676	0.0789	145.951	0.0069	11.5139	0.0869	9.6737
34	13.690	0.0731	158.627	0.0063	11.5869	0.0863	9.8208
35	14.785	0.0676	172.317	0.0058	11.6546	0.0858	9.9611
40	21.725	0.0460	259.057	0.0039	11.9246	0.0839	10.5699
45	31.920	0.0313	386.506	0.0026	12.1084	0.0826	11.0447
50	46.902	0.0213	573.770	0.0018	12.2335	0.0818	11.4107
55	68.914	0.0145	848.923	0.0012	12.3186	0.0812	11.6902
60	101.257	0.0099	1253.213	0.0008	12.3766	0.0808	11.9015
65	148.780	0.0067	1847.248	0.0006	12.4160	0.0806	12.0602
70	218.606	0.0046	2720.080	0.0004	12.4428	0.0804	12.1783
75	321.205	0.0031	4002.557	0.0003	12.4611	0.0803	12.2658
80	471.955	0.0021	5886.935	0.0002	12.4735	0.0802	12.3301
85	693.456	0.0015	8655.706	0.0001	12.4820	0.0801	12.3773
90	1018.915	0.0010	12723.939	0.0001	12.4877	0.0801	12.4116
95	1497.121	0.0007	18701.507	0.0001	12.4917	0.0801	12.4365
100	2199.761	0.0005	27484.516	0.0001	12.4943	0.0800	12.4545

TABLE D.2 9% Interest Factors

	Single payment		Equal-payment series				Uniform gradient-series factor
n	Compound-amount factor	Present-worth factor	Compound-amount factor	Sinking-fund factor	Present-worth factor	Capital-recovery factor	
	To find *F* given *P* *F*/*P*, *i*, *n*	To find *P* given *F* *P*/*F*, *i*, *n*	To find *F* given *A* *F*/*A*, *i*, *n*	To find *A* given *F* *A*/*F*, *i*, *n*	To find *P* given *A* *P*/*A*, *i*, *n*	To find *A* given *P* *A*/*P*, *i*, *n*	To find *A* given *G* *A*/*G*, *i*, *n*
1	1.090	0.9174	1.000	1.0000	0.9174	1.0900	0.0000
2	1.188	0.8417	2.090	0.4785	1.7591	0.5685	0.4785
3	1.295	0.7722	3.278	0.3051	2.5313	0.3951	0.9426
4	1.412	0.7084	4.573	0.2187	3.2397	0.3087	1.3925
5	1.539	0.6499	5.985	0.1671	3.8897	0.2571	1.8282
6	1.677	0.5963	7.523	0.1329	4.4859	0.2229	2.2498
7	1.828	0.5470	9.200	0.1087	5.0330	0.1987	2.6574
8	1.993	0.5019	11.028	0.0907	5.5348	0.1807	3.0512
9	2.172	0.4604	13.021	0.0768	5.9953	0.1668	3.4312
10	2.367	0.4224	15.193	0.0658	6.4177	0.1558	3.7978
11	2.580	0.3875	17.560	0.0570	6.8052	0.1470	4.1510
12	2.813	0.3555	20.141	0.0497	7.1607	0.1397	4.4910
13	3.066	0.3262	22.953	0.0436	7.4869	0.1336	4.8182
14	3.342	0.2993	26.019	0.0384	7.7862	0.1284	5.1326
15	3.642	0.2745	29.361	0.0341	8.0607	0.1241	5.4346
16	3.970	0.2519	33.003	0.0303	8.3126	0.1203	5.7245
17	4.328	0.2311	36.974	0.0271	8.5436	0.1171	6.0024
18	4.717	0.2120	41.301	0.0242	8.7556	0.1142	6.2687
19	5.142	0.1945	46.018	0.0217	8.9501	0.1117	6.5236
20	5.604	0.1784	51.160	0.0196	9.1286	0.1096	6.7675
21	6.109	0.1637	56.765	0.0176	9.2923	0.1076	7.0006
22	6.659	0.1502	62.873	0.0159	9.4424	0.1059	7.2232
23	7.258	0.1378	69.532	0.0144	9.5802	0.1044	7.4358
24	7.911	0.1264	76.790	0.0130	9.7066	0.1030	7.6384
25	8.623	0.1160	84.701	0.0118	9.8226	0.1018	7.8316
26	9.399	0.1064	93.324	0.0107	9.9290	0.1007	8.0156
27	10.245	0.0976	102.723	0.0097	10.0266	0.0997	8.1906
28	11.167	0.0896	112.968	0.0089	10.1161	0.0989	8.3572
29	12.172	0.0822	124.135	0.0081	10.1983	0.0981	8.5154
30	13.268	0.0754	136.308	0.0073	10.2737	0.0973	8.6657
31	14.462	0.0692	149.575	0.0067	10.3428	0.0967	8.8083
32	15.763	0.0634	164.037	0.0061	10.4063	0.0961	8.9436
33	17.182	0.0582	179.800	0.0056	10.4645	0.0956	9.0718
34	18.728	0.0534	196.982	0.0051	10.5178	0.0951	9.1933
35	20.414	0.0490	215.711	0.0046	10.5668	0.0946	9.3083
40	31.409	0.0318	337.882	0.0030	10.7574	0.0930	9.7957
45	48.327	0.0207	525.859	0.0019	10.8812	0.0919	10.1603
50	74.358	0.0135	815.084	0.0012	10.9617	0.0912	10.4295
55	114.408	0.0088	1260.092	0.0008	11.0140	0.0908	10.6261
60	176.031	0.0057	1944.792	0.0005	11.0480	0.0905	10.7683
65	270.846	0.0037	2998.288	0.0003	11.0701	0.0903	10.8702
70	416.730	0.0024	4619.223	0.0002	11.0845	0.0902	10.9427
75	641.191	0.0016	7113.232	0.0002	11.0938	0.0902	10.9940
80	986.552	0.0010	10950.574	0.0001	11.0999	0.0901	11.0299
85	1517.932	0.0007	16854.800	0.0001	11.1038	0.0901	11.0551
90	2335.527	0.0004	25939.184	0.0001	11.1064	0.0900	11.0726
95	3593.497	0.0003	39916.635	0.0000	11.1080	0.0900	11.0847
100	5529.041	0.0002	61422.675	0.0000	11.1091	0.0900	11.0930

TABLE D.3 10% Interest Factors

	Single payment		Equal-payment series				Uniform gradient-series factor
n	Compound-amount factor	Present-worth factor	Compound-amount factor	Sinking-fund factor	Present-worth factor	Capital-recovery factor	
	To find F given P $F/P, i, n$	To find P given F $P/F, i, n$	To find F given A $F/A, i, n$	To find A given F $A/F, i, n$	To find P given A $P/A, i, n$	To find A given P $A/P, i, n$	To find A given G $A/G, i, n$
1	1.100	0.9091	1.000	1.0000	0.9091	1.1000	0.0000
2	1.210	0.8265	2.100	0.4762	1.7355	0.5762	0.4762
3	1.331	0.7513	3.310	0.3021	2.4869	0.4021	0.9366
4	1.464	0.6830	4.641	0.2155	3.1699	0.3155	1.3812
5	1.611	0.6209	6.105	0.1638	3.7908	0.2638	1.8101
6	1.772	0.5645	7.716	0.1296	4.3553	0.2296	2.2236
7	1.949	0.5132	9.487	0.1054	4.8684	0.2054	2.6216
8	2.144	0.4665	11.436	0.0875	5.3349	0.1875	3.0045
9	2.358	0.4241	13.579	0.0737	5.7950	0.1737	3.3724
10	2.594	0.3856	15.937	0.0628	6.1446	0.1628	3.7255
11	2.853	0.3505	18.531	0.0540	6.4951	0.1540	4.0641
12	3.138	0.3186	21.384	0.0468	6.8137	0.1468	4.3884
13	3.452	0.2897	24.523	0.0408	7.1034	0.1408	4.6988
14	3.798	0.2633	27.975	0.0358	7.3667	0.1358	4.9955
15	4.177	0.2394	31.772	0.0315	7.6061	0.1315	5.2789
16	4.595	0.2176	35.950	0.0278	7.8237	0.1278	5.5493
17	5.054	0.1979	40.545	0.0247	8.0216	0.1247	5.8071
18	5.560	0.1799	45.599	0.0219	8.2014	0.1219	6.0526
19	6.116	0.1635	51.159	0.0196	8.3649	0.1196	6.2861
20	6.728	0.1487	57.275	0.0175	8.5136	0.1175	6.5081
21	7.400	0.1351	64.003	0.0156	8.6487	0.1156	6.7189
22	8.140	0.1229	71.403	0.0140	8.7716	0.1140	6.9189
23	8.953	0.1117	79.543	0.0126	8.8832	0.1126	7.1085
24	9.850	0.1015	88.497	0.0113	8.9848	0.1113	7.2881
25	10.835	0.0923	98.347	0.0102	9.0771	0.1102	7.4580
26	11.918	0.0839	109.182	0.0092	9.1610	0.1092	7.6187
27	13.110	0.0763	121.100	0.0083	9.2372	0.1083	7.7704
28	14.421	0.0694	134.210	0.0075	9.3066	0.1075	7.9137
29	15.863	0.0630	148.631	0.0067	9.3696	0.1067	8.0489
30	17.449	0.0573	164.494	0.0061	9.4269	0.1061	8.1762
31	19.194	0.0521	181.943	0.0055	9.4790	0.1055	8.2962
32	21.114	0.0474	201.138	0.0050	9.5264	0.1050	8.4091
33	23.225	0.0431	222.252	0.0045	9.5694	0.1045	8.5152
34	25.548	0.0392	245.477	0.0041	9.6086	0.1041	8.6149
35	28.102	0.0356	271.024	0.0037	9.6442	0.1037	8.7086
40	45.259	0.0221	442.593	0.0023	9.7791	0.1023	9.0962
45	72.890	0.0137	718.905	0.0014	9.8628	0.1014	9.3741
50	117.391	0.0085	1163.909	0.0009	9.9148	0.1009	9.5704
55	189.059	0.0053	1880.591	0.0005	9.9471	0.1005	9.7075
60	304.482	0.0033	3034.816	0.0003	9.9672	0.1003	9.8023
65	490.371	0.0020	4893.707	0.0002	9.9796	0.1002	9.8672
70	789.747	0.0013	7887.470	0.0001	9.9873	0.1001	9.9113
75	1271.895	0.0008	12708.954	0.0001	9.9921	0.1001	9.9410
80	2048.400	0.0005	20474.002	0.0001	9.9951	0.1001	9.9609
85	3298.969	0.0003	32979.690	0.0000	9.9970	0.1000	9.9742
90	5313.023	0.0002	53120.226	0.0000	9.9981	0.1000	9.9831
95	8556.676	0.0001	85556.760	0.0000	9.9988	0.1000	9.9889
100	13780.612	0.0001	137796.123	0.0000	9.9993	0.1000	9.9928

TABLE D.4 12% Interest Factors

n	Single payment		Equal-payment series				Uniform gradient-series factor
	Compound-amount factor	Present-worth factor	Compound-amount factor	Sinking-fund factor	Present-worth factor	Capital-recovery factor	
	To find F given P $F/P, i, n$	To find P given F $P/F, i, n$	To find F given A $F/A, i, n$	To find A given F $A/F, i, n$	To find P given A $P/A, i, n$	To find A given P $A/P, i, n$	To find A given G $A/G, i, n$
1	1.120	0.8929	1.000	1.0000	0.8929	1.1200	0.0000
2	1.254	0.7972	2.120	0.4717	1.6901	0.5917	0.4717
3	1.405	0.7118	3.374	0.2964	2.4018	0.4164	0.9246
4	1.574	0.6355	4.779	0.2092	3.0374	0.3292	1.3589
5	1.762	0.5674	6.353	0.1574	3.6048	0.2774	1.7746
6	1.974	0.5066	8.115	0.1232	4.1114	0.2432	2.1721
7	2.211	0.4524	10.089	0.0991	4.5638	0.2191	2.5515
8	2.476	0.4039	12.300	0.0813	4.9676	0.2013	2.9132
9	2.773	0.3606	14.776	0.0677	5.3283	0.1877	3.2574
10	3.106	0.3220	17.549	0.0570	5.6502	0.1770	3.5847
11	3.479	0.2875	20.655	0.0484	5.9377	0.1684	3.8953
12	3.896	0.2567	24.133	0.0414	6.1944	0.1614	4.1897
13	4.364	0.2292	28.029	0.0357	6.4236	0.1557	4.4683
14	4.887	0.2046	32.393	0.0309	6.6282	0.1509	4.7317
15	5.474	0.1827	37.280	0.0268	6.8109	0.1468	4.9803
16	6.130	0.1631	42.753	0.0234	6.9740	0.1434	5.2147
17	6.866	0.1457	48.884	0.0205	7.1196	0.1405	5.4353
18	7.690	0.1300	55.750	0.0179	7.2497	0.1379	5.6427
19	8.613	0.1161	63.440	0.0158	7.3658	0.1358	5.8375
20	9.646	0.1037	72.052	0.0139	7.4695	0.1339	6.0202
21	10.804	0.0926	81.699	0.0123	7.5620	0.1323	6.1913
22	12.100	0.0827	92.503	0.0108	7.6447	0.1308	6.3514
23	13.552	0.0738	104.603	0.0096	7.7184	0.1296	6.5010
24	15.179	0.0659	118.155	0.0085	7.7843	0.1285	6.6407
25	17.000	0.0588	133.334	0.0075	7.8431	0.1275	6.7708
26	19.040	0.0525	150.334	0.0067	7.8957	0.1267	6.8921
27	21.325	0.0469	169.374	0.0059	7.9426	0.1259	7.0049
28	23.884	0.0419	190.699	0.0053	7.9844	0.1253	7.1098
29	26.750	0.0374	214.583	0.0047	8.0218	0.1247	7.2071
30	29.960	0.0334	241.333	0.0042	8.0552	0.1242	7.2974
31	33.555	0.0298	271.293	0.0037	8.0850	0.1237	7.3811
32	37.582	0.0266	304.848	0.0033	8.1116	0.1233	7.4586
33	42.092	0.0238	342.429	0.0029	8.1354	0.1229	7.5303
34	47.143	0.0212	384.521	0.0026	8.1566	0.1226	7.5965
35	52.800	0.0189	431.664	0.0023	8.1755	0.1223	7.6577
40	93.051	0.0108	767.091	0.0013	8.2438	0.1213	7.8988
45	163.988	0.0061	1358.230	0.0007	8.2825	0.1207	8.0572
50	289.002	0.0035	2400.018	0.0004	8.3045	0.1204	8.1597

TABLE D.5 15% Interest Factors

	Single payment		Equal-payment series				Uniform gradient-series factor
n	Compound-amount factor	Present-worth factor	Compound-amount factor	Sinking-fund factor	Present-worth factor	Capital-recovery factor	
	To find F given P $F/P, i, n$	To find P given F $P/F, i, n$	To find F given A $F/A, i, n$	To find A given F $A/F, i, n$	To find P given A $P/A, i, n$	To find A given P $A/P, i, n$	To find A given G $A/G, i, n$
1	1.150	0.8696	1.000	1.0000	0.8696	1.1500	0.0000
2	1.323	0.7562	2.150	0.4651	1.6257	0.6151	0.4651
3	1.521	0.6575	3.473	0.2880	2.2832	0.4380	0.9071
4	1.749	0.5718	4.993	0.2003	2.8550	0.3503	1.3263
5	2.011	0.4972	6.742	0.1483	3.3522	0.2983	1.7228
6	2.313	0.4323	8.754	0.1142	3.7845	0.2642	2.0972
7	2.660	0.3759	11.067	0.0904	4.1604	0.2404	2.4499
8	3.059	0.3269	13.727	0.0729	4.4873	0.2229	2.7813
9	3.518	0.2843	16.786	0.0596	4.7716	0.2096	3.0922
10	4.046	0.2472	20.304	0.0493	5.0188	0.1993	3.3832
11	4.652	0.2150	24.349	0.0411	5.2337	0.1911	3.6550
12	5.350	0.1869	29.002	0.0345	5.4206	0.1845	3.9082
13	6.153	0.1625	34.352	0.0291	5.5832	0.1791	4.1438
14	7.076	0.1413	40.505	0.0247	5.7245	0.1747	4.3624
15	8.137	0.1229	47.580	0.0210	5.8474	0.1710	4.5650
16	9.358	0.1069	55.717	0.0180	5.9542	0.1680	4.7523
17	10.761	0.0929	65.075	0.0154	6.0472	0.1654	4.9251
18	12.375	0.0808	75.836	0.0132	6.1280	0.1632	5.0843
19	14.232	0.0703	88.212	0.0113	6.1982	0.1613	5.2307
20	16.367	0.0611	102.444	0.0098	6.2593	0.1598	5.3651
21	18.822	0.0531	118.810	0.0084	6.3125	0.1584	5.4883
22	21.645	0.0462	137.632	0.0073	6.3587	0.1573	5.6010
23	24.891	0.0402	159.276	0.0063	6.3988	0.1563	5.7040
24	28.625	0.0349	184.168	0.0054	6.4338	0.1554	5.7979
25	32.919	0.0304	212.793	0.0047	6.4642	0.1547	5.8834
26	37.857	0.0264	245.712	0.0041	6.4906	0.1541	5.9612
27	43.535	0.0230	283.569	0.0035	6.5135	0.1535	6.0319
28	50.066	0.0200	327.104	0.0031	6.5335	0.1531	6.0960
29	57.575	0.0174	377.170	0.0027	6.5509	0.1527	6.1541
30	66.212	0.0151	434.745	0.0023	6.5660	0.1523	6.2066
31	76.144	0.0131	500.957	0.0020	6.5791	0.1520	6.2541
32	87.565	0.0114	577.100	0.0017	6.5905	0.1517	6.2970
33	100.700	0.0099	664.666	0.0015	6.6005	0.1515	6.3357
34	115.805	0.0086	765.365	0.0013	6.6091	0.1513	6.3705
35	133.176	0.0075	881.170	0.0011	6.6166	0.1511	6.4019
40	267.864	0.0037	1779.090	0.0006	6.6418	0.1506	6.5168
45	538.769	0.0019	3585.128	0.0003	6.6543	0.1503	6.5830
50	1083.657	0.0009	7217.716	0.0002	6.6605	0.1501	6.6205

TABLE D.6 20% Interest Factors

	Single payment		Equal-payment series				Uniform gradient-series factor
n	Compound-amount factor	Present-worth factor	Compound-amount factor	Sinking-fund factor	Present-worth factor	Capital-recovery factor	Uniform gradient-series factor
	To find *F* given *P* *F/P, i, n*	To find *P* given *F* *P/F, i, n*	To find *F* given *A* *F/A, i, n*	To find *A* given *F* *A/F, i, n*	To find *P* given *A* *P/A, i, n*	To find *A* given *P* *A/P, i, n*	To find *A* given *G* *A/G, i, n*
1	1.200	0.8333	1.000	1.0000	0.8333	1.2000	0.0000
2	1.440	0.6945	2.200	0.4546	1.5278	0,6546	0.4546
3	1.728	0.5787	3.640	0.2747	2.1065	0.4747	0.8791
4	2.074	0.4823	5.368	0.1863	2.5887	0.3863	1.2742
5	2.488	0.4019	7.442	0.1344	2.9906	0.3344	1.6405
6	2.986	0.3349	9.930	0.1007	3.3255	0.3007	1.9788
7	3.583	0.2791	12.916	0.0774	3.6046	0.2774	2.2902
8	4.300	0.2326	16.499	0.0606	3.8372	0.2606	2.5756
9	5.160	0.1938	20.799	0.0481	4.0310	0.2481	2.8364
10	6.192	0.1615	25.959	0.0385	4.1925	0.2385	3.0739
11	7.430	0.1346	32.150	0.0311	4.3271	0.2311	3.2893
12	8.916	0.1122	39.581	0.0253	4.4392	0.2253	3.4841
13	10.699	0.0935	48.497	0.0206	4.5327	0.2206	3.6597
14	12.839	0.0779	59.196	0.0169	4.6106	0.2169	3.8175
15	15.407	0.0649	72.035	0.0139	4.6755	0.2139	3.9589
16	18.488	0.0541	87.442	0.0114	4.7296	0.2114	4.0851
17	22.186	0.0451	105.931	0.0095	4.7746	0.2095	4.1976
18	26.623	0.0376	128.117	0.0078	4.8122	0.2078	4.2975
19	31.948	0.0313	154.740	0.0065	4.8435	0.2065	4.3861
20	38.338	0.0261	186.688	0.0054	4.8696	0.2054	4.4644
21	46.005	0.0217	225.026	0.0045	4.8913	0.2045	4.5334
22	55.206	0.0181	271.031	0.0037	4.9094	0.2037	4.5942
23	66.247	0.0151	326.237	0.0031	4.9245	0.2031	4.6475
24	79.497	0.0126	392.484	0.0026	4.9371	0.2026	4.6943
25	95.396	0.0105	471.981	0.0021	4.9476	0.2021	4.7352
26	114.475	0.0087	567.377	0.0018	4.9563	0.2018	4.7709
27	137.371	0.0073	681.853	0.0015	4.9636	0.2015	4.8020
28	164.845	0.0061	819.223	0.0012	4.9697	0.2012	4.8291
29	197.814	0.0051	984.068	0.0010	4.9747	0.2010	4.8527
30	237.376	0.0042	1181.882	0.0009	4.9789	0.2009	4.8731
31	284.852	0.0035	1419.258	0.0007	4.9825	0.2007	4.8908
32	341.822	0.0029	1704.109	0.0006	4.9854	0.2006	4.9061
33	410.186	0.0024	2045.931	0.0005	4.9878	0.2005	4.9194
34	492.224	0.0020	2456.118	0.0004	4.9899	0.2004	4.9308
35	590.668	0.0017	2948.341	0.0003	4.9915	0.2003	4.9407
40	1469.772	0.0007	7343.858	0.0002	4.9966	0.2001	4.9728
45	3657.262	0.0003	18281.310	0.0001	4.9986	0.2001	4.9877
50	9100.438	0.0001	45497.191	0.0000	4.9995	0.2000	4.9945

TABLE D.7 25% Interest Factors

	Single payment		Equal-payment series				Uniform gradient-series factor
n	Compound-amount factor	Present-worth factor	Compound-amount factor	Sinking-fund factor	Present-worth factor	Capital-recovery factor	
	To find *F* given *P* $F/P, i, n$	To find *P* given *F* $P/F, i, n$	To find *F* given *A* $F/A, i, n$	To find *A* given *F* $A/F, i, n$	To find *P* given *A* $P/A, i, n$	To find *A* given *P* $A/P, i, n$	To find *A* given *G* $A/G, i, n$
1	1.250	0.8000	1.000	1.0000	0.8000	1.2500	0.0000
2	1.563	0.6400	2.250	0.4445	1.4400	0.6945	0.4445
3	1.953	0.5120	3.813	0.2623	1.9520	0.5123	0.8525
4	2.441	0.4096	5.766	0.1735	2.3616	0.4235	1.2249
5	3.052	0.3277	8.207	0.1219	2.6893	0.3719	1.5631
6	3.815	0.2622	11.259	0.0888	2.9514	0.3388	1.8683
7	4.768	0.2097	15.073	0.0664	3.1611	0.3164	2.1424
8	5.960	0.1678	19.842	0.0504	3.3289	0.3004	2.3873
9	7.451	0.1342	25.802	0.0388	3.4631	0.2888	2.6048
10	9.313	0.1074	33.253	0.0301	3.5705	0.2801	2.7971
11	11.642	0.0859	42.566	0.0235	3.6564	0.2735	2.9663
12	14.552	0.0687	54.208	0.0185	3.7251	0.2685	3.1145
13	18.190	0.0550	68.760	0.0146	3.7801	0.2646	3.2438
14	22.737	0.0440	86.949	0.0115	3.8241	0.2615	3.3560
15	28.422	0.0352	109.687	0.0091	3.8593	0.2591	3.4530
16	35.527	0.0282	138.109	0.0073	3.8874	0.2573	3.5366
17	44.409	0.0225	173.636	0.0058	3.9099	0.2558	3.6084
18	55.511	0.0180	218.045	0.0046	3.9280	0.2546	3.6698
19	69.389	0.0144	273.556	0.0037	3.9424	0.2537	3.7222
20	86.736	0.0115	342.945	0.0029	3.9539	0.2529	3.7667
21	108.420	0.0092	429.681	0.0023	3.9631	0.2523	3.8045
22	135.525	0.0074	538.101	0.0019	3.9705	0.2519	3.8365
23	169.407	0.0059	673.626	0.0015	3.9764	0.2515	3.8634
24	211.758	0.0047	843.033	0.0012	3.9811	0.2512	3.8861
25	264.698	0.0038	1054.791	0.0010	3.9849	0.2510	3.9052
26	330.872	0.0030	1319.489	0.0008	3.9879	0.2508	3.9212
27	413.590	0.0024	1650.361	0.0006	3.9903	0.2506	3.9346
28	516.988	0.0019	2063.952	0.0005	3.9923	0.2505	3.9457
29	646.235	0.0016	2580.939	0.0004	3.9938	0.2504	3.9551
30	807.794	0.0012	3227.174	0.0003	3.9951	0.2503	3.9628
31	1009.742	0.0010	4034.968	0.0003	3.9960	0.2503	3.9693
32	1262.177	0.0008	5044.710	0.0002	3.9968	0.2502	3.9746
33	1577.722	0.0006	6306.887	0.0002	3.9975	0.2502	3.9791
34	1972.152	0.0005	7884.609	0.0001	3.9980	0.2501	3.9828
35	2465.190	0.0004	9856.761	0.0001	3.9984	0.2501	3.9858

TABLE D.8 30% Interest Factors

	Single payment		Equal-payment series				Uniform gradient-series factor
n	Compound-amount factor	Present-worth factor	Compound-amount factor	Sinking-fund factor	Present-worth factor	Capital-recovery factor	
	To find F given P $F/P, i, n$	To find P given F $P/F, i, n$	To find F given A $F/A, i, n$	To find A given F $A/F, i, n$	To find P given A $P/A, i, n$	To find A given P $A/P, i, n$	To find A given G $A/G, i, n$
1	1.300	0.7692	1.000	1.0000	0.7692	1.3000	0.0000
2	1.690	0.5917	2.300	0.4348	1.3610	0.7348	0.4348
3	2.197	0.4552	3.990	0.2506	1.8161	0.5506	0.8271
4	2.856	0.3501	6.187	0.1616	2.1663	0.4616	1.1783
5	3.713	0.2693	9.043	0.1106	2.4356	0.4106	1.4903
6	4.827	0.2072	12.756	0.0784	2.6428	0.3784	1.7655
7	6.275	0.1594	17.583	0.0569	2.8021	0.3569	2.0063
8	8.157	0.1226	23.858	0.0419	2.9247	0.3419	2.2156
9	10.605	0.0943	32.015	0.0312	3.0190	0.3312	2.3963
10	13.786	0.0725	42.620	0.0235	3.0915	0.3235	2.5512
11	17.922	0.0558	56.405	0.0177	3.1473	0.3177	2.6833
12	23.298	0.0429	74.327	0.0135	3.1903	0.3135	2.7952
13	30.288	0.0330	97.625	0.0103	3.2233	0.3103	2.8895
14	39.374	0.0254	127.913	0.0078	3.2487	0.3078	2.9685
15	51.186	0.0195	167.286	0.0060	3.2682	0.3060	3.0345
16	66.542	0.0150	218.472	0.0046	3.2832	0.3046	3.0892
17	86.504	0.0116	285.014	0.0035	3.2948	0.3035	3.1345
18	112.455	0.0089	371.518	0.0027	3.3037	0.3027	3.1718
19	146.192	0.0069	483.973	0.0021	3.3105	0.3021	3.2025
20	190.050	0.0053	630.165	0.0016	3.3158	0.3016	3.2276
21	247.065	0.0041	820.215	0.0012	3.3199	0.3012	3.2480
22	321.184	0.0031	1067.280	0.0009	3.3230	0.3009	3.2646
23	417.539	0.0024	1388.464	0.0007	3.3254	0.3007	3.2781
24	542.801	0.0019	1806.003	0.0006	3.3272	0.3006	3.2890
25	705.641	0.0014	2348.803	0.0004	3.3286	0.3004	3.2979
26	917.333	0.0011	3054.444	0.0003	3.3297	0.3003	3.3050
27	1192.533	0.0008	3971.778	0.0003	3.3305	0.3003	3.3107
28	1550.293	0.0007	5164.311	0.0002	3.3312	0.3002	3.3153
29	2015.381	0.0005	6714.604	0.0002	3.3317	0.3002	3.3189
30	2619.996	0.0004	8729.985	0.0001	3.3321	0.3001	3.3219
31	3405.994	0.0003	11349.981	0.0001	3.3324	0.3001	3.3242
32	4427.793	0.0002	14755.975	0.0001	3.3326	0.3001	3.3261
33	5756.130	0.0002	19183.768	0.0001	3.3328	0.3001	3.3276
34	7482.970	0.0001	24939.899	0.0001	3.3329	0.3001	3.3288
35	9727.860	0.0001	32422.868	0.0000	3.3330	0.3000	3.3297

E FINITE QUEUING TABLES

TABLE E.1 TO E.3 FINITE QUEUING FACTORS

The probability of a delay, D, and the efficiency factor, F, are given for populations of 10, 20, and 30 units. Each set of values is keyed to the service factor, X, and the number of channels, M. (These tabular values are adapted with permission from L. G. Peck and R. N. Hazelwood, *Finite Queuing Tables* [New York: John Wiley & Sons, Inc., 1958].)

TABLE E.1 Finite Queuing Factors—Population 10

X	*M*	*D*	*F*	*X*	*M*	*D*	*F*	*X*	*M*	*D*	*F*
0.008	1	0.072	0.999		2	0.177	0.990		3	0.182	0.986
0.013	1	0.117	0.998		1	0.660	0.899		2	0.528	0.921
0.016	1	0.144	0.997	0.085	3	0.037	0.999		1	0.954	0.610
0.019	1	0.170	0.996		2	0.196	0.988	0.165	4	0.049	0.997
0.021	1	0.188	0.995		1	0.692	0.883		3	0.195	0.984
0.023	1	0.206	0.994	0.090	3	0.043	0.998		2	0.550	0.914
0.025	1	0.224	0.993		2	0.216	0.986		1	0.961	0.594
0.026	1	0.232	0.992		1	0.722	0.867	0.170	4	0.054	0.997
0.028	1	0.250	0.991	0.095	3	0.049	0.998		3	0.209	0.982
0.030	1	0.268	0.990		2	0.237	0.984		2	0.571	0.906
0.032	2	0.033	0.999		1	0.750	0.850		1	0.966	0.579
	1	0.285	0.988	0.100	3	0.056	0.998	0.180	5	0.013	0.999
0.034	2	0.037	0.999		2	0.258	0.981		4	0.066	0.996
	1	0.302	0.986		1	0.776	0.832		3	0.238	0.978
0.036	2	0.041	0.999	0.105	3	0.064	0.997		2	0.614	0.890
	1	0.320	0.984		2	0.279	0.978		1	0.975	0.549
0.038	2	0.046	0.999		1	0.800	0.814	0.190	5	0.016	0.999
	1	0.337	0.982	0.110	3	0.072	0.997		4	0.078	0.995
0.040	2	0.050	0.999		2	0.301	0.974		3	0.269	0.973
	1	0.354	0.980		1	0.822	0.795		2	0.654	0.873
0.042	2	0.055	0.999	0.115	3	0.081	0.996		1	0.982	0.522
	1	0.371	0.978		2	0.324	0.971	0.200	5	0.020	0.999
0.044	2	0.060	0.998		1	0.843	0.776		4	0.092	0.994
	1	0.388	0.975	0.120	4	0.016	0.999		3	0.300	0.968
0.046	2	0.065	0.998		3	0.090	0.995		2	0.692	0.854
	1	0.404	0.973		2	0.346	0.967		1	0.987	0.497
0.048	2	0.071	0.998		1	0.861	0.756	0.210	5	0.025	0.999
	1	0.421	0.970	0.125	4	0.019	0.999		4	0.108	0.992
0.050	2	0.076	0.998		3	0.100	0.994		3	0.333	0.961
	1	0.437	0.967		2	0.369	0.962		2	0.728	0.835
0.052	2	0.082	0.997		1	0.878	0.737		1	0.990	0.474
	1	0.454	0.963	0.130	4	0.022	0.999	0.220	5	0.030	0.998
0.054	2	0.088	0.997		3	0.110	0.994		4	0.124	0.990
	1	0.470	0.960		2	0.392	0.958		3	0.366	0.954
0.056	2	0.094	0.997		1	0.893	0.718		2	0.761	0.815
	1	0.486	0.956	0.135	4	0.025	0.999		1	0.993	0.453
0.058	2	0.100	0.996		3	0.121	0.993	0.230	5	0.037	0.998
	1	0.501	0.953		2	0.415	0.952		4	0.142	0.988
0.060	2	0.106	0.996		1	0.907	0.699		3	0.400	0.947
	1	0.517	0.949	0.140	4	0.028	0.999		2	0.791	0.794
0.062	2	0.113	0.996		3	0.132	0.991		1	0.995	0.434
	1	0.532	0.945		2	0.437	0.947	0.240	5	0.044	0.997
0.064	2	0.119	0.995		1	0.919	0.680		4	0.162	0.986
	1	0.547	0.940	0.145	4	0.032	0.999		3	0.434	0.938
0.066	2	0.126	0.995		3	0.144	0.990		2	0.819	0.774
	1	0.562	0.936		2	0.460	0.941		1	0.996	0.416
0.068	3	0.020	0.999		1	0.929	0.662	0.250	6	0.010	0.999
	2	0.133	0.994	0.150	4	0.036	0.998		5	0.052	0.997
	1	0.577	0.931		3	0.156	0.989		4	0.183	0.983
0.070	3	0.022	0.999		2	0.483	0.935		3	0.469	0.929
	2	0.140	0.994		1	0.939	0.644		2	0.844	0.753
	1	0.591	0.926	0.155	4	0.040	0.998		1	0.997	0.400
0.075	3	0.026	0.999		3	0.169	0.987	0.260	6	0.013	0.999
	2	0.158	0.992		2	0.505	0.928		5	0.060	0.996
	1	0.627	0.913		1	0.947	0.627		4	0.205	0.980
0.080	3	0.031	0.999	0.160	4	0.044	0.998		3	0.503	0.919

TABLE E.1 *(continued)* Finite Queuing Factors—Population 10

X	M	D	F	X	M	D	F	X	M	D	F
	2	0.866	0.732		4	0.533	0.906		7	0.171	0.982
	1	0.998	0.384		3	0.840	0.758		6	0.413	0.939
0.270	6	0.015	0.999		2	0.986	0.525		5	0.707	0.848
	5	0.070	0.995	0.400	7	0.026	0.998		4	0.917	0.706
	4	0.228	0.976		6	0.105	0.991		3	0.991	0.535
	3	0.537	0.908		5	0.292	0.963	0.580	8	0.057	0.995
	2	0.886	0.712		4	0.591	0.887		7	0.204	0.977
	1	0.999	0.370		3	0.875	0.728		6	0.465	0.927
0.280	6	0.018	0.999		2	0.991	0.499		5	0.753	0.829
	5	0.081	0.994	0.420	7	0.034	0.998		4	0.937	0.684
	4	0.252	0.972		6	0.130	0.987		3	0.994	0.517
	3	0.571	0.896		5	0.341	0.954	0.600	9	0.010	0.999
	2	0.903	0.692		4	0.646	0.866		8	0.072	0.994
	1	0.999	0.357		3	0.905	0.700		7	0.242	0.972
0.290	6	0.022	0.999		2	0.994	0.476		6	0.518	0.915
	5	0.093	0.993	0.440	7	0.045	0.997		5	0.795	0.809
	4	0.278	0.968		6	0.160	0.984		4	0.953	0.663
	3	0.603	0.884		5	0.392	0.943		3	0.996	0.500
	2	0.918	0.672		4	0.698	0.845	0.650	9	0.021	0.999
	1	0.999	0.345		3	0.928	0.672		8	0.123	0.988
0.300	6	0.026	0.998		2	0.996	0.454		7	0.353	0.954
	5	0.106	0.991	0.460	8	0.011	0.999		6	0.651	0.878
	4	0.304	0.963		7	0.058	0.995		5	0.882	0.759
	3	0.635	0.872		6	0.193	0.979		4	0.980	0.614
	2	0.932	0.653		5	0.445	0.930		3	0.999	0.461
	1	0.999	0.333		4	0.747	0.822	0.700	9	0.040	0.997
0.310	6	0.031	0.998		3	0.947	0.646		8	0.200	0.979
	5	0.120	0.990		2	0.998	0.435		7	0.484	0.929
	4	0.331	0.957	0.480	8	0.015	0.999		6	0.772	0.836
	3	0.666	0.858		7	0.074	0.994		5	0.940	0.711
	2	0.943	0.635		6	0.230	0.973		4	0.992	0.571
0.320	6	0.036	0.998		5	0.499	0.916	0.750	9	0.075	0.994
	5	0.135	0.988		4	0.791	0.799		8	0.307	0.965
	4	0.359	0.952		3	0.961	0.621		7	0.626	0.897
	3	0.695	0.845		2	0.998	0.417		6	0.870	0.792
	2	0.952	0.617	0.500	8	0.020	0.999		5	0.975	0.666
0.330	6	0.042	0.997		7	0.093	0.992		4	0.998	0.533
	5	0.151	0.986		6	0.271	0.966	0.800	9	0.134	0.988
	4	0.387	0.945		5	0.553	0.901		8	0.446	0.944
	3	0.723	0.831		4	0.830	0.775		7	0.763	0.859
	2	0.961	0.600		3	0.972	0.598		6	0.939	0.747
0.340	7	0.010	0.999		2	0.999	0.400		5	0.991	0.625
	6	0.049	0.997	0.520	8	0.026	0.998		4	0.999	0.500
	5	0.168	0.983		7	0.115	0.989	0.850	9	0.232	0.979
	4	0.416	0.938		6	0.316	0.958		8	0.611	0.916
	3	0.750	0.816		5	0.606	0.884		7	0.879	0.818
	2	0.968	0.584		4	0.864	0.752		6	0.978	0.705
0.360	7	0.014	0.999		3	0.980	0.575		5	0.998	0.588
	6	0.064	0.995		2	0.999	0.385	0.900	9	0.387	0.963
	5	0.205	0.978	0.540	8	0.034	0.997		8	0.785	0.881
	4	0.474	0.923		7	0.141	0.986		7	0.957	0.777
	3	0.798	0.787		6	0.363	0.949		6	0.995	0.667
	2	0.978	0.553		5	0.658	0.867	0.950	9	0.630	0.938
0.380	7	0.019	0.999		4	0.893	0.729		8	0.934	0.841
	6	0.083	0.993		3	0.986	0.555		7	0.994	0.737
	5	0.247	0.971	0.560	8	0.044	0.996				

TABLE E.2 Finite Queuing Factors—Population 20

X	M	D	F	X	M	D	F	X	M	D	F
0.005	1	0.095	0.999		1	0.837	0.866		3	0.326	0.980
0.009	1	0.171	0.998	0.052	3	0.080	0.998		2	0.733	0.896
0.011	1	0.208	0.997		2	0.312	0.986		1	0.998	0.526
0.013	1	0 246	0.996		1	0.858	0.851	0.100	5	0.038	0.999
0.014	1	0.265	0.995	0.054	3	0.088	0.998		4	0.131	0.995
0.015	1	0.283	0.994		2	0.332	0.984		3	0.363	0.975
0.016	1	0.302	0.993		1	0.876	0.835		2	0.773	0.878
0.017	1	0.321	0.992	0.056	3	0.097	0.997		1	0.999	0.500
0.018	2	0.048	0.999		2	0.352	0.982	0.110	5	0.055	0.998
	1	0.339	0.991		1	0.893	0.819		4	0.172	0.992
0.019	2	0.053	0.999	0.058	3	0.105	0.997		3	0.438	0.964
	1	0.358	0.990		2	0.372	0.980		2	0.842	0.837
0.020	2	0.058	0.999		1	0.908	0.802	0.120	6	0.022	0.999
	1	0.376	0.989	0.060	4	0.026	0.999		5	0.076	0.997
0.021	2	0.064	0.999		3	0.115	0.997		4	0.219	0.988
	1	0.394	0.987		2	0.392	0.978		3	0.514	0.950
0.022	2	0.070	0.999		1	0.922	0.785		2	0.895	0.793
	1	0.412	0.986	0.062	4	0.029	0.999	0.130	6	0.031	0.999
0.023	2	0.075	0.999		3	0.124	0.996		5	0.101	0.996
	1	0.431	0.984		2	0.413	0.975		4	0.271	0.983
0.024	2	0.082	0.999		1	0.934	0.768		3	0.589	0.933
	1	0.449	0.982	0.064	4	0.032	0.999		2	0.934	0.748
0.025	2	0.088	0.999		3	0.134	0.996	0.140	6	0.043	0.998
	1	0.466	0.980		2	0.433	0.972		5	0.131	0.994
0.026	2	0.094	0.998		1	0.944	0.751		4	0.328	0.976
	1	0.484	0.978	0.066	4	0.036	0.999		3	0.661	0.912
0.028	2	0.108	0.998		3	0.144	0.995		2	0.960	0.703
	1	0.519	0.973		2	0.454	0.969	0.150	7	0.017	0.999
0.030	2	0.122	0.998		1	0.953	0.733		6	0.059	0.998
	1	0.553	0.968	0.068	4	0.039	0.999		5	0.166	0.991
0.032	2	0.137	0.997		3	0.155	0.995		4	0.388	0.968
	1	0.587	0.962		2	0.474	0.966		3	0.728	0.887
0.034	2	0.152	0.996		1	0.961	0.716		2	0.976	0.661
	1	0.620	0.955	0.070	4	0.043	0.999	0.160	7	0.024	0.999
0.036	2	0.168	0.996		3	0.165	0.994		6	0.077	0.997
	1	0.651	0.947		2	0.495	0.962		5	0.205	0.988
0.038	3	0.036	0.999		1	0.967	0.699		4	0.450	0.957
	2	0.185	0.995	0.075	4	0.054	0.999		3	0.787	0.860
	1	0.682	0.938		3	0.194	0.992		2	0.987	0.622
0.040	3	0.041	0.999		2	0.545	0.953	0.180	7	0.044	0.998
	2	0.202	0.994		1	0.980	0.659		6	0.125	0.994
	1	0.712	0.929	0.080	4	0.066	0.998		5	0.295	0.978
0.042	3	0.047	0.999		3	0.225	0.990		4	0.575	0.930
	2	0.219	0.993		2	0.595	0.941		3	0.879	0.799
	1	0.740	0.918		1	0.988	0.621		2	0.996	0.555
0.044	3	0.053	0.999	0.085	4	0.080	0.997	0.200	8	0.025	0.999
	2	0.237	0.992		3	0.257	0.987		7	0.074	0.997
	1	0.767	0.906		2	0.643	0.928		6	0.187	0.988
0.046	3	0.059	0.999		1	0.993	0.586		5	0.397	0.963
	2	0.255	0.991	0.090	5	0.025	0.999		4	0.693	0.895
	1	0.792	0.894		4	0.095	0.997		3	0.938	0.736
0.048	3	0.066	0.999		3	0.291	0.984		2	0.999	0.500
	2	0.274	0.989		2	0.689	0.913	0.220	8	0.043	0.998
	1	0.815	0.881		1	0.996	0.554		7	0.115	0.994
0.050	3	0.073	0.998	0.095	5	0.031	0.999		6	0.263	0.980
	2	0.293	0.988		4	0.112	0.996		5	0.505	0.943

TABLE E.2 *(continued)* Finite Queuing Factors—Population 20

X	M	D	F	X	M	D	F	X	M	D	F
	4	0.793	0.852		4	0.998	0.555	0.500	14	0.033	0.998
	3	0.971	0.677	0.380	12	0.024	0.999		13	0.088	0.995
0.240	9	0.024	0.999		11	0.067	0.996		12	0.194	0.985
	8	0.068	0.997		10	0.154	0.989		11	0.358	0.965
	7	0.168	0.989		9	0.305	0.973		10	0.563	0.929
	6	0.351	0.969		8	0.513	0.938		9	0.764	0.870
	5	0.613	0.917		7	0.739	0.874		8	0.908	0.791
	4	0.870	0.804		6	0.909	0.777		7	0.977	0.698
	3	0.988	0.623		5	0.984	0.656		6	0.997	0.600
0.260	9	0.039	0.998		4	0.999	0.526	0.540	15	0.023	0.999
	8	0.104	0.994	0.400	13	0.012	0.999		14	0.069	0.996
	7	0.233	0.983		12	0.037	0.998		13	0.161	0.988
	6	0.446	0.953		11	0.095	0.994		12	0.311	0.972
	5	0.712	0.884		10	0.205	0.984		11	0.509	0.941
	4	0.924	0.755		9	0.379	0.962		10	0.713	0.891
	3	0.995	0.576		8	0.598	0.918		9	0.873	0.821
0.280	10	0.021	0.999		7	0.807	0.845		8	0.961	0.738
	9	0.061	0.997		6	0.942	0.744		7	0.993	0.648
	8	0.149	0.990		5	0.992	0.624		6	0.999	0.556
	7	0.309	0.973	0.420	13	0.019	0.999	0.600	16	0.023	0.999
	6	0.544	0.932		12	0.055	0.997		15	0.072	0.996
	5	0.797	0.848		11	0.131	0.991		14	0.171	0.988
	4	0.958	0.708		10	0.265	0.977		13	0.331	0.970
	3	0.998	0.536		9	0.458	0.949		12	0.532	0.938
0.300	10	0.034	0.998		8	0.678	0.896		11	0.732	0.889
	9	0.091	0.995		7	0.863	0.815		10	0.882	0.824
	8	0.205	0.985		6	0.965	0.711		9	0.962	0.748
	7	0.394	0.961		5	0.996	0.595		8	0.992	0.666
	6	0.639	0.907	0.440	13	0.029	0.999		7	0.999	0.583
	5	0.865	0.808		12	0.078	0.995	0.700	17	0.047	0.998
	4	0.978	0.664		11	0.175	0.987		16	0.137	0.991
	3	0.999	0.500		10	0.333	0.969		15	0.295	0.976
0.320	11	0.018	0.999		9	0.540	0.933		14	0.503	0.948
	10	0.053	0.997		8	0.751	0.872		13	0.710	0.905
	9	0.130	0.992		7	0.907	0.785		12	0.866	0.849
	8	0.272	0.977		6	0.980	0.680		11	0.953	0.783
	7	0.483	0.944		5	0.998	0.568		10	0.988	0.714
	6	0.727	0.878	0.460	14	0.014	0.999		9	0.998	0.643
	5	0.915	0.768		13	0.043	0.998	0.800	19	0.014	0.999
	4	0.989	0.624		12	0.109	0.993		18	0.084	0.996
0.340	11	0.029	0.999		11	0.228	0.982		17	0.242	0.984
	10	0.079	0.996		10	0.407	0.958		16	0.470	0.959
	9	0.179	0.987		9	0.620	0.914		15	0.700	0.920
	8	0.347	0.967		8	0.815	0.846		14	0.867	0.869
	7	0.573	0.924		7	0.939	0.755		13	0.955	0.811
	6	0.802	0.846		6	0.989	0.651		12	0.989	0.750
	5	0.949	0.729		5	0.999	0.543		11	0.998	0.687
	4	0.995	0.588	0.480	14	0.022	0.999	0.900	19	0.135	0.994
0.360	12	0.015	0.999		13	0.063	0.996		18	0.425	0.972
	11	0.045	0.998		12	0.147	0.990		17	0.717	0.935
	10	0.112	0.993		11	0.289	0.974		16	0.898	0.886
	9	0.237	0.981		10	0.484	0.944		15	0.973	0.833
	8	0.429	0.954		9	0.695	0.893		14	0.995	0.778
	7	0.660	0.901		8	0.867	0.819		13	0.999	0.722
	6	0.863	0.812		7	0.962	0.726	0.950	19	0.377	0.981
	5	0.971	0.691		6	0.994	0.625		18	0.760	0.943

TABLE E.3 Finite Queuing Factors—Population 30

X	M	D	F	X	M	D	F	X	M	D	F
0.004	1	0.116	0.999		1	0.963	0.772		3	0.426	0.976
0.007	1	0.203	0.998	0.044	4	0.040	0.999		2	0.847	0.873
0.009	1	0.260	0.997		3	0.154	0.996	0.075	5	0.069	0.998
0.010	1	0.289	0.996		2	0.474	0.977		4	0.201	0.993
0.011	1	0.317	0.995		1	0.974	0.744		3	0.486	0.969
0.012	1	0.346	0.994	0.046	4	0.046	0.999		2	0.893	0.840
0.013	1	0.374	0.993		3	0.171	0.996	0.080	6	0.027	0.999
0.014	2	0.067	0.999		2	0.506	0.972		5	0.088	0.998
	1	0.403	0.991		1	0.982	0.716		4	0.240	0.990
0.015	2	0.076	0.999	0.048	4	0.053	0.999		3	0.547	0.959
	1	0.431	0.989		3	0.189	0.995		2	0.929	0.805
0.016	2	0.085	0.999		2	0.539	0.968	0.085	6	0.036	0.999
	1	0.458	0.987		1	0.988	0.689		5	0.108	0.997
0.017	2	0.095	0.999	0.050	4	0.060	0.999		4	0.282	0.987
	1	0.486	0.985		3	0.208	0.994		3	0.607	0.948
0.018	2	0.105	0.999		2	0.571	0.963		2	0.955	0.768
	1	0.513	0.983		1	0.992	0.663	0.090	6	0.046	0.999
0.019	2	0.116	0.999	0.052	4	0.068	0.999		5	0.132	0.996
	1	0.541	0.980		3	0.227	0.993		4	0.326	0.984
0.020	2	0.127	0.998		2	0.603	0.957		3	0.665	0.934
	1	0.567	0.976		1	0.995	0.639		2	0.972	0.732
0.021	2	0.139	0.998	0.054	4	0.077	0.998	0.095	6	0.057	0.999
	1	0.594	0.973		3	0.247	0.992		5	0.158	0.994
0.022	2	0.151	0.998		2	0.634	0.951		4	0.372	0.979
	1	0.620	0.969		1	0.997	0.616		3	0.720	0.918
0.023	2	0.163	0.997	0.056	4	0.086	0.998		2	0.984	0.697
	1	0.645	0.965		3	0.267	0.991	0.100	6	0.071	0.998
0.024	2	0.175	0.997		2	0.665	0.944		5	0.187	0.993
	1	0.670	0.960		1	0.998	0.595		4	0.421	0.973
0.025	2	0.188	0.996	0.058	4	0.096	0.998		3	0.771	0.899
	1	0.694	0.954		3	0.288	0.989		2	0.991	0.664
0.026	2	0.201	0.996		2	0.695	0.936	0.110	7	0.038	0.999
	1	0.718	0.948		1	0.999	0.574		6	0.105	0.997
0.028	3	0.051	0.999	0.060	5	0.030	0.999		5	0.253	0.988
	2	0.229	0.995		4	0.106	0.997		4	0.520	0.959
	1	0.763	0.935		3	0.310	0.987		3	0.856	0.857
0.030	3	0.060	0.999		2	0.723	0.927		2	0.997	0.605
	2	0.257	0.994		1	0.999	0.555	0.120	7	0.057	0.998
	1	0.805	0.918	0.062	5	0.034	0.999		6	0.147	0.994
0.032	3	0.071	0.999		4	0.117	0.997		5	0.327	0.981
	2	0.286	0.992		3	0.332	0.986		4	0.619	0.939
	1	0.843	0.899		2	0.751	0.918		3	0.918	0.808
0.034	3	0.083	0.999	0.064	5	0.038	0.999		2	0.999	0.555
	2	0.316	0.990		4	0.128	0.997	0.130	8	0.030	0.999
	1	0.876	0.877		3	0.355	0.984		7	0.083	0.997
0.036	3	0.095	0.998		2	0.777	0.908		6	0.197	0.991
	2	0.347	0.988	0.066	5	0.043	0.999		5	0.409	0.972
	1	0.905	0.853		4	0.140	0.996		4	0.712	0.914
0.038	3	0.109	0.998		3	0.378	0.982		3	0.957	0.758
	2	0.378	0.986		2	0.802	0.897	0.140	8	0.045	0.999
	1	0.929	0.827	0.068	5	0.048	0.999		7	0.115	0.996
0.040	3	0.123	0.997		4	0.153	0.995		6	0.256	0.987
	2	0.410	0.983		3	0.402	0.979		5	0.494	0.960
	1	0.948	0.800		2	0.825	0.885		4	0.793	0.884
0.042	3	0.138	0.997	0.070	5	0.054	0.999		3	0.979	0.710
	2	0.442	0.980		4	0.166	0.995	0.150	9	0.024	0.999

TABLE E.3 *(continued)* Finite Queuing Factors—Population 30

X	*M*	*D*	*F*	*X*	*M*	*D*	*F*	*X*	*M*	*D*	*F*
	8	0.065	0.998		7	0.585	0.938		7	0.901	0.818
	7	0.155	0.993		6	0.816	0.868		6	0.981	0.712
	6	0.322	0.980		5	0.961	0.751		5	0.999	0.595
	5	0.580	0.944		4	0.998	0.606	0.290	14	0.023	0.999
	4	0.860	0.849	0.230	12	0.023	0.999		13	0.055	0.998
	3	0.991	0.665		11	0.056	0.998		12	0.117	0.994
0.160	9	0.036	0.999		10	0.123	0.994		11	0.223	0.986
	8	0.090	0.997		9	0.242	0.985		10	0.382	0.969
	7	0.201	0.990		8	0.423	0.965		9	0.582	0.937
	6	0.394	0.972		7	0.652	0.923		8	0.785	0.880
	5	0.663	0.924		6	0.864	0.842		7	0.929	0.795
	4	0.910	0.811		5	0.976	0.721		6	0.988	0.688
	3	0.996	0.624		4	0.999	0.580		5	0.999	0.575
0.170	10	0.019	0.999	0.240	12	0.031	0.999	0.300	14	0.031	0.999
	9	0.051	0.998		11	0.074	0.997		13	0.071	0.997
	8	0.121	0.995		10	0.155	0.992		12	0.145	0.992
	7	0.254	0.986		9	0.291	0.981		11	0.266	0.982
	6	0.469	0.961		8	0.487	0.955		10	0.437	0.962
	5	0.739	0.901		7	0.715	0.905		9	0.641	0.924
	4	0.946	0.773		6	0.902	0.816		8	0.830	0.861
	3	0.998	0.588		5	0.986	0.693		7	0.950	0.771
0.180	10	0.028	0.999		4	0.999	0.556		6	0.993	0.666
	9	0.070	0.997	0.250	13	0.017	0.999	0.320	15	0.023	0.999
	8	0.158	0.993		12	0.042	0.998		14	0.054	0.998
	7	0.313	0.980		11	0.095	0.996		13	0.113	0.994
	6	0.546	0.948		10	0.192	0.989		12	0.213	0.987
	5	0.806	0.874		9	0.345	0.975		11	0.362	0.971
	4	0.969	0.735		8	0.552	0.944		10	0.552	0.943
	3	0.999	0.555		7	0.773	0.885		9	0.748	0.893
0.190	10	0.039	0.999		6	0.932	0.789		8	0.901	0.820
	9	0.094	0.996		5	0.992	0.666		7	0.977	0.727
	8	0.200	0.990	0.260	13	0.023	0.999		6	0.997	0.625
	7	0.378	0.973		12	0.056	0.998	0.340	16	0.016	0.999
	6	0.621	0.932		11	0.121	0.994		15	0.040	0.998
	5	0.862	0.845		10	0.233	0.986		14	0.086	0.996
	4	0.983	0.699		9	0.402	0.967		13	0.169	0.990
0.200	11	0.021	0.999		8	0.616	0.930		12	0.296	0.979
	10	0.054	0.998		7	0.823	0.864		11	0.468	0.957
	9	0.123	0.995		6	0.954	0.763		10	0.663	0.918
	8	0.249	0.985		5	0.995	0.641		9	0.836	0.858
	7	0.446	0.963	0.270	13	0.032	0.999		8	0.947	0.778
	6	0.693	0.913		12	0.073	0.997		7	0.990	0.685
	5	0.905	0.814		11	0.151	0.992		6	0.999	0.588
	4	0.991	0.665		10	0.279	0.981	0.360	16	0.029	0.999
0.210	11	0.030	0.999		9	0.462	0.959		15	0.065	0.997
	10	0.073	0.997		8	0.676	0.915		14	0.132	0.993
	9	0.157	0.992		7	0.866	0.841		13	0.240	0.984
	8	0.303	0.980		6	0.970	0.737		12	0.392	0.967
	7	0.515	0.952		5	0.997	0 617		11	0.578	0.937
	6	0.758	0.892	0.280	14	0.017	0.999		10	0.762	0.889
	5	0.938	0.782		13	0.042	0.998		9	0.902	0.821
	4	0.995	0.634		12	0.093	0.996		8	0.974	0.738
0.220	11	0.041	0.999		11	0.185	0.989		7	0.996	0.648
	10	0.095	0.996		10	0.329	0.976	0.380	17	0.020	0.999
	9	0.197	0.989		9	0.522	0.949		16	0.048	0.998
	8	0.361	0.974		8	0.733	0.898		15	0.101	0.995

TABLE E.3 *(continued)* Finite Queuing Factors—Population 30

X	M	D	F	X	M	D	F	X	M	D	F
	14	0.191	0.988		16	0.310	0.977		22	0.038	0.998
	13	0.324	0.975		15	0.470	0.957		21	0.085	0.996
	12	0.496	0.952		14	0.643	0.926		20	0.167	0.990
	11	0.682	0.914		13	0.799	0.881		19	0.288	0.980
	10	0.843	0.857		12	0.910	0.826		18	0.443	0.963
	9	0.945	0.784		11	0.970	0.762		17	0.612	0.936
	8	0.988	0.701		10	0.993	0.694		16	0.766	0.899
	7	0.999	0.614		9	0.999	0.625		15	0.883	0.854
0.400	17	0.035	0.999	0.500	20	0.032	0.999		14	0.953	0.802
	16	0.076	0.996		19	0.072	0.997		13	0.985	0.746
	15	0.150	0.992		18	0.143	0.992		12	0.997	0.690
	14	0.264	0.982		17	0.252	0.983		11	0.999	0.632
	13	0.420	0.964		16	0.398	0.967	0.600	23	0.024	0.999
	12	0.601	0.933		15	0.568	0.941		22	0.059	0.997
	11	0.775	0.886		14	0.733	0.904		21	0.125	0.993
	10	0.903	0.823		13	0.865	0.854		20	0.230	0.986
	9	0.972	0.748		12	0.947	0.796		19	0.372	0.972
	8	0.995	0.666		11	0.985	0.732		18	0.538	0.949
0.420	18	0.024	0.999		10	0.997	0.667		17	0.702	0.918
	17	0.056	0.997	0.520	21	0.021	0.999		16	0.837	0.877
	16	0.116	0.994		20	0.051	0.998		15	0.927	0.829
	15	0.212	0.986		19	0.108	0.994		14	0.974	0.776
	14	0.350	0.972		18	0.200	0.988		13	0.993	0.722
	13	0.521	0.948		17	0.331	0.975		12	0.999	0.667
	12	0.700	0.910		16	0.493	0.954	0.700	25	0.039	0.998
	11	0.850	0.856		15	0.663	0.923		24	0.096	0.995
	10	0.945	0.789		14	0.811	0.880		23	0.196	0.989
	9	0.986	0.713		13	0.915	0.827		22	0.339	0.977
	8	0.998	0.635		12	0.971	0.767		21	0.511	0.958
0.440	19	0.017	0.999		11	0.993	0.705		20	0.681	0.930
	18	0.041	0.998		10	0.999	0.641		19	0.821	0.894
	17	0.087	0.996	0.540	21	0.035	0.999		18	0.916	0.853
	16	0.167	0.990		20	0.079	0.996		17	0.967	0.808
	15	0.288	0.979		19	0.155	0.991		16	0.990	0.762
	14	0.446	0.960		18	0.270	0.981		15	0.997	0.714
	13	0.623	0.929		17	0.421	0.965	0.800	27	0.053	0.998
	12	0.787	0.883		16	0.590	0.938		26	0.143	0.993
	11	0.906	0.824		15	0.750	0.901		25	0.292	0.984
	10	0.970	0.755		14	0.874	0.854		24	0.481	0.966
	9	0.994	0.681		13	0.949	0.799		23	0.670	0.941
	8	0.999	0.606		12	0.985	0.740		22	0.822	0.909
0.460	19	0.028	0.999		11	0.997	0.679		21	0.919	0.872
	18	0.064	0.997		10	0.999	0.617		20	0.970	0.832
	17	0.129	0.993	0.560	22	0.023	0.999		19	0.991	0.791
	16	0.232	0.985		21	0.056	0.997		18	0.998	0.750
	15	0.375	0.970		20	0.117	0.994	0.900	29	0.047	0.999
	14	0.545	0.944		19	0.215	0.986		28	0.200	0.992
	13	0.717	0.906		18	0.352	0.973		27	0.441	0.977
	12	0.857	0.855		17	0.516	0.952		26	0.683	0.953
	11	0.945	0.793		16	0.683	0.920		25	0.856	0.923
	10	0.985	0.724		15	0.824	0.878		24	0.947	0.888
	9	0.997	0.652		14	0.920	0.828		23	0.985	0.852
0.480	20	0.019	0.999		13	0.972	0.772		22	0.996	0.815
	19	0.046	0.998		12	0.993	0.714		21	0.999	0.778
	18	0.098	0.995		11	0.999	0.655	0.950	29	0.226	0.993
	17	0.184	0.989	0.580	23	0.014	0.999		28	0.574	0.973

F SELECTED BIBLIOGRAPHY

A selected bibliography has been compiled to include the references noted in the footnotes throughout this text, and some additional references that are specific to systems engineering and related fields. Although this material is certainly not to be considered all-inclusive, the references noted should aid the reader in gaining some additional insight into the systems engineering approach for the design and development of new systems or the re-engineering of existing systems. Specific areas covered include the following:

F.1 Systems, Systems Theory, Systems Engineering, and Systems Analysis
F.2 Concurrent/Simultaneous Engineering
F.3 Software/Computer-Aided Systems
F.4 Reliability, Maintainability, and Maintenance Engineering
F.5 Human Factors and Safety Engineering
F.6 Logistics and Logistics Engineering
F.7 Quality, Quality Engineering, and Quality Assurance
F.8 Production, Producibility, and Environmental Engineering
F.9 Operations Research and Operations Analysis
F.10 Engineering Economy, Economic Analysis, and Cost Estimating
F.11 Management and Supporting Areas

F.1 SYSTEMS, SYSTEMS THEORY, SYSTEMS ENGINEERING, AND SYSTEMS ANALYSIS

1. Ackoff, R. L. *Redesigning the Future*, John Wiley & Sons, Inc., New York, 1974.
2. Beam, W. R., *Systems Engineering: Architecture and Design*, McGraw-Hill Book Co., New York, 1990.
3. Belcher, R., and E. Aslaksen, *Systems Engineering*, Prentice Hall of Australia, Sydney, Australia, 1992.
4. Blanchard, B. S., *System Engineering Management*, 2nd ed., John Wiley & Sons, Inc., New York, 1998.
5. Blanchard, B. S., "The Systems Engineering Process: An Application for the Identification of Resource Requirements," *Systems Engineering:* Journal of the International Council on Systems Engineering (INCOSE), Seattle, Wash., Volume 1, Number 1, July/September 1994.
6. Blanchard, B. S., and W. J. Fabrycky, *Systems Engineering and Analysis*, 3rd ed., Prentice Hall, Inc., Upper Saddle River, N.J., 1998.
7. Blanchard, B. S., W. J. Fabrycly, and D. Verma (Eds.), *Application of the Systems Engineering Process to Define Requirements for Computer-Based Design Tools*, Monograph, International Society of Logistics (SOLE), Hyattsville, Md., 1994.
8. Boulding, K., "General Systems Theory: The Skeleton of Science," *Management Science*, April 1956.
9. DOD Regulation 5000.2, "Mandatory Procedures For Major Defense Acquisition Programs (MDAPs) And Major Automated Information System (MAIS) Acquisition Programs," Department of Defense, Washington, D.C., 1996.
10. DSMC, *Systems Engineering Management Guide*, Defense Systems Management College, Fort Belvoir, VA, latest ed.
11. EIA/IS-632, "Systems Engineering," Electronic Industries Association (EIA), 2001 Pennsylvania Avenue, N.W., Washington, D.C., 1994.
12. Fabrycky, W. J., "Modeling and Indirect Experimentation in System Design Evaluation," *Systems Engineering*, Journal of the International Council on Systems Engineering (INCOSE), Seattle, Wash., Volume 1, Number 1, July/September 1994.
13. Forrester, J. W., *Principles of Systems*, The MIT Press, Cambridge, MA, 1968.
14. Grady, J. O., *Systems Integration*, CRC Press, Boca Raton, FL, 1994.
15. Grady, J. O., *Systems Requirements Analysis*, McGraw-Hill Book Co., New York, 1993.
16. Grady, J. O., *System Engineering Planning and Enterprise Identity*, CRC Press, Boca Raton, FL, 1995.
17. Hall A. D., *A Methodology for Systems Engineering*, D. Van Nostrand Co., Ltd., Princeton, N.J. 1962.
18. Hubka, V., and W. E. Eder, *Theory of Technical Systems*, Springer-Verlag, Berlin, 1988.
19. IEEE-STD-1220-1994, "IEEE Trial Use Standard for Application and Management of the System Engineering Process," Institute of Electrical and Electronics Engineers (IEEE), 345 East 47th Street, New York 1994.
20. INCOSE, *Journal of the International Council on Systems Engineering*, United Airlines Bldg., Suite 804, 2033 Sixth Avenue, Seattle, WA 98121.
21. INCOSE, *Annual Conference*, Proceedings, International Council on Systems Engineering, United Airlines Bldg., Suite 804, 2033 Sixth Avenue, Seattle, WA 98121.

22. Lacy, J. A. *Systems Engineering Management: Achieving Total Quality*, McGraw-Hill Book Co., New York, 1992.
23. Martin, J. N., *Systems Engineering Guidebook: A Process for Developing Systems and Products*, CRC Press Boca Raton, FL 1997.
24. Rechtin, E., *System Architecting: Creating and Building Complex Systems*, Prentice Hall, Inc., Upper Saddle River, N.J., 1991.
25. Rechtin, E., and M. Maier, *The Art of Systems Architecting*, CRC Press, Inc., Boca Raton, FL, 1996.
26. Reilly, Norman B., *Successful Systems Engineering for Engineers and Managers*, Van Nostrand Reinhold, New York, 1993.
27. Sage, A. P, *Decision Support Systems Engineering*, John Wiley & Sons, Inc., New York, 1991.
28. Sage, A. P., *Systems Engineering*, John Wiley & Sons, Inc., New York, 1992.
29. Sage, A. P., *Systems Management for Information Technology and Software Engineering*, John Wiley & Sons, Inc., New York, 1995.
30. SECMM-95-01, "A Systems Engineering Capability Maturity Model (SE-CMM)," Version 1.1, Software Engineering Institute (SEI), Carnegie Melon University, Pittsburgh, PA, 1995.
31. Shishko, R., *NASA Systems Engineering Handbook*, NASA, Washington, D.C., 1995.
32. SPC-95075-CMC, "Systems Engineering Maturity and Benchmarking," Version 01.00.06, Software Productivity Consortium, SPC Building, 2214 Rock Hill Road, Herndon, VA, 1996.
33. Thome, Bernhard (Ed.), *Systems Engineering: Principles and Practice of Computer-Based Systems Engineering*, John Wiley & Sons, Inc., New York, 1993.
34. Truxal, J. G., *Introductory System Engineering*, McGraw-Hill Book Co., New York, 1972.
35. Von Bertalanffy, L., "General Systems Theory: A New Approach to Unity of Science," *Human Biology*, December 1951.
36. Von Bertalanffy, L., *General Systems Theory*, George Braziller Press, New York, 1968.
37. Wiener, N., *Cybernetics*, John Wiley & Sons, Inc., New York, 1948.
38. Wymore, A. W., *Model-Based Systems Engineering*, CRC Press, Inc., Boca Raton, FL, 1993.

F.2 CONCURRENT/SIMULTANEOUS ENGINEERING

1. Kusiak, A., *Concurrent Engineering: Automation, Tools, and Techniques*, John Wiley & Sons, Inc., New York, 1992.
2. Miller, L. C. G., *Concurrent Engineering Design: Integrating the Best Practices for Process Improvement*, Society of Manufacturing Engineers, Dearborn, MI, 1993.
3. Prasad, B., *Concurrent Engineering Fundamentals: Integrated Product and Process Organization*, Prentice Hall, Inc., Upper Saddle River, N.J., 1996.
4. Shina, S. G. (ed.), *Successful Implementation of Concurrent Engineering Products and Processes*, Van Nostrand Reinhold, New York, 1994.
5. Winner, R. I., J. P. Pennell, H. E. Bertrand, and M. M. G. Slusarczuk, *The Role of Concurrent Engineering in Weapons Systems Acquisition*, Report R-338, Institute for Defense Analysis, Arlington, VA, 1988.

F.3 SOFTWARE/COMPUTER-AIDED SYSTEMS

1. Boehm, B. W., *Software Engineering Economics*, Prentice Hall, Inc., Upper Saddle River, N.J., 1981.
2. Eisner, H., *Computer-Aided Systems Engineering*, Upper Saddle River, Prentice Hall, Inc., N.J., 1988.
3. Humphrey, W. S., *A Discipline for Software Engineering*, Addison-Wesley Publishing Co., Reading, MA, 1995.
4. Krouse, J. K., *What Every Engineer Should Know about Computer-Aided Design and Computer-Aided Manufacturing*, Marcel Dekker, Inc., New York, 1982.
5. MIL-STD-498, *Software Development and Documentation*, Department of Defense, Washington, D.C., 1994.
6. Pressman, R. S., *Software Engineering: A Practitioner's Approach*, 3rd ed., McGraw-Hill Book Co., New York, 1992.
7. Sage, A. P., and J. D. Palmer, *Software Systems Engineering*, John Wiley & Sons, Inc., New York, 1990.
8. Zeid, I., CAD/CAM Theory and Practice, McGraw-Hill Book Co., New York, 1991.

F.4 RELIABILITY, MAINTAINABILITY, AND MAINTENANCE ENGINEERING

1. *Annual Reliability and Maintainability Symposium*, Proceedings, Evans Associates, Durham, N.C.
2. Barlow, R. E., *Mathematical Theory of Reliability*, John Wiley & Sons, New York, 1965.
3. Blanchard, B. S., D. Verma, and E. L. Peterson, *Maintainability: A Key to Effective Serviceability and Maintenance Management*, John Wiley & Sons, Inc., New York, 1995.
4. EPRD-97, "Electronic Parts Reliability Data," Vols. 1 and 2, Reliability Analysis Center, Rome, NY, 1997.
5. Instruction Manual, *Potential Failure Mode and Effects Analysis*, Chrysler, Ford, and General Motors Corp., rev. 1993.
6. Ireson, W. G., and C. F. Coombs (Eds.), *Handbook of Reliability Engineering and Management*, McGraw-Hill Book Co., New York, 1988.
7. Kececioglu, D., *Reliability Engineering Handbook*, Vol. 1 and 2, Prentice Hall, Inc., Upper Saddle River, N.J., 1991.
8. Knezevic, J., *Reliability, Maintainability, and Supportability: A Probabilistic Approach*, McGraw-Hill Book Co., New York, 1993.
9. MIL-HDBK-217, Military Handbook, *Reliability Prediction of Electronic Equipment*, Department of Defense, Washington, D.C., latest ed.
10. MIL-STD-470B, Military Standard, *Maintainability Program for Systems and Equipment*, Department of Defense, Washington, D.C.
11. MIL-STD-785B, Military Standard, *Reliability Program for Systems and Equipment Development and Production*, Department of Defense, Washington, D.C.
12. Mobley, R. K., *An Introduction to Predictive Maintenance*, Van Nostrand Reinhold, New York, 1990.
13. Moubray, J., *Reliability-Centered Maintenance*, Butterworth-Heinemann Ltd., Boston, MA, 1992.

14. Nakajima, S. (ed.), *TPM Development Program: Implementing Total Productive Maintenance*, Productivity Press, Portland, OR, 1989.
15. NPRD-95, "Nonelectronic Parts Reliability Data," Reliability Analysis Center, Rome, NY, 1995.
16. O'Connor, P. D. T., *Practical Reliability Engineering*, 3rd ed., John Wiley & Sons, Inc., New York, 1991.
17. RAC, *Failure Mode, Effects and Criticality Analysis* (FMECA), Reliability Analysis Center, Rome, N.Y., 1992.
18. Raheja, D. G., *Assurance Technologies: Principles and Practices*, McGraw-Hill Book Co., New York, 1991.
19. Shooman, M. L., *Probabilistic Reliability: An Engineering Approach*, 2nd ed., Krieger, 1990.
20. Smith, A. M., *Reliability-Centered Maintenance*, McGraw-Hill Book Co., New York, 1993.
21. Wireman, T., *World Class Maintenance Management*, Industrial Press, New York, 1990.

F.5 HUMAN FACTORS AND SAFETY ENGINEERING

1. Hammer, W., *Occupational Safety Management and Engineering*, 4th ed., Prentice Hall, Inc., Upper Saddle River, N.J., 1988.
2. Kroemer, K., H. Kroemer, and K. Kroemer-Elbert, *Ergonomics: How to Design for Ease and Efficiency*, Prentice Hall, Inc., Upper Saddle River, N.J., 1994.
3. MIL-STD-1472, Military Standard, "Human Engineering Design Criteria For Military Systems, Equipment, And Facilities," Department of Defense, Washington, D.C., latest ed.
4. Roland, H. E., and B. Moriarty, *System Safety Engineering and Management*, 2nd ed., John-Wiley & Sons, Inc., New York, 1990.
5. Sanders, M. S., and E. J. McCormick, *Human Factors in Engineering and Design*, 7th ed., McGraw-Hill Co., New York, 1992.
6. Van Cott, H. P., and R. G. Kincade (eds.), *Human Engineering Guide to Equipment Design*, U.S. Government Printing Office, Washington, D.C., 1972.
7. Woodson, W. E., B. Tillman, and P. Tillman, *Human Factors Design*, 2nd ed., McGraw-Hill Book Co., New York, 1992.

F.6 LOGISTICS AND LOGISTICS ENGINEERING

1. Banks, J., and W. J. Fabrycky, *Procurement And Inventory Systems Analysis*, Prentice Hall, Inc., Upper Saddle River, N.J., 1987.
2. Blanchard, B. S., *Logistics Engineering and Management*, 5th ed., Prentice Hall, Inc., Upper Saddle River, N.J., 1998.
3. Coyle, J. J., E. J. Bardi, and J. L. Cavinato, *Transportation*, 4th ed., West Publishing, St. Paul, MN, 1992.
4. DSMC, *Integrated Logistic Support Guide*, Defense Systems Management College, Fort Belvoir, VA, latest ed.
5. Glaskowsky, N. A., D. R. Hudson, and R. M. Ivie, *Business Logistics*, 3rd ed., Dryden Press, Harcourt Brace Jovanovich Co., Orlando, FL, 1992.
6. International Society of Logistics, *Logistic Spectrum,* 8100 Professional Place, Suite 211, Hyattsville, MD 20785.

7. International Society of Logistics, Annual Symposium Proceedings, 8100 Professional Place, Suite 211, Hyattsville, MD 20785.
8. Jones, J. V., *Integrated Logistics Support Handbook*, Tab Books, Inc., Blue Ridge Summit, PA, 1987.
9. Langford, J. W., *Logistics Principles and Practices*, McGraw-Hill Book Co., New York, 1995.
10. MIL-HDBK-59A, Military Handbook "Computer-Aided Acquisition and Logistic Support (CALS)," Department of Defense, Washington, D.C.
11. MIL-HDBK-502, "Department of Defense Handbook—Acquisition Logistics," Department of Defense Washington, D.C., May 1997.
12. MIL-PRF-49506, "Performance Specification—Logistics Management Information," Department of Defense, Washington, D.C., November 1996.
13. MIL-STD-1388-2B, Military Standard, "DOD Requirements for a Logistic Support Analysis Record (LSAR)," Department of Defense, Washington, D.C., 1983.
14. Council of Logistics Management, *Journal of Business Logistics*, 2803 Butterfield Road, Oak Brook, IL 60521.

F.7 QUALITY, QUALITY ENGINEERING, AND QUALITY ASSURANCE

1. Akao, Y., *Quality Function Deployment: Integrating Customer Requirements into Product Design*, Productivity Press, Portland, OR, 1990.
2. Cohen, L., *Quality Function Deployment: How to Make QFD Work for You*, Addison-Wesley Publishing Co., Reading, MA, 1995.
3. Deming, W. E., *Out of the Crisis*, Massachusetts Institute of Technology Press, Cambridge, MA, 1986.
4. Grant, E. L., and R. S. Leavenworth, *Statistical Quality Control*, 6th ed., McGraw-Hill Book Co., New York, 1988.
5. Hauser, J. R., and D. Clausing, "The House of Quality," *Harvard Business Review*, May-June 1988.
6. Ishikawa, K., *Introduction to Quality Control*, Chapman & Hall, London, England, 1991.
7. ISO 9000-9004, International Standards Series, "Quality Standards," International Standards Organization.
8. Juran, J. M., and F. M., Gryna, *Quality Planning and Analysis*, 3rd ed., McGraw-Hill Book Co., New York, 1993.
9. Taguchi, G., E. A. Elsayed, and T. C. Hsiang, *Quality Engineering in Production Systems*, McGraw-Hill Book Co., New York, 1989.

F.8 PRODUCTION, PRODUCIBILITY, AND ENVIRONMENTAL, ENGINEERING

1. Defense Systems Management College (DSMC), *Manufacturing Management: Guide for Program Managers*, DSMC, For Belvoir, VA, latest ed.
2. Graedel, T. E., and B. R. Allenby, *Industrial Ecology*, Prentice Hall, Inc., Upper Saddle River, N.J., 1995.

3. ISO 14001, International Standard, "Environmental Management Systems (EMS)," International Standards Organization.
4. Kidd, P. T., *Agile Manufacturing: Forging New Frontiers*, Addison-Wesley Publishing Co., Reading, MA, 1994.

F.9 OPERATIONS RESEARCH, AND OPERATIONS ANALYSIS

1. Ackoff, R. L., *The Art of Problem Solving*, John Wiley & Sons, Inc., New York, 1978.
2. Churchman, C. W., R. L. Ackoff, and E. L. Arnoff, *Introduction to Operations Research*, John Wiley & Sons, Inc., New York, 1957.
3. Fabrycky, W. J., P. M. Ghare, and P. E. Torgersen, *Applied Operations Research and Management Science*, Prentice Hall, Inc., Upper Saddle River, N.J., 1984.
4. Hillier, F. S., and G. J. Lieberman, *Introduction to Operations Research*, 6th ed., McGraw-Hill Book Co., New York, 1995.
5. Taha, H. A., *Operations Research: An Introduction*, 5th ed., Macmillan Publishing Co., New York, 1992.

F.10 ENGINEERING ECONOMY, ECONOMIC ANALYSIS, AND COST ESTIMATING

1. Canada, J. R., W. G. Sullivan, and J. A. White, *Capital Investment Analysis for Engineering and Management*, 2nd ed., Prentice Hall, Inc., Upper Saddle River, N.J., 1996.
2. Dhillon, B. S., *Life Cycle Costing: Techniques, Models and Applications*, Gordon and Breach Science Publishers, New York, 1989.
3. Fabrycky, W. J., and B. S. Blanchard, *Life-Cycle Cost and Economic Analysis*, Prentice Hall, Inc., Upper Saddle River, N.J., 1991.
4. Fabrycky, W. J., G. J. Thuesen, and D. Verma, *Economic Decision Analysis*, 3rd ed., Prentice Hall, Inc., Upper Saddle River, N.J., 1998.
5. Ostwald, P. F., *Engineering Cost Estimating*, 3rd ed., Prentice Hall, Inc., Upper Saddle River, N.J., 1992.
6. Stewart, R. D., and R. M. Wyskida, *Cost Estimator's Reference Manual*, 2nd ed., John Wiley & Sons, Inc., New York, 1995.
7. Thuesen, G. J., and W. J. Fabrycky, *Engineering Economy*, 8th ed., Prentice Hall, Inc., Upper Saddle River, N.J., 1993.
8. White, J. A., M. H. Agee, and K. E. Case, *Principles of Engineering Economic Analysis*, 3rd ed., John Wiley & Sons, Inc., New York, 1989.

F.11 MANAGEMENT AND SUPPORTING AREAS

1. American Productivity and Quality Center, *The Benchmarking Management Guide*, Productivity Press, Portland, OR, 1993.
2. Balm G. J., *Benchmarking: A Practitioner's Guide for Becoming and Staying the Best of the Best*, QPMA Press, Schaumburg, IL, 1992.

3. Camp, R. C., *Benchmarking: The Search for Industry Best Practices that Lead to Superior Performance*, ASQC Press, Milwaukee, WI, 1989.

4. Kerzner, H., *Project Management: A Systems Approach to Planning, Scheduling, and Controlling*, 5th ed., Van Nostrand Reinhold, New York, 1995.

5. Koontz, H., C. O'Donnell, and H. Weihrich, *Essentials of Management*, 5th ed., McGraw-Hill Book Co., New York, 1990.

6. Stewart, R. D., and A. L. Stewart, *Proposal Preparation*, 2nd ed., John Wiley & Sons, Inc., New York, NY, 1992.

INDEX

F

G

H

I

N

O

P

Q

R

S